APPLIED CALCULUS

Second edition

Dennis D. Berkey

Boston University

SAUNDERS COLLEGE PUBLISHING
Philadelphia   Fort Worth   Chicago
San Francisco   Montreal   Toronto
London   Sydney   Tokyo

Text Typeface: Times Roman
Compositor: York Graphic Services, Inc.
Acquisitions Editor: Robert B. Stern
Developmental Editor: Alexa Barnes
Production Service: York Production Services
Text Design: York Production Services
Manager, Art and Design: Carol C. Bleistine
Art Coordinator: Doris Bruey
Production Manager: Joanne Cassetti
Director of EDP: Tim Frelick

Cover credit: York Production Services

Library of Congress Catalog Card Number: 89-24069

Applied Calculus, 2/e

ISBN 0-03-031267-1

Printed in the United States of America

012   039   98765432

19. $\displaystyle\int e^u \, du = e^u + C$

20. $\displaystyle\int a^u \, du = \frac{1}{\ln a} a^u + C, \; a \neq 1$

21. $\displaystyle\int u e^{au} \, du = \frac{e^{au}}{a^2} (au - 1) + C$

22. $\displaystyle\int \frac{du}{1 + e^u} = \ln\left(\frac{e^u}{1 + e^u}\right) + C$

23. $\displaystyle\int \frac{du}{a + be^{cu}} = \frac{u}{a} - \frac{1}{ac} \ln |a + be^{cu}| + C$

24. $\displaystyle\int \ln u \, du = u \ln u - u + C$

25. $\displaystyle\int u \ln u \, du = \frac{u^2}{2} \ln u - \frac{u^2}{4} + C$

26. $\displaystyle\int \frac{du}{u \ln u} = \ln |\ln u| + C$

27. $\displaystyle\int u^m \ln u \, du = u^{m+1}\left[\frac{\ln u}{m + 1} - \frac{1}{(m + 1)^2}\right] + C$

28. $\displaystyle\int \sin^2 u \, du = \frac{u}{2} - \frac{1}{4} \sin 2u + C$

29. $\displaystyle\int \sin^3 u \, du = -\frac{1}{3} \cos u(\sin^2 u + 2) + C$

30. $\displaystyle\int \cos^2 u \, du = \frac{1}{2} u + \frac{1}{4} \sin 2u + C$

31. $\displaystyle\int \cos^3 u \, du = \frac{1}{3} \sin u \, (\cos^2 u + 2) + C$

32. $\displaystyle\int \frac{du}{1 \pm \sin u} = \mp \tan\left(\frac{\pi}{4} \mp \frac{u}{2}\right) + C$

33. $\displaystyle\int \frac{du}{1 + \cos u} = \tan\left(\frac{u}{2}\right) + C$

34. $\displaystyle\int \frac{du}{1 - \cos u} = -\cot\left(\frac{u}{2}\right) + C$

APPLIED CALCULUS

Second edition

Preface

This text is intended for use in a full-year course in calculus for business, economics, and social science students. It assumes only a basic competence with college algebra.

EMPHASIS: My goal is to enable the student to understand the fundamental themes of the differential and integral calculus, to master the principal techniques associated with these results, and to apply these techniques and theories to the solution of real problems. Students who have mastered this text should be able to recognize and exploit the notions of function, graph, rate of change, optimization, and area of a region bounded by a curve when they arise in later courses or in practice. To accomplish this goal, I have incorporated these features:

- Relevant theorems are clearly stated, and immediately interpreted by examples and illustrations. Proofs are given only when they are instructive and accessible to students with the stated background.

- I have made every effort to write with mathematical honesty and precision, but the emphasis is primarily on the applications of the mathematical ideas, rather than on their justifications.

- The principal applications are in business and economic theory (revenue, cost, marginal analysis), although applications in the life and social sciences are also discussed (human growth, population growth, marginal utility, spread of epidemics). Because of the simplicity and precision with which the basic models of economic theory may be stated, they allow a very direct application of the calculus.

ORGANIZATION: This text follows a common organization of topics. An introductory chapter reviews the essentials of high school algebra that are required in what follows. Limits are then introduced in the context of finding the slope of a curve, and the rudiments of differentiation are developed (Chapter 2). This is followed immediately by applications of differentiation to finding relative extrema, and to max-min and related rates problems, including word problems (Chapter 3). The goal here is to provide students with an immediate payoff for the theory of the derivative, and to get them working with word problems as soon as possible. Chapter 3 also includes a section on curve sketching, to emphasize the relationship between the graph of a function and its derivatives.

The logarithm and exponential functions are introduced in Chapter 4, where further applications of the derivative are developed. Chapter 5 begins with a discussion of the antiderivative, its applications, and the method of integration by substitutions. Next, the definite integral is introduced as the limit of an approximation scheme for areas of regions in the plane, and the relationship between definite integrals and antiderivatives is established. Chapter 6 introduces the trigonometric functions and discusses their derivatives and integrals.

Chapter 7 treats multivariable topics, including partial differentiation, Lagrange multipliers, and multiple integrals. Chapter 8 concerns techniques of integration and improper integrals. The theory of infinite sequences and series, including Taylor polynomials and series, is presented in Chapter 9. Chapter 10 is a brief discussion of differential equations. A final chapter (Chapter 11) presents the theory of probability for discrete and continuous random variables, with particular attention to the normal and exponential probability distributions.

DISTINGUISHING FEATURES: Several features should help students master the techniques of the calculus:

- Many of the over 400 worked examples are presented in the split strategy/solution format, in which a summary of the important steps in the solution appears in the "strategy" column. This helps students to see *why* as well as how the steps are performed, and to understand what is required in developing problem-solving strategies of their own.

- Over 2700 exercises are presented, coordinated carefully with the worked examples and gradually increasing in difficulty within each set. Word problems are included in nearly every section.

- Special care has been taken to introduce and illustrate key concepts with applied and illustrative examples. Proofs, when given, are placed at the ends of discussions so that they may be omitted if the instructor wishes.

- Each chapter ends with a comprehensive Summary Outline of all of the important ideas and techniques in the chapter, and a set of Review Exercises.

- A clear, understandable writing style and conversational tone have been used throughout.

- Some reference is made to the use of computers, both in the text and in the exercises, and seven BASIC programs are included in Appendix I as models. The purpose is to indicate how computers can be used to illustrate and apply the calculus. This material is brief and can easily be omitted.

CHANGES TO THE SECOND EDITION: In addition to a careful reworking of the first edition, several significant changes were made in response to requests and advice from users of the first edition.

- Chapter 1 now includes the concept of *range* of a function.

- Among the applications of the derivative in Chapter 3, more discussion has been added on elasticity of demand—a topic which is vital to business students in their study of economics.

- In Chapter 4 the coverage of logarithmic differentiation has been expanded to make the discussion of logarithmic functions complete.

- A section on Newton's Method for approximating zeros has been added to Chapter 9, on infinite series. This topic is an application of Taylor series as well as an important numerical method.

- Chapter 10 has been expanded significantly by adding sections on applications to population growth and on the qualitative theory of differential equations. This material strengthens considerably the treatment of differential equations.

- Numerous examples and over 100 new exercises have been added throughout the text to both new and existing material.

- All the illustrations have been improved dramatically through the use of computer-generated art.

SUPPLEMENTS:

- A Student Solutions Manual, written by Fred Wright of Iowa State University of Science and Technology, is available for purchase: it contains chapter summaries, detailed solutions to all the odd-numbered exercises, and 10–15 additional practice problems per chapter. (The answers to the odd-numbered exercises are also found at the back of the text.) In effect, this ancillary increases the number of worked examples available for study and practice.

Instructors who adopt this text may receive, free of charge, the following items:

- Instructor's Manual—Also written by Fred Wright, this manual contains detailed solutions to all the exercises to assist the instructor in the classroom and in grading assignments.

- Prepared Tests—The Prepared Tests were written by Jack Porter and Jack Porter II of The University of Kansas and contain five tests for each chapter and two final exams. Three of the five tests per chapter are free-response and two are multiple-choice. Answers to all the test questions are also included.

- Computerized Test Bank (IBM version)—The computerized test bank, written and programmed by ipsTest, contains chapter tests modeled on the examples and exercises in the text. Each test has a mixture of 5 multiple-choice and 15 free-response questions. The free-response questions are generated algorithmically, so the instructor can create a large number of unique tests. The software solves each problem and provides an answer key. A printed version of this test bank is also available.

- Graph Tutor (IBM version)—George Bergeman of Northern Virginia Community College has written an interactive tutorial software package which provides graphical support for solving exercises in the text. Students can also use the computational capabilities of this package as a "super calculator" to perform otherwise cumbersome operations, thus permitting concentration on concepts and underlying theory.

ACKNOWLEDGMENTS: For the first edition the following mathematicians were kind enough to review various drafts of the manuscript. Their comments and suggestions were a major positive force in the development of the text.

Michael Bleicher, University of Wisconsin, Madison
Philip L. Bowers, Florida State University, Tallahassee
C. Kenneth Bradshaw, San Jose State University
Fred Brauer, University of Wisconsin
Edward A. Conners, University of Massachusetts, Amherst
Raymond Coughlin, Temple University
Robert W. Deming, State University of New York, Oswego
Garrett Etgen, University of Houston
Alec Ingraham, New Hampshire College
Edward M. Landesman, University of California, Santa Cruz
James W. Maxwell, Oklahoma State University
Eldon L. Miller, University of Mississippi
Mary Ellen O'Leary, University of South Carolina
William Perry, Texas A & M University
Donald R. Sherbert, University of Illinois
Henry A. Warchall, University of Texas, Austin
Fred Wright, Iowa State University

The following mathematicians served as readers for the second edition:

Alfred Bachman, California Polytechnic State University
Howard Beckwith, California State University
Fred Brauer, University of Wisconsin, Madison
John Busovicki, Indiana University of Pennsylvania
Jim Cribbs, City College, San Francisco
Ronald W. Dickey, University of Wisconsin, Madison
Paul Krajkiewcs, University of Nebraska
Jack Porter, University of Kansas

Professors Alfred Bachman, of California Polytechnic State University and Steve Reyner of SUNY, Oswego, read all galley proofs and page proofs to insure accuracy. The accuracy of this text results in large measure from their efforts.

The Saunders College Publishing team has continued to provide me with excellent support, especially Mathematics Editor Robert Stern, Developmental Editor Alexa Barnes, Production Coordinator Joanne Cassetti, and Publisher Liz Widdicombe.

Finally, I am again privileged to acknowledge the continuing warm and strong support of my students, my colleagues, and most especially, my family.

Dennis D. Berkey

Contents

Index of Applications

Biological Sciences

Chapter 1

Numbers, Functions, and Graphs

1.1 WORKING WITH REAL NUMBERS

One of the principal goals of this text is to help you use calculus to measure rates at which quantities change. For example, in studying a production model in economics, you might wish to know if a certain component of cost is increasing or decreasing. Or a psychology student armed with data from an experiment might wish to know how response time varies with experience.

Before proceeding to develop techniques for measuring changes among quantities, we need to review briefly several properties of the number system used to measure these quantities and to agree on the meaning of certain mathematical notation. This should prevent confusion later on and allow us to take full advantage of the power of the language of mathematics.

Integers and Real Numbers

The real numbers may be identified with points on the **number line,** as illustrated in Figure 1.1. The **integers** are the special subset of the real numbers $\{. . . -3, -2, -1, 0, 1, 2, 3, . . .\}$ as illustrated in Figure 1.2.

Figure 1.1 All locations on the number line correspond to real numbers.

Figure 1.2 The integers form a special subset of the real numbers.

We emphasize the distinction between integers and other real numbers because certain quantities are measured only as integers. For example, Figure 1.3 illustrates a typical relationship between the time following the injection of a certain drug in a person's bloodstream and the concentration of that drug in the person's blood. In this example both time, t, and concentration, c, are represented by all real numbers within certain ranges. Compare this example with Figure 1.4, which represents a typical relationship between the number, x, of personal computers produced per day by a manufacturer and the total daily cost, C, of manufacturing these computers. Here only integer values of x make sense.

Figure 1.3 Both drug concentration and time are represented by real numbers.

Figure 1.4 Production level x is measured only in integers.

In general, the theory of the calculus applies only to variables that can equal any real number (within certain ranges, sometimes). In order to deal with variables that can equal only integer values, such as production levels, we shall often *assume* that such variables can equal any real number. For example, Figure 1.5 illustrates how we might think about the relationship between output and cost in Figure 1.4 in this way. In such situations we shall be careful to note this assumption and the effect it might have on our calculations.

Figure 1.5 Effect of assuming that production level can equal any real number between 0 and 10. (Compare with Figure 1.4.)

Inequalities

The statement "$a < b$," read "a is less than b," means that a lies to the left of b on the number line (see Figure 1.6). This can also be written "$b > a$" and read "b is greater than a." The statement "$a \le b$" means that a is less than or equal to b.

Figure 1.6 The statement $a < b$ means a lies to the left of b.

Example 1 Each of the following inequalities is true. Note especially e and f.

(a) $2 < 6$ (b) $-3 < 5$ (c) $x^2 \ge 0$

(d) $\pi > 3$ (e) $-10 < -5$ (f) $-2 > -4$ □

Interval Notation

You are probably familiar with **set builder notation** used to specify certain sets of real numbers. For example, the set S_1 of all real numbers between 2 and 7 (but not including 2 or 7) can be written as

$$S_1 = \{x \mid 2 < x < 7\}$$

while the set S_2 consisting of the numbers 2, 7, and all real numbers between 2 and 7 is written as

$$S_2 = \{x \mid 2 \le x \le 7\}.$$

Such sets of real numbers, which include all real numbers between two given numbers, are called **intervals.** It is important to have a notation for intervals that indicates precisely whether or not each of the endpoints is included. We shall use the following notation for intervals:

$[a, b]$ means $\{x \mid a \le x \le b\}$ (both endpoints included)

$[a, b)$ means $\{x \mid a \le x < b\}$ (only left endpoint included)

$(a, b]$ means $\{x \mid a < x \le b\}$ (only right endpoint included)

(a, b) means $\{x \mid a < x < b\}.$ (neither endpoint included)

The conventions for graphing intervals on the number line are illustrated in Figure 1.7.

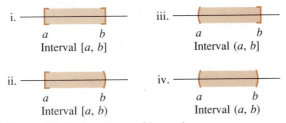

i. Interval $[a, b]$ iii. Interval $(a, b]$

ii. Interval $[a, b)$ iv. Interval (a, b)

Figure 1.7 Various types of intervals.

Intervals of the form $[a, b]$, which include both endpoints, are called *closed* intervals. Those of the form (a, b), which include neither endpoint, are called *open*.

Example 2 (a) A rain gauge has the shape of a cup 6 inches deep. If x represents the depth of water in the gauge after one month, the closed interval $[0, 6]$ represents the possible values of x.
(b) Let r be the radius of a piston that is to slide inside a cylinder of radius 10 cm. The possible values for r lie in the open interval $(0, 10)$. Why are 0 and 10 excluded? □

Infinite Intervals

Often we will need to specify sets consisting of all numbers larger or smaller than a given number. We do so by using the infinity symbol, ∞, as follows:

$$[a, \infty) \quad \text{means} \quad \{x \mid x \geq a\}$$
$$(a, \infty) \quad \text{means} \quad \{x \mid x > a\}$$
$$(-\infty, a] \quad \text{means} \quad \{x \mid x \leq a\}$$
$$(-\infty, a) \quad \text{means} \quad \{x \mid x < a\}.$$

It is important to note that the symbol ∞ is *not* a number. Rather, it is just a symbol used to denote all numbers to the right or left of a given number on the number line.

Example 3 The negative real numbers are those in the interval $(-\infty, 0)$, while the nonnegative real numbers constitute the interval $[0, \infty)$. □

Properties of Inequalities

We will frequently need to "solve" inequalities in one or more variables. The following theorem summarizes the relevant properties of inequalities.

THEOREM 1

Let a, b, and c be real numbers. Then

(i) If $a < b$, then $a + c < b + c$ for any number c.
(ii) If $a < b$ and $c > 0$, then $ac < bc$.
(iii) If $a < b$ and $c < 0$, then $ac > bc$.
(iv) If $a < b$ and $b < c$, then $a < c$.

Be especially careful when multiplying (or dividing) both sides of an inequality by a negative number. As part (iii) states, this reverses the sense of the inequality.

Example 4 Solve the inequality $3x - 4 < 8$

Strategy
Isolate x by

(i) adding 4 to both sides, and

Solution
Using Theorem 1, we have

$$3x - 4 < 8$$
$$3x - 4 + 4 < 8 + 4$$
$$3x < 12$$

(ii) multiplying both sides by $\frac{1}{3}$.

$$\frac{1}{3}(3x) < \frac{1}{3}(12)$$

$$x < 4$$

The solution set is therefore

$$\{x \mid x < 4\} \qquad \text{or} \qquad (-\infty, 4).$$

(See Figure 1.8.)   □

Figure 1.8 Solution of Example 4.   **Figure 1.9** Solution of Example 5.

Example 5 Solve the inequality $10 - 3x \leq 37$.

Strategy

Solve inequality by

(i) adding -10 to both sides

(ii) multiplying both sides by $-\frac{1}{3}$ *and reversing the inequality*.

Solution

Proceeding as before, we have

$$10 - 3x \leq 37$$
$$10 - 3x + (-10) \leq 37 + (-10)$$
$$-3x \leq 27$$

$$\left(-\frac{1}{3}\right)(-3x) \geq \left(-\frac{1}{3}\right)(27) \qquad \text{(Note reverse of inequality)}$$

$$x \geq -9.$$

The solution set is $[-9, \infty)$.
(See Figure 1.9.)   □

Example 6 A television set retailer wishes to stock up on sets costing $400 wholesale. The dealer must pay a $100 delivery charge on the order plus 5% sales tax on each set ordered. How many sets can the dealer order if he has $5000 available for the purchase?

Strategy

Label the unknown quantity as the variable x.
Find an expression involving x for total cost.

Write the inequality
Total cost $\leq$ 5000

Solve inequality by

(i) subtracting 100 from both sides

Solution

Let x be the number of sets purchased. The cost of each set is $(1.05)(\$400) = \$420$, including tax, so the total cost for the order will be $\$420x + \$100$, including shipping. The condition that must be satisfied is therefore

$$420x + 100 \leq 5000.$$

Thus

$$420x + 100 - 100 \leq 5000 - 100$$

so

$$420x \leq 4900$$

and

(ii) dividing both sides by 420.

$$x \le \frac{4900}{420} = 11\frac{2}{3}.$$

Note that the number of sets ordered must be an integer.

The solution $x \le 11\frac{2}{3}$ means that the dealer can order at most 11 sets at a total cost of ($420)(11) + $100 = $4620 + $100 = $4720. □

Exercise Set 1.1

In each of Exercises 1–10 determine whether the quantity described is measured by integers alone or by any real number.

1. Rainfall on a particular day

2. Height of a building, in feet

3. Pages in a book

4. Age

5. Inventory in a bookstore

6. Drug concentration in the bloodstream

7. U.S. auto production

8. Altitude

9. Temperature

10. Census

In each of Exercises 11–20 match the interval illustrated with one of the intervals (a)–(j) below.

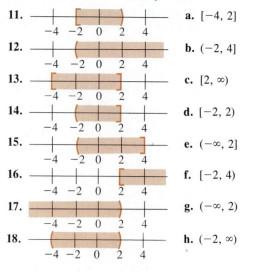

11.

a. [−4, 2]

12.

b. (−2, 4]

13.

c. [2, ∞)

14.

d. [−2, 2)

15.

e. (−∞, 2]

16.

f. [−2, 4)

17.

g. (−∞, 2)

18.

h. (−2, ∞)

19.

i. (−2, 2]

20.

j. (−4, 2)

21. True or false? Every integer is also a real number.

22. True or false? There is no largest real number.

23. For how many integers m is $\frac{1}{m}$ also an integer?

In Exercises 24–33 solve the inequality for x, expressing the solution set as an interval.

24. $2x > 10$

25. $6x - 2 \le 16$

26. $6 - 5x \ge 16$

27. $4 - 2x \le 12$

28. $4 - 3x > 5$

29. $6 + x \le 2x - 5$

30. $x - 7 \ge 3 - x$

31. $x(6 + x) \ge 3 + x^2$

32. $5 - 3x < 7 - 2x$

33. $(x - 4)^2 \ge x^2 - x + 12$

34. Restrictions in some U.S. Post Offices limit the dimensions of a parcel that can be mailed according to the rule "length plus girth not to exceed 108 inches." Write an inequality giving the possible girths g for a package 28 inches long.

35. A manufacturer of dishwashers produces a portable model and a built-in model. The portable costs $200 to produce, and the built-in model costs $250. Write an inequality involving the number of portables P and the number of built in models B in a production run showing that the total cost of production cannot exceed $10,000.

36. Refer to Exercise 35. If the production run must include at least 15 portables, what is the maximum number of built-in models that can be produced?

37. A caterer charges $12 per person at a party, plus a $50 delivery charge per party. How many guests can a host

afford, including the host, if $300 is available to pay the caterer?

38. A cylindrical container is to be made of two different kinds of materials. The material for the top and bottom costs $10 per square foot and the material for the side wall costs $30 per square foot. If the container has radius r feet and height h feet, write an inequality stating that the total cost of the material from which the container is made does not exceed $1000. (Ignore waste material.)

39. A rectangular window 54 inches high is to have between 1296 and 1620 square inches of area. Write an inequality that must be satisfied by its width w.

40. An automobile salesman earns a base salary of $900 per month plus 2% commission on each car that he sells. How many $10,000 cars must he sell in order to earn at least $2000 per month?

41. A computer saleswoman is offered the choice of two compensation plans. Plan A pays a monthly salary of $1100 plus 2% commission on all sales. Plan B pays only a commission of 3% on all sales. She is to market only one model of computer, which sells for $20,000.
   a. Let x be the number of computers that she sells in a given month. Write an inequality that must be satisfied for Plan A to provide higher compensation than Plan B.
   b. For which sales levels x will Plan B provide greater compensation?

42. Refer to Exercise 41. If compensation plan B is changed to a commission of 4% on all sales, for which sales levels x will Plan B provide greater compensation?

43. A bag of a dozen oranges weighs 3 pounds. Each orange weighs at least 3 ounces. What is the most any single orange can weigh?

1.2 GRAPHS OF EQUATIONS

In order to represent graphically a relationship between two variables, we make use of the **Cartesian coordinate plane.** For example, Figure 2.1 illustrates the number q of television sets that a manufacturer can sell per month at selling price p if the variables p and q are related by the equation $p + 2q = 1000$. The significance of the point $P = (500, 250)$ is that the manufacturer will be able to sell 250 sets per month if the price per set is $500. Similarly, the interpretation of the point $(1000, 0)$ is that no one will buy the manufacturer's television sets if the price is increased to $1000. The purpose of this section is to describe the coordinate plane and its use in constructing graphs of equations.

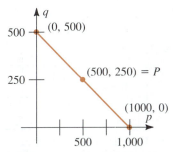

Figure 2.1 Graph of relationship $p + 2q = 1000$ between price p and monthly sales q.

The Coordinate Plane

Figure 2.2 shows that the coordinate plane is constructed by superimposing two number lines at right angles with the zero marks coincident at a point labeled (0, 0) called the **origin.** We have labeled the horizontal number line the x-axis and the vertical number line the y-axis.

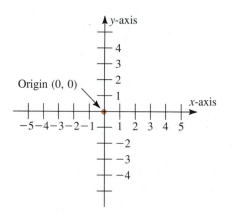

Figure 2.2 The coordinate plane.

With these axes in place we assign to each point P in the plane a unique ordered pair of numbers (a, b), called the **coordinates** of P. (See Figure 2.3.) The first number, a, is called the x-coordinate of P. It is the location where a vertical line through P meets the x-axis. Similarly, the y-coordinate, b, is the location where a horizontal line through P meets the y-axis. Figure 2.4 shows the coordinates of several particular points in the plane.

To find the point Q in the xy-plane corresponding to the given coordinates $Q = (c, d)$, we reverse the procedure described above, constructing a vertical line through $x = c$ on the x-axis and a horizontal line through $y = d$ on the y-axis. The two lines meet at the desired point $Q = (c, d)$.

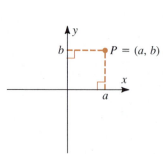

Figure 2.3 Method for assigning coordinates (a, b) to P.

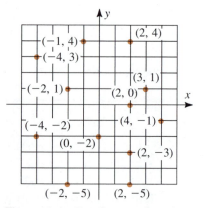

Figure 2.4 Coordinates of various points in the plane.

It is not necessary for the coordinate axes to be labeled x and y. Any two variables may be used. For example, economists often use the price p and demand q for a particular good or service, as in Figure 2.1. However, it is essential that the first number, a, in the coordinate pair (a, b) refer to the horizontal axis and that the second number, b, refer to the vertical axis.

Example 1 Table 2.1 shows, for each of the past five census years, the approximate percentage of U.S. citizens in the 25–29 age group who had completed four years of high school. Plot this data on a coordinate plane.

Table 2.1

t = years	1940	1950	1960	1970	1980
p = percentage	38	52	61	76	84

Solution: If we let n be the number of years after 1940, we can restate the data in Table 2.1 as in Table 2.2

Table 2.2

n = years after 1940	0	10	20	30	40
p = percentage	38	52	61	76	84

Then, using n to represent time on the horizontal axis and p to represent the percentage of citizens completing high school on the vertical axis, we plot the points

$P_1 = (0, 38)$
$P_2 = (10, 52)$
$P_3 = (20, 61)$
$P_4 = (30, 76)$
$P_5 = (40, 84)$

as illustrated in Figure 2.5. □

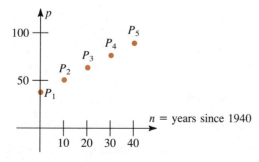

n = years since 1940

Figure 2.5 Percent of 25–29 year olds completing high school versus time.

Graphs of Equations

Example 1 shows how the coordinate plane can be used to illustrate relationships between two variables from a set of data. We shall be concerned primarily with using the coordinate plane to represent equations involving two variables. We do this by means of the *graph* of an equation.

DEFINITION 1

The **graph** of an equation involving the variables x and y is the set of all points $P = (x, y)$ whose coordinates satisfy the given equation.

One method of attempting to sketch the graph of an equation is to find and plot several points on the graph by the following method. First, select a particular number for one variable. Then insert this number into the equation and attempt to solve for the other variable. If successful, you will have determined a point on the graph. After finding several such points, you can attempt to sketch a curve through these points.

Example 2 Sketch the graph of the equation $xy = 1$.

Strategy

Solve the equation for y by multiplying both sides by $\dfrac{1}{x}$.

Solution

Beginning with the equation $xy = 1$, we find that

$$\left(\frac{1}{x}\right)xy = \left(\frac{1}{x}\right)1$$

so

$$y = \frac{1}{x}.$$

Select several numbers x. Find corresponding values of y.

Then if $x = 1$, $\quad y = \dfrac{1}{1} = 1$

if $x = 2$, $\quad y = \dfrac{1}{2}$

if $x = 3$, $\quad y = \dfrac{1}{3}$, etc.

Similarly if $x = -1$, $\quad y = \dfrac{1}{-1} = -1$

if $x = -2$, $\quad y = \dfrac{1}{-2} = -\dfrac{1}{2}$

if $x = -3$, $\quad y = \dfrac{1}{-3} = -\dfrac{1}{3}$, etc.

Plot the points (x, y) obtained, and sketch in the curve.

Thus for each number x the corresponding value of y is just the reciprocal, $\dfrac{1}{x}$. Finally, note that no y corresponds to $x = 0$ since division by zero is not defined. Figure 2.6 shows these points and the graph of the equation $xy = 1$. □

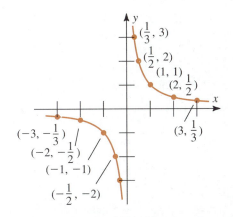

Figure 2.6 Graph of the equation $xy = 1$.

There are two difficulties with the method of plotting points to sketch graphs of equations. One is that it can be quite time consuming (although those with access to computers can write simple programs to plot large numbers of points quickly). The other difficulty is that the method is simply not very reliable. How do you know whether you have sketched the curve correctly between successive points?

One of the goals of the calculus is to develop more accurate techniques for analyzing graphs of equations. An important tool in doing this is an understanding of the equations associated with lines in the plane.

Equations for Lines

Suppose that ℓ is a nonvertical line in the plane. If $P_1 = (x_1, y_1)$ and $P_2 = (x_2, y_2)$ are distinct points on ℓ, the **slope** of ℓ is defined to be the ratio

$$m = \frac{y_2 - y_1}{x_2 - x_1} \tag{1}$$

(See Figure 2.7.)

For a given line the value of m is independent of the particular choices for (x_1, y_1) and (x_2, y_2). Notice that expression (1) for slope is not defined if $x_1 = x_2$. For this reason we say that *a vertical line has undefined slope*. Figure 2.8 shows the slopes of several lines through the origin.

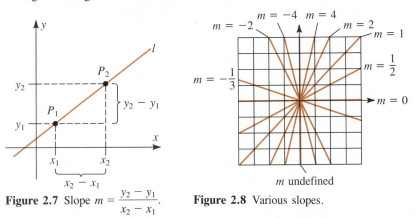

Figure 2.7 Slope $m = \dfrac{y_2 - y_1}{x_2 - x_1}$.

Figure 2.8 Various slopes.

Example 3 Find the slope of the line through the points

(a) $(-7, 2)$ and $(3, -3)$,   and
(b) $(-1, -4)$ and $(7, 16)$

Solution: From equation (1) for slope we have

(a) $m = \dfrac{-3 - 2}{3 - (-7)} = \dfrac{-5}{10} = -\dfrac{1}{2},$ and

(b) $m = \dfrac{16 - (-4)}{7 - (-1)} = \dfrac{20}{8} = \dfrac{5}{2}.$ □

Now suppose ℓ is a line with slope m that contains the point (x_1, y_1). To find an equation for ℓ, we let (x, y) denote the coordinate of any other point on ℓ. Then, by equation (1)

$$m = \frac{y - y_1}{x - x_1}.$$

Multiplying both sides of this equation by $x - x_1$ gives

$$y - y_1 = m(x - x_1) \qquad\qquad (2)$$

Equation (2) is called the **point-slope** form of the equation for ℓ since the slope m and the coordinates of one point on ℓ are required to write this equation.

Example 4 Find an equation for the line with slope 3 containing the point $(2, 4)$.

Solution: Using equation (2) with $m = 3$, $x_1 = 2$, and $y_1 = 4$, we obtain

$$y - 4 = 3(x - 2)$$

so

$$y - 4 = 3x - 6$$

or

$$y = 3x - 2.$$ □

(See Figure 2.9.)

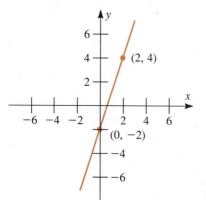

Figure 2.9 Line $y = 3x - 2$.

Example 5 Find an equation for the line through $(-3, 4)$ and $(1, -2)$.

Strategy

First, find the slope.

Solution

Since the slope is not given, we calculate it from the two given points using equation (1):

$$m = \frac{-2 - 4}{1 - (-3)} = \frac{-6}{4} = -\frac{3}{2}.$$

Then, use point-slope form of the line to find an equation.

Then, using the point $(x_1, y_1) = (-3, 4)$ and the point-slope equation (2), we have

$$y - 4 = \left(-\frac{3}{2}\right)(x - (-3))$$

or

$$y - 4 = -\frac{3}{2}(x + 3)$$

so

$$y - 4 = -\frac{3}{2}x - \frac{9}{2}.$$

Thus

$$y = -\frac{3}{2}x - \frac{1}{2}.$$

(See Figure 2.10.) □

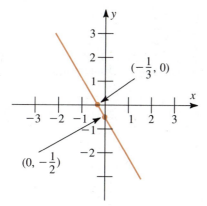

Figure 2.10 Line $y = -\frac{3}{2}x - \frac{1}{2}$.

If we solve equation (2) for y, we obtain

$$y = m(x - x_1) + y_1$$
$$= mx - mx_1 + y_1$$

or

$$y = mx + b \tag{3}$$

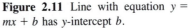

Figure 2.11 Line with equation $y = mx + b$ has y-intercept b.

where $b = -mx_1 + y_1$. Equation (3) is called the **slope-intercept form** of the equation for ℓ. This is because the number b is the **y-intercept,** or the y-coordinate of the point on ℓ with x-coordinate zero. (See Figure 2.11.)

Any equation of the form $Ax + By = C$ with $B \neq 0$ can be brought to the slope-intercept form (3), so it follows that its graph must be a line. Equations of this type are called **linear equations.**

Example 6 Graph the linear equation $6x + 2y = 8$.

Strategy	*Solution*
Solve for y to obtain slope-intercept form.	Solving for y, we obtain

$$6x + 2y = 8$$
$$2y = -6x + 8$$
$$y = -3x + 4.$$

Read off slope m and y-intercept b by comparing with equation (3).

This is a line with slope -3 and y-intercept 4. (See Figure 2.12.) $\qquad\square$

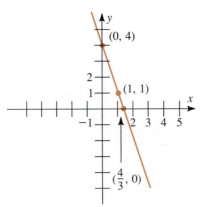

Figure 2.12 Graph of $y = -3x + 4$.

The *y*-intercept of a graph frequently has special meaning. The following example from economics shows how it may be interpreted as a "fixed cost."

Example 7 A manufacturer of garage door openers finds that the total cost C in dollars of producing x openers per week is given by the equation

$$C = 500 + 60x.$$

What is the significance of the constant 500?

Solution: If *no* openers are made, that is, if $x = 0$, the result is $C = 500$. Thus $500 is the fixed cost per week of being in business (having a factory, electricity, equipment, etc.) as opposed to the cost of materials and labor that go into the manufacture of a particular opener. Note in Figure 2.13 that only nonnegative values of x make sense. □

Figure 2.13 *y*-intercept is fixed cost.

Equations are often used as mathematical models to predict future values of a variable that changes with time. If the variable y satisfies an equation of the form $y = mx + b$, we say that it is a **linear model** for the variable y.

Example 8 The average weekly order for durable goods in the U.S. economy in 1967 was $26 billion. By 1982 this average had increased to $72 billion.

(a) Use these data to find a linear model for orders for durable goods in terms of time.
(b) What level of average weekly durable goods orders does this model predict for the year 1987?

Strategy
Name the variables.

Summarize data.

Solution

(a) We begin by letting y denote the average weekly order for durable goods and letting t denote time in years *since the year 1967*. We may then summarize the given data as follows:

t	0	15
y	26	72
(t, y)	(0, 26)	(15, 72)

Find the slope

$$m = \frac{y_2 - y_1}{t_2 - t_1}$$

and the y-intercept.

Write the linear equation in slope-intercept form.

A linear model is a straight line. From the data $(t_1, y_1) = (0, 26)$ and $(t_2, y_2) = (15, 72)$ we find that

$$\text{slope} = m = \frac{72 - 26}{15 - 0} = \frac{46}{15}$$

$$y\text{-intercept} = b = 26.$$

The linear model therefore has the equation

$$y = \frac{46}{15}t + 26.$$

(See Figure 2.14.)

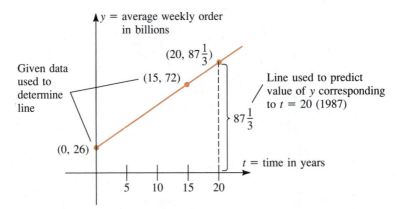

Figure 2.14 Linear model for average durable goods order level in U.S. economy.

Find the predicted value for

$t = 1987 - 1967$
$= 20$

by substituting into the equation.

(b) To find the value that this model predicts for the year 1987, we first note that this year corresponds to time $t_3 = 1987 - 1967 = 20$. We then substitute $t = 20$ into the model to obtain

$$y = \frac{46}{15}(20) + 26$$

$$= \frac{262}{3}$$

$$= 87\frac{1}{3} \text{ billion dollars.} \qquad \square$$

Of course, the method of Example 8 should be used only when the desired model is assumed to be linear, that is, when the data points lie on a straight line. In practice the actual model is often nonlinear (meaning a more general curve), or more than two data points are given, requiring the "best fitting" straight line to somehow be determined. We shall address both situations later in this text.

**Vertical and
Horizontal
Lines**

A vertical line has the property that all x-coordinates of points on the line are the same. It has an equation of the form $x = a$.

On a horizontal line all points have the same y-coordinates. A horizontal line has equation $y = b$. (It's just $y = mx + b$ with $m = 0$.)

Figure 2.15 shows the graphs of the vertical line $x = 2$ and the horizontal line $y = 1$.

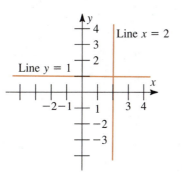

Figure 2.15 Horizontal and vertical lines.

**Parallel and
Perpendicular
Lines**

There are special relationships between the slopes of parallel or perpendicular lines.

THEOREM 2

Let $\ell_1: y = m_1x + b_1$ and $\ell_2: y = m_2x + b_2$ be equations for lines where neither m_1 nor m_2 is zero. Then

(i) ℓ_1 and ℓ_2 are *parallel* if and only if $m_1 = m_2$.
(ii) ℓ_1 and ℓ_2 are *perpendicular* if and only if $m_1m_2 = -1$.

Theorem 2 says that parallel lines have the same slope, while perpendicular lines have slopes that are negative reciprocals of each other.

Example 9 For the line $\ell_1: 3y - 9x = 12$ find an equation for

(a) the line ℓ_2 that is parallel to ℓ_1 and contains the point $(2, 0)$.
(b) the line ℓ_3 that is perpendicular to ℓ_1 and contains the point $(3, -2)$.

Strategy

First, find the slope of ℓ_1 by putting equation in slope-intercept form.

Take $m_2 = m_1$ by Theorem 1.

Solution

(a) Solving the ℓ_1 equation for y gives

$$3y - 9x = 12$$
$$3y = 9x + 12$$
$$y = 3x + 4.$$

The slope of ℓ_1 is therefore $m_1 = 3$. The slope of ℓ_2 must also be $m_2 = 3$ since the lines are parallel. Since $(2, 0)$ is on ℓ_2, an equation for ℓ_2 is

Find equation for ℓ_2 using point-slope form.

$$y - 0 = 3(x - 2)$$

or

$$y = 3x - 6.$$

Find m_3 from equation $m_1m_3 = -1$.

(b) Since ℓ_1 and ℓ_3 are perpendicular, we must have

$$m_3 = -\frac{1}{m_1} = -\frac{1}{3}.$$

Find equation for ℓ_3 using point-slope form.

Using this slope and the point $(3, -2)$, we find an equation for ℓ_3 to be

$$y - (-2) = -\frac{1}{3}(x - 3)$$

$$y + 2 = -\frac{1}{3}x + 1$$

$$y = -\frac{1}{3}x - 1.$$

Each of these lines is sketched in Figure 2.16. □

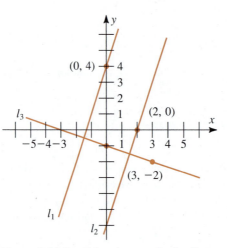

Figure 2.16 Parallel and perpendicular lines.

Exercise Set 1.2

1. Plot the following points in the coordinate plane.
   a. $(4, 2)$ **b.** $(-6, 3)$ **c.** $(-2, 0)$
   d. $(1, -3)$ **e.** $(-3, -3)$ **f.** $(6, -2)$

2. Plot the following points in the coordinate plane.
   a. $(1, -3)$ **b.** $(4, -2)$ **c.** $(-3, 1)$
   d. $(-1, -3)$ **e.** $(5, 1)$ **f.** $(1, 5)$

In each of Exercises 3–12 draw a rough sketch of the graph of the given equation by first plotting several points.

3. $3x - y = 5$ **4.** $y = x^2$

5. $y = \sqrt{x}$ **6.** $y - x^2 = 4$

7. $x - y = 1$ **8.** $x = y^2$

9. $y + \sqrt{x} = 4$

10. $y = \dfrac{4}{x}$

11. $y = \dfrac{1}{x - 2}$

12. $2x^2 + y = 3$

13. True or false? Every line has a slope.

14. True or false? Every real number can be the slope of a line.

15. Find a if the line through $(2, 4)$ and $(-2, a)$ has slope 3.

16. Find b if the line through $(b, 1)$ and $(1, -5)$ has slope 2.

In Exercises 17–31 find an equation for the line determined by the given information.

17. Slope 4 and y-intercept -2

18. Slope -2 and y-intercept 5

19. Slope zero and y-intercept -5

20. Through $(-1, 6)$ and $(4, 12)$

21. Through $(4, 1)$ and with slope 7

22. Through $(1, 3)$ and with slope -3

23. x-intercept -3 and y-intercept 6

24. Through $(-3, 5)$ and vertical

25. Through $(-3, 5)$ and horizontal

26. Through $(-2, 4)$ and $(-6, 8)$

27. Through $(0, 2)$ and $(-1, -4)$

28. Through $(1, 4)$ and parallel to the line with equation $2x - 6y + 5 = 0$.

29. Through $(5, -2)$ and parallel to the line with equation $x - y = 2$.

30. Through $(1, 3)$ and perpendicular to the line with equation $3x + y = 7$

31. Through $(4, -1)$ and perpendicular to the line through the points $(-2, 5)$ and $(-1, 9)$.

In Exercises 32–41 find the slope and y-intercept of the line determined by the given equation. Graph the line.

32. $x = 7 - y$

33. $3x - 2y = 6$

34. $x + y + 3 = 0$

35. $2x = 10 - 3y$

36. $y = 5$

37. $y - 2x = 9$

38. $x = 4$

39. $y = x$

40. $3x - y = 9$

41. $9x - 9y = 27$

42. A state has an income tax of 5% on all income over $5000.
   a. Write an equation that determines a person's tax T in terms of income i for a person earning more than $5000.
   b. At what income level will an individual owe $800 in state income taxes?

43. The cost C per unit time of producing flashlights for a particular manufacturer is known to be linearly related to the production level. That is, $C = mx + b$, where x is the number of flashlights produced per week and m and b are constants. The cost of producing 20 flashlights per week is $200, while the cost of producing 100 flashlights per week is $600.
   a. Find m and b.
   b. Find the cost of producing 150 flashlights per week.
   c. What is the manufacturer's fixed weekly cost?

44. Average per capita income in the United States in 1961 was $9400. By 1971 it had increased to $12,800. Assuming per capita income I to be linearly related to time, find the projected per capita income in
   a. 1991   **b.** 1996.

45. The consumer price index rose from a level of $C_1 = 100$ dollars in 1967 to a level of $C_2 = 293$ dollars in 1982.
   a. Based on these data find a linear model for C in terms of time. (Let 1967 be $t = 0$.)
   b. What value of the consumer price index does this model predict for the year 1987?

46. Incidence of severe disabilities in the U.S. population rose from 21 per 1000 in 1966 to a level of 36 per 1000 in 1976.
   a. Based on these data find a linear model for severe disabilities D in terms of time. (Let 1966 be $t = 0$.)
   b. What level of severe disabilities does this model predict for the year 1986?

47. Demand for leaded gasoline in the United States fell from 70 billion gallons in 1977 to 43 billion gallons in 1983.
   a. Based on these data find a linear model for demand D for leaded gasoline in terms of time. (Let 1977 be $t = 0$.)
   b. What demand for leaded gasoline does this model predict for 1987?
   c. When does this model predict that the demand for leaded gasoline will reach zero?

1.3 DISTANCE IN THE PLANE; EQUATIONS FOR CIRCLES

Figure 3.1 The Pythagorean theorem: $c^2 = a^2 + b^2$.

We will sometimes need to calculate the distance between two points P and Q in the coordinate plane. By this distance we mean the length of the line segment ℓ joining the points P and Q. Figure 3.2 illustrates the fact that if P has coordinates $P = (x_1, y_1)$ and Q has coordinates $Q = (x_2, y_2)$, then the line segment ℓ is the hypotenuse of the right triangle with vertices P, Q, and $R = (x_2, y_1)$. By the **Pythagorean theorem** (see Figure 3.1) the lengths of the sides of this triangle satisfy the equation

$$D^2 = (x_2 - x_1)^2 + (y_2 - y_1)^2.$$

Taking square roots of both sides of this equation gives the desired expression for D. (See Figure 3.2.)

DEFINITION 2

The distance between the two points (x_1, y_1) and (x_2, y_2) in the coordinate plane is

$$D = \sqrt{(x_2 - x_1)^2 + (y_2 - y_1)^2}. \tag{1}$$

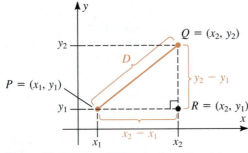

Figure 3.2 Distance D between (x_1, y_2) and (x_2, y_2) is $D = \sqrt{(x_2 - x_1)^2 + (y_2 - y_1)^2}$.

Example 1 Find the distance between the points $P = (-3, 2)$ and $Q = (5, -4)$.

Solution: Using equation (1) with $(x_1, y_1) = (-3, 2)$ and $(x_2, y_2) = (5, -4)$, we obtain

$$D = \sqrt{(5 - (-3))^2 + (-4 - 2)^2}$$
$$= \sqrt{8^2 + (-6)^2}$$
$$= \sqrt{100}$$
$$= 10. \qquad \Box$$

Note that the distance D between any two points will always be a nonnegative number.

Equations for Circles

The definition of distance in the plane gives a definition for circles that allows us to find their corresponding equations.

DEFINITION 3

A **circle** is the set of all points in the plane lying at a fixed distance r (called the **radius**) from a fixed point P (called the **center**).

Using the distance formula in equation (1) we can find an equation for the circle with center (h, k) and radius r. If we let $P = (x, y)$ be any point on the circle, the condition of Definition 3 that the distance between the point $P = (x, y)$ and the center (h, k) equal r can be written, using equation (1), as

$$\sqrt{(x - h)^2 + (y - k)^2} = r.$$

Squaring both sides of this equation then gives

$$(x - h)^2 + (y - k)^2 = r^2.$$

For easy reference we restate this result as a theorem.

THEOREM 3

The circle with center (h, k) and radius r has equation

$$(x - h)^2 + (y - k)^2 = r^2. \tag{2}$$

Equation (2) is called the **standard form** for the equation of a circle, since the radius r and the coordinates (h, k) of the center are easy to identify from this equation. (See Figure 3.3.)

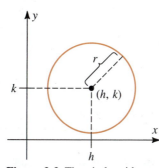

Figure 3.3 The circle with center (h, k) and radius r has equation

$$(x - h)^2 + (y - k)^2 = r^2.$$

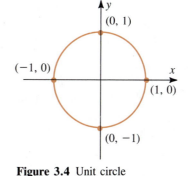

Figure 3.4 Unit circle

$$x^2 + y^2 = 1.$$

Example 2 The circle with center $(h, k) = (0, 0)$ and radius $r = 1$ has equation

$$x^2 + y^2 = 1$$

since $h = 0$, $k = 0$, and $r = 1$ in equation (2) gives this result. This is called the **unit circle.** (See Figure 3.4.) □

Example 3 An equation for the circle with center $(-4, 3)$ and radius 5 is found by setting $h = -4$, $k = 3$, and $r = 5$ in equation (2). We obtain

$$(x - (-4))^2 + (y - 3)^2 = 5^2$$

or

$$(x + 4)^2 + (y - 3)^2 = 25$$

which we can write as

$$(x^2 + 8x + 16) + (y^2 - 6y + 9) = 25$$

or

$$x^2 + 8x + y^2 - 6y = 0. \qquad \square$$

Completing the Square

In order to determine whether an equation involving quadratic (i.e., second degree) expressions in both x and y corresponds to a circle, we must attempt to bring the equation into the form of equation (2). Doing so requires the technique of **completing the square,** which goes as follows.

Recall that the result of squaring the binomial $(x + a)$ is the quadratic expression

$$(x + a)^2 = x^2 + 2ax + a^2. \tag{3}$$

The technique of completing the square answers just the opposite question: What constant, when added to the expression $x^2 + bx$, will produce a perfect square, as in equation (3)? By examining the right side of equation (3), you can see that *the constant term, a^2, in a perfect square is the square of half the coefficient of x.* We therefore "complete the square on $x^2 + bx$" as follows:

$$x^2 + bx = \left[x^2 + bx + \left(\frac{b}{2} \right)^2 \right] - \left(\frac{b}{2} \right)^2$$

$$= \left(x + \frac{b}{2} \right)^2 - \frac{b^2}{4}.$$

It is important to note that whatever constant is added to the expression $x^2 + bx$ must also be subtracted from the result, so that the values of the expression are not changed.

Example 4 Here are three examples of completing the square.

(a) $x^2 + 6x = (x^2 + 6x + 3^2) - 3^2$
$= (x + 3)^2 - 9$

(b) $x^2 - 8x = [x^2 - 8x + (-4)^2] - (-4)^2$
$= (x - 4)^2 - 16$

(c) $x^2 + \dfrac{3}{4}x = \left[x^2 + \dfrac{3}{4}x + \left(\dfrac{3}{8} \right)^2 \right] - \left(\dfrac{3}{8} \right)^2$

$= \left(x + \dfrac{3}{8} \right)^2 - \dfrac{9}{64}. \qquad \square$

The next example shows how the technique of completing the square is used to identify the equation for a circle.

Example 5 Describe the graph of the equation

$$x^2 + y^2 + 6x - 2y + 6 = 0.$$

Strategy	*Solution*
Complete the square on x and y terms and try to bring equation into form of equation (2).	We write the given equation as

$$(x^2 + 6x) + (y^2 - 2y) + 6 = 0. \tag{4}$$

To complete the square on the first term, we add the square of half the coefficient of x, that is, we add $(\frac{6}{2})^2 = 9$. This gives

Completing the square in x.

$$(x^2 + 6x) = (x^2 + 6x + 9) - 9 = (x + 3)^2 - 9. \tag{5}$$

Similarly, we complete the square on the second term as

Completing the square in y.

$$(y^2 - 2y) = (y^2 - 2y + 1) - 1 = (y - 1)^2 - 1. \tag{6}$$

Substituting the expressions in equations (5) and (6) into equation (4) gives the equation

Combining the results.

$$[(x + 3)^2 - 9] + [(y - 1)^2 - 1] + 6 = 0$$

or

$$(x + 3)^2 + (y - 1)^2 = 4.$$

Compare with equation (2).

The graph is therefore a circle with center $(-3, 1)$ and radius $r = \sqrt{4} = 2$. □

Exercise Set 1.3

In each of Exercises 1–8 find the distance between the given points.

1. $(2, -1)$ and $(0, 2)$ **2.** $(3, 1)$ and $(1, 3)$

3. $(0, 2)$ and $(1, -9)$ **4.** $(1, -3)$ and $(6, 6)$

5. $(1, 1)$ and $(-1, -1)$ **6.** $(-2, -2)$ and $(1, 2)$

7. $(-4, 2)$ and $(-3, -5)$ **8.** $(0, -9)$ and $(-6, 0)$

9. Verify that the **midpoint** of the line segment joining the points (x_1, y_1) and (x_2, y_2) has coordinates $\left(\dfrac{x_1 + x_2}{2}, \dfrac{y_1 + y_2}{2}\right)$. (*Hint:* Use the distance formula.)

10. Use the result of Exercise 9 to find the midpoints of the line joining the following pairs of points.
   a. $(0, 0)$ and $(0, 6)$
   b. $(1, 3)$ and $(4, 7)$
   c. $(-1, 2)$ and $(7, 4)$

In Exercises 11–15 find an equation for the circle with the stated properties.

11. Center $(0, 0)$ and radius 3

12. Center $(2, 4)$ and radius 6

13. Center $(4, 3)$ and radius 2

14. Center $(-2, 4)$ and radius 3

15. Center $(-6, -4)$ and radius 10

In Exercises 16–21 complete the square on the given expression.

16. $x^2 + 6x$ **17.** $x^2 - 10x$

18. $t - t^2$ **19.** $8x - x^2$

20. $x^2 - 6x + 5$ **21.** $3 - 2t - 2t^2$

In each of Exercises 22–27 find the center and radius of the circle whose equation is given.

22. $x^2 - 2x + y^2 - 8 = 0$

23. $x^2 - 2x + y^2 + 6y - 12 = 0$

24. $x^2 + y^2 + 4x + 2y - 11 = 0$

25. $x^2 + 14x + y^2 - 10y + 70 = 0$

26. $x^2 - 6x + y^2 - 4y + 8 = 0$

27. $x^2 + y^2 - 2x - 6y + 3 = 0$

28. Does the point $(1, 1)$ lie inside or outside the circle with equation $x^2 - 4x + y^2 + 6y = 3$? Why or why not?

29. Does the point $(2, 1)$ lie inside or outside the circle with equation $x^2 - 2x + y^2 - 4y + 1 = 0$?

30. Find an equation for the line tangent to the circle with equation $x^2 + y^2 - 6x - 8y = 0$ at the origin. (*Hint:* The tangent to a circle at point P is perpendicular to the radius of the circle at P.)

1.4 FUNCTIONS AND THEIR GRAPHS

The linear equation $y = mx + b$ is one example of how two variables can be related. The concept of a *function* is helpful in studying more general relationships between quantities.

A function is any rule between two kinds of objects that assigns one (and *only* one) object of the second kind to each object of the first kind. Figure 4.1 illustrates a rule that

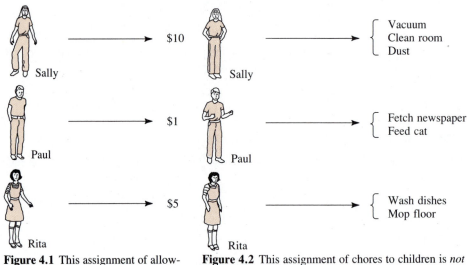

Figure 4.1 This assignment of allowances to children is a function.

Figure 4.2 This assignment of chores to children is *not* a function.

assigns a weekly allowance to each child. It is a function since exactly one amount is assigned to each child. Figure 4.2 describes a rule that is *not* a function, since it assigns *more* than one task to each child.

Here are some additional examples to illustrate when rules between quantities are functions.

Example 1 The rule that determines the *age* of each *student* in your calculus class is a function. It produces one age for each student. □

Example 2 The list that identifies the senators for each state in the United States is *not* a function since it identifies *two* senators for each state. (However, a list that assigns a state to each senator *is* a function, since a senator represents only one state.) □

Example 3 The rule that assigns a sales tax to each purchase made in the state of Ohio is a function. It assigns exactly one tax to each sales amount. □

Example 4 The list that identifies the airlines servicing the major American cities is *not* a function. It identifies more than one airline for each city. □

These very simple examples convey the reason why we say that a linear equation of the form $y = mx + b$ is a function (from x's to y's). That is because once we have selected a particular number x, the right-hand side of the equation produces one and only one value y.

For example, in the linear equation $y = 2x - 1$ we find that

$$\text{if } x = 0, \qquad y = 2(0) - 1 = -1$$
$$\text{if } x = 2, \qquad y = 2(2) - 1 = 3$$
$$\text{if } x = -1, \qquad y = 2(-1) - 1 = -3.$$

(See Figure 4.3.)

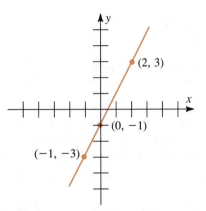

Figure 4.3 The linear equation $y = 2x - 1$ is a function. Each x corresponds to exactly one y.

However, not all equations involving two variables are functions. For example, in the equation

$$x^2 + y^2 = 4$$

we observe that

$$\text{if } x = 0, y^2 = 4, \quad \text{so} \quad y = 2 \ \ or \ \ y = -2$$
$$\text{if } x = 1, y^2 = 4 - 1, \quad \text{so} \quad y = \sqrt{3} \ \ or \ \ y = -\sqrt{3}.$$

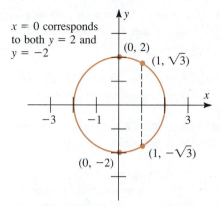

$x = 0$ corresponds to both $y = 2$ and $y = -2$

Figure 4.4 The equation $x^2 + y^2 = 4$ is *not* a function. Some x's correspond to *two* y's.

Figure 4.4 illustrates what is going wrong here. The graph of the equation $x^2 + y^2 = 4$ is a circle. It has the property that any x in the interval $(-2, 2)$ is associated with *two* values of y.

At this point a formal definition of function will help make our discussion of functions more precise.

DEFINITION 4   A **function** from set A to set B is a rule that assigns to each element of set A one and only one element of set B. Set A is called the **domain** of the function.

In other words, a function is an input-output device. For each input from the domain, the function produces a unique output. The terminology ''domain'' helps to distinguish the input set from the outputs. Referring to our previous examples, we see that

 (i) The domain of the allowance function in Figure 4.1 is the set of children, {Sally, Paul, Rita}.
 (ii) The domain of the age function in Example 1 is the set of students in your calculus class.
(iii) The domain of the sales tax function in Example 3 is the set of all possible purchase totals, $[0, \infty)$.
(iv) The domain of the linear function $y = 2x - 1$ is the set of all real numbers, $(-\infty, \infty)$.

Notation for Functions

We usually denote a function by a letter, such as f, and we write $f(x)$ to mean the *value* (output) that the function f assigns to the input x. Thus for the linear equation $y = 2x - 1$ we write

$$f(x) = 2x - 1$$

to emphasize that the numbers y are indeed values of a function, and we obtain particular values of this function as follows:

$$f(0) = 2(0) - 1 = -1$$
$$f(6) = 2(6) - 1 = 11$$
$$f(-3) = 2(-3) - 1 = -7$$

etc.

Another example of a function is the reciprocal function $f(x) = \dfrac{1}{x}$. $\left(\text{The graph of } y = \dfrac{1}{x} \text{ was sketched in Figure 2.6.}\right)$ Note that $x = 0$ is not in the domain of this function since division by zero is not defined. The domain of the function $f(x) = \dfrac{1}{x}$ consists of the two intervals $(-\infty, 0)$ and $(0, \infty)$.

Equations for Functions

Most of the functions with which we shall be concerned will be specified by equations such as

$$f(x) = x^3 - 6x + 6 \qquad (f \text{ as a function of } x)$$

or

$$f(t) = 32 - 16t^2 \qquad (f \text{ as a function of } t)$$

or

$$f(s) = \frac{s^2 - 1}{s + 1}. \qquad (f \text{ as a function of } s)$$

In such cases the variable appearing on the right-hand side of the equation is referred to as the **independent** variable. The advantage of using the function notation $f(\,)$ is that it specifies clearly the independent variable.

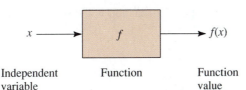

| Independent variable | Function | Function value |

Figure 4.5 A function produces a unique output $f(x)$ from an input x. $f(x)$ is the *value* that the function assigns to the number x.

Figure 4.5 illustrates the function concept and terminology.

When a function is specified by an equation, the domain, unless otherwise specified, is assumed to be all numbers for which the expression defining the function makes sense.

Example 5 For the functions $f(x) = x^2 - 2x + 1$ and $g(x) = \sqrt{x + 3}$ we have

(a) $f(5) = 5^2 - 2 \cdot 5 + 1 = 25 - 10 + 1 = 16$

(b) $f(-1) = (-1)^2 - 2(-1) + 1 = 1 + 2 + 1 = 4$

(c) $g(6) = \sqrt{6 + 3} = \sqrt{9} = 3$

(d) $g(-1) = \sqrt{-1 + 3} = \sqrt{2}$.

However, $g(-5) = \sqrt{-5 + 3} = \sqrt{-2}$ is not defined. □

Example 6 The domain of the function $f(x) = 1 + 2x - x^2$ is all real numbers, $(-\infty, \infty)$, since the right-hand side is defined for all x. □

Example 7 Find the domain of the function $f(x) = \sqrt{x - 2}$.

Solution: Since square roots of negative numbers are not defined, we must have $x - 2 \geq 0$, or $x \geq 2$. The domain is $[2, \infty)$. □

Example 8 Find the domain of the function $f(t) = \dfrac{1}{t^2 - t - 2}$.

Solution: The denominator can be factored as

$$t^2 - t - 2 = (t - 2)(t + 1)$$

so we can write

$$f(t) = \frac{1}{(t - 2)(t + 1)}.$$

The right-hand side is not defined when $t = 2$ or $t = -1$, since these numbers produce a factor of zero in the denominator. The domain consists of all other numbers t. That is, the intervals $(-\infty, -1)$, $(-1, 2)$, and $(2, \infty)$. □

Range of a Function

Generally speaking, the range of a function is the set of all its possible values. A more precise definition is the following:

DEFINITION 5   If f is a function with domain A, the *range* of f is the set of values $\{f(x) \mid x \in A\}$.

Determining the range of a function is sometimes a matter of "inspection," taking into account what values can be produced by selecting various numbers in the domain.

Example 9 The range of the squaring function $f(x) = x^2$ is $[0, \infty)$, since all nonnegative numbers are squares (of their square roots).

Example 10 The range of the function $f(x) = \dfrac{1}{x}$ is the pair of intervals $(-\infty, 0)$ and $(0, \infty)$. We leave the explanation to you.

Example 11 Find the range of the function $f(x) = x^2 - 4x + 1$.

Strategy

Complete the square to analyze f.

Solution

We find that

$$
\begin{aligned}
f(x) &= x^2 - 4x + 1 \\
&= [x^2 - 4x + 4] - 4 + 1 \\
&= (x - 2)^2 - 3.
\end{aligned}
$$

Determine the largest and smallest values of the resulting expression.

Since the smallest the expression $(x - 2)^2$ can be is zero, the smallest value of this function is $f(2) = 0 - 3 = -3$. Since the factor $(x - 2)^2$ has arbitrarily large values, the range of f is therefore $[-3, \infty)$.

Graphs of Functions

When functions are specified by equations, such as $f(x) = x^2 - 4$, we can graph the function on the coordinate plane by plotting the independent variable on the horizontal axis and values of the function on the vertical axis. That is, we plot the points $(x, f(x))$. Figures 4.6 and 4.7 show the graphs of the functions in Example 5.

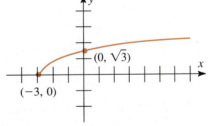

Figure 4.6 Graph of $f(x) = x^2 - 2x + 1$.

Figure 4.7 Graph of $f(x) = \sqrt{x + 3}$.

Since a function assigns only one value to each x in its domain, the graph of a function can have only one point corresponding to each x-coordinate. That is why a circle cannot be the graph of a function. (See Figure 4.8.) Since all points on a vertical line $x = a$ have the *same* x-coordinates, we have the following property (see Figure 4.9):

> **Vertical Line Property:** The graph of a function must have the property that any vertical line can intersect the graph at most once.

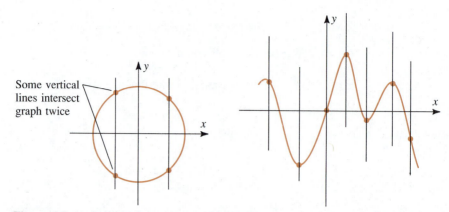

Figure 4.8 A circle does not have the vertical line property.

Figure 4.9 Graph of a function must have the vertical line property.

Applications to Business and Economics

Several types of functions are useful in modelling various aspects of economic activity. For example, suppose that the independent variable x represents the number of items of a particular kind that a manufacturing company produces per week. The variable x is then referred to as the *production level* for that item. Here are three important functions of production level:

(i) The cost function, $y = C(x)$, gives the total cost of producing and marketing x items per unit of time.

(ii) The revenue function, $y = R(x)$, gives the total amount of money received from the sale of x items.

(iii) The profit function, $y = P(x)$, gives the profit for the company in producing and selling x items per unit time.

We will assume that all items manufactured by the company in a given unit of time are sold during that same unit of time, so that

$$\text{profit} = \text{revenue} - \text{cost}$$

or

$$P(x) = R(x) - C(x)$$

(1)

The following examples involve these three functions.

Example 12 The cost to a company of producing goods usually consists of two types of costs. The *fixed costs* are the ongoing costs of maintaining a manufacturing facility (cost of building, electricity, managers' salaries, advertising, etc.). The *variable costs* are the costs directly assignable to the production of a particular item (cost of materials, direct cost of labor, shipping costs, etc.). Thus

$$\text{total costs} = \text{fixed costs} + \text{variable costs}.$$

(2)

If a bicycle manufacturer has fixed costs of $500 per week and variable costs of $30 per bicycle, find an equation for the total cost of producing x bicycles per week.

Solution: Since the variable costs are $30 per bicycle, the total variable costs in producing x bicycles are $30x$. With fixed costs of $500 per week, equation (2) gives total costs as

$$C(x) = 500 + 30x, \qquad x \geq 0 \tag{3}$$

dollars per week at production level x. This cost function is graphed in Figure 4.10. Note that the domain of C is $[0, \infty)$, since negative production levels are not defined. This is the meaning of the inequality $x \geq 0$ in line (3). □

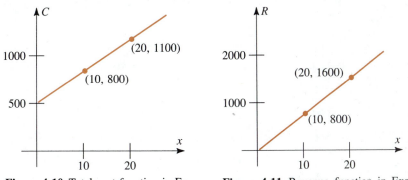

Figure 4.10 Total cost function in Example 12.

Figure 4.11 Revenue function in Example 13.

Example 13 Suppose the bicycle manufacturer in Example 12 sells the bicycles for $80 each. Find the revenue function R associated with output level x.

Solution: Total revenue is simply price per bicycle ($80) multiplied by the number of bicycles, x. Thus

$$R(x) = 80x, \qquad x \geq 0$$

dollars. The graph of the revenue function appears in Figure 4.11. □

Example 14 The bicycle manufacturer in Examples 12 and 13 wants to make a profit, of course.

(a) Find the profit function P.
(b) Find the output levels x at which $P(x)$ is positive.

Solution: Using Equation (1) and the results of Examples 12 and 13, we find that

$$\begin{aligned} P(x) &= R(x) - C(x) \\ &= 80x - (500 + 30x) \\ &= 50x - 500, \qquad x \geq 0. \end{aligned}$$

Thus $P(x)$ will be positive when

$$50x - 500 > 0$$

or

$$50x > 500$$

or

$$x > \frac{500}{50} = 10 \text{ bicycles per week.}$$

That is, the bicycle manufacturer must produce and sell 11 or more bicycles per week in order to show a profit. All three functions $C, R,$ and P are graphed on the same coordinate axes in Figure 4.12. □

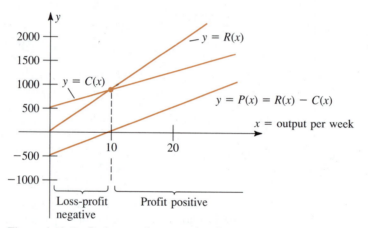

Figure 4.12 Profit, loss, and revenue functions.

Exercise Set 1.4

In Exercises 1–8 identify the given rule as either a function or not necessarily a function, and explain your reasoning.

1. The assigning of postage charges to packages

2. The calculation of federal income tax

3. Listing the names of your professors' children

4. The matching of companies with the products they produce

5. The calculation of average daily temperature

6. Measuring monthly rainfall

7. A listing of textbooks required for various college courses

8. A monthly electric bill for your home

9. For the function $f(x) = 7 - 3x$ find
   a. $f(0)$ **b.** $f(3)$
   c. $f(-4)$ **d.** $f(a)$

10. For the function $f(x) = \dfrac{1}{x - 7}$ find
   a. $f(8)$ **b.** $f(6)$
   c. $f(-1)$ **d.** $f(b)$

11. For the function $f(x) = \dfrac{x + 3}{x^2 - x + 2}$ find

   a. $f(0)$ **b.** $f(-3)$
   c. $f(1)$ **d.** $f(-1)$

12. For the function $f(x) = \dfrac{\sqrt{6 - x}}{1 + x^2}$ find

   a. $f(2)$ **b.** $f(6)$
   c. $f(-3)$ **d.** $f(-10)$

In Exercises 13–20 label the graph as either a function or not a function. (Use x as the independent variable.)

13.

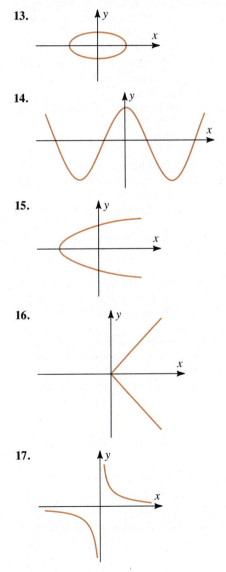

14.

15.

16.

17.

18.

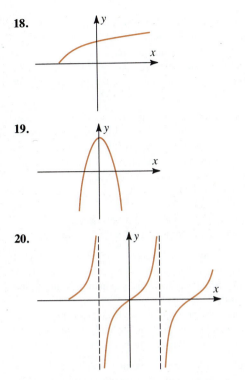

19.

20.

In Exercises 21–28 state the domain of the given function.

21. $f(x) = \dfrac{1}{x - 7}$ **22.** $f(x) = x - \sqrt{x}$

23. $f(s) = \dfrac{s - 1}{s + 1}$ **24.** $f(t) = \dfrac{t}{1 + \sqrt{t}}$

25. $g(x) = \sqrt{1 - x^2}$ **26.** $f(x) = \dfrac{1}{x^2 - 3x - 4}$

27. $f(x) = x^3 - x$ **28.** $f(x) = \sqrt{(x - 1)(x + 1)}$

29. What is the range of the function $f(x) = x^2 + 6x + 7$?

30. What is the range of the function in Exercise 27?

31. The cab fare from an airport to a nearby city is $2 per mile plus 50¢ for a bridge toll. Write the cost C as a function of the mileage x.

32. An automobile rents for $10 per day plus 20¢ per mile. Write a function $y = C(x)$ for the cost per day as a function of mileage.

33. A producer of lawnmowers has fixed costs of $1000 per week. The producer experiences variable costs of $70 per lawnmower. Express the producer's total weekly costs C as a function of the weekly production level x.

34. Refer to Exercise 33. If the producer of lawnmowers can sell his mowers for $150 each,
  a. Find the revenue function R.
  b. Find the manufacturer's weekly profit function P.
  c. How many lawnmowers must the producer sell per week in order to show a profit?

35. Market research indicates that the demand D for a certain type of umbrella is related to its selling price by the function $D(p) = 800 - 40p$, $0 \le p \le 20$. Here p is the selling price in dollars and $D(p)$ is the number of umbrellas that will be sold per week at price p.
  a. How many umbrellas will be sold per week at the price of $5?
  b. At what price will demand be greatest?
  c. At what price will demand have become zero?

36. A tool rental agency can rent 500 jackhammers per year at a daily rental of $30. For each one dollar increase in the daily rental fee 20 fewer jackhammers are rented per year.
  a. Express the yearly revenues from jackhammer rentals, R, as a function of the daily rental price p.
  b. At what price will revenues reach zero?

37. A hotel has 400 identical rooms. At daily rates below $20 per room all rooms will be filled. For each dollar increase in daily rate above $20, ten rooms will remain vacant. Express the daily revenues R as a function of the daily room rent p.

38. A supermarket manager discovers that after x hours of experience, a new employee at the checkout register can handle $N(x)$ customers per hour where

$$N(x) = 32 - \frac{24}{\sqrt{x + 1}}, \qquad x \ge 0.$$

  a. How many customers can a new employee (no experience) handle per hour?
  b. According to this model, does the employee's speed ever stop improving?

39. A typing instructor observes that after n weeks of instruction a student can, on average, type $W(n)$ words per minute where

$$W(n) = \sqrt{400 + 80n}.$$

  a. Graph this "learning curve" by plotting points.
  b. How many words per minute can a new student type on the first day ($n = 0$) of class?

40. A bank pays 8% interest on savings accounts, compounded once per year. A depositor places $1000 in a savings account and makes no further deposits. Express the amount in the account, P, as a function of the number of years that the funds have been on deposit.

41. Ajax Company, an underwriter of homeowners' insurance charges an annual premium of $75 plus 0.5% of the face amount of the policy.
  a. Express the annual premium P as a function of the face amount of the policy, x.
  b. What is the premium for a $40,000 policy?

42. Refer to Exercise 41. Bjax Insurance Company, a competing underwriter, charges an annual premium of $200 plus 0.3% of the face amount of the policy.
  a. Which company has a lower premium for a $40,000 policy?
  b. For what policy face amount are the two companies' premiums equal?

43. Carbon dioxide levels at the South Pole are reported to have risen from 315 parts per million in 1958 to 322 parts per million in 1966 and to 329 parts per million in 1974.
  a. Using $t = 0$ to correspond to the year 1958, find a linear function giving this carbon dioxide level as a function of time.
  b. What level does this function predict for the year 1990?

1.5 SOME SPECIAL TYPES OF FUNCTIONS

In Section 1.4 we presented several economic applications of linear functions—those of the form $f(x) = mx + b$. Many other types of functions are also important in applications. For example, a cost function $y = C(x)$ for a producer of a good or service is often nonlinear. Figure 5.1 shows one typical cost function. Note that following a short startup period, costs rise slowly as production increases, due primarily to efficiencies that can be

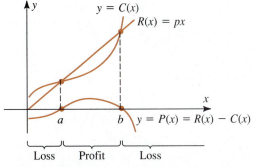

Figure 5.1 A nonlinear function $C(x)$ giving the cost of producing x items per unit time.

Figure 5.2 The profit function $P(x) = R(x) - C(x)$ is positive only in the interval (a, b). Losses result for production levels in $[0, a)$ or in (b, ∞).

achieved. But eventually costs rise rapidly as production capacity becomes strained (additional equipment is required, overtime wages are required, etc.).

Figure 5.2 shows the result of combining the cost function in Figure 5.1 with a linear revenue function $R(x) = px$ to form the profit function $y = P(x) = R(x) - C(x)$. Unlike the linear profit function of Example 14, Section 1.4, this profit function actually becomes negative (i.e., shows a loss) for higher production levels.

The purpose of this section is to review several types of nonlinear functions that we will encounter in various applications later on.

Power Functions

Among the simplest types of functions are the power functions of the form

$$f(x) = x^n$$

where n is a positive integer. Figures 5.3 and 5.4 show how we can sketch the graphs of the power functions $f(x) = x^2$ and $g(x) = x^3$ by first plotting particular points.

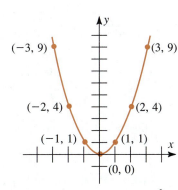

Figure 5.3 Graph of $f(x) = x^2$.

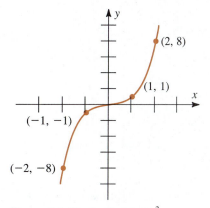

Figure 5.4 Graph of $g(x) = x^3$.

Figures 5.5 and 5.6 show the graphs of the power function $f(x) = x^4$ and $g(x) = x^5$. (What can you conjecture about the shape of the graph of the function $h(x) = x^n$ for n even versus n odd?)

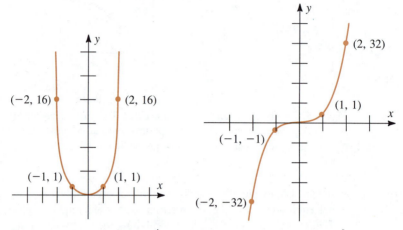

Figure 5.5 Graph of $f(x) = x^4$.   **Figure 5.6** Graph of $f(x) = x^5$.

Figure 5.7 shows the graphs of several functions of the form $f(x) = ax^2$. Note that each of these resembles the graph of $f(x) = x^2$, except that the y-coordinates have each been multiplied by the constant a. Note that the graph of $f(x) = ax^2$ opens upward when $a > 0$ and downward when $a < 0$.

x	0	1	−1	2	−2	3	−3
$f(x) = x^2$	0	1	1	4	4	9	9

x	0	1	−1	2	−2
$g(x) = x^3$	0	1	−1	8	−8

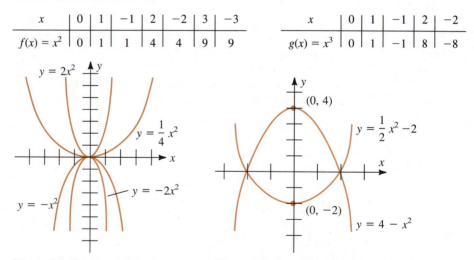

Figure 5.7 Functions of the form $f(x) = ax^2$.

Figure 5.8 Functions of the form $g(x) = ax^2 + b$.

Quadratic Functions

Each of the graphs in Figure 5.7 is a **parabola.** In general, a parabola is a curve that, when properly positioned in the coordinate plane, is the graph of a function of the form $f(x) = ax^2$. We shall not make a detailed study of parabolas, but later on we shall be able to show that the graph of any function of the form $f(x) = Ax^2 + Bx + C$ is a parabola. (Figure 5.8.) Such functions are called **quadratic** functions.

Polynomials

Quadratic functions are just one type of more general functions called **polynomials.** These are functions of the form

$$f(x) = a_n x^n + a_{n-1} x^{n-1} + \cdots + a_1 x + a_0$$

where $a_0, a_1, \ldots, a_n$ are constants, $a_n \neq 0$, and n is a positive integer. Like quadratic functions, polynomials are defined for all numbers x. The integer n is called the **degree** of the polynomial. Here are some examples:

$$\begin{array}{ll} f(x) = x^3 - 6x^2 + 3x - 4 & \text{(degree 3)} \\ f(x) = x^5 - 6x^2 + 14 & \text{(degree 5)} \\ f(x) = 9 - 6x^2 + x^7. & \text{(degree 7)} \end{array}$$

Figure 5.9 shows the graph of the polynomial $f(x) = x^3 - 4x + 1$. In general, the higher the degree of the polynomial, the greater the number of "turns" in its graph. We will see later how calculus locates these turns rather easily.

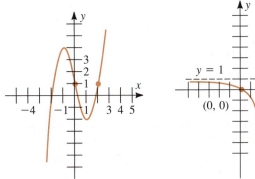

Figure 5.9 Graph of polynomial $f(x) = x^3 - 4x + 1$.

Figure 5.10 Graph of the rational function $f(x) = \dfrac{x}{x - 2}$.

Rational Functions

Rational functions are quotients of polynomials. That is, they are functions of the form

$$f(x) = \frac{p(x)}{q(x)}$$

where $p(x)$ and $q(x)$ are polynomials. It is important to note that the domain of a rational function excludes all numbers for which the denominator equals zero, since division by zero is not defined. Figure 5.10 shows the graph of the rational function $f(x) = \dfrac{x}{x - 2}$.

Note that $x = 2$ is excluded from the domain of this function.

Other Power Functions; Laws of Exponents

Finally, we note that the power function $f(x) = x^r$ can be defined for numbers r other than just positive integers. The following definition and theorem review the meaning and properties of $f(x) = x^r$ when r is a rational number. (A rational number is any quotient of integers: $r = n/m$, where n and m are integers and $m \neq 0$.)

DEFINITION 6

Let x and y be real numbers and let n and m be positive integers. Then

(a) x^n means $x \cdot x \cdot x \cdot \cdots \cdot x$ (n factors)

(b) x^{-n} means $\dfrac{1}{x^n}$ whenever $x \neq 0$.

(c) $x^{1/n} = y$ means $y^n = x$. (Here $x \geq 0$ if n is even.)

(d) $x^{m/n}$ means $(x^{1/n})^m$.

(e) x^0 means 1 whenever $x \neq 0$.

Example 1

(a) $2^5 = 2 \cdot 2 \cdot 2 \cdot 2 \cdot 2 = 32$

(b) $6^{-2} = \dfrac{1}{6^2} = \dfrac{1}{36}$

(c) $16^{3/4} = (16^{1/4})^3 = 2^3 = 8$

(d) $(-27)^{2/3} = [(-27)^{1/3}]^2 = (-3)^2 = 9$

But note that $(-27)^{1/2}$ is not defined since $(-27)^{1/2} = y$ would mean $y^2 = -27$, which has no solution. □

Example 1 points out the fact that power functions of the form $f(x) = x^{m/n} = (x^{1/n})^m$ are not defined for negative x when n is an even integer. For example, compare the graphs of $f(x) = \sqrt{x} = x^{1/2}$ and $g(x) = x^{-3/2}$ in Figure 5.11 with the graphs of $f(x) = x^{1/3}$ and $g(x) = x^{-2/3}$ in Figure 5.12.

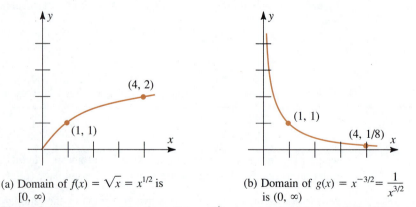

(a) Domain of $f(x) = \sqrt{x} = x^{1/2}$ is $[0, \infty)$

(b) Domain of $g(x) = x^{-3/2} = \dfrac{1}{x^{3/2}}$ is $(0, \infty)$

Figure 5.11 Functions of the form $f(x) = x^{m/n}$ are not defined for $x < 0$ if n is an even integer.

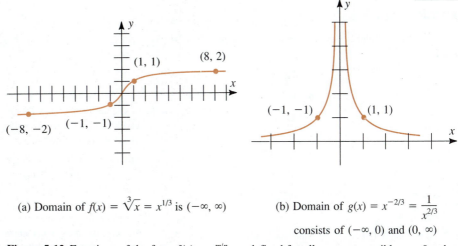

(a) Domain of $f(x) = \sqrt[3]{x} = x^{1/3}$ is $(-\infty, \infty)$ (b) Domain of $g(x) = x^{-2/3} = \dfrac{1}{x^{2/3}}$

consists of $(-\infty, 0)$ and $(0, \infty)$

Figure 5.12 Functions of the form $f(x) = x^{m/n}$ are defined for all x except possibly $x = 0$, when n is an odd integer.

The following theorem recalls the *laws of exponents,* which we must use in working with power functions.

THEOREM 4

(Laws of Exponents) Let x and y be real numbers, and let n and m be integers. Then

(a) $x^n \cdot x^m = x^{n+m}$.

(b) $\dfrac{x^n}{x^m} = x^{n-m}$, $x \neq 0$.

(c) $(x^n)^m = x^{nm}$.

(d) $x^{m/n} = (x^{1/n})^m = (x^m)^{1/n}$, $x \geq 0$ if n even.

(e) $(xy)^n = x^n y^n$.

Example 2 Using the Laws of Exponents the expression

$\dfrac{(4x^2 y)^3}{8xy^{-2}}$ may be simplified as follows:

$$\frac{(4x^2 y)^3}{8xy^{-2}} = \frac{4^3 \cdot (x^2)^3 \cdot y^3}{8xy^{-2}} = \frac{4^3 x^6 y^3}{8 \cdot x \cdot y^{-2}}$$

$$= \left(\frac{64}{8}\right)\left(\frac{x^6}{x}\right)\left(\frac{y^3}{y^{-2}}\right) = 8x^5 y^5. \qquad \square$$

Exercise Set 1.5

In Exercises 1–10 simplify the given expression as much as possible.

1. $27^{2/3}$

2. $81^{3/4}$

3. $4^{-3/2}$

4. $9^{-3/2}$

5. $\dfrac{2(3x)^4}{6x^2}$

6. $\dfrac{4x^3\sqrt{y}}{(x^2y)^{2/3}}$

7. $\dfrac{(3x^2y^{-1})^3}{\sqrt{xy^{-2}}}$

8. $\dfrac{\sqrt{4x^4y^2}}{\sqrt{y}}$

9. $\dfrac{6(xy)^3}{x^{2/3}y^{-3/2}}$

10. $\dfrac{(27x^6y^2)^{-1/3}}{3x^{-2}y^{4/3}}$

11. True or false? Every linear function is a polynomial.

12. True or false? Some polynomial functions are quadratic functions.

In Exercises 13–20 graph the given power function.

13. $f(x) = \dfrac{1}{2}x^2$

14. $f(x) = -x^3$

15. $f(x) = 3x^{2/3}$

16. $f(x) = 2x^{3/2}$

17. $f(x) = -4x^{1/3}$

18. $y = x^{-1/2}$

19. $y = -x^{-2/3}$

20. $f(x) = 2x^{-2}$

In Exercises 21–26 graph the given quadratic function by plotting points.

21. $f(x) = (x - 3)^2 - 4$

22. $f(x) = \dfrac{1}{2}(x + 1)^2 - 2$

23. $f(x) = x^2 - 2x + 1$

24. $f(x) = x^2 + 4x + 6$

25. $f(x) = -2x^2 - 4x + 1$

26. $f(x) = -\dfrac{3}{2}x^2 - 6x - 3$

In Exercises 27–32 graph the given polynomial function by plotting points.

27. $f(x) = x^3 + 4$

28. $f(x) = x^3 - x$

29. $f(x) = 4 - x^4$

30. $f(x) = x^3 - 4x + 2$

31. $f(x) = x^3 - x^2 - 2x + 2$

32. $f(x) = 1 - 3x + x^3$

In Exercises 33–38 graph the given rational function by plotting points and noting where the denominator equals zero.

33. $f(x) = \dfrac{1}{x - 3}$

34. $f(x) = \dfrac{3}{2 + x}$

35. $f(x) = \dfrac{x}{1 + x}$

36. $f(x) = \dfrac{x}{1 + x^2}$

37. $f(x) = \dfrac{x^3}{1 + x}$

38. $f(x) = \dfrac{x + 1}{1 - x^2}$

39. A rectangle has area 80 cm. Write a function w that gives the width of the rectangle in terms of its length. What is the domain of this function?

40. The statement that "the function f is *proportional* to the function g," means that $f(x) = kg(x)$ for some constant k and for all x common to the domains of f and g. Examples of one quantity being proportional to the *square* of another arise frequently in nature. For example, empirical evidence shows that the air resistance R for a moving automobile is proportional to the *square* of its speed, v. Write this statement in function form and discuss the nature of the associated graphs.

41. Your *utility function* associated with a certain commodity may be described as a measurement of your level of satisfaction associated with the possession of x units of the given commodity. For example, if $U(x)$ is your utility function for your consumption of x slices of pepperoni pizza, then $U(1)$ is the satisfaction you get from eating one slice, $U(2)$ the satisfaction from eating 2 slices, etc.

   a. Sketch the utility function $U_1(x) = \sqrt{x + 2}$.
   b. Sketch the utility function $U_2(x) = 6x - x^2$.
   c. In your opinion, which utility function seems more appropriate for the example of consuming x slices of pizza? Why?
   d. In your opinion, which utility function seems more appropriate for winning x thousands of dollars? Why?

42. The demand for a certain model of sports car is determined to be $D(p) = \sqrt{400 - p^2}$ cars per week where p is the selling price of the car in thousands of dollars.

   a. What is the domain of D?
   b. What is the demand per week at price $p = \$12,000$ per car?
   c. Sketch the graph of this demand function.

43. After x repetitions, an assembly worker can perform a certain task in $T(x) = 30\left(1 + \dfrac{2}{\sqrt{x + 1}}\right)$ seconds.

   a. How quickly will the employee perform the task on the fourth try?
   b. Graph this "learning curve."

44. In foreign language instruction it is observed that a person will have learned an average of $N(t) = 10t - 0.5t^2$ vocabulary terms in t continuous hours of study. Graph this quadratic "learning curve."

45. A manufacturer of garage doors finds that the total cost of manufacturing x garage doors per week is approximately $C(x) = 400 + 40x + 0.2x^2$.
  a. Sketch the graph of this cost function.
  b. What are the manufacturer's fixed costs?

46. The manufacturer in Exercise 45 can sell his garage doors for $100 each.
  a. Find the revenue function $y = R(x)$, which gives the amount received by the manufacturer from the sale of x items.
  b. Find the manufacturer's weekly profit function $y = P(x) = R(x) - C(x)$.
  c. Sketch the graphs for all three functions $C, R,$ and P on the same set of axes. Estimate the production levels x between which profit is positive.

1.6 FINDING ZEROS OF FUNCTIONS

If $P(x)$ represents a manufacturer's weekly profit from the sale of x items, the values of the function P may be either positive or negative, depending on the output level x. An output level x_0 for which $P(x_0) = 0$ is called a *break-even* level, for obvious reasons. The problem of finding break-even levels is just one example of the more general problem of finding the **zeros** of a given function f, that is, of finding the numbers x_0 for which $f(x_0) = 0$. (See Figure 6.1.)

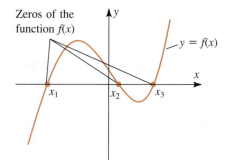

Figure 6.1 Zeros of the function f are numbers x for which $f(x) = 0$.

Finding Zeros by Factoring

When the equation for a function can be factored, the zeros of the function are just the zeros of the individual factors. For example, the zeros of the function $f(x) = (x - 2)(x + 1)$ are $x_1 = 2$ and $x_2 = -1$, since $x - 2 = 0$ when $x = 2$, and $x + 1 = 0$ when $x = -1$. These are *all* the zeros of f since $f(x)$ cannot equal zero unless one of its factors does. (See Figure 6.2.)

The following are additional examples of finding zeros of functions by factoring.

Example 1 Find the zeros of the function $f(x) = x^2 - 3x - 4$.

Strategy

Factor the right-hand side of the equation for f.

Solution

Since $f(x) = x^2 - 3x - 4$
$$= (x + 1)(x - 4)$$

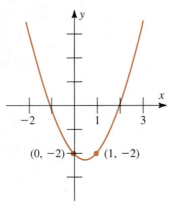

Figure 6.2 For $f(x) = x^2 - x - 2 = (x - 2)(x + 1)$ the zeros are $x_1 = -1$ and $x_2 = 2$.

Set each factor equal to zero and solve for x.

we will have $f(x) = (x + 1)(x - 4) = 0$ when $x + 1 = 0$ or when $x - 4 = 0$. Now

$$x + 1 = 0 \text{ gives } x = -1$$

and

$$x - 4 = 0 \text{ gives } x = 4.$$

The zeros of f are therefore $x_1 = -1$ and $x_2 = 4$.

Check answers by substituting x in $f(x)$ and verifying that $f(x) = 0$.

To check this answer, we substitute these numbers into the expression $f(x)$:

$$f(-1) = (-1)^2 - 3(-1) - 4 = 1 + 3 - 4 = 0$$
$$f(4) = 4^2 - 3 \cdot 4 - 4 = 16 - 12 - 4 = 0$$

so both answers are correct. ☐

Example 2 A manufacturer of central air-conditioning units finds that the total cost of producing x units per week is $C(x) = 10{,}000 + 350x + 25x^2$ dollars. If the manufacturer can sell each unit for 1600 dollars, what are the break-even level(s) of production?

Strategy

Find the weekly profit function $P = R - C$.

Solution

Since the revenue received from the sale of x units is $R(x) = 1600x$ dollars, the manufacturer's weekly profit is

$$P(x) = R(x) - C(x)$$
$$= 1600x - [10{,}000 + 350x + 25x^2]$$
$$= -25x^2 + 1250x - 10{,}000.$$

Factor $P(x)$.

Since

$$P(x) = -25(x^2 - 50x + 400)$$
$$= -25(x - 10)(x - 40)$$

we shall have $P(x) = 0$ when $x - 10 = 0$ or $x - 40 = 0$.

Set factors of $P(x)$ equal to zero and solve. Solutions are zeros of P.

The equation

$$x - 10 = 0 \text{ gives } x = 10$$

and the equation

$$x - 40 = 0 \text{ gives } x = 40.$$

The break-even levels are therefore $x_1 = 10$ units per week and $x_2 = 40$ units per week.

Figure 6.3 shows that the manufacturer will experience a true profit ($P(x) > 0$) for $10 < x < 40$, and losses ($P(x) < 0$) for $0 \le x < 10$ or $x > 40$. □

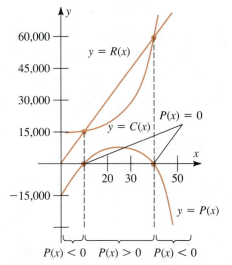

Figure 6.3 Zeros for profit function $P(x) =$ $-25x^2 + 1250x - 10{,}000$ give break-even points $x_1 = 10$ and $x_2 = 40$.

Example 3 Find the zeros of the function $f(x) = x^{5/3} - 4x$.

Solution: We factor the right-hand side as

$$f(x) = x^{5/3} - 4x$$
$$= x(x^{2/3} - 4).$$

Then $f(x) = 0$ if $x = 0$ or if $x^{2/3} - 4 = 0$. The equation

$$x^{2/3} - 4 = 0$$

gives

$$x^{2/3} = 4.$$

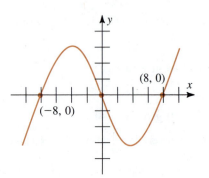

Figure 6.4 Zeros of $f(x) = x^{5/3} - 4x$ are $x_1 = -8$, $x_2 = 0$, and $x_3 = 8$.

Cubing both sides gives $x^2 = 4^3 = 64$, which has solutions $x = \pm 8$. The zeros of f are therefore $x_1 = -8$, $x_2 = 0$, and $x_3 = 8$. (See Figure 6.4.) $\square$

The Quadratic Formula

For the special case of quadratic functions of the form

$$f(x) = ax^2 + bx + c \tag{1}$$

the zeros are given by the **quadratic formula**

$$x = \frac{-b \pm \sqrt{b^2 - 4ac}}{2a} \tag{2}$$

The quadratic formula is particularly useful when factors of a quadratic expression are not easily recognized.

Example 4 Find the zeros of the function $f(x) = 2x^2 + x - 3$.

Solution: This function has the form of the quadratic function in equation (1) with $a = 2$, $b = 1$, and $c = -3$. Thus by the quadratic formula (2) the zeros are

$$x = \frac{-1 \pm \sqrt{1 - 4(2)(-3)}}{2 \cdot 2}$$

$$= \frac{-1 \pm \sqrt{1 + 24}}{4}$$

$$= -\frac{1}{4} \pm \frac{\sqrt{25}}{4}.$$

The zeros are therefore $x_1 = -\frac{1}{4} - \frac{5}{4} = -\frac{3}{2}$ and $x_2 = -\frac{1}{4} + \frac{5}{4} = 1$. $\square$

Equality of Function Values

Closely related to the problem of finding the zeros of a single function is the problem of finding the numbers for which two given functions have the same values. Note that finding the numbers x for which $f(x) = g(x)$ is the same as finding the zeros of the function $h = g - f$. The following example requires just this observation.

Example 5 The population of Town A is currently 20 thousand people. It is projected by demographers that the population of Town A will have grown to $P(t) = \sqrt{400 + 200t}$ thousand after t years. The neighboring Town B currently has 10 thousand people and is projected to have $Q(t) = (10 + 8t)$ thousand after t years. When will the towns have equal populations?

Strategy

Set $P = Q$.

Eliminate radical by squaring both sides.

Collect all terms on one side of the equation.

Find the zeros of this function using the quadratic formula.

Since time must be positive, disregard any negative values for t.

Verify that $P(t) = Q(t)$ at the time $t = \frac{5}{2}$ years.

Solution

The two towns have equal populations if $P(t) = Q(t)$, or

$$\sqrt{400 + 200t} = 10 + 8t.$$

Squaring both sides of this equation gives

$$400 + 200t = 100 + 160t + 64t^2$$

so

$$64t^2 - 40t - 300 = 0.$$

By the quadratic formula the solutions are

$$t = \frac{40 \pm \sqrt{(40)^2 - 4(64)(-300)}}{2(64)}$$

$$= \frac{40 \pm \sqrt{1600 + 76{,}800}}{128}$$

$$= \frac{40 \pm 280}{128}.$$

Now the time $t_1 = \dfrac{40 - 280}{128} = -\dfrac{15}{8}$ is meaningless in this application, so the only time at which the two towns will have equal populations is

$$t_2 = \frac{40 + 280}{128} = \frac{320}{128} = \frac{5}{2}$$

years from the present. At this time the populations will be

$$P\left(\frac{5}{2}\right) = \sqrt{400 + 200\left(\frac{5}{2}\right)} = \sqrt{900} = 30 \text{ thousand}$$

which is the same as

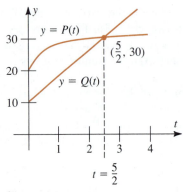

Figure 6.5 Population curves in Example 5.

$$Q\left(\frac{5}{2}\right) = 10 + 8\left(\frac{5}{2}\right) = 10 + 20 = 30 \text{ thousand.}$$

These population curves are graphed in Figure 6.5. □

Demand and Supply Functions

Under certain market conditions the selling price p of a particular item (Brand X dishwashers, for example) determines the number of items that consumers will purchase in a given unit of time. If we let $D(p)$ denote the number of items that will be sold at price p during the unit of time, then D is called a **demand function.** Figure 6.6 shows the graph of a typical demand function. The property that higher prices correspond to lower demands is called the *law of demand*.

Similarly, the function S giving the number $S(p)$ of items that suppliers are willing to produce at selling price p in a unit of time is called the **supply function** for that item. In general, higher selling prices result in higher levels of supply, as reflected by the graph of a typical supply function in Figure 6.6.

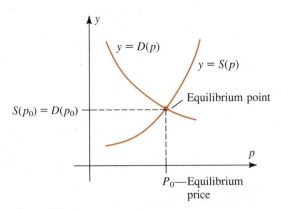

Figure 6.6 Graphs of typical demand and supply curves.

For a given item and market the point where the graphs of the demand and supply functions intersect is called the *equilibrium* point. It indicates the price at which all who wish to purchase the item will be able to do so, and no surplus items will exist. The techniques of this section are useful in locating equilibrium points.

Example 6 Market research indicates that when a certain type of alloy tennis racket is priced at p dollars per racket, consumers will purchase $D(p) = 50 - \frac{1}{2}p$ rackets per week and suppliers will produce $S(p) = \frac{2}{5}p - 4$ rackets per week. Find the equilibrium point for these demand and supply functions.

Solution: Setting the demand and supply functions equal to each other, we obtain the equation

$$\frac{2}{5}p - 4 = 50 - \frac{1}{2}p.$$

Thus

$$\left(\frac{2}{5} + \frac{1}{2}\right)p = 50 + 4$$

or

$$\frac{9}{10}p = 54.$$

Thus

$$p = \left(\frac{10}{9}\right)54$$

$$= 60.$$

The equilibrium price is therefore $60, corresponding to a demand and supply of $D(60) = S(60) = 20$ rackets per week. The equilibrium point is therefore $(60, 20)$. (See Figure 6.7.) □

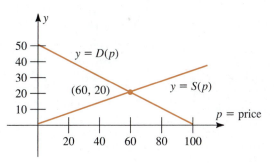

Figure 6.7 Demand and supply functions in Example 6.

Exercise Set 1.6

In Exercises 1–10 find the zero(s) of the given function by factoring, if necessary.

1. $f(x) = x^2 - 4$

2. $f(x) = x - x^3$

3. $f(x) = x^2 + x - 6$

4. $f(x) = x^2 - 9x + 20$

5. $f(x) = x^2 + 2x - 15$

6. $f(x) = x^3 + 2x^2 + x$

7. $f(x) = \dfrac{x - 2}{x + 1}$

8. $f(x) = \dfrac{x^2 - 3}{x^2 + 1}$

9. $f(x) = x^{3/2} - 3\sqrt{x}$

10. $f(x) = x - 2\sqrt{x} + 1$

In Exercises 11–18 find the zeros of the given function by using the quadratic formula.

11. $f(x) = x^2 + x - 1$

12. $f(x) = x^2 + 3x + 1$

13. $f(x) = 2x^2 + 3x - 1$

14. $f(x) = 1 + 4x - x^2$

15. $f(x) = 3 + 2x - x^2$

16. $f(x) = 2x^2 - 2 - 2x$

17. $f(x) = x - 3x^2 + 4$

18. $f(x) = 3 - 3x - 2x^2$

In Exercises 19–28 find the point(s) of intersection of the graphs of f and g.

19. $f(x) = 2x + 1$

$g(x) = 7 - x$

20. $f(x) = 1 - \dfrac{x}{3}$

$g(x) = 8 + 2x$

21. $f(x) = x^2$

$g(x) = x + 2$

22. $f(x) = 4 - x^2$

$g(x) = 2 - x$

23. $f(x) = \sqrt{x}$

$g(x) = \dfrac{1}{2}x$

24. $f(x) = 7 - x^2$

$g(x) = x^2 - 1$

25. $f(x) = x^3 + 2$

$g(x) = x + 2$

26. $f(x) = \sqrt{x + 2}$

$g(x) = \dfrac{1}{3}x + \dfrac{4}{3}$

27. $f(x) = 3x^2 - 12x$

$g(x) = 5x - 20$

28. $f(x) = \sqrt{x} + 1$

$g(x) = \dfrac{1}{3}x + 1.$

29. Figure 6.8 shows typical demand and supply curves for an item selling at price p. Let p_0 denote the equilibrium price as indicated. Explain why a shortage of this item will exist if the selling price is set at price $p_1 < p_0$.

30. Refer to Exercise 29 and Figure 6.8. Explain why a surplus of the item will exist if the selling price is set at price $p_2 > p_0$.

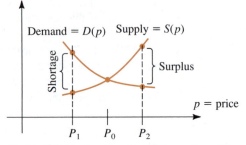

Figure 6.8 At other than equilibrium price, either a shortage or surplus will exist.

31. A consumer's utility function (a measure of satisfaction) for dollars spent annually on leisure activities is $L(x) = 100\sqrt{x}$ while the same consumer's utility function for dollars spent annually on investments is $S(x) = \frac{5}{3}x$. Find the expenditure level x for which these two utility functions are equal.

32. A manufacturer's cost of producing x pairs of roller skates per week is $C(x) = 400 + 15x$. If the roller skates can be sold for $25 per pair, what is the manufacturer's break-even production level?

33. The weekly cost of manufacturing x telephones per week is found by a manufacturer to be $C(x) = 500 + 20x + x^2$ dollars. The telephones can be sold at a price of $p = \$80$ each.

a. Find the manufacturer's break-even production level(s).

b. For what production levels will the manufacturer experience a profit?

34. The demand for a certain type of T-shirt is determined to be $D(p) = 20 - p$ T-shirts per week at price p. At this price suppliers are willing to produce $S(p) = p^2$ T-shirts per week. Find the equilibrium price.

35. Deaths from trauma after a serious injury are known to be of two types. Type A is due to severe damage to a major organ (heart, brain, etc.). Type B is due to blood loss, either through internal hemorrhages or through loss of blood from the body. Figure 6.9 shows typical graphs for the frequencies of each type of death from trauma. If

$$f(t) = \frac{25}{t} \quad \text{and} \quad g(t) = \frac{50}{1 + (t - 2)^2}$$

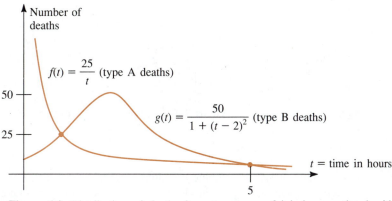

Figure 6.9 Distribution of deaths from two types of injuries associated with trauma. (See Exercise 35.)

find the times t at which these functions have equal values.

36. The percentage of the U.S. labor force involved in farming fell linearly from 70% in 1830 to 20% in 1930. According to these data, when was it 40%?

37. If $C(x)$ denotes the total cost of producing x items, the *average cost* per item is $c(x) = \dfrac{C(x)}{x}$. Suppose that a

manufacturer of picnic tables finds that its total cost of producing x tables is given by the function $C(x) = 500 + 10x + x^2$ dollars. If the tables can be sold for $70 each, at what production level(s) does the selling price equal the average cost?

38. Derive the quadratic formula by completing the square on the expression $ax^2 + bx + c$.

1.7 COMBINATIONS OF FUNCTIONS

We have already seen several examples of ways in which certain functions are combined to form other functions of interest. Two such examples involve the revenue function $y = R(x)$ and the total cost function $y = C(x)$ per unit time associated with the manufacture and sale of x items. They are the profit and average cost functions:

$$P(x) = R(x) - C(x) \qquad \text{(profit function)}$$

$$c(x) = \frac{C(x)}{x}. \qquad \text{(average cost function—see Exercise 37, Section 1.6)}$$

Figures 7.1 to 7.3 describe another such example. They are what industrial psychologists refer to as *work curves*. In such curves the horizontal axis represents time and the vertical axis represents productivity or efficiency of a worker performing a certain task. The curve in Figure 7.1 represents what is called the *warm-up effect*. This is the phenomenon of a worker's efficiency initially rising during the early part of the work period. Figure 7.2 shows the *fatigue effect*—as the work period drags on, the worker tires and becomes less efficient. Figure 7.3 shows how the warm-up and fatigue effects combine through addition to produce an *overall* work curve.

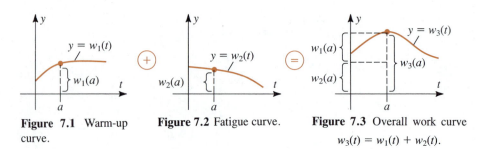

Figure 7.1 Warm-up curve.

Figure 7.2 Fatigue curve.

Figure 7.3 Overall work curve $w_3(t) = w_1(t) + w_2(t)$.

The Algebra of Functions

It is a simple matter to combine functions algebraically to form new functions that are sums, differences, multiples, products, or quotients of the given functions. It is done by calculating the values of the individual terms in the combination separately and then forming the given combination using those values.

DEFINITION 7

Given the functions f and g, and the number c, the functions $f + g$, $f - g$, cf, fg, and $\dfrac{f}{g}$ are defined by the following equations:

$$(f + g)(x) = f(x) + g(x)$$
$$(f - g)(x) = f(x) - g(x)$$
$$(cf)(x) = c[f(x)]$$
$$(fg)(x) = f(x)g(x)$$
$$(f/g)(x) = f(x)/g(x). \qquad \text{(provided } g(x) \neq 0)$$

Of course, these combinations can be formed only for numbers x that are in the domains of both functions. Although these definitions may seem obvious, we will need a clear understanding of these definitions in studying limits of functions in Chapter 2.

Example 1 Let $f(x) = x^3 + 2$ and $g(x) = \sqrt{x + 1}$. Then

(a) for $h_1 = f + g$, $\quad h_1(x) = x^3 + 2 + \sqrt{x + 1}$

so

$$h_1(3) = f(3) + g(3) = [3^3 + 2] + \sqrt{3 + 1} = 29 + 2 = 31$$

and

$$h_1(1) = f(1) + g(1) = [1^3 + 2] + \sqrt{1 + 1} = 3 + \sqrt{2}$$

but

$h_1(-2)$ is undefined since $g(-2) = \sqrt{1 - 2}$ is undefined.

(b) for $h_2 = \dfrac{f}{3g}$, $\quad h_2(x) = \dfrac{x^3 + 2}{3\sqrt{x + 1}}$

so

$$h_2(3) = \frac{f(3)}{3g(3)} = \frac{3^3 + 2}{3\sqrt{3} + 1} = \frac{27 + 2}{3\sqrt{4}} = \frac{29}{6}$$

and

$$h_2(0) = \frac{f(0)}{3g(0)} = \frac{0^3 + 2}{3\sqrt{0} + 1} = \frac{2}{3}.$$ □

Composite Functions

Often two functions are combined not by an algebraic operation such as addition, but by letting the second function act on the result (output) of the first. For example, Figure 7.4 shows how one would calculate the distance between an air controller at ground level and an airplane flying at an altitude of 2 miles:

(i) find the distance $d_1(t)$ from the control tower to the point on the ground directly beneath the plane;
(ii) use the Pythagorean theorem to find the actual distance $d_2(t)$ between the controller and the airplane as

$$d_2(t) = \sqrt{[d_1(t)]^2 + 2^2}. \tag{1}$$

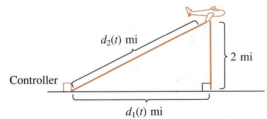

$d_2(t)$ mi

2 mi

Controller

$d_1(t)$ mi

Figure 7.4 Distance between controller and airplane is $d_2(t) = \sqrt{[d_1(t)]^2 + 2^2}$.

In this example we have written both distances $d_1(t)$ and $d_2(t)$ as functions of t (time), since both change in time (hopefully!) with the location of the airplane. Note in equation (1) that the square root function acts on the function $f(t) = [d_1(t)]^2 + 4$ to produce the value of the function d_2. This is an example of a *composite* function.

DEFINITION 8

Given two functions f and g the **composite** function $g \circ f$ is the result of the function g acting on the values of the function f. That is

$$(g \circ f)(x) = g(f(x))$$

as illustrated in Figure 7.5.

The difference between composite functions, or functions formed by *composition,* and algebraic combinations of functions has to do with the order in which the operations are performed. In an algebraic combination the individual function values are found first and

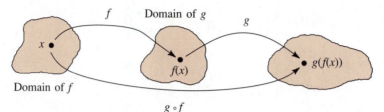

Figure 7.5 The composite function $g \circ f$ is the result of the function g acting on values of the function f.

then combined. In the composite function $g \circ f$ we first find the value of the *inside* function $f(x)$ and then insert this number into the function g. Obviously, the number $f(x)$ must lie in the domain of the function g for the value $g(f(x))$ to be defined.

Example 2 Let $f(x) = x^2 + 3$ and $g(x) = \sqrt{x}$. Then

(a) $(g \circ f)(x) = g(f(x)) = g(x^2 + 3) = \sqrt{x^2 + 3}$

(b) $(f \circ g)(x) = f(g(x)) = f(\sqrt{x}) = (\sqrt{x})^2 + 3 = x + 3$

whenever $x \geq 0$. □

Example 3 Let $f(x) = (3x + 1)^2$ and $g(x) = x - 2$. Then

(a) $(g \circ f)(x) = g((3x + 1)^2) = (3x + 1)^2 - 2$
$= (9x^2 + 6x + 1) - 2 = 9x^2 + 6x - 1$

(b) $(f \circ g)(x) = f(x - 2) = [3(x - 2) + 1]^2 = (3x - 5)^2 = 9x^2 - 30x + 25.$ □

Notice in both Example 2 and Example 3 that $g(f(x)) \neq f(g(x))$ in general. The next example shows how we must be careful about order in determining the domain of a composite function.

Example 4 For $f(x) = \sqrt{x}$ and $g(x) = x + 4$ find

(a) the composite function $g \circ f$ and its domain.
(b) the composite function $f \circ g$ and its domain.

Solution: In (a) we have

$$g(f(x)) = g(\sqrt{x}) = \sqrt{x} + 4$$

and the domain of $g \circ f$ is $[0, \infty)$ since we must have $x \geq 0$. In part (b) we have

$$f(g(x)) = f(x + 4) = \sqrt{x + 4}.$$

The domain of $f \circ g$ is therefore all x with $x + 4 \geq 0$. This requires $x \geq -4$, so the domain is $[-4, \infty)$. □

Split Functions

In certain applications we will need to work with functions that are defined differently for various numbers x. This occurs frequently in pricing structures where manufacturers wish to encourage volume purchases.

For example, suppose that the producer of the Easy-as-Pie software product, which retails for $500 a copy, offers discounts of 20% on orders for 10 or more copies, and an additional 20% on orders for 25 or more copies. The total cost $C(x)$ of purchasing x copies of this software would then be:

(i) $500x$ if $1 \leq x < 10$

(ii) $(.8)(500x) = 400x$ if $10 \leq x < 25$

(iii) $(.6)(500x) = 300x$ if $25 \leq x.$

In this case we would write the *split function C* as

$$C(x) = \begin{cases} 500x & \text{if} \quad 1 \leq x < 10 \\ 400x & \text{if} \quad 10 \leq x < 25 \\ 300x & \text{if} \quad 25 \leq x. \end{cases} \tag{2}$$

To find the value $C(x)$ for a particular number x, we would first determine which of the three inequalities in statement (2) the number x satisfies and then use the corresponding equation for the function. The graph of $y = C(x)$ appears in Figure 7.6.

Example 5 For the function $f(x) = \begin{cases} 4 - x^2 & \text{if} \quad -\infty < x \leq 1 \\ x + 2 & \text{if} \quad 1 < x < \infty \end{cases}$

(a) find $f(-2)$
(b) find $f(3)$
(c) graph f.

Solution:

(a) Since $x = -2$ satisfies $-\infty < x \leq 1$, we have

$$f(-2) = 4 - (-2)^2 = 4 - 4 = 0.$$

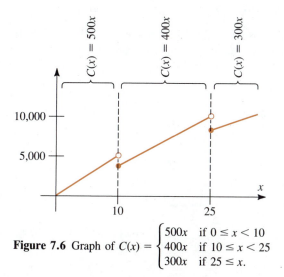

Figure 7.6 Graph of $C(x) = \begin{cases} 500x & \text{if } 0 \leq x < 10 \\ 400x & \text{if } 10 \leq x < 25 \\ 300x & \text{if } 25 \leq x. \end{cases}$

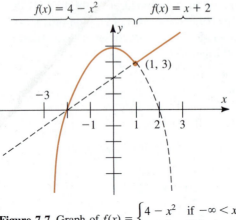

$f(x) = 4 - x^2$ $f(x) = x + 2$

(1, 3)

Figure 7.7 Graph of $f(x) = \begin{cases} 4 - x^2 & \text{if } -\infty < x \leq 1 \\ x + 2 & \text{if } 1 < x < \infty. \end{cases}$

(b) Since $x = 3$ satisfies $1 < x < \infty$, we have

$$f(3) = 3 + 2 = 5.$$

(c) The graph of f appears in Figure 7.7.   □

The Absolute Value Function

An important split function is the **absolute value** function. It is defined as follows:

$$|x| = \begin{cases} x & \text{if } x \geq 0 \\ -x & \text{if } x < 0. \end{cases}$$

In words, the absolute value of a number is just the number itself if the number is nonnegative. If the number is negative, its absolute value is found by changing its sign. Thus

$$|3| = 3, \ |0| = 0, \ |-7| = 7, \quad \text{and} \quad |-\sqrt{2}| = \sqrt{2}.$$

The graph of the absolute value function appears in Figure 7.8. In working with absolute values in composite functions, we must be careful to note the sign of the expression inside the absolute value signs as the following example shows.

$y = |x|$

$y = -x$ $y = x$

Figure 7.8 Graph of the absolute value function.

Example 6 Graph the function $f(x) = |4 - x^2|$

Strategy

Determine where the expression inside absolute value signs is nonnegative.

Rewrite f as a split function.

Graph each branch corresponding to one part of the split function separately.

Solution

We shall have $4 - x^2 \geq 0$ if $4 \geq x^2$, which requires $-2 \leq x \leq 2$. Thus

$$f(x) = \begin{cases} -(4 - x^2) = x^2 - 4 & \text{if} \quad -\infty < x < -2 \\ 4 - x^2 & \text{if} \quad -2 \leq x \leq 2 \\ -(4 - x^2) = x^2 - 4 & \text{if} \quad 2 < x < \infty. \end{cases}$$

The graph appears in Figure 7.9. Note that the effect of the absolute value signs is to turn the "legs" of the graph of $4 - x^2$ that extend below the x-axis upward. □

$f(x) = x^2 - 4$ $f(x) = 4 - x^2$ $f(x) = x^2 - 4$

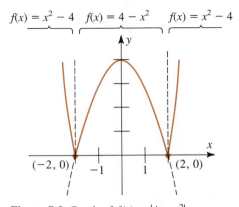

Figure 7.9 Graph of $f(x) = |4 - x^2|$.

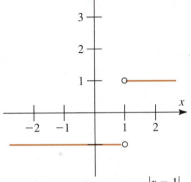

Figure 7.10 Graph of $f(x) = \dfrac{|x - 1|}{x - 1}$.

Example 7 Graph the function $f(x) = \dfrac{|x - 1|}{x - 1}$.

Solution: Since the numerator $|x - 1|$ equals zero when $x = 1$, we know that

$$|x - 1| = \begin{cases} x - 1 & \text{if} \quad x \geq 1 \\ 1 - x & \text{if} \quad x < 1 \end{cases}$$

by definition of the absolute value of the quantity $x - 1$. We can therefore write f as the split function

$$f(x) = \begin{cases} \dfrac{x - 1}{x - 1} = 1 & \text{if} \quad x > 1 \\ \dfrac{1 - x}{x - 1} = -1 & \text{if} \quad x < 1. \end{cases}$$

The graph of f appears in Figure 7.10. Note that $f(1)$ is not defined. □

Exercise Set 1.7

In Exercises 1–6 let $f(x) = x - 7$, $g(x) = x^2 + 2$, $h_1 = f + g$, and $h_2 = \dfrac{f}{g}$. Find

1. $h_1(x)$ **2.** $h_2(x)$

3. $h_1(2)$ **4.** $h_2(-1)$

5. $h_2(-4)$ **6.** $h_1(0)$

In Exercises 7–12 let $f(x) = \dfrac{1}{x + 2}$, $g(x) = \sqrt{3 + x}$, $h_1 = f - 2g$, and $h_2 = \dfrac{g}{f}$. Find

7. $h_1(x)$ **8.** $h_2(x)$

9. $h_2(1)$ **10.** $h_1(-3)$

11. $h_1(1)$ **12.** $h_2(6)$

13. Let $f(x) = x^2 - x$. Find g so that $(f + g)(x) = 4 - x$.

14. Let $f(x) = \sqrt{x} - 2$. Find g so that $(fg)(x) = x - 4$.

In Exercises 15–22 use the functions

$$f(x) = 3x + 1 \qquad g(x) = x^3$$

$$h(x) = \frac{1}{x + 1} \qquad u(x) = \sqrt{x}$$

to find equations for the indicated composite function.

15. $f \circ u$ **16.** $u \circ h$

17. $g \circ f$ **18.** $u \circ g$

19. $h \circ u$ **20.** $h \circ f \circ u$

21. $f \circ g \circ u$ **22.** $g \circ h \circ u$

23. Let $u(x) = x - 3$. Find a function f so that $f(u(x)) = x^2$.

24. Let $f(x) = x^2$. Find a function u so that $f(u(x)) = x^2 + 8x + 16$.

25. Let $f(x) = x + 4$ and $g(x) = (x - 2)^2$. Find a function u so that $f(g(u(x))) = 4x^2 - 8x + 8$.

In each of Exercises 26–31 match the split function with the correct graph among figures a–f.

26. $f(x) = \begin{cases} x + 1, & x \le 2 \\ 9 - 3x, & x > 2 \end{cases}$ **a.**

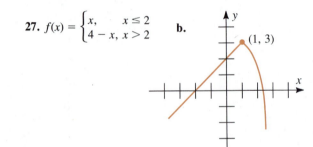

27. $f(x) = \begin{cases} x, & x \le 2 \\ 4 - x, & x > 2 \end{cases}$ **b.**

28. $f(x) = \begin{cases} 4 - x^2, & x \le 1 \\ 4 - x, & x > 1 \end{cases}$ **c.**

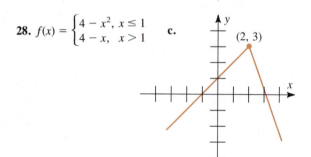

29. $f(x) = \begin{cases} x + 2, & x \le 1 \\ 4 - x^2, & x > 1 \end{cases}$ **d.**

30. $f(x) = \begin{cases} 4 - 2x, & x \le 1 \\ \dfrac{2}{x}, & x \ge 1 \end{cases}$ **e.**

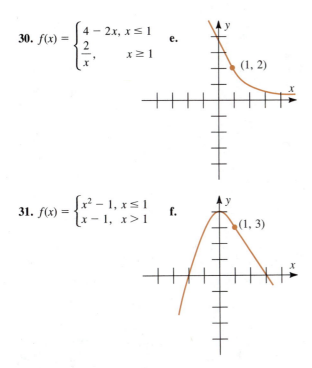

(1, 2)

31. $f(x) = \begin{cases} x^2 - 1, & x \le 1 \\ x - 1, & x > 1 \end{cases}$ **f.**

(1, 3)

In Exercises 32–35 match the function written with absolute value notation with its equivalent split function form a–d.

32. $f(x) = |2x - 4|$

a. $f(x) = \begin{cases} -4 - 2x, & x < -2 \\ 4 + 2x, & x \ge -2 \end{cases}$

33. $f(x) = |4 + 2x|$

b. $f(x) = \begin{cases} 4 - 2x, & x < 2 \\ 2x - 4, & x \ge 2 \end{cases}$

34. $f(x) = |4 - 2x|$

c. $f(x) = \begin{cases} 4 - 2x, & x \le 2 \\ 2x - 4, & x > 2 \end{cases}$

35. $f(x) = |4x - 2|$

d. $f(x) = \begin{cases} 2 - 4x, & x < \dfrac{1}{2} \\ 4x - 2, & x \ge \dfrac{1}{2} \end{cases}$

In Exercises 36–41 graph the given function.

36. $f(x) = \begin{cases} x^2 - 2, & x \le 2 \\ 4 - x, & x > 2 \end{cases}$

37. $f(x) = \begin{cases} \sqrt{4 - x}, & x \le 0 \\ 2 - x^2, & x > 0 \end{cases}$

38. $f(x) = |x + 2|$

39. $f(x) = x + |x|$

40. $f(x) = |x^2 - 4|$

41. $f(x) = |x^2 - x - 2|$

42. A manufacturing plant can produce one dishwasher every 30 minutes. The total variable costs associated with the production of x dishwashers are $C(x) = 30x + 0.02x^2$. Write a composite function C giving the total variable costs resulting from t hours of dishwasher production.

43. A producer of a certain software product sells the product according to a price structure that encourages volume purchases. For orders of up to five copies the price is $500 per copy. On orders of more than five copies the price per copy is reduced by $10 for each copy in excess of five, except that all orders for more than 25 copies are filled at $300 per copy.
 a. Write a function p giving the price per copy in purchasing x copies of this software.
 b. Graph the function p.

44. Find the function R giving the total revenue received from a single order of x copies of the software described in Exercise 43.

45. Foxes and rabbits coexist in a certain ecological niche. The number of rabbits r is determined by the number of carrots according to the function $r(c) = 10 + 0.2c + 0.01c^2$. The number of foxes, which prey on the rabbits, is given by the function $f(r) = \sqrt{r} + 5$. Find the composite function f giving the number of foxes as a function of the number of carrots.

46. A delivery truck depreciates in value at the rate of 30% per year. Comprehensive theft and damage insurance on the truck costs 8% of its value per year. Find a function C giving the cost of the comprehensive insurance policy on the truck after t years if the original cost of the truck is $20,000.

47. A lawnmower manufacturer sells mowers to a certain retailer for $200 each, plus a handling charge of $75 on each order. The retailer applies a markup of 40% to the total price paid to the manufacturer.
 a. Find the function C giving the retailer's total cost of a single order of x lawnmowers.
 b. Find the function R giving the retailer's total revenues from the sale of these x lawnmowers.
 c. If the retailer sells all these lawnmowers at the same price, what is the retail price per lawnmower?

SUMMARY OUTLINE OF CHAPTER 1

■ The **inequality** $a < b$ means a lies to the left of b on the number line. (Page 3)

■ *Interval notation* (Page 3)

$[a, b]$ means $\{x \mid a \leq x \leq b\}$
(a, b) means $\{x \mid a < x < b\}$
$[a, b)$ means $\{x \mid a \leq x < b\}$
$(a, b]$ means $\{x \mid a < x \leq b\}$
(a, ∞) means $\{x \mid a < x\}$, etc.

■ *Theorem (Properties of Inequalities):*   Let a, b, and c be real numbers. (Page 4)

  (i) If $a < b$ then $a + c < b + c$.
  (ii) If $a < b$ and $c > 0$,   then $ac < bc$.
  (iii) If $a < b$ and $c < 0$,   then $ac > bc$.
  (iv) If $a < b$ and $b < c$,   then $a < c$.

■ The **slope** of the line through the points (x_1, y_1) and (x_2, y_2) is the number (Page 11)

$$m = \frac{y_2 - y_1}{x_2 - x_1}.$$

■ The **point-slope** form of the equation for the line with slope m containing the point (x_1, y_1) is (Page 12)

$$y - y_1 = m(x - x_1).$$

■ The **slope-intercept** form of the equation for the line with slope m and y-intercept b is (Page 14)

$$y = mx + b.$$

■ An equation for the horizontal line through the point (a, b) is $y = b$. (Page 17)

■ An equation for the vertical line through the point (a, b) is $x = a$. (Page 17)

■ *Theorem:*   Let $\ell_1: y = m_1 x + b_1$ and $\ell_2: y = m_2 x + b_2$ be equations for lines ℓ_1 and ℓ_2. Then (Page 17)

  (i) ℓ_1 and ℓ_2 are **parallel** if $m_1 = m_2$.
  (ii) ℓ_1 and ℓ_2 are **perpendicular** if $m_1 m_2 = -1$.

■ *Definition:*   A **function** from set A to set B is a rule that assigns to each element of set A one and only one element of set B. Set A is called the **domain** of the function. (Page 26)

■ If, for a given manufacturer, C, R, and P denote the total cost, revenues, and profit from the production and sale of x items, then (Page 30)

$$P(x) = R(x) - C(x) \qquad \text{(profit equation)}.$$

■ *Definition:*   Let x and y be real numbers and let n and m be positive integers. Then (Page 38)

  (a) x^n means $x \cdot x \cdot x \cdot \cdots \cdot x$ (n factors)

  (b) x^{-n} means $\dfrac{1}{x^n}$ whenever $x \neq 0$.

  (c) $x^{1/n} = y$ means $y^n = x$.   ($x \geq 0$ if n is even.)

  (d) $x^{m/n}$ means $(x^{1/n})^m$.   ($x \geq 0$ if n is even.)

  (e) x^0 means   1 whenever $x \neq 0$.

■ **Theorem (Laws of Exponents):** Let x and y be real numbers and let n and m be integers. (Page 39)
Then

(a) $x^n x^m = x^{n+m}$
(b) $x^n/x^m = x^{n-m}, \quad x \neq 0$
(c) $(x^n)^m = x^{nm}$
(d) $x^{m/n} = (x^{1/n})^m = (x^m)^{1/n}$
(e) $(xy)^n = x^n y^n$.

■ The **quadratic formula** gives the zeros of the function $f(x) = ax^2 + bx + c$: (Page 44)

$$x = \frac{-b \pm \sqrt{b^2 - 4ac}}{2a}.$$

■ The **demand function** D gives the number of items that consumers will purchase in a unit of (Page 46)
time at price p. The **supply function** S gives the number of items that producers will supply in
a unit of time at price p. The **equilibrium price** p_0 is the price for which

$D(p_0) = S(p_0)$ (equilibrium price equation).

If $C(x)$ gives the total cost of producing x items, the **average cost** per item is

$$c(x) = \frac{C(x)}{x}.$$

■ The **composite function** $g \circ f$ has values $(g \circ f)(x) = g(f(x))$. (Page 51)

■ The **absolute value** function $f(x) = |x|$ is defined by (Page 54)

$$|x| = \begin{cases} x & \text{if } x \geq 0 \\ -x & \text{if } x < 0. \end{cases}$$

REVIEW EXERCISES—CHAPTER 1

1. Write each of the following sets of real numbers as an interval.
   a. $\{x \mid -6 < x \leq 3\}$
   b. $\{x \mid x < 4\}$
   c. $\{x \mid 2 \leq x\}$

In Exercises 2–5 solve the given inequality

2. $2x - 6 \leq 4$

3. $3 - x \geq 4$

4. $6x - 6 < x + 3$

5. $2 - x \leq 4 - 4x$

In Exercises 6–9 sketch the graph of the given function by plotting points.

6. $f(x) = x^3 - 2$

7. $f(x) = 2\sqrt{x} + 1$

8. $f(x) = \dfrac{1}{x - 3}$

9. $f(x) = \dfrac{x}{x^2 + 1}$

10. Find the slope of the line through the points $(-3, -2)$ and $(4, 1)$.

11. Find the slope of a line parallel to the line with equation $x - 6y = 24$.

12. Find an equation of the line perpendicular to the line $x - 6y = 24$ that contains the point $(2, -3)$.

13. Find an equation for the horizontal line containing the point $(-3, 5)$.

14. Find an equation for the vertical line containing the point $(-6, 2)$.

15. Find b if the point $(b, 4)$ lies on the line with no slope that contains the point $(7, -2)$.

16. Find an equation for the line containing the points $(2, -6)$ and $(-3, 4)$.

In Exercises 17–20 state the slope and y-intercept for the line determined by the given equation. Graph the line.

17. $x - 2y = 6$

18. $3x - y = -4$

19. $2x + 2y = 8$

20. $3 - y = 4x + 2$

In Exercises 21–24 state the domain of the given function.

21. $f(x) = \dfrac{x - 7}{x}$

22. $f(x) = \dfrac{x - 2}{x^2 - 4}$

23. $f(x) = \sqrt{9 - x^2}$

24. $f(x) = \dfrac{x + 3}{x^2 - 2x - 15}$

25. For the function $f(x) = \dfrac{\sqrt{x - 2}}{x + 3}$ find

    a. $f(6)$ **b.** $f(11)$

26. For the function $f(x) = x^{3/2} + 2x$ find
    a. $f(4)$ **b.** $f(9)$

In Exercises 27–30 simplify the given expression as much as possible.

27. $16^{3/4}$

28. $27^{-2/3}$

29. $\dfrac{(x^2 y^4)^2}{xy^2}$

30. $\dfrac{3\sqrt{xy}}{x^2\sqrt{y}}$

In Exercises 31–36 graph the circle whose equation is given.

31. $x^2 + y^2 = 9$

32. $x^2 + y^2 = 16$

33. $x^2 + 2x + y^2 = 5$

34. $x^2 + y^2 + 6y = 13$

35. $x^2 + 4x + y^2 + 2y = 14$

36. $x^2 - 2x + y^2 - 9y = 22$

In Exercises 37–42 find all zeros of the given function.

37. $f(x) = x^2 - 3x + 2$

38. $f(x) = x^2 + 4x + 4$

39. $f(x) = \dfrac{x - 3}{x + 2}$

40. $f(x) = \sqrt{x} - x^{3/2}$

41. $f(x) = \dfrac{x^2 - 9}{x + 2}$

42. $f(x) = 6 - 2x - x^2$

In Exercises 43–46 find the point(s) of intersection of the graphs of f and g.

43. $f(x) = x - 6$
    $g(x) = 2x + 2$

44. $f(x) = \sqrt{x + 2}$
    $g(x) = x$

45. $f(x) = 2 - x^2$

    $g(x) = 4x^2 - 3$

46. $f(x) = \dfrac{1}{x + 2}$

    $g(x) = 4$

In Exercises 47–50 let $f(x) = x^3 + 2$, $g(x) = x + 4$, $h_1 = fg$, and $h_2 = f/g$. Find

47. $h_1(x)$

48. $h_2(x)$

49. $h_1(-2)$

50. $h_2(-2)$

In Exercises 51–54 let $f(x) = x^2 - 7$, $g(x) = \sqrt{x + 3}$ and $u(x) = 4x$. Find equations for the indicated composite functions.

51. $g \circ f$

52. $g \circ u$

53. $u \circ f$

54. $g \circ f \circ u$

In Exercises 55–58 graph the given function.

55. $f(x) = |x + 3|$

56. $f(x) = |x^2 - 2x - 3|$

57. $f(x) = \begin{cases} x - 4, & x < 2 \\ -\frac{1}{2}x^2, & x \geq 2 \end{cases}$

58. $f(x) = \begin{cases} \sqrt{2 - x}, & x \leq 1 \\ 2x - 1, & x > 1 \end{cases}$

59. Market research indicates that the demand for 12″ color television sets is $D(p) = 2000 - 6p$ sets per month if the sets sell for p dollars each. Producers are willing to supply $S(p) = 4p$ sets per month at price p. What is the equilibrium price $[D(p_0) = S(p_0)]$ for this market?

60. After x repetitions a worker can perform a task in

$$T(x) = 10\left(1 + \frac{4}{\sqrt{x + 2}}\right) \text{ seconds. Graph this learn-}$$

ing curve.

61. Which of Figures a–d are graphs of functions?

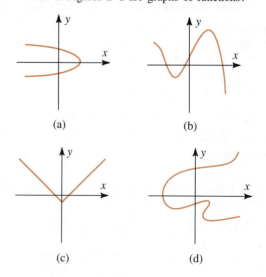

(a)

(b)

(c)

(d)

62. Find the lowest point on the graph of $f(x) = 3(x - 2)^2 + 4$.

63. True or false? The graph of $y = 3x^2 + 2x - 1$ opens downward.

64. Find an equation for the line with slope $m = -4$ and containing the point $(2, -3)$.

65. A restaurant agrees to serve a banquet at a cost of $20 per person for the first 50 people and $15 per person for each additional person. Find the function C giving the total cost of the meal.

66. The rate of personal savings in the United States rose from 5.6% of disposable income in the first quarter of 1981 to 6.0% in the first quarter of 1982. If the personal savings rate is a linear function of time, what did these data predict for a personal savings rate in the first quarter of 1986?

67. The price of copper in the United States rose from 66¢ per pound in 1982 to 74¢ per pound in 1984. If the price of copper is a linear function of time, what price do these data predict for 1990?

68. Air quality engineers predict that the level of air pollutants in the center of a particular city will be $c(n) = \sqrt{1 + 0.2n}$ parts per million, where n is the average number of automobiles in the center city area. Traffic engineers predict that the average number of automobiles in the center of the city after t years will be $n(t) = 10{,}000 + 50t^{2/3}$. Write a composite function $c \circ n$ giving the level of air pollutants after t years.

69. The average daily temperature in a certain city rose from 58°F in 1940 to 59.2°F in 1980. If the relationship between temperature and time is assumed to be linear, what temperature do these data project for this city's average daily temperature in 1990?

70. Cab fare in a certain city is $2 per ride plus $1.20 per mile. How far can you ride for $10?

71. The *disposable annual income* for an individual is defined by the equation $D = (1 - r)T$ where T is total income and r is the net tax rate applied to his income.
   a. Find the disposable income for an individual with a total income of $30,000 subject to a net tax rate of $r = 30\% = 0.30$.

b. What net tax rate gives a disposable income of $16,000 from a total income of $20,000?
   c. Under a net tax rate of 40%, what total income is required to yield a disposable income of $30,000?

72. If a piece of equipment originally valued at V_0 dollars is assumed to have a useful life of N years, after which it will have a salvage value of V_s dollars, the *straight line* (constant rate) method of depreciation gives its value after n years as $V(n) = V_s + (V_0 - V_s)\left(1 - \dfrac{n}{N}\right)$. Find m and b so that this linear function can be written in the form $V(n) = mn + b$.

73. Refer to Exercise 72. If a computer originally valued at $30,000 is to be depreciated by the straight line method to a salvage value of $5,000 over a period of five years, find its assumed value at the end of three years.

74. Market research shows that the demand for a particular type of mechanical pencil will be $D(p) = \dfrac{400}{p^2 + 2p + 5}$ thousand pencils per year at a selling price of p dollars per pencil. At what price will demand equal 20 thousand pencils per year?

75. Agricultural research shows that a certain type of apple tree will yield 400 pounds of apples per year minus 2 pounds for each tree planted per acre. Write a function Y giving the annual yield in pounds of a one-acre plot containing x such apple trees.

76. A 6-inch thick blanket of glass fibre insulation has an R-value of $R(6) = 19$. If R-value is a linear function of the thickness w of the insulation, what R-value is produced by 15 inches of glass fibre insulation? (*Hint:* $R(0) = 0$.)

77. A manufacturer of smoke detectors finds that its total cost in producing x units per day is $C(x) = C_0 + 16x + 0.02x^2$ dollars where C_0 represents fixed daily costs. The alarms are sold to retailers for $24 each. Find C_0 if the manufacturer's break-even production level (total costs equal total revenue) is to be $x = 100$ units per day.

78. Write a linear function giving the total cost C of leasing a car for x months if a driver travels 800 miles per month and the leasing fees are $100 per month plus 20¢ per mile for each mile over 500 per month.

79. For tax purposes a building that originally cost $100,000 is to be fully depreciated by the straight line method over a period of $N = 20$ years. This means that its value after n years is assumed to be $V(n) = (100,000)\left(1 - \dfrac{n}{20}\right)$.

a. Find $V(5)$, its value after 5 years.

b. Find an expression for $V(n)$ if the depreciation period is changed to $N = 30$ years.

c. Find $V(5)$ if $N = 30$ years.

80. Demand for a certain type of dishwasher is known to be

$$D(p) = \frac{154}{p}$$

units per month at selling price p hundred dollars. If producers are willing to supply $S(p) = 2p - 8$ units per month at selling price p hundred dollars $(p \geq 4)$, find the equilibrium price p_0 at which $S(p_0) = D(p_0)$.

Chapter 2 — *The Derivative*

Much of both natural and social science has to do with the study of change. How does climatic fluctuation affect the growth of vegetation in an ecological niche? How will an increase in excise taxes affect the consumption of gasoline? Or, what increase in sales will result from a 20% reduction in the price of a personal computer?

The **derivative** is a mathematical tool that we can use to measure rates of change between and among such variables. Although a powerful tool, this concept can best be understood by beginning with the slope of a straight line as a simple example of a rate of change and then extending the notion to more general types of functions.

2.1 THE SLOPE OF A CURVE

Figure 1.1 shows how the slope of a straight line may be interpreted as a **rate of change.** Since the definition of slope is

$$m = \frac{y_2 - y_1}{x_2 - x_1} \qquad \left(= \frac{\text{change in } y\text{-coordinate}}{\text{change in } x\text{-coordinate}} \right) \tag{1}$$

when m is known, we may choose $x_2 - x_1 = 1$ and use equation (1) to conclude that the y-coordinate of a point on ℓ changes by $m = \dfrac{y_2 - y_1}{1} = y_2 - y_1$ units whenever the x-coordinate increases by 1 unit. (See Figure 1.2.) This is what is meant by the *rate* at which the y-coordinate changes with respect to change in the x-coordinate.

For example, if the linear demand function $D(p) = 40 - 2p$ gives the number of hair dryers that consumers will purchase per week at price p, the slope $m = -2$ signifies that demand will decrease by two hair dryers per week for each one dollar increase in price. (See Figure 1.3.) That is, demand will decrease at a *rate* of two items per week for each one dollar increase in price.

But many of the functions of interest in business and other applications are not linear functions. As Figure 1.4 suggests, we would like to be able to determine the rate at which

Figure 1.1 Slope of ℓ is $m = \dfrac{y_2 - y_1}{x_2 - x_1}$.

Figure 1.2 When $x_2 - x_1 = 1$, we have $m = \dfrac{y_2 - y_1}{1} = y_2 - y_1$, the rate at which y changes with respect to change in x.

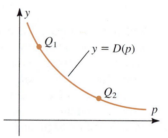

Figure 1.3 For $D(p) = 40 - 2p$ demand changes by $m = -2$ for each one dollar increase in price.

Figure 1.4 At what rate is demand changing at points Q_1 or Q_2?

values of a demand function, for example, change with respect to change in price even though its graph is not a straight line. Since the graph is not a straight line, we should expect to find different rates of change associated with different prices.

We shall answer this question in terms of *tangents* to a given curve. After resolving the question in this geometric setting, we shall return to the discussion of rates of change and the economic and physical interpretations of this concept.

The Tangent to a Curve

Intuitively, the line **tangent** to a curve C at a point P is the straight line that most resembles the curve at P. Figure 1.5 shows one way to visualize this concept. By repeatedly magnifying a small region surrounding the point P, we should observe the part of the curve "near" P flattening out and resembling a straight line. This line is what we mean by the tangent to the curve C at P.

Figure 1.6 shows both the curve C and its tangent at point P. The reason why we are so interested in this tangent is that its slope, which we can compute because a tangent is a straight line, is what we shall mean by the slope of the curve C at point P.

The difficulty with the preceding discussion is that it does not provide a precise meaning for the concept of a line tangent to a curve. Here is a more careful description of what

Figure 1.5 By magnifying an increasingly small region near P we observe the curve C looking more and more like a straight line. This straight line is the *tangent* to the curve C at P.

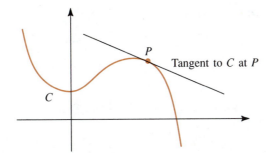

Figure 1.6 The slope of a curve C at point P is the slope of the line *tangent* to the curve C at P.

a tangent really is, one that also provides a means for actually calculating the slope of the tangent that is defined.

We begin by assuming that the curve C is actually the graph of a function $y = f(x)$ and that the point P has coordinates $P = (x_0, y_0)$ with $y_0 = f(x_0)$. If we let $Q = (x_1, y_1)$ be a second point on this graph, then the line ℓ_{sec} through P and Q is called a **secant** to the graph of $y = f(x)$ at P. (See Figure 1.7.)

Figure 1.8 shows how these secants yield a tangent at point P. By moving point Q toward point P along the graph of $y = f(x)$, we generate an entire family of secant lines through P, which approach a "limiting position" as Q approaches P. This "limiting position" is our tangent, ℓ_{tan}, to the graph of $y = f(x)$ at point $P = (x_0, y_0)$.

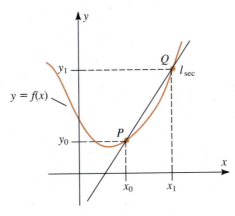

Figure 1.7 Secant ℓ_{sec} through $P = (x_0, y_0)$ and $Q = (x_1, y_1)$ on the graph of $y = f(x)$.

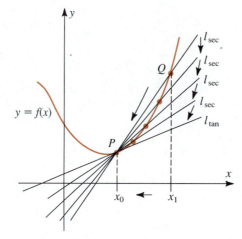

Figure 1.8 Tangent $\ell_{\tan}$ is limiting position of secants as Q approaches P.

The Slope of a Tangent to a Curve

By examining this process a bit more critically, we can see how to find the slope of $\ell_{\tan}$. With $h = x_1 - x_0$ we can write $x_1 = x_0 + h$ and $f(x_1) = f(x_0 + h)$. Then, as Figure 1.9 shows, the slope of the secant through P and Q is

$$\text{slope of } \ell_{\sec} = \frac{f(x_0 + h) - f(x_0)}{h} \qquad h \neq 0.$$

Since the tangent $\ell_{\tan}$ is the limiting position of $\ell_{\sec}$ as Q approaches P (that is, as h approaches zero), we say that

$$\text{slope of } \ell_{\tan} = \{\text{``limit'' of slope of } m_{\sec} \text{ as } h \to 0\}$$

$$= \lim_{h \to 0} \frac{f(x_0 + h) - f(x_0)}{h}.$$

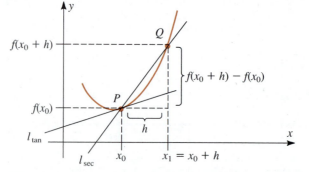

Figure 1.9 Slope of tangent to graph of $y = f(x)$ at the point $P = (x_0, y_0)$ is

$$m = \lim_{h \to 0} \frac{f(x_0 + h) - f(x_0)}{h}.$$

in the sense that the number $\dfrac{f(x_0 + h) - f(x_0)}{h}$ can be made as close as desired to the slope of ℓ_{tan} simply by taking h sufficiently close to 0.

Combining this observation with the point-slope form for the equation of a line, we may summarize our findings about ℓ_{tan} as follows:

If ℓ is the line tangent to the graph of $y = f(x)$ at the point $(x_0, f(x_0))$, then the slope of ℓ is

$$m = \lim_{h \to 0} \frac{f(x_0 + h) - f(x_0)}{h} \qquad (2)$$

and an equation for ℓ is

$$y - y_0 = m(x - x_0) \qquad (3)$$

where $y_0 = f(x_0)$.

The symbol "$\lim\limits_{h \to 0}$" in equation (2) is read, "the *limit* as h approaches zero of." The meaning of this, roughly speaking, is the result of allowing the number h to "shrink to zero" in the expression that follows. For example, we would write

$$\lim_{h \to 0} (2 + h) = 2$$

read, "the limit of the expression $2 + h$ as h approaches zero is 2," since the number $2 + h$ "approaches" the number 2 as the number h "approaches" zero. Similarly, we would write

$$\lim_{h \to 0} (x + 4h) = x$$

since, as h approaches zero, the number $4h$ approaches zero but the number x remains unchanged. We shall discuss the concept of *limit* more generally in Section 2.2.

The following examples show how equation (2) may be applied to actually calculate the slope of a tangent to a particular curve at a given point.

Example 1 Use equation (2) to find the slope of the line tangent to the graph of the function $f(x) = x^2$ at the point $(2, 4)$.

Strategy	*Solution*
Identify the number x_0.	Here the specified value of x_0 is

$$x_0 = 2$$

so a nearby value of x is

$$x_0 + h = 2 + h.$$

From the equation for $f(x)$ determine $f(x_0)$ and $f(x_0 + h)$.

Since $f(x) = x^2$, we have

$$f(x_0) = f(2) = 2^2 = 4$$

and

$$f(x_0 + h) = f(2 + h) = (2 + h)^2.$$

Set up the expression for m from equation (2).

Substituting each of these expressions in equation (2), we find that

Simplify this expression as much as possible.

$$m = \lim_{h \to 0} \frac{f(2 + h) - f(2)}{h} \qquad (4)$$

$$= \lim_{h \to 0} \frac{(2 + h)^2 - 2^2}{h}$$

$$= \lim_{h \to 0} \frac{[4 + 4h + h^2] - 4}{h}$$

$$= \lim_{h \to 0} \frac{4h + h^2}{h}$$

Factor h from each term in the numerator.

$$= \lim_{h \to 0} \frac{h(4 + h)}{h}$$

$$= \lim_{h \to 0} \left(\frac{h}{h}\right)(4 + h)$$

"Cancel" common factor of $h \neq 0$. Evaluate the limit by letting h approach zero.

$$= \lim_{h \to 0} (4 + h) \qquad \left(\frac{h}{h} = 1, \text{ provided } h \neq 0\right)$$

$$= 4.$$

Thus $m = 4$ at the point $(2, 4)$. (See Figure 1.10.) □

It is important to note that we cannot evaluate the expression for m in equation (2) by simply setting $h = 0$, since this will produce a factor of zero in its denominator. In particular, setting $h = 0$ in line (4) of Example 1 would have given the meaningless expression

$$\frac{f(2 + 0) - f(2)}{0}. \qquad \text{(undefined)}$$

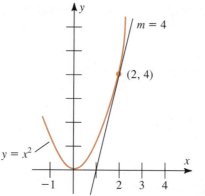

Figure 1.10 Slope of tangent to graph of $f(x) = x^2$ at $(2, 4)$ is $m = 4$.

To evaluate properly the expression for m, we must follow the steps outlined in Example 1.

Method for finding m, the slope of $\ell_{\tan}$ at $x = x_0$:

(1) Set up the expression
$$m = \lim_{h \to 0} \frac{f(x_0 + h) - f(x_0)}{h}.$$

(2) Using algebra, simplify the numerator and factor h from each of its terms.

(3) "Cancel" the common factor of h from both numerator and denominator.
$$\left(\text{That is, use the fact that } \frac{h}{h} = 1 \text{ when } h \neq 0. \right)$$

(4) Finally, determine the limit by allowing the remaining factors of h to approach zero.

Here is another example of how these steps are followed in calculating the slope of a curve.

Example 2 Find the slope of the line tangent to the graph of the function $f(x) = 1 + 2x - x^2$ at the point where $x = 1$.

Strategy

Identify x_0.

Set up expression for m.

Substitute expressions for $f(x_0 + h)$, $f(x_0)$.

Simplify by algebra.

"Cancel" common factor of h.

Let h approach zero.

Solution

With $x_0 = 1$ we have

$$m = \lim_{h \to 0} \frac{f(1 + h) - f(1)}{h}$$

$$= \lim_{h \to 0} \frac{[1 + 2(1 + h) - (1 + h)^2] - [1 + 2 \cdot 1 - 1^2]}{h}$$

$$= \lim_{h \to 0} \frac{[1 + 2(1 + h) - (1 + 2h + h^2)] - 2}{h}$$

$$= \lim_{h \to 0} \frac{(2 - h^2) - 2}{h}$$

$$= \lim_{h \to 0} \frac{-h^2}{h}$$

$$= \lim_{h \to 0} - \left(\frac{h}{h} \right) h$$

$$= \lim_{h \to 0} - h$$

$$= 0.$$

Thus $m = 0$ when $x = 1$.
(See Figure 1.11.)

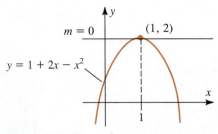

Figure 1.11 Slope of tangent to graph of $f(x) = 1 + 2x - x^2$ at $(1, 2)$ is $m = 0$.

The following examples show how equations (2) and (3) can be used together to find an *equation* for the line tangent to the graph of $y = f(x)$ at the point $P = (x_0, y_0)$.

Example 3 Find an equation for the line tangent to the graph of $f(x) = x^2$ at the point $(2, 4)$.

Solution: The slope of this tangent line was found in Example 1 to be $m = 4$. Using equation (3) with $x_0 = 2$ and $y_0 = 4$, we obtain the equation

$$y - 4 = 4(x - 2)$$

or

$$y = 4x - 4.$$ □

Example 4 Find an equation for the line tangent to the graph of the function $f(x) = \dfrac{3}{x + 1}$ at the point where $x = 2$. (See Figure 1.12.)

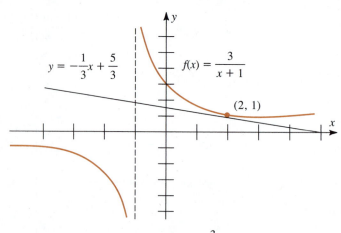

Figure 1.12 Tangent to graph of $f(x) = \dfrac{3}{x + 1}$ at $(2, 1)$ has equation $y = -\dfrac{1}{3}x + \dfrac{5}{3}.$

Strategy	**Solution**

Strategy

Set up the expression for m.

Solution

Using $x_0 = 2$, we first calculate the slope of this tangent line as

$$m = \lim_{h \to 0} \frac{f(2 + h) - f(2)}{h}$$

Substitute for $f(x_0 + h)$, $f(x_0)$.

$$= \lim_{h \to 0} \frac{\dfrac{3}{(2 + h) + 1} - \dfrac{3}{2 + 1}}{h}$$

Simplify the expression by

$$= \lim_{h \to 0} \left(\frac{1}{h}\right)\left[\frac{3}{3 + h} - 1\right]$$

1. finding a common denominator

$$= \lim_{h \to 0} \left(\frac{1}{h}\right)\left[\frac{3}{3 + h} - \left(\frac{3 + h}{3 + h}\right)\right]$$

2. combining terms

$$= \lim_{h \to 0} \left(\frac{1}{h}\right)\left[\frac{3 - (3 + h)}{3 + h}\right]$$

3. factoring h from the numerator

$$= \lim_{h \to 0} \left(\frac{1}{h}\right)\left(\frac{-h}{3 + h}\right)$$

4. canceling the common factor of h

$$= \lim_{h \to 0} \left(\frac{h}{h}\right)\left(\frac{-1}{3 + h}\right)$$

$$= \lim_{h \to 0} \left(\frac{-1}{3 + h}\right)$$

Evaluate the limit.

$$= -\frac{1}{3}.$$

The desired slope is therefore $m = -\dfrac{1}{3}$.

Use equation (3) to write the equation for the tangent line.

Now when $x_0 = 2$, $y_0 = f(2) = \dfrac{3}{2 + 1} = 1$.

Equation (3) then gives the equation for the desired tangent line as

$$y - 1 = -\frac{1}{3}(x - 2)$$

or

$$y = -\frac{1}{3}x + \frac{5}{3}.$$

(See Figure 1.12.) □

In this section we have defined the slope of a curve at a point P to be the slope of the line *tangent* to the curve at P. To find the slope of this tangent, we have introduced the concept of *limit*. The notion of the slope of a tangent will lead us in Section 4 to the more general concept of the *derivative* of a function, one of the principal objects of interest in the calculus. Before taking up the study of derivatives, we need to develop a better understanding of the idea of the *limit* of a function. This is the agenda for Sections 2.2 and 2.3.

Exercise Set 2.1

In Exercises 1–6 estimate the slope m_{tan} of the line tangent to the graph of the given function at point P by positioning a straight edge in the position of the tangent and determining the slope by inspection.

1. $f(x) = 9 - x^2$

2. $f(x) = x^3$

3. $f(x) = \sqrt{x}$

4. $f(x) = \dfrac{1}{x - 2}$

5. $f(x) = \dfrac{1}{1 + x^2}$

6. $f(x) = x^2 + 2x - 3$

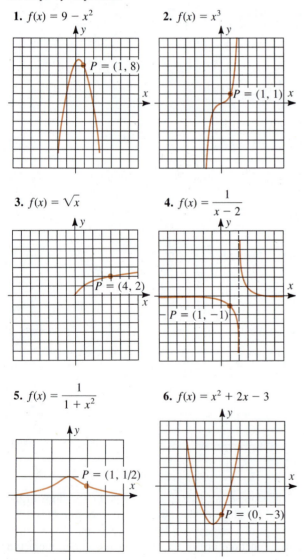

In Exercises 7–16 use the method of this section to find the slope of the line tangent to the graph of the given function at point P.

7. $f(x) = x^2 + 2$, $P = (1, 3)$

8. $f(x) = 3x^2 - 1$, $P = (2, 11)$

9. $f(x) = 2 - x^2$, $P = (1, 1)$

10. $f(x) = x^2 + x$, $P = (-1, 0)$

11. $f(x) = x^3$, $P = (1, 1)$

12. $f(x) = x^2 + x - 2$, $P = (2, 4)$

13. $f(x) = x^3 + x$, $P = (-1, -2)$

14. $f(x) = \dfrac{1}{x}$, $P = (1, 1)$

15. $f(x) = \dfrac{1}{x + 3}$, $P = (-2, 1)$

16. $f(x) = \dfrac{x}{1 + x}$, $P = (2, \frac{2}{3})$

17. Find an equation for the line tangent to the graph of $f(x) = 4x^2 - 3$ at the point $(1, 1)$.

18. Find an equation for the line tangent to the graph of $f(x) = 9 - x^2$ at the point where $x = -2$.

19. Find an equation for the line tangent to the graph of $f(x) = \dfrac{2}{3 + x}$ at the point $(1, \frac{1}{2})$.

20. Use the method of this section to show that the line tangent to the graph of $f(x) = mx + b$ at any point P on the graph is just the line $y = mx + b$.

21. Find the slope of the supply curve $S(p) = \frac{1}{4}p^2 + 4$, $p > 0$, at the point where $p = 4$.

22. What is the slope of the demand curve $D(p) = \dfrac{10}{p + 4}$, $p > 0$, at the point where $p = 6$?

23. Find the slope of the graph of
$$f(x) = \begin{cases} 4 - x^2, & x \leq 1 \\ 2x + 1, & x > 1 \end{cases} \quad \text{at the point}$$
  a. $P = (-1, 3)$.
  b. $P = (2, 5)$.

24. Find the slope of the graph of the function $f(x) = |x - 3|$ at the point $P = (a, f(a))$ for
  a. $a < 3$.
  b. $a > 3$.

25. For the supply function $S(p) = \frac{1}{4}p^2$ and the demand function $D(p) = 12 - 2p$, find
  a. the equilibrium price p_0, where $S(p_0) = D(p_0)$.
  b. the slope of the supply curve at the equilibrium point $(p_0, S(p_0))$.
  c. the slope of the demand curve at the point $(p_0, D(p_0))$.

26. The quadratic function $y = a + bt + ct^2$ is used to model human length during certain periods of growth. Find the slope of the line tangent to the graph of this function at the point where $t = 1$.

27. During the first 50 days following the outbreak of influenza on a college campus the number $N(t)$ of students infected after t days is given by the function $N(t) = 25t - (\frac{1}{2})t^2$, $\quad 0 \leq t \leq 50$.
  a. Graph the function $y = N(t)$ for $0 \leq t \leq 50$.
  b. Find the slope of the line tangent to this graph at the point where $t = 1$.
  c. Find the slope of the line tangent to this graph at the point where
  $$t = a, \quad 0 \leq a \leq 50.$$
  d. At what time t, between 0 and 50, is the slope of the tangent to this graph horizontal?

2.2 LIMITS OF FUNCTIONS

In Section 2.1 we determined that the slope of the line tangent to the graph of $y = f(x)$ at the point $(x_0, f(x_0))$ is

$$m = \lim_{h \to 0} \frac{f(x_0 + h) - f(x_0)}{h}. \tag{1}$$

The purpose of this section is to take a closer look at the meaning of the *limit* concept used in equation (1). You will discover that the notion of limit is a central theme in the calculus, one that is used in studying many ideas in addition to the slope of a tangent line.

Before addressing the more general notion of the limit of a function, let's take one additional look at the limit that arises in calculating the slope of a tangent to a curve. Figure 2.1 shows the graph of the function $f(x) = x^2$ and one of the secant lines used in defining the tangent at the point $(1, 1)$. According to the way we have labeled Figure 2.1, the slope of this secant is

$$m_{\text{sec}} = \frac{f(x) - f(1)}{x - 1} = \frac{x^2 - 1}{x - 1}. \tag{2}$$

Since $h = x - 1$ in this example, to find the slope of the tangent at $(1, 1)$ we want to find the limit of the expression for m_{sec} in equation (2) as x approaches 1 (which is the same as h approaching zero):

$$m = \lim_{x \to 1} \frac{x^2 - 1}{x - 1}.$$

Figure 2.2 shows the graph of this quotient $g(x) = \dfrac{x^2 - 1}{x - 1}$ and what happens as we

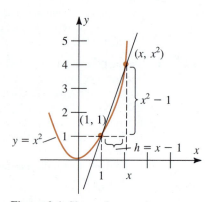

Figure 2.1 Slope of secant is $m = \dfrac{x^2 - 1}{x - 1}$.

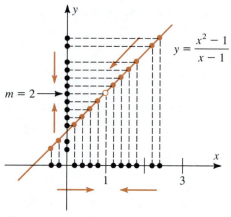

Figure 2.2 As the numbers x approach 1, values of $\dfrac{x^2 - 1}{x - 1}$ approach 2. Thus

$$\lim_{x \to 1} \frac{x^2 - 1}{x - 1} = 2.$$

choose x "closer and closer" to $x = 1$. Some examples of such choices are listed in Table 2.1.

Table 2.1

Number x	2.0	1.5	1.1	1.01	1.001	0.999	0.99	0.9	0.5	0
Value of $g(x) = \dfrac{x^2 - 1}{x - 1}$	3.0	2.5	2.1	2.01	2.001	1.999	1.99	1.9	1.5	1

Note that as x approaches the number 1 from either direction, the values $g(x)$ approach the number $m = 2$. That is why we conclude that

$$m = \lim_{x \to 1} g(x) = \lim_{x \to 1} \frac{x^2 - 1}{x - 1} = 2.$$

There are two important observations concerning this result:

1. We *cannot* determine the limit $\lim_{x \to 1} g(x)$ by simply setting $x = 1$ since $g(1)$ is not defined. [Indeed, setting $x = 1$ in $g(x) = \dfrac{x^2 - 1}{x - 1}$ gives $\dfrac{1^2 - 1}{1 - 1} = \dfrac{0}{0}$, which is not defined. That is why there is a "hole" in the graph of $y = g(x)$ at location $(1, 2)$ in Figure 2.2.]

2. In this example, however, we could have used algebra to factor the numerator and simplify, thus allowing the limit to be obtained "by inspection," as in Section 2.1, as follows:

$$m = \lim_{x \to 1} \frac{x^2 - 1}{x - 1}$$

$$= \lim_{x \to 1} \frac{(x - 1)(x + 1)}{x - 1}$$

$$= \lim_{x \to 1} \left(\frac{x - 1}{x - 1} \right)(x + 1)$$

$$= \lim_{x \to 1} (x + 1) \qquad \left(\left(\frac{x - 1}{x - 1} \right) = 1, \text{ provided } x \neq 1 \right)$$

$$= (1 + 1)$$

$$= 2.$$

This is an example of the more general notion of the limit of a function.

The Limit of a Function

The example of finding the slope of a tangent to a curve is one case of needing to know what happens to the values of a function, say, $f(x)$, as x approaches a particular number, say, $x = a$, *even though $f(a)$ itself may not be defined*. Figure 2.3 shows such a situation, one where the values $f(x)$ approach the number $y = L$ as x approaches the number $x = a$. This is what we mean by writing $L = \lim_{x \to a} f(x)$.

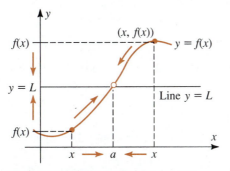

Figure 2.3 $L = \lim_{x \to a} f(x)$; values $f(x)$ approach the number L as x approaches a.

DEFINITION 1

We say that the number L is the *limit of the function f as x approaches a*, written

$$L = \lim_{x \to a} f(x)$$

if the values $f(x)$ approach the unique number L as x approaches a from either direction. (See Figure 2.3.)

The following examples illustrate various typical situations in which limits of functions are determined.

Example 1 Since we have already mastered the techniques for graphing quadratic functions, it is clear from the graph of a quadratic function $f(x) = Ax^2 + Bx + C$ that

$$\lim_{x \to a} (Ax^2 + Bx + C) = Aa^2 + Ba + C.$$

That is, the limit as x approaches a of a quadratic function f is just the value $f(a)$. Figure 2.4 shows the particular case

$$\lim_{x \to 3} (4 + 3x - x^2) = 4.$$

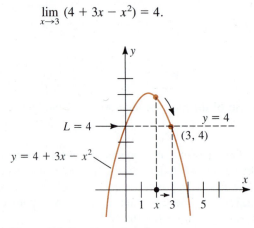

Figure 2.4 $\lim_{x \to 3} (4 + 3x - x^2) = 4.$

Functions that have the property that $\lim_{x \to a} f(x) = f(a)$ are called *continuous* at the number $x = a$. We will have more to say about continuous functions in the next section. □

Example 2 The limit $\lim_{x \to 2} \dfrac{x^2 + x - 6}{x - 2}$ cannot be evaluated by simply letting $x = 2$ since the rational function $f(x) = \dfrac{x^2 + x - 6}{x - 2}$ is not defined for $x = 2$. We must apply the method of Section 2.1. That is, we must try to factor the "offending" term $(x - 2)$ from the numerator as follows:

$$\lim_{x \to 2} \frac{x^2 + x - 6}{x - 2} = \lim_{x \to 2} \frac{(x - 2)(x + 3)}{x - 2}$$

$$= \lim_{x \to 2} \left(\frac{x - 2}{x - 2}\right)(x + 3)$$

$$= \lim_{x \to 2} (x + 3)$$

$$= 2 + 3$$

$$= 5.$$

In this case we have used the fact that values of the rational function $f(x) = \dfrac{x^2 + x - 6}{x - 2}$ equal those of the linear function $g(x) = x + 3$ except when $x = 2$, because $f(2)$ is undefined. (See Figure 2.5.) □

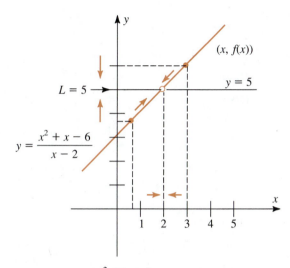

Figure 2.5 $\lim\limits_{x \to 2} \dfrac{x^2 + x - 6}{x - 2} = 5.$

Example 3 If we change the numerator in the function in Example 2 to $x^2 + x + 6$, we find that the limit

$$\lim_{x \to 2} \frac{x^2 + x + 6}{x - 2}$$

does not exist. This is because $(x - 2)$ is no longer a factor of the numerator, so it cannot be "canceled." As x approaches 2, the numerator approaches $2^2 + 2 + 6 = 12$ while the denominator approaches zero. Thus the quotient "blows up." Table 2.2 shows the result of using BASIC Program 1 in Appendix I to calculate values $f(x) = \dfrac{x^2 + x + 6}{x - 2}$ for various numbers x near $x_0 = 2$. You can see the phenomenon described here in these data. Figure 2.6, the graph of $y = f(x)$, illustrates graphically why this limit does not exist. □

Table 2.2. Values $f(x) = \dfrac{x^2 + x + 6}{x - 2}$

near $x_0 = 2$

x	$f(x) = \dfrac{x^2 + x + 6}{x - 2}$
3	18
2.5	29.5
2.1	125
2.01	1205
2.001	12005
2.0001	120005
1	-8
1.5	-19.5
1.9	-115
1.99	-1195
1.999	-11995
1.9999	-119995

Figure 2.6 $\displaystyle\lim_{x\to 2} \dfrac{x^2 + x + 6}{x - 2}$ does not exist.

Properties of Limits

In Section 1.7 we reviewed the ways in which functions may be combined using algebra to form new functions. When limits of each of these individual functions exist, the limits of the resulting combinations of functions can be calculated from the limits of the original functions. The following theorem shows how this is done.

THEOREM 1

Let f and g be functions for which

$$\lim_{x\to a} f(x) = L, \quad \text{and} \quad \lim_{x\to a} g(x) = M$$

and let c be any real number. Then

(i) $\displaystyle\lim_{x\to a} (f + g)(x) = \lim_{x\to a} f(x) + \lim_{x\to a} g(x) = L + M$

(ii) $\displaystyle\lim_{x\to a} (cf)(x) = c[\lim_{x\to a} f(x)] = cL$

(iii) $\displaystyle\lim_{x\to a} (fg)(x) = [\lim_{x\to a} f(x)] \cdot [\lim_{x\to a} g(x)] = L \cdot M$

(iv) $\displaystyle\lim_{x\to a} (f/g)(x) = [\lim_{x\to a} f(x)]/[\lim_{x\to a} g(x)] = L/M$, provided $M \neq 0$.

The following examples show how Theorem 1 may be used.

Example 4 Find $\lim_{x \to 3} (2x^2 + 4x + 9)$.

Strategy   *Solution*

Apply Theorem 1, part (i).

$$\lim_{x \to 3} (2x^2 + 4x + 9) = \lim_{x \to 3} (2x^2) + \lim_{x \to 3} (4x) + \lim_{x \to 3} (9)$$

Apply Theorem 1, part (ii).

$$= 2 \lim_{x \to 3} (x^2) + 4 \lim_{x \to 3} (x) + \lim_{x \to 3} (9)$$

$$= 2 \cdot 3^2 + 4 \cdot 3 + 9$$

$$= 39. \qquad \square$$

Example 5 Find $\lim_{x \to 2} \dfrac{x^2 - 4x}{3x + 6}$.

Strategy   *Solution*

Apply Theorem 1, part (iv).

$$\lim_{x \to 2} \frac{x^2 - 4x}{3x + 6} = \frac{\lim_{x \to 2} (x^2 - 4x)}{\lim_{x \to 2} (3x + 6)}$$

Apply Theorem 1, part (i).

$$= \frac{\lim_{x \to 2} (x^2) + \lim_{x \to 2} (-4x)}{\lim_{x \to 2} (3x) + \lim_{x \to 2} (6)}$$

Apply Theorem 1, part (ii).

$$= \frac{\lim_{x \to 2} (x^2) - 4 \lim_{x \to 2} (x)}{3 \lim_{x \to 2} (x) + \lim_{x \to 2} (6)}$$

$$= \frac{2^2 - 4 \cdot 2}{3 \cdot 2 + 6}$$

$$= -\frac{4}{12}$$

$$= -\frac{1}{3}. \qquad \square$$

Limits at Infinity

We can use the notion of limit to describe the behavior of a function as x becomes very large rather than as x approaches a particular number. Figure 2.7 shows the graph of the function

$$f(x) = \frac{x + 2}{x}.$$

Notice that as x increases in the positive direction, the values of the function $f(x)$ approach the number 1. We can see why this happens by writing $f(x)$ as

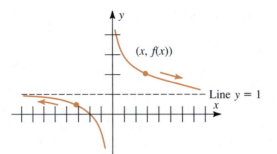

Figure 2.7 $\lim\limits_{x\to\infty} \dfrac{x+2}{x} = 1$; $\lim\limits_{x\to-\infty} \dfrac{x+2}{x} = 1$.

$$f(x) = \frac{x+2}{x}$$

$$= \frac{x}{x} + \frac{2}{x}$$

$$= 1 + \frac{2}{x}.$$

Since the number $\dfrac{2}{x}$ "shrinks toward zero" as x becomes large, the values $f(x)$ approach the number $1 + 0 = 1$. In this case we would write

$$\lim_{x\to\infty} \frac{x+2}{x} = 1.$$

The meaning of this *limit at infinity* is the following.

DEFINITION 2

We say that the number L is the *limit of the function f as x approaches infinity*, written

$$L = \lim_{x\to\infty} f(x)$$

if the values $f(x)$ approach the unique number L as x increases without bound.

    We say that the number M is the *limit of the function f as x approaches negative infinity*, written

$$M = \lim_{x\to-\infty} f(x)$$

if the values $f(x)$ approach the number M as x decreases without bound.

Figure 2.7 also shows that

$$\lim_{x\to-\infty} \frac{x+2}{x} = 1.$$

That is, the values of the function $f(x) = \dfrac{x + 2}{x}$ approach the number 1 as x decreases without bound through negative numbers.

The following example shows how we can use the technique of *dividing the numerator and denominator by the highest power of x present to evaluate a limit at infinity.*

Example 6 To evaluate the limit

$$\lim_{x \to \infty} \frac{3x^2 + 2x + 1}{x^2 + 4x + 5}$$

we divide both numerator and denominator by x^2, the highest power of x in both the numerator and denominator, to obtain

$$\lim_{x \to \infty} \frac{3x^2 + 2x + 1}{x^2 + 4x + 5} = \lim_{x \to \infty} \frac{3x^2 + 2x + 1}{x^2 + 4x + 5} \left(\frac{\dfrac{1}{x^2}}{\dfrac{1}{x^2}} \right)$$

$$= \lim_{x \to \infty} \left(\frac{\dfrac{3x^2}{x^2} + \dfrac{2x}{x^2} + \dfrac{1}{x^2}}{\dfrac{x^2}{x^2} + \dfrac{4x}{x^2} + \dfrac{5}{x^2}} \right)$$

$$= \lim_{x \to \infty} \left(\frac{3 + \dfrac{2}{x} + \dfrac{1}{x^2}}{1 + \dfrac{4}{x} + \dfrac{5}{x^2}} \right)$$

$$= \frac{3 + 0 + 0}{1 + 0 + 0}$$

$$= 3$$

since $\lim\limits_{x \to \infty} \dfrac{2}{x} = 0$, $\lim\limits_{x \to \infty} \dfrac{1}{x^2} = 0$, $\lim\limits_{x \to \infty} \dfrac{4}{x} = 0$, and $\lim\limits_{x \to \infty} \dfrac{5}{x^2} = 0$. □

The idea behind the technique of Example 6 is that once we have divided all terms by the highest power of x present, we will be left with only constants and terms of the form $\dfrac{c}{x^n}$, $n > 0$, for which

$$\lim_{x \to \infty} \frac{c}{x^n} = 0$$

since the denominator "blows up" while the numerator remains constant.

Example 7 Using this technique, we find that

$$\lim_{x \to \infty} \frac{x-3}{2x+x^3} = \lim_{x \to \infty} \frac{x-3}{2x+x^3} \left(\frac{\dfrac{1}{x^3}}{\dfrac{1}{x^3}} \right)$$

$$= \lim_{x \to \infty} \left(\frac{\dfrac{1}{x^2} - \dfrac{3}{x^3}}{\dfrac{2}{x^2} + 1} \right)$$

$$= \frac{0-0}{0+1}$$

$$= 0. \qquad \square$$

Not all functions have limits at infinity. For example, the quadratic function $f(x) = x^2$ has the property that

$$\lim_{x \to \infty} x^2 = +\infty; \qquad \lim_{x \to -\infty} x^2 = +\infty. \tag{3}$$

The meaning of the statements in line (3) is that as x increases without bound, so do the values of the function $f(x) = x^2$. (See Figure 2.8.) In such cases we say that the limit does not exist since the symbol ∞ is not a number.

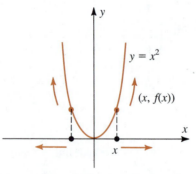

Figure 2.8 $\lim\limits_{x \to \infty} x^2 = \infty$ and $\lim\limits_{x \to -\infty} x^2 = \infty$.

Example 8 Find $\lim\limits_{x \to \infty} \dfrac{4x^3 + 3x + 3}{2x^2 + x}$.

Solution: Here we divide numerator and denominator by x^3 to obtain

$$\lim_{x \to \infty} \frac{4x^3 + 3x + 3}{2x^2 + x} = \lim_{x \to \infty} \left(\frac{4 + \dfrac{3}{x^2} + \dfrac{3}{x^3}}{\dfrac{2}{x} + \dfrac{1}{x^2}} \right).$$

In the numerator we obtain

$$\lim_{x \to \infty} \left(4 + \frac{3}{x^2} + \frac{3}{x^3} \right) = 4 + 0 + 0 = 4$$

while in the denominator

$$\lim_{x \to \infty} \left(\frac{2}{x} + \frac{1}{x^2} \right) = 0 + 0 = 0.$$

Since the denominator "shrinks" to zero while the numerator approaches 4, the quotient "blows up," that is, increases without bound through positive numbers. Thus

$$\lim_{x \to \infty} \frac{4x^3 + 3x + 3}{2x^2 + x} = +\infty. \qquad \square$$

Exercise Set 2.2

In Exercises 1–18 find the indicated limit if it exists.

1. $\lim\limits_{x \to 3} (2x - 6)$

2. $\lim\limits_{x \to 3} \dfrac{2x - 1}{x + 5}$

3. $\lim\limits_{x \to 1} (x^2 - 4x)$

4. $\lim\limits_{x \to 2} \dfrac{x^2 - 4}{x - 2}$

5. $\lim\limits_{x \to -3} \dfrac{x^2 - 9}{x + 3}$

6. $\lim\limits_{x \to 0} \dfrac{x^3 - x}{x}$

7. $\lim\limits_{x \to 4} \dfrac{x - 4}{x^2 - x - 12}$

8. $\lim\limits_{x \to 6} \dfrac{36 - x^2}{x - 6}$

9. $\lim\limits_{x \to 1} \dfrac{x^3 - 1}{x - 1}$

10. $\lim\limits_{x \to 0} |x|$

11. $\lim\limits_{h \to 0} \dfrac{(4 + h)^2 - 16}{h}$

12. $\lim\limits_{x \to -1} \dfrac{x^2 - 2x - 3}{x + 1}$

13. $\lim\limits_{x \to 4} \dfrac{x - 4}{\sqrt{x} - 2}$

14. $\lim\limits_{x \to -3} \dfrac{x^2 - 6x - 27}{x^2 + 3x}$

15. $\lim\limits_{x \to -3} \dfrac{x^2 - 4x - 21}{x + 3}$

16. $\lim\limits_{x \to 4} \dfrac{|x - 4| + 2x}{x + 3}$

17. $\lim\limits_{x \to 1} \dfrac{1 - x^4}{x^2 - 1}$

18. $\lim\limits_{x \to -1} \dfrac{x^3 + 2x^2 - 5x - 6}{x + 1}$

In Exercises 19–30 find the indicated limit at infinity if it exists.

19. $\lim\limits_{x \to \infty} \dfrac{3x^2 + 2}{10x^2 - 3x}$

20. $\lim\limits_{x \to \infty} \dfrac{6x^4 - 8x}{7 - 3x^4}$

21. $\lim\limits_{x \to \infty} \dfrac{x(4 - x^3)}{3x^4 + 2x^2}$

22. $\lim\limits_{x \to \infty} \dfrac{(x - 3)(x + 4)}{2x^2 + 2}$

23. $\lim\limits_{x \to -\infty} \dfrac{x^2 - 7}{3 + 4x^2}$

24. $\lim\limits_{x \to -\infty} \dfrac{2x + 6}{x^2 + 1}$

25. $\lim\limits_{x \to \infty} \dfrac{3x^2 + 7x}{1 - x^4}$

26. $\lim\limits_{x \to \infty} \dfrac{x^4 - 4}{x^3 + 7x^2}$

27. $\lim\limits_{x \to \infty} \dfrac{\sqrt{x} - 1}{x^2}$

28. $\lim\limits_{x \to \infty} \dfrac{x^{2/3} + x^{4/3}}{x^2}$

29. $\lim\limits_{x \to \infty} \dfrac{x + 7 - x^3}{10x^2 + 18}$

30. $\lim\limits_{x \to \infty} \dfrac{3x^{2/3} - x^{5/2}}{6 + x^{3/2}}$

2.3 ONE-SIDED LIMITS AND CONTINUITY

Figure 3.1 shows a situation in which $\lim\limits_{x \to a} f(x)$ fails to exist. It is the graph of the function

$$f(x) = \frac{|x - 1|}{x - 1}.$$

According to the definition of absolute value, we may rewrite this function as

$$f(x) = \begin{cases} \dfrac{x - 1}{x - 1} = 1 & \text{if } x > 1 \\[2mm] \dfrac{-(x - 1)}{x - 1} = -1 & \text{if } x < 1. \end{cases}$$

The reason that $\lim\limits_{x \to 1} f(x)$ fails to exist is that, as x approaches 1 *from the right* all values $f(x)$ are $f(x) = 1$, while as x approaches 1 *from the left* all values $f(x)$ are $f(x) = -1$. Since $1 \neq -1$, $f(x)$ does not approach a *unique* number L as x approaches 1 as required by the definition of limit. (See Figure 3.1.)

Figure 3.1 $\lim\limits_{x \to 1} \dfrac{|x - 1|}{x - 1}$ does not exist.

However, when we restrict the graph of $y = f(x)$ in Figure 3.1 to just one side of $x = 1$, either the right or the left, we do observe a certain behavior in the values $f(x)$ that we can describe with a more restricted type of limit. It is the *one-sided* limit defined as follows.

DEFINITION 3   We say that L is the **left-hand limit** of f as x approaches a, written

$$L = \lim\limits_{x \to a^-} f(x)$$

if the values $f(x)$ approach the unique number L as x approaches a *from the left only.* Similarly, we say that M is the **right-hand limit** of f as x approaches a, written

$$M = \lim\limits_{x \to a^+} f(x)$$

if the values $f(x)$ approach the unique number M as x approaches a *from the right only.* (See Figure 3.2.)

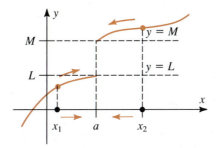

Figure 3.2 $\lim\limits_{x \to a^-} f(x)$ does not exist, but

$$\lim_{x \to a^-} f(x) = L$$

$$\lim_{x \to a^+} f(x) = M.$$

With these one-sided limits we may refer to the function $f(x) = \dfrac{|x - 1|}{x - 1}$ in Figure 3.1 by saying that

$$\lim_{x \to 1^-} \frac{|x - 1|}{x - 1} = -1; \qquad \lim_{x \to 1^+} \frac{|x - 1|}{x - 1} = 1.$$

Finally, we note that our original (two-sided) limit, $\lim\limits_{x \to a} f(x)$, can exist if and only if both one-sided limits exist and are equal. That is

$$L = \lim_{x \to a} f(x) \quad \text{if and only if} \quad \lim_{x \to a^-} f(x) = L = \lim_{x \to a^+} f(x).$$

Thus, for example, $\lim\limits_{x \to 1} \dfrac{|x - 1|}{x - 1}$ does not exist.

Example 1 The *greatest integer* function is defined by

$$[x] = \text{the largest integer } n \text{ with } n \leq x.$$

(See Figure 3.3.) We have

$$\lim_{x \to 2^+} [x] = 2; \qquad \lim_{x \to 2^-} [x] = 1$$

$$\lim_{x \to -2^-} [x] = -3; \qquad \lim_{x \to -2^+} [x] = -2.$$

But neither $\lim\limits_{x \to 2} [x]$ nor $\lim\limits_{x \to -2} [x]$ exist. □

Example 2 For the split function $f(x) = \begin{cases} \dfrac{1}{x} & \text{if } x > 0 \\ -x^2 & \text{if } x \le 0 \end{cases}$

the left-hand limit at $a = 0$ is

$$\lim_{x \to 0^-} f(x) = \lim_{x \to 0^-} (-x^2) = 0.$$

However, the right-hand limit fails to exist since the values $f(x)$ become increasingly large as x approaches zero from the right. (See Figure 3.4.) □

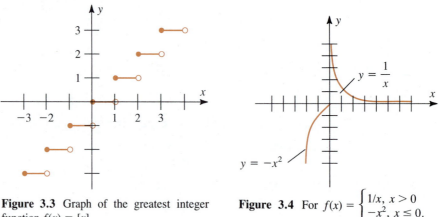

Figure 3.3 Graph of the greatest integer function $f(x) = [x]$.

Figure 3.4 For $f(x) = \begin{cases} 1/x, & x > 0 \\ -x^2, & x \le 0. \end{cases}$

Continuity

Figure 3.2 shows that when the two one-sided limits for f at $x = a$ exist but are not equal, the graph of f has a "jump" or a "tear" at $x = a$. Such phenomena are referred to as *discontinuities* of f. Other ways in which discontinuities can occur are that $\lim_{x \to a} f(x)$ fails to exist (Figure 3.5), that $f(a)$ fails to exist (Figure 3.6), or that $f(a)$ does not equal $\lim_{x \to a} f(x)$ (Figure 3.7). The property of *continuity*, defined here, rules out all such "tears" or "holes" in the graph of f.

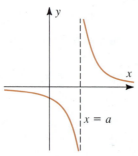

Figure 3.5 $\lim_{x \to a} f(x)$ does not exist.

Figure 3.6 $f(a)$ does not exist.

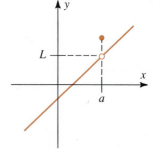

Figure 3.7 $f(a) \ne \lim_{x \to a} f(x)$.

| DEFINITION 4 | The function f is said to be **continuous** at $x = a$ if each of the following conditions holds: |

(i) $\lim\limits_{x \to a} f(x)$ exists.

(ii) $f(a)$ is defined.

(iii) $f(a) = \lim\limits_{x \to a} f(x)$.

The function f is said to be continuous on the interval (a, b) if it is continuous at each x in (a, b). It is said to be **discontinuous** at $x = c$ if it is not continuous at $x = c$.

From our experience in graphing lines and parabolas we know that constant, linear, and quadratic functions are continuous for all x. More generally, all polynomial functions are continuous for all x. Discontinuities almost always occur because a factor in the denominator of a function becomes zero.

Example 3 Find the intervals on which the function

$$f(x) = \frac{x}{x^2 - x - 2}$$

is continuous.

Strategy

Factor the denominator and find its zeros.

Solution

The denominator of this function is

$$x^2 - x - 2 = (x + 1)(x - 2).$$

It equals zero for $x = -1$ and $x = 2$. The function

Discontinuities occur at each zero of the denominator.

f is continuous at all other x.

$$f(x) = \frac{x}{x^2 - x - 2} = \frac{x}{(x + 1)(x - 2)}$$

will therefore be discontinuous at $x = -1$ and $x = 2$. There are no other zeros for the denominator. Thus, f is continuous on the intervals $(-\infty, -1)$, $(-1, 2)$ and $(2, \infty)$. (See Figure 3.8.) □

A final remark: Many of the functions that we shall work with in this text are inherently discontinuous because they are defined only for integers, strictly speaking. Here we refer primarily to the functions of business and economics that involve the cost, price, revenues, profits, and losses associated with the production and sale of certain numbers (that is, integers) of items. However, unless we state otherwise we shall assume that these functions are defined for all real numbers (even though manufacturing $7\frac{1}{2}$ bicycles is not possible). This affords the advantage of working with continuous functions for the most part and allows us to associate discontinuities with truly dramatic behavior on the part of these models.

Figure 3.8 $f(x) = \dfrac{x}{(x + 1)(x - 2)}$ is discontinuous at $x = -1, 2$.

Figure 3.9 Tax schedule in Example 4 is discontinuous at income levels $x_1 = \$15,000$ and $x_2 = \$20,000$.

Example 4 A state income tax schedule shows the following taxes due on these ranges of adjusted gross income for single taxpayers:

Income Range	Tax
$\$10,000 \le x < \$15,000$	$\$500 + 5\%$ of amount over $\$10,000$
$\$15,000 \le x < \$20,000$	$\$800 + 6\%$ of amount over $\$15,000$
$\$20,000 \le x < \$25,000$	$\$1200 + 7\%$ of amount over $\$20,000$.

If $T(x)$ denotes the tax due on income level x, is T a continuous function for all x? Why or why not?

Solution: The answer is no. The problem is that T is discontinuous at $x = \$15,000$ and at $x = \$20,000$. To see this, note that

$$T(x) = 500 + 0.05(x - 10,000) \quad \text{if} \quad 10,000 \le x < 15,000.$$

Thus

$$\lim_{x \to 15,000^-} T(x) = \lim_{x \to 15,000^-} [500 + 0.05(x - 10,000)]$$

$$= 500 + 0.05(15,000 - 10,000)$$

$$= 750.$$

But $T(15,000) = 800$ according to the second line of the tax table. Thus $\lim\limits_{x \to 15,000^-} T(x) \ne T(15,000)$, so T is discontinuous at $x = 15,000$. A similar calculation shows that T is discontinuous at $x = \$20,000$. The graph of $y = T(x)$ appears in Figure 3.9. $\quad\square$

Exercise Set 2.3

1. For the function $y = f(x)$ in Figure 3.10 find

   a. $f(1)$

   b. $\lim_{x \to 1^-} f(x)$

   c. $\lim_{x \to 1} f(x)$

   d. $\lim_{x \to 0} f(x)$

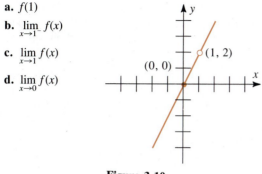

Figure 3.10

2. For the function $y = f(x)$ in Figure 3.11 find

   a. $f(1)$

   b. $\lim_{x \to 1^-} f(x)$

   c. $\lim_{x \to 1^+} f(x)$

   d. $\lim_{x \to 1} f(x)$

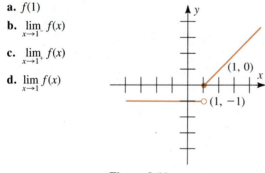

Figure 3.11

3. For the function $y = f(x)$ in Figure 3.12 find

   a. $\lim_{x \to 0^-} f(x)$

   b. $\lim_{x \to 0^+} f(x)$

   c. $\lim_{x \to -2} f(x)$

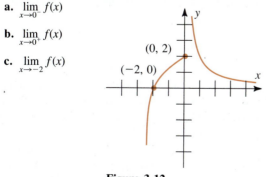

Figure 3.12

4. For the function $y = f(x)$ in Figure 3.13 find

   a. $\lim_{x \to -2^+} f(x)$

   b. $\lim_{x \to 2^-} f(x)$

   c. $\lim_{x \to 2^+} f(x)$

   d. $f(2)$

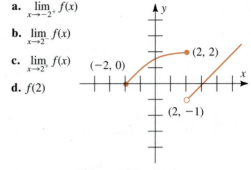

Figure 3.13

In Exercises 5–10 find the indicated one-sided limit.

5. $\lim_{x \to 0^-} \dfrac{|x|}{x}$

6. $\lim_{x \to 0^+} \dfrac{|x|}{x}$

7. $\lim_{x \to 0^+} \sqrt{x}$

8. $\lim_{x \to 4^-} \sqrt{4 - x}$

9. $\lim_{x \to 2^+} \dfrac{|x + 2|}{x + 2}$

10. $\lim_{x \to 6^+} \dfrac{\sqrt{x^2 - 5x - 6}}{x - 3}$

In Exercises 11–18 find the numbers x at which the given function is discontinuous.

11. $f(x) = \dfrac{1}{1 - x^2}$

12. $f(x) = \dfrac{x + 2}{x^2 - x - 2}$

13. $f(x) = \dfrac{x^2 - 9}{x + 3}$

14. $f(x) = \dfrac{x^3 + 4x^2 - 7x - 10}{x - 2}$

15. $f(x) = \dfrac{x + 3}{x^3 + 3x^2 - x - 3}$

16. $f(x) = \dfrac{|x + 2|}{x + 2}$

17. $f(x) = \begin{cases} 1 - x, & x \le 2 \\ x - 1, & x > 2 \end{cases}$

18. $f(x) = \begin{cases} x^2, & x < 0 \\ 3x, & x \ge 0 \end{cases}$

In Exercises 19–21 find the constant k that makes the function continuous at $x = a$.

19. $f(x) = \begin{cases} x^k, & x \le 2 \\ 10 - x, & x > 2 \end{cases} \quad a = 2$

20. $f(x) = \begin{cases} k, & x \ge 1 \\ \dfrac{1}{\sqrt{kx^2 + k}}, & x < 1 \end{cases} \quad a = 1$

21. $f(x) = \begin{cases} (x - k)(x + k), & x \le 2 \\ kx + 5, & x > 2 \end{cases} \quad a = 2$

22. The supply $S(p)$ of electric razors that manufacturers will supply weekly at price p is given by the function

$$S(p) = \begin{cases} \dfrac{1}{2}p & 0 \le p < 20 \\ \dfrac{2}{3}p - 10 & 20 \le p < 40 \\ \dfrac{4}{3}p - \dfrac{110}{3} & 40 \le p \end{cases}$$

For what values of p is this function discontinuous?

23. A manufacturer of electric hedge trimmers offers reduced prices (that is, discounts) for volume purchases according to the following schedule.

Quantity purchased	Price per trimmer $p(q)$
$0 \le q < 50$	$p(q) = 40$
$50 \le q < 100$	$p(q) = 35$
$100 \le q$	$p(q) = 30$

For what values of q is this price function p discontinuous?

24. Graph the total revenue function $R(q) = q \cdot p(q)$ where $p(q)$ is the price per trimmer in Exercise 23. Is the revenue function R discontinuous at $q = 50$? Why or why not?

25. The 1983 Federal Income Tax Schedule provided the following taxes for single taxpayers, in part.

Income Range	Tax
$15,000–18,200	$2097 + 24% of amount over $15,000
$18,200–23,500	A + 28% of amount over $18,200

Find A if the tax T was a continuous function of income at level $x = \$18,200$.

26. A television salesman receives a monthly salary and commission payment $S(x)$, in dollars, according to the following schedule, where x represents the number of television sets sold:

$$S(x) = \begin{cases} 500 + 150x, & 0 \le x \le 5 \\ 1230 + 100(x - 5), & 5 < x \le 10 \\ 1800 + 50(x - 10), & 10 < x. \end{cases}$$

Assuming the function S to be defined for all nonnegative real numbers x, find any numbers where S is discontinuous.

27. The function f defined by

$$f(x) = \begin{cases} ax^2 + 2, & 0 \le x \le 3 \\ 3x + 11, & x > 3 \end{cases}$$

is continuous at $x = 3$. Find a.

2.4 THE DERIVATIVE

We are now prepared to generalize the result of Section 2.1, that the slope of the line tangent to the graph of $y = f(x)$ at $P = (x_0, y_0)$ is

$$m = \lim_{h \to 0} \frac{f(x_0 + h) - f(x_0)}{h}. \tag{1}$$

Figure 4.1 illustrates the generalization we have in mind. It is simply that since equation (1) assigns a slope m to each number $x = x_0$, it actually defines a new *function* (from x's to m's). This function is called the *derivative* of f, since it is derived from f via the slope calculation.

The derivative of f is denoted by f'. According to what we have just said, $f'(x_0)$ is the slope of the line tangent to the graph of $y = f(x)$ at $P = (x_0, y_0)$. That is, $f'(x_0) = m$. For this reason the derivative f' is sometimes called the *slope function* for f. (See Figure 4.2.) The following definition formalizes these observations.

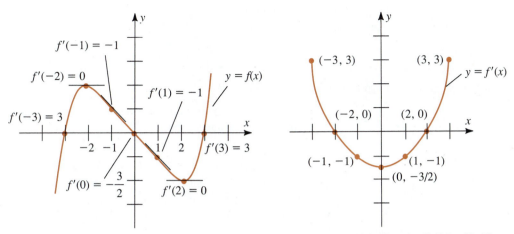

Figure 4.1 The derivative $f'(x_0)$ gives the slope of the graph of $y = f(x)$ at (x_0, y_0).

Figure 4.2 "Slope function" f' for f in Figure 4.1. Slopes for f are values of the function f'.

DEFINITION 5

The **derivative** of the function f with respect to x is the function f' whose value at x is

$$f'(x) = \lim_{h \to 0} \frac{f(x + h) - f(x)}{h}.$$

provided this limit exists.
The domain of f' is the set of all x for which this limit exists.

In Section 2.1 we determined that the slope of the line tangent to the graph of $f(x) = x^2$ at the point $P = (2, 4)$ was $m = 4$. In the notation of Definition 5 we would write this conclusion, "$f'(2) = 4$ if $f(x) = x^2$." The following example shows how to find the derivative of $f(x) = x^2$ for all x. (The process of finding a derivative is called differentiation.)

Example 1 Find the derivative of the function $f(x) = x^2$.

Strategy

Solution

If $f(x) = x^2$, then $f(x + h) = (x + h)^2$. Thus the definition of $f'(x)$ gives

Set up expression for $f'(x)$ using Definition 5.

$$f'(x) = \lim_{h \to 0} \frac{f(x + h) - f(x)}{h}$$

$$= \lim_{h \to 0} \frac{(x + h)^2 - x^2}{h}$$

Substitute expressions for $f(x + h)$, $f(x)$, and simplify, using algebra.

$$= \lim_{h \to 0} \frac{(x^2 + 2xh + h^2) - x^2}{h}$$

$$= \lim_{h \to 0} \frac{2xh + h^2}{h}$$

Factor h from terms of numerator and use $\dfrac{h}{h} = 1,\ h \neq 0$.

$$= \lim_{h \to 0} \left(\frac{h}{h}\right)(2x + h)$$

$$= \lim_{h \to 0} (2x + h)$$

Evaluate limit by letting h approach zero.

$$= 2x$$

Thus $f'(x) = 2x$. In terms of the slope of the graph of $f(x) = x^2$ this means, for instance, that

$$
\begin{aligned}
\text{if } x = -2, \quad & m = f'(-2) = 2(-2) = -4 \\
\text{if } x = -1, \quad & m = f'(-1) = 2(-1) = -2 \\
\text{if } x = 0, \quad & m = f'(0) \quad = 2(0) \quad = \quad 0 \\
\text{if } x = 1, \quad & m = f'(1) \quad = 2(1) \quad = \quad 2 \\
\text{if } x = 2, \quad & m = f'(2) \quad = 2(2) \quad = \quad 4
\end{aligned}
$$

etc. (See Figure 4.3.) □

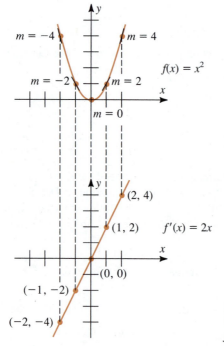

Figure 4.3 Graph of $f(x) = x^2$ (upper) and its "slope function" $f'(x) = 2x$ (lower).

Example 2 Find the derivative of the function $f(x) = \dfrac{1}{x}$.

Solution: Following the same steps as in Example 1, we find that

$$f'(x) = \lim_{h \to 0} \frac{f(x + h) - f(x)}{h}$$

$$= \lim_{h \to 0} \frac{\dfrac{1}{x+h} - \dfrac{1}{x}}{h}$$

$$= \lim_{h \to 0} \left(\frac{1}{h}\right)\left[\frac{x - (x+h)}{x(x+h)}\right]$$

$$= \lim_{h \to 0} \left(\frac{h}{h}\right)\left[\frac{-1}{x(x+h)}\right]$$

$$= \lim_{h \to 0} \frac{-1}{x(x+h)}$$

$$= -\frac{1}{x^2}.$$

Thus $f'(x) = -\dfrac{1}{x^2}$. Note that neither $f(x)$ nor $f'(x)$ is defined if $x = 0$. Graphs of both functions appear in Figure 4.4. □

Other Notation for Derivatives Another notation that is often used to denote the derivative of the function f is the symbol $\dfrac{dy}{dx}$. This notation is due to the German mathematician Gottfried Leibniz:

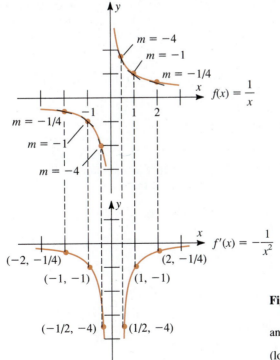

Figure 4.4 Graph of $f(x) = \dfrac{1}{x}$ (upper) and its "slope function" $f'(x) = -\dfrac{1}{x^2}$ (lower).

$$\frac{dy}{dx} = \lim_{h \to 0} \frac{f(x + h) - f(x)}{h} = f'(x).$$

Abstracting just a bit further, we interpret the symbol "$\frac{d}{dx}$" as meaning "the derivative with respect to x of." Thus for $y = f(x)$

$$\frac{dy}{dx}, \quad \frac{d}{dx}f(x), \quad \text{and} \quad \frac{df}{dx} \quad \text{all mean } f'.$$

Using this notation, we can write the results of Examples 1 and 2 as

(1) $\dfrac{d}{dx}(x^2) = 2x;$ or $\dfrac{dy}{dx} = 2x$ if $y = x^2$

(2) $\dfrac{d}{dx}\left(\dfrac{1}{x}\right) = -\dfrac{1}{x^2};$ or $\dfrac{dy}{dx} = -\dfrac{1}{x^2}$ if $y = \dfrac{1}{x}.$

Rules for Calculating Derivatives

Our next objective is to free ourselves from the tedium of using the formal definition in calculating the derivative of each function with which we intend to work. We do so by trying to determine derivatives for entire classes of functions of similar types.

The simplest functions are the constant functions, $f(x) \equiv c$. (The meaning of the symbol "$\equiv$" is that $f(x)$ equals c for every x.) Since the graph of any constant function is a horizontal line (with slope zero), the derivative of any constant function should be zero. This is our first rule.

> **Rule 1:** If f is a constant function $f(x) \equiv c$, then $f'(x) \equiv 0$. That is
>
> $$\frac{d}{dx}(c) = 0.$$

The proof of this is almost as easy as the intuition. Since $f(x) = c$ for all x, we have from the definition of the derivative that

$$f'(x) = \lim_{h \to 0} \frac{f(x + h) - f(x)}{h}$$

$$= \lim_{h \to 0} \frac{c - c}{h}$$

$$= \lim_{h \to 0} \left(\frac{0}{h}\right)$$

$$= \lim_{h \to 0} (0)$$

$$= 0.$$

Again, drawing on the fact that $f'(x)$ is the slope of the graph of $y = f(x)$, we would expect the derivative of a linear function $f(x) = mx + b$ to be its slope, $f'(x) = m$. This is our next rule.

Rule 2: If $f(x) = mx + b$, then $f'(x) \equiv m$.
That is

$$\frac{d}{dx}(mx + b) = m.$$

Example 3 Using Rule 2, we find that

$$\frac{d}{dx}(3x + 6) = 3$$

$$\frac{d}{dx}(4 - 2x) = -2,$$

etc.

The proof of Rule 2 is straightforward. With $f(x) = mx + b$ the definition of $f'(x)$ gives

$$f'(x) = \lim_{h \to 0} \frac{f(x + h) - f(x)}{h}$$

$$= \lim_{h \to 0} \frac{[m(x + h) + b] - [mx + b]}{h}$$

$$= \lim_{h \to 0} \frac{mx + mh + b - mx - b}{h}$$

$$= \lim_{h \to 0} m\left(\frac{h}{h}\right)$$

$$= \lim_{h \to 0} m$$

$$= m$$

as required.

Powers of x

From Example 1, Example 2, Rule 2 (with $m = 1$, $b = 0$), and Exercises 4 and 5 in this section the following derivatives of powers of x are obtained:

$$
\begin{array}{lll}
(n = -1) & \text{if } f(x) = x^{-1}, & f'(x) = -x^{-2} \\
(n = 1) & \text{if } f(x) = x, & f'(x) = 1 \ (= x^0) \\
(n = 2) & \text{if } f(x) = x^2, & f'(x) = 2x \\
(n = 3) & \text{if } f(x) = x^3, & f'(x) = 3x^2 \\
(n = 4) & \text{if } f(x) = x^4, & f'(x) = 4x^3.
\end{array}
$$

Indeed, this pattern holds for all powers of x, as the following rule states.

> **Rule 3 (Power Rule):** Let n be any real number with $n \neq 0$.
> If $f(x) = x^n$, then $f'(x) = nx^{n-1}$
> That is
>
> $$\frac{d}{dx}(x^n) = nx^{n-1}. \qquad \text{(provided } x \neq 0 \text{ if } n < 0)$$

A proof of the power rule for the case where n is a positive integer is outlined in Exercise 38. Ideas developed later in the text will allow us to prove this rule for all real powers of x.

According to the power rule, we have the following examples:

$$\frac{d}{dx}(x^7) = 7x^6 \qquad (n = 7)$$

$$\frac{d}{dx}(x^{-2}) = -2x^{-3} \qquad (n = -2)$$

$$\frac{d}{dx}(x^{1/2}) = \frac{1}{2}x^{-1/2} \qquad \left(n = \frac{1}{2}\right).$$

Example 4 Find an equation for the line tangent to the graph of $f(x) = \sqrt[3]{x}$ at the point where $x = 8$.

Solution: Since $\sqrt[3]{x}$ means $x^{1/3}$, we have $f'(x) = \frac{1}{3}x^{-2/3}$ by the power rule. The slope of the tangent is therefore

$$m = f'(8) = \frac{1}{3}(8^{-2/3}) = \frac{1}{3(8^{2/3})} = \frac{1}{3(8^{1/3})^2} = \frac{1}{3 \cdot 2^2} = \frac{1}{12}.$$

Since $(8, f(8)) = (8, \sqrt[3]{8}) = (8, 2)$, the line has equation

$$y - 2 = \frac{1}{12}(x - 8), \quad \text{or} \quad 12y - x = 16. \qquad \square$$

Properties of the Derivative

Two important properties of the derivative enable us to use Rules 1 to 3 to differentiate polynomials. In stating them, we use the terminology "f is differentiable on an interval" to mean that $f'(x)$ exists for each x in that interval.

THEOREM 2
Properties of the Derivative

Let the functions f and g be differentiable on an interval I, and let c be a constant. Then, on I

(i) the function $h = f + g$ is differentiable, and

$$h'(x) = f'(x) + g'(x).$$

That is,

$$\frac{d}{dx}(f + g) = \frac{df}{dx} + \frac{dg}{dx}.$$ (2)

(ii) the function $r = cf$ is differentiable, and

$$r'(x) = cf'(x).$$

That is

$$\frac{d}{dx}(cf) = c\frac{df}{dx}.$$ (3)

Here are some examples of how these rules are used.

Example 5 Find $f'(x)$ for $f(x) = x^3 + x^{2/3}$.

Solution: Using equation (2), we find

$$\frac{d}{dx}(x^3 + x^{2/3}) = \frac{d}{dx}(x^3) + \frac{d}{dx}(x^{2/3})$$

$$= 3x^2 + \frac{2}{3}x^{-1/3}. \qquad \Box$$

Example 6 For $y = 3x^5$ find $\frac{dy}{dx}$.

Solution: By equation (3) we obtain

$$\frac{d}{dx}(3x^5) = 3\frac{d}{dx}(x^5) = 3(5x^4) = 15x^4. \qquad \Box$$

Example 7 Find $f'(x)$ for $f(x) = 6x^3 - 2x^{-3/4} + 8$.

Strategy   *Solution*

Apply equation (2). $$\frac{d}{dx}(6x^3 - 2x^{-3/4} + 8) = \frac{d}{dx}(6x^3) + \frac{d}{dx}(-2x^{-3/4}) + \frac{d}{dx}(8)$$

Apply equation (3). $$= 6 \cdot \frac{d}{dx}(x^3) + (-2)\frac{d}{dx}(x^{-3/4}) + \frac{d}{dx}(8)$$

Apply Rules 1 and 3. $$= 6(3x^2) + (-2)\left(-\frac{3}{4}x^{-7/4}\right) + 0$$

Simplify. $= 18x^2 + \dfrac{3}{2}x^{-7/4}.$ □

Example 8 Here are two additional examples whose solutions are more typical of how you will actually apply these rules and theorems:

(i) $\dfrac{d}{dx}[3x^5 - 7x^2 + x^{-2}] = 3(5x^4) - 7(2x) + (-2x^{-3})$

$$= 15x^4 - 14x - 2x^{-3}.$$

(ii) $\dfrac{d}{dx}[4x^{5/3} + 6x^{-1/4} + 5] = 4\left(\dfrac{5}{3}x^{2/3}\right) + 6\left(-\dfrac{1}{4}x^{-5/4}\right) + 0$

$$= \dfrac{20}{3}x^{2/3} - \dfrac{3}{2}x^{-5/4}.$$ □

Proof of Theorem 2: The first part of Theorem 2 is proved by using the basic definition of the derivative and the first part of Theorem 1 (the limit of a sum is the sum of the limits):

$$\dfrac{d}{dx}(f + g)(x) = \lim_{h\to 0} \dfrac{(f + g)(x + h) - (f + g)(x)}{h}$$

$$= \lim_{h\to 0} \dfrac{[f(x + h) + g(x + h)] - [f(x) + g(x)]}{h}$$

$$= \lim_{h\to 0} \dfrac{[f(x + h) - f(x)] + [g(x + h) - g(x)]}{h}$$

$$= \lim_{h\to 0} \left\{\dfrac{f(x + h) - f(x)}{h} + \dfrac{g(x + h) - g(x)}{h}\right\}$$

$$= \lim_{h\to 0} \dfrac{f(x + h) - f(x)}{h} + \lim_{h\to 0} \dfrac{g(x + h) - g(x)}{h}$$

$$= f'(x) + g'(x).$$

The second part of Theorem 2 is proved in a similar way and is left for you as Exercise 37.

□

Exercise Set 2.4

In Exercises 1–6 use the basic definition of the derivative to find $f'(x)$.

1. $f(x) = 2x + 5$ **2.** $f(x) = 3 - x$

3. $f(x) = x^2 + 4$ **4.** $f(x) = x^3$

5. $f(x) = x^4$ **6.** $f(x) = \sqrt{x}$

In Exercises 7–24 use Rules 1 to 3 and properties of the derivative to find $f'(x)$ or $\dfrac{dy}{dx}$.

7. $f(x) = 3x^2 + 4x + 2$

8. $y = 4x^3 - x + 5$

9. $y = 3x^4 + 2x^2 - 4x$

10. $f(x) = 6 - 3x + 4x^2 + x^5$

11. $f(x) = x^3 - x^5$

12. $f(x) = x^2 + x^{-2} + 7$

13. $f(x) = 3x^4 - 6x - 1$

14. $y = 4x^{-2} - 6x^3 + 4$

15. $y = 2x^5 - 3x^{-2}$

16. $f(x) = 4 + \sqrt{x}$

17. $f(x) = 2x^{2/3} + 3x^{5/3}$

18. $y = x^{-1/2} - x^{-2/3}$

19. $y = 9x^{-2} + 3\sqrt{x} - 6$

20. $f(x) = 9 - 4x^{-2} + 10x^{2/3}$

21. $y = 20x^9 + 3\sqrt[3]{x} - 6$

22. $f(x) = (x - 3)^2$ (*Hint:* square first)

23. $f(x) = 4x^{-2/3} - 7x^{-1/4}$

24. $f(x) = 6 + 3\sqrt{x} - 5x^{-4/3}$

In Exercises 25–30 find $f'(a)$ for the given function $f(x)$ and number $x = a$.

25. $f(x) = 4x^3 - 2x^2, \quad a = 2$

26. $f(x) = x^3 - \sqrt{x}, \quad a = 4$

27. $f(x) = 3x^4 - 6x^2 + 4x + 9, \quad a = -3$

28. $f(x) = 4x^3 - 3x^4, \quad a = -1$

29. $f(x) = 7 - 6\sqrt{x} + 5x^{2/3}, \quad a = 8$

30. $f(x) = 4x^3 + 5x^{-2}, \quad a = -2$

31. Find an equation for the line tangent to the graph of $f(x) = 4x^2 - 2x + 1$ at the point $(1, 3)$.

32. Find an equation for the line tangent to the graph of $y = \sqrt{x}$ at the point $(4, 2)$.

33. Find the slope of the graph of $f(x) = x^{1/2} - x^{-1/2}$ at the point $(4, 3/2)$.

34. Find the slope of the graph of $y = 3x^4 - 4x^3$ at the point $(1, -1)$.

35. Find a if the line tangent to the graph of $f(x) = x^2 + ax - 6$ at the point where $x = 2$ has slope $m = 9$.

36. Find b if the graph of $y = \dfrac{b}{x^2}$ has tangent

$$4y - bx - 21 = 0 \text{ when } x = -2.$$

37. Prove statement (ii) of Theorem 2.

38. Prove the power rule as follows:
  a. For $f(x) = x^n$, n a positive integer, show that

$$f'(x) = \lim_{h \to 0} \frac{(x + h)^n - x^n}{h}$$

  b. Use the Binomial Theorem

$$(a + b)^n = a^n + na^{n-1}b + \cdots + nab^{n-1} + b^n$$

  to write

$$(x + h)^n = x^n + nx^{n-1}h + \cdots + nxh^{n-1} + h^n.$$

  c. Use part (b) to show that $f'(x)$ in (a) can be written

$$f'(x) = \lim_{h \to 0} \{nx^{n-1} + \text{terms involving factors of } h\}$$

  d. Conclude that $f'(x) = nx^{n-1}$.

2.5 APPLICATIONS OF THE DERIVATIVE: MARGINAL ANALYSIS AND VELOCITY

For a function f the difference quotient

$$\frac{f(x_0 + h) - f(x_0)}{h} \tag{1}$$

is a ratio of the change in the values $f(x)$ divided by the corresponding change in x from $x = x_0$ to $x = x_0 + h$. We refer to expression (1) as the *average* rate of change of $f(x)$ for x between x_0 and $x_0 + h$. (See Figure 5.1.) By letting h approach zero in this expression,

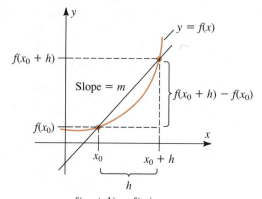

Figure 5.1 $\dfrac{f(x_0 + h) - f(x_0)}{h}$ is the average rate of change of $f(x)$ between x_0 and $x_0 + h$.

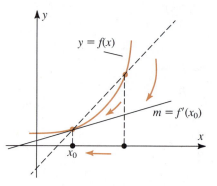

Figure 5.2 $m = f'(x_0)$ is the rate at which $f(x)$ is changing at $x = x_0$.

we effectively shorten the length of the interval over which this average rate of change is computed to just the single number x_0. That is why we interpret the derivative

$$f'(x_0) = \lim_{h \to 0} \frac{f(x_0 + h) - f(x_0)}{h}$$

as simply the *rate* at which $f(x)$ is changing at $x = x_0$. (See Figure 5.2.)

The objective of this section is to introduce several applications of this interpretation of the derivative as a rate of change.

Marginal Cost

Let $C(x)$ be a manufacturer's total cost in producing x items per unit time (such as, say, per week). Then the change in total cost between production level x_0 and production level $x_0 + h$ is $C(x_0 + h) - C(x_0)$. Thus the ratio

$$\frac{C(x_0 + h) - C(x_0)}{h} \qquad \text{(average rate of change of cost)}$$

is the average rate of change of total cost between production levels x_0 and $x_0 + h$. Letting h approach zero, and assuming C to be a differentiable function, we conclude that the derivative

$$MC(x_0) = C'(x_0) = \lim_{h \to 0} \frac{C(x_0 + h) - C(x_0)}{h} \qquad \text{(marginal cost)}$$

is the rate at which total cost will change with change in x at production level x_0. The derivative C' is called the **marginal cost** function associated with the total cost function C.

When the production level x is restricted to integer values only (such as 2 cars or 3 cars, but not 2.7 cars), economists use the term "marginal cost at production level x_0" to mean the change in total cost caused by an increase in production of one unit to level $x_0 + 1$. For a differentiable cost function C this is not quite the same as the marginal cost $C'(x_0)$, although the approximation is usually quite close. For the purposes of this section we will work exclusively with differentiable cost functions and marginal cost as defined by the derivative C'.

Example 1 A manufacturer of leather billfolds determines that the total cost of producing x billfolds per week is

$$C(x) = 40 + 5x + \frac{1}{4}x^2. \quad \text{(See Figure 5.3.)}$$

In this case the manufacturer's marginal cost function is

$$MC(x) = C'(x) = 5 + \frac{1}{2}x. \quad \text{(See Figure 5.4.)}$$

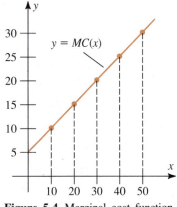

Figure 5.3 Total cost function $C(x) = 40 + 5x + \frac{1}{4}x^2$ and various marginal costs (slopes).

Figure 5.4 Marginal cost function $MC(x) = C'(x) = 5 + \frac{1}{2}x$ for $y = C(x)$ in Figure 5.3.

Thus at production level $x_0 = 10$ billfolds per week the marginal cost $C'(10) = 10$ means that total cost will increase at a rate of \$10 per additional billfold. Compare this with production level $x_1 = 40$ billfolds per week, where total cost will increase at a rate of $C'(40) = 25$ dollars per additional billfold. $\square$

Marginal Revenue

If $R(x)$ represents the total revenue received by a merchant for the sale of x items in a unit of time, then the quotient

$$\frac{R(x_0 + h) - R(x_0)}{h} \quad \text{(average rate of change of revenue)}$$

gives the average rate of change of total revenues between sales levels $x = x_0$ and $x = x_0 + h$. If we assume R to be differentiable and let the change in sales level h approach zero, we obtain the derivative

$$MR(x_0) = R'(x_0) = \lim_{h \to 0} \frac{R(x_0 + h) - R(x_0)}{h} \quad \text{(marginal revenue)}$$

as the rate at which total revenue will change with change in x at sales level x_0. As in the case of marginal cost, we refer to this rate of change of revenue as the **marginal revenue function.** The value $MR(x_0) = R'(x_0)$ provides a close approximation to the increase (or decrease) in total revenue resulting from the sale of one additional item at sales level x_0.

Revenue and Demand

When the number of items that a particular merchant or company sells in a unit of time is small relative to the total consumer demand, the selling price per item is often a fixed price p regardless of the number x of items sold. In this case the total revenue function is simply

$$R(x) = px$$

and the marginal revenue function is just

$$MR(x) = R'(x) = p$$

the selling price. However, when a company's position in a market approaches a monopoly, the selling price p and sales level x are often related to each other via a *demand equation.* Among the simplest of these is the linear demand equation of the form $p(x) = a - bx$. The next example shows how in such cases an increase in sales volume can actually correspond to a decrease in total revenue.

Example 2 Market research by the manufacturer in Example 1 shows that the selling price p and expected weekly sales levels x for billfolds are related by the demand equation

$$p(x) = 60 - x. \qquad \text{(See Figure 5.5.)}$$

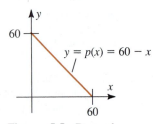

Figure 5.5 Demand curve $p(x) = 60 - x$.

Thus total revenue from the sale of x billfolds per week will be

$$\begin{aligned} R(x) &= xp(x) \\ &= x(60 - x) \\ &= 60x - x^2 \text{ dollars.} \end{aligned}$$

In this case the marginal revenue function is

$$MR(x) = R'(x) = 60 - 2x$$

dollars per billfold at sales level x. Thus

(a) at sales level $x = 10$ the marginal revenue is $MR(10) = 60 - 2(10) = 40$ dollars per billfold. This means that the additional revenue resulting from the sale of the eleventh billfold per week will be *approximately* $40. Note that we can use the expression for $R(x)$ to calculate the *actual* additional revenue as

$$R(11) - R(10) = [60(11) - 11^2] - [60(10) - 10^2]$$
$$= 39 \text{ dollars.}$$

(b) At sales level $x = 30$ we have $MR(30) = 60 - 2(30) = 0$. This means that the additional revenue resulting from the sale of the 31st billfold per week will be *approximately* zero. The *actual* additional revenue, according to the equation $R(x) = 60x - x^2$, will be

$$R(31) - R(30) = [60(31) - 31^2] - [60(30) - 30^2]$$
$$= -1 \text{ dollar.}$$

Note from Figures 5.6 and 5.7 that total revenue actually *decreases* as sales levels increase beyond 30 billfolds per week. This happens, for example, when a market becomes so "flooded" with a product that to sell additional quantities requires that the price be lowered so far as to more than offset the effect of additional sales.

□

Figure 5.6 Revenue function $R(x) = 60x - x^2$ and various marginal revenues (slopes).

Figure 5.7 Marginal revenue function $MR(x) = R'(x) = 60 - 2x$.

REMARK: It is important to note that the marginal rates $MR(x)$ provide only *approximations* to the actual change in revenue as sales levels are increased from x to $x + 1$. However, since "models" for revenue functions themselves are usually only approximations, these marginal rates provide valuable insight into how values of the revenue function change over small intervals of the form $[x, x + 1]$. The same observation holds for the other economic models for cost, profit, etc.

The Profit Equation

The profit from the production and sale of x items in a unit of time is given by the equation

$$P(x) = R(x) - C(x) \qquad \text{(profit equation)}$$

where $R(x)$ and $C(x)$ are the total revenue and total cost, respectively, at sales level x. Thus the rate at which profits are changing at production level x is given by the derivative

$$MP(x) = P'(x) = R'(x) - C'(x).$$

That is

$$MP(x) = MR(x) - MC(x)$$

Thus, **marginal profit**, $MP(x)$, is simply the difference between marginal revenue and marginal cost.

Example 3 For the manufacturer of billfolds in Examples 1 and 2 the total revenue and costs are

$$R(x) = 60x - x^2 \quad \text{and} \quad C(x) = 40 + 5x + \frac{1}{4}x^2.$$

The profit function P is therefore

$$P(x) = (60x - x^2) - \left(40 + 5x + \frac{1}{4}x^2\right)$$

$$= -40 + 55x - \frac{5}{4}x^2 \text{ dollars.}$$

The marginal profit at production level x is

$$MP(x) = P'(x) = 55 - \frac{5}{2}x \text{ dollars per billfold.}$$

Thus at production level $x = 20$ profits are increasing at the rate of $MP(20) = 55 - \frac{5}{2}(20) = 5$ dollars per additional billfold, while at production level $x = 30$ profits change by $MR(30) = 55 - \frac{5}{2}(30) = -20$ dollars per additional billfold. The two functions R and C are graphed on a common set of axes in Figure 5.8. Their difference, P, is graphed in Figure 5.9. □

Distance and Velocity

A physical application of the derivative as a rate has to do with motion along a line (called rectilinear motion). If $s(t)$ gives the location of an object on a line (such as an automobile on a straight section of highway) at time t, then the difference $s(t_0 + h) - s(t_0)$ gives the change in the object's location between times $t = t_0$ and $t = t_0 + h$. The ratio

$$\frac{s(t_0 + h) - s(t_0)}{h} \qquad \left(\text{average velocity} = \frac{\text{change in distance}}{\text{change in time}}\right)$$

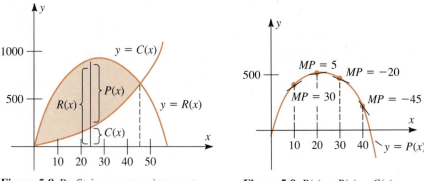

Figure 5.8 Profit is revenue minus cost.

Figure 5.9 $P(x) = R(x) - C(x)$
$$= -40 + 55x - \tfrac{5}{4}x^2.$$

is therefore the average rate of change of location between times t_0 and $t_0 + h$, which we refer to as **average velocity.**

For example, a basic result from elementary physics is that a freely falling body will have fallen

$$s(t) = 16t^2$$

feet t seconds after it is released with zero initial velocity. Thus, between times $t = 1$ second and $t = 4$ seconds it falls through a distance of

$$s(4) - s(1) = 16 \cdot 4^2 - 16 \cdot 1^2 = 240 \text{ feet.} \qquad \text{(See Figure 5.10.)}$$

During this interval of time its average velocity is

$$\frac{s(4) - s(1)}{4 - 1} = \frac{240 \text{ feet}}{3 \text{ seconds}} = 80 \text{ feet per second.}$$

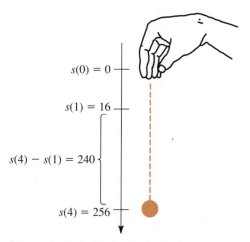

Figure 5.10 A freely falling body falls $s(t) = 16t^2$ feet in t seconds.

But we know that 80 feet per second is only the *average* velocity for the object during this time period, and that except for a single instant, the object is falling either slower or faster than 80 feet per second.

To determine the velocity of the object *at time* t_0, we allow h to approach zero in the ratio $\dfrac{s(t_0 + h) - s(t_0)}{h}$, obtaining

$$v(t_0) = s'(t_0) = \lim_{h \to 0} \frac{s(t_0 + h) - s(t_0)}{h}.$$

This equation gives the velocity $v(t_0)$ of an object when its position along a line at time t is given by the differentiable function s. It is simply the derivative $v = \dfrac{ds}{dt}$.

Example 4 Starting at time $t = 0$ a particle moves along a line so that its position after t seconds is $s(t) = t^2 - 6t + 8$. The unit of measurement along the line is feet.

(a) What is its velocity at time t?
(b) When is its velocity zero?

Strategy

Use equation for $v(t)$ to find the velocity.

Solution

Since the position function is

$$s(t) = t^2 - 6t + 8$$

the velocity function is

$$v(t) = s'(t)$$
$$= 2t - 6 \text{ feet per second.}$$

Set $v(t) = 0$ and solve to find time when velocity equals zero.

Setting the velocity function equal to zero gives

$$v(t) = 2t - 6 = 0$$
$$2t = 6$$
$$t = 3 \text{ seconds.} \qquad \square$$

Figure 5.11 is the graph of the position function in Example 4. Notice that the distance $s(t)$ decreases from time $t = 0$ until time $t = 3$, which is reflected in the sign of the velocity function $v(t) = 2t - 6$: for $0 \le t \le 3$, $-6 \le 2t - 6 \le 0$ so $v(t) \le 0$. After $t = 3$ seconds the position function is increasing, and this corresponds to positive velocity: $t > 3$ implies $2t - 6 > 0$, so $v(t) > 0$. Figure 5.12 shows another way of thinking about motion along a line and its relation to velocity. The point here is that *velocity indicates both a speed and a direction*. The *sign* of $v(t)$ determines the direction in which the object is moving.

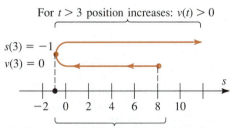

Figure 5.11 Position function $s(t) = t^2 - 6t + 8$ and several values of velocity $v(t) = s'(t)$.

Figure 5.12 Position of an object at various times with s as in Figure 5.11.

Example 5 When an object is launched vertically upward from ground level with an initial velocity of 72 feet per second, its location after t seconds will be

$$s(t) = 72t - 16t^2$$

feet above ground level.

(a) When does the object stop rising?
(b) What is its maximum height?

Solution: The velocity function is

$$v(t) = \frac{d}{dt}(72t - 16t^2)$$

$$= 72 - 32t.$$

The object stops rising when $v(t_0) = 0$. That is, when

$$72 - 32t_0 = 0$$

or

$$t_0 = \frac{72}{32} = \frac{9}{4} \text{ seconds.}$$

At this time its height is

$$s(t_0) = 72\left(\frac{9}{4}\right) - 16\left(\frac{9}{4}\right)^2$$

$$= 81 \text{ feet.}$$

(See Figure 5.13.)   □

Figure 5.13 Object in Example 5.

Exercise Set 2.5

In Exercises 1–8 a revenue function R and a total cost function C are given. In each exercise find

a. the marginal revenue function R'

b. the marginal cost function C'

c. the profit function $P = R - C$, and the marginal profit function P'.

1. $R(x) = 10x$
$C(x) = 50 + 6x$

2. $R(x) = 100x$
$C(x) = 400 + 40x + x^2$

3. $R(x) = 100x - 2x^2$
$C(x) = 20x + x^3 + 400$

4. $R(x) = 20\sqrt{x}$
$C(x) = 10x + \sqrt{x} + 30$

5. $R(x) = 40x + 50\sqrt{x}$
$C(x) = 150 + 20x + x^{2/3}$

6. $R(x) = 40(20 - x)$
$C(x) = 250 + 20x + x^{1/3}$

7. $R(x) = 400x - 10x^{2/3}$
$C(x) = \dfrac{5000}{x} + 40$

8. $R(x) = 100(x - 40) - \dfrac{30}{x}$
$C(x) = 200 + 3x^{2/3} - \dfrac{10}{x^2}$

In Exercises 9–15 a function s giving the position of an object along a line (in feet) at time t (in seconds) is given.

Find

a. its velocity function v, and

b. its velocity at time t_0.

9. $s(t) = t^2 - 6t + 7$; $t_0 = 2$

10. $s(t) = 4t^3 - 6t^2 + 3t - 5$; $t_0 = 3$

11. $s(t) = 20 - \sqrt{t} + 3t^{3/2}$; $t_0 = 4$

12. $s(t) = 3t^{2/3} - t^{1/3}$; $t_0 = 8$

13. $s(t) = 6t^{1/2} + 3t^3 - t^{-1/2}$; $t_0 = 4$

14. $s(t) = \dfrac{t^2 + 4t}{t}$; $t_0 = 6$

15. $s(t) = \dfrac{t^2 + 9t + 20}{t + 4}$; $t_0 = 5$.

16. A manufacturer of soccer balls finds that the profit from the sale of x balls per week is given by

$$P(x) = 160x - x^2$$

dollars.

a. Find the marginal profit at production level $x = 40$ balls per week.

b. For what sales levels of x is marginal profit positive?

17. A manufacturing company determines that the revenue obtained from the sale of x items will be $R(x) = 200x - 4x^2$ dollars for $0 \le x \le 25$ while the cost of producing these x items will be $C(x) = 900 + 40x$. Find the output level x for which marginal revenue equals marginal cost.

18. A manufacturer of dishwashers can produce up to 100 dishwashers per week. Sales experience indicates that the manufacturer can sell x dishwashers per week at price p where $p + 3x = 600$ dollars. Production records show that the cost of producing x dishwashers per week is

$$C(x) = 4000 + 150x + 0.5x^2.$$

a. Find the weekly revenue function $R(x) = xp(x)$. (*Hint:* Here $p(x) = 600 - 3x$.)
b. Find the weekly profit function.
c. Find the marginal cost and marginal revenue functions.
d. Find the weekly production level x for which marginal cost equals marginal revenue.

19. A manufacturer of cameras finds that the price at which it can sell x cameras per week is roughly $p(x) = (500 - x)$ dollars. Furthermore, the total cost of producing x cameras per week is $C(x) = 150 + 4x + x^2$ dollars.
a. Find the weekly revenue function $R(x) = xp(x)$.
b. Find the profit $P(x)$ obtained from selling x items per week.
c. Find the marginal revenue and marginal cost functions.
d. Find the production level x for which marginal cost will equal marginal revenue.
e. If the manufacturer is currently producing $x = 125$ cameras per week, should the level of production be increased or decreased? Why?

20. Suppose that a foreign language student has learned $N(t) = 20t - t^2$ vocabulary terms after t hours of uninterrupted study.
a. How many terms are learned between times $t = 2$ and $t = 3$ hours?
b. What is the rate in terms per hour at which the student is learning at time $t = 2$ hours?

21. A student can master $5\sqrt{t} - \frac{1}{4}t$ basic skills after t continuous hours of practice.
a. What is the rate of basic skills per hour at which the student is learning?
b. Does this rate ever become negative? If so, when?

22. A city's population t years after 1980 is predicted to be $P(t) = 40{,}000 + 2000\sqrt{t}$.
a. Find the rate, in terms of people per year, at which the city is growing.
b. Does the population of the city in 1980 affect its growth rate?

23. A manufacturer's total weekly cost in producing x items can be written $C(x) = F + V(x)$ where F, a constant, represents fixed costs (space, light, heat, etc.) and $V(x)$ represents the variable costs that depend on the production level x. Show that the marginal cost is independent of fixed costs.

24. Suppose that the demand for an item is such that its selling price p and weekly sales level x are related by the equation $p(x) = a - bx$ where a and b are constants. Show that the marginal revenue function is a linear function with slope $m = -2b$ (which is twice the slope of the demand curve $p = a - bx$).

25. The *total utility* $U(x)$ attained by a consumer from a commodity is the amount of satisfaction the consumer receives by consuming x units of the commodity in a unit of time. Marginal utility, $U'(x)$, is the rate of change of utility. Figure 5.14 shows a consumer's utility function $U(x) = x(6 - x)$ associated with the consumption of x ice cream cones per day.
a. Find the consumer's marginal utility function $MU(x) = U'(x)$.
b. For what value of x is $MU(x) = 0$?
c. According to this model, would the person wish to consume another ice cream cone if she had already eaten four cones that day?

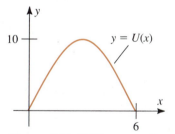

Figure 5.14 Utility curve for daily consumption of ice cream cones.

26. A particle moves along a line so that after t seconds its position is $s(t) = 6 + 5t - t^2$. (Distance unit is feet.) Find its maximum distance from the origin during the time interval $[0, 6]$.

27. An object moves along a line so that at time t its location is $s(t) = 6t - t^2$. (Time is in seconds, distance is in feet.)
a. What is its velocity at time $t = 0$?
b. When does it change direction?
c. How fast is it moving when it reaches the origin a second time?

28. The quadratic function $y = At^2 + Bt + C$ is sometimes used to model human growth during certain time intervals, with t representing time since birth, in years, and with A, B, and C being constants. In such growth models the *velocity* of the growth curve is the derivative $v(t) = 2At + B$. For this growth model, find the constants A, B, and C if $y(0) = 40$, $y(5) = 315$, and $v(5) = 105$.

29. A manufacturer of men's dress shirts finds that its cost in producing x shirts per week is $C(x) = 400 + 10x +$ $20\sqrt{x}$ dollars, while its revenue from the sale of x shirts is $R(x) = 30x$ dollars.

a. Does the manufacturer earn a profit at production level $x_0 = 16$ shirts per week?

b. Does the manufacturer earn a profit at production level $x_0 = 100$?

c. At what production level x_0 does cost equal revenue? (*Hint:* Obtain a quadratic equation in x_0.)

d. Find the profit function P and the values $P(16)$, $P(25)$, $P(49)$, and $P(100)$.

2.6 THE PRODUCT AND QUOTIENT RULES

We can paraphrase Theorem 2 of Section 2.4 by saying, "The derivative of a sum is the sum of the derivatives." This simple relationship may lead you to suspect a similar rule for products. However, the following example shows that the derivative of a product is *not* simply the product of the individual derivatives, in general. Let

$$f(x) = x \quad \text{and} \quad g(x) = x^3.$$

Then

$$f'(x) = 1 \quad \text{and} \quad g'(x) = 3x^2.$$

Now the product $p(x) = (fg)(x) = x \cdot x^3 = x^4$ has derivative

$$p'(x) = \frac{d}{dx}(x^4) = 4x^3$$

while the product of the individual derivatives is

$$f'(x)g'(x) = (1)(3x^2) = 3x^2.$$

Since $4x^3 \neq 3x^2$, we see that $\dfrac{d}{dx}(fg) \neq \left(\dfrac{df}{dx}\right)\left(\dfrac{dg}{dx}\right)$.

The correct rule for differentiating a product is the following:

Rule 4 (Product Rule): Let f and g be differentiable at x. Then the product function $p = fg$ is differentiable at x, and

$$p'(x) = f'(x)g(x) + f(x)g'(x).$$

That is

$$\frac{d}{dx}(fg) = \left(\frac{df}{dx}\right)g + f\left(\frac{dg}{dx}\right).$$

We shall defer the proof of the Product Rule to the end of this section. The following examples illustrate the use of this rule.

Example 1 For $f(x) = (2x + 7)(x - 9)$ find $f'(x)$.

Solution: One way to find $f'(x)$ is to first multiply the two binomial factors in $f(x)$ to find that

$$f(x) = (2x + 7)(x - 9)$$
$$= 2x^2 - 11x - 63.$$

Then

$$f'(x) = 4x - 11.$$

Now let's rework the problem using the Product Rule:

$$f(x) = (2x + 7)(x - 9)$$

so

$$f'(x) = \left[\frac{d}{dx}(2x + 7)\right](x - 9) + (2x + 7)\left[\frac{d}{dx}(x - 9)\right]$$

$$= 2(x - 9) + (2x + 7)(1)$$
$$= 2x - 18 + 2x + 7$$
$$= 4x - 11.$$

Of course, the result must be the same by either method. □

Example 2 For $f(x) = (3x^3 - 6x)(9x^{-2} + x^{1/2})$ find $f'(x)$.

Solution: This time we proceed directly by the Product Rule:

$$f'(x) = \left[\frac{d}{dx}(3x^3 - 6x)\right](9x^{-2} + x^{1/2}) + (3x^3 - 6x)\left[\frac{d}{dx}(9x^{-2} + x^{1/2})\right]$$

$$= (9x^2 - 6)(9x^{-2} + x^{1/2}) + (3x^3 - 6x)\left(-18x^{-3} + \frac{1}{2}x^{-1/2}\right)$$

$$= (81 - 54x^{-2} + 9x^{5/2} - 6x^{1/2}) + \left(-54 + 108x^{-2} + \frac{3}{2}x^{5/2} - 3x^{1/2}\right)$$

$$= \frac{21}{2}x^{5/2} - 9x^{1/2} + 54x^{-2} + 27.$$ □

Example 3 Find the slope of the line tangent to the graph of the function $f(x) = (\sqrt{x} - 2)(x^3 - 6)$ at the point $(4, 0)$.

Solution: By the Product Rule we have

$$f'(x) = \left[\frac{d}{dx}(x^{1/2} - 2)\right](x^3 - 6) + (x^{1/2} - 2)\left[\frac{d}{dx}(x^3 - 6)\right]$$

$$= \frac{1}{2}x^{-1/2}(x^3 - 6) + (x^{1/2} - 2)(3x^2)$$

$$= \frac{7}{2}x^{5/2} - 6x^2 - 3x^{-1/2}.$$

Thus

$$m = f'(4) = \frac{7}{2}(4)^{5/2} - 6(4)^2 - 3(4)^{-1/2}$$

$$= \frac{7}{2}(32) - 6(16) - 3\left(\frac{1}{2}\right)$$

$$= \frac{29}{2}. \qquad \square$$

The Quotient Rule

The rule for differentiating the quotient of two functions is even more surprising than the Product Rule.

Rule 5 (Quotient Rule): Let f and g be differentiable at x with $g(x) \neq 0$. Then the quotient $q = f/g$ is differentiable at x, and

$$q'(x) = \frac{g(x)f'(x) - f(x)g'(x)}{[g(x)]^2}.$$

That is

$$\frac{d}{dx}\left(\frac{f}{g}\right) = \frac{g \cdot \dfrac{df}{dx} - f \cdot \dfrac{dg}{dx}}{g^2}.$$

Example 4 For $f(x) = \dfrac{2x}{x+3}$ find $f'(x)$.

Solution: Using the Quotient Rule, we obtain

$$f'(x) = \frac{(x+3)\left[\dfrac{d}{dx}(2x)\right] - (2x)\left[\dfrac{d}{dx}(x+3)\right]}{(x+3)^2}$$

$$= \frac{(x+3)(2) - (2x)(1)}{(x+3)^2}$$

$$= \frac{6}{(x+3)^2}. \qquad \square$$

Example 5 Find $f'(x)$ for $f(x) = \dfrac{3x^2 + 7x + 1}{9 - x^3}.$

Solution: By the Quotient Rule we have

$$f'(x) = \frac{(9 - x^3)\left[\dfrac{d}{dx}(3x^2 + 7x + 1)\right] - (3x^2 + 7x + 1)\left[\dfrac{d}{dx}(9 - x^3)\right]}{(9 - x^3)^2}$$

$$= \frac{(9 - x^3)(6x + 7) - (3x^2 + 7x + 1)(-3x^2)}{(9 - x^3)^2}$$

$$= \frac{3x^4 + 14x^3 + 3x^2 + 54x + 63}{(9 - x^3)^2}. \qquad \text{(as above)} \qquad \square$$

Example 6 For $f(x) = \dfrac{x^{-2} + 3\sqrt{x}}{4 - \sqrt{x}}$ find $f'(x)$.

Solution: Again applying the Quotient Rule, we obtain

$$f'(x) = \frac{(4 - x^{1/2})\left[\dfrac{d}{dx}(x^{-2} + 3x^{1/2})\right] - (x^{-2} + 3x^{1/2})\left[\dfrac{d}{dx}(4 - x^{1/2})\right]}{(4 - x^{1/2})^2}$$

$$= \frac{(4 - x^{1/2})\left(-2x^{-3} + \dfrac{3}{2}x^{-1/2}\right) - (x^{-2} + 3x^{1/2})\left(-\dfrac{1}{2}x^{-1/2}\right)}{(4 - x^{1/2})^2}$$

$$= \frac{\left(-8x^{-3} + 2x^{-5/2} + 6x^{-1/2} - \dfrac{3}{2}\right) - \left(-\dfrac{1}{2}x^{-5/2} - \dfrac{3}{2}\right)}{(4 - x^{1/2})^2}$$

$$= \frac{-8x^{-3} + \dfrac{5}{2}x^{-5/2} + 6x^{-1/2}}{(4 - x^{1/2})^2} \qquad \square$$

Differentiability and Continuity

In order to prove the Product and Quotient rules we need to use the following theorem which we can paraphrase by saying that, "If f is differentiable at x then f must be continuous at x."

THEOREM 3

If f is differentiable at x, then

$$\lim_{h \to 0} f(x + h) = f(x).$$

Proof of Theorem 3: Since f is differentiable at x the limit

$$\lim_{h \to 0} \frac{f(x + h) - f(x)}{h} \tag{1}$$

exists and equals $f'(x)$. Since $\lim\limits_{h\to 0} h = 0$ in the denominator of (1), we must also have $\lim\limits_{h\to 0} [f(x + h) - f(x)] = 0$ in the numerator; otherwise the limit (1) would not exist. Thus

$$0 = \lim_{h\to 0} [f(x + h) - f(x)]$$

$$= [\lim_{h\to 0} f(x + h)] - f(x)$$

so

$$\lim_{h\to 0} f(x + h) = f(x). \qquad\blacksquare$$

Proof of the Product Rule: By the definition of the derivative we have, for $p = fg$, that

$$p'(x) = \lim_{h\to 0} \frac{f(x + h)g(x + h) - f(x)g(x)}{h}.$$

In order to factor the numerator, we introduce the term

$$-f(x)g(x + h) + f(x)g(x + h)$$

which is simply zero, and use the properties of limits as follows:

$$p'(x) = \lim_{h\to 0} \frac{[f(x + h)g(x + h) - f(x)g(x + h)] + [f(x)g(x + h) - f(x)g(x)]}{h}$$

$$= \lim_{h\to 0} \left[\left(\frac{f(x + h) - f(x)}{h}\right)g(x + h) + f(x)\left(\frac{g(x + h) - g(x)}{h}\right)\right]$$

$$= \left[\lim_{h\to 0} \frac{f(x + h) - f(x)}{h}\right]\left[\lim_{h\to 0} g(x + h)\right] + f(x)\left[\lim_{h\to 0} \frac{g(x + h) - g(x)}{h}\right]$$

$$= f'(x)g(x) + f(x)g'(x). \tag{2}$$

In line (2) we have used the definition of $f'(x)$ and $g'(x)$ and Theorem 3 in the form $\lim\limits_{h\to 0} g(x + h) = g(x)$. $\qquad\blacksquare$

Proof of the Quotient Rule: The technique is the same as in the proof of the Product Rule except that the algebra is a bit trickier. For $q = f/g$ we have

$$q'(x) = \lim_{h\to 0} \frac{\dfrac{f(x + h)}{g(x + h)} - \dfrac{f(x)}{g(x)}}{h}$$

$$= \lim_{h\to 0} \frac{f(x + h)g(x) - f(x)g(x + h)}{hg(x + h)g(x)}$$

$$= \lim_{h\to 0} \frac{[f(x + h)g(x) - f(x)g(x)] + [f(x)g(x) - f(x)g(x + h)]}{hg(x + h)g(x)}$$

$$= \lim_{h \to 0} \frac{\left[\dfrac{f(x+h) - f(x)}{h}\right] g(x) - f(x)\left[\dfrac{g(x+h) - g(x)}{h}\right]}{g(x+h)g(x)}$$

$$= \frac{\left[\displaystyle\lim_{h \to 0} \dfrac{f(x+h) - f(x)}{h}\right] g(x) - f(x)\left[\displaystyle\lim_{h \to 0} \dfrac{g(x+h) - g(x)}{h}\right]}{\displaystyle\lim_{h \to 0} g(x+h)g(x)}$$

$$= \frac{g(x)f'(x) - f(x)g'(x)}{[g(x)]^2} \qquad\qquad (3)$$

We have again used Theorem 3 in line (3) to conclude that $\lim_{h \to 0} g(x+h) = g(x)$. ∎

Exercise Set 2.6

In Exercises 1–10 use the Product Rule to find $f'(x)$.

1. $f(x) = (x - 1)(x + 1)$

2. $f(x) = (x^2 - 1)(2 - x)$

3. $f(x) = (3x^2 - 8x)(x^2 + 2)$

4. $f(x) = (x^2 + x + 1)(x + 1)$

5. $f(x) = (x^3 - x)^2$

6. $f(x) = \left(x^2 - \dfrac{3}{x^2}\right)^2$

7. $f(x) = \sqrt{x}(x^3 + x^{-1})$

8. $f(x) = (x^2 - x)(x^{2/3} + 2x^{1/3})$

9. $f(x) = (3x^{-2} + x^{-1})(x - 4)$

10. $f(x) = (\sqrt{x} - x^{-2})^2$

In Exercises 11–20 use the Quotient Rule to find $f'(x)$.

11. $f(x) = \dfrac{x + 2}{x - 2}$

12. $f(x) = \dfrac{x^2 - 6}{3 - x}$

13. $f(x) = \dfrac{(8x + 2)(x + 1)}{x - 3}$

14. $f(x) = \dfrac{x^4 + 4x + 4}{1 - x^3}$

15. $f(x) = \dfrac{x^2 - 4}{x + 2}$

16. $f(x) = \dfrac{(x^2 + 7)(x + 2)}{x(x - 3)}$

17. $f(x) = \dfrac{1 - x}{(1 + x)^2}$

18. $f(x) = \dfrac{\sqrt{x}}{1 + x^2}$

19. $f(x) = \dfrac{x^2 - x^{2/3}}{\sqrt{x} + 1}$

20. $f(x) = \dfrac{x - 4 + x^2}{x^2 - \sqrt{x} + 1}$

In Exercises 21–26 find the derivative.

21. $f(x) = x^{1/3}(x + 2)$

22. $f(x) = \dfrac{x^2}{1 - x^4}$

23. $f(x) = \left(\dfrac{1}{x + 1}\right)\left(\dfrac{x - 3}{x}\right)$

24. $f(x) = \sqrt{x}\left(\dfrac{1 - x^3}{1 + x}\right)$

25. $f(x) = (ax + b)(cx^2 + d)$

26. $f(x) = \dfrac{C(x)}{x}$ \qquad (assume $C(x)$ exists)

27. Find the points at which the tangent to the graph of
$$f(x) = \frac{3x}{2x - 4} \text{ has slope } m = -3.$$

28. Find the constant a so that the graph of $y = \dfrac{1}{ax + 2}$ has tangent $4y + 3x - 2 = 0$ at $(0, 1/2)$.

29. Verify the Product Rule for three functions

$$\frac{d}{dx}(fgh) = \left(\frac{df}{dx}\right)gh + f\left(\frac{dg}{dx}\right)h + fg\left(\frac{dh}{dx}\right)$$

by applying the Product Rule for two functions to the function fv where $v = gh$.

In Exercises 30–33 use the result of Exercise 29 to find $f'(x)$.

30. $f(x) = x(x + 1)(x + 2)$

31. $f(x) = (x - 3)(1 - x)(x^3 - x)$

32. $f(x) = \sqrt{x}(1 + x)(1 - x)$

33. $f(x) = (x - x^2)(x^3 - x^{-2})(x^{-1} - x^{-3})$

34. A manufacturer's total weekly cost in producing x bicycles is $C(x) = 500 + (1 + \sqrt{x})(20x + x^2)$.
  a. Find the marginal cost $MC(x) = C'(x)$.
  b. Find $MC(16)$.

35. An automobile dealer finds that the total monthly revenue from the sale of x automobiles of a certain model is
$$R(x) = \frac{10{,}000x^{3/2}}{2 + \sqrt{x}}.$$
  a. Find the marginal revenue $MR(x) = R'(x)$.
  b. Find $MR(9)$.

36. Find an equation for the line tangent to the graph of
$$f(x) = \frac{x}{1 + x^2} \text{ at the point } (2, 2/5).$$

37. If $C(x)$ gives a manufacturer's total weekly cost for the production of x items, the *average* weekly cost per item

is defined to be the ratio $c(x) = \dfrac{C(x)}{x}$.
The *marginal average cost* per item is defined to be the derivative $c'(x)$.
If $C(x) = 400 + 20x + x^2$, find
  a. the marginal cost $MC(x) = C'(x)$
  b. the average cost $c(x)$
  c. the marginal average cost $c'(x)$.

38. Refer to Exercise 37. Show that marginal average cost can be written as
$$c'(x) = \frac{xMC(x) - C(x)}{x^2}.$$

39. A typing instructor finds that after t hours of instruction a typical student can type
$$W(t) = \frac{50t^2}{10 + t^2}$$
words per minute. Find $W'(t)$, the rate at which the student's skill is improving after t hours.

40. An object moves along a line so that after t seconds its position is
$$s(t) = \frac{t^2 - 6t + 4}{t + 3}.$$
  a. Find the object's velocity $v(t) = s'(t)$.
  b. Find $v(3)$.

41. For the utility curve $U(x) = \dfrac{x}{x^2 + 9}$ find
  a. the marginal utility function $MU = \dfrac{dU}{dx}$
  b. the value of $x > 0$ for which $MU(x) = 0$.

2.7 THE CHAIN RULE

When a given function is the square of another function, $f(x) = [u(x)]^2$, we can find $f'(x)$ using the Product Rule:

$$\frac{d}{dx}[u(x)]^2 = \frac{d}{dx}[u(x)u(x)]$$

$$= u'(x)u(x) + u(x)u'(x)$$

$$= 2u(x)u'(x).$$

Similarly, we can use the Product Rule for three functions (See Exercise 29, Section 2.6) to find the derivative of $f(x) = [u(x)]^3$.

$$\frac{d}{dx}[u(x)]^3 = \frac{d}{dx}[u(x)u(x)u(x)]$$

$$= u'(x)u(x)u(x) + u(x)u'(x)u(x) + u(x)u(x)u'(x)$$

$$= 3[u(x)]^2u'(x).$$

Each of these results is a special case of the rule for differentiating a power of a function.

> **Rule 6 (General Power Rule):** Let n be any nonzero real number. If the function u is differentiable at x and if both $f(x) = [u(x)]^n$ and $[u(x)]^{n-1}$ exist, then for
>
> $$f(x) = [u(x)]^n, \qquad f'(x) = n[u(x)]^{n-1} \cdot u'(x)$$
>
> That is
>
> $$\frac{d}{dx}[u(x)]^n = n[u(x)]^{n-1} \cdot u'(x).$$
>
> (In stating the Power Rule, we assume that $u(x) \neq 0$ if $n < 0$.)

Rule 6 states that the function $f(x) = [u(x)]^n$ is differentiated just as if the function u itself were the independent variable (multiply by n and reduce the exponent by one) except that an additional factor $u'(x)$ appears. When $u(x) = x$ Rule 6 reduces to our earlier Power Rule for differentiating $f(x) = x^n$.

The General Power Rule is itself a special case of the Chain Rule, which we will discuss later in this section.

Example 1 For $f(x) = (x^2 - 6x + 2)^5$ find $f'(x)$.

Solution: Here $f(x)$ has the form $f(x) = [u(x)]^5$ where the "inside" function is $u(x) = x^2 - 6x + 2$. Thus by the General Power Rule

$$\frac{d}{dx}(x^2 - 6x + 2)^5 = 5(x^2 - 6x + 2)^4 \cdot \frac{d}{dx}(x^2 - 6x + 2)$$

$$= 5(x^2 - 6x + 2)^4(2x - 6)$$

$$= 10(x - 3)(x^2 - 6x + 2)^4. \qquad \square$$

Example 2 For $f(x) = \sqrt{9 - x^3}$ find $f'(x)$.

Strategy

Identify the inside function u.

Solution

In this case the inside function is

$$u(x) = 9 - x^3$$

so

Express $f(x)$ as $f(x) = [u(x)]^n$.

$$f(x) = \sqrt{9 - x^3} = \sqrt{u(x)} = [u(x)]^{1/2}.$$

Thus

Apply General Power Rule.

$$f'(x) = \frac{1}{2}[u(x)]^{-1/2} \cdot u'(x)$$

$$= \frac{1}{2}(9 - x^3)^{-1/2} \cdot \frac{d}{dx}(9 - x^3)$$

$$= \frac{1}{2}(9 - x^3)^{-1/2}(-3x^2)$$

$$= -\frac{3}{2}x^2(9 - x^3)^{-1/2}. \qquad \square$$

Example 3 For $f(x) = \left(\dfrac{x - 3}{x^2 + 7}\right)^9$ find $f'(x)$.

Strategy

Identify the inside function u.

Solution

In this case we let

$$u(x) = \frac{x - 3}{x^2 + 7}.$$

Since $u(x)$ is somewhat complicated, calculate $u'(x)$ before applying the Power Rule to $f(x)$.

Using the Quotient Rule we find that

$$u'(x) = \frac{(x^2 + 7)(1) - (x - 3)(2x)}{(x^2 + 7)^2}$$

$$= \frac{7 + 6x - x^2}{(x^2 + 7)^2}.$$

Write $f(x)$ as $f(x) = [u(x)]^n$.

We can now apply the Power Rule. Since

$$f(x) = \left(\frac{x - 3}{x^2 + 7}\right)^9 = [u(x)]^9$$

Apply the General Power Rule.

we have

$$f'(x) = 9[u(x)]^8 \cdot u'(x)$$

$$= 9\left(\frac{x - 3}{x^2 + 7}\right)^8\left[\frac{7 + 6x - x^2}{(x^2 + 7)^2}\right]$$

$$= \frac{9(7 + 6x - x^2)(x - 3)^8}{(x^2 + 7)^{10}}. \qquad \square$$

Composite Functions: The Chain Rule

A power function $f(x) = [u(x)]^n$ is one type of *composite* function of the form $f(x) = (g \circ u)(x) = g(u(x))$. In this case the "outside" function is $g(u) = u^n$. Figure 7.1 shows how we represented composite functions in Section 1.7.

The **Chain Rule** specifies the derivative of a composite function whether or not the outside function g is a power function. Thus it is a more general differentiation rule than the General Power Rule. Although we will not encounter outside functions other than

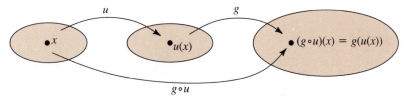

Figure 7.1 The composite function $(g \circ u)(x) = g(u(x))$.

power functions until Chapter 4, it is important to begin thinking of power functions as composite functions now so that the transition in Chapter 4 will be less abrupt.

> **Rule 7 (Chain Rule):** If the function u is differentiable at x, and if the function $g(u)$ is differentiable at $u(x)$, then the composite function $f(x) = g(u(x))$ is differentiable at x, and for
>
> $$f(x) = g(u(x)), \qquad f'(x) = g'(u(x)) \cdot u'(x). \tag{1}$$
>
> That is
>
> $$\frac{d}{dx}[g(u(x))] = g'(u(x)) \cdot u'(x). \tag{2}$$

The Chain Rule says this: "To differentiate the composite function $f(x) = g(u(x))$, first differentiate g as a function of u (as if u were the independent variable). Then multiply the result by the derivative of u."

Before looking at why the Chain Rule is true, we apply it in several examples.

Example 4 For $f(x) = (x^2 - 3x)^{2/3}$ find $f'(x)$.

Solution: Here the inside function is

$$u(x) = x^2 - 3x, \qquad \text{with } u'(x) = 2x - 3$$

and the outside function is

$$g(u) = u^{2/3}, \qquad \text{with } g'(u) = \frac{2}{3}u^{-1/3}.$$

Thus

$$g'(u(x)) = \frac{2}{3}[u(x)]^{-1/3} = \frac{2}{3}(x^2 - 3x)^{-1/3}.$$

Thus by the Chain Rule, as stated in Equation (2), we obtain

$$\frac{d}{dx}(x^2 - 3x)^{2/3} = \frac{d}{dx}g(u(x))$$

$$= g'(u(x)) \cdot u'(x)$$

$$= \frac{2}{3}(x^2 - 3x)^{-1/3} \cdot (2x - 3)$$

$$= \frac{2(2x - 3)}{3(x^2 - 3x)^{1/3}}. \qquad \square$$

Example 5 For $f(x) = \dfrac{1}{(6x^3 - x)^4}$ find $f'(x)$.

Solution: In this example the inside function is

$$u = 6x^3 - x, \qquad \text{with } u'(x) = 18x^2 - 1$$

and the outside function is

$$g(u) = \frac{1}{u^4} = u^{-4}, \qquad \text{with } g'(u) = -4u^{-5},$$

so

$$g'(u(x)) = -4[u(x)]^{-5} = -4(6x^3 - x)^{-5}.$$

Using the Chain Rule, as stated in equation (2), we have

$$\frac{d}{dx}\left[\frac{1}{(6x^3 - x)^4} \right] = \frac{d}{dx}[g(u(x))]$$

$$= g'(u(x)) \cdot u'(x)$$

$$= -4(6x^3 - x)^{-5}(18x^2 - 1)$$

$$= \frac{-4(18x^2 - 1)}{(6x^3 - x)^5}. \qquad \square$$

Example 6 For $f(x) = \sqrt{\dfrac{x^2 - 1}{x^2 + 1}}$ find $f'(x)$.

Strategy	*Solution*
Identify the inside function u.	We let $u(x) = \dfrac{x^2 - 1}{x^2 + 1}$ be the inside function u. Then, using the Quotient Rule, we have that
Find $u'(x)$.	$$u'(x) = \frac{(x^2 + 1)(2x) - (x^2 - 1)(2x)}{(x^2 + 1)^2}$$ $$= \frac{4x}{(x^2 + 1)^2}.$$
Identify the outside function g.	The outside function is $g(u) = \sqrt{u} = u^{1/2}$, so
Find $g'(u)$.	$$g'(u) = \frac{1}{2}u^{-1/2}.$$

Apply the Chain Rule:

$$f'(x) = g'(u(x)) \cdot u'(x).$$

By the Chain Rule we have, for $f(x) = g(u(x))$, that

$$f'(x) = g'(u(x)) \cdot u'(x)$$

$$= \frac{1}{2}\left[\frac{x^2 - 1}{x^2 + 1}\right]^{-1/2} \cdot \left[\frac{4x}{(x^2 + 1)^2}\right]$$

$$= \frac{2x}{(x^2 + 1)^{3/2}(x^2 - 1)^{1/2}}. \qquad \square$$

Justifying the Chain Rule

Here is an informal explanation of why the Chain Rule works. According to the definition of the derivative, we have

$$\frac{d}{dx}g(u(x)) = \lim_{h \to 0} \frac{g(u(x + h)) - g(u(x))}{h}.$$

If we assume that $u(x + h) - u(x) \neq 0$, we may rewrite this limit as

$$\frac{d}{dx}g(u(x)) = \lim_{h \to 0}\left[\frac{g(u(x + h)) - g(u(x))}{u(x + h) - u(x)}\right] \cdot \left[\frac{u(x + h) - u(x)}{h}\right]. \qquad (3)$$

Now if we let $v = u(x + h) - u(x)$, we can write the first factor on the right-hand side of equation (3) as

$$\frac{g(u(x + h)) - g(u(x))}{u(x + h) - u(x)} = \frac{g(u + v) - g(u)}{v}. \qquad (4)$$

Also, since u is differentiable at x, Theorem 3 guarantees that

$$\lim_{h \to 0}[u(x + h) - u(x)] = 0.$$

That is, $v \to 0$ as $h \to 0$. Thus we may combine equations (3) and (4) and use properties of limits to write

$$\frac{d}{dx}g(u(x)) = \left[\lim_{v \to 0}\frac{g(u + v) - g(u)}{v}\right] \cdot \left[\lim_{h \to 0}\frac{u(x + h) - u(x)}{h}\right]$$

$$= \frac{dg}{du} \cdot \frac{du}{dx}.$$

as desired. However, this argument is not an entirely rigorous proof because we have not insured that the denominator $v = u(x + h) - u(x)$ is nonzero for $h \neq 0$. A rigorous proof that treats this difficulty can be given, but it is beyond the scope of this text.

Exercise Set 2.7

In each of Exercises 1–30 find $f'(x)$.

1. $f(x) = (x + 4)^3$

2. $f(x) = (3x - 2)^4$

3. $f(x) = (x^2 - 7x)^6$

4. $f(x) = (x^2 + 8x + 8)^9$

5. $f(x) = x(x^4 - 5)^3$

6. $f(x) = (x^2 - 7)^3(5 - x^3)^2$

7. $f(x) = \dfrac{1}{(x^2 - 9)^3}$

8. $f(x) = \dfrac{x + 3}{(x^2 - 6x + 3)^2}$

9. $f(x) = (3\sqrt{x} - 2)^4$ **10.** $f(x) = (x^6 - x^2 + 2)^{-3}$

11. $f(x) = (x^{2/3} - x^{1/4})^4$

12. $f(x) = \sqrt{x^2 - 6x + 6}$

13. $f(x) = \left(\dfrac{x-3}{x+3}\right)^4$

14. $f(x) = \left(\dfrac{1+\sqrt{x}}{1-\sqrt{x}}\right)^6$

15. $f(x) = \left(\dfrac{x+2}{\sqrt[3]{x}}\right)^3$

16. $f(x) = (x^{-2/3} - 2x^{1/3})^{1/4}$

17. $f(x) = (4x^{-1/4} + 6)^{-3}$

18. $f(x) = (3 - 2x - x^4)^{1/4}$

19. $f(x) = \dfrac{(x^3 + 1)^3 + 1}{1 - x}$

20. $f(x) = \dfrac{x+2}{3 + (x^2 + 1)^3}$

21. $f(x) = \dfrac{1}{(x^2 + x + 1)^6}$

22. $f(x) = \left(\dfrac{1+x}{1-x}\right)^6$

23. $f(x) = \left(\dfrac{x^{2/3}}{1 + \sqrt{x}}\right)^3$

24. $f(x) = \dfrac{1}{(1 + \sqrt{x})^4}$

25. $f(x) = \left(\sqrt{x} - \dfrac{1}{\sqrt{x}}\right)^3$

26. $f(x) = \dfrac{\sqrt{x^2 + 1}}{(x + 5)^3}$

27. $f(x) = \dfrac{(x-4)^5}{\sqrt{x^3 + 6}}$

28. $f(x) = (x^{1/4} - x^{3/4})^3(x^{-2/3} + x^{-4/3})^5$

29. $f(x) = (x^{1/3} - 4)^3(x - \sqrt{x})^{-2/3}$

30. $f(x) = \dfrac{\sqrt{1 - x^2}}{\sqrt{1 + x^2}}$

In Exercises 31–34 find an equation for the line tangent to the graph of $y = f(x)$ at point P.

31. $f(x) = (1 - x^2)^3$; $P = (1, 0)$

32. $f(x) = \left(\dfrac{x}{x+1}\right)^2$; $P = (0, 0)$

33. $f(x) = \sqrt{1 + x^3}$; $P = (2, 3)$

34. $f(x) = \sqrt{\dfrac{1-x}{1+x}}$; $P = (0, 1)$

35. A manufacturer's total monthly cost in producing x items is $C(x) = 500 + \sqrt{40 + 16x^2}$. Find the marginal cost $MC(x)$.

36. A manufacturer's monthly revenue from the sale of x items is $R(x) = 20\sqrt{100x - x^2}$.
   a. Find the marginal revenue $MR(x)$.
   b. For which values of $x \geq 0$ is $MR(x) \geq 0$?

37. Ajax Corporation determines that its weekly profit from the sale of x dishwashers is
$P(x) = (x^3 + 10x + 125)^{1/3} - 15$.
   a. Find its marginal weekly profit.
   b. Do increasing sales always correspond to increasing profit in this example? Why or why not?

38. An object moves along a line so that after t seconds its position is $s(t) = (t^3 - 9t + 2)^3$.
   a. Find its velocity $v(t)$.
   b. For which values of $t \geq 0$ is $v(t)$ negative?

39. A grocery store determines that after t hours on the job a new cashier can ring up $N(t) = 20 - \dfrac{30}{\sqrt{9 + t^2}}$ items per minute.
   a. Find $N'(t)$, the rate at which the cashier's speed is increasing.
   b. According to this model, does the cashier ever stop improving? Why?

40. The demand for a new type of toothbrush is predicted to be $D(p) = \left(\dfrac{20}{2 + p}\right)^{2/3}$ toothbrushes per week at price p, while producers are predicted to be willing to supply $S(p) = (p^2 + 1)^{2/3}$ toothbrushes per week at price p.
   a. Verify that $p = 2$ is the equilibrium price for this model.
   b. Find $D'(2)$, the rate at which demand is decreasing at equilibrium.
   c. Find $S'(2)$, the rate at which supply is increasing at equilibrium.

41. Find the marginal average monthly cost for the manufacturer in Exercise 35.

SUMMARY OUTLINE OF CHAPTER 2

■ The **derivative** of the function f at $x = x_0$ is $f'(x_0) = \lim\limits_{h \to 0} \dfrac{f(x_0 + h) - f(x_0)}{h}$. (Page 91)

■ The **slope** of the line tangent to the graph of $y = f(x)$ at the point $(x_0, f(x_0))$ is the derivative $f'(x_0)$. (Page 67)

■ An **equation** for the line tangent to the graph of $y = f(x)$ at the point $(x_0, f(x_0))$ is (Page 67)

$$y - f(x_0) = f'(x_0)(x - x_0)$$

■ The statement $L = \lim\limits_{x \to a} f(x)$, read, "the **limit** of f as x approaches a is L," means that the values $f(x)$ approach the unique number L as x approaches the number a. (Page 75)

■ *Theorem* (Properties of Limits) (Page 78)

$$\lim_{x \to a} (f + g)(x) = \lim_{x \to a} f(x) + \lim_{x \to a} g(x)$$

$$\lim_{x \to a} (cf)(x) = c \cdot \lim_{x \to a} f(x)$$

$$\lim_{x \to a} (fg)(x) = [\lim_{x \to a} f(x)] \cdot [\lim_{x \to a} g(x)]$$

$$\lim_{x \to a} (f/g)(x) = [\lim_{x \to a} f(x)]/[\lim_{x \to a} g(x)] \qquad \text{if}$$

$$\lim_{x \to a} g(x) \neq 0$$

■ The **one-sided limit,** $\lim\limits_{x \to a^+} f(x) = L$, means that $f(x)$ approaches L as x approaches a from the *right only*. The one-sided limit, $\lim\limits_{x \to a^-} f(x) = L$, means that $f(x)$ approaches L as x approaches a from the left only. (Page 84)

■ The function f is continuous at $x = a$ if both $f(a)$ and $\lim\limits_{x \to a} f(x)$ exist, and if $f(a) = \lim\limits_{x \to a} f(x)$. (Page 87)

■ *Rules for Calculating Derivatives* (Page 94)

$$\frac{d}{dx}(c) = 0$$

$$\frac{d}{dx}(mx + b) = m$$

$$\frac{d}{dx}(x^n) = nx^{n-1} \qquad \text{(Power Rule)}$$

$$\frac{d}{dx}[f(x) + g(x)] = \frac{d}{dx}f(x) + \frac{d}{dx}g(x)$$

$$\frac{d}{dx}[cf(x)] = c\,\frac{d}{dx}f(x)$$

$$\frac{d}{dx}[f(x)g(x)] = \left[\frac{d}{dx}f(x)\right]g(x) + f(x)\left[\frac{d}{dx}g(x)\right] \qquad \text{(Product Rule)}$$

$$\frac{d}{dx}[f(x)/g(x)] = \frac{g(x)\left[\dfrac{d}{dx}f(x)\right] - f(x)\left[\dfrac{d}{dx}g(x)\right]}{[g(x)]^2}$$ (Quotient Rule)

$$\frac{d}{dx}[u(x)]^n = n \cdot u(x)^{n-1}u'(x)$$ (Power Rule)

$$\frac{d}{dx}g(u(x)) = g'(u(x))u'(x)$$ (Chain Rule)

■ **Applications** (Page 100)

If $C(x)$ = total cost, then $MC(x) = C'(x)$ = marginal cost.
If $R(x)$ = total revenue, then $MR(x) = R'(x)$ = marginal revenue.
If $P(x)$ = total profit, then $MP(x) = P'(x)$ = marginal profit.
$MP(x) = MR(x) - MC(x)$

■ If $s(t)$ is the position at time t of an object moving along a line and if $v(t)$ is the object's veloc- (Page 106)
ity at time t, then

$$v(t) = s'(t).$$

■ If $U(x)$ is the utility associated with the consumption of x units of a quantity per unit time, (Page 109)
then $MU(x) = U'(x)$ is the marginal utility function.

REVIEW EXERCISES—CHAPTER 2

In Exercises 1–14 find the indicated limit.

1. $\lim\limits_{x \to 2} (3x - 1)$

2. $\lim\limits_{x \to 2} \dfrac{2x - 1}{x + 6}$

3. $\lim\limits_{x \to 3} (x^4 - 4)$

4. $\lim\limits_{x \to -5} \dfrac{x^2 - 25}{x - 5}$

5. $\lim\limits_{x \to 3/2} \dfrac{4x^2 - 9}{2x - 3}$

6. $\lim\limits_{x \to 3} \dfrac{x - 3}{x^2 - 9}$

7. $\lim\limits_{x \to -2} \dfrac{x^2 + x - 2}{x + 2}$

8. $\lim\limits_{x \to 1} \dfrac{x^2 + 6x - 7}{x - 1}$

9. $\lim\limits_{x \to 1} \dfrac{x^2 + 2x - 3}{x^2 + x - 2}$

10. $\lim\limits_{x \to -2} \dfrac{x^2 + 2x - 3}{x^2 + x - 2}$

11. $\lim\limits_{t \to -1} \sqrt{\dfrac{1 - t^2}{1 + t}}$

12. $\lim\limits_{x \to 8} \sqrt{\dfrac{x - 7}{x + 2}}$

13. $\lim\limits_{x \to 0} \dfrac{(2 + x)^2 - 4}{x}$

14. $\lim\limits_{x \to 0} \dfrac{3x + 5x^2}{x}$

In Exercises 15–20 find those values of x for which $f(x)$ is discontinuous.

15. $f(x) = \dfrac{x^2 - 4}{x - 2}$

16. $f(x) = \dfrac{x^2 - 7}{x - 3}$

17. $f(x) = \dfrac{x + 2}{x^2 - x - 6}$

18. $f(x) = \dfrac{x}{|x|}$

19. $f(x) = \begin{cases} x - x^2, & x \le 2 \\ -x & x > 2 \end{cases}$

20. $f(x) = \begin{cases} \sqrt{x}, & x \le 4 \\ x - 1, & x > 4 \end{cases}$

In Exercises 21–46 find the derivative of the given function.

21. $f(x) = x^2 - x + 3$

22. $f(x) = 9 - x^3$

23. $f(x) = x^{2/3} + 4$

24. $f(x) = \sqrt{x} - 3$

25. $f(x) = \dfrac{1}{x^2}$

26. $f(x) = \dfrac{9}{x^3}$

27. $f(x) = \dfrac{1}{x-2}$

28. $f(x) = x^{1/3} - x^{-1/3}$

29. $f(x) = mx + b$

30. $f(x) = \sqrt{1-x}$

31. $f(x) = \dfrac{x}{3x-7}$

32. $f(x) = \dfrac{x}{1+\sqrt{x}}$

33. $f(x) = \dfrac{x-1}{x+3}$

34. $f(x) = \dfrac{\sqrt{x}}{1+\sqrt{x}}$

35. $f(x) = \dfrac{1}{x^2+6}$

36. $f(x) = \dfrac{1}{(x+3)^3}$

37. $f(x) = (x^2 + 4x + 4)^3$

38. $f(x) = \sqrt{9+x^2}$

39. $f(x) = (\sqrt{x}+3)^{-2/3}$

40. $f(x) = x^{-4} + (x^3-6)^{-3/4}$

41. $f(x) = \dfrac{1}{\sqrt{1+x^2}}$

42. $f(x) = \dfrac{(x+1)^2}{x-3}$

43. $f(x) = x\sqrt{1+x^2}$

44. $f(x) = (x-2)^3(x^2+9)^4$

45. $f(x) = (1-x^2)^3(6+2x)^{-3}$

46. $f(x) = (x^2-9)\sqrt{3-x}$

47. Find the slope of the demand curve $D(p) = \dfrac{20}{\sqrt{p-1}}$, $p > 1$ at the point $(5, 10)$.

48. Find the slope of the line tangent to the graph of $f(x) = (9 - x^2)^{2/3}$ at the point $(1, 4)$.

49. Find an equation for the line in Exercise 48.

50. The graph of the function $f(x) = x^2 - 1$ has two tangents that pass through the point $(0, -2)$. Find equations for these lines.

51. For the demand curve $D(p) = \dfrac{12}{p-1}$, $p > 1$ and the supply curve $S(p) = p - 5$, $p > 5$, find
a. the equilibrium price p_0 where $S(p_0) = D(p_0)$
b. the slope of the demand curve at $(p_0, D(p_0))$
c. the slope of the supply curve at $(p_0, S(p_0))$.

52. Excise taxes on automobiles in a certain state are levied according to the following schedule:

Value of Automobile	Excise Tax
$0 \le x < 2000$	$(10 + 0.02x)$ dollars
$2000 \le x < 5000$	$(A + 0.03(x - 2000))$ dollars
$5000 \le x$	$(B + 0.04(x - 5000))$ dollars

Find A and B if the function $y = T(x)$, given that the tax as a function of x is continuous at $x_1 = 2000$ and $x_2 = 5000$.

53. The demand for lawnmowers in a certain market sector is determined to be

$$D(p) = \begin{cases} 2p - 5, & 50 \le p < 200 \\ p + 400, & 200 \le p < 400 \\ \dfrac{1}{2}p + 800, & 400 \le p \end{cases}$$

where $D(p)$ is the number of lawnmowers sold per week during the month of April at price p. For which prices p is D discontinuous?

54. a. Graph the revenue function

$$R(x) = \begin{cases} 4x, & 0 \le x < 10 \\ 140 - (x - 20)^2, & 10 \le x < 20 \\ 160 - x, & 20 \le x \end{cases}$$

b. For what value of x is marginal revenue equal to 10?
c. For which value(s) of x is marginal revenue equal to 4?

55. A manufacturer finds that the cost of producing x tennis rackets per week is $C(x) = 500 + 30x + \frac{1}{2}x^2$.
a. Find the marginal cost $MC(x)$.
b. Find the production level x for which $MC(x) = 60$.

56. A rug weaver makes a weekly revenue from the sale of x rugs of a certain type to be $R(x) = 450x - \dfrac{x^2}{4}$.
a. Find the marginal revenue $MR(x)$.
b. Find $MR(40)$.
c. For which values of $x > 0$ is $MR(x) > 0$?
d. For which values of $x > 0$ is $MR(x) < 0$?

57. An athletic equipment supplier experiences weekly costs of $C(x) = 700 + 40x + \frac{1}{3}x^3$ in producing x baseball gloves per week.
a. Find the marginal cost $MC(x)$.
b. Find the production level x for which marginal cost is \$76 per glove.

58. A manufacturer of kitchen appliances experiences revenues from the sale of x refrigerators per month of
$$R(x) = 750x - \frac{x^2}{6} - \frac{2x^3}{3} \text{ dollars.}$$
   a. Find the marginal revenue $MR(x)$.
   b. Find the marginal revenue at sales level $x = 10$ refrigerators per month.

59. A manufacturer of cameras finds that the number x of cameras that can be sold per month at price p is given by the demand equation $x = 500 - 2p$.
   a. Find price $p(x)$ as a function of x.
   b. Find the monthly revenue function $R(x) = xp(x)$.
   c. Find the marginal revenue function $MR = \dfrac{dR}{dx}$.
   d. For what production level x is $MR(x) = 0$?

60. The manufacturer in Exercise 59 finds that the total cost of producing x cameras per month is $C(x) = 40 + 50x$ dollars.
   a. Find the marginal cost $MC(x)$.
   b. Find the monthly profit $P(x) = R(x) - C(x)$.
   c. Find the marginal profit $MP(x)$.
   d. Find the production level at which $MP(x) = 0$.

61. A manufacturer experiences total costs of producing x items per day of $C(x) = 400 + 50x$ dollars.
   a. Find the average daily cost per item $c(x) = \dfrac{C(x)}{x}$.
   b. Find the marginal average daily cost per item $c'(x)$.

62. For the utility curve $U(x) = 20\sqrt{x^2 + 9}$ find
   a. marginal utility $MU(x) = U'(x)$
   b. marginal utility for $x = 4$, $MU(4)$.

Chapter **3**

Applications of the Derivative

3.1 INCREASING AND DECREASING FUNCTIONS

Figure 1.1 shows the graph of a manufacturer's weekly profit function $y = P(x)$, which is typical of many of the profit functions we have encountered thus far. Note that profit rises as production x increases up to production level $x = x_0$. Beyond $x = x_0$, profit falls as production increases. This happens, for example, when increased production floods a market, causing price to fall so far that profit actually decreases as production increases.

In this chapter we shall use the derivative to determine the numbers x for which a function is increasing or decreasing and to find the maximum or minimum values of a function, such as $P_0 = P(x_0)$ in Figure 1.1. We begin by giving a precise meaning to the terms increasing and decreasing.

DEFINITION 1   Let I be an open interval. Then

(i) The function f is said to be *increasing* on I if
$$f(x_2) > f(x_1) \quad \text{whenever} \quad x_1 < x_2$$
for any numbers x_1 and x_2 in I.

(ii) The function f is said to be *decreasing* on I if
$$f(x_2) < f(x_1) \quad \text{whenever} \quad x_1 < x_2$$
for any numbers x_1 and x_2 in I.

Definition 1 says that f is increasing if the values $f(x)$ increase when x does. Thus the graph of an increasing function "runs uphill." (See Figure 1.2.) And a decreasing function has the property that its values $f(x)$ *decrease* when x increases. Graphs of decreasing functions "run downhill." (See Figure 1.3.)

Figure 1.4 indicates how we can use the derivative to determine the intervals on which a function is either increasing or decreasing. Since the derivative $f'(x)$ measures the slope of the graph of f at x, f is increasing (running uphill) when $f'(x)$ is positive, and f is

Figure 1.1 Typical profit function.

Figure 1.2 f is increasing if $f(x_2) > f(x_1)$ whenever $x_1 < x_2$.

Figure 1.3 f is decreasing if $f(x_2) < f(x_1)$ whenever $x_1 < x_2$.

Figure 1.4 Sign of $f'(x)$ determines whether f is increasing or decreasing at x.

decreasing (running downhill) when $f'(x)$ is negative. Theorem 1 formalizes this observation. While its truth should seem obvious from the geometric interpretation of the derivative as a slope, we shall actually prove this theorem in Section 3.9, using the Mean Value Theorem.

THEOREM 1

Let f be differentiable on an open interval I.

(i) If $f'(x) > 0$ for all x in I, then f is increasing on I.
(ii) If $f'(x) < 0$ for all x in I, then f is decreasing on I.

How can we use Theorem 1 to find the intervals on which a function is either increasing or decreasing? The idea is really very simple. Notice in Figure 1.4 that f changes from increasing to decreasing at $x = a$ and $f'(a)$ *equals zero*. Also, f changes from decreasing to increasing at $x = b$ *and* $f'(b)$ *equals zero*. More generally, when f' is a continuous function, f can change from increasing to decreasing (or vice versa) only where $f'(x) = 0$. (This is because the *continuous* function f' cannot "skip" the value zero in changing sign from positive to negative, or vice versa, without actually equaling zero for some number x.) So we simply find the zeros of f' and check the sign of f' on the resulting intervals between the zeros.

Example 1 For the function $f(x) = x(20 - x)$ find the intervals on which f is (a) increasing, and (b) decreasing.

Strategy

Find f'.

Set $f'(x) = 0$ and solve to find the zero(s) of f'.

Identify the intervals determined by the zeros of f'.

Determine the sign of f' on each interval by checking its sign at any particular number of x.

Apply Theorem 1: f is increasing if $f'(x) > 0$; decreasing if $f'(x) < 0$.

Solution

For the function $f(x) = x(20 - x) = 20x - x^2$ we have

$$f'(x) = 20 - 2x.$$

Setting $f'(x) = 0$ gives

$$20 - 2x = 0$$

or

$$x = 10.$$

This single zero of f' gives the two intervals $(-\infty, 10)$ and $(10, \infty)$ on which we must determine the sign of f'. We do this by checking the sign of $f'(x)$ at an arbitrarily chosen number x in each interval:

(i) In the interval $(-\infty, 10)$ we pick, say, $x_1 = 5$. Since

$$f'(5) = 20 - 2(5) = 10 > 0$$

we conclude that f is *increasing* on $(-\infty, 10)$.

(ii) In the interval $(10, \infty)$ we pick, say, $x_2 = 15$. Since

$$f'(15) = 20 - 2(15) = -10 < 0$$

we conclude that f is *decreasing* on $(10, \infty)$.

(See Table 1.1 and Figure 1.5.)   □

Table 1.1 Sign of $f'(x)$ determines whether f is increasing or decreasing.

Interval I	Test number x in I	Sign of $f'(x)$	Conclusion
$(-\infty, 10)$	$x_1 = 5$	$f'(5) = 10 > 0$	f is increasing
$(10, \infty)$	$x_2 = 15$	$f'(15) = -10 < 0$	f is decreasing

Example 2 Find the intervals on which the function $f(x) = 2x^3 + 3x^2 - 12x + 1$ is (a) increasing and (b) decreasing.

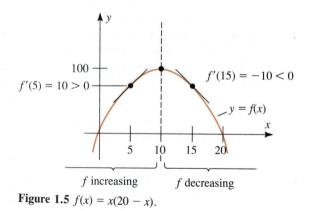

Figure 1.5 $f(x) = x(20 - x)$.

Strategy

Find f'.

Set $f'(x) = 0$ and solve to find the zeros of f.

Check the sign of $f'(x)$ on each of the resulting intervals and apply Theorem 1.

Solution

The derivative is

$$f'(x) = 6x^2 + 6x - 12$$
$$= 6(x^2 + x - 2)$$
$$= 6(x + 2)(x - 1).$$

Setting $f'(x) = 0$ gives $x = -2$ and $x = 1$. We must therefore check the sign of f' on each of the intervals $(-\infty, -2)$, $(-2, 1)$ and $(1, \infty)$.

(i) Choosing $x_1 = -3$ in the interval $(-\infty, -2)$, we find

$$f'(-3) = 6(-3)^2 + 6(-3) - 12 = 24 > 0.$$

Thus f is increasing on $(-\infty, -2)$.

(ii) Choosing $x_2 = 0$ in the interval $(-2, 1)$ gives

$$f'(0) = 6(0)^2 + 6(0) - 12 = -12 < 0$$

so f is decreasing on $(-2, 1)$.

(iii) Choosing $x_3 = 2$ in the interval $(1, \infty)$ gives

$$f'(2) = 6(2)^2 + 6(2) - 12 = 24 > 0$$

so f is increasing on $(1, \infty)$. (See Table 1.2 and Figure 1.6.) □

Table 1.2 Analysis of $f(x) = 2x^3 + 3x^2 - 12x + 1$.

Interval I	Test Number x in I	Sign of $f'(x)$	Conclusion
$(-\infty, -2)$	$x_1 = -3$	$f'(-3) = 24 > 0$	f increasing
$(-2, 1)$	$x_2 = 0$	$f'(0) = -12 < 0$	f decreasing
$(1, \infty)$	$x_3 = 2$	$f'(2) = 24 > 0$	f increasing

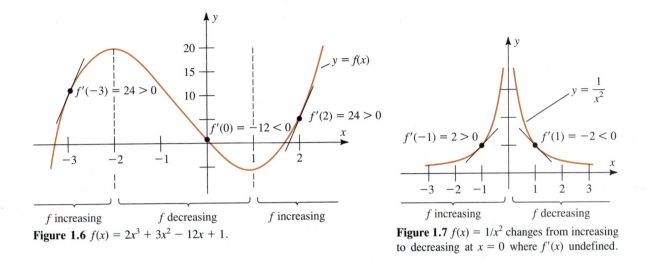

f increasing f decreasing f increasing

Figure 1.6 $f(x) = 2x^3 + 3x^2 - 12x + 1$.

f increasing f decreasing

Figure 1.7 $f(x) = 1/x^2$ changes from increasing to decreasing at $x = 0$ where $f'(x)$ undefined.

When $f'(x)$
Fails to Exist

In Examples 1 and 2 the functions f each had a continuous derivative. Thus the only place at which f could change from increasing ($f'(x) > 0$) to decreasing ($f'(x) < 0$), or vice versa, was at a zero of f'. However, when the derivative $f'(x)$ fails to exist at a number $x = x_0$ it is possible that f may change from increasing to decreasing, or vice versa, at $x = x_0$. The next two examples show that this can happen when $f(x_0)$ itself is undefined, or when $f'(x_0)$ fails to exist because of a factor of zero in its denominator.

Example 3 For the function $f(x) = \dfrac{1}{x^2} = x^{-2}$, the derivative is $f'(x) = -2x^{-3} = \dfrac{-2}{x^3}$.

Note that

(i) $f'(0)$ is undefined, as is $f(0)$.

(ii) If $x < 0$, $f'(x) = \dfrac{-2}{x^3}$ is positive, so f is increasing on $(-\infty, 0)$.

(iii) But if $x > 0$, $f'(x) = \dfrac{-2}{x^3}$ is negative, so f is decreasing on $(0, \infty)$.

Thus f changes from increasing to decreasing at $x_0 = 0$ where $f'(x_0)$ is undefined. (See Figure 1.7.) □

Example 4 The function $f(x) = x^{2/3}$ is defined for all x. But the derivative $f'(x) = \dfrac{2}{3}x^{-1/3} = \dfrac{2}{3\sqrt[3]{x}}$ fails to exist at $x = 0$. Moreover

(i) If $x < 0$, $f'(x) = \dfrac{2}{3\sqrt[3]{x}}$ is negative, so f is decreasing on $(-\infty, 0)$.

(ii) If $x > 0$, $f'(x) = \dfrac{2}{3\sqrt[3]{x}}$ is positive, so f is increasing on $(0, \infty)$.

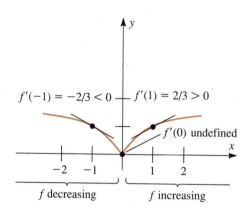

Figure 1.8 $f(x) = x^{2/3}$ changes from decreasing to increasing where $f'(x)$ undefined.

Here f changes from decreasing to increasing at $x_0 = 0$ where $f'(x_0)$ is undefined. (See Figure 1.8.)

The following procedure summarizes our findings.

> **Procedure for Determining the Intervals on Which *f* Is Either Increasing or Decreasing:**
>
> (1) Find all numbers x for which $f'(x) = 0$.
> (2) Find all numbers x for which $f'(x)$ fails to exist.
> (3) Identify all intervals I in the domain of f determined by the numbers in steps (1) and (2).
> (4) In each interval I select one "test number" x_0 and determine the sign of $f'(x_0)$:
> (a) If $f'(x_0) > 0$, f is increasing on I.
> (b) If $f'(x_0) < 0$, f is decreasing on I.

Example 5 A manufacturer of motorcycles finds that its weekly revenue from the sale of x motorcycles is approximately

$$R(x) = 1440x - 54x^2 - 3x^3 \text{ dollars.}$$

At which production levels x will revenue increase as production increases?

Solution: Here the derivative is

$$R'(x) = MR(x) = 1440 - 108x - 9x^2$$
$$= (20 + x)(72 - 9x).$$

Thus $R'(x) = 0$ when $(20 + x) = 0$ or when $(72 - 9x) = 0$. Since only nonnegative levels of production make sense, the domain of R is $[0, \infty)$ and the only zero of R' that lies in this domain occurs when $72 - 9x = 0$, that is, at $x = 8$. Also, there are no numbers x for which $R'(x)$ is undefined. We must therefore examine the sign of R' on the intervals $(0, 8)$ and $(8, \infty)$.

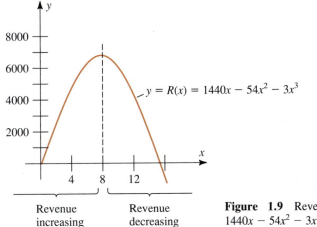

Figure 1.9 Revenue function $R(x) = 1440x - 54x^2 - 3x^3$ in Example 5.

(i) On the interval $(0, 8)$ we choose the test number $x_1 = 4$, and we find

$$R'(4) = 1440 - 108(4) - 9(4^2) = 864 > 0.$$

Since $R'(4) > 0$, revenue is increasing for $0 < x < 8$.

(ii) On the interval $(8, \infty)$ we choose the test number $x_2 = 10$, and we find

$$R'(10) = 1440 - 108(10) - 9(10^2) = -540 < 0.$$

Since $R'(10) < 0$, revenue is decreasing for $x > 8$.

Figure 1.9 shows the graph of the revenue function $y = R(x)$. □

Exercise Set 3.1

In Exercises 1–20 find the intervals on which f is (a) increasing and (b) decreasing.

1. $f(x) = 7 + 3x$

2. $f(x) = 9 - x$

3. $f(x) = 9 - x^2$

4. $f(x) = x^2 - 7x + 6$

5. $f(x) = x - x^3$

6. $f(x) = \sqrt{x + 2}$

7. $f(x) = x^3 - 3x + 6$

8. $f(x) = 2x^3 - 3x^2 - 12x + 6$

9. $f(x) = x^3 - 12x - 6$

10. $f(x) = x^3 + 3x^2 - 9x + 6$

11. $f(x) = 2x^3 - 3x^2 - 36x + 6$

12. $f(x) = x^4 - 2x^2 + 8$

13. $f(x) = 3x^4 + 4x^3 - 12x^2 + 12$

14. $f(x) = \dfrac{1}{x - 3}$

15. $f(x) = x^{4/3} + 3$

16. $f(x) = \sqrt{x^2 + 2}$

17. $f(x) = \dfrac{x}{x + 1}$

18. $f(x) = (x + 3)^{2/3}$

19. $f(x) = \sqrt[3]{8 - x^3}$

20. $f(x) = |9 - x^2|$

21. The function $f(x) = 2x^3 - 3ax^2 + 6$ decreases only on the interval $(0, 3)$. Find a.

22. Profits for the XYZ Corporation during the 10 years of its existence were given by the function $P(t) = \dfrac{50}{1 + (t - 6)^2}$. During which periods were profits (a) increasing, (b) decreasing?

23. The sum of two numbers is 50. If x denotes one of these numbers, for which numbers x is the product of the two numbers increasing?

24. The total cost of producing x reading lamps per week is determined by a manufacturer to be $C(x) = 100 + 5x + 0.02x^2$.
    a. For which numbers $x > 0$ is total cost $C(x)$ increasing?
    b. For which numbers $x > 0$ is average per unit cost $c(x) = \dfrac{C(x)}{x}$ increasing?

25. A manufacturer of sofas finds that its total cost in producing x sofas per week is $C(x) = 20 + 200x + 0.01x^3$. For which numbers x is average weekly per item cost $c(x) = \dfrac{C(x)}{x}$
    a. increasing?
    b. decreasing?

26. A producer of graphics software finds that the selling price p for its software is related to its annual sales level x by the demand equation $x = 10,000 - 200p$, while its total annual cost of producing x copies of its software is given by the function $C(x) = 450,000 + 5x$.
    a. Find $C(p)$, its total annual cost as a function of selling price p.
    b. Is $C(p)$ an increasing or decreasing function of p? Why?
    c. Find $R(p) = xp$, its annual revenue as a function of p.

    d. For which numbers $p > 0$ is $R(p)$ an increasing function of p?
    e. Find $P(p) = R(p) - C(p)$, its annual profit as a function of p.
    f. For which numbers $p > 0$ is P increasing?
    g. For which numbers $p > 0$ is P decreasing?

27. For the revenue function $R(x) = 1500x - 60x^2 - x^3$
    a. For which numbers $x > 0$ is R increasing?
    b. For which numbers $x > 0$ is marginal revenue decreasing?

28. For the utility function $U(x) = \sqrt{40 - 10x + x^2}$
    a. Find the marginal utility function $MU(x) = U'(x)$.
    b. For which numbers $x > 0$ is utility increasing?
    c. For which numbers $x > 0$ is utility decreasing?

29. A producer of art prints finds that it can sell $x = 1000 - 2p$ prints monthly at retail price p. The total monthly cost of producing these x prints is $C(x) = 400 + 2x$ dollars.
    a. Find the producer's monthly revenue $R(x) = xp$.
    b. For which numbers $x > 0$ is monthly revenue increasing?
    c. Find the producer's monthly profit $P(x) = R(x) - C(x)$.
    d. For which numbers $x > 0$ is P increasing?
    e. For which numbers $x > 0$ is P decreasing?

30. A tool rental company can achieve 500 daily rentals of jackhammers per month at a daily rate of $30 per jackhammer. For each $1 increase in price 10 fewer daily rentals are achieved.
    a. Find $R(p)$, the company's monthly revenue from renting jackhammers at p dollars per day.
    b. For which numbers $p > 0$ is R increasing as a function of p?
    c. For which numbers $p > 0$ is R decreasing as a function of p?

3.2 RELATIVE MAXIMA AND MINIMA

We can use the basic facts about increasing and decreasing functions to determine the high and low points on the graph of a function. Figure 2.1 shows one reason why we would want to do this: The profit function $y = P(x)$ has its largest value at the sales level x_0 for which $(x_0, P(x_0))$ is the highest point on the graph of $y = P(x)$.

Figure 2.2 shows a more general situation for the graph of a function $y = f(x)$ with several "relative" high and low points. We refer to these values as *relative maxima and minima*, which we define as follows.

Figure 2.1 Profit $P(x)$ is a maximum at sales level $x = x_0$.

Figure 2.2 Several relative maxima and minima for $y = f(x)$.

DEFINITION 2

The value $M = f(c)$ is a **relative maximum** for the function f if there exists a number $h > 0$ so that

$$f(c) \geq f(x)$$

for all x in the interval $(c - h, c + h)$.

The value $m = f(d)$ is a **relative minimum** for the function f if there exists a number $h > 0$ so that

$$f(d) \leq f(x)$$

for all x in the interval $(d - h, d + h)$.

Figure 2.3 $M = f(c)$ is a relative maximum if $f(c) \geq f(x)$ for all x near c.

Figure 2.4 $m = f(d)$ is a relative minimum if $f(d) \leq f(x)$ for all x near d.

In other words, $M = f(c)$ is a relative maximum if $f(c)$ is the largest value of f for all x "near" c. Similarly, $m = f(d)$ is a relative minimum if $f(d)$ is the smallest value of f for x "near" d. (See Figures 2.3 and 2.4.) We use the term **relative extremum** to refer to either a relative maximum or a relative minimum.

The following theorem is our basic tool in finding relative extrema.

THEOREM 2

Let f be a continuous function defined on some open interval containing the number c. If $f(c)$ is a relative maximum or minimum value of the function f, then either

(i) $f'(c) = 0$, or else
(ii) $f'(c)$ fails to exist.

Theorem 2 is proved using Theorem 1 in an argument that we shall only sketch. The idea is simply to rule out the possibilities that $f'(c) > 0$ or $f'(c) < 0$ when $f(c)$ is a relative extremum. This will leave only the possibilities (i) $f'(c) = 0$, or (ii) $f'(c)$ fails to exist.

The possibility $f'(c) > 0$ is ruled out by showing that if $f'(c) > 0$, then $f(x) > f(c)$ for sufficiently small x with $x > c$. But this would mean that $f(c)$ could not be a relative maximum. Also, we would have $f(x) < f(c)$ for sufficiently large x with $x < c$, so $f(c)$ could not be a relative minimum. Thus the possibility $f'(c) > 0$ leads to the contradiction that $f(c)$ is not a relative extremum, so it cannot be true and must be ruled out. A similar argument rules out the possibility $f'(c) < 0$.

Example 1 Figure 2.5 shows one example of the condition $f'(c) = 0$ at a relative maximum. From our study of the quadratic function $f(x) = 2 - (x - 1)^2$ we know that its graph is a parabola whose highest point occurs at the point $(1, 2)$. Since $f'(x) = -2(x - 1)$, we have $f'(x) = 0$ when $x = 1$, the x-coordinate of the "high point." □

Example 2 Figure 2.6 shows an example of the condition of $f'(c)$ failing to exist at a relative minimum. The function $f(x) = (x - 2)^{2/3}$ has derivative $f'(x) = \dfrac{2}{3}(x - 2)^{-1/3} = \dfrac{2}{3(x - 2)^{1/3}}$. This derivative fails to exist at $x = 2$, which corresponds to the relative minimum value $f(2) = 0$. Note the "cusp" or "point" on the graph of $y = f(x)$ at $(2, 0)$. □

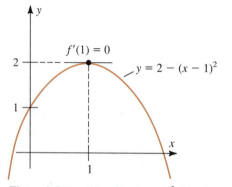

Figure 2.5 For $f(x) = 2 - (x - 1)^2$ the relative maximum is $f(1) = 2$, the "high point" of the parabola, and $f'(1) = 0$.

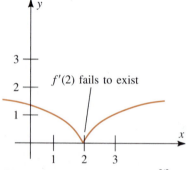

Figure 2.6 For $f(x) = (x - 2)^{2/3}$ the relative minimum is $f(2) = 0$ where $f'(x)$ is undefined.

Critical Numbers

Since the only numbers x for which $f(x)$ can be a relative extremum are those for which $f'(x) = 0$ or $f'(x)$ is undefined, we give these numbers a special name.

DEFINITION 3

The **critical numbers** for the function f are those numbers x for which $f(x)$ exists and either

(i) $f'(x) = 0$, or
(ii) $f'(x)$ fails to exist.

Using this terminology we may paraphrase Theorem 2 by stating that *relative extrema can occur only at critical numbers*. This is the point of Figure 2.7.

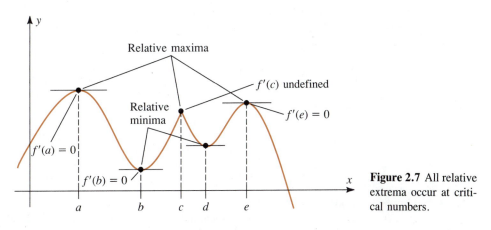

Figure 2.7 All relative extrema occur at critical numbers.

Identifying Relative Extrema

To find the relative maxima and minima for f, we obviously begin by finding the critical numbers for f. But if c is such a critical number, how can we determine if $f(c)$ is indeed a relative maximum or minimum? The answer is suggested by Figures 2.8 and 2.9:

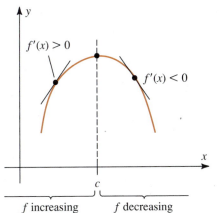

Figure 2.8 If $f'(x) > 0$ for $x < c$ and $f'(x) < 0$ for $x > c$, then $f(c)$ is a relative maximum.

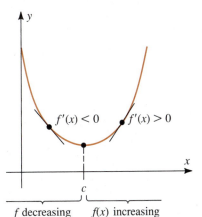

Figure 2.9 If $f'(x) < 0$ for $x < c$ and $f'(x) > 0$ for $x > c$, then $f(c)$ is a relative minimum.

(i) If f is increasing for $x < c$ and decreasing for $x > c$, then $f(c)$ is a relative maximum (i.e., a high point on the graph of $y = f(x)$).

(ii) If f is decreasing for $x < c$ and increasing for $x > c$, then $f(c)$ is a relative minimum (i.e., a low point on the graph of $y = f(x)$).

Since the sign of f' determines whether f is increasing or decreasing, the following theorem formalizes these observations. It is called the *First Derivative Test for relative extrema*.

THEOREM 3
First Derivative Test

Let c be a critical number for f and let I be an interval containing c but no other critical numbers for f. Then, for x in the interval I

(i) If $f'(x) > 0$ for $x < c$ and $f'(x) < 0$ for $x > c$, then $f(c)$ is a relative maximum.

(ii) If $f'(x) < 0$ for $x < c$ and $f'(x) > 0$ for $x > c$, then $f(c)$ is a relative minimum.

(iii) If f' does not change sign at $x = c$, then $f(c)$ is neither a relative maximum nor a relative minimum.

Example 3 Find the relative extrema for the function $f(x) = 2x^3 - 3x^2 - 12x + 1$.

Strategy

Find f'.

Solution

The derivative is

$$f'(x) = 6x^2 - 6x - 12$$
$$= 6(x^2 - x - 2)$$
$$= 6(x + 1)(x - 2).$$

Set $f'(x) = 0$ to find the critical numbers.

Thus $f'(x) = 0$ if $x = -1$ or $x = 2$. Since $f'(x)$ is defined for all x, the only critical numbers are $c_1 = -1$ and $c_2 = 2$. We must therefore check the sign of f' on the intervals $(-\infty, -1)$, $(-1, 2)$, and $(2, \infty)$. We do so by selecting one "test number" in each interval. Table 2.1 shows the results.

Check the sign of f' on each of the intervals determined by the critical numbers.

Table 2.1

Interval I	Test number t	Value of $f'(t)$	Sign of $f'(t)$
$(-\infty, -1)$	$t = -2$	$f'(-2) = 24$	$+$
$(-1, 2)$	$t = 0$	$f'(0) = -12$	$-$
$(2, \infty)$	$t = 3$	$f'(3) = 24$	$+$

Figure 2.10 shows the results of Table 2.1 as a "sign analysis" of f' on the number line.

Figure 2.10 Sign of f'.

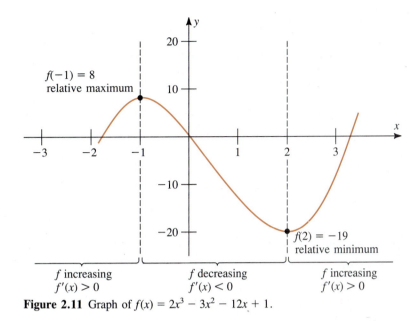

$f(-1) = 8$
relative maximum

$f(2) = -19$
relative minimum

| f increasing | f decreasing | f increasing |
| $f'(x) > 0$ | $f'(x) < 0$ | $f'(x) > 0$ |

Figure 2.11 Graph of $f(x) = 2x^3 - 3x^2 - 12x + 1$.

Apply the First Derivative Test to the information about the sign of f'.

According to the First Derivative Test

$$f(-1) = 8 \text{ is a relative maximum.}$$
$$f(2) = -19 \text{ is a relative minimum.}$$

The graph of $y = f(x)$ appears in Figure 2.11. □

Example 4 The function $f(x) = x^3$ provides an example of a critical number that yields neither a relative maximum nor a relative minimum. Here $f'(x) = 3x^2$, so setting $f'(x) = 0$ gives the single critical number $c = 0$. But since $f'(x) = 3x^2$ is positive on both the intervals $(-\infty, 0)$ and $(0, \infty)$, the derivative does not change sign at the critical number $c = 0$. In fact, $f(x) = x^3$ is increasing for all x. (See Figure 2.12.) □

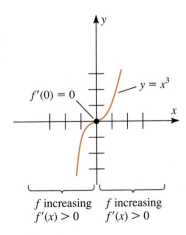

$f'(0) = 0$

$y = x^3$

| f increasing | f increasing |
| $f'(x) > 0$ | $f'(x) > 0$ |

Figure 2.12 $f(x) = x^3$ has a critical number at $c = 0$ but no relative extrema.

Example 5 For $f(x) = 4x^2(1 - x^2)$, find all relative extrema.

Solution: Since $f(x) = 4x^2 - 4x^4$, we have

$$f'(x) = 8x - 16x^3$$
$$= 8x(1 - 2x^2).$$

Setting $f'(x) = 0$ gives $8x(1 - 2x^2) = 0$, so either

$$8x = 0 \quad \text{or} \quad 1 - 2x^2 = 0.$$

Thus either

$$x = 0 \quad \text{or} \quad x^2 = \frac{1}{2}.$$

The three critical numbers are therefore $c_1 = 0$, $c_2 = -\dfrac{\sqrt{2}}{2}$, and $c_3 = \dfrac{\sqrt{2}}{2}$. Table 2.2 shows the result of checking the sign of $f'(x)$ on each of the resulting intervals.

Table 2.2

Interval I	Test number t	Value $f'(t)$	Sign of $f'(t)$
$\left(-\infty, -\dfrac{\sqrt{2}}{2}\right)$	$t = -1$	$f'(-1) = 8$	$+$
$\left(-\dfrac{\sqrt{2}}{2}, 0\right)$	$t = -\dfrac{1}{2}$	$f'\left(-\dfrac{1}{2}\right) = -2$	$-$
$\left(0, \dfrac{\sqrt{2}}{2}\right)$	$t = \dfrac{1}{2}$	$f'\left(\dfrac{1}{2}\right) = 2$	$+$
$\left(\dfrac{\sqrt{2}}{2}, \infty\right)$	$t = 1$	$f'(1) = -8$	$-$

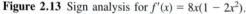

Figure 2.13 Sign analysis for $f'(x) = 8x(1 - 2x^2)$.

Figure 2.14 Relative extrema for $f(x) = 4x^2(1 - x^2)$.

Figure 2.13 shows the result of this sign analysis on the number line. From the First Derivative Test we conclude that

$$f\left(-\frac{\sqrt{2}}{2}\right) = 1 \text{ is a relative maximum.}$$

$$f(0) = 0 \text{ is a relative minimum.}$$

$$f\left(\frac{\sqrt{2}}{2}\right) = 1 \text{ is a relative maximum.}$$

The graph of $f(x) = 4x^2(1 - x^2)$ appears in Figure 2.14. □

Exercise Set 3.2

In Exercises 1–20 find all critical numbers for f. Then classify each critical number as corresponding to a relative maximum, a relative minimum, or neither. Use this information to sketch the graph of f, where possible.

1. $f(x) = x^2 - 2x$

2. $f(x) = x(x - 3)$

3. $f(x) = 3x^2 - 6x + 3$

4. $f(x) = x^2 - x - 6$

5. $f(x) = 9 - 4x - x^2$

6. $f(x) = (x - 3)(2 + x)$

7. $f(x) = 6\sqrt[3]{x} - 2$

8. $f(x) = x^3 - 3x^2 + 6$

9. $f(x) = x^3 - 3x + 7$

10. $f(x) = 2x^3 + 3x^2 - 36x + 4$

11. $f(x) = x^{3/2} - 3x + 7$

12. $f(x) = x^{4/3} - 4x - 3$

13. $f(x) = 4x^3 + 9x^2 - 12x + 7$

14. $f(x) = x^4 + 4x^3 - 8x^2 - 48x + 9$

15. $f(x) = \dfrac{x + 1}{x}$

16. $f(x) = \dfrac{x - 1}{x + 1}$

17. $f(x) = 9x - x^{-1}$

18. $f(x) = x^{1/3}(x - 7)^2$

19. $f(x) = x^{5/3} - 5x^{2/3} + 3$

20. $f(x) = \dfrac{x - 1}{x^2 + 2}$

21. A company determines that its profit resulting from the sale of x items per week is $P(x) = 160 - 96x + 18x^2 - x^3$ thousand dollars. Find all relative extrema for this profit function.

22. Find all relative extrema for the profit function $P(x) = 500 - 60(3x - 27)^{2/3}$.

23. Explain why a function of the form $f(x) = ax + b$ has no relative extrema.

24. Explain why a function of the form $f(x) = x^2 + bx + c$ always has precisely one relative minimum and no relative maximum.

25. The function $f(x) = x^2 - ax + b$ has a relative minimum at $x = 2$. Find a.

3.3 CONCAVITY AND THE SECOND DERIVATIVE

The following example shows that in analyzing the behavior of a function we often need to know more than simply whether the function is increasing or decreasing on a particular interval.

Figure 3.1 Total cost function for Alphaware increases slowly as production increases.

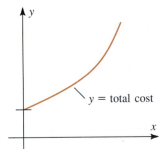

Figure 3.2 Total cost function for Betaware increases rapidly as production increases.

Example 1 Figures 3.1 and 3.2 show total monthly cost functions for two computer software companies, Alphaware and Betaware. Although both cost functions are increasing functions of production, these functions are very different for reasons we shall now describe.

Alphaware produces only one software product, its Allpurpose software, so its costs associated with producing x software systems per month are primarily just the costs of copying and distributing its single product. Thus total costs rise slowly as production increases, as indicated in Figure 3.1.

Betaware, on the other hand, produces only custom software. Every software system that it ships is individually written to a customer's specifications. Since the manpower required to produce software is known to increase rapidly as the amount of software increases, total costs for Betaware rise rapidly as the monthly production level increases. (See Figure 3.2.) □

Note that the graph of the slowly increasing cost function in Figure 3.1 has a "cupped down" shape, while the graph of the rapidly increasing cost function in Figure 3.2 is "cupped up." Figures 3.3 and 3.4 explain in more detail why this happens. Figure 3.3 shows both the cost function $y = C_A(x)$ and its derivative, the marginal cost function $MC_A = C'_A$. Since marginal cost is just the slope of the cost curve, *the graph of $y = C_A(x)$ is cupped down because marginal cost MC_A is a decreasing function.* Similarly, *the graph of the cost function $y = C_B(x)$ in Figure 3.4 is cupped up because the marginal cost function $MC_B = C'_B$ is an increasing function.*

The Second Derivative

This simple example of cost curves suggests that whether the graph of a function f is cupped up or cupped down is determined by whether the derivative (i.e., slope) is an increasing or a decreasing function. To determine whether f' itself is increasing or decreasing, we must examine the derivative of f', if it exists, which we denote by f''. That is,

$$f''(x) \quad \text{means} \quad \frac{d}{dx} f'(x).$$

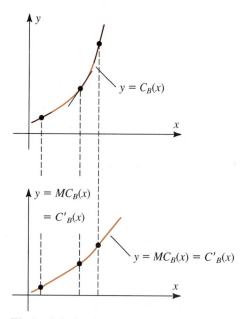

Figure 3.3 Graph of $y = C_A(x)$ is cupped *down* and derivative $C'_A = MC_A$ is *decreasing*.

Figure 3.4 Graph of $y = C_B(x)$ is cupped *up* and derivative $C'_B = MC_B$ is *increasing*.

We refer to f'' as the *second derivative* of f since it is obtained by differentiating f twice. Another notation for the second derivative is $\dfrac{d^2f}{dx^2}$. That is

$$\frac{d^2}{dx^2}f(x) \quad \text{means} \quad f''(x), \quad \text{or} \quad \frac{d}{dx}f'(x).$$

Example 2 For $f(x) = 3x^4 - x^2 - x^{-1}$ we have

$$f'(x) = 12x^3 - 2x + x^{-2}$$

so

$$f''(x) = \frac{d}{dx}(12x^3 - 2x + x^{-2})$$

$$= 36x^2 - 2 - 2x^{-3}.$$ □

Example 3 Since $\dfrac{d}{dx}\sqrt{9 - x^2} = \dfrac{1}{2}(9 - x^2)^{-1/2}(-2x)$

$$= -x(9 - x^2)^{-1/2}$$

we have

$$\frac{d^2}{dx^2}\sqrt{9-x^2} = \frac{d}{dx}[-x(9-x^2)^{-1/2}]$$

$$= -(9-x^2)^{-1/2} - x\left[-\frac{1}{2}(9-x^2)^{-3/2}(-2x)\right]$$

$$= -(9-x^2)^{-1/2} - x^2(9-x^2)^{-3/2}. \qquad \square$$

Concavity and the Second Derivative

Mathematicians prefer the terminology *concave up* when referring to graphs that we have called "cupped up" and *concave down* for graphs that are "cupped down." Since the function f' is increasing when $\frac{d}{dx}(f'(x)) = f''(x)$ is positive, it follows that the graph of f is concave up when $f''(x)$ is positive. Similarly, the graph of f is concave down when $f''(x)$ is negative. (See Figures 3.5 and 3.6.) This observation is the basis for our formal definition of concavity.

Figure 3.5 The graph of f is concave up if f' is an increasing function on (a, b), that is, if $f''(x) > 0$ for all x in (a, b).

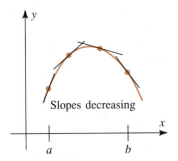

Figure 3.6 The graph of f is concave down if $f'(x)$ is a decreasing function on (a, b), that is, if $f''(x) < 0$ for all x in (a, b).

DEFINITION 4

Let f be twice differentiable (i.e., let $f'(x)$ and $f''(x)$ exist) for all x in (a, b). Then

(i) The graph of f is said to be **concave up** on (a, b) if $f''(x) > 0$ for all x in (a, b).
(ii) The graph of f is said to be **concave down** on (a, b) if $f''(x) < 0$ for all x in (a, b).

Points on the graph of f that separate arcs of opposite concavity are called **inflection points**.

Definition 4 gives *a procedure for determining the intervals on which the graph of $y = f$ is concave up or concave down:*

(i) Find all numbers x for which $f''(x) = 0$ or $f''(x)$ fails to exist.
(ii) Check the sign of f'' on each of the resulting intervals to determine concavity.

Example 4 Determine the concavity for the graph of the function $f(x) = x^4 - 6x^2 + 2$.

Strategy

Find $f''(x)$.

Solution

For $f(x) = x^4 - 6x^2 + 2$ we have

$$f'(x) = 4x^3 - 12x$$

and

$$f''(x) = 12x^2 - 12.$$

Find numbers x for which $f''(x) = 0$ or $f''(x)$ undefined.

Setting $f''(x) = 0$ gives $x^2 = 1$, so the zeros of the second derivative are $x_1 = -1$ and $x_2 = 1$. There are no numbers x for which $f''(x)$ is undefined.

Check the sign of f'' on each of the resulting intervals.

We therefore proceed to determine the sign of f'' on each of the intervals $(-\infty, -1)$, $(-1, 1)$, and $(1, \infty)$ by choosing one "test number" t in each interval and calculating $f''(t)$. The results are given in Table 3.1.

Table 3.1

Interval	Test number t	$f''(t)$	Sign of $f''(t)$
$(-\infty, -1)$	$t = -2$	$f''(-2) = 36$	$+$
$(-1, 1)$	$t = 0$	$f''(0) = -12$	$-$
$(1, \infty)$	$t = 2$	$f''(2) = 36$	$+$

Apply Definition 4.

Applying Definition 4 to the results in the last column of Table 3.1, we conclude that the graph of $f(x) = x^4 - 6x^2 + 2$ is

  (i) concave up on $(-\infty, -1)$.
  (ii) concave down on $(-1, 1)$.
  (iii) concave up on $(1, \infty)$.

Identify inflection points where $f''(x)$ changes sign.

Since the concavity of the graph changes both at $(-1, f(-1)) = (-1, -3)$ and $(1, f(1)) = (1, -3)$, these are inflection points. The graph appears in Figure 3.7. $\square$

Example 5 Determine the concavity and find the inflection points for the graph of $f(x) = x^{2/3} - \frac{1}{5}x^{5/3}$.

Solution: Here $f'(x) = \dfrac{2}{3}x^{-1/3} - \dfrac{1}{3}x^{2/3}$, so

$$f''(x) = -\frac{2}{9}x^{-4/3} - \frac{2}{9}x^{-1/3}$$

$$= -\frac{2(1 + x)}{9x^{4/3}}.$$

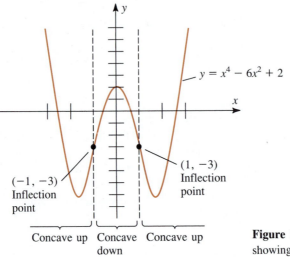

$y = x^4 - 6x^2 + 2$

$(1, -3)$
Inflection
point

$(-1, -3)$
Inflection
point

Concave up   Concave   Concave up
             down

Figure 3.7 Graph of $f(x) = x^4 - 6x^2 + 2$ showing intervals of concavity.

Thus $f''(x) = 0$ for $x = -1$ and $f''(x)$ is undefined for $x = 0$. We must therefore check the sign of f'' on each of the intervals $(-\infty, -1)$, $(-1, 0)$, and $(0, \infty)$ to determine concavity. The results of doing so are given in Table 3.2.

Table 3.2

Interval	Test number t	$f''(t)$	Sign of $f''(t)$
$(-\infty, -1)$	$t = -2$	$f''(-2) = \dfrac{2}{9}$	$+$
$(-1, 0)$	$t = -\dfrac{1}{2}$	$f''\left(-\dfrac{1}{2}\right) = -\dfrac{2^{4/3}}{9}$	$-$
$(0, \infty)$	$t = 1$	$f''(1) = -\dfrac{4}{9}$	$-$

From Table 3.2 and Theorem 4 we conclude that the graph of $f(x) = x^{2/3} - \frac{1}{5}x^{5/3}$ is

(i) concave up on $(-\infty, -1)$.
(ii) concave down on $(-1, 0)$.
(iii) concave down on $(0, \infty)$.

Thus the point $(-1, f(-1)) = (-1, \frac{6}{5})$ is an inflection point but the point $(0, f(0)) = (0, 0)$ is not. (See Figure 3.8.) □

Example 6 Figure 3.9 shows a typical example of a "political supply curve." The values of the function represented by this graph are the percentages of voters supporting various levels of proposed expenditures in support of a "public good." For example, this

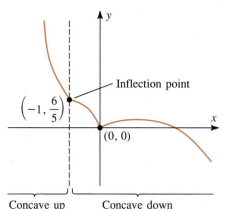

Concave up   Concave down

Figure 3.8 Graph of $f(x) = x^{2/3} - \frac{1}{5}x^{5/3}$ showing intervals of concavity.

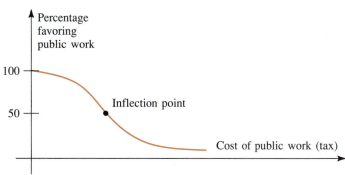

Figure 3.9 A typical ''political supply curve'' showing percentage of voters favoring a public work at various levels of required tax.

graph might represent the percentage of voters favoring various tax increases proposed to support the construction of a new public library. The fact that $P(0) = 100$ means that everyone would support construction of a new library if it cost the taxpayers nothing. But as the size of the required tax increases, the percentage of voters favoring construction of the new library decreases. Typically such curves change from concave down to concave up at an inflection point somewhere in the vicinity of $P = 50\%$. Can you think of reasons why such a political supply curve might have this shape?   □

The Second Derivative Test for Extrema

The second derivative provides us with a very straightforward test for determining whether certain critical numbers for the function f yield relative extrema. It is often easier to apply than the First Derivative Test.

THEOREM 4
Second Derivative Test

Let f be differentiable on an open interval containing the critical number $x = c$ with $f'(c) = 0$. Suppose also that $f''(x)$ exists throughout this interval.

(i) If $f''(c) < 0$, then $f(c)$ is a relative maximum.
(ii) If $f''(c) > 0$, then $f(c)$ is a relative minimum.
(iii) If $f''(c) = 0$ there is no conclusion from this test.

Although we shall not provide a formal proof of Theorem 4, Figures 3.10 and 3.11 indicate why it is true. If $f''(x) > 0$ near $x = c$, then the point $(c, f(c))$ must be a ''low point,'' since the graph is concave up. Similarly, if $f''(x) < 0$ near $x = c$, the graph is concave down, so $(c, f(c))$ must be a high point.

In applying the Second Derivative Test, be sure to note statement (iii): if $f''(c) = 0$, we obtain no conclusion as to whether $f(c)$ is a relative extremum. Exercise 40 presents several examples of just this situation. Also, note that the Second Derivative Test applies only to critical numbers c with $f'(c) = 0$. It does *not* apply to critical numbers c for which $f'(c)$ fails to exist. In either case we must resort to the First Derivative Test to classify $f(c)$.

JOHN DRURY

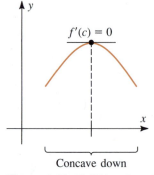

Concave up

Concave down

Figure 3.10 If $f'(c) = 0$ and $f''(c) > 0$, then $f(c)$ is a relative minimum.

Figure 3.11 If $f'(c) = 0$ and $f''(c) < 0$, then $f(c)$ is a relative maximum.

Example 7 Find all relative extrema for the function $f(x) = x^4 - 8x^2 + 2$.

Strategy

Set $f'(x) = 0$ to find the critical numbers.

Solution

Setting $f'(x) = 4x^3 - 16x = 0$, we obtain the equation

$$4x(x^2 - 4) = 0$$

so $f'(x) = 0$ if $x = 0$ or if $x^2 = 4$. The three critical numbers are therefore $x = 0$, $x = -2$, and $x = 2$. Since the second derivative is

Find f''.

$$f''(x) = 12x^2 - 16$$

we may apply the Second Derivative Test to each of the three critical numbers as follows:

Check the sign of $f''(c)$ for each critical number c and apply the Second Derivative Test.

(i) $f''(0) = 12(0)^2 - 16 = -16 < 0$
  so $(0, f(0)) = (0, 2)$ is a relative maximum.
(ii) $f''(-2) = 12(-2)^2 - 16 = 32 > 0$
  so $(-2, f(-2)) = (-2, -14)$ is a relative minimum.
(iii) $f''(2) = 12(2)^2 - 16 = 32 > 0$,
  so $(2, f(2)) = (2, -14)$ is a relative minimum.

The graph of $f(x) = x^4 - 8x^2 + 2$ appears in Figure 3.12. □

Higher Order Derivatives

There is no reason why we cannot differentiate a given function more than twice, although we shall have little practical reason to do so in this text. The notation for such "higher order" derivatives is this:

$$f'''(x), \quad \text{or} \quad \frac{d^3}{dx^3}f(x), \quad \text{means} \quad \frac{d}{dx}(f''(x))$$

$$f^{(4)}(x), \quad \text{or} \quad \frac{d^4}{dx^4}f(x), \quad \text{means} \quad \frac{d}{dx}(f'''(x))$$

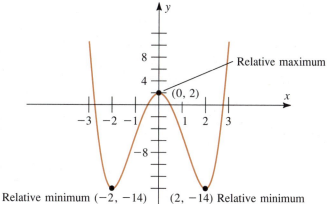

Relative maximum

(0, 2)

Relative minimum (−2, −14) (2, −14) Relative minimum

Figure 3.12 Graph of $f(x) = x^4 - 8x^2 + 2$.

and, in general

$$f^{(n)}(x), \quad \text{or} \quad \frac{d^n}{dx^n}f(x), \quad \text{means} \quad \frac{d}{dx}(f^{(n-1)}(x)), \qquad n = 4, 5, 6, \ldots$$

Example 8 For $f(x) = x^5 - 2x^3 + x^{2/3} - 6$ we have

$$f'(x) = 5x^4 - 6x^2 + \frac{2}{3}x^{-1/3}$$

$$f''(x) = 20x^3 - 12x - \frac{2}{9}x^{-4/3}$$

$$f'''(x) = 60x^2 - 12 + \frac{8}{27}x^{-7/3}$$

$$f^{(4)}(x) = 120x - \frac{56}{81}x^{-10/3}$$

etc.   □

Exercise Set 3.3

In Exercises 1–14 find $f'(x)$ and $f''(x)$.

1. $f(x) = x - x^3 + 6$

2. $f(x) = x^4 - 4x^2 + 3$

3. $f(x) = 9 - x^6$

4. $f(x) = x^{-2} - 2x^3$

5. $f(x) = 3x^6 - 6x^4 + 3$

6. $f(x) = 4x - x^{-1} + 3x^{-2}$

7. $f(x) = \dfrac{1}{x - 1}$

8. $f(x) = \dfrac{x}{x + 3}$

9. $f(x) = \sqrt{x + 2}$

10. $f(x) = (x - 6)^{2/3}$

11. $f(x) = x(x - 1)^{2/3}$

12. $f(x) = \dfrac{x}{1 + x^2}$

13. $f(x) = (1 - x^2)^3$

14. $f(x) = \sqrt{1 + x^3}$

In Exercises 15–30 describe the concavity of the graph of f and find the inflection points. Use this information to sketch the graph of f, where possible.

15. $f(x) = x^2 - 3x + 2$

16. $f(x) = 9 - x^3$

17. $f(x) = x^3 - 9x^2 + 12x - 6$

18. $f(x) = x^2 - x^3 + 3x - 6$

19. $f(x) = \dfrac{1}{x}$

20. $f(x) = \dfrac{1}{x - 4}$

21. $f(x) = \dfrac{x}{x + 1}$

22. $f(x) = (x + 2)^{1/3}$

23. $f(x) = (2x + 1)^3$

24. $f(x) = \sqrt{x + 4}$

25. $f(x) = 2x^3 - 3x^2 + 18x - 12$

26. $f(x) = \dfrac{x + 4}{x - 4}$

27. $f(x) = (x - 2)^{1/3}$

28. $f(x) = |4 - x^2|$

29. $f(x) = \dfrac{x}{1 - x}$

30. $f(x) = x^{5/3} - 5x^{2/3} + 3$

In Exercises 31–35 use the Second Derivative Test to determine whether f has a relative extremum at the given value of x. If so, identify the relative extremum as a relative maximum or a relative minimum.

31. $f(x) = x^2 + \dfrac{2}{x}; \qquad x = 1$

32. $f(x) = 2x^3 - 3x^2; \qquad x = 1$

33. $f'(x) = \dfrac{x - 1}{x + 1}; \qquad x = 1$

34. $f'(x) = (x - 1)(x + 2); \qquad x = -2$

35. $f'(x) = (x^3 - 4x)^{4/3}; \qquad x = 2$

36. What can you say about the concavity of the graph of the total revenue function $y = R(x)$ if marginal revenue is decreasing for all $x > 0$?

37. What can you say about the concavity of the graph of the total cost function $y = C(x)$ if marginal cost is increasing for all $x > 0$?

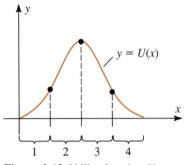

Figure 3.13 Utility function U.

38. Figure 3.13 is a graph of a function $y = U(x)$ giving total utility U as a function of quantity x. On which interval is
  a. utility U increasing but marginal utility $MU = U'$ decreasing?
  b. both utility and marginal utility increasing?
  c. both utility and marginal utility decreasing?
  d. utility decreasing but marginal utility increasing?

39. Figure 3.14 shows the graphs of two utility curves $y = U_1(x)$ and $y = U_2(x)$.
  a. For which curve is marginal utility an increasing function of x?
  b. If the two utility curves represent the amount of annual returns that two individual investors can achieve investing x dollars, which investor "knows how to make money work for him"? Why?
  c. With assumptions as in part b, which investor would be more inclined to risk an additional dollar on his investments
    (i) at $x = a$?
    (ii) at $x = b$?

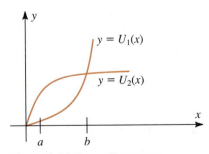

Figure 3.14 Two utility curves.

40. Verify for each of the following functions that the number $a = 0$ is a critical number for f, that $f''(a) = 0$, and that the value $f(a)$ satisfies the stated property:

a. $f(x) = x^3$; $f(0) = 0$ is neither a relative maximum nor a relative minimum.
b. $f(x) = x^4$; $f(0) = 0$ is a relative minimum.
c. $f(x) = -x^4$; $f(0) = 0$ is a relative maximum.

3.4 CURVE SKETCHING

The purpose of this section is to note how several of the concepts we have developed can be used in sketching the graph of a function. The basic tools that we shall use are primarily the facts about derivatives developed in Sections 3.1 through 3.3. Before outlining this program, however, we need to develop the techniques for finding *asymptotes* of graphs, the one graphing technique that does not depend on the derivative.

Asymptotes

Up to this point we have encountered several types of functions having the property that their graphs "approach straight lines" in one way or another. For example

(a) The graph of the function $f(x) = \dfrac{1}{x^2}$ in Example 3, Section 3.1, approaches the y-axis as x approaches zero from either direction (see Figure 1.7).

(b) The graph of the function $f(x) = \dfrac{1}{x^2}$ also approaches the x-axis as x becomes very large (see Figure 1.7).

(c) Branches of the graph of the function $f(x) = \dfrac{1 - x}{1 + x}$ approach the vertical line $x = -1$ as x approaches -1 from either direction. Moreover, the branches of this same graph also approach the horizontal line $y = -1$ as $|x|$ becomes very large, either through positive or through negative values (see Figure 4.1).

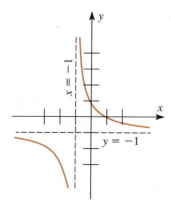

Figure 4.1 Graph of $y = \dfrac{1 - x}{1 + x}$ approaches line $x = -1$ as x approaches -1, and line $y = -1$ as $|x|$ becomes large.

Each of these is an example of an **asymptote,** which is just any line in the plane approached by the graph of f. We restrict our consideration of asymptotes to two types: vertical and horizontal. Since asymptotes tell us something about the nature of the graphs of functions, we need to be able to recognize when they occur, and to know how to find them.

Vertical Asymptotes

A **vertical asymptote** occurs in the graph of f at $x = a$ when the values $f(x)$ become infinite (with either positive or negative sign) as x approaches a. This phenomenon may occur as x approaches a from one side only or as x approaches a from either direction (as in Figure 4.1).

Vertical asymptotes are fairly easy to spot. Many functions that have vertical asymptotes do so because their denominator approaches zero as x approaches a, while their numerator does not. Consider the following examples.

Example 1 The graph of $f(x) = \dfrac{1}{x - 2}$ has a vertical asymptote at $x = 2$ because as x approaches 2 the denominator $(x - 2)$ approaches zero. Since the numerator (1) is fixed, this causes the values $f(x)$ to "blow up" as x approaches 2, for they are the result of dividing the constant 1 by increasingly tiny numbers. (See Figure 4.2.) □

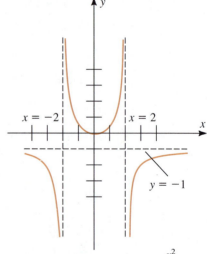

Figure 4.2 Graph of $f(x) = \dfrac{1}{x - 2}$ has a vertical asymptote at $x = 2$.

Figure 4.3 Graph of $f(x) = \dfrac{x^2}{4 - x^2}$ has vertical asymptotes at $x = 2$ and $x = -2$.

Example 2 To find the vertical asymptotes for the graph of $f(x) = \dfrac{x^2}{4 - x^2}$, we set the denominator equal to zero and solve for x: $4 - x^2 = 0$ gives $x^2 = 4$ so $x = \pm 2$. This graph has vertical asymptotes at both $x = -2$ and $x = 2$ since the denominator approaches zero as x approaches either of these numbers but the numerator approaches $x^2 = 4$. (See Figure 4.3.) □

Example 3 By way of comparison with Examples 1 and 2, we recall that the graph of $f(x) = \dfrac{x^2 - 4}{x - 2}$ does *not* have a vertical asymptote at $x = 2$ since both its numerator and denominator approach zero as x approaches 2. In fact, we can write f as

$$f(x) = \frac{x^2 - 4}{x - 2} = \frac{(x - 2)(x + 2)}{(x - 2)} = x + 2, \qquad x \neq 2$$

so the graph of f is the line with equation $y = x + 2$, minus the single point $(2, 4)$. (See Figure 4.4.) □

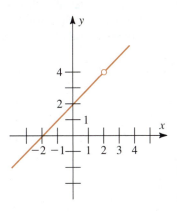

Figure 4.4 The graph of $f(x) = \dfrac{x^2 - 4}{x - 2}$ has no asymptote at $x = 2$, even though $f(2)$ is undefined.

Once a vertical asymptote for f has been located at $x = a$, sketching the graph of f in the vicinity of the asymptote requires knowledge of the one-sided limits (defined in Section 2.3)

$$\lim_{x \to a^+} f(x) \qquad \text{and} \qquad \lim_{x \to a^-} f(x).$$

Recall that the statement "$\lim\limits_{x \to a^+} f(x) = +\infty$" means that the values $f(x)$ "increase without bound" as x approaches a from the right. This is the case in Example 1 where

$$\lim_{x \to 2^+} \frac{1}{x - 2} = +\infty. \tag{1}$$

In comparison, the statement "$\lim\limits_{x \to a^-} f(x) = -\infty$" means that the values $f(x)$ decrease through negative values without bound as x approaches a from the left. Again referring to Example 1, we find that

$$\lim_{x \to 2^-} \frac{1}{x - 2} = -\infty. \tag{2}$$

Taken together, statements (1) and (2) tell us how to sketch the graph of $f(x) = 1/(x - 2)$ near the vertical asymptote $x = 2$. In general, the sign of these one-sided limits will indicate how the graph of f approaches an asymptote.

Example 4 Find the vertical asymptotes and corresponding one-sided limits for the function $f(x) = \dfrac{x + 1}{9 - x^2}$.

Solution: By factoring the denominator as $9 - x^2 = (3 - x)(3 + x)$, we can write f as

$$f(x) = \frac{x + 1}{(3 - x)(3 + x)}.$$

Clearly, the denominator has zeros at $x = \pm 3$. Since the numerator is not zero at either of these numbers, the graph of f has vertical asymptotes at both $x = -3$ and $x = 3$. Figure 4.5 shows a method of keeping track of the sign of each factor of $f(x)$, from which we can determine the sign of f on each of the three intervals $(-\infty, -3)$, $(-3, 3)$, and $(3, \infty)$:

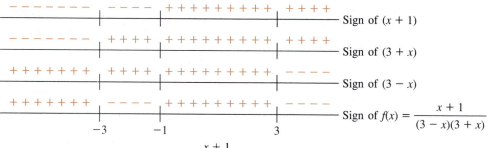

Figure 4.5 Sign analysis for $f(x) = \dfrac{x + 1}{9 - x^2}$.

From the sign of $f(x)$ as illustrated in Figure 4.5 we conclude that

$$\lim_{x \to -3^-} \frac{x + 1}{9 - x^2} = +\infty$$

$$\lim_{x \to -3^+} \frac{x + 1}{9 - x^2} = -\infty$$

$$\lim_{x \to 3^-} \frac{x + 1}{9 - x^2} = +\infty$$

$$\lim_{x \to 3^+} \frac{x + 1}{9 - x^2} = -\infty$$

The graph of f in Figure 4.6 reflects these results. □

Horizontal Asymptotes

We say that the line $y = L$ is a **horizontal asymptote** for the graph of f if the values $f(x)$ approach the number L as x increases without bound or as x decreases without bound. That is, $y = L$ is a horizontal asymptote for the graph of f if either

$$\lim_{x \to \infty} f(x) = L \quad \text{or} \quad \lim_{x \to -\infty} f(x) = L. \tag{3}$$

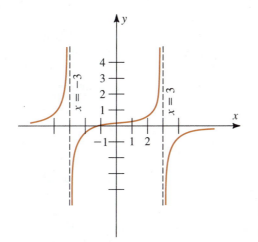

Figure 4.6 The graph of $f(x) = \dfrac{x+1}{9-x^2}$ has vertical asymptotes at $x = -3$ and $x = 3$.

Thus the problem of finding horizontal asymptotes boils down to that of evaluating the *limits at infinity* appearing in equation (3). (Recall, these limits at infinity were discussed in Section 2.2.)

Example 5 For the function $y = \dfrac{1-x}{1+x}$ the limit

$$\lim_{x \to \infty} \frac{1-x}{1+x}$$

is not easy to evaluate since x appears in both the numerator and denominator. However, using the technique of Section 2.2, we divide both the numerator and denominator by x $\left(\text{i.e., multiply both by } \dfrac{1}{x}\right)$ and use the fact that $\lim\limits_{x \to \infty} \dfrac{1}{x} = 0$ to find that

$$\lim_{x \to \infty} \left(\frac{1-x}{1+x}\right) = \lim_{x \to \infty} \left(\frac{1-x}{1+x}\right)\left(\frac{1/x}{1/x}\right)$$

$$= \lim_{x \to \infty} \frac{\dfrac{1}{x} - 1}{\dfrac{1}{x} + 1}$$

$$= \frac{0-1}{0+1}$$

$$= -1.$$

Thus the line $y = -1$ is a horizontal asymptote. Examining the other limit at infinity $\lim\limits_{x \to -\infty} \left(\dfrac{1-x}{1+x}\right)$ by this technique leads to the same result. (See Figure 4.1.) □

Example 6 Find the horizontal asymptotes for the graph of $f(x) = \dfrac{x^2-9}{x^2}$.

Strategy

Identify highest power of x.

Divide numerator and denominator by this power of x.

Evaluate limit using fact that
$$\lim_{x \to \infty} \frac{1}{x^2} = 0.$$

Horizontal asymptote is $y = \lim_{x \to \infty} f(x)$.

Solution

The highest power of x present is x^2. We therefore divide numerator and denominator by x^2 to evaluate the limit at infinity:

$$\lim_{x \to \infty} \left(\frac{x^2 - 9}{x^2} \right) \left(\frac{1/x^2}{1/x^2} \right) = \lim_{x \to \infty} \frac{1 - 9/x^2}{1}$$

$$= \frac{1 - 0}{1}$$

$$= 1.$$

Thus the graph of f has the horizontal asymptote $y = 1$. (See Figure 4.7.) □

Horizontal asymptotes often occur in economic models when a variable such as total cost has a fixed component and a component that varies proportionally with an independent variable such as production or sales level. In the "long run" the average value of such a variable will sometimes approach a constant value, which is then a horizontal asymptote for its graph.

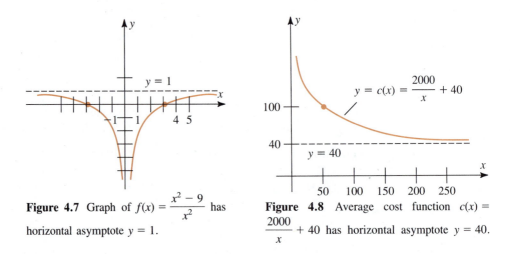

Figure 4.7 Graph of $f(x) = \dfrac{x^2 - 9}{x^2}$ has horizontal asymptote $y = 1$.

Figure 4.8 Average cost function $c(x) = \dfrac{2000}{x} + 40$ has horizontal asymptote $y = 40$.

Example 7 A manufacturer of bicycles experiences fixed weekly costs of $2000 plus variable costs of $40 per bicycle. Find

(a) the manufacturer's *average* cost per bicycle $c(x)$ for the production of x bicycles per week.

(b) the "long run" value of $c(x)$, that is, $\lim_{x \to \infty} c(x)$.

Solution: We are given that the total weekly costs are

$$C(x) = 2000 + 40x \text{ dollars}$$

so the *average* weekly cost per bicycle is

$$c(x) = \frac{2000 + 40x}{x} = \frac{2000}{x} + 40 \text{ dollars per bicycle.}$$

The horizontal asymptote $y = 40$ arises from the limit

$$\lim_{x \to \infty} c(x) = \lim_{x \to \infty} \left(\frac{2000}{x} + 40 \right)$$
$$= 0 + 40$$
$$= 40.$$

(See Figure 4.8.) ☐

Curve Sketching

We now have available to us the following steps to use in sketching the graph of a function. The examples that follow illustrate how these steps are executed.

Steps to Follow in Sketching the Graph of f

(1) Determine the domain of f.

(2) If possible, locate the zeros of f by solving the equation $f(x) = 0$.

(3) Find all vertical asymptotes by finding the numbers a for which $\lim_{x \to a^+} f(x)$

 or $\lim_{x \to a^-} f(x)$ is infinite.

(4) Find all horizontal asymptotes by examining $\lim_{x \to \pm\infty} f(x)$.

(5) Find all critical numbers, determine whether f is increasing or decreasing on the resulting intervals, and classify the extrema. (Sections 3.1, 3.2.)

(6) Determine the concavity and locate the inflection points. (Section 3.3.)

(7) Calculate the values of the function at a few convenient numbers and locate the corresponding points on the graph. Then sketch the graph according to the above information.

Example 8 Sketch the graph of the function $f(x) = 2x^3 + 3x^2 - 12x + 3$.

Solution: We follow the steps outlined above.

(1)   The domain of f is all real numbers since f is a polynomial.

(2)   Since f is a polynomial of degree 3, it is not easy to find its zeros. We therefore skip this step since the point here is to obtain a quick sketch.

(3,4) The graph of f has no vertical or horizontal asymptotes since f is a polynomial.

(5)   To find the critical numbers, we set $f'(x) = 0$:

$$f'(x) = 6x^2 + 6x - 12$$
$$= 6(x^2 + x - 2)$$
$$= 6(x + 2)(x - 1)$$

so the equation $f'(x) = 0$ has solutions $x = -2$ and $x = 1$. These are the only critical numbers. Table 4.1 shows the results of testing the sign of $f'(x)$ on each of the resulting intervals.

Table 4.1

Interval	Test Number t	$f'(t)$	Conclusion
$(-\infty, -2)$	$t = -3$	$f'(-3) = 24 > 0$	f increasing
$(-2, 1)$	$t = 0$	$f'(0) = -12 < 0$	f decreasing
$(1, \infty)$	$t = 2$	$f'(2) = 24 > 0$	f increasing

From these conclusions we know that

$$f(-2) = 2(-2)^3 + 3(-2)^2 - 12(-2) + 3 = 23$$

is a relative maximum for f, and that

$$f(1) = 2(1)^3 + 3(1)^2 - 12(1) + 3 = -4$$

is a relative minimum.

(6)   The second derivative of f is

$$f''(x) = 12x + 6$$

so setting $f''(x) = 0$ gives $12x = -6$, or $x = -\frac{1}{2}$. We test the sign of $f''(x)$ on the resulting intervals to determine the concavity as indicated in Table 4.2.

Table 4.2

Interval	Test Number t	$f''(t)$	Conclusion
$\left(-\infty, -\dfrac{1}{2}\right)$	$t = -1$	$f''(-1) = -6 < 0$	f concave down
$\left(-\dfrac{1}{2}, \infty\right)$	$t = 0$	$f''(0) = 6 > 0$	f concave up

Thus the point $(-\frac{1}{2}, f(-\frac{1}{2})) = (-\frac{1}{2}, \frac{19}{2})$ is an inflection point.

(7)   Table 4.3 shows several points $(x, f(x))$, which we use to plot the graph of f in Figure 4.9.   □

Table 4.3

x	-2	-1	0	1	2
$y = f(x)$	23	16	3	-4	7
(x, y)	$(-2, 23)$	$(-1, 16)$	$(0, 3)$	$(1, -4)$	$(2, 7)$

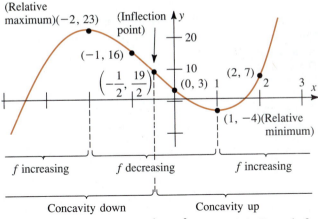

(Relative maximum)$(-2, 23)$

(Inflection point)

$(-1, 16)$

$\left(-\dfrac{1}{2}, \dfrac{19}{2}\right)$

$(0, 3)$

$(2, 7)$

$(1, -4)$(Relative minimum)

f increasing *f* decreasing *f* increasing

Concavity down Concavity up

Figure 4.9 Graph of $f(x) = 2x^3 + 3x^2 - 12x + 3$ in Example 8.

Example 9 Sketch the graph of $f(x) = \dfrac{x}{1 - x^2}$.

Solution: Following the outline, we find that

(1) The *domain* of f is all real numbers except $x = -1$ and $x = 1$. That is because the zeros of the denominator are $x = \pm 1$.

(2) The equation $f(x) = 0$ has the single solution $x = 0$. Thus the point $(0, 0)$ is on the graph.

(3) Since the zeros of the denominator are $x = \pm 1$, we check for vertical asymptotes at these numbers. Since

$$\lim_{x \to -1^-} \left(\frac{x}{1 - x^2} \right) = \infty, \qquad \lim_{x \to -1^+} \left(\frac{x}{1 - x^2} \right) = -\infty$$

$$\lim_{x \to 1^-} \left(\frac{x}{1 - x^2} \right) = \infty, \qquad \text{and} \qquad \lim_{x \to 1^+} \left(\frac{x}{1 - x^2} \right) = -\infty$$

both $x = -1$ and $x = 1$ are vertical asymptotes.

(4) Checking for horizontal asymptotes, we find that

$$\lim_{x \to \infty} \left(\frac{x}{1 - x^2} \right) = \lim_{x \to \infty} \left(\frac{x}{1 - x^2} \right) \left(\frac{1/x^2}{1/x^2} \right) = \lim_{x \to \infty} \left[\frac{1/x}{(1/x^2) - 1} \right] = \frac{0}{0 - 1} = 0$$

and

$$\lim_{x \to -\infty} \left(\frac{x}{1 - x^2} \right) = \lim_{x \to -\infty} \left(\frac{x}{1 - x^2} \right) \left(\frac{1/x^2}{1/x^2} \right) = \lim_{x \to -\infty} \left[\frac{1/x}{(1/x^2) - 1} \right] = \frac{0}{0 - 1} = 0.$$

Thus the line $y = 0$ (i.e., the *x*-axis) is a horizontal asymptote.

(5) To find the critical numbers, we calculate the derivative using the Quotient Rule:

$$f'(x) = \frac{d}{dx} \left(\frac{x}{1 - x^2} \right) = \frac{(1 - x^2)(1) - x(-2x)}{(1 - x^2)^2} = \frac{1 + x^2}{(1 - x^2)^2}.$$

Since $1 + x^2 \geq 1$ for all x, this derivative is never zero. However, $f'(x)$ is undefined for $x = \pm 1$, but so is $f(x)$. Thus there are no critical numbers. Finally, since both the numerator and denominator of f' are positive for all $x \neq \pm 1$, f is increasing on each of the intervals $(-\infty, -1)$, $(-1, 1)$, and $(1, \infty)$.

(6) The second derivative of $f(x)$ is

$$f''(x) = \frac{d}{dx}\left[\frac{1 + x^2}{(1 - x^2)^2}\right] = \frac{(1 - x^2)^2(2x) - (1 + x^2) \cdot 2(1 - x^2)(-2x)}{(1 - x^2)^4}$$

$$= \frac{2x^3 + 6x}{(1 - x^2)^3}$$

$$= \frac{2x(x^2 + 3)}{(1 - x^2)^3}.$$

Thus the concavity of the graph of f may change at $x = 0$ ($f''(x) = 0$) and $x = \pm 1$ ($f''(x)$ undefined). The sign of $f''(x)$ on each of the resulting intervals is shown in Figure 4.10, which also shows the sign of each factor of $f''(x)$.

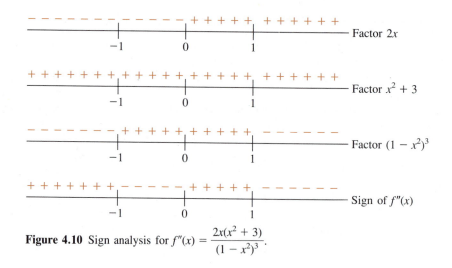

Figure 4.10 Sign analysis for $f''(x) = \dfrac{2x(x^2 + 3)}{(1 - x^2)^3}$.

From the sign of $f''(x)$ we conclude that the graph of f is

concave up on	$(-\infty, -1)$,
concave down on	$(-1, 0)$,
concave up on	$(0, 1)$, and
concave down on	$(1, \infty)$.

The point $(0, f(0)) = (0, 0)$ is therefore an inflection point.

(7) Since $f(-2) = 2/3$, $f(0) = 0$, and $f(2) = -2/3$, the points $(-2, 2/3)$, $(0, 0)$, and $(2, -2/3)$ are on the graph. The graph is sketched in Figure 4.11. □

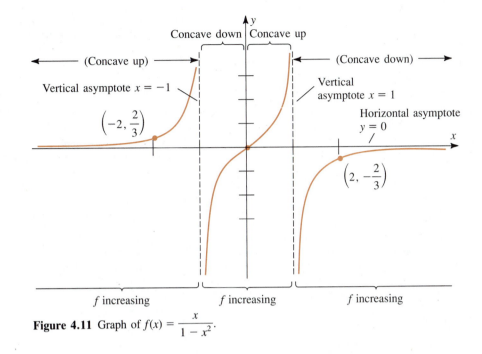

Figure 4.11 Graph of $f(x) = \dfrac{x}{1 - x^2}$.

Exercise Set 3.4

In Exercises 1–20 find the indicated limit.

1. $\displaystyle \lim_{x \to \infty} \frac{3x + 5}{4 - x}$

2. $\displaystyle \lim_{x \to \infty} \frac{3x^2 + 2}{10x^2 - 3x}$

3. $\displaystyle \lim_{x \to \infty} \frac{x(9 - x)}{x^2 + 6}$

4. $\displaystyle \lim_{x \to \infty} \frac{x^2}{16 + x^2}$

5. $\displaystyle \lim_{x \to \infty} \frac{1}{7 + x^{-2}}$

6. $\displaystyle \lim_{x \to \infty} \frac{3}{1 + \sqrt{x}}$

7. $\displaystyle \lim_{x \to \infty} \frac{x^4 - 3x^2}{5x^2 + x^5}$

8. $\displaystyle \lim_{x \to -\infty} \frac{2x + 6}{x^2 + 1}$

9. $\displaystyle \lim_{x \to \infty} \frac{3x^2 + 4x}{1 - x^4}$

10. $\displaystyle \lim_{x \to \infty} \frac{(x - 3)(x + 4)}{3x - 7}$

11. $\displaystyle \lim_{x \to \infty} \frac{4}{x + \sqrt{x}}$

12. $\displaystyle \lim_{x \to \infty} \frac{\sqrt{x} + 7}{1 - \sqrt[3]{x}}$

13. $\displaystyle \lim_{x \to 2^+} \frac{3}{x - 2}$

14. $\displaystyle \lim_{x \to -1^-} \frac{4}{x + 1}$

15. $\displaystyle \lim_{x \to -2^-} \frac{1}{(x + 2)^2}$

16. $\displaystyle \lim_{x \to \infty} \frac{\sqrt{x} - 1}{x^2}$

17. $\displaystyle \lim_{x \to 2^+} \frac{x^2 + 1}{x^2 - 4}$

18. $\displaystyle \lim_{x \to 2^-} \frac{x^2 + 1}{x^2 - 4}$

19. $\displaystyle \lim_{x \to 3^-} \frac{2}{x^2 - 9}$

20. $\displaystyle \lim_{x \to 0^+} x + \frac{1}{x}$

In Exercises 21–28 find the horizontal asymptotes for the given function.

21. $y = \dfrac{1}{1 + x}$

22. $y = \dfrac{1 + x}{3 - x}$

23. $y = \dfrac{2x^2}{x^2 + 1}$

24. $y = \dfrac{x^2}{1 - x}$

25. $y = \dfrac{2x^2}{(x^2 + 1)^2}$

26. $y = \dfrac{x^2 + 3}{1 - 3x^2}$

27. $y = 7 + 3x^{-2}$

28. $y = \dfrac{4x - \sqrt{x}}{x^{2/3} + x}$

In Exercises 29–38 find the vertical asymptotes.

29. $y = \dfrac{1}{x - 4}$

30. $f(x) = \dfrac{1}{(x + 2)^2}$

31. $f(x) = \dfrac{x^2 + 4}{x^2 - 4}$

32. $y = \dfrac{x^2 + 1}{x^2 - 1}$

33. $y = \dfrac{4}{4 + 5x + x^2}$

34. $f(x) = \dfrac{x + 3}{x^2 - 4x + 3}$

35. $f(x) = \dfrac{x + 3}{x^2 + x - 6}$

36. $f(x) = \dfrac{x^2 + x - 6}{x^2 + 3x - 10}$

37. $f(x) = \dfrac{x^2 + 5x + 6}{x^2 - 5x + 6}$

38. $f(x) =$

$$\dfrac{x^3 + 2x^2 - 5x - 6}{x^2 - x - 2}$$

39. A manufacturing company finds that its weekly fixed costs are $2500 and that its variable costs of production for its single product are $80 per item. Find
   a. $C(x)$, its total weekly cost in producing x items.
   b. $c(x) = \dfrac{C(x)}{x}$, its average cost per item.
   c. $\lim\limits_{x \to \infty} c(x)$.

40. A manufacturer of picnic tables has total cost $C(x) = 500 + 20\sqrt{x} + 40x$ in producing x picnic tables per week. Find the limit $\lim\limits_{x \to \infty} c(x)$ of the average cost function $c(x) = \dfrac{C(x)}{x}$.

41. A manufacturer has total cost function $C(x) = 2000 + 500x^{1/3} + Ax$. Find A if, for the average cost function $c(x) = \dfrac{C(x)}{x}$, $\lim\limits_{x \to \infty} c(x) = 35$.

42. A firm determines that the average cost per apple of supplying x apples per week to a certain grocery chain is $c(x) = \dfrac{22 + 10x + 20x^2}{1 + x^2}$ cents. Does this average cost function have a horizontal asymptote? If so, what?

43. A public relations firm determines that the demand for a certain product, as a function of the amount x spent per month on advertising the product, is

$$d(x) = \dfrac{72 + 96x^{1/3}}{10x^{1/5} + 6x^{1/3}}$$

thousand sales per month. Does this demand function have a horizontal asymptote? If so, what?

44. Suppose that the supply of uranium available on world markets is related to the price per pound by the function

$$S(p) = \dfrac{100}{200 - p}$$

where p is the price per pound, $0 \le p < 200$. Sketch the graph of $S(p)$ and identify the vertical asymptote.

In each of Exercises 45–63 sketch the graph of f according to the outline described in this section.

45. $f(x) = x^2 - 2x - 8$

46. $f(x) = 2x^3 - 3x^2$

47. $f(x) = x^3 + x^2 - 8x + 8$

48. $f(x) = x^3 - 12x + 6$

49. $y = \dfrac{x + 4}{x - 4}$

50. $f(x) = 9x - \dfrac{1}{x}$

51. $f(x) = |4 - x^2|$

52. $f(x) = x^{5/3} - x^{2/3}$

53. $y = \dfrac{1}{x(x - 4)}$

54. $y = x^2 - \dfrac{9}{x^2}$

55. $f(x) = \dfrac{x}{(2x + 1)^2}$

56. $f(x) = (x - 3)^{2/3} + 1$

57. $y = \dfrac{1}{3}x^3 - x^2 - 3x + 4$

58. $f(x) = \dfrac{x^2 - 4x + 5}{x - 2}$

59. $y = 4x^2(1 - x^2)$

60. $f(x) = \dfrac{x^2}{9 - x^2}$

61. $y = 16 - 20x^3 + 3x^5$

62. $f(x) = x^4 - 8x^2 + 10$

63. $f(x) = 8 - 12x^2 + x^4$

3.5 FINDING ABSOLUTE EXTREMA

Many applications of the derivative have to do with finding the *absolute maximum* or *absolute minimum* value of a given function. For example, we may wish to find the maximum value of a company's profits, a college's enrollment, or a crop's yield, or the minimum value of a company's total costs, a college's attrition rate, or a farmer's crop loss due to a plant disease.

By the **absolute maximum** value of a function f we simply mean the largest value of $f(x)$ for all x in the domain of f. Similarly, the **absolute minimum** value of f is just the smallest value of $f(x)$ for all x in the domain of f. We use the term **absolute extremum** to refer to either an absolute maximum or an absolute minimum. (When we use the terms maximum or minimum value we are referring to an *absolute* maximum or minimum, rather than a *relative* maximum or minimum.)

While a relative maximum or minimum value of a function may also be an absolute maximum or minimum value for that function, the two concepts are not the same, and the procedure for finding absolute extrema is somewhat different from that for finding relative extrema. In this section we highlight the differences between these two types of extrema and develop a procedure for finding absolute extrema.

We begin with two examples in which absolute extrema either occur at relative extrema or fail to exist.

Example 1 For the function $f(x) = 4x^2 - 2x^4$ graphed in Figure 5.1 the absolute maximum value $f(x) = 2$ occurs both at $x = -1$ and at $x = 1$. Both of these numbers x are critical numbers since the derivative is

$$f'(x) = 8x - 8x^3 = 8x(1 - x^2)$$

which equals zero for $x = \pm 1$. Also, since

$$f''(x) = 8 - 24x^2$$

we have $f''(-1) = f''(1) = -16 < 0$, so both $f(-1)$ and $f(1)$ are relative maxima by the Second Derivative Test.

However, f has no absolute minimum value since this function takes on increasingly large negative values as $|x|$ increases. (Note that a relative minimum exists at $x = 0$, but this is *not* an absolute minimum.) □

Example 2 The function $f(x) = \sqrt{x}$ has an absolute minimum value $f(x) = 0$ at $x = 0$. However, since $f(x) = \sqrt{x}$ is an increasing function for all $x > 0$, this function has no absolute maximum value. (See Figure 5.2.) □

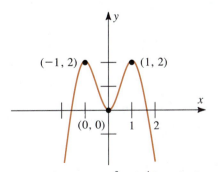

Figure 5.1 $f(x) = 4x^2 - 2x^4$ has absolute and relative maxima $f(-1) = f(1) = 2$, but no absolute minimum value. Relative minimum at $(0, 0)$ is not an absolute minimum.

Figure 5.2 $f(x) = \sqrt{x}$ has an absolute minimum at $x = 0$ but no absolute maximum.

**Endpoint
Extrema**

The point of these first two examples is that absolute extrema may exist at relative extrema, but that an absolute maximum or minimum may not exist even though relative extrema are present. Figures 5.3 and 5.4 show another way in which absolute extrema can occur—at endpoints of the domain of a function. It is important to note that when the domain of a function is restricted to a closed interval $[a, b]$, the endpoint values $f(a)$ and $f(b)$ may become absolute extrema, even though they are not relative extrema when the domain of f is unrestricted.

The following theorem states that when a continuous function is restricted to a closed bounded interval $[a, b]$, there will always exist precisely one maximum and one minimum value for the function (although either of them may occur at more than one point in the interval). Its proof is beyond the scope of this text.

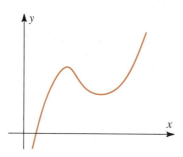

Figure 5.3 Function f has no absolute extrema.

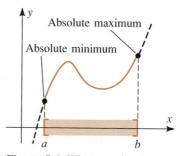

Figure 5.4 When restricted to the domain $[a, b]$, the function f has an absolute minimum at endpoint $x = a$ and an absolute maximum at endpoint $x = b$.

THEOREM 5

If the function f is continuous on the closed interval $[a, b]$, then f has both a maximum and a minimum value on $[a, b]$.

Most problems in which we shall need to find maximum or minimum values of a function will involve closed intervals of the form $[a, b]$ as the domain of the function. (When this is not the case, we shall need to keep in mind the fact that extrema need not exist, as in Examples 1 and 2.)

When the domain of f is restricted to the closed interval $[a, b]$, we have seen that maximum or minimum values can exist

(a) at relative extrema, or
(b) at the endpoints $x = a$ and $x = b$.

These are the only possibilities. Since relative extrema can exist only at critical numbers, we have the following procedure for finding absolute extrema.

> **Procedure for Finding the Maximum and Minimum Values of a Continuous Function f on a Closed Interval [a, b]:**
>
> (1) Find all critical numbers c in $[a, b]$. (These are the numbers c for which $f'(c) = 0$ or $f'(c)$ is undefined.)
> (2) Compute the endpoint values $f(a)$ and $f(b)$ and all values $f(c)$ where c is a critical number in $[a, b]$.
> (3) The largest of the values found in Step 2 is the absolute maximum value of f on $[a, b]$. The smallest is the absolute minimum value of f on $[a, b]$.

Example 3 Find the maximum and minimum values for the function $f(x) = x^3 - 3x^2 - 24x + 5$ for x on the interval $[-3, 8]$.

Strategy

Find f'.

Solution

For $f(x) = x^3 - 3x^2 - 24x + 5$ the derivative is

$$f'(x) = 3x^2 - 6x - 24$$
$$= (3x + 6)(x - 4).$$

Set $f'(x) = 0$ to find the critical numbers.

Thus $f'(x) = 0$ if $3x + 6 = 0$ or if $x - 4 = 0$. These two equations give the critical numbers $x = -2$ and $x = 4$. Since $f'(x)$ exists for all x, there are no other critical numbers.

Calculate $f(x)$ for each critical number and endpoint of the interval $[-3, 8]$.

Calculating $f(x)$ for each critical number and both endpoints of the interval $[-3, 8]$, we find

$$f(-3) = (-3)^3 - 3(-3)^2 - 24(-3) + 5 = 23$$
$$f(-2) = (-2)^3 - 3(-2)^2 - 24(-2) + 5 = 33$$
$$f(4) = 4^3 - 3(4)^2 - 24(4) + 5 = -75$$
$$f(8) = 8^3 - 3(8)^2 - 24(8) + 5 = 133.$$

Select the largest and smallest of these values.

The maximum value of f is therefore $f(8) = 133$ and the minimum value is $f(4) = -75$. Note from Figure 5.5 that the maximum value of f occurs at an endpoint while the minimum value of f occurs at a critical number. □

Figure 5.5 For $f(x) = x^3 - 3x^2 - 24x + 5$ the maximum value is $f(8) = 133$ and the minimum value is $f(4) = -75$ on the interval $[-3, 8]$.

Example 4 Find the maximum and minimum values for the function $f(x) = x^{2/3}$ for $-1 \le x \le 8$.

Strategy

Find f'.

Solution

For the function $f(x) = x^{2/3}$ the derivative is

$$f'(x) = \frac{2}{3}x^{-1/3} = \frac{2}{3x^{1/3}}.$$

Find all critical numbers.

Calculate $f(x)$ for all critical numbers and endpoints.

Now $f'(x)$ is nonzero for all x, but $f'(x)$ fails to exist for $x = 0$. Thus the only critical number for f is $x = 0$. Checking the values $f(x)$ at this critical number and at both endpoints of $[-1, 8]$ gives

$$f(-1) = (-1)^{2/3} = 1$$
$$f(0) = (0)^{2/3} = 0$$
$$f(8) = 8^{2/3} = (8^{1/3})^2 = 2^2 = 4.$$

Select the largest and smallest of these values.

Thus the maximum value $f(8) = 4$ occurs at an endpoint and the minimum value $f(0) = 0$ occurs at the critical number. (See Figure 5.6.)

□

Our last example involves finding the maximum value of a function defined on the interval $[0, \infty)$. Although this interval is not of the form $[a, b]$, very large numbers x often do not make sense (for instance, a factory may not be able to produce a million units per day) and can be ruled out. Thus the technique for finding the desired maximum is to check the single endpoint $x = 0$ plus any relative extrema at critical numbers lying in the interval $(0, \infty)$.

Example 5 A manufacturer of pocket cameras finds that the total cost of producing and marketing x cameras per month is $C(x) = 400 + 10x + \frac{1}{2}x^2$ dollars. If the manufacturer can sell all cameras produced for \$50 each, what monthly production level yields the maximum profit for the manufacturer?

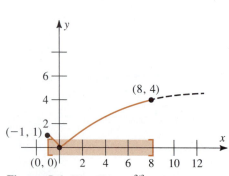

Figure 5.6 For $f(x) = x^{2/3}$ on the interval $[-1, 8]$ the maximum is $f(8) = 4$ and the minimum is $f(0) = 0$.

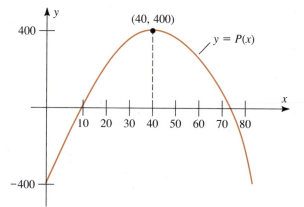

Figure 5.7 Maximum value of the profit function $P(x) = 40x - 400 - \frac{1}{2}x^2$ is $P(40) = 400$.

Solution: The revenue from the sale of x cameras is

$$R(x) = 50x$$

so the monthly profit to the manufacturer is

$$P(x) = R(x) - C(x)$$

$$= 50x - \left(400 + 10x + \frac{1}{2}x^2\right)$$

$$= 40x - 400 - \frac{1}{2}x^2.$$

Since negative production levels do not make sense, the problem is to find the maximum value of the profit function

$$P(x) = 40x - 400 - \frac{1}{2}x^2$$

for x in the interval $[0, \infty)$. Setting $P'(x) = 0$ gives

$$P'(x) = 40 - x = 0$$

so $x = 40$ is the only critical number. Since

$$P''(x) = -1 < 0$$

for all x, the Second Derivative Test verifies that

$$P(40) = 40(40) - 400 - \frac{1}{2}(40)^2 = 400$$

is a relative maximum value for P. Since

$$P(0) = 40(0) - 400 - \frac{1}{2}(0)^2 = -400$$

the relative maximum yields a larger value $P(x)$ than does the endpoint $x = 0$. Thus the maximum profit of $P(40) = 400$ dollars per month occurs at the production level of 40 cameras per month. (See Figure 5.7.) $\square$

Exercise Set 3.5

In each of Exercises 1–20 find the maximum and minimum values of the given function on the given interval.

1. $f(x) = 9 - x^2$, x in $[-3, 3]$

2. $f(x) = 9 - x^2$, x in $[-4, 1]$

3. $f(x) = x^2 - 2x + 3$, x in $[-1, 3]$

4. $f(x) = 7 - x^3$, x in $[-2, 3]$

5. $f(x) = 3 - x + x^2$, x in $(-\infty, \infty)$

6. $f(x) = 3x - x^3$, x in $[-1, 1]$

7. $f(x) = x^2(x - 1)$, x in $[0, 3]$

8. $f(x) = |x - 2|$, x in $[0, 5]$

9. $f(x) = x^3 - 2x^2$, x in $[-1, 2]$

10. $f(x) = x^2(x - 3)$, x in $[-1, 2]$

11. $f(x) = x + \frac{1}{x}$, x in $[1/2, 2]$

12. $f(x) = (\sqrt{x} + x)^2$, x in $[0, 4]$

13. $f(x) = x^{2/3} + 2$, x in $[0, \infty)$

14. $f(x) = 3 - x^{2/3}$, x in $[-1, 8]$

15. $f(x) = \dfrac{1}{x(x - 4)}$, x in $[1, 3]$

16. $f(x) = \dfrac{(x + 1)}{(x - 1)}$, x in $[-3, 0]$

17. $f(x) = \dfrac{(x + 1)}{(x - 1)}$, x in $[2, \infty)$

18. $f(x) = 8x^{1/3} - 2x^{4/3}$, x in $[-1, 8]$

19. $f(x) = \sqrt{x}(1 - x)$, x in $[0, 4]$

20. $f(x) = 3x^5 - 5x^3$, x in $[-2, 2]$

21. The population of a certain city is predicted to be $P(t) = 100{,}000 + 48t^{3/2} - 4t^2$ people t years after 1985. In what year will the population be a maximum according to this prediction?

22. A manufacturer of men's shoes experiences a total cost of $C(x) = 400 + 20x + x^2$ dollars in producing x pairs of shoes per week. If the manufacturer receives $60 for each pair of shoes produced, what is the weekly production level that maximizes profit?

23. A company determines that the profit resulting from the sale of x of its products per week is $P(x) = 160 - 96x + 18x^2 - x^3$ thousand dollars. Find the maximum and minimum values for this function on the interval $[0, 10]$ and the sales level x at which each occurs.

24. Find the maximum and minimum values of the profit function

$$P(x) = 500 - 60(3x - 27)^{2/3}$$

on the interval $[0, 15]$.

25. A producer of picture frames finds that x frames can be sold per week at price p where $x = 400 - 4p$ for $0 \le p \le 100$. At what price will revenue be a maximum? $[R(x) = xp(x).]$

26. A large suburban community receives its electrical power supply from an urban power plant. A local engineer, after studying data on energy usage, determines that a good model for daily energy usage between 6:00 a.m. and 8:00 p.m. is given by the function

$$E(t) = 4 - 2\left(\frac{t - 13}{7}\right)^4, \qquad 6 \le t \le 20$$

where t represents hours after midnight. The engineer argues that through the use of solar generators the community could reduce its energy demand from the urban power plant by the amount $S(t)$ where

$$S(t) = 1 - \left(\frac{t - 13}{7}\right)^2, \qquad 6 \le t \le 20.$$

(See Figure 5.8.)

a. Find the peak demand for energy according to the engineer's model for present usage and the time when it occurs.

b. Find the resulting peak demand according to the model for the contribution due to solar power and the time(s) at which it occurs.

c. Calculate the amount by which peak demand from the power plant can be reduced according to this model. (See Figure 5.9.)

Figure 5.8

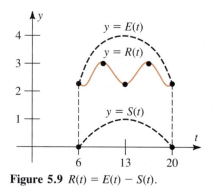

Figure 5.9 $R(t) = E(t) - S(t)$.

3.6 APPLIED MAXIMUM–MINIMUM PROBLEMS

It is frequently the case that problems to which we would like to apply the techniques of the calculus are not stated for us mathematically, but rather, they are stated in the language of a particular field of application—business, the social sciences, and so on. It is important to develop the ability to reformulate such problems mathematically, so that appropriate mathematical techniques can be applied to solve the problem. This act is called *mathematical modelling*. The following diagram indicates the three basic steps in solving applied problems by this approach.

This section contains many applied problems. Each is stated in the language of some area of application, and the mathematical solution of each involves using the derivative to find the maximum or minimum value of a certain function on a particular interval. In each case you will need to begin by reformulating the given problem into a mathematical problem of finding extrema, as in Section 3.5, and in each case you will be executing the three basic steps in Diagram 6.1.

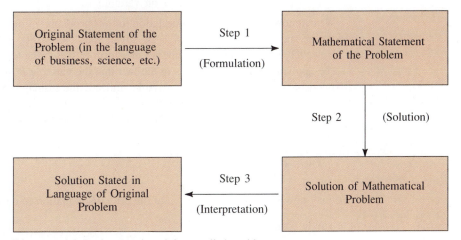

Diagram 6.1 Basic steps in solving applied problems.

We begin with a typical example, following which we shall describe a general procedure for solving additional problems of this type.

Example 1 An individual is interested in building a rectangular one-story house containing 180 square meters of floor space. Due to energy conservation considerations, the perimeter of the house is to be as small as possible. Also, for aesthetic reasons neither the length nor the width of the house is to be less than 10 meters. What dimensions for the house meet all these requirements? (See Figure 6.1.)

Solution: We begin by identifying and naming the relevant variables. We let

ℓ = length of the house
w = width of the house
P = perimeter of the house
A = area of the floor space.

Figure 6.1 House in Example 1.

We next identify all known relationships among the variables as well as any information concerning these variables. We have

$P = 2\ell + 2w$ (principal equation defining P)
$A = \ell w = 180$ (auxiliary equation giving condition on A)
$\ell \geq 10$ (constraint on variable ℓ)
$w \geq 10$ (constraint on variable w)

We can now state a precise formulation for our problem: "Find values for ℓ and w that minimize the expression $P = 2\ell + 2w$ given that $\ell w = 180$, $\ell \geq 10$, and $w \geq 10$."

However, this formulation is not yet in a form to which the techniques of Section 3.5 can be applied. The difficulty lies in the fact that the variable P to be minimized is not a function of a single independent variable. Indeed, the principal equation $P = 2\ell + 2w$ gives P as a function of *two* independent variables, ℓ and w. To overcome this difficulty, we use the auxiliary equation $\ell w = 180$ to solve for one of these variables in terms of the other:

$$\ell w = 180 \quad \text{gives} \quad w = \frac{180}{\ell}.$$

We may now substitute $w = \dfrac{180}{\ell}$ into the principal equation to obtain P as a function of the single variable ℓ:

$$P(\ell) = 2\ell + 2\left(\frac{180}{\ell}\right)$$
$$= 2\ell + \frac{360}{\ell}.$$

Finally, since it is given that neither ℓ nor w can be less than 10, and since $\ell w = 180$, it follows that ℓ cannot be greater than $\dfrac{180}{10} = 18$. That is, we must have $10 \leq \ell \leq 18$. We can therefore state our problem as follows:

"Find the minimum value of the function

$$P(\ell) = 2\ell + \frac{360}{\ell} \quad \text{for } \ell \text{ in } [10, 18]."$$

To solve this problem, we proceed as in Section 3.5. The derivative of P is

$$P'(\ell) = 2 - \frac{360}{\ell^2}.$$

Setting $P'(\ell) = 0$ gives the equation $\dfrac{360}{\ell^2} = 2$, so $\ell^2 = 180$. Thus $\ell = \pm\sqrt{180} = \pm 6\sqrt{5}$.

Now $P'(\ell)$ is undefined for $\ell = 0$, but 0 lies outside the interval [10, 18] as does the critical number $-6\sqrt{5}$. Thus the only critical number for P in the interval [10, 18] is $\ell = 6\sqrt{5}$. The minimum value for P must therefore lie among the three values

$$P(10) = 2(10) + \frac{360}{10} = 56 \text{ meters} \qquad \text{(endpoint)}$$

$$P(6\sqrt{5}) = 2(6\sqrt{5}) + \frac{360}{6\sqrt{5}} \approx 53.67 \text{ meters} \qquad \text{(critical number)}$$

$$P(18) = 2(18) + \frac{360}{18} = 56 \text{ meters} \qquad \text{(endpoint)}$$

The dimensions that minimize the perimeter are therefore $\ell = 6\sqrt{5}$ and $w = \dfrac{180}{\ell} = \dfrac{180}{6\sqrt{5}} = 6\sqrt{5}$. In other words, the house should be built in the shape of a square $6\sqrt{5}$ meters on each side. □

At this point you should look back over the solution of Example 1 to identify where each of the three basic steps (formulation, solution, interpretation) occurred. The following procedure breaks each of these three conceptual steps down into finer detail, corresponding more closely to the actual steps that you will follow in working the remaining problems in this section.

Procedure for Solving Applied Max–Min Problems

Step 1: Formulation

a. Draw a sketch, if appropriate, illustrating all variables and relevant constant quantities or dimensions.
b. Identify and label all variables.
c. Identify the variable for which the extremum is to be found.
d. Find an equation expressing this variable in terms of other variables and constants (principal equation).
e. Find all other equations among variables (auxiliary equations).
f. Identify any other constraints (e.g., intervals within which variables must lie).

Step 2: Mathematical Solution

a. Use auxiliary equations, if necessary, to substitute for variables in the principal equation until the principal equation expresses the variable for which the extremum is sought as a function of a single independent variable.
b. Determine the interval within which the independent variable lies.
c. Solve the resulting problem by the methods of Section 3.5.

Step 3: Interpretation

a. From the auxiliary equation(s), find the values of all remaining variables corresponding to the solution found above.
b. Describe the solution in the language of the original statement of the problem.

In the examples that follow the comments in the *Strategy* column will point out which of the steps in the procedure are being applied. Use these comments as a guide to help you think through the basic steps required to solve a problem before actually beginning any of the calculations.

Example 2 A highway engineer has sampled the speed and density of automobiles along a particular section of highway and determined that a good model of the relationship between velocity and density is

$$v(\rho) = \frac{100}{1 + \rho^2} \text{ km/hr}, \qquad 0 \le \rho \le 3.$$

(See Figure 6.2.)

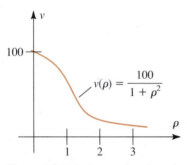

Figure 6.2 Velocity as a function of density.

Here ρ represents density in units of 100 automobiles per kilometer, and v represents velocity. Find the density at which *traffic flow* $q = \rho v$ will be a maximum according to this model.

Strategy	*Solution*
Identify the relevant variables.	The relevant variables are

$$q = \text{traffic flow}$$
$$v = \text{traffic velocity}$$
$$\rho = \text{traffic density}.$$

Find the principal equation.

The principal equation for flow is given as

$$q = \rho v \qquad (1)$$

Identify the auxiliary equation.

and the auxiliary equation for velocity is

$$v(\rho) = \frac{100}{1 + \rho^2}. \qquad (2)$$

Use constraints to find the interval in which the independent variable lies.

Finally, we have the constraint that density must satisfy the inequality

$$0 \le \rho \le 3. \qquad (3)$$

Use the auxiliary equation to eliminate one of the two variables on the right-hand side of the principal equation.

If we are to maximize flow as a function of density, we must work with ρ as the independent variable. Thus we use auxiliary equation (2) to substitute for v in principal equation (1):

$$q = \rho v$$

$$= \rho\left(\frac{100}{1 + \rho^2}\right).$$

The principal equation now gives flow q as a function of ρ alone:

$$q(\rho) = \frac{100\rho}{1 + \rho^2} \text{ hundred automobiles per hour.}$$

State the mathematical problem to be solved.

Our mathematical problem is therefore to find the maximum value of this function for ρ in the closed interval $[0, 3]$. (See Figure 6.3.)

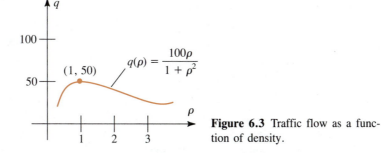

Figure 6.3 Traffic flow as a function of density.

Set the derivative equal to zero and solve to find critical numbers.

To solve this problem, we begin by calculating the derivative of q using the Quotient Rule:

$$q'(\rho) = \frac{100(1 + \rho^2) - 2\rho(100\rho)}{(1 + \rho^2)^2}$$

$$= \frac{100(1 - \rho^2)}{(1 + \rho^2)^2}.$$

Now $q'(\rho) = 0$ only if $1 - \rho^2 = 0$. Thus the critical numbers for q are $\rho = -1$ and $\rho = 1$. Since only $\rho = 1$ lies within the interval $[0, 3]$, the maximum flow lies among the values

Calculate values of flow at endpoints and critical numbers.

$$q(0) = \frac{100(0)}{1 + 0^2} = 0 \qquad \text{(endpoint)}$$

$$q(1) = \frac{100}{1 + 1^2} = 50 \qquad \text{(critical number)}$$

$$q(3) = \frac{100(3)}{1 + 3^2} = 30 \qquad \text{(endpoint)}$$

Select maximum value.
Interpret result.

The maximum flow is therefore $50 \times 100 = 5000$ automobiles per hour, which corresponds to a density of $\rho = 1$ hundred automobiles per kilometer and a velocity of $v(1) = 50$ kilometers per hour. ☐

Example 3 A producer of computer graphics software finds that the selling price p of its software is related to the number x of copies of its software sold annually by the demand equation

$$x = 10,000 - 200p \tag{4}$$

while its total cost in producing and marketing these x copies is given by the function

$$C(x) = 50,000 + 5x. \tag{5}$$

Find the price p for which profits will be a maximum.

Strategy	*Solution*

Strategy

Identify all relevant variables.

Solution

Since profits depend on price, quantity, costs, and revenues, the relevant variables are

$$P = \text{annual profits}$$
$$R = \text{annual revenues}$$
$$C = \text{annual costs}$$
$$p = \text{selling price}$$
$$x = \text{production level.} \qquad \text{(copies produced and sold annually)}$$

Find principal equation for profit, the variable to be maximized.

We have seen earlier that profit is determined from revenues and costs by the equation

$$P = R - C. \tag{6}$$

Find auxiliary equation for revenue.

Also, we know that revenues are determined from price and production level as $R = xp$. Using equation (4), we can express revenues as a function of price alone as

$$\begin{aligned} R(p) &= xp \\ &= (10,000 - 200p)p \\ &= 10,000p - 200p^2. \end{aligned} \tag{7}$$

Identify constraints on independent variables. Use to determine interval for independent variable p.

Since neither price nor production level can be negative, we have the constraints

$$p \geq 0 \quad \text{and} \quad x \geq 0.$$

The constraint $x \geq 0$, together with equation (4), gives the inequality

$$10,000 - 200p \geq 0.$$

Thus

$$200p \leq 10,000$$

so

$$0 \leq p \leq 50.$$

(See Figure 6.4.)

Use auxiliary equations to substitute for variables in the principal equation.

Using equations (5), (6), and (7) we can now express profit P as

$$\begin{aligned} P &= R - C \\ &= (10,000p - 200p^2) - (50,000 + 5x) \\ &= 10,000p - 200p^2 - 50,000 - 5x. \end{aligned}$$

Figure 6.4 Demand equation $x = 10,000 - 200p$ gives the constraint $0 \le p \le 50$.

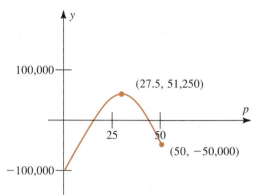

Figure 6.5 Maximum profit of $P = 51,250$ dollars occurs at selling price $p = \$27.50$.

Using equation (4), we can express profit P as a function of price alone as

Use auxiliary equation (4) to eliminate second independent variable in principal equation.

$$P(p) = 10,000p - 200p^2 - 50,000 - 5(10,000 - 200p)$$
$$= 11,000p - 200p^2 - 100,000. \tag{8}$$

This is the function for which we seek a maximum value.

State the mathematical problem to be solved.

The mathematical problem is therefore to find the maximum value of the function

$$P(p) = 11,000p - 200p^2 - 100,000$$

for p in the closed interval $[0, 50]$.

Find critical numbers.

To do so, we set $P'(p) = 0$ and find the critical numbers

$$P'(p) = 11,000 - 400p = 0$$

gives

$$p = \frac{11,000}{400} = 27.5.$$

This is the only critical number, and it does lie in the interval $[0, 50]$. Using equation (8), we calculate the values

Calculate the values $P(p)$ at endpoints and critical numbers.

$$P(0) = 11,000(0) - 200(0)^2 - 100,000$$
$$= -100,000 \qquad \text{(endpoint)}$$
$$P(27.5) = 11,000(27.5) - 200(27.5)^2 - 100,000$$
$$= 51,250 \qquad \text{(critical number)}$$

Select maximum.

$$P(50) = 11,000(50) - 200(50)^2 - 100,000$$
$$= -50,000. \qquad \text{(endpoint)}$$

Interpret result.

The maximum profit of \$51,250 will therefore occur at selling price $p = \$27.50$ per copy. (See Figure 6.5.) □

Example 4 An orchard currently has 25 trees per acre. The average yield has been found to be 495 apples per tree. It is predicted that for each additional tree planted per acre the average yield will be reduced by 15 apples per tree. According to this information, how many additional trees per acre should be planted in order to maximize total yield per acre?

Solution: Let x be the number of additional trees to be planted per acre, N the total number of trees per acre, and Y the total yield per acre. Since

$$\left\{ \begin{array}{c} \text{Total yield} \\ \text{per acre} \end{array} \right\} \text{ equals } \left\{ \begin{array}{c} \text{Number of trees} \\ \text{per acre} \end{array} \right\} \text{ times } \{\text{Yield per tree}\}$$

we have the principal equation

$$Y = N(495 - 15x).$$

The auxiliary equation $N = 25 + x$ may be combined with the principal equation to give yield as a function of x alone:

$$Y = (25 + x)(495 - 15x)$$
$$= 12{,}375 + 120x - 15x^2.$$

Finally, we note that the yield per tree will reach zero when $495 - 15x = 0$, or when $x = 33$. Thus our problem is to find the maximum value of the function

$$Y(x) = 12{,}375 + 120x - 15x^2$$

for x in the interval $[0, 33]$. Since

$$Y'(x) = 120 - 30x$$

we shall have $Y'(x) = 0$ when $x = 120/30 = 4$. This is the only critical number for Y. Since

$$\begin{array}{ll} Y(0) = 12{,}375 + 120(0) - 15(0)^2 = 12{,}375 & \text{(endpoint)} \\ Y(4) = 12{,}375 + 120(4) - 15(4)^2 = 12{,}615 & \text{(critical number)} \\ Y(33) = 12{,}375 + 120(33) - 15(33)^2 = 0 & \text{(endpoint)} \end{array}$$

the maximum yield per acre of 12,615 apples will occur when $x = 4$ additional trees are planted per acre. □

Exercise Set 3.6

1. The percentage of school age children attending private schools in a certain city is predicted to be $p(t) = \sqrt{24 + 10t - t^2}$ where t represents years after 1985. When will this percentage be the largest between 1985 and 1995?

2. A manufacturer finds that its total monthly costs in producing x hairdryers is $C(x) = 100 + 5x + 0.02x^2$. For what production level x is *average cost per item* a mini-

mum? $\left(\text{Recall, average cost } c(x) \text{ is defined by } c(x) = \dfrac{C(x)}{x}. \right)$

3. Efficiency specialists determine that t hours after beginning a shift a plant worker's efficiency on the job is approximately $R(t) = 0.92 + 0.08t - 0.02t^2$ percent. When is efficiency a maximum?

4. The sum of two nonnegative numbers is 10. Find these numbers if their product is as large as possible.

5. The sum of two nonnegative numbers is 36. Find these numbers if the first plus the square of the second is a maximum.

6. A rectangular yard is to be constructed along the side of a house by erecting a fence on three sides, using the house as the fourth wall of the play yard. Find the dimensions that produce the play yard of maximum area if 20 meters of fence is available for the project.

7. A model for the spread of disease assumes that the rate at which disease spreads is proportional to the product of the number of people infected and the number not infected. Assume the size of the population to be a constant N. When is the disease spreading most rapidly? *Hint:* Use the equation $R(t) = I(t)[N - I(t)]$.

8. A farmer has 120 meters of fencing with which to make a rectangular pen. The pen is to have one internal fence running parallel to the end fence which divides the pen into two sections. Find the dimensions that produce the pen of maximum area if the length of the larger section is to be twice the length of the smaller section.

9. Find the minimum and maximum values of the slopes of the lines tangent to the graph of

$$f(x) = x^3 - 9x^2 + 7x - 6, \qquad 1 \le x \le 4$$

and the points where these slopes occur.

10. For the total cost function $C(x) = 16 + 4x + x^2$, find the production level x for which average cost per item, $\dfrac{C(x)}{x}$, is a minimum.

11. Demand for a certain type of electric appliance is related to its selling price p by the equation $x = 2{,}000 - 100p$. Here x is the number of appliances that can be

sold per month at price p. Find the selling price for which revenues received from sales will be a maximum.

12. For the appliance manufacturer in Exercise 11 the total cost of producing x appliances per month is $C(x) = 500 + 10x$. Find the selling price p at which profits are a maximum.

13. A tool rental company determines that it can achieve 500 daily rentals of jackhammers per year at a daily rental fee of $30. For each $1 increase in rental price 10 fewer jackhammers will be rented. What rental price maximizes revenue?

14. An open box is to be made from a square sheet of cardboard by cutting out squares of equal size from each of the four corners and bending up the flaps. The sheet of cardboard measures 20 cm on each side. Find the dimensions of the box of maximum volume that can be made in this way.

15. A rectangle is inscribed in a triangle with sides of length 6 cm, 8 cm, and 10 cm, respectively. Find the dimensions of the rectangle of maximum area if two sides of the rectangle lie along two sides of the triangle.

16. A pattern for a rectangular box with a top is cut from a sheet of cardboard measuring 10 cm by 16 cm. Find the dimensions of the box for which volume is a maximum.

17. A window has the shape of a rectangle surmounted by a semicircle. Find the dimensions that provide maximum area if the perimeter of the window is 10 meters.

18. A hotel finds that it can rent 200 rooms per day if it charges $40 per room. For each $1 increase in rental rate 4 fewer rooms will be rented per day. What room rate maximizes revenues?

19. The level of antacid in a person's stomach t minutes after a certain brand of antacid tablet is taken is found to be $f(t) = \dfrac{6t}{t^2 + 2t + 1}$. When is the antacid level a maximum?

20. Utility is a measure of an individual's satisfaction due to the consumption of x units of a given quantity or commodity during a unit of time. For a utility function $U(x)$ the *saturation* quantity x_0 is the number for which $U(x_0)$ is the maximum value of the utility function.

Find the saturation quantity for the utility function

$$U(x) = \frac{6x - x^2}{10x^2 - 60x + 100}.$$

21. For the utility function $U(x) = 2 - (x - 8)^{2/3}$, $x \geq 0$, for which x is utility, $U(x)$, a maximum?

22. The cost per hour of operating a truck is $C(v) = kv^{3/2}$ where v is velocity and k is a constant. If the cost per hour for the driver is a constant A, what is the most economical speed at which to operate the truck over a route of fixed distance?

23. An orchard presently has 30 trees per acre. The average yield per tree has been found to be 300 apples. It is predicted that for each additional tree planted per acre the yield will be reduced by 5 apples per tree. According to this information, how many additional trees per acre should be planted in order to maximize yield per acre?

24. A rectangular beam is to be cut from a round log 20 cm in diameter. If the strength of the beam is proportional to the product of its width and the square of its depth, find the dimensions of the cross-section for the beam of maximum strength.

25. Suppose that the density of automobiles along a section of highway between the hours of $t = 3$ p.m. and $t = 7$ p.m. is given by the function

$$\rho(t) = -10(t - 3)(t - 7) \text{ automobiles per kilometer.}$$

If the velocity $v(\rho)$ as a function of density is as given in Example 2, find the maximum and minimum velocities and the times at which they occur.

26. A wire 50 cm long is to be cut into two pieces, one of which is to be bent into the shape of a circle and the other of which is to be bent into the shape of a square. Find where the wire should be cut so that the combined area of the resulting figures is a maximum.

27. A supervisor has the option of scheduling a laborer for up to 4 hours per day of overtime work in addition to the regular 8-hour workday. For a certain job requiring 200 man-hours to complete, the supervisor has available only one laborer. The laborer is paid $10 per hour during regular hours and $15 per hour during overtime. In addition, the company incurs daily costs of $200 for each workday in which the job remains incomplete. How should the supervisor schedule the laborer's time in order to minimize costs? (That is, should overtime work be scheduled and, if so, how much?)

28. A swimmer is in the ocean 100 meters from a straight shoreline. A person in distress is located on the shoreline 300 meters from the point on the shoreline closest to the swimmer. If the swimmer can swim 3 meters per second and run 5 meters per second, what path should the swimmer follow in order to reach the distressed person as quickly as possible?

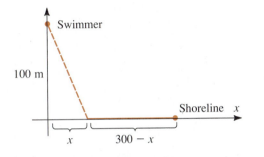

29. An underground telephone cable is to be laid between two boat docks on opposite sides of a straight river. One boat-house is 600 meters downstream from the other. The river is 200 meters wide. If the cost of laying the cable is $50 per meter under water and $30 per meter on land, how should the cable be laid to minimize cost?

30. Observations by plant biologists support the thesis that the survival rate for seedlings in the vicinity of a parent tree is proportional to the product of the density of the seeds on the ground and their probability of survival against herbivores. (The density of the herbivores tends to decrease as distance from the tree increases since the density of the food supply also decreases.) Let x denote the distance in meters from the trunk of the tree. The results of sampling indicate that the density of seeds on the ground for $0 \le x \le 10$ is given by

$$d(x) = \frac{1}{1 + (0.2x)^2}$$

and the probability of survival against herbivores is

$$p(x) = (0.1)x.$$

Find the distance, according to the model proposed here, at which the survival rate is a maximum.

31. In Exercise 30, suppose that the density of seeds for $1 \le x \le 9$ is given by the function

$$d(x) = \frac{1}{x}$$

and the probability of a seed surviving against herbivores is

$$p(x) = \frac{1}{(x - 10)^2}.$$

Find the distance from the parent tree at which seed survival will be a minimum.

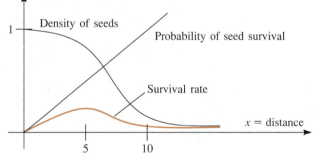

3.7 APPLICATIONS TO ECONOMICS AND BUSINESS

Our preceding examples have made frequent use of the concepts of production, cost, revenue, supply, demand, and profit. We briefly revisit these relationships here and consider several additional examples of how the derivative is useful in treating problems in business and economic theory.

Maximizing Profit

If $P(x)$, $R(x)$, and $C(x)$ represent the profit, revenue, and total cost associated with the production and sale of x items in a given period of time, the basic relationship among these three quantities is

$$P(x) = R(x) - C(x). \tag{1}$$

That is, profit equals revenues minus costs.

We have also made use of the rates at which revenue and cost change with respect to production level x. We have defined these marginal quantities as

$$\text{marginal revenue} = MR(x) = R'(x)$$

and

$$\text{marginal cost} = MC(x) = C'(x).$$

Of course, in defining marginal cost and marginal revenue we have assumed that the revenue and cost functions are differentiable functions of x. When this is the case, there exists an important relationship among the production level x_0 that maximizes profit and these two marginal quantities. To obtain this relationship, we differentiate both sides of equation (1) to obtain

$$P'(x) = R'(x) - C'(x) \tag{2}$$
$$= MR(x) - MC(x).$$

When profit is a maximum, we have, in general, $P'(x_0) = 0$. Setting $P'(x_0) = 0$ in equation (2) gives $MR(x_0) - MC(x_0) = 0$, or $MR(x_0) = MC(x_0)$. That is, *marginal revenue equals marginal cost when profit is a maximum*. (The exception to this principle is that x_0 may be an endpoint, rather than a critical number.)

Example 1 A monopoly exists when a single firm is the sole producer of a particular product or service. In this situation there is a direct relationship between the price the monopolist charges for the item and the number of items the public will purchase in a given period of time. Decreasing the price increases sales, and conversely. A typical demand curve appears in Figure 7.1.

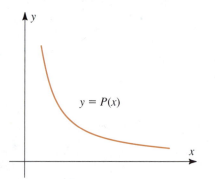

Figure 7.1 Typical demand curve.

Suppose that a monopolist can manufacture at most 175 items per week and that market research shows that the price at which the monopolist can sell x items per week is approximately $p(x) = (500 - x)$ dollars. Furthermore, suppose that the monopolist estimates that the cost of producing x items per week is $C(x) = 150 + 4x + x^2$ dollars. Then

(a) the revenue obtained from the sale of x items per week is

$$R(x) = xp(x) = x(500 - x) = 500x - x^2$$

(b) the profit obtained from selling x items per week is

$$P(x) = R(x) - C(x)$$
$$= (500x - x^2) - (150 + 4x + x^2)$$
$$= -2x^2 + 496x - 150 \text{ dollars}$$

(c) the maximum weekly profit corresponds to the maximum value of the function $P(x)$. To find this maximum, we set

$$P'(x) = -4x + 496 = 0$$

and obtain the single critical number $x_0 = 496/4 = 124$. Since the weekly production must satisfy the inequality $0 \leq x \leq 175$, the maximum profit must lie among the numbers

$$
\begin{aligned}
P(0) &= -150 &&\text{(endpoint)} \\
P(124) &= 30{,}602 &&\text{(critical number)} \\
P(175) &= 25{,}400 &&\text{(endpoint)}
\end{aligned}
$$

The maximum profit of \$30,602 is therefore achieved at the production level $x_0 = 124$ items per week.

(d) The marginal revenue and cost functions are

$$MR(x) = R'(x) = \frac{d}{dx}(500x - x^2) = 500 - 2x$$

and

$$MC(x) = C'(x) = \frac{d}{dx}(150 + 4x + x^2) = 4 + 2x.$$

At the production level $x_0 = 124$ for which profit is maximized we have that

$$MR(124) = 500 - 2(124) = 252$$

and

$$MC(124) = 4 + 2(124) = 252$$

which shows that *marginal revenue equals marginal cost when profit is maximized.*

□

Monopolists Versus Perfect Competitors

Example 1 concerned a monopoly. This is a market in which only one producer of a particular item exists, so there is no competition among sellers. In such situations the number of items the monopolist can sell is related to the price that is charged by a demand curve such as Figure 7.1. Since higher sales levels (values of x) correspond to lower prices, the corresponding revenue function eventually decreases to zero. Figure 7.2 shows the graphs of the revenue, cost, and profit functions for the monopoly in Example 1. Note that revenue increases to a maximum value (at $x = 250$) and then decreases to zero. Figure 7.2(a) also illustrates the principle that marginal revenue equals marginal cost when profit is a maximum. This translates into the fact that the line tangents to the revenue and cost curves are parallel for $x = 124$, the production level for which profit is a maximum.

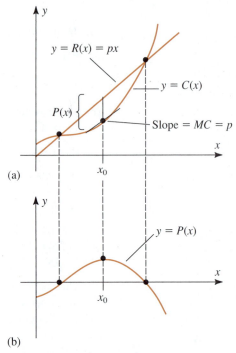

Figure 7.2 (a) shows revenue and cost curves in Example 1. Note $MR(x) = MC(x)$ when $P(x)$ is a maximum.
(b) shows profit function $P(x) = R(x) - C(x)$.

Figure 7.3 (a) shows revenue curve for a "perfect competitor." Note that revenue is a linear function. $MC(x)$ equals the slope of the revenue line when profit is a maximum.
(b) shows the corresponding profit function.

At the opposite extreme from a monopoly is a purely competitive market. In this situation many suppliers are present, with none commanding a large enough share of the market to be able to affect price by increasing or decreasing production. The selling price p for the item is constant, so the revenue function for any supplier is just the linear function $R(x) = px$. This situation is illustrated in Figure 7.3. Since marginal revenue is just $MR(x) = R'(x) = p$, we must have $MC(x_0) = p = MR(x_0)$ at the output level x_0 for which profit is a maximum.

Minimizing Costs

Since a perfect competitor cannot affect the price at which an item will sell, minimizing costs is doubly important. The techniques used to do this are just those of Sections 3.5 and 3.6 if we can find an expression for the total cost function C. The following example describes one such situation.

Example 2 A merchant sells brooms rather uniformly at a rate of about 1600 brooms a year. Accounting procedures determine that the cost of carrying a broom in stock is $2 per year. When ordering brooms, the merchant experiences a fixed order cost of $25 plus a variable order cost of 10¢ per broom. Assuming that orders can be placed so that delivery occurs precisely when stock is depleted, how often should the merchant order brooms so as to minimize yearly costs?

Strategy

Write equation for cost in words. Then translate words into mathematical expressions.

Solution

Our merchant faces a tradeoff between ordering brooms frequently, thereby reducing storage costs but increasing ordering costs, or ordering infrequently with the opposite consequences. To analyze this cost problem, we first identify the contributing factors of cost, as described above. We find that

$$\text{Yearly cost} = (\text{storage costs}) \qquad (3)$$
$$+ (\text{fixed ordering costs})$$
$$+ (\text{variable ordering costs}).$$

Label independent variable.

The only independent variable present is the number of times the merchant orders brooms per year, which we denote by x. Then the fraction of a year each order lasts in stock is $1/x$. Since brooms are sold at a uniform rate, individual brooms remain in stock half this long, $1/2x$, on the average. The merchant's annual storage costs are therefore as follows:

Write equation for storage costs in words. Then convert to mathematical expressions.

$$\text{storage costs} = (\text{number of brooms})$$
$$\cdot (\text{average storage time per broom})$$
$$\cdot (\text{annual storage cost per broom})$$
$$= (1600)\left(\frac{1}{2x}\right)(2)$$
$$= \frac{1600}{x} \text{ dollars per year.}$$

Write equation for fixed ordering costs in words. Then convert to mathematical expressions.

The fixed ordering costs will be

$$\text{fixed ordering costs} = (\text{orders per year})$$
$$\cdot (\$25 \text{ per order})$$
$$= 25x \text{ dollars per year.}$$

Find equation for variable ordering costs.

Obtain equation (3) in mathematical terms.

Finally, the variable ordering costs will be $(1600) \cdot (0.10) = 160$ dollars per year regardless of the frequency with which orders are placed. With these observations we can return to equation (3) and write down the annual cost $C(x)$ as

$$C(x) = \frac{1600}{x} + 25x + 160 \text{ dollars per year.}$$

Now x will be limited to an interval, say, $[1, 52]$. Then the problem of minimizing cost is simply the problem of finding the minimum value for the function C on the interval $[1, 52]$. We find that

$$C'(x) = -\frac{1600}{x^2} + 25$$

Set $C'(x) = 0$ to find critical numbers.

so setting $C'(x) = 0$ gives the equation

$$x^2 = \frac{1600}{25} = \left(\frac{40}{5}\right)^2 = 8^2$$

Remaining critical numbers occur where $C'(x)$ is undefined.

which yields the critical numbers $x = \pm 8$. In addition, $x = 0$ is also a critical number since $C'(0)$ is undefined. However, among these critical numbers only $x = 8$ lies within the interval of feasible solutions $[1, 52]$. Since

$$C(1) = 1600 + 25 + 160 = 1785 \text{ dollars}$$
$$C(8) = 200 + 200 + 160 = 560 \text{ dollars, and}$$
$$C(52) \approx 31 + 1300 + 160 = 1491 \text{ dollars}$$

Examine $C(x)$ at critical points and endpoints to find minimum.

the minimum cost occurs when brooms are ordered eight times per year.

□

Taxation and Other Policy Questions

In addition to serving the fundamental capitalist objective of maximizing profits, models such as those of the preceding examples can be helpful in analyzing policy questions such as the effect on output of taxation. For instance, if a government decides to impose a tax of t dollars on each item produced, the basic profit equation (1) becomes

$$P(x) = R(x) - C(x) - tx, \qquad x = \text{output.} \tag{4}$$

If, on the other hand, the tax is imposed as a percentage α of profits, the profit equation (1) becomes

$$P(x) = (1 - \alpha)[R(x) - C(x)]. \tag{5}$$

(In practice, a tax on profits rarely occurs, mainly because of the difficulty in establishing accurate figures on profit levels.)

Example 3 Suppose that the city in which the monopolist in Example 1 is located is contemplating a tax of \$20 to be imposed on each item manufactured by the monopolist. How will the monopolist respond, given the objective of profit maximization?

Solution: Applying the taxed profit equation (4) to the calculations of Example 1 gives the monopolist's revised profit equation as

$$\begin{aligned} P(x) &= R(x) - C(x) - 20x \\ &= (-2x^2 + 496x - 150) - 20x \\ &= -2x^2 + 476x - 150 \text{ dollars.} \end{aligned}$$

The new maximum is obtained by setting

$$P'(x) = -4x + 476 = 0$$

which gives the new critical number $x = 119$. Inspection of endpoint values $P(0)$ and $P(175)$ shows that, indeed, the new maximum value of profit is $P(119) = 28,172$ dollars, a reduction in profit of \$2430 per week. Notice that the sale price corresponding to the new output level is $p = (500 - 119) = 381$ dollars per item, as opposed to $p = 376$ dollars per item at the original output level of 124 items per week. The results of the proposed tax will therefore be that:

(a) The monopolist will reduce output from 124 to 119 items per week, achieving a maximum profit of $28,172 weekly.

(b) The monopolist will sell the items at a price of $381, thus passing $5 of the $20 tax along to the consumer and absorbing the remaining $15.

(c) The tax will generate a weekly revenue of 119 × 20 = 2380 dollars to the city government. □

Elasticity of Demand

Figure 7.4 shows two different demand curves relating the price p at which an item sells to the quantity $Q(p)$ of that item that consumers will purchase in a given period of time. Note that because of the steepness of the graph of Q_1, a small change in price will result in a substantial change in demand. Economists refer to such demand curves as *elastic*. Because the slope of the demand curve Q_2 is relatively small, a small change in p will cause only a small change in demand. Demand curves with this property are called *inelastic*.

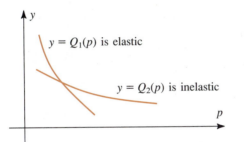

Figure 7.4 Two demand curves.

We can make this concept more precise by defining the **elasticity,** $E(p)$, of the demand curve for Q at price level p to be the negative of the limit of the relative change in demand divided by the relative change in price as $h \to 0$, that is

$$E(p) = -\lim_{h \to 0} \frac{\left[\dfrac{Q(p+h) - Q(p)}{Q(p)}\right]}{\dfrac{h}{p}}.$$

If $Q(p)$ is a differentiable function of p, this definition becomes

$$E(p) = -\lim_{h \to 0} \left[\frac{Q(p+h) - Q(p)}{h}\right]\left(\frac{p}{Q(p)}\right) = -\frac{pQ'(p)}{Q(p)}. \tag{6}$$

The definition of $E(p)$ is written with a negative sign by convention so that $E(p)$ is always a positive quantity. (Both p and $Q(p)$ are positive, and $Q'(p)$ is negative since demand decreases as price increases.)

When $E(p) > 1$, the demand curve is called **elastic** at p. This is because the relative change in demand is larger than the relative change in price $\left(\dfrac{h}{p}\right)$. When $E(p) < 1$ the demand curve is called **inelastic** at p because just the opposite is happening.

Example 4 Find the elasticity of the demand curve with equation

$$Q(p) = \frac{500}{(p+1)^2}$$

at price level $p = 4$.

Solution With $Q(p) = 500(p+1)^{-2}$ we have

$$Q'(p) = -1000(p+1)^{-3}$$

so

$$
\begin{aligned}
E(p) &= \frac{-pQ'(p)}{Q(p)} \\
&= \frac{-p[-1000(p+1)^{-3}]}{500(p+1)^{-2}} \\
&= \frac{2p}{p+1}.
\end{aligned}
$$

When $p = 4$, the elasticity is

$$E(4) = \frac{(2)(4)}{4+1} = \frac{8}{5}.$$

Since $\dfrac{8}{5} > 1$, we say that this demand curve is *elastic* at price $p = 4$. □

Example 5 The demand for an item is found to be roughly

$$Q(p) = 200 - p$$

for prices between $p = 25$ and $p = 200$ dollars. For which prices is this demand curve (a) elastic? (b) inelastic?

Solution: Here $Q'(p) = \dfrac{d}{dp}(200 - p) = -1$, so we have from equation (6) that

$$E(p) = \frac{-(-1)}{\left(\dfrac{200-p}{p}\right)} = \frac{p}{200-p}.$$

To determine for which p we shall have $E(p) > 1$ or $E(p) < 1$, we begin by finding the price for which $E(p) = 1$. Setting $E(p) = 1$ gives

$$\frac{p}{200-p} = 1, \quad \text{or} \quad p = 200 - p.$$

Thus $2p = 200$, so $p = 100$. We must therefore check the magnitude of $E(p)$ on the intervals $[25, 100)$ and $(100, 200]$.

(i) Choosing the test point $p_1 = 50$ in $[25, 100)$, we find

$$E(50) = \frac{50}{200 - 50} = \frac{50}{150} = \frac{1}{3} < 1.$$

Thus $E(p) < 1$ and the demand curve is inelastic for $25 \leq p < 100$.

(ii) Choosing the test point $p_2 = 150$ in $(100, 200]$, we find

$$E(150) = \frac{150}{200 - 150} = \frac{150}{50} = 3 > 1.$$

Thus $E(p) > 1$ and the demand curve is elastic for $100 < p \leq 200$. ☐

Interest in the concept of elasticity has to do in part with its relation to marginal revenue. If $Q(p)$ denotes the demand for an item at price p, then the revenue function is $R(p) = pQ(p)$. The marginal revenue is therefore

$$MR(p) = R'(p) = \frac{d}{dp}[pQ(p)]$$

$$= Q(p) + pQ'(p)$$

$$= Q(p)\left[1 + p\frac{Q'(p)}{Q(p)}\right]$$

$$= Q(p)[1 - E(p)].$$

From this calculation we can conclude:

(1) If the demand curve is *elastic* at p (i.e., if $E(p) > 1$), revenue will decrease if price is increased since in this case $MR(p) = R'(p) < 0$.

(2) If the demand curve is *inelastic* at p, meaning $E(p) < 1$, then we shall have $MR(p) = R'(p) > 0$, so revenue will *increase* if price is increased.

These observations illustrate the type of conclusion that can be drawn from general models such as the profit equation (1) and the preceding definition of elasticity, even though explicit equations for the demand function and its derivative are not known. The determining factor in these observations is the magnitude of elasticity, $E(p)$, which in turn depends on the relationship between the variables p and Q and the derivative $Q'(p)$. The study of relationships between marginal rates constitutes a significant part of modern economic theory, and the concept of the derivative clearly plays a central role in this subject.

Exercise Set 3.7

In Exercises 1–4 the cost $C(x)$ and revenue $R(x)$ from the manufacture and sale of x items per month for a company is given along with the range of possible production levels. In each case find the production level that most nearly maximizes profits, $P(x) = R(x) - C(x)$.

1. $C(x) = 500 + 4x$
$R(x) = 100x - \frac{1}{4}x^2$
$0 \leq x \leq 200$

2. $C(x) = 1000 + 30x + \frac{1}{2}x^2$
$R(x) = 80x$
$0 \leq x \leq 500$

3. $C(x) = 500 + 20x + x^2$
$R(x) = 30x$
$0 \le x \le 100$

4. $C(x) = 200 + 50x + \frac{1}{10}x^2$
$R(x) = 100x - \frac{1}{10}x^2$
$0 \le x \le 40$

5. Not all of the companies in Exercises 1–4 are profitable, even at the output level that maximizes profit. Which are profitable $(P(x_0) > 0)$?

6. The cost of operating a small aircraft is calculated to be $200 per hour plus the cost of fuel. Fuel costs are $s/100$ dollars per kilometer, where s represents speed in kilometers per hour. Find the speed that minimizes the cost per kilometer.

7. An automobile parts store sells 3000 headlight bulbs per year at a uniform rate. Orders can be placed from the distributor so that shipments arrive just as supplies run out. In ordering bulbs the parts store encounters costs of $20 per order plus 5¢ per bulb. How frequently should the parts store order bulbs if the cost of maintaining one bulb in stock for one year is estimated to be 48¢?

8. A chartered cruise requires a minimum of 100 persons. If 100 people sign up, the cost is $400 per person. For each additional person the cost per person decreases by $1.50. Find the number of passengers that maximizes revenue.

9. A merchant sells shirts uniformly at a rate of about 2000 per year. The cost of carrying a shirt in stock is $3 per year. The cost of ordering shirts is $50 per order plus 20¢ per shirt. Approximately how many times per year should the merchant order shirts to minimize his total yearly costs? (Assume that orders can be placed so that shipment arrives precisely when stock is depleted.)

10. Find the effect on the broom merchant's reorder scheme in Example 2 if sales fall to 1200 brooms per year and reorder costs increase by 20%.

11. A manufacturing company determines that the revenue obtained from the sale of x items will be $R(x) = 200x - 4x^2$ dollars for $0 \le x \le 25$ and that the cost of producing these x items will be $C(x) = 900 + 40x$. Find the output x at which profits will be a maximum.

12. A publisher can produce a certain book at a net cost of $4 per book. Market research indicates that 20,000 copies of the book can be sold at a price of $16 per

book and that sales can be increased by 2000 copies for each $1 reduction in price.
a. What price maximizes revenue?
b. What is the marginal cost per book?
c. What price maximizes profit?

13. A manufacturer of dishwashers can produce up to 100 dishwashers per week. Sales experience indicates that the manufacturer can sell x dishwashers per week at price p where $p + 3x = 600$ dollars. Production records show that the cost of producing x dishwashers per week is

$$C(x) = 400 + 150x + 0.5x^2.$$

a. Find the weekly revenue function.
b. Find the weekly profit function.
c. Find the marginal cost and marginal revenue functions.
d. Find the weekly production level for which profit is a maximum.
e. What is the relationship between marginal cost and marginal revenue for the production level found in part **d**?

14. Suppose that the supply of uranium available on world markets is related to the price per pound by the function

$$S(p) = \frac{100}{200 - p}$$

where p is the price per pound, $0 \le p \le 180$.
a. Find the price at which the marginal supply equals 0.25 ton.
b. Explain why S is an increasing function.
c. What would be unrealistic about this model if p were to have the range $0 \le p < 200$?

15. Let $C(x)$ denote the cost of producing x items. Then the *average cost per item* of production for these items is $c(x) = C(x)/x$. Assuming C and c to be differentiable functions of x, find the relationship between average cost and marginal cost when average cost is a minimum.

16. From equation (4), find the relationship between marginal cost and marginal revenue, in the presence of a tax of t dollars per item, when profit is a maximum.

17. A firm is called a *perfect competitor* if the price it can receive for its items is independent of production level (in other words, price is a fixed constant p beyond the control of the firm). In this case the revenue obtained from the sale of x items is simply $R(x) = px$. Show that for a perfect competitor subject to a per item tax, an increase in the tax rate will cause a decrease in output level.

18. Suppose a tax of $30 per dishwasher is imposed on the manufacturer in Exercise 13. Find
  a. the output level that maximizes profit.
  b. the resulting change in price per dishwasher.
  c. the amount of tax absorbed by the manufacturer.

19. In Exercise 13, find the resulting production level, assuming the manufacturer will always schedule production so as to maximize profit, of the granting of a tax credit of $35 per dishwasher to the manufacturer.

In Exercises 20–25 find the elasticity of the given demand function Q at the given price level p_0.

20. $Q(p) = \dfrac{1}{p^2}$, $\qquad p_0 = 5$

21. $Q(p) = \dfrac{1}{\sqrt{p}}$, $\qquad p_0 = 3$

22. $Q(p) = \dfrac{40}{(1 + p)^2}$, $\qquad p_0 = 5$

23. $Q(p) = \dfrac{10}{(p + 2)^2}$, $\qquad p_0 = 2$

24. $Q(p) = 500 - \dfrac{p}{2}$, $\qquad p_0 = 500$

25. $Q(p) = 200 - 5p$, $\qquad p_0 = 10$

26. For the demand curve $Q = 40 + 6p - p^2$, $0 < p < 10$, find
  a. the elasticity of demand $E(p)$.
  b. $E(4)$.

27. A producer in a monopoly faces a demand curve $Q(p) = 200 - p$ where p is the price per item of the manufactured item. Suppose that the cost of manufacturing Q items is given by the function $C(Q) = 2Q + 100$.
  a. Find the output Q that maximizes profits and the corresponding price.
  b. Find the elasticity of demand at the output that maximizes profit.

For each of Exercises 28–31, determine for which numbers p the given demand curve is elastic or inelastic.

28. $Q(p) = \dfrac{1}{\sqrt{p}}$

29. $Q(p) = \dfrac{1}{p^2}$

30. $Q(p) = 500 - \tfrac{1}{2}p$

31. $Q(p) = \dfrac{\sqrt{p}}{1 + p}$, $p > 1$

3.8 RELATED RATES AND IMPLICIT DIFFERENTIATION

Often one encounters problems in which two or more variables are each functions of time. For example, when an ice cube melts, its volume, its weight, and each of its dimensions change continuously as time passes. For any such variable the rate at which it changes is given by its derivative with respect to time. A *related rates* problem is one in which we are asked to find the rate at which one variable is changing in terms of the rate(s) of change of the other variable(s).

Example 1 Suppose that a child is inflating a spherical balloon. If V and r represent the volume and radius of the balloon, these two variables are related by the familiar formula for the volume of a sphere

$$V = \frac{4}{3}\pi r^3. \tag{1}$$

Now suppose we know that the balloon is being inflated at a rate of 100 cm³ per second. Since this is the rate of change of volume, this means that

$$\frac{dV}{dt} = 100 \text{ cm}^3/\text{sec}. \tag{2}$$

Can we use this information to determine the rate at which the radius r is increasing?

We begin by taking the point of view that equation (1) states that V and $4/3(\pi r^3)$ are *the same function* of time. Thus the derivatives of these two functions must be the same. Now the derivative of the left-hand side of equation (1) with respect to t is just $\frac{dV}{dt}$. To find the derivative of the right-hand side, we must remember that r is a function of t and apply the Chain Rule:

$$\frac{d}{dt}\left(\frac{4}{3}\pi r^3\right) = \frac{4}{3}\pi \cdot \frac{d}{dt}r^3 = \frac{4}{3}\pi \cdot 3r^2 \cdot \frac{dr}{dt} = 4\pi r^2 \cdot \frac{dr}{dt}.$$

Thus by differentiating both sides of the equation

$$V = \frac{4}{3}\pi r^3$$

with respect to t we find that

$$\frac{dV}{dt} = 4\pi r^2 \cdot \frac{dr}{dt}. \tag{3}$$

We can now solve equation (3) for $\frac{dr}{dt}$:

$$\frac{dr}{dt} = \frac{1}{4\pi r^2} \cdot \frac{dV}{dt}. \tag{4}$$

Equation (4) gives the rate $\frac{dr}{dt}$ at which the radius is increasing for particular values of r and $\frac{dV}{dt}$. For example, when $r = 9$ cm and $\frac{dV}{dt} = 100$ cm³/sec, we have

$$\frac{dr}{dt} = \frac{1}{4\pi \cdot 9^2}(100)$$

$$= \frac{25}{81\pi} \approx 0.098 \text{ cm/sec}.$$

In general, when two or more functions of time are related by a single equation, we can find the relationship between the *rates* at which these variables are changing by simply differentiating both sides of the given equation with respect to t. Unless the basic equation is quite simple, this will require a careful application of the Chain Rule. □

Example 2 A manufacturer of insulated steel doors finds that the revenue received from the production and sale of x doors per month is

$$R(x) = 200x - 3x^{2/3}$$

dollars per month. If the production level x is increasing at the rate of 10 doors per month, at what rate is revenue changing at production level $x = 1000$ doors per month?

Solution: Since the production level x is changing in time, both x and the revenue function R are functions of $t = $ time. The principal equation relating revenue and production is given as

$$R = 200x - 3x^{2/3}. \tag{5}$$

Differentiating both sides of this equation with respect to t gives

$$\frac{dR}{dt} = \frac{d}{dt}[200x - 3x^{2/3}]$$

$$= 200\frac{d}{dt}(x) - 3\frac{d}{dt}(x^{2/3})$$

$$= 200\frac{dx}{dt} - 3\left(\frac{2}{3}x^{-1/3} \cdot \frac{dx}{dt}\right)$$

$$= (200 - 2x^{-1/3})\frac{dx}{dt}.$$

When $x = 1000$ and $\dfrac{dx}{dt} = 10$, this gives the rate

$$\frac{dR}{dt} = [200 - 2(1000)^{-1/3}](10)$$

$$= \left(200 - \frac{2}{\sqrt[3]{1000}}\right)(10)$$

$$= 1998 \text{ dollars per month.} \qquad \square$$

In this example it is important to note that the particular value $x = 1000$ could not be substituted in equation (5) until the differentiations had been performed. That is because once we have substituted a particular value for the variable x, differentiation will simply produce zero. We must differentiate both sides of the basic equation *before* substituting particular values for the variables so that the desired rates (derivatives) will appear.

Example 3   At noon a truck leaves a depot in Columbus, Ohio, traveling east at a rate of 40 miles per hour. An hour later a second truck leaves the depot traveling north at a rate of 60 miles per hour. At what rate is the distance between the trucks increasing at 2:00 p.m. that day?

Strategy
From a sketch name the variables.

Solution
Figure 8.1 shows that we may think of the trucks as moving along the x and y axes of a coordinate plane with the depot located at the origin. If we let x be the distance of the eastbound truck from the depot and y be the distance of the northbound truck from the depot, the Pythagorean Theorem gives the distance between the trucks as

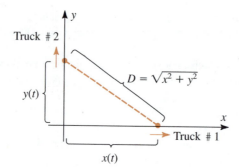

Figure 8.1 Distance between trucks in Example 3.

Find an equation relating the variables.

$$D = \sqrt{x^2 + y^2}. \tag{6}$$

Since each of the variables D, x and y are functions of t we differentiate both sides of equation (6) with respect to t to obtain

Differentiate both sides of the principal equation to obtain equation relating the rates.

$$\frac{dD}{dt} = \frac{d}{dt}\sqrt{x^2 + y^2}$$

$$= \frac{1}{2}(x^2 + y^2)^{-1/2} \cdot \frac{d}{dt}(x^2 + y^2)$$

$$= \frac{1}{2}(x^2 + y^2)^{-1/2}\left(2x \cdot \frac{dx}{dt} + 2y\,\frac{dy}{dt}\right).$$

Determine values for all variables and rates on right-hand side from given data.

At 2:00 p.m. the eastbound truck has been traveling at 40 mph for two hours, so

$$x = 80 \qquad \text{and} \qquad \frac{dx}{dt} = 40$$

while the northbound truck has been traveling at 60 mph for one hour, so

$$y = 60 \quad \text{and} \quad \frac{dy}{dt} = 60.$$

Substitute known values into rate equation to determine desired rate.

Substituting these values into the expression for $\dfrac{dD}{dt}$ gives

$$\frac{dD}{dt} = \frac{1}{2}(80^2 + 60^2)^{-1/2}[2(80)(40) + 2(60)(60)]$$

$$= \frac{6800}{\sqrt{10,000}}$$

$$= 68 \text{ miles per hour.} \qquad \square$$

Implicit Differentiation

The basic idea of differentiating both sides of an equation in order to find a relationship between the derivatives can also be applied to find the derivative $\dfrac{dy}{dx}$ when y is *implicitly* defined as a function of x by an equation in those two variables. For example, the equation

$$x^2 + y^2 = 25 \tag{7}$$

does not specify y as an *explicit* function of x of the form $y = f(x)$. (This can be most easily seen from the graph of Figure 8.2, a circle that fails to have the vertical line property.) However, as Figure 8.3 illustrates, in the region of the plane near the point $(3, 4)$ the graph of equation (7) does appear to be the graph of a function. To find the

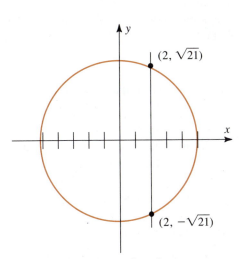

Figure 8.2 Graph of $x^2 + y^2 = 25$ does not have the vertical line property. For each x with $-5 < x < 5$ there are *two* points on the graph with this x-coordinate.

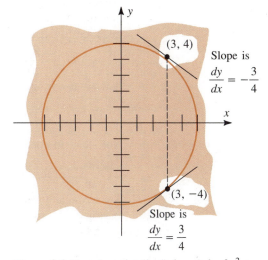

Figure 8.3 Near the point $(3, 4)$ the graph of $x^2 + y^2 = 25$ represents a function. The slope of the tangent is $\dfrac{dy}{dx} = -\dfrac{x}{y} = -\dfrac{3}{4}$.

derivative $\dfrac{dy}{dx}$ of this function, we simply differentiate both sides of equation (7) with respect to x, thinking of y as $f(x)$:

$$\frac{d}{dx}(x^2 + y^2) = \frac{d}{dx}(25)$$

gives $2x + 2y\,\dfrac{dy}{dx} = 0$

so

$$2y\,\frac{dy}{dx} = -2x$$

and

$$\frac{dy}{dx} = -\frac{2x}{2y}$$

$$= -\frac{x}{y}.$$

To find the value of this derivative at the point $(3, 4)$, we must substitute *both* coordinates $x = 3$ and $y = 4$ to obtain $\frac{dy}{dx} = -\frac{3}{4}$, the slope of the line tangent to the graph of equation (7). (See Figure 8.3.)

There are two important points to note concerning this example:

1. Since we differentiated both sides of equation (6) *with respect to x* we were *assuming* that the dependent variable y represented a differentiable function of x. Thus we were required to use the Chain Rule in differentiating the term y^2 to obtain

$$\frac{d}{dx}(y^2) = 2y \cdot \frac{dy}{dx}$$

just as we do when the function y is written in the form $y = f(x)$:

$$\frac{d}{dx}[f(x)]^2 = 2f(x) \cdot f'(x).$$

Note that, in contrast, $\frac{d}{dx}(x^2) = 2x$, as usual; you could think of this as $\frac{d}{dx}(x^2) = 2x \cdot \frac{dx}{dx}$, because $\frac{dx}{dx} = 1$.

2. Since for many values of x the graph of equation (6) contains more than one point with the same x-coordinate, we must specify *both* the x and y coordinate of a particular point to find the value of the derivative at that point. For example, the point $(3, -4)$ is also on the graph of $x^2 + y^2 = 25$, and the derivative at that point is $\frac{dy}{dx} = -\frac{x}{y} = -\frac{3}{(-4)} = \frac{3}{4}$. (See Figure 8.3.)

In general, when an equation *implicitly* defines y as a function of x, we find the derivative $\frac{dy}{dx}$ by *implicit differentiation,* which involves the following steps:

(i) Differentiate all terms on both sides of the equation remembering that the derivative of any term involving y must include the factor $\frac{dy}{dx}$, according to the Chain Rule.

(ii) Solve the resulting equation for the term $\frac{dy}{dx}$.

(iii) To find the value of $\frac{dy}{dx}$ at a particular point (a, b), substitute the numbers $x = a$ and $y = b$ into the expression for $\frac{dy}{dx}$.

Example 4 Assuming that the equation

$$y^2 + x^2y = 3x^2$$

defines y as a differentiable function of x near the point $(2, 2)$, find an expression for $\dfrac{dy}{dx}$ near $(2, 2)$ and the slope of the line tangent to the graph at this point.

Strategy

Differentiate all terms with respect to x.

Solution

Differentiating both sides of the equation with respect to x gives

$$\underbrace{2y\frac{dy}{dx}}_{\frac{d}{dx}(y^2)} + \underbrace{2xy + x^2\frac{dy}{dx}}_{\frac{d}{dx}(x^2y)} = \underbrace{6x}_{\frac{d}{dx}(3x^2)}$$

Collect all terms involving $\dfrac{dy}{dx}$.

so

$$(x^2 + 2y)\frac{dy}{dx} = 6x - 2xy$$

Solve for $\dfrac{dy}{dx}$.

and

$$\frac{dy}{dx} = \frac{6x - 2xy}{x^2 + 2y}.$$

Substitute $x = 2$ and $y = 2$ to find value of $\dfrac{dy}{dx}$ at $(2, 2)$.

At the point $(2, 2)$ the value of this derivative is

$$\frac{dy}{dx} = \frac{6(2) - 2(2)(2)}{2^2 + 2(2)} = \frac{4}{8} = \frac{1}{2}$$

which is the desired slope. □

Example 5 Find $\dfrac{dy}{dx}$ if $\sqrt[3]{x} + \sqrt[3]{y} = 1$.

Solution: Differentiating all terms in the equation

$$x^{1/3} + y^{1/3} = 1$$

gives

$$\frac{1}{3}x^{-2/3} + \frac{1}{3}y^{-2/3} \cdot \frac{dy}{dx} = 0$$

so

$$\frac{dy}{dx} = \frac{-\dfrac{1}{3}x^{-2/3}}{\dfrac{1}{3}y^{-2/3}} = -\left(\frac{y}{x}\right)^{2/3}.$$ □

Proof of the Power Rule

Up to this point we have been able to prove the Power Rule

$$\frac{d}{dx}(x^r) = rx^{r-1}$$

(8)

only for the case where r is an integer. Assuming x^r to be differentiable, we can use the technique of implicit differentiation to verify the Power Rule for all rational exponents r as follows. First, we write $r = p/q$ where p and q are integers with $q \neq 0$. Then if $y = x^r$ with $x \neq 0$, we have, $y = x^{p/q} \neq 0$, so

$$y^q = (x^{p/q})^q = x^p.$$

Differentiating both sides of this equation gives

$$qy^{q-1} \cdot \frac{dy}{dx} = px^{p-1}.$$

Recalling that $y = x^{p/q} \neq 0$, we can now solve for $\dfrac{dy}{dx}$:

$$\frac{dy}{dx} = px^{p-1}\left(\frac{1}{q}\right)y^{1-q}$$

$$= \left(\frac{p}{q}\right)x^{p-1}(x^{p/q})^{1-q}$$

$$= \left(\frac{p}{q}\right)x^{p-1} \cdot x^{\frac{p}{q}-p}$$

$$= \left(\frac{p}{q}\right)x^{(p-1+\frac{p}{q}-p)}$$

$$= \left(\frac{p}{q}\right)x^{\frac{p}{q}-1}$$

which is equation (8), the Power Rule, with $r = p/q$. ∎

Exercise Set 3.8

In Exercises 1–12 assume y to be a differentiable function of x and find an expression for $\dfrac{dy}{dx}$.

1. $x^2 + y^2 = 9$ **2.** $x^2 - y^2 = 16$

3. $xy^2 = 6$ **4.** $4x^2 + 2y^2 = 4$

5. $x^2 + 2xy + y^2 = 8$ **6.** $\sqrt{x} + \sqrt{y} = 4$

7. $x = y(y - 1)$ **8.** $(xy)^{1/2} = x - y$

9. $y^4 = x^5$ **10.** $\sqrt{x + y} = y$

11. $x^{1/2} - xy = y^{1/3}$ **12.** $x^3 - xy + 2y^2 = 4$

In Exercises 13–18 find the value of the derivative $\dfrac{dy}{dx}$ at the indicated point.

13. $xy = 9$ at $(3, 3)$

14. $x^2 + y^2 = 4$ at $(\sqrt{2}, \sqrt{2})$

15. $x^3 + y^3 = 16$ at $(2, 2)$

16. $x^2y^2 = 16$ at $(-1, 4)$

17. $\dfrac{x + y}{x - y} = 4$ at $(5, 3)$

18. $(y - x)^2 = x$ at $(9, 12)$

19. Find an equation for the line tangent to the curve $\sqrt{x} + \sqrt{y} = 4$ at the point $(4, 4)$.

20. Find an equation for the line tangent to the graph of the equation $x^2 + y^2 = 8$ at the point $(-2, 2)$.

21. Find the slope of the line tangent to the graph of $y^3 - x^2 = 7$ at the point $(1, 2)$.

22. Find the slope of the line tangent to the graph of the ellipse $\dfrac{x^2}{16} + \dfrac{y^2}{9} = 1$ at the point $\left(2, \dfrac{3\sqrt{3}}{2}\right)$.

23. When a pebble is tossed into a still pond, ripples move out in the shape of concentric circles from the point where the stone hits. Find the rate at which the area of the disturbed water is increasing when the radius of the outermost circle equals 10 meters if the radius is increasing at a rate of 2 meters per second.

24. A radio transmitter is located 3 kilometers from a straight section of interstate highway. A truck is traveling away from the transmitter along the highway at a speed of 80 kilometers per hour. How fast is the distance between the truck and the transmitter increasing when they are 5 kilometers apart?

25. Following the outbreak of an epidemic, a population of $N(t)$ people can be regarded as being made up of immunes, $I(t)$, and susceptibles, $S(t)$. That is, $N(t) = I(t) + S(t)$. $I(t)$ includes both those who have contracted the disease and those who cannot contract the disease. If the rate of decrease of susceptibles is 20 persons per day, and the rate of increase of immunes is 24 persons per day, how fast is the population growing?

26. An orchard has 100 apple trees. Currently the average yield per tree is 200 pounds of apples per year. This yield is increasing by 20 pounds per year. If the price of apples is expected to increase by 10 cents per pound per year for the next few years and the current price of apples is 60 cents per pound, find the rate at which the orchard's annual revenue from the sale of its apples will be increasing after one year.

27. The population of a certain city is predicted to be $P(t) = \sqrt{2 + t}$ million people t years after 1985. If the per capita income is expected to be $I(t) = 9 + 0.2t$ thousand dollars at this same time, and if the city has a tax on personal income of 5%, find the rate at which revenues from the city's personal income tax will be increasing in 1992.

28. Tuition at a small college is currently $8000 per year. The college currently has 1000 students. If tuition is increasing at the rate of $400 per year and enrollment is declining at the rate of 20 students per year, find the rate at which annual revenues from tuition will be changing two years from now.

29. A company estimates that it experiences a cost of $40 per day for each employee who misses work due to illness. It also estimates that t days after the outbreak of a certain type of flu among its employees $I(t) = \sqrt{4 + t^2}$ employees will be ill. At what rate is the company's total cost due to illness increasing 6 days after the outbreak?

30. The velocity at which a viscous fluid flows through a circular tube is not the same at all points of a cross-section. The flow is a maximum at the center of the cross-section and decreases to zero at the walls. For a point at a radial distance r from the center of the tube, the velocity of the flow is

$$v = \frac{\alpha}{L}(R^2 - r^2)$$

where R is the radius of the tube, L is its length and α is a constant. Find the acceleration of the fluid moving at the center of the tube if $L = 25$ cm is fixed and R is increasing at a rate of 0.02 cm/min at the instant when $R = 10$ cm.

3.9 THE MEAN VALUE THEOREM

The final theorem of this chapter has a very simple geometric interpretation, yet it is an important and powerful tool in establishing key theorems in both this chapter and others that follow.

Figure 9.1 shows a portion of the graph of the function f corresponding to an interval $[a, b]$. There is also shown the secant line containing the points $(a, f(a))$ and $(b, f(b))$. The **Mean Value Theorem** guarantees that, under certain conditions on f, there will always exist at least one number c in the interval (a, b) for which the line tangent to the

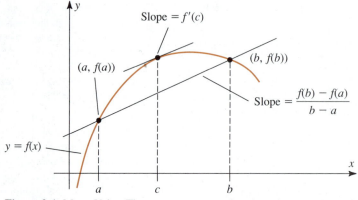

Figure 9.1 Mean Value Theorem guarantees that there is a number c in (a, b) for which $f'(c) = \dfrac{f(b) - f(a)}{b - a}$.

graph of f at $(c, f(c))$ is parallel to the secant through $(a, f(a))$ and $(b, f(b))$.
The precise statement of this result is the following.

THEOREM 6 **(Mean Value** **Theorem)**	If the function f is continuous on the closed interval $[a, b]$ and differentiable on the open interval (a, b), there exists at least one number c in (a, b) for which $$f'(c) = \frac{f(b) - f(a)}{b - a}. \qquad (1)$$

Comparing equation (1) with Figure 9.1, you will note that the left-hand side of equation (1) gives the slope of the tangent to the graph of f at $(c, f(c))$, while the right-hand side of equation (1) is the slope of the secant through $(a, f(a))$ and $(b, f(b))$. The terminology "Mean Value" refers to the slope of this secant, which we have referred to earlier as the average (mean) rate of change of the function f over the interval $[a, b]$. The Mean Value Theorem states that the derivative $f'(x)$ must equal this average rate of change for at least one number $x = c$ in (a, b).

A proof of the Mean Value Theorem is sketched in Exercise 12. Before giving an economic interpretation of this theorem and using it to prove an earlier result, we give one example showing how the number c in equation (1) may be found.

Example 1 Find the number c satisfying the Mean Value Theorem for the function $f(x) = \sqrt{x}$ on the interval $[0, 4]$.

Solution: Since $f(x) = \sqrt{x}$ is continuous on $[0, 4]$ and $f'(x) = \dfrac{1}{2\sqrt{x}}$ exists for all x in $(0, 4)$, the Mean Value Theorem applies. We therefore seek a number c for which

$$f'(c) = \frac{f(4) - f(0)}{4 - 0}$$

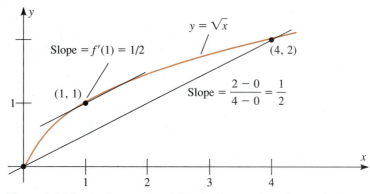

Figure 9.2 The number $c = 1$ satisfies the Mean Value Theorem for $f(x) = \sqrt{x}$ on $[0, 4]$.

that is, for which

$$\frac{1}{2\sqrt{c}} = \frac{\sqrt{4} - \sqrt{0}}{4 - 0} = \frac{2}{4} = \frac{1}{2}.$$

This requires $2\sqrt{c} = 2$, or $\sqrt{c} = 1$. Thus $c = 1$ satisfies the Mean Value Theorem. (See Figure 9.2.) ◻

An Economic Interpretation of the Mean Value Theorem

Figure 9.3 shows the graphs of a typical total cost function $y = C(x)$ and a total revenue function of the form $R(x) = px$, p constant. If we assume that these graphs intersect at $x = a$ and $x = b$, and that $R(x) > C(x)$ for $a < x < b$, then the profit function $P(x) = R(x) - C(x)$ is nonnegative for $a \le x \le b$. Thus if P is differentiable (it will be if C is), there must exist a number c in (a, b) for which $P(c)$ is the maximum value of P on $[a, b]$. This is because the differentiable, hence continuous, function P must have a maximum value on a closed, bounded interval, according to Theorem 5. This number c is shown in Figure 9.3.

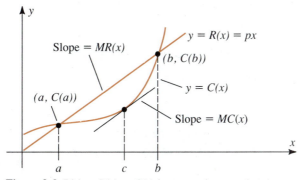

Figure 9.3 $P(x) = R(x) - C(x)$ has a maximum value at $x = c$ where $MR(c) = MC(c)$.

So far we are just reviewing the theory of extrema for the function P on the interval $[a, b]$, according to Sections 3.5 and 3.6. But now let us combine these observations with the principle from Section 3.7 that *marginal cost equals marginal revenue when profit is a maximum*. This means that

$$MC(c) = MR(c). \tag{2}$$

Now since the revenue function in Figure 9.3 is the linear function $R(x) = px$, the marginal revenue function is the constant function $MR(x) \equiv p$, *which is the slope of the straight line through* $(a, C(a))$ *and* $(b, C(b))$. That is

$$MR(x) \equiv p = \frac{C(b) - C(a)}{b - a}, \qquad x \text{ in } [a, b]. \tag{3}$$

Since the marginal cost function is just the derivative $MC(x) = C'(x)$, we can use equation (3) to rewrite equation (2) as

$$C'(c) = \frac{C(b) - C(a)}{b - a}. \tag{4}$$

What we have discovered in equation (4) may be stated as follows: "Given a differentiable cost function C on a closed interval $[a, b]$, there exists a number c in (a, b) where the slope of the *curve* $y = C(x)$ equals the slope of the *line* through $(a, C(a))$ and $(b, C(b))$." But this "discovery" is, of course, just the statement of the Mean Value Theorem applied to the cost function C.

Proof of Theorem 1, Chapter 3

Using the Mean Value Theorem we can prove one of our earlier results. We shall prove the first part of this theorem, which states that "if $f'(x) > 0$ for all x in (a, b), then f is increasing on (a, b)."

To show that f is increasing on (a, b), we let x_1 and x_2 be numbers in (a, b) with $x_1 < x_2$. We must show that $f(x_2) > f(x_1)$.

Since x_1 and x_2 are in (a, b), the closed interval $[x_1, x_2]$ is in (a, b). Thus f is differentiable, hence also continuous, on $[x_1, x_2]$. Then, by the Mean Value Theorem, there is a number c in (x_1, x_2) with

$$f'(c) = \frac{f(x_2) - f(x_1)}{x_2 - x_1}.$$

Now since $f'(x) > 0$ for *all* x in (a, b), we must have $f'(c) > 0$. Thus

$$\frac{f(x_2) - f(x_1)}{x_2 - x_1} > 0. \tag{5}$$

But since $x_2 > x_1$, the denominator $x_2 - x_1$ is positive. Thus for inequality (5) to hold the numerator $f(x_2) - f(x_1)$ must also be positive. But $f(x_2) - f(x_1) > 0$ means that $f(x_2) > f(x_1)$, which is what we needed to show.

The other statement of Theorem 1, that "if $f'(x) < 0$ for all x in (a, b), then f is decreasing on (a, b)" is proved in a similar way. This task is left for you as an exercise. ■

Exercise Set 3.9

1. Sketch the function $f(x) = |x|$ for $-1 \le x \le 1$. Does the Mean Value Theorem hold for this example? Why or why not?

In each of Exercises 2–8 find the value of c that satisfies the Mean Value Theorem for the given function and interval.

2. $f(x) = 2x - 7$, x in $[-1, 5]$

3. $f(x) = 4 - x^2$, x in $[0, 2]$

4. $f(x) = x^2 + 2x$, x in $[0, 4]$

5. $f(x) = \sqrt{x}$, x in $[0, 1]$

6. $f(x) = \dfrac{1}{x}$, x in $[1, 3]$

7. $f(x) = x^2 + 2x - 3$, x in $[-3, 0]$

8. $f(x) = x^3 - 2x + 4$, x in $[0, 2]$

9. If an automobile travels 120 miles in two hours, at varying speeds, why can you conclude that the driver violated the 55 mph speed limit at least once?

10. A monopolist determines that its total revenue from the sale of 100 items per month is the same as its total revenue from the sale of 300 items per month.
   a. Does this necessarily mean that marginal revenue equals zero for all sales levels x with $100 < x < 300$? Why?
   b. Assuming the monopolist's total revenue function $y = R(x)$ to be differentiable, is there a sales level x with $100 \le x \le 300$ for which marginal revenue equals zero? Why?

11. Use the Mean Value Theorem to prove that if $f'(x) < 0$ for all x in (a, b) then f is decreasing on (a, b).

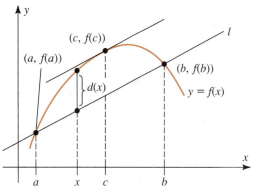

Figure 9.4

12. Prove the Mean Value Theorem as follows, referring to Figure 9.4.
   a. Let d be the function defined by
   $$d(x) = f(x) - \left[f(a) + \frac{f(b) - f(a)}{b - a}(x - a) \right].$$
   b. Explain why d is differentiable on (a, b).
   c. Show that $d(a) = d(b) = 0$.
   d. Explain why either $d(x) = 0$ for all x in $[a, b]$ or else d has a relative extremum at some number c in (a, b).
   e. Using part **d**, explain why there must be a number c in (a, b) for which $d'(c) = 0$.
   f. Show that the equation $d'(c) = 0$ gives
   $$f'(c) = \frac{f(b) - f(a)}{b - a}.$$

SUMMARY OUTLINE OF CHAPTER 3

■ **Theorem:** If $f'(x) > 0$ for all x in I, then f is increasing on the open interval I.   (Page 127)
If $f'(x) < 0$ for all x in I, then f is decreasing on I.

■ $M = f(c)$ is a **relative maximum** for f if $f(c) \ge f(x)$ for all x near c.   (Page 135)

■ $m = f(d)$ is a **relative minimum** for f if $f(d) \le f(x)$ for all x near d.   (Page 135)

■ **Theorem:** If $f(c)$ is a relative maximum or minimum, then either $f'(c) = 0$ or else $f'(c)$ fails   (Page 136)
to exist.

■ A **critical number** for f is a number x in the domain of f for which either $f'(x) = 0$ or else $f'(x)$ fails to exist. (Page 137)

■ **Theorem:** If c is a critical number for f, then (Page 138)

   (i) If $f'(x) > 0$ for $x < c$ and $f'(x) < 0$ for $x > c$, then $f(c)$ is a relative maximum.
   (ii) If $f'(x) < 0$ for $x < $ c and $f'(x) > 0$ for $x > c$, then $f(c)$ is a relative minimum.

■ $f''(x)$, the **second derivative** of $f(x)$, is $\dfrac{d}{dx} f'(x)$. (Page 142)

■ The graph of f is (Page 144)

   (i) **concave up** on (a, b) if $f''(x) > 0$ for all x in (a, b).
   (ii) **concave down** on (a, b) if $f''(x) < 0$ for all x in (a, b).

■ Points on the graph of f that separate arcs of opposite concavity are called **inflection points.** (Page 144)

■ **Theorem:** Let c be a critical number for f with $f'(c) = 0$. (Page 147)

   (i) If $f''(c) < 0$, then $f(c)$ is a relative maximum.
   (ii) If $f''(c) > 0$, then $f(c)$ is a relative minimum.
   (iii) If $f''(c) = 0$, there is no conclusion.

■ $f'''(x)$ means $\dfrac{d}{dx} f''(x)$, $f^{(4)}(x)$ means $\dfrac{d}{dx} f'''(x)$, etc. (Page 148)

■ The line $x = a$ is a **vertical asymptote** for the graph of $y = f(x)$ if $\lim\limits_{x \to a^-} f(x) = \pm\infty$ or if $\lim\limits_{x \to a^+} f(x) = \pm\infty$. (Page 152)

■ The line $y = a$ is a **horizontal asymptote** for the graph of $y = f(x)$ if $\lim\limits_{x \to \infty} f(x) = a$ or if $\lim\limits_{x \to -\infty} f(x) = a$. (Page 154)

■ $m = f(c)$ is the **absolute maximum** value of f for x in $[a, b]$ if $f(c) \geq f(x)$ for all x in $[a, b]$. The number $m = f(d)$ is the **absolute minimum** for x in $[a, b]$ if $f(d) \leq f(x)$ for all x in $[a, b]$. (Page 163)

■ The **elasticity** $E(p)$ of the demand curve $Q(p)$ at price level p is (Page 185)

$$E(p) = -\frac{pQ'(p)}{Q(p)}.$$

■ **Theorem (Mean Value Theorem):** If f is differentiable on (a, b) and continuous on $[a, b]$, there exists at least one number c in (a, b) for which (Page 198)

$$f'(c) = \frac{f(b) - f(a)}{b - a}.$$

REVIEW EXERCISES—CHAPTER 3

In Exercises 1–10 find the intervals on which f is increasing or decreasing and all relative extrema.

1. $f(x) = 4x - x^2$

2. $f(x) = x^2 - 6x + 6$

3. $f(x) = x^2 - 2x + 3$

4. $f(x) = \dfrac{1 - x}{x}$

5. $f(x) = \dfrac{x - 3}{x + 3}$

6. $f(x) = \dfrac{x^2}{x^2 - 1}$

7. $f(x) = x\sqrt{16 - x^2}$

8. $f(x) = x^4 - 2x^2$

9. $f(x) = x^3 - 3x^2 + 2$

10. $f(x) = x^2 + \dfrac{2}{x}$

In Exercises 11–20 find the maximum and minimum values of the function on the given interval and the corresponding values of x.

11. $f(x) = \dfrac{1}{x^2 - 4}$, x in $[-1, 1]$

12. $f(x) = x^3 - 3x^2 + 1$, x in $[-1, 1]$

13. $f(x) = x\sqrt{1 - x}$, x in $[-3, 0]$

14. $f(x) = x + x^{2/3}$, x in $[-1, 1]$

15. $f(x) = x - \sqrt{1 - x^2}$, x in $[-1, 1]$

16. $f(x) = \sqrt[3]{x} - x$, x in $[-1, 1]$

17. $f(x) = x - 4x^{-2}$, x in $[-3, -1]$

18. $f(x) = \dfrac{x}{1 + x^2}$, x in $[-2, 2]$

19. $f(x) = x^4 - 2x^2$, x in $[-2, 2]$

20. $f(x) = \dfrac{x - 1}{x^2 + 3}$, x in $[-2, 4]$

In Exercises 21–30 find all horizontal and vertical asymptotes.

21. $f(x) = \dfrac{1}{7 - x}$

22. $f(x) = \dfrac{x - 2}{x + 2}$

23. $y = \dfrac{x^2}{1 + x^2}$

24. $f(x) = \dfrac{x^2}{3 - x^2}$

25. $y = \dfrac{x^2 + 9}{x^2 - 9}$

26. $y = \dfrac{x - 1}{x^2 + x - 2}$

27. $y = \dfrac{x}{|x|}$

28. $f(x) = \dfrac{9}{x^2 + 3x + 2}$

29. $y = 4 + 6x^{-3}$

30. $f(x) = \dfrac{x^3}{(x^2 + 6)^2}$

In Exercises 31–40 find the indicated limit.

31. $\displaystyle\lim_{x \to \infty} \dfrac{2x + 6}{9 - x}$

32. $\displaystyle\lim_{x \to \infty} \dfrac{9 + x^2}{3x^2 + 6}$

33. $\displaystyle\lim_{x \to \infty} \dfrac{7}{3 - x^{-3}}$

34. $\displaystyle\lim_{x \to \infty} \dfrac{7}{3 - x^{-3}}$

35. $\displaystyle\lim_{x \to 4^-} \dfrac{x^2 + 5}{x^2 - 16}$

36. $\displaystyle\lim_{x \to 0^+} \dfrac{|x|}{x}$

37. $\displaystyle\lim_{x \to 2^-} \dfrac{6}{x^2 - 4}$

38. $\displaystyle\lim_{x \to \infty} \dfrac{(x - 6)(x + 3)}{2x + 8}$

39. $\displaystyle\lim_{x \to 1^-} \dfrac{5}{1 - x}$

40. $\displaystyle\lim_{x \to \infty} \dfrac{1 - x^{-2}}{2 + x^{-2}}$

In Exercises 41–50 sketch the graph of the given function.

41. $y = 4x - x^2$

42. $y = x(x - 1)(x + 3)$

43. $f(x) = x^2 - 2x - 3$

44. $f(x) = \dfrac{1 - x}{x}$

45. $y = \dfrac{x - 3}{x + 3}$

46. $f(t) = \dfrac{t^2}{t^2 - 1}$

47. $f(x) = x\sqrt{16 - x^2}$

48. $y = \dfrac{\sqrt{x}}{1 + \sqrt{x}}$

49. $y = \dfrac{x^2}{x^2 + 9}$

50. $f(x) = x^2 + \dfrac{2}{x}$

51. The sides of a square are increasing at a rate of 2 cm/sec. At what rate is the area increasing when the length of a side equals 10 cm?

52. A peach orchard has 25 trees per acre, and the average yield is 300 peaches per tree. For each additional tree planted per acre the average yield per tree will be reduced by 10 peaches. How many trees per acre give the largest peach crop?

53. A volume V of oil is spilled at sea. The spill takes the shape of a disc whose radius is increasing at a rate of $a/\sqrt{t}$ m/sec where a is a constant. Find the rate at which the thickness of the layer of oil is decreasing after 9 seconds if the radius at that time is $r(9) = 6a$.

54. A fertilizer company can sell x pounds of fertilizer per week at a price of $p(x)$ cents per pound where

$$p(x) = 90 - \dfrac{x}{500}.$$

  a. Find the total weekly revenue.
  b. Find the marginal revenue function.
  c. If weekly production is limited to 50 tons, what sales level maximizes revenue?

55. The cost of producing x wastebaskets per week is found by a manufacturer to be $C(x) = 5000 + 2x$ dollars. Market research shows that the relationship between the number x of wastebaskets that can be sold and the selling price p is $p(x) = 5 + \dfrac{200}{x}$. If production is limited to 1000 wastebaskets per week, what production level maximizes profit?

56. The cost per hour of driving a particular automobile at speed v is calculated to be $C(v) = 10 + 0.004v^2$ dollars per hour. What is the most economical speed at which to operate this automobile? (Assume v to be in units of miles per hour.)

57. A truck traveling 80 kilometers per hour is heading north on highway X, which crosses east–west highway Y at point P. A car traveling east on highway Y at 60 kilometers per hour passed point P when the truck was 20 kilometers south of point P. At what rate is the distance between the truck and the car increasing when the truck passes point P?

58. A student 1.6 meters tall walks directly away from a street light 8 meters above the ground at a rate of 1.2 meters per second. Find the rate at which the student's shadow is increasing when the student is 20 meters from the point directly beneath the light.

59. An appliance dealer sells 150 dishwashers per year at a uniform rate. The storage costs are $15 per year per dishwasher. The ordering costs are $2 per dishwasher, plus an additional $25 per order. How many times per year should the store order dishwashers so as to minimize costs?

60. A monopolist can manufacture at most 200 items per week. The price at which the monopolist can sell x items per week is $p(x) = 800 - 2x$ dollars, while the cost of producing x items per week is $C(x) = 100 + 6x + x^2$ dollars. Find the output level x_0 that maximizes profits.

61. How would the imposition of a $30 per item manufacturing tax change the optimal output level in Exercise 60?

62. A publisher currently sells 2000 subscriptions annually for a professional journal at a rate of $40 per year. Its marketing department predicts that for each one dollar reduction in subscription price an additional 50 subscriptions can be sold. Under this assumption what subscription rate maximizes revenues?

63. The price p and demand x for a certain commodity are related by the equation $p = \sqrt{100 - x^2}$, $0 \le x \le 10$. Find the price that maximizes revenue.

64. Find the slope of the line tangent to the circle with equation $(x + 2)^2 + (y - 3)^2 = 4$ at the point $(-1, 3 + \sqrt{3})$.

65. A pebble that is thrown vertically upward from ground level is at a height of $s(t) = 96t - 16t^2$ feet t seconds after it is thrown.
  a. Find its velocity function v.
  b. When is height $s(t)$ a maximum? At what height?
  c. When is velocity a maximum? At what height?
  d. What is its speed when it strikes the ground?

66. A lawn care service experiences the following profit function in contracting to service x lawns per week:

$$P(x) = 40x - 300 - 0.5x^2.$$

  a. Find the interval on which P is increasing.
  b. How many lawns should the service handle per week in order to maximize profit?

67. A tire manufacturer finds that its total cost of producing x tires per day is

$$C(x) = 30x + 200 + 0.5x^2 \text{ dollars.}$$

  a. Find its average daily cost per tire, $c(x) = \dfrac{C(x)}{x}$.
  b. For what production level x will average cost be a minimum?

68. A bicycle merchant finds that the combined annual cost of ordering and storing x bicycles at a time is

$$C(x) = 8x + \dfrac{3200}{x} \text{ dollars.}$$

Find the lot size x that minimizes this cost.

In Exercises 69 and 70 find $\dfrac{dy}{dx}$.

69. $x^2y + xy^2 = -4$ **70.** $x^3y - xy^3 = 10$

In Exercises 71 and 72 find the value of the derivative $\dfrac{dy}{dx}$ at the indicated point.

71. $\sqrt[3]{xy} - y = -6$ at $(3, 9)$

72. $\sqrt{x} - \sqrt{y} = 1$ at $(25, 16)$

Chapter **4**

Exponential and Logarithmic Functions

4.1 EXPONENTIAL FUNCTIONS

Many models of interest in economics and business, as well as in the life sciences, involve quantities whose rate of growth depends on their current size. In business the amount of interest that a sum of money will earn depends, in part, on the size of the sum. And in the biology lab the rate at which a population of bacteria increases depends on the number of bacteria present. Such situations are often modelled using **exponential functions.**

Exponential Functions

If n is a positive integer and a is a nonzero number we define the expression a^n by

$$a^n = \overbrace{a \cdot a \cdot a \cdot \cdots \cdot a}^{n \text{ factors}}.$$

Further, we define $a^0 = 1$ for all $a \neq 0$, and we define

$$a^{-n} = \frac{1}{a^n}, \qquad a \neq 0.$$

These definitions determine the values of the *exponential function*

$$f(n) = a^n \tag{1}$$

for all integers n. This function has the following properties:

(E1) $\quad a^n a^m = a^{n+m}$
(E2) $\quad a^n/a^m = a^{n-m}$
(E3) $\quad [a^n]^m = a^{nm}.$

Properties (E1)–(E3) hold for all integers n and m. Proofs of these properties are direct applications of the above definitions. For example, to prove property (E1), we note that

$$a^n a^m = \overbrace{a \cdot a \cdot a \cdot \cdots \cdot a}^{n \text{ factors}} \overbrace{a \cdot a \cdot a \cdot \cdots \cdot a}^{m \text{ factors}}$$

$$\overbrace{= a \cdot a \cdot a \cdot \cdots \cdot a}^{(n + m) \text{ factors}}$$
$$= a^{n+m}.$$

Figure 1.1 shows the graph of the exponential function $f(n) = 2^n$.

We have already seen in Chapter 1 that the domain of the exponential function $f(n) = a^n$ can be extended to include rational numbers. We do so in two steps.

(i) We define the nth root function by saying that

$$a^{1/n} = y \quad \text{means} \quad a = y^n. \tag{2}$$

For example, $9^{1/2} = 3$ since $3^2 = 9$, and $(-8)^{1/3} = -2$ since $(-2)^3 = -8$. (Recall however, that we must require $a > 0$ if n is even. That is, even roots of negative numbers are not defined.)

(ii) Then, we use composition of functions to define the exponential function for rational exponents by saying that

$$a^{n/m} \quad \text{means} \quad (a^{1/m})^n. \tag{3}$$

For example, $8^{2/3} = (8^{1/3})^2 = 2^2 = 4$, and $(-27)^{4/3} = [(-27)^{1/3}]^4 = (-3)^4 = 81$. (Again, we must require that $a > 0$ if the denominator m of the exponent is even.)

Equations (2) and (3) together allow us to define the exponential function $f(x) = a^x$ for all rational numbers x when $a > 0$. Using these definitions, we can show that properties (E1)–(E3) hold for rational exponents. The following example shows how these properties are used.

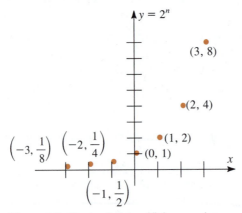

Figure 1.1 Graph of $f(n) = 2^n$ for n an integer.

Example 1 According to the above definitions and properties (E1)–(E3), we have:

(i) $2^5 = 2 \cdot 2 \cdot 2 \cdot 2 \cdot 2 = 32$.
(ii) $3^2/3^5 = 3^{2-5} = 3^{-3} = 1/3^3 = 1/27$.
(iii) $27^{1/3} = 3$ since $3^3 = 27$.
(iv) $64^{-5/6} = 1/(64^{1/6})^5 = 1/2^5 = 1/32$. □

Graphs of Exponential Functions

Figure 1.2 shows the graph of $f(x) = 2^x$. Figure 1.3 shows the graph of the exponential functions $f(x) = 3^x$ and $g(x) = -2^x$. Compare these graphs with those of Figure 1.4 for the functions $f(x) = 2^{-x}$ and $g(x) = 10^x$.

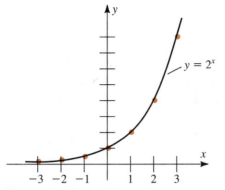

Figure 1.2 Graph of $f(x) = 2^x$ for x rational.

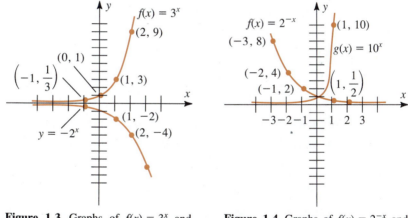

Figure 1.3 Graphs of $f(x) = 3^x$ and $g(x) = -2^x$.

Figure 1.4 Graphs of $f(x) = 2^{-x}$ and $g(x) = 10^x$.

Irrational Exponents

Although the graphs of the exponential functions in Figures 1.2–1.4 suggest that these functions have been defined for all x, this is not the case, because we have defined the expression a^x only when x is a rational number. Thus when x is an irrational number, such as $\sqrt{2}$ or π, the expression a^x is as yet undefined.

Since, for $a > 0$, we would like our exponential functions $f(x) = a^x$ to be not only defined but also continuous and differentiable for all values of x, we wish to extend the domain of our exponential functions to include all real numbers, both irrational and rational. We shall do so here quite informally by saying that we define the values of $f(x) = a^x$ for x irrational "by continuity" from the values of $f(x) = a^x$ for x rational. That is, if x_0 is an irrational number, we define $f(x_0) = a^{x_0}$ to be the number that makes $f(x) = a^x$ continuous at x_0 given the values of a^x for x rational. Geometrically, this simply says that we draw the graph of $f(x) = a^x$ by sketching a smooth curve through the points (x, a^x) for x rational, thus producing the values of a^x for x irrational "by continuity."

More advanced mathematical texts present a formal extension of the domain of the exponential function $f(x) = a^x$ to irrational numbers. We shall not pursue such formalities here. Rather, we shall henceforth assume that, for $a > 0$, the exponential function $f(x) = a^x$ is defined, continuous, and differentiable for all values of x.

Compound Interest: An Application of Exponential Functions

Exponential functions occur naturally in formulas for the periodic compounding of interest. For example, if an annual rate of interest of r percent is applied to a principal amount P_0 placed in a savings account, the amount $P(1)$ on deposit at the end of one year is

$$P(1) = P_0 + rP_0$$

interest
original deposit

$$= (1 + r)P_0.$$

At the end of the second year the interest rate is again applied to this amount giving a total $P(2)$ on deposit after 2 years of

$$P(2) = (1 + r)P_0 + r[(1 + r)P_0]$$

interest
amount on deposit after one year

$$= (1 + r)[(1 + r)P_0]$$
$$= (1 + r)^2 P_0.$$

Continuing in this manner, we find that the amount on deposit after 3 years is

$$P(3) = (1 + r)^2 P_0 + r[(1 + r)^2 P_0]$$
$$= (1 + r)^3 P_0$$

and, after n years, the amount on deposit is

$$P(n) = (1 + r)^n P_0. \tag{4}$$

Equation (4) involves the exponential function $f(n) = a^n$ with $a = (1 + r)$.

Example 2 A student deposits \$100 in a savings account that pays 8% interest compounded annually. No additional deposits or withdrawals are made. Find the amount in this account after 1, 2, and 4 years.

Solution: With $r = 0.08$ equation (4) gives

$$P(1) = (1.08)(100) = \$108$$
$$P(2) = (1.08)^2(100)$$
$$= \$116.64$$
$$P(4) = (1.08)^4(100)$$
$$= \$136.05.$$

The Number e   For reasons that will soon become apparent, there is a very special number that we prefer to use as the base for our exponential function. It is the irrational number that we denote by the letter e whose first 12 digits are

$$e = 2.71828182845 \ldots .$$

There is a very precise way to define the number e. It is the "limit at infinity" shown in equation (5).

$$e = \lim_{x \to \infty} \left(1 + \frac{1}{x}\right)^x. \tag{5}$$

It is a straightforward exercise to calculate values of the expression on the right-hand side of equation (5) using a hand calculator or a computer for increasingly large values of x. Doing so one obtains values of the expression $(1 + 1/x)^x$, which approach the number e "in the limit." Table 1.1 shows the result of doing just this, using BASIC Program 2 in Appendix I on a personal computer.

Table 1.1. Approximation to $e = \lim_{x \to \infty} \left(1 + \frac{1}{x}\right)^x$ using Program 2, Appendix I.

x	$(1 + 1/x)^x$
10	2.593742
20	2.653297
100	2.704813
200	2.711517
1000	2.716925
2000	2.717605
10,000	2.718159

Values of the exponential function $f(x) = e^x$ have been calculated to great precision. Table 1 in Appendix II gives values of this function to four decimal place accuracy for various values of x. And almost every scientific calculator has a key marked "e^x" or "exp," which gives values of the exponential function $f(x) = e^x$ for any specified value of x.

Using the approximation $e \approx 2.718 \ldots$, we can plot various points (x, e^x) and sketch the graph of the exponential function $f(x) = e^x$ as in Figure 1.5.

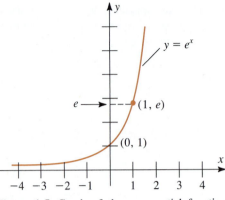

Figure 1.5 Graph of the exponential function $f(x) = e^x$.

Continuous Compounding of Interest: An Application of $f(x) = e^x$

Returning to the example of compound interest, we find the first application of the exponential function e^x.

If a *nominal* interest rate r is used by a bank that compounds interest k times per year, rather than once per year, this means that on each of k days equally spaced throughout the year an interest payment of r/k percent is added to the current balance in an account. Thus if interest is compounded semiannually ($k = 2$), at the end of 6 months ($\frac{1}{2}$ year) an original deposit of P_0 dollars will have grown to

$$P\left(\frac{1}{2}\right) = \left(1 + \frac{r}{2}\right)P_0 \text{ dollars}$$

and at the end of the first year the balance in the account will be

$$P(1) = \left(1 + \frac{r}{2}\right)P\left(\frac{1}{2}\right) = \left(1 + \frac{r}{2}\right)\left[\left(1 + \frac{r}{2}\right)P_0\right] = \left(1 + \frac{r}{2}\right)^2 P_0 \text{ dollars.}$$

Similarly, if interest is compounded quarterly ($k = 4$), the balances on deposit at the end of the four quarters of the first year would be

$$P\left(\frac{1}{4}\right) = \left(1 + \frac{r}{4}\right)P_0$$

$$P\left(\frac{1}{2}\right) = \left(1 + \frac{r}{4}\right)^2 P_0$$

$$P\left(\frac{3}{4}\right) = \left(1 + \frac{r}{4}\right)^3 P_0$$

$$P(1) = \left(1 + \frac{r}{4}\right)^4 P_0.$$

In general, it follows that compounding k times per year at a nominal interest rate r gives an amount on deposit at the end of one year of

$$P(1) = \left(1 + \frac{r}{k}\right)^k P_0. \tag{6}$$

Applying this result again, we find that the amount on deposit after 2 years is

$$P(2) = \underbrace{\left(1 + \frac{r}{k}\right)^k}_{} \underbrace{\left[\left(1 + \frac{r}{k}\right)^k P_0\right]}_{}$$

—— amount on deposit after 1 year

—— annual compounding factor from equation (6)

$$= \left(1 + \frac{r}{k}\right)^{2k} P_0$$

and more generally, the amount on deposit after n years is

$$P(n) = \left(1 + \frac{r}{k}\right)^{nk} P_0. \tag{7}$$

By experimenting with various values of k using a calculator or a computer, you can obtain results such as those in Table 1.2. Notice that the amount $P(1)$ increases as k increases, but that the rate at which $P(1)$ increases seems to slow considerably as k increases. This raises the question of whether there exists a "limiting value" for $P(1)$ as the number of compoundings per year becomes infinitely large.

Table 1.2. Annual growth due to interest using Program 3, Appendix I.

k = number of compoundings per year.	$P(1) = \left(1 + \frac{r}{k}\right)^k P_0 =$ amount on deposit after one year ($P_0 = 100$, $r = 0.08$)
$k = 1$ (annual)	108.00
$k = 2$ (semiannual)	108.16
$k = 4$ (quarterly)	108.24
$k = 12$ (monthly)	108.30
$k = 365$ (daily)	108.33

If we were to allow the number of compoundings in equation (7) to become infinite, we would find the amount $P(1)$ on deposit after one year to be

$$P(1) = \lim_{k \to \infty} \left(1 + \frac{r}{k}\right)^k P_0. \tag{8}$$

To analyze the limit in equation (8), we let $x = k/r$. Then $x \to \infty$ as $k \to \infty$ since r is fixed, and we can write

$$P(1) = \lim_{k \to \infty} \left(1 + \frac{r}{k}\right)^k P_0$$

$$= \lim_{k \to \infty} \left(1 + \frac{r}{k}\right)^{(k/r)r} P_0 \qquad \left[\left(\frac{k}{r}\right)r = k\right]$$

$$= \lim_{x \to \infty} \left[\left(1 + \frac{1}{x}\right)^x\right]^r P_0 \qquad \left[x = \frac{k}{r}\right]$$

$$= e^r \cdot P_0 \tag{9}$$

since $e = \lim\limits_{x \to \infty} \left(1 + \dfrac{1}{x}\right)^x$. That is, the amount on deposit after 1 year, under "infinitely many" compoundings, is just the value of the exponential function $P(1) = e^r$. Such compounding is called "continuous" compounding of interest, reflecting the notion that interest is being compounded at each instant of time.

Extending this result, we find that the amount $P(2)$ on deposit after 2 years resulting from continuous compounding of interest at a nominal rate of r percent on an initial deposit of P_0 dollars would be

$$P(2) = e^r[e^r P_0]$$

— amount $P(1)$ on deposit after 1 year

— annual compounding factor from equation (9)

$$= e^{2r} P_0$$

and, in general, that the amount on deposit after t years will be

$$P(t) = e^{rt} P_0. \tag{10}$$

Equation (10) is the formula used by banks and other institutions in calculating the value of an initial investment of P_0 dollars compounded continuously at a nominal rate of interest of r percent for a period of t years.

Example 3 Find the amount that an investor will have on deposit 5 years after making a single deposit of $1000 in a savings account paying 8% nominal interest if interest is compounded:

(a) quarterly
(b) continuously.

Solution:

(a) Under quarterly compounding of interest we have $k = 4$ compoundings per year. Using equation (7) with $r = 0.08$, $n = 5$, and $P_0 = 1000$, we obtain

$$P(5) = \left(1 + \frac{0.08}{4}\right)^{5 \cdot 4}(1000)$$

$$= (1.02)^{20}(1000)$$

$$= \$1485.95.$$

(The last step requires use of a calculator or computer to evaluate the expression a^x with $a = 1.02$ and $x = 20$, in order to avoid a very lengthy calculation.)

(b) Under continuous compounding of interest we apply equation (10) with $t = 5$, $r = 0.08$, and $P_0 = 1000$ to obtain

$$P(5) = e^{5(0.08)}(1000)$$

$$= 1000e^{0.4}$$

$$= 1000(1.49182)$$

$$= \$1491.82.$$

(The value $e^{0.4} = 1.49182$ was obtained from the table of values for e^x in Appendix II.)

Note in this example that continuous compounding of interest produced a yield $\$1491.82 - \$1485.95 = \$5.87$ larger than quarterly compounding of interest. □

Exercise Set 4.1

1. Simplify the following expressions.
   a. $9^{3/2}$ **b.** $16^{1/4}$
   c. $49^{3/2}$ **d.** $4^{-3/2}$

2. Simplify the following expressions.
   a. $36^{-3/2}$ **b.** $81^{-3/4}$
   c. $8^{5/3}$ **d.** $125^{-2/3}$

3. Simplify the following expressions.
   a. $(1/4)^{3/2}$ **b.** $(27/8)^{2/3}$
   c. $(1/16)^{-3/4}$ **d.** $(81/36)^{-3/2}$

In Exercises 4–11 sketch the graph of the given exponential function.

4. $f(x) = 4^x$ 5. $f(x) = 3^{-x}$

6. $f(x) = (1/2)^x$ 7. $f(x) = 4^{-x}$

8. $f(x) = 2^{2x}$ 9. $f(x) = e^{-x}$

10. $f(x) = -2e^x$ 11. $f(x) = -3e^{-x}$

In Exercises 12–21 use properties (E1)–(E3) of exponents to simplify the given expression.

12. $2^3 \cdot 4^{-2}$

13. $\dfrac{(27)^{1/3} \cdot 4^2}{48}$

14. $\dfrac{3^{-6} \cdot 3^5}{3^2}$

15. $\dfrac{[2^3 \cdot 3^{-2}]^2}{\frac{1}{3}(8)}$

16. $e^2 e^{-2}$

17. $e^3 \cdot e^x$

18. $\dfrac{e^3 e^{-2x}}{e^{-x}}$

19. $\sqrt{16e^{4x}}$

20. $(27e^{-6}e^{3x})^{1/3}$

21. $\dfrac{(4e^{3x})^{2/3}}{\sqrt{8e^4}}$

22. If $1000 is invested at 10% nominal interest, find the value of this investment after 3 years under
    a. annual compounding of interest.
    b. continuous compounding of interest.

23. Find the value of an investment of $500 if it is deposited in a savings account paying 8% nominal interest compounded semiannually
    a. after one year.
    b. after 2 years.
    c. after 18 months.

24. Find the value of the investment in Exercise 23 for each of the stated periods of time if the interest is compounded continuously.

25. If we multiply both sides of the equation $P(t) = e^{rt}P_0$ by e^{-rt}, we obtain the equation

$$P_0 = e^{-rt}P(t) = \frac{P(t)}{e^{rt}}.$$

This equation gives the initial amount P_0 that must be invested in order that an amount $P(t)$ will result from the continuous compounding of interest at a nominal rate r on the original amount P_0. For this reason P_0 is called the *present value* of the amount $P(t)$ that will be available in t years. Find the present value of an investment that will be worth $1000 after continuous compounding for 4 years at 10% annual interest.

26. Use the method of Exercise 25 to determine what amount of money should be invested at 8% interest compounded continuously so as to produce a total of $20,000 after 10 years.

27. In the presence of a nominal inflation rate of 6%, compounded continuously, what annual salary will be required in 5 years to match the purchasing power of a current annual salary of $20,000?

28. Find the size of the deposit that, when compounded continuously for 12 years at 12% interest, will yield $10,000.

29. What rate of interest, compounded continuously, will produce an *effective annual yield* of 8%? (Effective annual yield is the rate required under *annual* compounding of interest to obtain the specified yield. *Hint:* Do this by trial and error using a calculator.)

30. What is the present value of an investment that will be worth $1,000 5 years from now if the prevailing rate of interest, compounded continuously, is 10%?

31. Find the effective annual yield (see Exercise 29) on
a. an interest rate of 10% compounded semiannually.
b. an interest rate of 10% compounded continuously.

4.2 LOGARITHM FUNCTIONS

If b and y are positive numbers, we define the **logarithm to the base b of y** by the statement

$$\log_b y = x \quad \text{if and only if} \quad y = b^x. \tag{1}$$

Thus

$$\log_{10} 100 = 2 \quad \text{since} \quad 10^2 = 100$$
$$\log_2 8 \quad = 3 \quad \text{since} \quad 2^3 = 8$$
$$\log_{27} 3 \quad = 1/3 \quad \text{since} \quad 27^{1/3} = 3.$$

In other words, $\log_b y$ is just the *exponent* that we must apply to the *base b* in order to obtain y.

For any base b we say that the logarithm function $g(y) = \log_b y$ is the *inverse* of the exponential function $f(x) = b^x$, because if $y = b^x$ then $x = \log_b y$ and vice versa. That is, the logarithm function reverses the effect of the exponential function and vice versa, as illustrated in Figure 2.1.

Figure 2.1 The logarithm function $x = \log_b y$ is the inverse of the function $y = b^x$.

Two important identities summarize the fact that the logarithm and exponential functions are inverses of each other:

$$\log_b(b^y) = y, \qquad b^{\log_b x} = x. \tag{2}$$

Figure 2.2 shows the graphs of the functions $f(x) = 2^x$ and $g(x) = \log_2 x$. Since the graph of $f(x) = 2^x$ consists of points of the form (x, y) with $y = 2^x$, the graph of $g(x) = \log_2 x$ consists of points of the form $(y, x) = (2^x, x)$. That is, the graph of $g(x) = \log_2 x$ is obtained from the graph of $f(x) = 2^x$ by interchanging x and y coordinates, or by *reflecting* the graph of $f(x) = 2^x$ about the line $y = x$. (Note that since the exponential function b^x can have only positive values, $\log_b x$ is defined only for positive numbers.)

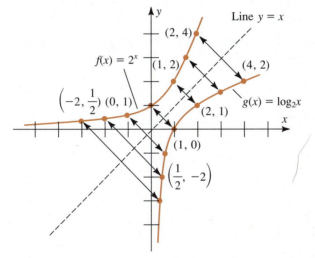

Figure 2.2 Graphs of $f(x) = 2^x$ and $g(x) = \log_2 x$.

Example 1 Here are some values of various logarithm functions.

$$\begin{array}{lll} \log_2 16 = 4 & \text{since} & 2^4 = 16 \\ \log_5 25 = 2 & \text{since} & 5^2 = 25 \\ \log_{81} 3 = \tfrac{1}{4} & \text{since} & 81^{1/4} = 3 \\ \log_2(\tfrac{1}{8}) = -3 & \text{since} & 2^{-3} = \tfrac{1}{8} \\ \log_{16}(\tfrac{1}{4}) = -\tfrac{1}{2} & \text{since} & 16^{-1/2} = \tfrac{1}{4}. \end{array}$$

□

Example 2 The Beer–Lambert law relates the absorption of light traveling in a material to the concentration and thickness of the material. If I_0 and I denote the intensities of light of a particular wavelength before and after passing through the material, respectively, and if x denotes the length of the path followed by the beam of light passing through the material, then

$$\log_{10}\left(\frac{I}{I_0}\right) = kx$$

where k is a constant depending on the material. To express I as a function of x, we use equation (1):

$$\log_{10}\left(\frac{I}{I_0}\right) = kx \quad \text{if and only if} \quad 10^{kx} = \frac{I}{I_0}.$$

Thus $I = I_0 \cdot 10^{kx}$.   □

Properties of Logarithms

Logarithms have three algebraic properties that are due to properties (E1)–(E3) of exponential functions. Here x, y, and b are positive numbers and r is any real number:

(L1) $\log_b(xy) = \log_b x + \log_b y$

(L2) $\log_b\left(\dfrac{x}{y}\right) = \log_b x - \log_b y$

(L3) $\log_b x^r = r \log_b x.$

Proofs of these properties follow from properties (E1)–(E3) for exponential functions. (See Exercises 27–29.)

Example 3 Simplify the following expressions.

(a) $\log_2\left(\dfrac{16}{4^x}\right)$ (b) $\log_3(x^2 \cdot 3^x)$

Strategy

Apply property (L2), property (L3), and equation (1).

Apply property (L1), property (L3), and equation (2).

Solution

(a) $\log_2\left(\dfrac{16}{4^x}\right) = \log_2 16 - \log_2 4^x$

$= 4 - x \log_2 4$
$= 4 - 2x.$

(b) $\log_3(x^2 \cdot 3^x) = \log_3 x^2 + \log_3 3^x$
$= 2 \log_3 x + x.$ □

Common Logarithms vs. Natural Logarithms

Logarithms to the base 10 are called *common* logarithms. Common logarithms once played an important role in mathematics and engineering. This was due to properties (L1)–(L3), which allowed multiplications and divisions involving very large or very small numbers to be carried out as additions or subtractions. The development of modern calculating devices has all but eliminated this role for the logarithm function. However, the *natural* logarithm function, which is the logarithm to the base e, plays an important role in modelling various phenomena that are characterized by a certain *rate of growth*.

We shall address the applications of the natural logarithm function in Section 4.5, after we have determined its derivative and related properties.

DEFINITION 1

The natural logarithm function, $y = \ln x$, is the logarithm function with base e. That is,

$$\ln x = \log_e x, \quad x > 0.$$

It is important to note that the natural logarithm function satisfies properties (L1)–(L3) of logarithms, even though the abbreviated notation ln x does not display the base e. For the natural logarithm function equation (1) becomes

$$\ln y = x \quad \text{if and only if} \quad y = e^x \tag{3}$$

and the identities in equation (2) become

$$\ln(e^x) = x, \quad \text{and} \quad e^{\ln x} = x \tag{4}$$

The following examples illustrate how the natural logarithm function is useful in simplifying expressions involving the exponential function. (Values of the natural logarithm function are listed in Appendix II and may be obtained on most hand calculators.)

Example 4 Solve the equation $e^{x+3} = 1$.

Solution: Taking the natural logarithm of both sides gives

$$\ln(e^{x+3}) = \ln(1).$$

Using the first identity in equation (4), we may simplify the left-hand side as

$$\ln(e^{x+3}) = x + 3.$$

Since $e^0 = 1$, we must have $\ln(1) = 0$ by equation (3). Thus we obtain

$$x + 3 = 0, \quad \text{or} \quad x = -3. \qquad \square$$

Example 5 Scientists who study human growth often model human body parameters, such as length, by functions involving polynomials and logarithmic and exponential functions. One such example would be

$$f(t) = a + 3t + 4e^{bt}$$

where a and b are constants. For this function, find a and b if it is known that

$$f(0) = 9 \quad \text{and} \quad f(2) = 23.$$

Solution: Substituting $t = 0$ in the equation for f and using the given information that $f(0) = 9$ gives

$$f(0) = a + (3)(0) + 4e^{(b)(0)}$$
$$= a + 4$$
$$= 9$$

so

$$a = 5.$$

Next, with $a = 5$ we substitute $t = 2$ in the equation for f and use the fact that $f(2) = 23$ to obtain

$$f(2) = 5 + (3)(2) + 4e^{2b}$$
$$= 11 + 4e^{2b}$$
$$= 23$$

so

$$4e^{2b} = 12$$

or

$$e^{2b} = 3.$$

To solve for b, we take natural logarithms of both sides and apply equation (4):

$$\ln (e^{2b}) = \ln 3$$
$$2b = \ln 3$$
$$b = \frac{1}{2} \ln 3 \cong 0.54930.$$ □

Example 6 Find the rate of interest that, when compounded continuously, produces an effective annual yield of 10%.

Solution: The effective annual yield of 10% on an investment of P dollars is just 110% of P, or $(1.1)P$ dollars. Continuous compounding of P dollars for one year at r percent interest produces the amount $e^{1 \cdot r}P = e^r P$. Thus we must solve the equation

$$e^r P = (1.1)P$$

for r. Dividing both sides by P gives

$$e^r = 1.1.$$

Taking the natural logarithm of both sides of this equation and using the first identity in equation (4) gives

$$\ln(e^r) = \ln(1.1)$$

or

$$r = \ln(1.1)$$
$$= 0.09531 \qquad \text{(from Table 2, Appendix II).}$$

Thus the rate 9.531%, when compounded continuously, gives an effective annual yield of 10%. □

The Natural Exponential Function

Using the natural logarithm function, we can express any exponential b^x in the form e^{ax} for some constant a. To do so, we begin with the identity $b = e^{\ln b}$ from equation (4). This gives

$$b^x = [e^{\ln b}]^x$$
$$= e^{x \ln b}$$ (5)

by property (E3) of exponents. Thus $b^x = e^{ax}$ with $a = \ln b$. For example

$$2^x = e^{x \ln 2}$$
$$10^x = e^{x \ln 10}$$

etc. For this reason we shall henceforth work exclusively with exponential functions with base e. These are sometimes referred to as *natural* exponential functions, although we shall simply refer to $f(x) = e^x$ as "the" exponential function in what follows.

Exercise Set 4.2

1. Use the definition of the logarithm $\log_b y$ in equation (1) to find the value of each of the following.
   a. $\log_{10} 100$
   b. $\log_{10} 10$
   c. $\log_2 16$
   d. $\log_4 64$
   e. $\log_3 81$
   f. $\ln e^2$

2. Use the definition of the logarithm in equation (1) to find the value of each of the following.
   a. $\log_8 2$
   b. $\log_2 \sqrt{2}$
   c. $\log_9 \left(\frac{1}{3}\right)$
   d. $\log_{10} \left(\frac{1}{100}\right)$
   e. $\log_{10}(0.001)$
   f. $\log_4 \left(\frac{1}{8}\right)$

3. Use Table 2, Appendix II, where necessary, to find the value of the following natural logarithms.
   a. $\ln e$
   b. $\ln 1$
   c. $\ln(2.2)$
   d. $\ln(0.3)$
   e. $\ln(e^3)$
   f. $\ln\left(\frac{1}{e^2}\right)$

4. Complete the following steps to establish the formulas

$$\log_a x = \frac{\ln x}{\ln a} \quad \text{and} \quad \ln x = (\ln a)\log_a x$$

for converting natural logarithms to logarithms to the base a and vice versa.
   a. Let $y = \log_a x$. Explain why $x = a^y$ follows.
   b. Take natural logs of both sides of the equation $x = a^y$ to show that

   $$\ln x = \ln(a^y) = y \ln a.$$

   c. Solve the equation in (b.) for y to obtain

   $$y = \frac{\ln x}{\ln a}.$$

   d. Use (a.) to show that both formulas follow from (c.).

5. Use the formulas in Exercise 4 to find formulas for converting between common and natural logarithms:

$$\log_{10} x = \frac{\ln x}{\ln 10} \quad \text{and} \quad \ln x = (\ln 10)\log_{10} x.$$

6. Use the results of Exercises 4 and 5 together with Table 2, Appendix II (or a hand calculator) to find values for the following logarithms.
   a. $\log_{10} 5$
   b. $\log_3 8$
   c. $\log_2 12$
   d. $\log_8 4$
   e. $\log_3 \frac{1}{2}$
   f. $\log_4 6$

In Exercises 7–16 use properties of logarithms and exponents to solve for x.

7. $e^x = 1$
8. $e^x = 3$
9. $e^{x^2-3} = e$
10. $\ln x^2 = 2$
11. $e^{x^2} = 1$
12. $2^{x-2} = 8$
13. $\ln x^2 = 8$
14. $e^{x^2-2x-3} = 1$
15. $e^{\ln(2x+3)} = 7$
16. $e^{2x} + e^x - 2 = 0$

17. The Antoine equation for the relationship between the vapor pressure P and temperature t of a pure liquid is

$$\log_{10} P = \frac{-A}{t + C} + B$$

where A, B, and C are empirically determined constants and t is in units of degrees centigrade. Use Exercise 4 to rewrite this equation in terms of $\ln P$.

18. Use Exercise 4 and the definition of the natural logarithm to solve the equation in Exercise 17 for P.

19. The selling price p and quantity sold per month x for a certain product are related by the equation $p = 100e^{-x}$. Solve this equation for x as a function of p.

20. Under continuous compounding of interest at a nominal rate r, an initial amount P_0 will grow to the amount $P_T = e^{rT}P_0$ after T years. Use the natural logarithm to show that this equation may be solved for T to give

$$T = \frac{1}{r} \ln\left(\frac{P_T}{P_0}\right).$$

21. Use the result of Exercise 20 to find the number of years required for an initial deposit of $1000 to grow to $1500 under continuous compounding of interest at nominal rate $r = 10\%$.

22. Repeat Exercise 21 with $r = 5\%$.

23. Use the result of Exercise 20 to find the number of years required for an investment to double under continuous compounding at a nominal rate of $r = 10\%$.

24. Suppose that the selling price p and the number x of items that can be sold per month at price p are related by the equation $p(x) = 100e^{-0.02x}$. What price will be required in order to sell 100 items per month?

25. For the price–demand equation in Exercise 24, find the number of items that can be sold per month at price $p = 50$ dollars.

26. For the growth model $f(t) = 10 + 3t + 6 \ln t$, find (a) $f(1)$, (b) $f(2)$, and (c) $f(e)$.

27. In the growth model $f(t) = a + b \ln t$ assume that $f(1) = 15$ and $f(e) = 20$. Find a and b.

28. The growth model $f(t) = a + b(1 - e^{2t})$ has the property that $f(0) = 10$ and $f(\ln 2) = 4$. Find the constants a and b.

29. True or false? Under continuous compounding of interest, doubling the interest rate will reduce the time required for an investment to double in value by one half. Why?

30. Prove property (L1) of logarithms as follows.
   a. Let $u = \log_b x$ and $v = \log_b y$. Then

   $$x = b^u \quad \text{and} \quad y = b^v. \quad \text{(Why?)}$$

   b. Then $\log_b(xy) = \log_b(b^u b^v) = \log_b b^{u+v} = u + v$. (Why?)

   c. Finally, since $u = \log_b x$ and $v = \log_b y$ it follows that $\log_b(xy) = \log_b x + \log_b y$.

31. Prove property (L2) of logarithms using the method of Exercise 30.

32. Prove property (L3) of logarithms using the method of Exercise 30.

4.3 DIFFERENTIATING THE NATURAL LOGARITHM FUNCTION

The natural logarithm function has a very simple derivative, one that is very important in applications. We shall show at the end of the section that it is simply the reciprocal function.

For $f(x) = \ln x$, $f'(x) = \dfrac{1}{x}$ (1)

or,

$$\frac{d}{dx} \ln x = \frac{1}{x}.$$

This result is not all that surprising when we think about the graph of $f(x) = \ln x$ (see Figure 3.1). Recall that the logarithm function $\ln x$ is defined only for $x > 0$, and that its graph rises quickly at first, increasing more slowly as x increases. This property is verified by its derivative $f'(x) = 1/x$, which measures the slope of the tangents to its graph (see

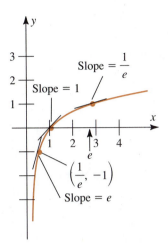

Figure 3.1 Graph of $y = \ln x$ is increasing and concave down for all $x > 0$. Slope decreases as x increases in the domain of $\ln x$.

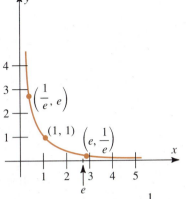

Figure 3.2 Graph of $f'(x) = \dfrac{1}{x}$. This derivative is positive and decreasing for all $x > 0$.

Figure 3.2), because $f'(x) = 1/x$ is *positive* for all x in the domain of $\ln x$. In the language of Section 3.1, this says that the function $f(x) = \ln x$ is increasing for all x.

But we can say more about the graph of $\ln x$. Since its second derivative is

$$\frac{d^2}{dx^2}(\ln x) = \frac{d}{dx}\left(\frac{1}{x}\right) = -\frac{1}{x^2}$$

which is negative for all x, the graph of $y = \ln x$ is concave down for all x the domain of $\ln x$.

Differentiating Composite Functions

Most logarithm functions that we encounter in applications are actually composite functions of the form $\ln u$ where u is a function of x. When u is differentiable, we can differentiate such functions using equation (1) together with the Chain Rule:

> If $f(x) = \ln[u(x)]$, $f'(x) = \dfrac{1}{u(x)} \cdot u'(x)$ (2)
>
> or, in Leibniz notation
>
> $\dfrac{d}{dx} \ln u = \dfrac{1}{u} \cdot \dfrac{du}{dx}.$

We shall establish the differentiation formula for $\ln x$ at the end of this section. Before doing so, we consider several examples involving equation (2).

Example 1 For $f(x) = \ln(3x + 5)$ find $f'(x)$.

Solution: Here $f(x) = \ln[u(x)]$ with $u(x) = 3x + 5$. Using equation (2), we obtain

$$f'(x) = \underbrace{\frac{1}{3x + 5}}_{\frac{1}{u(x)}} \cdot \underbrace{3}_{u'(x)} = \frac{3}{3x + 5}.$$

□

Example 2 For $f(x) = \ln(3x^{2/3} - 6x^2 + \sqrt{x})$, find $f'(x)$.

Strategy
Here $u(x) = 3x^{2/3} - 6x^2 + \sqrt{x}$.

Use equation (2).

Solution
By equation (2) we have

$$f'(x) = \frac{1}{3x^{2/3} - 6x^2 + \sqrt{x}}\left(2x^{-1/3} - 12x + \frac{1}{2}x^{-1/2}\right)$$

$$= \frac{2x^{-1/3} - 12x + \frac{1}{2}x^{-1/2}}{3x^{2/3} - 6x^2 + \sqrt{x}}.$$

□

Example 3 Find $\dfrac{dy}{dx}$ for $y = \sqrt{x + 3}\,\ln(4 + x^2)$.

Solution: Here we must use the Product Rule. We obtain

$$\frac{dy}{dx} = \left[\frac{d}{dx}(x + 3)^{1/2}\right]\ln(4 + x^2) + (x + 3)^{1/2}\left[\frac{d}{dx}\ln(4 + x^2)\right]$$

$$= \frac{1}{2}(x + 3)^{-1/2}\ln(4 + x^2) + (x + 3)^{1/2}\left(\frac{1}{4 + x^2}\right)(2x)$$

$$= \frac{\ln(4 + x^2)}{2\sqrt{x + 3}} + \frac{2x\sqrt{x + 3}}{4 + x^2}.$$

□

Example 4 For $y = \ln(x^2 + 2x + 1)^5$, find $\dfrac{dy}{dx}$.

Solution: This function has the form $y = \ln u$ with $u = (x^2 + 2x + 1)^5$. Since $\dfrac{du}{dx} = 5(x^2 + 2x + 1)^4(2x + 2)$, the derivative of y is

$$\frac{dy}{dx} = \underbrace{\frac{1}{(x^2 + 2x + 1)^5}}_{\frac{1}{u}}\underbrace{[5(x^2 + 2x + 1)^4](2x + 2)}_{\frac{du}{dx}}$$

$$= \frac{5(2x + 2)}{x^2 + 2x + 1}$$

$$= \frac{10x + 10}{x^2 + 2x + 1}$$

$$= \frac{10(x + 1)}{(x + 1)^2}$$

$$= \frac{10}{x + 1}. \qquad \square$$

Example 5 For $f(x) = \ln \dfrac{\sqrt{x + 3}}{x^2 - x + 2}$, find $f'(x)$.

Solution: Here it simplifies matters to use properties of logarithms to first rewrite $f(x)$ as

$$f(x) = \ln \frac{(x + 3)^{1/2}}{(x^2 - x + 2)} = \ln(x + 3)^{1/2} - \ln(x^2 - x + 2) \qquad \text{(Property L2)}$$

$$= \frac{1}{2} \ln(x + 3) - \ln(x^2 - x + 2). \qquad \text{(Property L3)}$$

Differentiation gives

$$f'(x) = \frac{1}{2} \left(\frac{1}{x + 3} \right) - \left(\frac{1}{x^2 - x + 2} \right)(2x - 1)$$

$$= \frac{1}{2x + 6} - \frac{2x - 1}{x^2 - x + 2}. \qquad \square$$

Logarithmic Differentiation

There are two general types of situations in which the derivative of a given function is more easily found by first differentiating the **natural logarithm** of the function and then solving the resulting expression for the desired derivative. This technique is called **logarithmic differentiation.**

The first of these situations involves a function that has the form of a variable base expression raised to a power that is also variable. Examples of such functions are x^x, $(x^2 + 4)^{1/x}$, and so on. Note, in the following example, how the natural logarithm of such a function will involve a **product** of such factors, rather than an exponentiation.

Example 6 Find $\dfrac{dy}{dx}$ for $y = x^{\sqrt{x}}, \quad x > 0$.

Solution: Note that we have no differentiation formula available to use in calculating $\dfrac{dy}{dx}$ directly because the exponent $\sqrt{x}$ is not a *fixed* real number. But if we first take natural logarithms of both sides of the equation

$$y = x^{\sqrt{x}}$$

we obtain

$$\ln y = \ln (x\sqrt{x})$$
$$= \sqrt{x} \ln x. \qquad \text{(by property (L3) of logarithms)}$$

Differentiating both sides of the equation $\ln y = \sqrt{x} \ln x$ and remembering that y is a function of x gives

$$\left(\frac{1}{y}\right)\frac{dy}{dx} = \left(\frac{d}{dx}\sqrt{x}\right)\ln x + \sqrt{x}\left(\frac{d}{dx}\ln x\right)$$

$$= \frac{1}{2\sqrt{x}} \ln x + \sqrt{x}\left(\frac{1}{x}\right)$$

$$= \frac{\ln x + 2}{2\sqrt{x}}.$$

Solving for $\dfrac{dy}{dx}$ now gives

$$\frac{dy}{dx} = y\left(\frac{\ln x + 2}{2\sqrt{x}}\right)$$

or, since $y = x^{\sqrt{x}}$,

$$\frac{dy}{dx} = x^{\sqrt{x}}\left(\frac{\ln x + 2}{2\sqrt{x}}\right)$$

□

Example 7 Find $\dfrac{dy}{dx}$ for $y = (2x^2 + 3)^x, \quad x > 0.$

Strategy

Take natural logs of both sides to eliminate the variable exponent.

Solution

Taking natural logs of both sides of the equation

$$y = (2x^2 + 3)^x$$

gives

Use Property (L3).

$$\ln y = \ln [(2x^2 + 3)^x]$$
$$= x \ln (2x^2 + 3).$$

Differentiate both sides

Differentiating both sides of this equation gives

$$\frac{1}{y}\frac{dy}{dx} = \ln (2x^2 + 3) + x\left[\frac{4x}{2x^2 + 3}\right]$$

$$= \ln (2x^2 + 3) + \frac{4x^2}{2x^2 + 3}$$

so

Solve for $\dfrac{dy}{dx}$.

$$\frac{dy}{dx} = y\left[\ln(2x^2 + 3) + \frac{4x^2}{2x^2 + 3}\right]$$

$$= (2x^2 + 3)^x\left[\ln(2x^2 + 3) + \frac{4x^2}{2x^2 + 3}\right]. \qquad \square$$

The other situation in which logarithmic differentiation is sometimes useful is in differentiating a rational expression involving products and exponentiations. Because logarithms can convert exponentiations, multiplications, and divisions into products, sums, and differences, it is sometimes useful to differentiate the **natural logarithm** of the given function, as the following example shows.

Example 8 Find the derivative of the function

$$y = \frac{x^3(x^4 + 6)^5}{\sqrt{1 + x^3}}.$$

Strategy

Because of the powers, products, and quotient, first take natural log and simplify.

Solution

Taking natural logs of both sides of the equation for y gives

$$\ln y = \ln \frac{x^3(x^4 + 6)^5}{\sqrt{1 + x^3}}$$

$$= \ln x^3 + \ln(x^4 + 6)^5 - \ln(1 + x^3)^{1/2}$$

$$= 3 \ln x + 5 \ln(x^4 + 6) - \left(\frac{1}{2}\right)\ln(1 + x^3)$$

Differentiate both sides.

Differentiating both sides, and recalling that y is a function of x, gives

$$\left(\frac{1}{y}\right)\frac{dy}{dx} = \frac{3}{x} + \frac{20x^3}{x^4 + 6} - \frac{3x^2}{2(1 + x^3)}$$

Solve for $\dfrac{dy}{dx}$ and substitute for y.

so

$$\frac{dy}{dx} = \frac{x^3(x^4 + 6)^5}{\sqrt{1 + x^3}}\left[\frac{3}{x} + \frac{20x^3}{x^4 + 6} - \frac{3x^2}{2(1 + x^3)}\right]. \qquad \square$$

The Rate of Growth of a Function

Economists define the **rate of growth** of a function as the ratio

$$G(x) = \frac{f'(x)}{f(x)}$$

of the rate of change of f (i.e., its derivative) to its current size $f(x)$. This ratio is sometimes called the *relative rate of change* of f.

For example, if a company's profit P as a function of its production level x is

$$P(x) = 400x + 5000 - x^2$$

then its marginal profit is

$$P'(x) = 400 - 2x$$

and at production level $x = 50$ its rate of increase in profits per unit increase in production is the marginal profit

$$P'(50) = 300 \text{ dollars.}$$

But what does this really mean? By calculating its *rate of growth* in profits

$$G(50) = \frac{P'(50)}{P(50)} = \frac{300}{22,500} \approx 0.0133$$

we can say that the *relative* rate of increase in profits is 0.0133, or $0.0133 \times 100 \approx 1.3\%$. Thus an increase in production of 2% (from $x = 50$ to $x = 51$) will result in an increase in profits of roughly 1.3%.

Notice that the growth function $G(x)$ can be written as a *logarithmic derivative*:

$$G(x) = \frac{f'(x)}{f(x)} = \frac{d}{dx} \ln[f(x)].$$

This interpretation sheds some light on the definition of elasticity of demand discussed in Section 3.7. Recall, if $Q(p)$ is the number of items that can be sold per unit time at price p, elasticity of demand is defined as the quantity

$$E(p) = -\frac{pQ'(p)}{Q(p)}.$$

Note that we can write elasticity as

$$E(p) = -\frac{\left(\dfrac{Q'(p)}{Q(p)}\right)}{\left(\dfrac{1}{p}\right)}$$

$$= -\frac{\dfrac{d}{dp} \ln[Q(p)]}{\dfrac{d}{dp} \ln p}$$

$$= -\frac{\text{rate of growth of } Q(p)}{\text{rate of growth of } p}.$$

We may therefore interpret elasticity as the (negative of the) ratio of the *relative* change in demand to the *relative* change in price at price level p.

Proof of the Differentiation Formula (1)

Assuming the function $y = \ln x$ to be differentiable for all $x > 0$, we establish the differentiation formula for $\ln x$ using the basic definition of the derivative

$$\frac{d}{dx} f(x) = \lim_{h \to 0} \frac{f(x + h) - f(x)}{h}.$$

We obtain

$$\frac{d}{dx}\ln x = \lim_{h\to 0}\frac{\ln(x+h)-\ln x}{h}$$

$$= \lim_{h\to 0}\left(\frac{1}{h}\right)\ln\left(\frac{x+h}{x}\right) \qquad \text{(by property (L2) of logarithms)}$$

$$= \lim_{h\to 0}\ln\left(1+\frac{h}{x}\right)^{1/h}. \qquad \text{(by property (L3) of logarithms)}$$

We now let $t = \dfrac{x}{h}$. Then since x is fixed, we have that $t\to\infty$ as $h\to 0^+$ and vice versa. (Similarly, $t\to -\infty$ as $h\to 0^-$, but we shall consider only the case of the right-hand limit.) The above limit can be written as

$$\frac{d}{dx}\ln x = \lim_{t\to\infty}\ln\left(1+\frac{1}{t}\right)^{t/x}$$

$$= \lim_{t\to\infty}\ln\left[\left(1+\frac{1}{t}\right)^{t}\right]^{1/x}$$

$$= \lim_{t\to\infty}\frac{1}{x}\cdot\ln\left(1+\frac{1}{t}\right)^{t} \qquad \text{(property (L3) of logarithms again)}$$

$$= \left(\frac{1}{x}\right)\left[\lim_{t\to\infty}\ln\left(1+\frac{1}{t}\right)^{t}\right].$$

But since $\lim\limits_{t\to\infty}\left(1+\dfrac{1}{t}\right)^{t} = e$, it follows from our assumption that $\ln x$ is a continuous function that

$$\lim_{t\to\infty}\ln\left(1+\frac{1}{t}\right)^{t} = \ln\left[\lim_{t\to\infty}\left(1+\frac{1}{t}\right)^{t}\right] = \ln e = 1.$$

Thus

$$\frac{d}{dx}\ln x = \left(\frac{1}{x}\right)(1) = \frac{1}{x}$$

as stated. ■

Exercise Set 4.3

In Exercises 1–20 find the derivative of the given function.

1. $y = \ln 2x$

2. $f(x) = \ln ax$

3. $f(x) = 6\ln(3x-2)$

4. $f(x) = 4\ln(9-x^2)$

5. $y = x\ln x$

6. $y = \ln\sqrt{x}$

7. $f(x) = \ln\sqrt{x^3 - x}$

8. $f(x) = \ln(x^2 - 2x)^3$

9. $f(x) = [\ln(x^2 - 2x)]^3$

10. $f(x) = \ln\sqrt{1+\sqrt{x}}$

11. $y = \ln(\ln(t))$

12. $y = \sqrt{1+\ln x}$

13. $f(x) = \dfrac{x}{1+\ln x}$

14. $f(x) = x^2\ln^2 x$ ($\ln^2 x$ means $[\ln x]^2$.)

15. $y = (3 \ln \sqrt{x})^4$

16. $f(x) = \dfrac{\ln(a + bx)}{\ln(c + dx)}$

17. $f(x) = x^2 \ln(3x - 6)$

18. $y = \ln\left(\dfrac{x^2 + 3}{\sqrt{x}}\right)$

19. $f(x) = \ln x \ln \sqrt{x}$

20. $f(x) = \dfrac{1 + \ln x}{1 - \ln x}$

In Exercises 21–24 use properties of logarithms to simplify as much as possible before finding $f'(x)$.

21. $f(x) = \ln(x^3 + 3)^{4/3}$

22. $f(x) = \ln[\sqrt{x}(x^2 + 3)]$

23. $f(x) = \ln \dfrac{(x - 6)^{2/3}}{\sqrt{1 + x}}$

24. $f(x) = \ln \dfrac{\sqrt{x}(x^2 + 2)}{\sqrt[3]{x}(1 + x^2)}$

In Exercises 25–32 find $\dfrac{dy}{dx}$ by logarithmic differentiation.

25. $y = (x + 3)^x$

26. $y = x^{\sqrt{x+2}}$

27. $y = x^{\sqrt{x+1}}$

28. $y = (x^2 + 3)^{\sqrt{x+1}}$

29. $y = \dfrac{x(x + 1)(x + 2)}{(x + 3)(x + 4)}$

30. $y = \dfrac{\sqrt{1 + x^2}}{x(1 - x^5)}$

31. $y = \dfrac{(x^2 + 3)(1 - x)^4}{x\sqrt{1 + x}}$

32. $y = \dfrac{x(x^3 + 6)^4}{\sqrt{1 + x^2}}$

In Exercises 33 and 34 find $\dfrac{dy}{dx}$ by implicit differentiation.

33. $\ln(xy) = x + y$

34. $x \ln y + y \ln x = 1$.

35. Find an equation for the line tangent to the graph of the equation $y = x(\ln x)^2 + \dfrac{x}{\ln x}$ at the point $(e, 2e)$.

In Exercises 36–39 find all relative extrema.

36. $y = x \ln x$

37. $y = x - \ln x$

38. $y = x^2 \ln x$

39. $y = \ln(x^2 - x)$

In Exercises 40 and 41 sketch the graph of the given function.

40. $y = \ln x^3$

41. $y = \ln(4 - x)$

42. A manufacturer of refrigerators has a total cost of producing x refrigerators per month of $C(x) = 2000 + 200x + x^2$.
   a. What is its marginal cost at production level $x = 20$?
   b. What is its growth in cost at production level $x = 20$?

43. A manufacturer of watches experiences a total cost of $C(x) = 20x + 300$ dollars in producing x watches per month. If its total revenues from the sale of x watches per month are $R(x) = 60x - x^2$, find
   a. its growth in costs at production level $x = 20$.
   b. its growth in revenues at production level $x = 20$.
   c. its growth in profits at production level $x = 20$.

44. A manufacturer experiences total monthly costs of $C(x) = 200 + 10x$ dollars and total monthly revenues of $R(x) = 50x$ dollars at production level x.
   a. Find the manufacturer's break-even point x_0.
   b. Show that growth in profits is increasing for $0 < x < x_0$ and decreasing for $x > x_0$.
   c. Does the result of part **b** mean that profits are decreasing for $x > x_0$? Why or why not?

45. For the utility function $U(x) = \ln \sqrt{x}$ show that marginal utility is positive but decreasing for all $x > 0$.

46. Complete the following argument to show that the graph of the natural logarithm function has no horizontal asymptote. (That is, $y = \ln x$ is increasing but unbounded for $x > 0$.)
   a. Recall that $\ln 2^n = n \ln 2$ for all $n > 0$.
   b. Note that $\ln 2 > 0$.
   c. Show that $\lim\limits_{n \to \infty} \ln 2^n = \lim\limits_{n \to \infty} n \cdot \ln 2 = +\infty$.
   d. Conclude that $y = \ln x$ has no horizontal asymptote.

47. Use an argument similar to that of Exercise 46 to show that $\lim\limits_{x \to 0^+} \ln x = -\infty$. That is, show that the y-axis is a vertical asymptote for $y = \ln x$.

4.4 DIFFERENTIATING THE EXPONENTIAL FUNCTION

The identity linking the natural logarithm and exponential functions

$$\ln(e^x) = x \tag{1}$$

provides us almost immediately with the derivative of the function $y = e^x$. Differentiating both sides of equation (1) with respect to x (using the Chain Rule formula $\dfrac{d}{dx} \ln u =$

$\dfrac{1}{u} \cdot \dfrac{du}{dx}$ with $u = e^x$ on the left-hand side) gives

$$\frac{1}{e^x} \cdot \frac{d}{dx} e^x = 1.$$

Multiplying both sides of this equation by e^x gives the surprisingly simple result that

$$\frac{d}{dx} e^x = e^x = e^x. \tag{2}$$

Thus the function $f(x) = e^x$ is its own derivative! It and its multiples are the only functions that have this property, which turns out to make them very useful in applications.

The Graph of $y = e^x$

Figure 4.1 shows the graph of $f(x) = e^x$. Since $f'(x) = e^x$ also, this means that the y-coordinate of any point on the graph of $y = e^x$ is the same as the slope of the line tangent to the graph at that point. Since the values of e^x are positive for all x, the slopes of these tangents are always positive. Moreover, these slopes increase as x increases. Finally, since

$$\frac{d^2}{dx^2} e^x = \frac{d}{dx} \left(\frac{d}{dx} e^x \right) = \frac{d}{dx} e^x = e^x > 0$$

the graph of $y = e^x$ is concave up for all x.

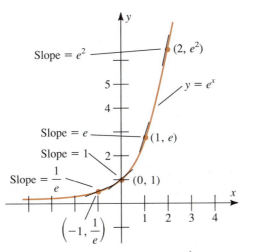

Figure 4.1 Graph of $y = e^x$. Since $\dfrac{d}{dx} e^x = e^x$,

slope at each point equals its y-coordinate.

Combining equation (2) with the Chain Rule, we obtain the following rule for differentiating composite functions of the form e^u where u is a differentiable function of x.

If $f(x) = e^{u(x)}$, $\quad f'(x) = e^{u(x)} \cdot u'(x)$. $\hspace{3cm}$ (3)

That is

$$\frac{d}{dx}e^u = e^u \cdot \frac{du}{dx}.$$

The following examples illustrate the use of equation (3).

Example 1 The function $f(x) = e^{3x}$ has the form $f(x) = e^u$ with $u = 3x$. Thus by equation (3)

$$f'(x) = \underbrace{e^{3x}}_{e^{u(x)}} \cdot \underbrace{3}_{u'(x)} = 3e^{3x}.$$

More generally, if a is any constant, we have

$$\frac{d}{dx}e^{ax} = e^{ax} \cdot \frac{d}{dx}(ax)$$

$$= e^{ax} \cdot a$$

$$= ae^{ax}. \qquad \square$$

Example 2 Find $f'(x)$ for $f(x) = e^{\sqrt{x}}$.

Strategy

Identify $u(x) = \sqrt{x}$.

Find $u'(x) = \dfrac{1}{2\sqrt{x}}$.

Solution

With $u(x) = \sqrt{x}$ we have, from equation (3), that

$$f'(x) = \underbrace{e^{\sqrt{x}}}_{e^{u(x)}} \cdot \underbrace{\frac{1}{2\sqrt{x}}}_{u'(x)} = \frac{1}{2\sqrt{x}}e^{\sqrt{x}}. \qquad \square$$

Example 3 For $f(x) = \dfrac{e^{x^2}}{1 + x}$, find $f'(x)$.

Solution: Since this function is a quotient, we must begin by applying the Quotient Rule:

$$\frac{d}{dx}\left(\frac{e^{x^2}}{1 + x}\right) = \frac{(1 + x)\left[\dfrac{d}{dx}e^{x^2}\right] - e^{x^2}\left[\dfrac{d}{dx}(1 + x)\right]}{(1 + x)^2}$$

$$= \frac{(1 + x)[e^{x^2} \cdot 2x] - e^{x^2}(1)}{(1 + x)^2}$$

$$= \frac{2xe^{x^2}(1+x) - e^{x^2}}{(1+x)^2}$$

$$= \frac{(2x^2 + 2x - 1)e^{x^2}}{(1+x)^2}. \qquad \square$$

Example 4 For the function $f(x) = xe^x$, find all relative extrema, determine the intervals on which f is increasing or decreasing, find all inflection points, and determine the concavity.

Strategy

Find f'.

Solution

Using the Product Rule, we find that

$$f'(x) = \left[\frac{d}{dx}(x)\right]e^x + x\left[\frac{d}{dx}e^x\right]$$

$$= (1)e^x + xe^x$$

$$= (1 + x)e^x.$$

Set $f'(x) = 0$ to find critical numbers.

Since e^x is never zero, the only zero of f' occurs when $x = -1$. This is the only critical number.

Test the critical number using the Second Derivative Test to determine whether it is a relative maximum or minimum.

To test this number, we find the second derivative:

$$f''(x) = \frac{d}{dx}[(1+x)e^x]$$

$$= \left[\frac{d}{dx}(1+x)\right]e^x + (1+x)\left[\frac{d}{dx}e^x\right]$$

$$= 1 \cdot e^x + (1+x)e^x$$

$$= (2 + x)e^x.$$

Since $f''(-1) = (2 - 1)e^{-1} = 1/e > 0$, the Second Derivative Test shows that the value $f(-1) = -1/e$ is a relative minimum. Thus f must be decreasing on $(-\infty, -1)$ and increasing on $(-1, \infty)$. (Why?)

Find the zeros of the second derivative.

The second derivative $f''(x) = (2 + x)e^x$ equals zero only when $x = -2$. In the interval $(-\infty, -2)$ we use the test point $t = -3$ for which

Check the sign of $f''(x)$ on each of the resulting intervals to determine the concavity.

$$f''(-3) = (2 - 3)e^{-3} = -1/e^3 < 0$$

so the graph is concave down on $(-\infty, -2)$. On the interval $(-2, \infty)$ we use the test point $t = 0$:

Identify inflection points where concavity changes.

$$f''(0) = (2 + 0)e^0 = 2 > 0$$

so the graph is concave up on $(-2, \infty)$. Since the concavity changes at $x = -2$, the point $(-2, -2e^{-2})$ is an inflection point. The graph appears in Figure 4.2. $\qquad \square$

Example 5 The owner of a stand of timber estimates that the timber is currently worth $5000 and that it will increase in value with time (as the trees continue to grow) according

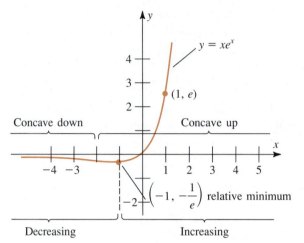

Figure 4.2 Graph of $f(x) = xe^x$.

to the formula

$$V(t) = 5000e^{\sqrt{t}/2}$$

where t is time in years. But prevailing interest rates are expected to remain at 12%, and the owner worries that by waiting longer to convert the timber to cash considerable interest income will be forgone. When should the owner sell the timber so as to maximize revenue?

Solution: Since dollars received in the future have different values today depending on when they are to be received, we must convert the revenue to be received in the future into its *present value* using the formula $P_0 = e^{-rt}P(t)$ developed in Exercise Set 4.1. With $P(t) = V(t)$ and $r = 0.12$ this gives the present value of the revenue received in selling the timber after t years as

$$R(t) = e^{-0.12t}V(t)$$
$$= (5000e^{\sqrt{t}/2})e^{-0.12t}$$
$$= 5000e^{\sqrt{t}/2 - 0.12t}.$$

To find the maximum value of R, we set $R'(t) = 0$:

$$R'(t) = \underbrace{5000e^{(\sqrt{t}/2) - 0.12t}}_{e^{u(t)}}\underbrace{\left[\frac{1}{4\sqrt{t}} - 0.12\right]}_{u'(t)} = 0.$$

Since the exponential function $e^{u(t)}$ is never zero, we must have

$$\frac{1}{4\sqrt{t}} - 0.12 = 0 \qquad\qquad (4)$$

or

$$\sqrt{t} = \frac{1}{4(0.12)} = \frac{1}{0.48}.$$

Thus

$$t = \left(\frac{1}{0.48}\right)^2 = 4.34 \text{ years}$$

or approximately 4 years and 4 months. (You can verify that this critical number actually yields the maximum value of $R(t)$ by applying the Second Derivative Test.) □

More on the Rate of Growth of a Function: Applications to Optimal Harvesting

The preceding example is typical of many situations in which the value of an asset $P(t)$ is growing in time. The question of when to sell the asset is called an *optimal harvest* problem because it applies to harvesting fish from a fishery, trees from a woodland, wine from an aging facility, and so on. What makes the question of when to harvest interesting is that while the value of the asset $P(t)$ is growing in time, the owner is missing the opportunity to invest these dollars at the prevailing interest rate r. This means that $P(t)$ must be *discounted* by the factor e^{-rt} used to calculate *present* value. As in Example 5, the present value of the revenue received from harvesting after t years is thus

$$R(t) = P(t)e^{-rt}.$$

To determine the *optimal* harvest time, we maximize $R(t)$ by setting $R'(t) = 0$. Since

$$R'(t) = P'(t)e^{-rt} + P(t)[e^{-rt}(-r)]$$
$$= [P'(t) - rP(t)]e^{-rt}$$

we will have $R'(t) = 0$ when

$$P'(t) - rP(t) = 0$$

or when

$$r = \frac{P'(t)}{P(t)}. \qquad \text{(Rate of growth of } P(t)\text{)} \tag{5}$$

Now recall that the ratio $\frac{P'(t)}{P(t)}$ is what we have defined as the *rate of growth* of the function P. Thus, under a constant prevailing rate of interest r, *the optimal time to harvest the asset whose value is given by the function P is when the rate of growth of P equals the prevailing interest rate*.

In Example 5 we had $P(t) = e^{\sqrt{t}/2}$ and $r = 0.12$, so this principle (equation (5)) would have required that we harvest when

$$0.12 = \frac{P'(t)}{P(t)} = \frac{e^{\sqrt{t}/2}\left(\frac{1}{4\sqrt{t}}\right)}{e^{\sqrt{t}/2}} = \frac{1}{4\sqrt{t}}.$$

This is precisely the condition in equation (4) that led to the solution $t = 4.34$ years.

Derivatives of Other Exponential and Logarithm Functions

In Section 4.2 we obtained a formula equivalent to

$$a^x = e^{x \ln a} \qquad \text{(Equation (5), Section 4.2)}$$

which shows how to express exponential functions in other bases in terms of the natural exponential function. Differentiating both sides of this equation using equation (3) gives

$$\frac{d}{dx} a^x = \frac{d}{dx} e^{x \ln a}$$

$$= e^{x \ln a} \cdot \left\{ \frac{d}{dx} (x \ln a) \right\}$$

$$= e^{x \ln a} \cdot \ln a$$

$$= a^x \ln a.$$

Combining this result with the Chain Rule gives the formula for differentiating a^u when u is a differentiable function of x:

If $f(x) = a^{u(x)}, \qquad f'(x) = a^{u(x)} \cdot u'(x) \cdot \ln a$ (6)

or

$$\frac{d}{dx} a^u = a^u \cdot \frac{du}{dx} \cdot \ln a.$$

Similarly, we may differentiate both sides of the equation

$$\log_a x = \frac{\ln x}{\ln a} \qquad \text{(Exercise 4, Section 4.2)}$$

to obtain the rule for differentiating logarithm functions to the base a:

$$\frac{d}{dx} \log_a x = \frac{d}{dx} \left(\frac{1}{\ln a} \cdot \ln x \right)$$

$$= \left(\frac{1}{\ln a} \right) \left(\frac{d}{dx} \ln x \right)$$

$$= \frac{1}{\ln a} \cdot \frac{1}{x}$$

$$= \frac{1}{x \ln a}.$$

If $u = u(x)$ is a differentiable function, the Chain Rule combines with the above result to give the following:

If $f(x) = \log_a u(x), \qquad f'(x) = \dfrac{1}{u(x) \ln a} \cdot u'(x)$ (7)

or

$$\frac{d}{dx} \log_a u = \frac{1}{u \ln a} \cdot \frac{du}{dx}.$$

Example 6 Using differentiation formulas (6) and (7), we obtain the following results:

(a) $\dfrac{d}{dx} 3^x = 3^x \ln 3$

(b) $\dfrac{d}{dx}(7^{x^2-\sqrt{x}}) = 7^{x^2-\sqrt{x}} \cdot \left(2x - \dfrac{1}{2\sqrt{x}}\right) \ln 7$

(c) $\dfrac{d}{dx}(\log_{10} x) = \dfrac{1}{x \ln 10}$

(d) $\dfrac{d}{dx}\{[\log_2(9 - x^3)]^4\} = 4[\log_2(9 - x^3)]^3 \cdot \dfrac{1}{(9 - x^3) \ln 2}(-3x^2)$

$$= \dfrac{-12x^2[\log_2(9 - x^3)]^3}{(9 - x^3) \ln 2}. \qquad \square$$

Exercise Set 4.4

In Exercises 1–20 find the derivative of the given function.

1. $f(x) = e^{6x}$

2. $f(x) = e^{9-x}$

3. $f(x) = e^{x^2-4}$

4. $f(x) = 3e^{2\sqrt{x}}$

5. $f(x) = \dfrac{e^x}{x}$

6. $f(x) = x^2 e^{-x}$

7. $f(x) = \sqrt{1 + e^{-x}}$

8. $f(x) = \dfrac{e^x + 1}{e^x}$

9. $f(x) = e^{x^2-\sqrt{x}}$

10. $f(x) = xe^{-2\ln x}$

11. $f(x) = \ln \dfrac{e^x + 1}{x + 1}$

12. $y = \dfrac{e^x - 1}{e^{-x} + 1}$

13. $y = \dfrac{1}{2}(e^x + e^{-x})$

14. $y = (x^2 + x - 1)e^{x^2+3}$

15. $f(x) = (x - e^{-2x})^4$

16. $y = e^{\sqrt{x}} \ln \sqrt{x}$

17. $f(x) = xe^{1/x^2}$

18. $y = \ln(x + e^{x^2})$

19. $f(x) = \dfrac{xe^x}{1 + x^2}$

20. $y = x^2 e^{\sqrt{1+x^2}}$

In Exercises 21 and 22 find $\dfrac{dy}{dx}$ by implicit differentiation.

21. $e^{xy} = x$

22. $e^{x-y} = ye^x$

In Exercises 23–28 find the derivative.

23. $f(x) = x2^x$

24. $f(x) = 7^{x^3-6x}$

25. $y = \sqrt{1 + 4^x}$

26. $f(x) = \log_{10}(x^2 - x^{2/3})$

27. $f(x) = [\log_2 \sqrt{x}]^2$

28. $y = \log_8(x^2 - 4)^3$

29. Find the maximum value of the function $f(x) = (3 - x^2)e^x$.

30. Find all relative extrema for the function $y = x^2 e^{1-x^2}$.

31. Find all relative extrema for the function $y = xe^{1-x^3}$.

32. The line $y = -\dfrac{1}{e}$ is tangent to the graph of $y = xe^x$ at the point P. Find P.

33. The selling price p and the number of items x that can be sold per month are related by the equation $p(x) = 100e^{-0.2x}$. Find the price p for which revenue is a maximum.

34. The population of a certain city is projected to be $P(t) = 100{,}000(1 + 5t)e^{-0.05t}$ in t years. When will the city's population be a maximum?

35. A manufacturer experiences weekly costs of $C(x) = 500 + 40x + e^{0.5x}$ in producing x hundred items. If the items can be sold for $100 each, find the weekly production level x that maximizes profits.

36. The owner of a valuable oil painting estimates that the value of the painting will be approximately $V(t) = 5000(2)^{\sqrt{t}}$ dollars t years from now. At what rate will the value of the painting be increasing in 4 years?

37. Find the rate of *growth* of the value of the painting in Exercise 36 after 4 years.

38. The value of a case of a certain French wine t years after being imported into the United States is $V(t) = 100(1.5)^{\sqrt{t}}$. If interest rates under continuous compounding are expected to remain constant at 12.5% nominally, when should the wine be sold so as to maximize profits? (Assume storage costs to be negligible.)

39. Land bought for speculation is expected to be worth $V(t) = 10,000(1.2)^{\sqrt{t}}$ dollars after t years. If the cost of money remains constant at 10% per year, when should the land be sold so as to maximize revenue to the owner?

40. Find the x-coordinate of the equilibrium point P for the supply curve $S(x) = 10e^{0.5x}$ and the demand curve $D(x) = 20e^{-0.5(x-10)}$. (P is where the curves intersect.)

41. Human growth in height from age 1 year to adulthood typically looks like the graph in Figure 4.3. A function that has been used to model this kind of growth is the *double-logistic*

$$y(t) = \frac{a}{1 + e^{-b_1(t-c_1)}} + \frac{f - a}{1 + e^{-b_2(t-c_2)}}$$

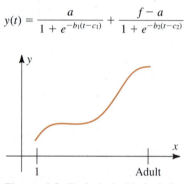

Figure 4.3 Typical double-logistic.

where the parameters a_1, b_1, a_2, b_2, c_1, c_2, and f are different for each individual.
a. Show that the parameter f is adult height. That is, show that

$$f = \lim_{t \to \infty} y(t).$$

b. If $b_1 = b_2$ and $c_1 = c_2 = c$, show that the time of most rapid growth is $t = c$.

42. The function $L(t) = a + b(1 - e^{-ct})$ is used to model the length of human infants as a function of time. Here a, b, and c are nonzero constants.
a. Find the velocity function $v(t) = L'(t)$ associated with this model by finding the first derivative of L.
b. Does the function L have a maximum value on $(0, \infty)$? Why or why not?
c. Find the acceleration function $a(t) = L''(t)$ associated with this model.
d. Show that the model has constant negative acceleration if $b > 0$.

43. Show that the function $f(x) = e^{-x^2/2}$ has a relative maximum when $x = 0$, inflection points when $x = \pm 1$, and a graph such as that in Figure 4.4.

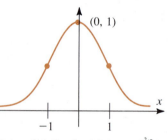

Figure 4.4 Graph of $f(x) = e^{-x^2/2}$ is a bell-shaped curve.

44. The function

$$f(x) = \frac{1}{\sigma\sqrt{2\pi}} e^{-x^2/2\sigma^2}$$

is called the normal probability density function. Comparing this function with that of Exercise 43, show that f has a maximum value at $x = 0$ and inflection points where $x = \pm\sigma$. (The constant σ is called the *standard deviation* for the normal curve.)

4.5 EXPONENTIAL GROWTH AND DECAY

Imagine the following experiment conducted in a biology course. On a certain day a number n of fruit flies is placed in an enclosed environment such as a large bell jar. If the environment is supportive (e.g., sufficient food supply and proper temperature), the number of flies will increase as time passes. The experiment is to record the number of fruit flies $N(t)$ present after t days, and to find a mathematical relationship between time and

Figure 5.1 Plot of typical data on fruit flies.

population size. In experiments of this kind, data such as those in Table 5.1 are often obtained. (These data are plotted in Figure 5.1.)

Table 5.1. Typical data on the growth of fruit flies

t (days)	0	4	8	12	16	20	24
$N(t)$ fruit flies	10	18	35	72	107	208	361

On examining these data, we observe that $N(t)$ increases more rapidly as the population size itself grows. That is, the larger the population the faster it grows. Biologists refer to this phenomenon as the **Law of Natural Growth,** which is often stated as

{rate of change of population size} $\propto$ {current size of population}.

The symbol "$\propto$" means "is proportional to". The statement of proportionality, $A \propto B$, is expressed mathematically by the equation $A = kB$ where k is a constant.

If we assume the population size N to be a differentiable function of t, the derivative $N'(t)$ is the rate of change of the population size. We may then write the *Law of Natural Growth* as the equation

$$N'(t) = kN(t) \tag{1}$$

or

$$\frac{dN}{dt} = kN.$$

The constant k in equation (1) is called the **growth constant.** Equation (1) itself is called a **differential equation** because it is an equation involving an unknown function N and its derivative N'. Solving equation (1) involves finding a differentiable function N for which equation (1) is true.

Fortunately, the solution of equation (1) is quite straightforward. If we let $N(t) = Ce^{kt}$ where C is any constant, then

$$N'(t) = \frac{d}{dt}Ce^{kt}$$

$$= C\left(\frac{d}{dt}e^{kt}\right)$$

$$= C(ke^{kt})$$

$$= k(Ce^{kt})$$

$$= kN(t)$$

so the function $N(t) = Ce^{kt}$ is a solution of equation (1). Moreover, it can be shown that *every* solution of equation (1) must be of this form.

For example, the differential equation

$$N'(t) = 2N(t) \tag{2}$$

has the form of equation (1) with $k = 2$, so it has solutions $N_1(t) = 3e^{2t}$, $N_2(t) = 40e^{2t}$, $N_3(t) = -6e^{2t}$, $N_4(t) = \pi e^{2t}$, and, in general, $N(t) = Ce^{2t}$ for any number C (including zero!).

Similarly, the differential equation

$$\frac{dy}{dt} = 6y \tag{3}$$

has the form of equation (1), so its solutions all have the form $y = Ce^{6t}$.

Finally, we can say a bit more about the constant C. Setting $t = 0$ in the function $N(t) = Ce^{kt}$ gives

$$N(0) = Ce^{k \cdot 0} = Ce^0 = C.$$

Thus, given any solution Ce^{kt} of equation (1), the constant C is just the value of this solution when $t = 0$. For this reason the constant C is called the **initial value** of the solution Ce^{kt}. If we specify a particular initial value $N(0) = N_0$ that a solution of equation (1) must have, we then obtain the *unique* solution $N(t) = N_0e^{kt}$ since both constants k and C are known. For example, if we specify $N(0) = 10$ in the differential equation (2), we obtain the unique solution $N(t) = 10e^{2t}$, since $k = 2$ and $C = 10$. Specifying the initial value $y(0) = -5$ in equation (3) gives the unique solution $y = -5e^{6t}$ since $k = 6$ and $C = -5$.

The following theorem summarizes our findings.

THEOREM 1
Law of Natural Growth

If a differentiable function of time, N, has the property that its rate of change is proportional to its present size at each instant of time, then it satisfies the differential equation

$$N'(t) = kN(t) \tag{4}$$

for some constant k. The *general solution* of equation (4) is

$$N(t) = Ce^{kt} \tag{5}$$

where C is an arbitrary constant. If, in addition, one specifies the *initial condition*

$$N(0) = N_0$$

for some number N_0, then equation (4) has the *unique* solution

$$N(t) = N_0 e^{kt}. \tag{6}$$

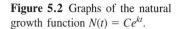

$N(t) = Ce^{kt}, \ k < 0$

$N(t) = Ce^{kt}, \ k > 0$

$(0, C) = (0, N(0))$

t

Figure 5.2 Graphs of the natural growth function $N(t) = Ce^{kt}$.

Figure 5.2 shows the graphs of two typical functions of the form $N(t) = Ce^{kt}$, one with $k > 0$ (exponential growth) and one with $k < 0$ (exponential decay). The advantage of having taken the trouble to develop the statement of Theorem 1 is that we now know that any quantity that grows or decays in time at a rate proportional to its present size can be represented by an exponential function as in equations (5) or (6).

Example 1 **(Population Growth)** Assume that the rate of increase of a population of fruit flies is proportional to the population size at each instant of time. If 100 fruit flies are present initially and 300 are present after 10 days, how many will be present after 15 days?

Strategy

Apply Law of Natural Growth to find the form of $N(t)$.

Solution

Since it is stated that the population function N satisfies the differential equation

$$N'(t) = kN(t)$$

the solution $N(t)$ must have the form

$$N(t) = Ce^{kt}.$$

Use the initial value $N(0) = 100$ to find C.

We are given that $N(0) = 100$, so $C = N(0) = 100$. The solution therefore has the form

$$N(t) = 100 e^{kt} \tag{7}$$

Substitute in other given data to obtain an equation in k alone.

but we do not as yet know k. To find k, we use the fact that $N(10) = 300$. Substituting $t = 10$ and $N(10) = 300$ in equation (7) gives

$$300 = 100 e^{k \cdot 10}$$

or

Take natural logs of both sides to solve for k.

$$e^{10k} = \frac{300}{100} = 3.$$

To solve for k we take natural logs of both sides:

$$\ln e^{10k} = \ln 3$$

or

$$10k = \ln 3 \text{ (since } \ln e^{10k} = 10k).$$

Write $N(t)$ using known values for C and k.

Thus $k = (1/10) \ln 3$, and our population function is

$$N(t) = 100e^{(1/10 \ln 3)t}.$$

Set $t = 15$ and solve to find the desired population size.

The size of the population after $t = 15$ days is therefore

$$\begin{aligned} N(15) &= 100e^{(1/10 \ln 3)(15)} \\ &= 100e^{(3/2) \ln 3} \\ &= 100e^{\ln(3^{3/2})} \\ &= 100 \cdot 3^{3/2} \\ &\approx 520 \text{ fruit flies.} \end{aligned}$$

□

Example 2 (**Radioactive Decay**) Certain radioactive isotopes, such as uranium, decay at a rate proportional to the amount present. If a block of 50 grams of such material decays to 40 grams in 5 days, what is the half-life of this material? (The half-life is the amount of time required for the material to decay to half the original amount.)

Strategy

Label variables. State form $A(t)$ using Theorem 1.

Solution

Let $A(t)$ denote the amount present after t days. Since $A(t)$ satisfies the Law of Natural Growth, we have

$$A(t) = Ce^{kt}.$$

Apply initial condition $A(0) = 50 = C$.

We are given that $A(0) = C = 50$ grams, so

$$A(t) = 50e^{kt}. \tag{8}$$

Substitute in other data to obtain equation for k.

To find k, we substitute the given data $A(5) = 40$ to obtain

$$40 = 50e^{k \cdot 5}.$$

Thus

Take ln's of both sides to solve for k.

$$e^{5k} = \frac{40}{50} = \frac{4}{5}$$

so

$$5k = \ln\left(\frac{4}{5}\right), \quad \text{or} \quad k = \frac{1}{5} \ln\left(\frac{4}{5}\right).$$

Write down half-life condition.

We seek a time T so that

$$A(T) = \frac{1}{2}A(0).$$

With $A(t)$ as in equation (8) this equation becomes

$$50e^{kT} = \frac{1}{2} \cdot 50e^0$$

Solve for T.   so

$$e^{kT} = \frac{1}{2}$$

or

$$kT = \ln\left(\frac{1}{2}\right) = -\ln 2.$$

Substitute known value for k.   Using the value for k found above, we have

$$T = \frac{1}{k}(-\ln 2) = \frac{-\ln 2}{\frac{1}{5}\ln\left(\frac{4}{5}\right)}$$

$$= \frac{-5\ln 2}{\ln 4 - \ln 5}$$

$$\approx 15.53 \text{ days.} \qquad \square$$

Example 3 (**Compound Interest**) We have previously seen that the value of an initial investment of $P(0) = P_0$ dollars compounded continuously for t years at a nominal interest rate of r percent is

$$P(t) = P_0 e^{rt}. \tag{9}$$

That this function has the form of the solution to the differential equation

$$P'(t) = rP(t)$$

with initial condition $P(0) = P_0$ confirms our earlier interpretation of continuous compounding of interest: the rate of growth of the investment (i.e., the addition of interest due to compounding) is the constant $k = r$ times the current size of the investment at each instant of time.

The formula for present value

$$P_0 = e^{-rt}P(t) \tag{10}$$

giving the current value P_0 of an investment that will have value $P(t)$ in t years under continuous compounding at rate r, is also a solution of an equation for natural growth. In this case $k = -r$. $\qquad \square$

Example 4 Find the value of an initial deposit of $500 after 4 years in a bank paying 10% nominal interest compounded continuously.

Solution: With $P_0 = 500$, $r = 0.10$, and $t = 4$ in equation (9) we obtain

$$P(4) = 500e^{(0.10)4} = 500e^{0.4} = 500(1.49182) = \$745.91.$$ □

Example 5 An undergraduate business student with $10,000 to invest wishes to have $20,000 available after 4 years with which to launch a small business. What annual rate of interest, compounded continuously, will produce this yield?

Solution: We again use equation (9), this time with $P_0 = 10{,}000$, $t = 4$, and $P(t) = P(4) = 20{,}000$. We obtain

$$P(4) = P_0 e^{r \cdot 4}$$

or

$$20{,}000 = 10{,}000 e^{4r}.$$

Thus

$$e^{4r} = \frac{20{,}000}{10{,}000} = 2$$

so

$$4r = \ln 2$$

and

$$r = \frac{\ln 2}{4} = \frac{0.69315}{4} = 0.17329.$$

Thus a rate of $r = 17.329\%$ is required. □

Example 6 It turns out that the best rate the undergraduate in Example 5 can find is $r = 10\%$. How much *additional* money must she add to her initial deposit of $10,000 in order to reach her goal via continuous compounding of interest?

Solution: This time we use equation (10) with $r = 0.10$, $t = 4$, and $P(t) = P(4) = 20{,}000$ to find the *present value*

$$\begin{aligned}
P_0 &= e^{-(0.10)(4)} 20{,}000 \\
&= e^{-0.40}(20{,}000) \\
&= (0.67032)(20{,}000) \\
&= 13{,}406.40.
\end{aligned}$$

Thus she will require an additional $3406.40. □

Example 7 All living matter contains two types of carbon, ^{14}C and ^{12}C in its molecules. While the organism is alive the ratio of ^{14}C to ^{12}C is constant, as the ^{14}C is interchanged with fixed levels of $^{14}CO_2$ in the atmosphere. However, when a plant or animal dies, this replenishment ceases and the radioactive ^{14}C present decays exponentially with a half-life of 5760 years. By examining the $^{14}C/^{12}C$ ratio, archaeologists can determine the percentage of the original ^{14}C level remaining and thereby date a once living fossil.

(a) What percentage of the original ^{14}C will exist after 2000 years?
(b) If a fossil contains 10% of its original amount, what is its age?

Solution: Let $A(t)$ be the percentage amount of ^{14}C remaining t years after the fossil's death. Since $A(t)$ decays exponentially, we know that $A(t)$ has the form

$$A(t) = A_0 e^{kt}.$$

Clearly $A(0) = 100\%$, so $A_0 = 100$. To find k, we use the fact that the half-life is 5760 years. That is

$$A(5760) = \frac{1}{2}A_0.$$

Thus

$$A_0 e^{5760k} = \frac{1}{2}A_0$$

so

$$e^{5760k} = \frac{1}{2}$$

$$5760k = \ln\left(\frac{1}{2}\right) = \ln 1 - \ln 2 = -\ln 2$$

and

$$k = \frac{-\ln 2}{5760}.$$

The function $A(t)$ is therefore

$$A(t) = 100e^{\left(\frac{-\ln 2}{5760}t\right)}.$$

(a) After $t = 2000$ years the percent of ^{14}C remaining will be

$$A(2000) = 100e^{\left(\frac{-\ln 2}{5760}\right)(2000)}$$
$$= 100e^{-0.24}$$
$$\approx 79\%.$$

(b) To find the number of years T after which 10% of the original ^{14}C will remain, we set up the equation

$$A(T) = 0.10A(0)$$

which gives

$$100e^{\left(\frac{-\ln 2}{5760}\right)T} = 0.10(100)$$

or

$$e^{\left(\frac{-\ln 2}{5760}\right)T} = 0.10.$$

Thus

$$\left(\frac{-\ln 2}{5760}\right)T = \ln 0.1$$

so

$$T = \frac{5760(\ln 0.1)}{-\ln 2}$$

$$\approx 19{,}134 \text{ years.}$$

□

Exercise Set 4.5

In Exercises 1–6 find the general solution of the differential equation.

1. $N'(t) = 4N(t)$

2. $\dfrac{dN}{dt} = -3N$

3. $f'(x) = -2f(x)$

4. $\dfrac{dy}{dt} = 9y$

5. $\dfrac{dy}{dt} + 3y = 0$

6. $f'(t) - 6f(t) = 0$

In Exercises 7–10 find the unique solution of the differential equation satisfying the given initial condition.

7. $N'(t) = 4N(t)$, $N(0) = 3$

8. $\dfrac{dy}{dx} = -2y$, $y(0) = -4$

9. $\dfrac{dN}{dt} + 3N = 0$, $N(0) = 6$

10. $\dfrac{dy}{dt} - 5y = 0$, $y(0) = 2$.

11. An annuity pays 10% interest, compounded continuously. What amount of money deposited today will have grown to $2500 in 8 years?

12. What rate of interest, compounded continuously, will produce an effective annual yield of 8%? (Effective annual yield is the rate required under *annual* compounding of interest to obtain the specified yield.)

13. Show that the effective annual rate of interest, i, for continuous compounding of interest, is $i = e^r - 1$.

14. True or false? If the population $P(t)$ of a city is increasing at a rate of 5% per year, then the function $P(t)$, if differentiable, satisfies the equation $P'(t) = 0.05P(t)$. If this is false, what is the correct equation?

15. How long does it take for a deposit of P_0 dollars to double at 10% interest compounded continuously?

16. True or false? The half-life of an isotope depends on the amount present. Why?

17. A radioactive isotope decays exponentially with a half-life of 20 days. Assume that 50 mg of the isotope remain after 10 days.
   a. How much of the isotope was present initially?
   b. How much will remain after 30 days?

18. For the isotope in Exercise 17, when will 90% of the initial amount have disintegrated?

19. The number of bacteria in a certain culture grows from 50 to 400 in 12 hours. Assume that the rate of increase is proportional to the number of bacteria present.
   a. How long does it take for the number of bacteria present to double?
   b. How many bacteria will be present after 16 hours?

20. Show that the relative rate of growth of an investment of P_0 dollars compounded continuously at a rate of r percent per year is r.

21. Bacteria grow rapidly in a certain rich culture, at a rate proportional to their current number. If 100,000 bacteria grow in number to 150,000 in 2 hours, how many bacteria will be present after 5 hours?

22. The half-life of radium is approximately 1600 years. How long will it take 10 grams of radium to decay to 6 grams?

23. After 3 years a deposit of $1000 has earned a total of $400 in accrued interest compounded continuously at a nominal rate of r percent. Find r.

24. How long will it be until a deposit of P dollars triples under continuous compounding of interest at a nominal interest rate of 10%?

25. When a certain drug is injected into a patient the amount of the drug present in the patient's bloodstream t hours after the injection satisfies the differential equation $A'(t) = -0.10A(t)$. Find the amount present 6 hours after an injection of 10 milligrams.

26. After 2 years, 80% of an original amount of a radioactive isotope remains. What is its half-life?

27. A startup company's sales increased from $10,000 in year one to $150,000 in year two. If the growth in sales is exponential, what sales level will result in year three?

28. In a simple model of the growth of a biological population the rate of increase of N, the number of individuals, is proportional to the difference between the birthrate, b, and the death rate, d. That is, the simple birth–death model of population growth is

$$\frac{dN}{dt} = bN - dN.$$

a. Find an expression for population size, N, as a function of the birthrate b, the death rate d, and the initial population size N_0.

b. In a population modelled by this process there are 20 births and 10 deaths per year per 100 individuals. Find the anticipated size of the population in 20 years if its current size is $N_0 = 200$ individuals.

29. When a foreign substance is introduced into the body, the body's defense mechanisms move to break down the substance and excrete it. The rate of excretion is usually proportional to the concentration in the body, and the half-life of the resulting exponential decay is referred to as the *biological half-life* of the substance. If, after 12 hours, 30% of a massive dosage of a substance has been excreted by the body, what is the biological half-life of the substance?

30. A fossil contains 80% of its original amount of ^{14}C. What is its age? (See Example 7.)

31. The body concentrates iodine in the thyroid gland. This observation leads to the treatment of thyroid cancer by the injection of radioactive iodine into the bloodstream. One isotope used has a half-life of approximately 8 days and decays exponentially in time. If 50 micrograms of this isotope are injected, what amount remains in the body after three weeks?

32. A chemical dissolves in water at a rate proportional to the amount still undissolved. If 20 grams of the chemical are placed in water and 10 grams remain undissolved 5 minutes later, when will 90% of the chemical be dissolved?

SUMMARY OUTLINE OF CHAPTER 4

■ **Definition:** $a^n = a \cdot a \cdot a \cdot \cdots \cdot a$ (*n* factors)   (Page 205)

$$a^{-n} = \frac{1}{a^n}, \qquad a \neq 0$$

$$a^0 = 1$$

■ **Properties of Exponents**   (Page 205)

(E1) $a^n a^m = a^{n+m}$

(E2) $a^n/a^m = a^{n-m}$

(E3) $[a^n]^m = a^{nm}$

■ *Definition:* $a^{1/n} = y$ means $y^n = a$ $(a > 0$ if n is even). (Page 206)

■ *Formula for Compound Interest:* (Page 211)

$$P(n) = \left(1 + \frac{r}{k}\right)^{nk} P_0$$

where P_0 = original amount, r = interest rate, k = number of compoundings annually, n = years.

■ *Definition:* $e = \lim_{x \to \infty} \left(1 + \frac{1}{x}\right)^x$ (Page 209)

$$e \approx 2.718281828. \ldots$$

■ *Formula for Continuous Compounding of Interest:* $P(t) = e^{rt}P_0$. (Page 212)

■ *Definition:* $\log_b y = x$ if and only if $y = b^x$. (Page 214)

■ *Properties of Logarithms* (Page 216)

(L1) $\log_b(xy) = \log_b x + \log_b y$
(L2) $\log_b(x/y) = \log_b x - \log_b y$
(L3) $\log_b x^r = r \log_b x$

■ *Definition:* $\ln x = \log_e x$. Thus $\ln y = x$ if and only if $y = e^x$. (Page 217)

■ *Definition:* $b^x = e^{x \ln b}$ (Page 218)

■ For $f(x) = \ln x,$ $f'(x) = \frac{1}{x}; \frac{d}{dx} \ln x = \frac{1}{x}.$ (Page 220)

■ For $f(x) = \ln u(x),$ $f'(x) = \frac{u'(x)}{u(x)}; \frac{d}{dx} \ln u = \frac{1}{u} \cdot \frac{du}{dx}.$ (Page 221)

■ The **rate of growth** of f at $x = x_0$ is $G(x_0) = \frac{f'(x_0)}{f(x_0)}.$ (Page 225)

■ For $f(x) = e^x,$ $f'(x) = e^x;$ $\frac{d}{dx} e^x = e^x.$ (Page 229)

■ For $f(x) = e^{u(x)},$ $f'(x) = e^{u(x)} u'(x);$ $\frac{d}{dx} e^u = e^u \cdot \frac{du}{dx}.$ (Page 230)

■ For $f(x) = a^x,$ $f'(x) = a^x \ln a;$ $\frac{d}{dx} a^x = a^x \ln a.$ (Page 234)

■ For $f(x) = \log_a x,$ $f'(x) = \frac{1}{x \ln a};$ $\frac{d}{dx} \log_a x = \frac{1}{x \ln a}.$ (Page 234)

■ A function that satisfies the **Law of Natural Growth** (Page 238)

$$P'(t) = kP(t)$$

must have the form $P(t) = P_0 e^{kt}$, where $P_0 = P(0)$.

REVIEW EXERCISES—CHAPTER 4

1. Simplify the following.

   a. $49^{-3/2}$ **b.** $8^{2/3}$

   c. $\left(\dfrac{8}{27}\right)^{2/3}$ **d.** $\left(\dfrac{36}{81}\right)^{-3/2}$

2. Simplify the given expression.

   a. $e^6 e^{-2}$ **b.** $\dfrac{(2^{-3} \cdot 3^{-2})^3}{\frac{1}{3}(16)}$

   c. $\dfrac{\{8e^{3x}\}^{-2/3}}{\sqrt{9e^2}}$ **d.** $\{16e^{-4}2^{8x}\}^{3/4}$

3. Find the value of a deposit of $400 after 3 years in a savings account paying 6% nominal interest compounded semiannually.

4. Find the value of the investment in Exercise 3 if the interest is compounded continuously.

5. Find the present value of an investment that will yield a sum of $10,000 in 5 years if the prevailing rate of interest is expected to be 10%.

6. Find $\log_b x$.

   a. $\log_{10} \dfrac{1}{100}$ **b.** $\log_3 27$

   c. $\log_2 \dfrac{1}{4}$ **d.** $\ln(e^{-3})$

7. Solve for x:

   a. $e^{3x} = 1$ **b.** $e^{x^2+x-2} = 1$

   c. $e^{\ln(3x-4)} = 2$ **d.** $e^{2x} - e^x - 2 = 0$

8. The selling price p and quantity sold, x, per month for a particular item are related by the equation $p = 50e^{-2x}$. Solve this equation for x as a function of p.

9. For the item in Exercise 8, find the selling price p for which revenue is a maximum.

10. For the price–demand model in Exercise 8, how many items will be sold per month at price $p = 2$ dollars?

In Exercises 11–30 find the derivative of the given function.

11. $y = \ln 6x$ **12.** $f(x) = e^{-3x}$

13. $y = e^{6-x^2}$ **14.** $f(x) = 4\ln(3x + 2)$

15. $f(t) = e^{t^3-3t+2}$ **16.** $y = te^{1-t}$

17. $f(x) = \ln\sqrt{x+1}$ **18.** $f(x) = \sqrt{1 - \ln x}$

19. $y = x^2 e^{\sqrt{x}}$ **20.** $f(x) = \ln(e^{2x} - 2)$

21. $y = \ln\left(\dfrac{x+2}{\sqrt{x}}\right)$ **22.** $y = e^{\sqrt{x}} \ln x$

23. $f(x) = \ln \ln \sqrt{x}$ **24.** $y = x \ln(\sqrt{x} - e^{-x})$

25. $y = xe^{x-\sqrt{x}}$ **26.** $y = e^{t+\ln t}$

27. $y = \ln(1 - e^{-2x})$ **28.** $y = \sqrt{e^x - e^{-x}}$

29. $f(t) = e^{\sqrt{t} - \ln t}$ **30.** $f(x) = x^2 e^{1-\sqrt{x}}$

In Exercises 31–34 find $\dfrac{dy}{dx}$ by implicit differentiation.

31. $x \ln y + y \ln x = 4$ **32.** $x^2 + e^{xy} - y = 2$

33. $e^{xy} = xy$ **34.** $\ln xy = x - y$

35. Find all relative extrema for the function $y = x^2 \ln x$.

36. Find the maximum and minimum values of the function $f(x) = e^{x^2/3}$ for x in the interval $[-1, \sqrt{8}]$.

37. On what intervals is the graph of the function $y = \ln(1 + x^2)$ concave up?

38. Find all relative extrema for the function $y = x^2 - \ln x^2$.

39. Find r if $2^x = e^{rx}$ for all x.

40. Find the equation of the line tangent to the graph of $y = xe^{2x}$ at the point $(\ln 2, 4 \ln 2)$.

41. A cylindrical jar 12 centimeters in diameter is leaking water at a rate of $\ln(t^2)$ cubic centimeters per minute. How fast is the water level in the jar falling?

Find the solutions of the differential equation.

42. $\dfrac{dy}{dx} = 2y$ **43.** $\dfrac{dy}{dx} + y = 0$

44. Find the solution of the initial value problem.

   $2y' + 4y = 0$

   $y(0) = \pi$

45. The population of the United States in 1970 was 203 million. By 1980 the population had grown to 227 million. Assuming exponential growth, what will the population be in 1990?

46. In the chemical reaction called the inversion of raw sugar, the inversion rate is proportional to the amount of raw sugar remaining. If 100 kilograms of raw sugar is reduced to 75 kilograms in 6 hours, how long will it be until

a. half the raw sugar has been inverted?

b. 90% of the raw sugar has been inverted?

47. A population of fruit flies grows exponentially. If initially there were 100 flies and if after 10 days there were 500 flies, how many flies were present after 4 days?

48. 100 grams of a radioactive substance is reduced to 40 grams in 6 hours. If the decay is exponential, what is the half-life?

49. Show that the exponential function $f(x) = Ce^{kx}$ satisfies the differential equation $f'(x + a) = kf(x + a)$ for every real number a.

5

Antiderivatives and the Definite Integral

In this chapter we develop two major themes—reversing the process of differentiation (antidifferentiation), and calculating the areas of certain types of regions in the plane. While these two issues may seem unrelated, we shall see, using the Fundamental Theorem of Calculus, that they are intimately related, and that both yield important applications in business and economics.

5.1 ANTIDERIVATIVES: REVERSING THE PROCESS OF DIFFERENTIATION

We have seen that the derivative of a function may be interpreted as the rate at which it changes. For example, if the function $y = C(x)$ gives a company's total cost in producing x lawnmowers per week, the derivative $C'(x)$ gives the rate at which total cost changes with respect to change in the production level x. In this case we have referred to the derivative C' as the *marginal cost* function. In short, marginal cost is obtained from total cost by the process of differentiation.

Now consider just the reverse situation. Suppose you were told that the company's marginal cost in producing lawnmowers was \$100 per lawnmower, regardless of current production level. That is, suppose

$$C'(x) = MC(x) = 100$$

is given. What can you say about the total cost function C? Obviously, it must be a function C whose derivative is $C'(x) = 100$. One such function that comes to mind immediately is

$$C_1(x) = 100x.$$

But there are others, including

$$C_2(x) = 500 + 100x$$
$$C_3(x) = 10,000 + 100x$$

and

$$C_4(x) = K + 100x$$

where K is any constant. Each of these functions has the property that its derivative is $MC(x) = 100$. For this reason each is called an *antiderivative* of the marginal cost function $MC(x)$.

DEFINITION 1

The function F is called an **antiderivative** for the function f if F is differentiable and if

$$F'(x) = f(x).$$

In other words, an antiderivative for f is just any function whose derivative is f. The following examples show that a given function has many antiderivatives.

Example 1 An antiderivative for $f(x) = 2x + 1$ is the function $F(x) = x^2 + x$, since

$$F'(x) = \frac{d}{dx}(x^2 + x) = 2x + 1 = f(x).$$

Other antiderivatives for f are

$$G(x) = x^2 + x + 7$$

and

$$H(x) = x^2 + x - 45. \qquad \square$$

Example 2 An antiderivative for $f(x) = e^{2x} + \dfrac{1}{x}$ is $F(x) = \dfrac{1}{2}e^{2x} + \ln x + 5$, since

$$F'(x) = \frac{d}{dx}\left(\frac{1}{2}e^{2x} + \ln x + 5\right) = e^{2x} + \frac{1}{x} = f(x).$$

Other antiderivatives for f are

$$G(x) = \frac{1}{2}e^{2x} + \ln x + 100$$

and

$$H(x) = \frac{1}{2}e^{2x} + \ln x - e. \qquad \square$$

Examples 1 and 2 suggest that whenever F is an antiderivative for f so is $F + C$ where C is any constant. This is true because the derivative of a constant is zero. That is

$$\frac{d}{dx}(F(x) + C) = \frac{d}{dx}F(x) + \frac{d}{dx}C$$

$$= \frac{d}{dx}F(x) + 0$$

$$= \frac{d}{dx}F(x).$$

However, this is the only complicating factor in the quest for the antiderivatives of f. The following theorem states that once we have found a particular antiderivative F, any other antiderivative for f has the form $F + C$ for some constant C.

THEOREM 1

Let F and G be antiderivatives for the function f. Then there exists a constant C so that

$$G(x) = F(x) + C.$$

We shall prove Theorem 1 at the end of this section using the Mean Value Theorem.

Notation for Antiderivatives

The notation used to denote the antiderivatives of f is

$$\int f(x)\, dx, \quad \text{or just} \quad \int f\, dx.$$

It consists of an elongated s preceding the function and the letters dx following $f(x)$. The symbol $\int$ is called the **integral sign,** and the function f is called the **integrand.** The symbol dx, called **the differential for x,** identifies the independent variable for f. We shall say more about the meanings associated with these symbols in Section 5.4. For now, we simply think of $\int$ and dx as pieces of notation used to denote the antiderivative of a given function.

According to Theorem 1, once we have found a particular antiderivative F for f, we can express all others as $F + C$. Accordingly, we write

$$\int f(x)\, dx = F(x) + C \tag{1}$$

as the **general form** of the antiderivative for f.

The term **indefinite integral** is synonymous with antiderivative, and we speak of the process of finding an antiderivative as **antidifferentiation** or **integration.** Finally, the constant C in equation (1) is called the **constant of integration.**

At this point the only technique available to you for finding antiderivatives is to try to recognize the given function f as the derivative of a familiar function—in other words, to "think backward" through the differentiation process. In Section 5.3 we shall begin to develop a systematic approach to antidifferentiation.

Example 3 Using the notation of equation (1), we write

$$\int \frac{1}{2\sqrt{x}}\, dx = \sqrt{x} + C$$

since $\dfrac{d}{dx}(\sqrt{x} + C) = \dfrac{1}{2\sqrt{x}}.$ □

Example 4 For $x > 0$ we have

$$\int \frac{1}{x}\, dx = \ln x + C$$

since $\dfrac{d}{dx}(\ln x + C) = \dfrac{1}{x}$, $\qquad x > 0$. ☐

Example 5 The rule for differentiating a polynomial tells us that

$$\int (5x^4 + 4x^3 + 3x^2 + 2x + 1)\, dx = x^5 + x^4 + x^3 + x^2 + x + C$$

as you can verify by differentiation. ☐

Example 6 Using equation (2) of Section 4.4, we have $\dfrac{d}{dx}(e^x + C) = e^x$, so

$$\int e^x\, dx = e^x + C$$ ☐

The Power Rule

One of the most basic differentiation rules is that for differentiating a power of x: $\dfrac{d}{dx} x^n = nx^{n-1}$. Reversing this operation gives the Power Rule for antiderivatives:

$$\int x^n\, dx = \frac{x^{n+1}}{n + 1} + C, \qquad n \neq -1.$$ (2)

Equation (2) is easy to verify:

$$\frac{d}{dx}\left[\frac{x^{n+1}}{n + 1} + C\right] = \frac{1}{n + 1}(n + 1)x^n + 0 = x^n.$$

Thus, for example

$$\int x^3\, dx = \frac{x^4}{4} + C$$

$$\int x^{3/2}\, dx = \frac{x^{5/2}}{5/2} + C = \frac{2}{5}x^{5/2} + C \qquad \text{(Here we must require } x \geq 0.)$$

$$\int x^{-2/3}\, dx = \frac{x^{1/3}}{1/3} + C = 3x^{1/3} + C \qquad \text{(Here we require } x \neq 0.)$$

and

$$\int x^{-5}\, dx = \frac{x^{-4}}{-4} + C = -\frac{1}{4}x^{-4} + C. \qquad \text{(Here we require } x \neq 0.)$$

It is important to note that we must *divide by the new exponent* as well as increase the old exponent by 1 in finding $\int x^n \, dx$. It is also important to note that *the Power Rule does not apply in the case $n = -1$*. This is because, as seen in Example 4

$$\int x^{-1} \, dx = \int \frac{1}{x} \, dx = \ln x + C, \qquad x > 0.$$

Properties of Antiderivatives

Since finding antiderivatives is just a matter of reversing the process of differentiation, the same algebraic laws govern both operations. Thus both equations

$$\int [f(x) + g(x)] \, dx = \int f(x) \, dx + \int g(x) \, dx \tag{3}$$

and

$$\int cf(x) \, dx = c \int f(x) \, dx \tag{4}$$

are true. They simply restate properties of the derivative: The derivative (antiderivative) of the sum of two functions equals the sum of the individual derivatives (antiderivatives); and constants may be "factored out" of the differentiation (antidifferentiation) process.

Equations (3) and (4), together with the Power Rule (2), allow us to find the antiderivative of any polynomial.

Example 7 Find $\int (6x^3 + 12x^2 - 4x + 5) \, dx$.

Strategy

Solution

Using equations (2), (3), and (4), we find that

$$\int (6x^3 + 12x^2 - 4x + 5) \, dx$$

Break integral into sum of simpler terms using equation (3).

$$= \int 6x^3 \, dx + \int 12x^2 \, dx + \int (-4x) \, dx + \int 5 \, dx$$

"Factor out" constants using equation (4).

$$= 6 \int x^3 \, dx + 12 \int x^2 \, dx - 4 \int x \, dx + 5 \int 1 \, dx$$

Apply Power Rule, equation (2), to each integral.

$$= 6 \left(\frac{x^4}{4} \right) + 12 \left(\frac{x^3}{3} \right) - 4 \left(\frac{x^2}{2} \right) + 5x + C$$

$$= \frac{3}{2} x^4 + 4x^3 - 2x^2 + 5x + C.$$

(Note that we write the constant of integration C only once, even though one could have been written for each integral. Since each is arbitrary, we have collected them all in one constant C.) □

Just as the Power Rule for differentiation applies to all real exponents, so does the Power Rule for antiderivatives. Pay careful attention to the treatment of fractional and negative exponents in the next example.

Example 8 Find $\int (2x^{3/2} - 4x^{1/2} + 2x^{-2/3} - 3x^{-7/3})\, dx$.

Solution: $\int (2x^{3/2} - 4x^{1/2} + 2x^{-2/3} - 3x^{-7/3})\, dx$

$$= \int 2x^{3/2}\, dx + \int (-4x^{1/2})\, dx + \int 2x^{-2/3}\, dx + \int (-3x^{-7/3})\, dx$$

$$= 2\int x^{3/2}\, dx - 4\int x^{1/2}\, dx + 2\int x^{-2/3}\, dx - 3\int x^{-7/3}\, dx$$

$$= 2\left(\frac{x^{5/2}}{5/2}\right) - 4\left(\frac{x^{3/2}}{3/2}\right) + 2\left(\frac{x^{1/3}}{1/3}\right) - 3\left(\frac{x^{-4/3}}{-4/3}\right) + C$$

$$= \frac{4}{5}x^{5/2} - \frac{8}{3}x^{3/2} + 6x^{1/3} + \frac{9}{4}x^{-4/3} + C.$$ □

Proof of Theorem 1

The main step in the proof of Theorem 1 involves the following fact, whose proof involves the Mean Value Theorem of Section 3.9.

LEMMA

The only antiderivatives for the zero function are the constant functions. That is, if $F'(x) = 0$ for all x, then $F(x) = C$ for some constant C and all x.

Proof of Lemma: Let x and a be any two numbers with $x > a$. Then by the Mean Value Theorem there is a number c in $[a, x]$ so that

$$F'(c) = \frac{F(x) - F(a)}{x - a}.$$

But $F'(c) = 0$ by assumption, so we must have $F(x) - F(a) = 0$. That is, $F(x) = F(a)$ whenever $x > a$. Taking $x < a$ and repeating the argument on $[x, a]$ gives the same conclusion. Thus, with $C = F(a)$, we have determined that $F(x) = C$ for all x. ■

Proof of Theorem 1: We now proceed to prove Theorem 1, "If F and G are antiderivatives for f, then there exists a constant C so that $G(x) = F(x) + C$ for all x."

Given antiderivatives F and G for f, we know that

$$F'(x) = G'(x) = f(x).$$

Thus if we let $H(x) = (G - F)(x) = G(x) - F(x)$, we have

$$H'(x) = G'(x) - F'(x) = 0$$

for all x. Since $H'(x) = 0$ for all x, the lemma guarantees that $H(x) = C$ for some constant C and all x. Since $H(x) = G(x) - F(x)$, this gives

$$G(x) - F(x) = C$$

or $G(x) = F(x) + C$ for all x, as desired. ∎

Exercise Set 5.1

In each of Exercises 1–20 find the general form of the anti-derivative.

1. $\int 5 \, dx$

2. $\int 3x^2 \, dx$

3. $\int (2x^3 + 5x) \, dx$

4. $\int (4 - 4x^3) \, dx$

5. $\int (x^3 - 6x^2 + 2x - 1) \, dx$

6. $\int (9 - x^3 + 5x^4) \, dx$

7. $\int (x^{2/3} - 3x^{1/2}) \, dx$

8. $\int e^{3x} \, dx$

9. $\int \frac{4}{x} \, dx$

10. $\int 2e^{-x} \, dx$

11. $\int (x^{1/3} - 2x^{-2/3} + x^{-5/3}) \, dx$

12. $\int (x^4 - 4x^{2/3} + 5) \, dx$

13. $\int \left(4e^{-x} - \frac{1}{x} + \sqrt{x} \right) dx$

14. $\int (x - 1)(x + 2) \, dx$

15. $\int (4x^{-6} - 6x^{-4}) \, dx$

16. $\int (\sqrt{x} - x^{1/3}) \, dx$

17. $\int (\sqrt{x} - x^{1/3})^2 \, dx$

18. $\int \frac{x^2 - 3x + 2}{x - 1} \, dx$

19. $\int \frac{1}{x + 3} \, dx$

20. $\int \sqrt{x}(x - 4) \, dx$

5.2 INTERPRETING ANTIDERIVATIVES

Figure 2.1 reminds us that we may interpret differentiation as the operation of beginning with the graph of a function and finding the *slope function* $f'(x)$. Antidifferentiation is just the reverse. We begin with the slope function g and find the function $G(x) = \int g(x) \, dx$ whose slope at each number x is $g(x)$. (See Figure 2.2.)

Figure 2.2 also shows the difficulty that arises in trying to pass from values $g(x)$ as slopes to the antiderivative $G(x) = \int g(x) \, dx$, which has g as its slope function. Specifying the slope of a graph at each value of x produces not one graph but an entire family of graphs, each with slope function g.

But this is not surprising since we already know that the antiderivative for g is actually a family of functions of the form

$$\int g(x) \, dx = G_1(x) + C \tag{1}$$

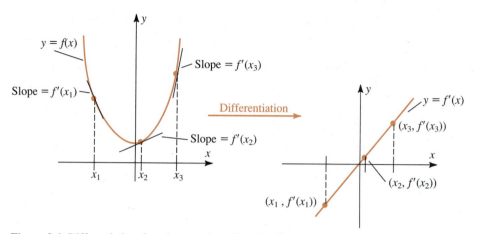

Figure 2.1 Differentiating f produces a slope function f'.

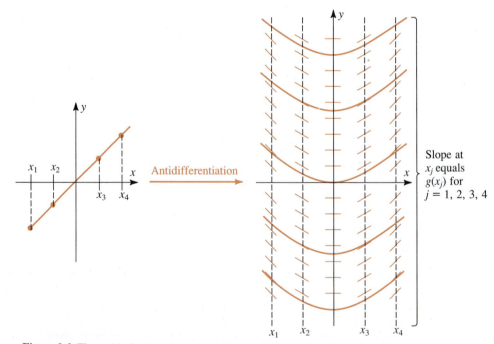

Figure 2.2 The antiderivative of g is actually an infinite family of functions $G(x) = \int g(x)\, dx$ each of which has slope function g.

where G_1 is any particular antiderivative of g. Figure 2.3 shows the interpretation of equation (1) for the particular case $g(x) = x$. Antidifferentiation produces a family of congruent curves, all of which have equation $y = \frac{1}{2}x^2 + C$ and slope function $g(x) = x$. The various values for the constant C are just the y-intercepts of these curves.

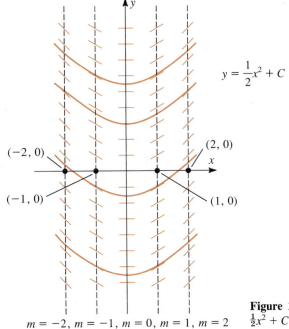

$$y = \frac{1}{2}x^2 + C$$

$(2, 0)$

$(-2, 0)$

$(-1, 0)$

$(1, 0)$

$m = -2, \; m = -1, \; m = 0, \; m = 1, \; m = 2$

Figure 2.3 Graphs of antiderivatives $G(x) = \frac{1}{2}x^2 + C$ of $g(x) = x$.

Example 1 Find the family of curves that have slopes given by the function $f(x) = e^{3x} + 4x^2 - \dfrac{3}{x^2}$ at each value of $x \neq 0$.

Solution: The solution is simply the antiderivative for f, which is

$$F(x) = \int (e^{3x} + 4x^2 - 3x^{-2}) \, dx$$

$$= \int e^{3x} \, dx + 4 \int x^2 \, dx - 3 \int x^{-2} \, dx$$

$$= \frac{1}{3}e^{3x} + \frac{4}{3}x^3 + 3x^{-1} + C.$$

(Note that the antiderivative of e^{3x} is $\frac{1}{3}e^{3x}$, since

$$\frac{d}{dx}\left(\frac{1}{3}e^{3x}\right) = \frac{1}{3} \cdot \frac{d}{dx}e^{3x} = \frac{1}{3}(e^{3x} \cdot 3) = e^{3x}).$$

□

Particular Antiderivatives

When a particular value of the antiderivative

$$\int f(x) \, dx = F(x) + C$$

is specified, we may substitute this information into the general form of the antiderivative to determine the constant C. The next example shows how this is done.

Example 2 Find the particular antiderivative F of the function $f(x) = 2x + e^{-x}$ that satisfies the condition $F(0) = 3$.

Solution: The general form of the antiderivative for f is

$$F(x) = \int (2x + e^{-x})\, dx$$

$$= x^2 - e^{-x} + C$$

as you can verify. Substituting $x = 0$ and using the condition that $F(0) = 3$ gives

$$3 = F(0) = 0^2 - e^{-0} + C = -1 + C.$$

Thus

$$3 = -1 + C$$

so

$$C = 4.$$

The desired antiderivative is $F(x) = x^2 - e^{-x} + 4$.

This solution can be easily verified. We note that

$$F'(x) = \frac{d}{dx}(x^2 - e^{-x} + 4)$$

$$= 2x - (-e^{-x}) + 0$$
$$= 2x + e^{-x}$$
$$= f(x)$$

and that

$$F(0) = 0^2 - e^{-0} + 4 = 0 - 1 + 4 = 3$$

as required. (It is a good idea to always check the results of antidifferentiation by differentiating the answer. This must bring you back to the function with which you began.)

☐

Marginal Analysis

Since marginal cost and marginal revenue have been defined as the derivatives of total cost and total revenue, respectively (that is, $MC(x) = C'(x)$; $MR(x) = R'(x)$), we can determine total cost and revenue from these marginal rates by antidifferentiation:

$$\int MC(x)\, dx = C(x) + A \qquad (A \text{ constant}) \qquad (2)$$

$$\int MR(x)\, dx = R(x) + B. \qquad (B \text{ constant}) \qquad (3)$$

Of course, equations (2) and (3) determine total cost and revenue only to within a constant of integration. In each case a particular value of the antiderivative is needed to determine the function completely.

Example 3 Find a manufacturer's total cost $C(x)$ in producing x items per week if its marginal cost per item is $MC(x) = 100$ dollars and if its fixed costs are $500 per week.

Solution: Since $MC(x) = C'(x)$, we find total costs $C(x)$ from $MC(x)$ by antidifferentiation:

$$C(x) = \int MC(x)\, dx = \int 100\, dx = 100x + K. \tag{4}$$

To determine the constant K, we recall that fixed costs are the costs that one incurs if no items are manufactured. The given information on fixed costs therefore means that $C(0) = 500$. Substituting this information into equation (4) gives

$$500 = C(0) = 100(0) + K = K.$$

Thus $K = 500$, and the total cost function is

$$C(x) = 100x + 500. \qquad \square$$

Example 4 Find a manufacturer's total cost in producing x items per month if its marginal cost per item is

$$MC(x) = 400 + 2x$$

and it is known that total monthly costs are $60,000 when $x = 100$ items are produced per month.

Solution: As in Example 3 we begin by finding the antiderivative for $MC(x)$:

$$C(x) = \int MC(x)\, dx = \int (400 + 2x)\, dx$$
$$= 400x + x^2 + K.$$

To find K, we insert the given data $C(100) = 60,000$ and obtain the equation

$$60,000 = C(100) = 400(100) + (100)^2 + K$$
$$= 50,000 + K.$$

Thus

$$K = 60,000 - 50,000$$
$$= 10,000$$

is the fixed monthly cost. Total monthly cost is therefore

$$C(x) = 400x + x^2 + 10,000 \text{ dollars.} \qquad \square$$

Exercise Set 5.2

In Exercises 1–4 match the given function with the correct slope portrait for its antiderivative.

1.

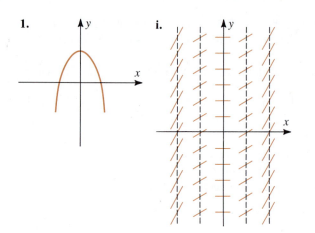

2.

3.

i.

ii.

iii.

4.

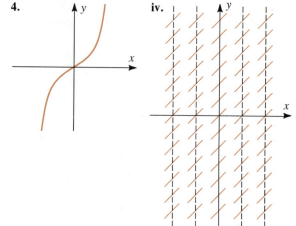

iv.

In Exercises 5–14 find the function F that has the given derivative and that has the specified value.

5. $F'(x) = 4$; $F(0) = 3$

6. $F'(x) = x - 2$; $F(0) = 6$

7. $F'(x) = x^2 - 2x$; $F(0) = -3$

8. $F'(x) = \sqrt{x}$; $F(4) = 7$

9. $F'(x) = \dfrac{1}{x}$; $F(e) = 5$

10. $F'(x) = 4e^x + 5$; $F(1) = 4e + 8$

11. $F'(x) = x^{2/3} - x^{-1/3}$; $F(0) = 5$

12. $F'(x) = x^2 - 4 + 3x^{-7/2}$; $F(0) = -4$

13. $F'(x) = 5e^{2x} + 4$; $F(0) = 10$

14. $F'(x) = \sqrt{x} - \dfrac{1}{x}$; $F(1) = 3$

15. A manufacturer of typewriters finds that its marginal cost per typewriter is \$250. Find its total weekly cost in producing x typewriters if its fixed costs are \$700 per week.

16. For the manufacturer in Exercise 15, suppose that fixed costs are unknown, but that the total cost in producing 20 typewriters per week is known to be \$7000.
 a. Find weekly fixed costs.
 b. Find total weekly cost, $C(x)$.

17. A manufacturer of dishwashers experiences marginal costs of

$$MC(x) = 400 + \frac{1}{4}x \text{ dollars}$$

at a production level of x dishwashers per month. If the total cost in producing ten dishwashers per month is known to be $5200.

a. Find the manufacturer's fixed monthly cost.

b. Find the total monthly cost function $C(x)$.

18. Marginal revenue from the sale of x cameras per month is known to be

$$MR(x) = 250 - \frac{1}{2}x \text{ dollars.}$$

Find the total revenue function R. (*Hint:* Use the fact that $R(0) = 0$ to find the value of the constant of integration.)

19. If a person has a marginal utility function of $MU(x) = \sqrt{x}$ for a particular commodity, find the utility function U. (Recall, $MU(x) = U'(x)$. Use the condition that $U(0) = 0$ to evaluate the constant.)

20. A manufacturer of a certain product finds that it has a marginal cost of $MC(x) = 40 + \frac{1}{2}x$ dollars, marginal revenue of $60, and a fixed cost of $800 per month, where x is the number of items produced per month.

a. Find the manufacturer's monthly cost, $C(x)$.

b. Find the manufacturer's monthly revenue, $R(x)$.

c. Find the manufacturer's monthly profit, $P(x)$.

d. How can marginal profit, $MP(x) = P'(x)$, be obtained from marginal cost and marginal revenue?

21. Recall that if an object moves along a line so that after t seconds its position on the line is given by the differen-

tiable function $s(t)$, then the velocity of the object at time t is the derivative $v(t) = s'(t)$ of the position function. Thus

$$\int v(t) \, dt = s(t) + C.$$

That is, we can recover $s(t)$ from $v(t)$ by integration.

Suppose that an object moves along a line with velocity $v(t) = 3t^2 + 6t + 2$ and that its location at time $t = 0$ is $s(0) = 4$.

a. Find its location after t seconds, $s(t)$.

b. Find its location after 4 seconds, $s(4)$.

22. An object dropped from rest achieves a velocity of $v(t) = -32t$ ft/sec t seconds after it is released. How far has such an object fallen

a. 4 seconds after release?

b. 8 seconds after release?

c. t seconds after release?

23. An epidemiologist estimates that the rate at which influenza is spreading throughout a city is

$$\frac{dN}{dt} = 20 + 24\sqrt{t}$$

persons per day t days after its outbreak. Assuming that $N(0) = 2$.

a. Find $N(t)$, the number of people who have contracted influenza t days after its outbreak.

b. Find $N(16)$.

5.3 FINDING ANTIDERIVATIVES BY SUBSTITUTION

Each differentiation rule that we develop gives a corresponding rule for antidifferentiation. Here is a list of the rules we have developed so far.

Differentiation Rule	Antidifferentiation Rule	
$\dfrac{d}{dx}(ax) = a$	$\displaystyle\int a \, dx = ax + C$	(1)
$\dfrac{d}{dx}(x^n) = nx^{n-1}$	$\displaystyle\int x^n \, dx = \frac{1}{n+1}x^{n+1} + C$	(2)
$\dfrac{d}{dx}\ln x = \dfrac{1}{x}, \quad x > 0$	$\displaystyle\int \frac{1}{x} \, dx = \ln x + C, \quad x > 0$	(3)
$\dfrac{d}{dx}e^x = e^x$	$\displaystyle\int e^x \, dx = e^x + C$	(4)

The goal of this section is to develop a general technique for finding antiderivatives of composite functions involving the functions in equations (1)–(4).

Inverting the Chain Rule

The Chain Rule for differentiating a composite function is

$$\frac{d}{dx} F(u(x)) = F'(u(x)) \cdot u'(x). \tag{5}$$

Now suppose f is the derivative of F. That is, let $f = F'$. Then $F'(u(x)) = f(u(x))$, and we can write equation (5) as

$$\frac{d}{dx} F(u(x)) = f(u(x)) \cdot u'(x). \tag{6}$$

Now let's interpret equation (6) as an antidifferentiation formula. We obtain

$$\int f(u(x)) \cdot u'(x) \, dx = F(u(x)) + C$$

where $F'(x) = f(x)$.

$$\tag{7}$$

Equation (7) is really just a reminder that the derivative of a composite function $F \circ u$ has two factors: $F'(u(x))$ and $u'(x)$. When we are faced with the problem of integrating a function that is the product of two factors, we should check to see if this product has the form $f(u(x)) \cdot u'(x)$. If so, an antiderivative is $F \circ u$, where F is an antiderivative of f.

A special case of equation (7) involving the power function $f(x) = x^n$ occurs frequently enough to warrant special attention. In this case we have $f(u(x)) = [u(x)]^n$ and $F(x) = \frac{1}{n+1} x^{n+1} + C$, according to equation (2). Thus

$$\int [u(x)]^n \, u'(x) \, dx = \frac{1}{n+1} [u(x)]^{n+1} + C. \qquad n \neq -1 \tag{8}$$

Example 1 Find $\displaystyle\int (x^2 + 1)^5 \cdot 2x \, dx$.

Solution: We let $u(x) = x^2 + 1$. Then $u'(x) = 2x$, so the integral has the form of that in equation (8) with $n = 5$. Thus

$$\int \underbrace{(x^2 + 1)^5}_{[u(x)]^5} \cdot \underbrace{2x}_{u'(x)} \cdot dx = \underbrace{\frac{1}{6}(x^2 + 1)^6}_{\frac{1}{6}[u(x)]^6} + C.$$

□

Example 2 Find $\displaystyle\int 2xe^{x^2 + 5} \, dx$.

Solution: We let $u(x) = x^2 + 5$. Then $u'(x) = 2x$, and the integral has the form $\int e^{u(x)} u'(x) \, dx$. Using equation (7) with $f(u) = e^u$, and equation (4), we obtain

$$\int \underbrace{e^{x^2+5}}_{e^{u(x)}} \cdot \underbrace{2x}_{u'(x)} \, dx = \underbrace{e^{x^2+5}}_{e^{u(x)}} + C.$$

□

Example 3 Find $\int \dfrac{3x^2 + 4}{x^3 + 4x} \, dx, \qquad x > 0.$

Solution: If we let $u(x) = x^3 + 4x$, we note that $u'(x) = 3x^2 + 4$, which is precisely the numerator of the integrand. Thus using equation (7), with $f(u) = 1/u$, and equation (3), we obtain

$$\int \frac{3x^2 + 4}{x^3 + 4x} \, dx = \int \underbrace{\frac{1}{x^3 + 4x}}_{\dfrac{1}{u(x)}} \underbrace{(3x^2 + 4)}_{u'(x)} \, dx = \underbrace{\ln(x^3 + 4x)}_{\ln u(x)} + C.$$

□

Example 4 Find $\int (x^2 + 6)^4 x \, dx.$

Solution: If we take $u(x) = x^2 + 6$, we shall need to have the factor $u'(x) = 2x$ present in the integral. But the integral contains only the factor of x, not the required $2x$. However, we can multiply x by $\frac{2}{2}(=1)$ and "pull the factor of $\frac{1}{2}$ in front of the integral sign" as follows:

$$\int (x^2 + 6)^4 x \, dx = \int (x^2 + 6)^4 \left(\frac{2}{2}\right) x \, dx$$

$$= \frac{1}{2} \int \underbrace{(x^2 + 6)^4 2x}_{[u(x)]^4 u'(x)} \, dx$$

$$= \frac{1}{2} \cdot \frac{1}{5}(x^2 + 6)^5 + C$$

$$= \frac{1}{10}(x^2 + 6)^5 + C.$$

□

Example 5 Find $\int (x^3 - 6x + 2)^3 (x^2 - 2) \, dx.$

Solution: With $u(x) = x^3 - 6x + 2$ we require $u'(x) = 3x^2 - 6 = 3(x^2 - 2)$. Since only the factor $(x^2 - 2)$ is present in the integrand, we multiply by $\frac{3}{3}$ and proceed as in Example 4:

$$\int (x^3 - 6x + 2)^3 (x^2 - 2) \, dx = \int (x^3 - 6x + 2)^3 \left(\frac{3}{3}\right)(x^2 - 2) \, dx$$

$$= \frac{1}{3} \int \underbrace{(x^3 - 6x + 2)^3 (3x^2 - 6)} \, dx$$

$$[u(x)]^3 u'(x)$$

$$= \frac{1}{3} \cdot \frac{1}{4}(x^3 - 6x + 2)^4 + C$$

$$= \frac{1}{12}(x^3 - 6x + 2)^4 + C. \qquad \square$$

Differential Notation

The notation of differentials allows us to formalize these techniques a bit further. We have already made reference to the symbol dx as the differential for the independent variable x. If u is a differentiable function of x, we define the differential du as

$$du = u'(x) \, dx \qquad (9)$$

or, in Leibniz notation

$$du = \frac{du}{dx} \cdot dx. \qquad (10)$$

For example, if $u = x^3 - 2x + 1$, then $du = (3x^2 - 2) \, dx$.

In other words, the differential for u is just its derivative multiplied by the differential dx. Note that equation (10) suggests that we may think of the derivative symbol $\dfrac{du}{dx}$ as a quotient of the differentials du and dx, whereby canceling common factors of dx on the right-hand side leaves us with the tautology $du = du$.

Differentials actually have a deeper meaning than we have described here, one we shall not pursue in this text. Their usefulness to us is revealed when we rewrite equation (7) using the definition of du in equation (9):

The equation

$$\int \underbrace{f(u(x))}_{f(u)} \cdot \underbrace{u'(x) \, dx}_{du} = \underbrace{F(u(x))}_{F(u)} + C \qquad (11)$$

becomes simply

$$\int f(u) \, du = F(u) + C. \qquad (12)$$

The Method of Substitution

The point of equations (11) and (12) is that through the use of the differential notation $du = u'(x)\,dx$ we can rewrite the composite integrand in equation (11) as the simple function $f(u)$ of the independent variable u in equation (12). This greatly simplifies the integration to be performed.

Example 6 Find $\displaystyle\int x\sqrt{x^2 + 4}\,dx$.

Strategy

Substitute u for the expression under the radical.

Find du by $du = u'(x)\,dx$.

Solve equation involving du for factors actually present in the integrand.

Substitute expressions for u and du:

$x^2 + 4 = u$;

$dx = \dfrac{1}{2}\,du$.

Integrate with respect to u.

Substitute back.

Solution

We let $u = x^2 + 4$. Then

$$du = \left[\frac{d}{dx}(x^2 + 4)\right]dx = 2x\,dx.$$

Since only the factors x and dx are present in the integrand, we solve this last equation for $x\,dx$:

$$x\,dx = \frac{1}{2}\,du.$$

Substituting these expressions involving u and du into the integrand gives

$$\int x\sqrt{x^2 + 4}\,dx = \int (x^2 + 4)^{1/2}x\,dx$$

$$= \int u^{1/2} \cdot \frac{1}{2}\,du$$

$$= \frac{1}{2}\int u^{1/2}\,du$$

$$= \frac{1}{2} \cdot \frac{2}{3}u^{3/2} + C$$

$$= \frac{1}{3}(x^2 + 4)^{3/2} + C. \qquad \square$$

Example 7 Find $\displaystyle\int \frac{e^{\sqrt{x}}}{\sqrt{x}}\,dx$.

Solution: The integral can be written

$$\int \frac{e^{\sqrt{x}}}{\sqrt{x}}\,dx = \int e^{\sqrt{x}} \cdot \frac{1}{\sqrt{x}}\,dx.$$

Letting $u = \sqrt{x}$, we have $du = \dfrac{1}{2\sqrt{x}}\,dx$. Solving this equation for the factor $\dfrac{1}{\sqrt{x}}\,dx$ in the integrand gives

$$\frac{1}{\sqrt{x}}\,dx = 2\,du.$$

With these substitutions the integral becomes

$$\int e^{\sqrt{x}} \cdot \frac{1}{\sqrt{x}} \, dx = \int e^u \cdot 2 \, du$$

$$= 2 \int e^u \, du$$

$$= 2e^u + C$$

$$= 2e^{\sqrt{x}} + C.$$

$\square$

Example 8 Find $\int \frac{x^2 + e^{3x}}{x^3 + 6 + e^{3x}} \, dx, \ x > 0.$

Solution: Since the numerator of the integrand resembles the derivative of the denominator, we suspect that the integral can be brought into the form $\int \frac{du}{u}$. We therefore try the substitution

$$u = x^3 + 6 + e^{3x}$$

for the denominator. Then

$$du = (3x^2 + 3e^{3x}) \, dx$$
$$= 3(x^2 + e^{3x}) \, dx.$$

Thus

$$(x^2 + e^{3x}) \, dx = \frac{1}{3} \, du.$$

With these substitutions the integral becomes

$$\int \frac{x^2 + e^{3x}}{x^3 + 6 + e^{3x}} \, dx = \int \frac{1}{x^3 + 6 + e^{3x}} (x^2 + e^{3x}) \, dx$$

$$= \int \frac{1}{u} \cdot \frac{1}{3} \, du$$

$$= \frac{1}{3} \int \frac{1}{u} \, du$$

$$= \frac{1}{3} \ln u + C$$

$$= \frac{1}{3} \ln(x^3 + 6 + e^{3x}) + C.$$

$\square$

More on

The integration formula

$$\int \frac{1}{x} \, dx = \ln x + C, \qquad x > 0 \tag{13}$$

has thus far been restricted to $x > 0$ since ln x is defined only for $x > 0$. We shall now show that this formula can be extended to the formula

$$\int \frac{1}{x}\, dx = \ln |x| + C, \qquad x \neq 0 \tag{14}$$

valid for all $x \neq 0$.

If $x > 0$ formula (14) is the same as formula (13) since $|x| = x$, so we need only address the case $x < 0$. But then $-x > 0$, so $|x| = -x$. Thus ln $|x|$ is defined, and

$$\frac{d}{dx} \ln |x| = \frac{d}{dx} \ln(-x) = \frac{1}{-x} \cdot \frac{d}{dx}(-x) = \frac{1}{-x}(-1) = \frac{1}{x} \qquad \text{for } x < 0.$$

Since $\dfrac{d}{dx} \ln |x| = \dfrac{1}{x}$, we have $\displaystyle \int \frac{1}{x}\, dx = \ln |x| + C$ as claimed.

Example 9 Find $\displaystyle \int \frac{x}{9 - 4x^2}\, dx$.

Solution: We let $u = 9 - 4x^2$. Then $du = -8x\, dx$, so $x\, dx = -\frac{1}{8}\, du$. Thus

$$\int \frac{x}{9 - 4x^2}\, dx = \int \frac{1}{9 - 4x^2} x\, dx = \int \frac{1}{u}\left(-\frac{1}{8}\right) du$$

$$= -\frac{1}{8} \int \frac{1}{u}\, du$$

$$= -\frac{1}{8} \ln |u| + C$$

$$= -\frac{1}{8} \ln |9 - 4x^2| + C. \qquad \square$$

Exercise Set 5.3

In Exercises 1–26 find the antiderivative.

1. $\displaystyle \int \sqrt{x + 2}\, dx$

2. $\displaystyle \int x\sqrt{x^2 + 1}\, dx$

9. $\displaystyle \int \frac{x}{1 - 3x^2}\, dx$

10. $\displaystyle \int \frac{\ln x}{x}\, dx$

3. $\displaystyle \int 2x\sqrt{9 + x^2}\, dx$

4. $\displaystyle \int x^2(4 + x^3)^{-1/2}\, dx$

11. $\displaystyle \int e^{2x}(1 + e^{2x})^3\, dx$

12. $\displaystyle \int \frac{e^x}{1 + e^x}\, dx$

5. $\displaystyle \int xe^{x^2}\, dx$

6. $\displaystyle \int \frac{x^2\, dx}{\sqrt{1 + x^3}}$

13. $\displaystyle \int \frac{e^x}{\sqrt{e^x + 1}}\, dx$

14. $\displaystyle \int \frac{1}{x \ln x}\, dx$

7. $\displaystyle \int (x^3 + 5)^4 x^2\, dx$

8. $\displaystyle \int \frac{x - 1}{x^2 - 2x}\, dx$

15. $\displaystyle \int \frac{2x + 3}{(x^2 + 3x + 6)^3}\, dx$

16. $\displaystyle \int \frac{3x + 3}{(x^2 + 2x - 3)^3}\, dx$

17. $\int \left(1 - \dfrac{1}{x}\right)^3 \left(\dfrac{1}{x^2}\right) dx$

18. $\int \dfrac{x^2 + 2x + 3}{\sqrt{x}} \, dx$

19. $\int \dfrac{x^3}{\sqrt{5 + x^4}} \, dx$

20. $\int \dfrac{x}{x^2 + 5} \, dx$

21. $\int \dfrac{4 \ln x}{x} \, dx$

22. $\int \dfrac{1}{\sqrt{x}(1 - \sqrt{x})} \, dx$

23. $\int \dfrac{(x^{2/3} - 5)^{2/3}}{\sqrt[3]{x}} \, dx$

24. $\int \dfrac{e^{-3x}}{1 + e^{-3x}} \, dx$

25. $\int \dfrac{(1 + e^{\sqrt{x}})e^{\sqrt{x}}}{\sqrt{x}} \, dx$

26. $\int (x^{5/3} - 2x)^{3/2}(5x^{2/3} - 6) \, dx$

27. A manufacturer finds that its marginal cost at production level x in the production of a certain product is

$$MC(x) = C'(x) = 40 + \dfrac{20x}{1 + x^2} \text{ dollars.}$$

a. Find $\lim_{x \to \infty} MC(x)$.

b. Find the total cost function C if $C(0) = 500$ dollars.

28. The manufacturer in Exercise 27 sells its products for $100 each.

a. Find the revenue function R giving the revenue from the sale of x items.

b. What is $R(0)$?

29. The marginal revenue from the sale of picnic tables at sales level x per month is

$$MR(x) = R'(x) = \dfrac{10,000x}{100 + 0.2x^2}.$$

a. Find $\lim_{x \to \infty} MR(x)$.

b. Find the total revenue function R.

30. An individual's marginal utility function for the consumption of x items per unit time is $MU(x) = U'(x) = 4x(4 + x^2)^{-2/3}$. Find the utility function $U(x)$ assuming that $U(0) = 0$.

31. A city's population is predicted to grow at a rate of $P'(t) = \dfrac{100e^{20t}}{1 + e^{20t}}$ people per year where t is time in years from the present. Find the city's population $P(t)$ t years from now if its current population is $P(0) = 40,000$ people.

5.4 THE AREA PROBLEM AND THE DEFINITE INTEGRAL

Almost all of what was developed in Chapters 1–4 was related, in one way or another, to the *tangent line problem*. We come now to the second principal problem of the calculus, the *area problem:*

> **Area Problem:** Given the continuous nonnegative function f, find the area of the region R bounded by the x-axis, the graph of $y = f(x)$, the vertical line $x = a$, and the vertical line $x = b$. (See Figure 4.1.)

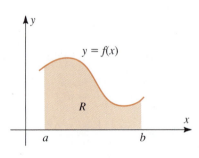

Figure 4.1 Area Problem: Find the area of region R.

Although the area problem is stated in purely geometrical terms, there are many practical situations in which we would like to be able to solve it. For example, Figure 4.2 shows the graphs of the interest paid by two money-market mutual funds over a common period of time. Although both funds experienced the same maximum and minimum interest rates, the fact that Fund B paid a higher rate of return over the period in question is reflected in the fact that the area of region R_B appears to be larger than the area of region R_A. This is because Fund B's rate of interest was higher than Fund A's during most of the period.

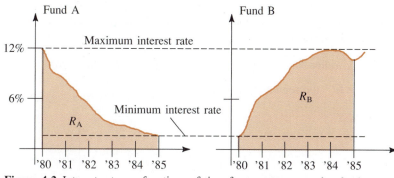

Figure 4.2 Interest rates as functions of time for two money-market funds.

Approximating the Area of R

Formulas from plane geometry enable us to calculate the area of a region R corresponding to a square, a circle, a triangle, or a trapezoid. But when R is bounded by the graph of an arbitrary continuous function f, as in Figure 4.1, no formula from geometry prescribes (or even defines) the area of R.

Figure 4.3 shows how we can use the formula for the area of a rectangle to *approximate* the area of such a region R when $f(x) \ge 0$ for all x in $[a, b]$. To do so, we

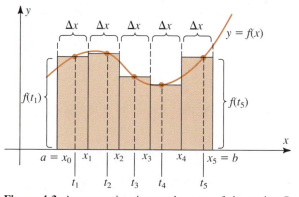

Figure 4.3 An approximation to the area of the region R bounded by the graph of $y = f(x)$ and the x-axis between $x = a$ and $x = b$ using five subintervals of equal length.

(i) Divide the interval $[a, b]$ into n subintervals of equal length $\Delta x = \dfrac{b - a}{n}$.

(ii) In each subinterval select one "test number" t; then approximate the area of the region lying over this subinterval by the area of the *rectangle* with width Δx and height $f(t)$.

(iii) Sum the approximations associated with the subintervals to obtain the approximation

$$\text{Area of } R \approx f(t_1)\, \Delta x + f(t_2)\, \Delta x + \cdots + f(t_n)\, \Delta x$$
$$= \{f(t_1) + f(t_2) + \cdots + f(t_n)\}\, \Delta x. \tag{1}$$

Of course, the approximation in equation (1) will not give the exact value of the area of R because the rectangles do not fit "exactly" under the graph of f. However, as we increase the number n of rectangles used in the approximation, thereby decreasing their widths $\Delta x = \dfrac{b - a}{n}$, we might expect the region determined by the rectangles to "better fit" the region R, as illustrated in Figure 4.4.

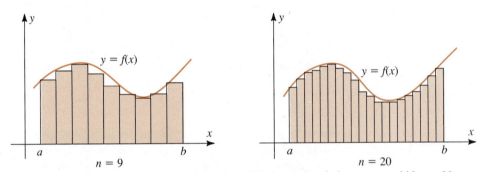

Figure 4.4 Increasing the number of rectangles while decreasing their common width provides an increasingly accurate approximation to the area of R, in general.

Example 1 Figure 4.5 shows how we can use this scheme to approximate the area of the region R bounded by the graph of $f(x) = x^2$ and the x-axis between $a = 0$ and $b = 4$.

(i) If we use $n = 4$ subintervals of equal width, we have

$$\Delta x = \frac{b - a}{n} = \frac{4 - 0}{4} = 1.$$

If we arbitrarily choose the "test point" t_j to be the left endpoint of each subinterval, we have

$$t_1 = 0, \quad t_2 = 1, \quad t_3 = 2, \quad \text{and} \quad t_4 = 3.$$

The approximation given by (1) is then

$$\text{Area of } R \approx \{f(0) + f(1) + f(2) + f(3)\}\, \Delta x$$
$$= \{0^2 + 1^2 + 2^2 + 3^2\}(1)$$
$$= 14.$$

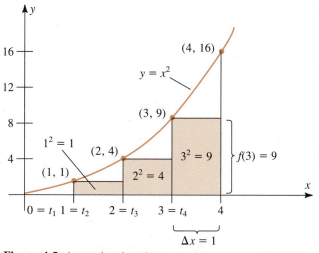

Figure 4.5 Approximating the area of the region bounded by graph of $y = x^2$ and x-axis, for $0 \le x \le 4$, by rectangles of width $\Delta x = 1$.

As you can see from Figure 4.5, this approximation is too small, since the corresponding rectangles lie within R but do not fill R entirely.

(ii) Figure 4.6 illustrates the approximation of the area of this same region using $n = 8$ subintervals rather than 4. In this case the width of each subinterval is

$$\Delta x = \frac{4 - 0}{8} = \frac{1}{2}.$$

If we again take left endpoints as our "test points" t_j, we have

$$t_1 = 0, \quad t_2 = \frac{1}{2}, \quad t_3 = 1, \quad t_4 = \frac{3}{2}, \quad t_5 = 2, \quad t_6 = \frac{5}{2}, \quad t_7 = 3, \quad t_8 = \frac{7}{2}.$$

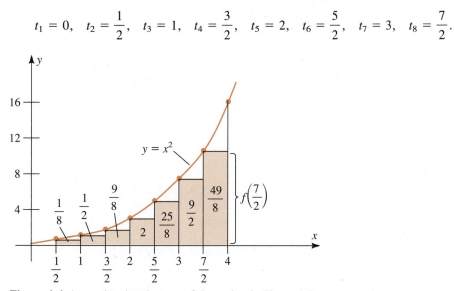

Figure 4.6 Approximating the area of the region in Figure 4.5 using $n = 8$ rectangles.

Our approximation is then

$$\text{Area of } R \approx \left\{ f(0) + f\left(\frac{1}{2}\right) + f(1) + \cdots + f\left(\frac{7}{2}\right) \right\}\left(\frac{1}{2}\right)$$

$$= \left\{ 0^2 + \left(\frac{1}{2}\right)^2 + 1^2 + \left(\frac{3}{2}\right)^2 + 2^2 + \left(\frac{5}{2}\right)^2 + 3^2 + \left(\frac{7}{2}\right)^2 \right\}\left(\frac{1}{2}\right)$$

$$= \left\{ \frac{1}{4} + 1 + \frac{9}{4} + 4 + \frac{25}{4} + 9 + \frac{49}{4} \right\}\left(\frac{1}{2}\right)$$

$$= \frac{140}{8}$$

$$= 17.5.$$

(iii) Continuing in this way we could increase the value of n and recalculate the approximation given in (1). Although a tedious chore by hand calculation, such approximations are easily carried out for large values of n by computers. Table 4.1 shows the result of using BASIC Program 4 in Appendix I to implement the approximation used in parts (i) and (ii) for various values of n.

n	Approximation
10	18.2400
20	19.7600
50	20.6976
100	21.0144
200	21.1736
500	21.2694

Table 4.1. Approximations to area of R in Example 1 using Program 4 in Appendix I. (The actual value is $\frac{64}{3} \approx 21.3333$.)

In Section 5.5 we shall learn a method for calculating the exact value of the area of R, which is $\frac{64}{3} = 21.333. \ldots$. Note that the approximations in Table 4.1 increase toward this exact value as n increases. ☐

Example 2 Let R be the region bounded by the x-axis and the graph of $f(x) = \dfrac{1}{x}$ between $a = 1$ and $b = 3$. Figure 4.7 shows the result of approximating the area of R using $n = 4$ subintervals of width $\Delta x = (3 - 1)/4 = 1/2$ and again using the left endpoint of each interval as the "test point" t_j. Approximation (1) becomes

$$\text{Area of } R \approx \left\{ f(1) + f\left(\frac{3}{2}\right) + f(2) + f\left(\frac{5}{2}\right) \right\}\left(\frac{1}{2}\right)$$

$$= \left\{ 1 + \frac{2}{3} + \frac{1}{2} + \frac{2}{5} \right\}\left(\frac{1}{2}\right)$$

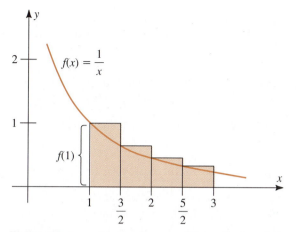

Figure 4.7 Approximating the area of the region bounded by the graph of $f(x) = 1/x$ and the x-axis for $1 \leq x \leq 3$ using $n = 4$ subintervals.

$$= \left(\frac{30 + 20 + 15 + 12}{30} \right) \left(\frac{1}{2} \right)$$

$$= \frac{77}{60}$$

$$\approx 1.2833. \ldots$$

The value of this approximation is *larger* than the actual area (which we shall later show is ln 3 ≈ 1.0986) since each of the approximating rectangles contains more than just the part of R over the corresponding subinterval. Table 4.2 shows the result of using BASIC Program 4 in Appendix I to approximate the area of R. Note that the approximations decrease toward the exact value ln 3 ≈ 1.0986 as n increases.   □

n	Approximation
10	1.1682
20	1.1327
50	1.1121
100	1.1053
200	1.1019
500	1.0999

Table 4.2. Approximations to the area of the region bounded by the graph of the function $f(x) = 1/x$ and the x-axis for $1 \leq x \leq 3$ using BASIC Program 4 in Appendix I. (The actual value, approximated to four decimal places, is 1.0986.)

Limits of Approximating Sums

The idea of approximating regions in the plane by rectangles leads to more than just a method for making "good" estimates for the areas of these regions. The following theorem guarantees that for a continuous function f and a closed finite interval $[a, b]$ all approximations by rectangles *will approach the same limit* as n increases without bound. The proof of this theorem is beyond the scope of this text.

THEOREM 2

Let f be a continuous function on the interval $[a, b]$. Let the n numbers $t_1, t_2, \ldots, t_n$ be chosen arbitrarily, one from each of the n subintervals of $[a, b]$ of equal length $\Delta x = \dfrac{b - a}{n}$ as previously described. Then the limit as n increases without bound of the approximating sum

$$S_n = \{f(t_1) + f(t_2) + \cdots + f(t_n)\}\,\Delta x \tag{2}$$

exists and is unique, regardless of how the numbers $t_1, t_2, \ldots, t_n$ are selected. When $f(x) \geq 0$ for all x in $[a, b]$, this limit is the area of the region bounded by the graph of $y = f(x)$ and the x-axis between $x = a$ and $x = b$. (See Figure 4.8.)

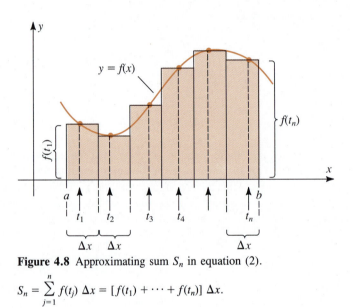

Figure 4.8 Approximating sum S_n in equation (2).

$$S_n = \sum_{j=1}^{n} f(t_j)\,\Delta x = [f(t_1) + \cdots + f(t_n)]\,\Delta x.$$

Sigma Notation

The symbol $\sum$, capital sigma, is often used as shorthand notation to indicate the sum of a number of similar terms. For example

$$\sum_{j=1}^{5} x_j \quad \text{means} \quad x_1 + x_2 + x_3 + x_4 + x_5$$

$$\sum_{j=1}^{4} j^2 \quad \text{means} \quad 1^2 + 2^2 + 3^2 + 4^2 = 30$$

and

$$\sum_{j=1}^{4} x^j \quad \text{means} \quad x + x^2 + x^3 + x^4.$$

In other words, $\sum_{j=1}^{n} f(x_j)$ means, "sum all expressions of the form $f(x_j)$ for j beginning at $j = 1$ and proceeding through all positive integers up to and including n."

Using sigma notation, we can write the approximating sum in equation (2) as

$$
\begin{aligned}
S_n &= \{f(t_1) + f(t_2) + \cdots + f(t_n)\} \, \Delta x \\
&= f(t_1) \, \Delta x + f(t_2) \, \Delta x + \cdots + f(t_n) \, \Delta x \\
&= \sum_{j=1}^{n} f(t_j) \, \Delta x
\end{aligned}
$$

since each term in the sum is the product of $f(t_j)$ (a height of a rectangle, when $f(t_j) \geq 0$) and Δx (the width of the rectangle).

When $f(x) \geq 0$ for all x in the interval $[a, b]$, we can summarize the conclusions of Theorem 2 by writing that

$$\text{Area of } R = \lim_{n \to \infty} S_n = \lim_{n \to \infty} \sum_{j=1}^{n} f(t_j) \, \Delta x \tag{3}$$

where Δx and t_j are as defined above.

Of course, the question remains of how to find the limit of the approximating sums in equation (3). This is the topic of Section 5.5.

The Definite Integral

A special symbol denotes the limit of the approximating sum in equation (3) for the continuous function f. It is

$$\int_{a}^{b} f(x) \, dx$$

which is called the **definite integral** of the function f on the interval $[a, b]$. That is, we *define*

$$\int_{a}^{b} f(x) \, dx = \lim_{n \to \infty} \sum_{j=1}^{n} f(t_j) \, \Delta x. \tag{4}$$

(Remember that the widths Δx and the numbers t_j change as n increases.) As with the indefinite integral, the elongated s, $\int$, is referred to as the **integral sign**. But the **lower limit** a and **upper limit** b identify the fact that the definite integral is not a function (as is an antiderivative). Rather, $\int_{a}^{b} f(x) \, dx$ is a *number* associated with the **integrand** f and the *interval* $[a, b]$. The symbol $\int_a^b f(x)\,dx$ is read, "the integral from a to b of f with respect to x."

When $f(x) \geq 0$ for all x in $[a, b]$, we may combine equations (3) and (4) by writing

$$\text{Area of } R = \int_a^b f(x)\,dx. \tag{5}$$

(See Figure 4.9.)

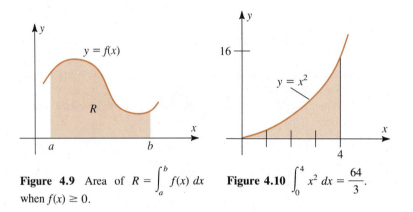

Figure 4.9 Area of $R = \int_a^b f(x)\,dx$ when $f(x) \geq 0$.　　**Figure 4.10** $\int_0^4 x^2\,dx = \dfrac{64}{3}$.

Example 3 As stated in part (iii) of Example 1, we shall show in Section 5.5 that

$$\int_0^4 x^2\,dx = \frac{64}{3}.$$

(See Figure 4.10.)　　　　　　　　　　　　　　　　　　　□

Example 4 According to Example 2, we have

$$\int_1^3 \frac{1}{x}\,dx = \ln 3 \approx 1.0986.$$

(See Figure 4.11.)　　　　　　　　　　　　　　　　　　　□

Example 5 Figure 4.12 shows the region R bounded above by the graph of $f(x) = 2x + 1$ and below by the x-axis for $1 \leq x \leq 3$. Since R is a trapezoid with bases of length $b = 3$ and $B = 7$, and height $h = 2$, the area of R is

$$A = \frac{1}{2}(B + b)h = \frac{1}{2}(3 + 7)(2) = 10.$$

Thus using equation (5), we can write that

$$\int_1^3 (2x + 1)\,dx = 10.$$
　　　　　　　　　　　　　　　　　　　　　　　　　　　□

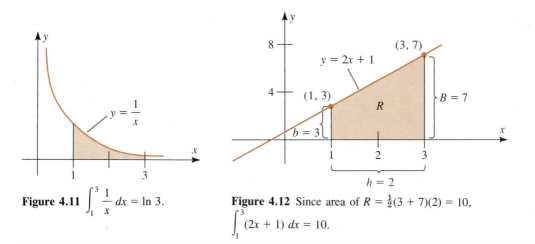

Figure 4.11 $\int_1^3 \frac{1}{x}\, dx = \ln 3.$

Figure 4.12 Since area of $R = \frac{1}{2}(3 + 7)(2) = 10,$

$$\int_1^3 (2x + 1)\, dx = 10.$$

Exercise Set 5.4

In each of Exercises 1–5 find the approximating sum, equation (1), for f on the interval $[a, b]$ using n subintervals of equal length and the left endpoint of each subinterval as the "test number" t.

1. $f(x) = 2x + 5,$ $a = 0,$ $b = 4,$ $n = 4$

2. $f(x) = x^2 + 3,$ $a = 0,$ $b = 2,$ $n = 4$

3. $f(x) = 4 - x^2,$ $a = -2,$ $b = 2,$ $n = 8$

4. $f(x) = \dfrac{1}{x + 2},$ $a = 0,$ $b = 3,$ $n = 6$

5. $f(x) = \dfrac{1}{1 + x^2},$ $a = -1,$ $b = 1,$ $n = 4$

In each of Exercises 6–10 apply the instructions for Exercises 1–5 except that the right endpoint of each subinterval should be used as the "test number."

6. $f(x) = 2x + 5,$ $a = 0,$ $b = 4,$ $n = 4$

7. $f(x) = x^2 + 3,$ $a = 0,$ $b = 2,$ $n = 4$

8. $f(x) = 9 - x^2,$ $a = -3,$ $b = 3,$ $n = 6$

9. $f(x) = \dfrac{1}{4 - x},$ $a = 0,$ $b = 2,$ $n = 4$

10. $f(x) = \dfrac{1}{2 + x^2},$ $a = -2,$ $b = 2,$ $n = 4$

In each of Exercises 11–20 use geometry to find the value of the definite integral.

11. $\int_0^2 x\, dx$

12. $\int_1^3 (x + 4)\, dx$

13. $\int_{-1}^3 4\, dx$

14. $\int_{-1}^2 (x + 1)\, dx$

15. $\int_0^9 (9 - x)\, dx$

16. $\int_3^6 3\, dx$

17. $\int_{-1}^2 |x|\, dx$

18. $\int_2^4 (3x - 6)\, dx$

19. $\int_0^4 |x - 2|\, dx$

20. $\int_0^5 |2x - 4|\, dx$

(Computer). In each of Exercises 21–26 use Program 4 in Appendix I to approximate the area of the region R.

21. R is the region bounded above by the graph of $f(x) = \sqrt{x + 2}$ and below by the x-axis for $0 \le x \le 2$.

22. R is the region bounded above by the graph of $f(x) = \dfrac{10}{1 + x^2}$ and below by the x-axis for $0 \le x \le 3$.

23. R is the region bounded above by the graph of $f(x) = \ln(x + 3)$ and below by the x-axis for $0 \le x \le 4$.

24. R is the region whose area is $\int_0^2 \sqrt{4 - x^2}\, dx$.

25. R is the region whose area is $\int_0^4 \sqrt{5 + x^2}\, dx$.

26. R is the region whose area is $\int_{-2}^2 \sqrt{9 - x^2}\, dx$.

5.5 THE FUNDAMENTAL THEOREM OF CALCULUS

In Section 5.4 we have defined the definite integral of the continuous function f on the interval $[a, b]$ as

$$\int_a^b f(x)\, dx = \lim_{n \to \infty} \sum_{j=1}^n f(t_j)\, \Delta x \qquad (1)$$

where $\Delta x = \dfrac{b - a}{n}$ and $t_1, t_2, \ldots, t_n$ are as in Figure 5.1. When $f(x) \geq 0$ for $a \leq x \leq b$, the definite integral gives the area of the region bounded by the graph of f and the x-axis for $a \leq x \leq b$.

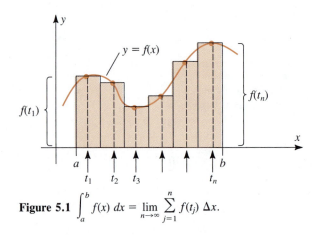

Figure 5.1 $\int_a^b f(x)\, dx = \lim_{n \to \infty} \sum_{j=1}^n f(t_j)\, \Delta x$.

Although the definite integral $\int_a^b f(x)\, dx$ gives useful information about the graph of $y = f(x)$, the procedure outlined in Section 5.4 for approximating its value is tedious, at best, and we do not yet have a method for determining the precise value for the limit in equation (1).

However, the following theorem shows that the exact value of the definite integral $\int_a^b f(x)\, dx$ can be computed very easily if we can find an *antiderivative* for the function f. This result is so central to our theory that it is called the ***Fundamental Theorem of Calculus***.

THEOREM 3
Fundamental
Theorem

Let f be continuous on $[a, b]$ and let F be any antiderivative for f. That is, let $F'(x) = f(x)$. Then

$$\int_a^b f(x)\, dx = F(b) - F(a). \qquad (2)$$

That is, the value of the definite integral $\int_a^b f(x)\, dx$ is found by

(i) Finding an antiderivative F for f, and
(ii) Finding the difference $F(b) - F(a)$ between its values at the endpoints of $[a, b]$.

In applying the Fundamental Theorem, we will make use of the notation

$$F(x)]_a^b = F(b) - F(a). \tag{3}$$

We shall defer the proof of the Fundamental Theorem to the end of this section.

Example 1 Since an antiderivative for $f(x) = x$ is $F(x) = x^2/2$, we have

$$\int_0^4 x \, dx = \frac{x^2}{2}\Big]_0^4 = \frac{4^2}{2} - \frac{0^2}{2} = 8.$$

This result is easy to verify geometrically since the value of the definite integral is the area of the triangle in Figure 5.2. □

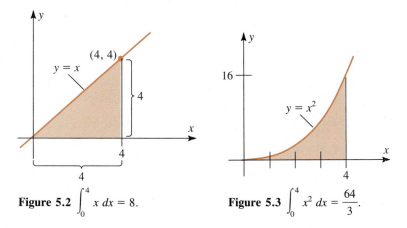

Figure 5.2 $\int_0^4 x \, dx = 8.$

Figure 5.3 $\int_0^4 x^2 \, dx = \dfrac{64}{3}.$

Example 2 In Example 1, Section 5.4, we claimed that, for the region bounded by the graph of $y = x^2$ and the x-axis for $0 \le x \le 4$, the exact value of the area is $\dfrac{64}{3}$. We can verify this claim using the Fundamental Theorem and the antiderivative $F(x) = \dfrac{x^3}{3}$ for $f(x) = x^2$:

$$\int_0^4 x^2 \, dx = \frac{x^3}{3}\Big]_0^4 = \frac{4^3}{3} - \frac{0^3}{3} = \frac{64}{3}.$$

(See Figure 5.3.) □

Example 3 $\displaystyle\int_1^3 \frac{1}{x} \, dx = \ln x \Big]_1^3 = \ln 3 - \ln 1 = \ln 3$

since

$$\frac{d}{dx} \ln x = \frac{1}{x}.$$ □

Example 4 $\displaystyle\int_{-1}^2 (x^3 - 3x^2 - 2x + 2) \, dx$

$$= \frac{x^4}{4} - x^3 - x^2 + 2x \bigg]_{-1}^{2}$$

$$= \left[\frac{2^4}{4} - 2^3 - 2^2 + 2(2) \right] - \left[\frac{(-1)^4}{4} - (-1)^3 - (-1)^2 + 2(-1) \right]$$

$$= (4 - 8 - 4 + 4) - \left(\frac{1}{4} - 1 - 1 - 2 \right)$$

$$= -\frac{9}{4}. \qquad \square$$

Example 5 $\displaystyle\int_{-2}^{2} \left(e^{-x} + \frac{1}{x^2} \right) dx = -e^{-x} - \frac{1}{x} \bigg]_{-2}^{2}$

$$= \left[-e^{-2} - \frac{1}{2} \right] - \left[-e^{-(-2)} - \frac{1}{(-2)} \right]$$

$$= e^2 - \frac{1}{e^2} - 1. \qquad \square$$

REMARK: Note that we do not include a constant of integration in writing an antiderivative for f when using the Fundamental Theorem to evaluate $\displaystyle\int_a^b f(x)\, dx$. That is because the Fundamental Theorem calls for *any* particular antiderivative of f, not the most general form of the antiderivative. Including an arbitrary constant would be inconsequential since

$$F(x) + C]_a^b = [F(b) + C] - [F(a) + C]$$
$$= F(b) + C - F(a) - C$$
$$= F(b) - F(a)$$
$$= F(x)]_a^b.$$

Example 6 $\displaystyle\int_1^4 \frac{x+1}{\sqrt{x}}\, dx = \int_1^4 \left(\sqrt{x} + \frac{1}{\sqrt{x}} \right) dx$

$$= \int_1^4 (x^{1/2} + x^{-1/2})\, dx$$

$$= \frac{2}{3} x^{3/2} + 2x^{1/2} \bigg]_1^4$$

$$= \left(\frac{2}{3} \cdot 4^{3/2} + 2 \cdot 4^{1/2} \right) - \left(\frac{2}{3} \cdot 1^{3/2} + 2 \cdot 1^{1/2} \right)$$

$$= \frac{20}{3}. \qquad \square$$

Substitutions in Definite Integrals

Often the method of substitution is required to find an antiderivative in evaluating a definite integral. The following example is one such situation. Note that we substitute *back* from u's to x's before the antiderivative is evaluated.

Example 7 Find $\displaystyle\int_{-1}^{3} \frac{x}{\sqrt{7 + x^2}}\, dx$.

Solution: To find an antiderivative $\displaystyle\int \frac{x\, dx}{\sqrt{7 + x^2}}$, we use the substitution

$$u = 7 + x^2, \qquad du = 2x\, dx.$$

Then $x\, dx = \tfrac{1}{2}\, du$, and we obtain

$$\int \frac{x\, dx}{\sqrt{7 + x^2}} = \int \frac{\tfrac{1}{2}\, du}{\sqrt{u}} = \tfrac{1}{2} \int u^{-1/2}\, du = u^{1/2} + C$$

$$= \sqrt{7 + x^2} + C.$$

We therefore use $F(x) = \sqrt{7 + x^2}$ in the Fundamental Theorem to find that

$$\int_{-1}^{3} \frac{x\, dx}{\sqrt{7 + x^2}}\, dx = \sqrt{7 + x^2}\,\Big]_{-1}^{3} = \sqrt{7 + 3^2} - \sqrt{7 + (-1)^2}$$

$$= \sqrt{16} - \sqrt{8}$$

$$= 4 - 2\sqrt{2}. \qquad \square$$

Example 8 Find $\displaystyle\int_{1}^{2} \frac{3x^2 + 6}{x^3 + 6x - 6}\, dx$.

Solution: This time we notice that the numerator of the integrand is the derivative of the denominator. This suggests that we find the antiderivative by using the substitution

$$u = x^3 + 6x - 6 \quad \text{and} \quad du = (3x^2 + 6)\, dx.$$

We obtain the antiderivative

$$\int \frac{3x^2 + 6}{x^3 + 6x - 6}\, dx = \int \frac{du}{u} = \ln|u| + C = \ln|x^3 + 6x - 6| + C.$$

Thus

$$\int_{1}^{2} \frac{3x^2 + 6}{x^3 + 6x - 6}\, dx = \ln|x^3 + 6x - 6|\,\Big]_{1}^{2}$$

$$= \ln(2^3 + 6(2) - 6) - \ln(1^3 + 6(1) - 6)$$

$$= \ln 14 - \ln 1$$

$$= \ln 14. \qquad \square$$

Properties of the Definite Integral

There are several properties of the definite integral that are useful in calculations. If f and g are continuous functions on the interval $[a, b]$

$$(\text{I1}) \quad \int_{a}^{b} [f(x) + g(x)]\, dx = \int_{a}^{b} f(x)\, dx + \int_{a}^{b} g(x)\, dx$$

(I2) $\displaystyle\int_a^b cf(x)\ dx = c \cdot \int_a^b f(x)\ dx$ (c is any constant)

(I3) $\displaystyle\int_a^a f(x)\ dx = 0$

(I4) $\displaystyle\int_a^b f(x)\ dx = \int_a^c f(x)\ dx + \int_c^b f(x)\ dx$ (c is any number between a and b)

Although each of these properties follows from the basic definition of the definite integral, it is particularly simple to demonstrate their validity using the Fundamental Theorem of Calculus when antiderivatives for f and g can be found. For example, if $F'(x) = f(x)$ and $G'(x) = g(x)$ for all x in $[a, b]$, then $F + G$ is an antiderivative for $f + g$, so

$$\int_a^b [f(x) + g(x)]\ dx = (F(x) + G(x))]_a^b$$
$$= [F(b) + G(b)] - [F(a) + G(a)]$$
$$= [F(b) - F(a)] + [G(b) - G(a)]$$
$$= \int_a^b f(x)\ dx + \int_a^b g(x)\ dx$$

which demonstrates property (I1). We leave it for you as an exercise to verify properties (I2)–(I4) in the same way.

Properties (I3) and (I4) have particularly simple geometric interpretations. Property (I3) means that the definite integral over an interval of length zero (i.e., a single point) must equal zero. This corresponds to the fact that a rectangle of height $f(a) \geq 0$ and width zero has area equal to $(0)f(a) = 0$.

Figure 5.4 gives the geometric interpretation of property (I4) when $f(x) \geq 0$ for all x in $[a, b]$. The area of the region bounded by the graph of $y = f(x)$ and the x-axis for $a \leq x \leq b$ equals the sum of the areas of the regions bounded by the graph of $y = f(x)$ and the x-axis for $a \leq x \leq c$ (region R_1) and for $c \leq x \leq b$ (region R_2).

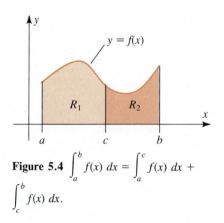

Figure 5.4 $\displaystyle\int_a^b f(x)\ dx = \int_a^c f(x)\ dx + \int_c^b f(x)\ dx.$

Proof of the Fundamental Theorem of Calculus

Since the definite integral is the limit

$$\int_a^b f(x)\,dx = \lim_{n\to\infty} \sum_{j=1}^n f(t_j)\,\Delta x \tag{4}$$

of an approximating sum, we begin by letting $x_0, x_1, x_2, \ldots, x_n$ denote the successive endpoints of the subintervals of $[a, b]$ of length $\Delta x = \dfrac{b-a}{n}$. (See Figure 5.5.) Also, since F is an antiderivative for f, the function F is differentiable and continuous on the interval $[a, b]$. Thus F satisfies the hypotheses of the Mean Value Theorem, which guarantees that in each subinterval $[x_{j-1}, x_j]$ there is a number t_j so that

$$\frac{F(x_j) - F(x_{j-1})}{x_j - x_{j-1}} = F'(t_j) = f(t_j). \tag{5}$$

This is the test number t_j that we shall use in each interval to form the approximating sum in equation (4). Using equation (5) and the fact that $\Delta x = x_j - x_{j-1}$ for each $j = 1, 2, \ldots, n$ gives

$$\int_a^b f(x)\,dx = \lim_{n\to\infty} \sum_{j=1}^n f(t_j)\,\Delta x$$

$$= \lim_{n\to\infty} \sum_{j=1}^n \left[\frac{F(x_j) - F(x_{j-1})}{x_j - x_{j-1}}\right](x_j - x_{j-1})$$

$$= \lim_{n\to\infty} \sum_{j=1}^n [F(x_j) - F(x_{j-1})].$$

Now the last line above contains a telescoping sum in which every term cancels except the first and the last:

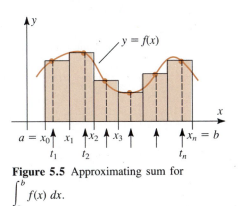

Figure 5.5 Approximating sum for $\displaystyle\int_a^b f(x)\,dx$.

$$\sum_{j=1}^{n} [F(x_j) - F(x_{j-1})] = \{[F(x_1) - F(x_0)] + [F(x_2) - F(x_1)]$$
$$+ [F(x_3) - F(x_2)] + [F(x_4) - F(x_3)]$$
$$+ \cdots$$
$$+ [F(x_{n-1}) - F(x_{n-2})] + [F(x_n) - F(x_{n-1})]\}$$
$$= F(x_n) - F(x_0)$$
$$= F(b) - F(a)$$

since $x_0 = a$ and $x_n = b$. (See Figure 5.5.) Thus

$$\int_a^b f(x)\,dx = \lim_{n \to \infty} \sum_{j=1}^{n} [F(x_j) - F(x_{j-1})]$$
$$= \lim_{n \to \infty} [F(b) - F(a)]$$
$$= F(b) - F(a)$$

as stated. ■

Exercise Set 5.5

In Exercises 1–24 use the Fundamental Theorem of Calculus to evaluate the definite integral.

1. $\displaystyle\int_0^2 (x + 3)\,dx$

2. $\displaystyle\int_1^3 (4 - 2x)\,dx$

3. $\displaystyle\int_1^5 (x^2 - 6)\,dx$

4. $\displaystyle\int_2^7 3\,dx$

5. $\displaystyle\int_{-3}^2 (9 + x^2)\,dx$

6. $\displaystyle\int_{-2}^3 \left(6 + \frac{1}{x^2}\right)dx$

7. $\displaystyle\int_0^1 e^{4x}\,dx$

8. $\displaystyle\int_1^5 \frac{1}{x + 3}\,dx$

9. $\displaystyle\int_1^4 \left(\sqrt{x} - \frac{3}{\sqrt{x}}\right)dx$

10. $\displaystyle\int_0^2 xe^{x^2}\,dx$

11. $\displaystyle\int_0^1 (x^4 - 6x^3 + x)\,dx$

12. $\displaystyle\int_0^2 \left(\frac{x - 1}{x + 1}\right)dx$

13. $\displaystyle\int_1^4 \left(\frac{x + 3}{\sqrt{x}}\right)dx$

14. $\displaystyle\int_0^2 (\sqrt{x} - 2)(x + 1)\,dx$

15. $\displaystyle\int_0^2 (x^3 - 6x^2 + 3x + 3)\,dx$

16. $\displaystyle\int_{-1}^1 (4x^3 - 3x^4)\,dx$

17. $\displaystyle\int_1^2 \frac{x + 6}{x^2 + 12x}\,dx$

18. $\displaystyle\int_0^1 xe^{3x^2}\,dx$

19. $\displaystyle\int_2^4 \frac{x - 1}{\sqrt{x} - 1}\,dx$

20. $\displaystyle\int_0^3 x(\sqrt[3]{x} - 2)\,dx$

21. $\displaystyle\int_0^1 (x^{3/5} - x^{5/3})\,dx$

22. $\displaystyle\int_1^2 \frac{1 - x}{x^3}\,dx$

23. $\displaystyle\int_0^4 \frac{dx}{\sqrt{2x + 1}}$

24. $\displaystyle\int_1^4 \frac{e^{\sqrt{x}}}{\sqrt{x}}\,dx$

In Exercises 25–30 use the method of substitution to evaluate the definite integral.

25. $\displaystyle\int_0^2 \frac{x}{\sqrt{16 + x^2}}\,dx$

26. $\displaystyle\int_0^1 \frac{x^2 + 1}{x^3 + 3x + 7}\,dx$

27. $\displaystyle\int_0^3 x\sqrt{9 - x^2}\,dx$

28. $\displaystyle\int_1^2 \frac{x}{(2x^2 - 1)^3}\,dx$

29. $\displaystyle\int_1^2 x(x^2 - 1)^{1/3}\,dx$

30. $\displaystyle\int_1^4 \left(1 - \frac{1}{2\sqrt{x}}\right)e^{x - \sqrt{x}}\,dx$

5.6 FINDING AREAS BY INTEGRATION

Figure 6.1 summarizes the fact that when $f(x) \geq 0$ for all x in $[a, b]$, the area of the region R bounded by the graph of the continuous function f and the x-axis between $x = a$ and $x = b$ is given by the definite integral $\int_a^b f(x) \, dx$:

$$\text{Area of } R = \int_a^b f(x) \, dx \tag{1}$$
$$\text{if} \quad f(x) \geq 0 \quad \text{for all } x \text{ in } [a, b].$$

That is because the area of R is approximated by the areas of rectangles whose heights are $f(t_j)$ and whose widths are Δx, as illustrated in Figure 6.2.

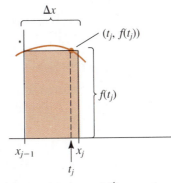

Figure 6.1 Area of $R = \int_a^b f(x) \, dx$

$$= \lim_{n \to \infty} \sum_{j=1}^n f(t_j) \, \Delta x.$$

Figure 6.2 Area of jth approximating rectangle is $f(t_j) \, \Delta x$.

Example 1 Find the area of the region R bounded by the graph of $f(x) = \sqrt{x} + 2$ and the x-axis between $x = 1$ and $x = 4$. (See Figure 6.3.)

Solution: Here $f(x) > 0$ for all x in $[1, 4]$, so we apply equation (1) with $a = 1$ and $b = 4$:

$$\text{Area of } R = \int_1^4 (\sqrt{x} + 2) \, dx$$
$$= \int_1^4 (x^{1/2} + 2) \, dx$$
$$= \frac{2}{3} x^{3/2} + 2x \Big]_1^4$$

$$= \left(\frac{2}{3} \cdot 4^{3/2} + 2 \cdot 4\right) - \left(\frac{2}{3} \cdot 1^{3/2} + 2 \cdot 1\right)$$

$$= \frac{2}{3}(8) + 8 - \frac{2}{3} - 2$$

$$= \frac{32}{3}. \qquad \qquad \square$$

In the next example the endpoints a and b defining the region R are not explicitly stated. They must be found by finding the *zeros* of the function f.

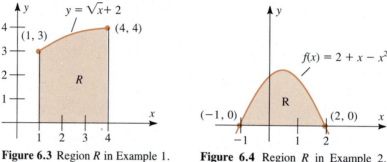

Figure 6.3 Region R in Example 1.   **Figure 6.4** Region R in Example 2.

Example 2 Find the area of the region bounded by the graph of $f(x) = 2 + x - x^2$ and the x-axis.

Strategy

Sketch the graph to locate the region R.

Determine the interval $[a, b]$ defining the region by finding the zeros of f.

Use equation (1) to find the area of R.

Solution

The region bounded by the graph of $f(x) = 2 + x - x^2$ and the x-axis lies above the x-axis, as shown in Figure 6.4. To find the largest and smallest values of x associated with this region, we locate the zeros of f by factoring:

$$f(x) = 2 + x - x^2$$
$$= (2 - x)(1 + x).$$

Thus $x = a = -1$ and $x = b = 2$ are the zeros of f. The area of R therefore is

$$\text{Area} = \int_{-1}^{2} (2 + x - x^2) \, dx$$

$$= 2x + \frac{1}{2}x^2 - \frac{1}{3}x^3 \Big]_{-1}^{2}$$

$$= \left[2 \cdot 2 + \frac{1}{2} \cdot 2^2 - \frac{1}{3} \cdot 2^3\right]$$

$$\quad - \left[2(-1) + \frac{1}{2}(-1)^2 - \frac{1}{3}(-1)^3\right]$$

$$= \frac{9}{2}. \qquad \qquad \square$$

Example 3 Find the area of the region R bounded above by the graph of $y = |x - 3|$ and below by the x-axis for $1 \leq x \leq 6$.

Strategy
Use definition of absolute value to rewrite $f(x) = |x - 3|$.

Solution
Since we do not have an antiderivative for $f(x) = |x - 3|$, we cannot proceed directly to apply equation (1). However, using the definition of absolute value, we may rewrite f as

$$f(x) = |x - 3| = \begin{cases} x - 3 & \text{if } x \geq 3 \\ 3 - x & \text{if } x < 3. \end{cases}$$

Calculate areas of regions over [1, 3] and [3, 6] separately, writing $f(x)$ as a simple linear function in each interval.

This suggests that we calculate separately the areas of regions R_1 and R_2 as illustrated in Figure 6.5. We obtain

$$\text{Area of } R = \text{area of } R_1 + \text{area of } R_2$$

$$= \int_1^3 (3 - x) \, dx + \int_3^6 (x - 3) \, dx$$

$$= \left(3x - \frac{1}{2}x^2 \right) \Big]_1^3 + \left(\frac{1}{2}x^2 - 3x \right) \Big]_3^6$$

$$= \left\{ \left(3 \cdot 3 - \frac{1}{2} \cdot 3^2 \right) - \left(3 \cdot 1 - \frac{1}{2} \cdot 1^2 \right) \right\}$$

$$+ \left\{ \left(\frac{1}{2} \cdot 6^2 - 3 \cdot 6 \right) - \left(\frac{1}{2} \cdot 3^2 - 3 \cdot 3 \right) \right\}$$

$$= \frac{13}{2}.$$

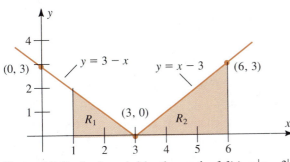

Figure 6.5 Region bounded by the graph of $f(x) = |x - 3|$ and the x-axis for $1 \leq x \leq 6$.

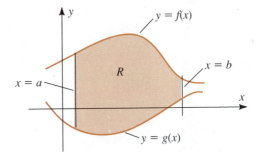

Figure 6.6 Region bounded by two graphs.

The Area of a Region Bounded by Two Curves

Figure 6.6 shows a region R bounded by the graphs of two continuous functions—above by the graph of f, and below by the graph of g—for values of x between a and b. We can use the definite integral to compute the area of such regions as follows. If we divide the interval $[a, b]$ into n equal subintervals of length $\Delta x = \dfrac{b - a}{n}$ and select one "test number" t_j in each subinterval, we can approximate the part of the region corresponding to that subinterval by the rectangle with width Δx and height $[f(t_j) - g(t_j)]$. (See Figures 6.7 and 6.8). Summing the individual approximations then gives the approximation to the area of R:

$$\text{Area of } R \approx \sum_{j=1}^{n} [f(t_j) - g(t_j)] \, \Delta x \tag{2}$$

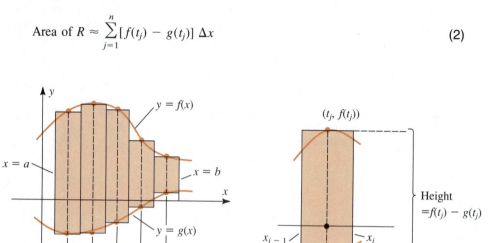

Figure 6.7 Approximating R by rectangles.

Figure 6.8 Area of jth rectangle is $[f(t_j) - g(t_j)] \, \Delta x$.

Now in the limit as n becomes infinitely large the approximating sum on the right-hand side of equation (2) becomes the definite integral

$$\int_{a}^{b} [f(x) - g(x)] \, dx = \lim_{n \to \infty} \sum_{j=1}^{n} [f(t_j) - g(t_j)] \, \Delta x \tag{3}$$

which we define to be the area of R:

> The area of the region R bounded above by the graph of the continuous $\quad$ (4)
> function f and below by the graph of the continuous function g, between
> $x = a$ and $x = b$, is
>
> $$\text{Area of } R = \int_{a}^{b} [f(x) - g(x)] \, dx.$$
>
> (Note that this requires $f(x) \geq g(x)$ for all x. See Figure 6.6.)

Note that equation (4) generalizes equation (1), since $g(x) = 0$ when the region R is bounded below by the x-axis. Notice also that there are no restrictions on the sign of $f(x)$ or $g(x)$ in equation (4)—either may be positive or negative. *It is essential that $f(x) \geq g(x)$, however, so that the integrand $[f(x) - g(x)]$ is nonnegative.*

Example 4 Find the area of the region R bounded above by the graph of $f(x) = e^x$ and below by the graph of $g(x) = \dfrac{1}{x}$ for $1 \le x \le 2$. (See Figure 6.9).

Solution: Since $f(x) = e^x > \dfrac{1}{x} = g(x)$ for all x with $1 \le x \le 2$, the top curve is indeed $f(x) = e^x$ for all such x. According to equation (4), the area of R is

$$\text{Area of } R = \int_1^2 \left[e^x - \frac{1}{x} \right] dx$$
$$= (e^x - \ln x)]_1^2$$
$$= (e^2 - \ln 2) - (e^1 - \ln 1)$$
$$= e^2 - e - \ln 2. \qquad (\ln 1 = 0)$$

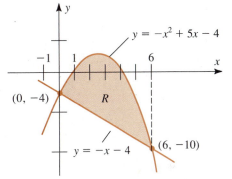

Figure 6.9 Region R in Example 4.   **Figure 6.10** Region R in Example 5.

Example 5 Find the area of the region R bounded by the graphs of the equations $y = -x^2 + 5x - 4$ and $y = -x - 4$ and the lines $x = 0$ and $x = 6$. (See Figure 6.10.)

Strategy

Determine which graph is on top.

(To find area, we must integrate top curve minus bottom.)

Apply equation (4).

Solution

By graphing the two equations (or by checking individual values), we find that the graph of $f(x) = -x^2 + 5x - 4$ forms the top boundary while the graph of $g(x) = -x - 4$ is the bottom boundary of the region R. Applying equation (4) gives

$$\text{Area of } R = \int_0^6 [(-x^2 + 5x - 4) - (-x - 4)]\, dx$$
$$= \int_0^6 (-x^2 + 6x)\, dx$$
$$= -\frac{1}{3}x^3 + 3x^2 \Big]_0^6$$
$$= \left(-\frac{1}{3}\right)6^3 + 3 \cdot 6^2$$
$$= 36.$$

In Example 5 the region R was completely determined by the graphs of the two functions f and g, even though the numbers $x = 0$ and $x = 6$ were provided in the statement of the problem. If these numbers are not provided in such problems, it is necessary to first solve the equation $f(x) = g(x)$ in order to determine the horizontal extremities of the region R.

Example 6 Find the area of the region R bounded by the graphs of the functions $y = x^4 + 1$ and $y = 2x^2$.

Strategy

Sketch the region.

Find the points of intersection by equating the two functions and factoring.

Determine which curve is on top.

Apply equation (4).

Solution

A rough sketch of the region shows that the graphs intersect at two points. To find these points, we equate the two functions, obtaining

$$x^4 + 1 = 2x^2$$

or

$$x^4 - 2x^2 + 1 = (x^2 - 1)^2 = 0.$$

Thus $x^2 - 1 = 0$, so $x = \pm 1$. The points of intersection are therefore $(-1, 2)$ and $(1, 2)$.

Checking function values for any x in $(-1, 1)$ shows that the graph of $f(x) = x^4 + 1$ is the upper boundary, while the graph of $g(x) = 2x^2$ is the lower boundary. Thus by equation (4)

$$\text{Area of } R = \int_{-1}^{1} [(x^4 + 1) - (2x^2)]\, dx$$

$$= \frac{1}{5}x^5 - \frac{2}{3}x^3 + x \bigg]_{-1}^{1}$$

$$= \left[\frac{1}{5}(1)^5 - \frac{2}{3}(1)^3 + 1 \right]$$

$$\quad - \left[\frac{1}{5}(-1)^5 - \frac{2}{3}(-1)^3 + (-1) \right]$$

$$= \frac{16}{15}.$$

(See Figure 6.11.) □

Sometimes a region bounded by two curves will actually consist of several parts with different "top" and "bottom" graphs in each part. In such cases you should calculate the area of each part separately, being careful to identify properly the upper and lower boundary for each part.

Example 7 Find the area of the region R bounded by the graphs of the functions $f(x) = x^3 - 4x$ and $g(x) = 5x$.

Solution: Equating the two functions, we find that

$$x^3 - 4x = 5x$$

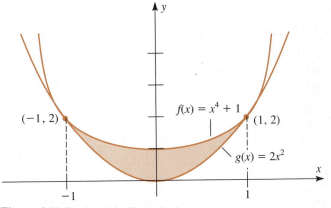

Figure 6.11 Region R in Example 6.

gives

$$x^3 - 9x = 0, \quad \text{or} \quad x(x^2 - 9) = 0.$$

Thus the two graphs intersect when $x = -3, 0,$ and 3. The region therefore consists of two subregions: R_1, corresponding to the interval $[-3, 0]$, and R_2, corresponding to the interval $[0, 3]$.

On the interval $[-3, 0]$ we have $x^3 - 4x \geq 5x$ for all x, so the area of R_1 is

$$\text{Area of } R_1 = \int_{-3}^{0} [(x^3 - 4x) - (5x)] \, dx$$

$$= \frac{1}{4}x^4 - \frac{9}{2}x^2 \bigg]_{-3}^{0}$$

$$= 0 - \left[\frac{1}{4}(-3)^4 - \frac{9}{2}(-3)^2 \right]$$

$$= \frac{81}{4}.$$

On the interval $[0, 3]$ we have $5x \geq x^3 - 4x$ for all x, so the area is

$$\text{Area of } R_2 = \int_{0}^{3} [5x - (x^3 - 4x)] \, dx$$

$$= \frac{9}{2}x^2 - \frac{1}{4}x^4 \bigg]_{0}^{3}$$

$$= \left[\frac{9}{2}3^2 - \frac{1}{4}3^4 \right] - 0$$

$$= \frac{81}{4}.$$

Thus

$$\text{Area of } R = \text{area of } R_1 + \text{area of } R_2$$

$$= \frac{81}{4} + \frac{81}{4}$$

$$= \frac{81}{2}.$$

(See Figure 6.12.)

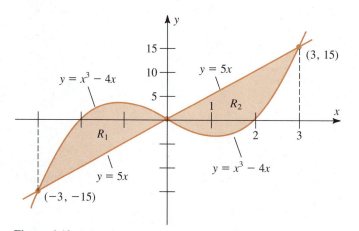

Figure 6.12 Area of R is sum of area of R_1 and area of R_2.

(Note that if we had mistakenly calculated only the integral $\displaystyle\int_{-3}^{3} (f(x) - g(x))\, dx$, we would have obtained a number that is *not* the total area of the region R. In fact, we would have obtained

$$\int_{-3}^{3} (f(x) - g(x))\, dx = \int_{-3}^{3} [(x^3 - 4x) - (5x)]\, dx$$

$$= \int_{-3}^{3} (x^3 - 9x)\, dx$$

$$= \frac{1}{4}x^4 - \frac{9}{2}x^2 \Big]_{-3}^{3}$$

$$= \left(\frac{81}{4} - \frac{81}{2}\right) - \left(\frac{81}{4} - \frac{81}{2}\right)$$

$$= 0.$$

The reason for this result is that the integrand $f(x) - g(x)$ is negative on $[-3, 0)$ and positive on $(0, 3]$, so the integrals corresponding to the two intervals "cancel.") ☐

Regions Bounded by Graphs of Negative Functions

The area formula

$$\text{Area of } R = \int_a^b [f(x) - g(x)] \, dx \tag{5}$$

allows us to resolve the problem of finding the area of a region R bounded by the graph of the function g and the x-axis when $g(x) \le 0$ for $a \le x \le b$ as in Figure 6.13. In this case the curve defining the upper boundary of R is simply the x-axis, which has equation $f(x) = 0$ for all x. Equation (5) then gives

$$
\begin{aligned}
\text{Area of } R &= \int_a^b [0 - g(x)] \, dx \\
&= \int_a^b [-g(x)] \, dx \\
&= -\int_a^b g(x) \, dx \qquad \text{(by Property (I2), Section 5.5)}.
\end{aligned}
$$

That is, when $g(x) \le 0$ for all x in $[a, b]$, the definite integral $\int_a^b g(x) \, dx$ gives the *negative* of the area of the region R. This is why we must be so careful in ensuring that the integrand $f(x) - g(x)$ is nonnegative for all values of x when calculating the area of a region using equation (5).

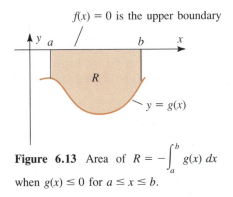

$f(x) = 0$ is the upper boundary

Figure 6.13 Area of $R = -\displaystyle\int_a^b g(x) \, dx$ when $g(x) \le 0$ for $a \le x \le b$.

Exercise Set 5.6

In each of Exercises 1–8 find the area of the region bounded above by the graph of $y = f(x)$ and below by the x-axis for $a \le x \le b$.

1. $f(x) = 2x + 5, \quad a = 0, \quad b = 2$

2. $f(x) = 9 - x^2, \quad a = -3, \quad b = 3$

3. $f(x) = \dfrac{1}{x - 2}, \quad a = 3, \quad b = 5$

4. $f(x) = e^{2x}, \quad a = 0, \quad b = \ln 2$

5. $f(x) = \dfrac{x + 1}{x^2 + 2x}, \quad a = 1, \quad b = 2$

6. $f(x) = xe^{1-x^2}, \quad a = 0, \quad b = 1$

7. $f(x) = \sqrt{5 + x}, \quad a = -4, \quad b = 4$

8. $f(x) = \dfrac{x}{\sqrt{9 + x^2}}, \quad a = 0, \quad b = 4$

In each of Exercises 9–16 find the area of the region bounded by the graphs of the given functions for $a \leq x \leq b$.

9. $f(x) = 9 - x^2$, $g(x) = -2$, $a = -2$, $b = 2$

10. $f(x) = -x - 1$, $g(x) = \sqrt{x}$, $a = 1$, $b = 4$

11. $f(x) = x + 1$, $g(x) = -2x + 1$, $a = 0$, $b = 2$

12. $f(x) = 2x + 3$, $g(x) = x^2 - 4$, $a = -1$, $b = 1$

13. $f(x) = \sqrt{x}$, $g(x) = -x^2$, $a = 0$, $b = 4$

14. $f(x) = \dfrac{1}{x^2}$, $g(x) = x^{2/3}$, $a = 1$, $b = 8$

15. $f(x) = x\sqrt{9 - x^2}$, $g(x) = -x$, $a = -3$, $b = 3$

16. $f(x) = |4 - x^2|$, $g(x) = 5$, $a = -3$, $b = 3$

In each of Exercises 17–24 find the area of the region bounded by the graphs of the given functions.

17. $y = 4 - x^2$, $y = x - 2$

18. $y = x^2$, $y = x^3$

19. $y = x^3$, $y = x$

20. $y = 9 - x^2$, $9y - x^2 + 9 = 0$

21. $y = x^2 - 4$, $y = 2 - x$

22. $y = \sqrt{x}$, $y = \sqrt[3]{x}$

23. $y = x^{2/3}$, $y = x^2$

24. $y = x^{2/3}$, $y = 2 - x^2$

25. Find the area of the region bounded by the graphs of $y = x^3$, $y = -x^3$, $y = 1$ and $y = -1$.

26. Find the area of the region bounded *below* by the graph of $y = x^2 - 9$ and above by the x-axis.

27. Find the number a so that the line with equation $x = a$ divides into two parts of equal area the region bounded by the graph of the equation $y = \sqrt{x}$ and the x-axis between $x = 0$ and $x = 4$.

5.7 APPLICATIONS OF THE DEFINITE INTEGRAL TO ECONOMICS AND BUSINESS

Total Functions from Marginal Functions

We have already seen that the derivative of the total cost function $y = C(x)$ is the marginal cost function $MC(x) = C'(x)$. Thus one antiderivative of marginal cost is $C(x)$ and, in general

$$\int MC(x)\, dx = C(x) + K \tag{1}$$

where K is an arbitrary constant. Because of equation (1) we can evaluate the definite integral of the marginal cost function MC between any two production levels a and b as

$$\int_a^b MC(x)\, dx = C(x)]_a^b = C(b) - C(a). \tag{2}$$

That is, the definite integral of the marginal cost function from $x = a$ to $x = b$ is the change in total cost resulting from a change in production from $x = a$ to $x = b$ units per unit time.

Rewritten in the following form, equation (2) shows how total cost at production level b can be determined from total cost at production level a and knowledge of the marginal cost function:

$$C(b) = C(a) + \int_a^b MC(x)\, dx. \tag{3}$$

Figure 7.1 $\int_a^b MC(x)\,dx = C(b) - C(a)$

= change in total cost.

Figure 7.1 shows that the area of the region between the graph of the marginal cost function and the x-axis may be interpreted as change in total cost.

Equations similar to (1)–(3) hold for marginal revenue functions. That is, if $R(x)$ is the revenue received from the sale of x items, and if $MR(x) = R'(x)$ is the corresponding marginal revenue function, then since

$$\int MR(x)\,dx = R(x) + K \tag{4}$$

we have

$$\int_a^b MR(x)\,dx = R(x)]_a^b = R(b) - R(a) \tag{5}$$

and

$$R(b) = R(a) + \int_a^b MR(x)\,dx. \tag{6}$$

Example 1 A manufacturer of pocket cameras finds that its total cost in producing 100 cameras per week is $C(100) = \$2150$ and that its marginal cost is $MC(x) = 10 + 0.2x$ dollars per camera at production level x cameras per week.

(a) Find the amount by which total cost will increase if production is increased from $a = 100$ to $b = 120$ cameras per week.
(b) Find the total cost in producing 120 cameras per week.

Solution:

(a) Using equation (2), we find that the change in total cost is

$$C(120) - C(100) = \int_{100}^{120} (10 + 0.2x)\,dx$$

$$= 10x + 0.1x^2]_{100}^{120}$$

$$= [10(120) + (0.1)(120)^2] - [10(100) + 0.1(100)^2]$$

$$= 640 \text{ dollars.}$$

(b) Equation (3) now gives the total cost in producing $b = 120$ cameras as

$$C(120) = C(100) + \int_{100}^{120} (10 + 0.2x)\, dx$$

$$= 2150 + 640$$

$$= 2790 \text{ dollars.} \qquad \square$$

Example 2 The camera manufacturer in Example 1 received total revenues of $R(100) = 3000$ dollars from the sale of 100 cameras per week. Also, it has determined that its marginal revenue function is $MR(x) = 50 - 0.4x$ at sales level x cameras per week.

(a) Find the change in total revenues if sales are increased from 100 to 120 cameras per week.
(b) Find $R(120)$.
(c) What is the change in profit as production and sales levels are increased from $x = 100$ to $x = 120$ cameras per week?

Solution:

(a) Using equation (5), we find that

$$R(120) - R(100) = \int_{100}^{120} (50 - 0.4x)\, dx$$

$$= 50x - 0.2x^2]_{100}^{120}$$

$$= [50(120) - 0.2(120)^2] - [50(100) - 0.2(100)^2]$$

$$= 120 \text{ dollars.}$$

(b) Equation (6) gives $R(120)$ as

$$R(120) = R(100) + \int_{100}^{120} (50 - 0.4x)\, dx$$

$$= 3000 + 120$$

$$= 3120 \text{ dollars.}$$

(c) Since profit is defined by the equation $P(x) = R(x) - C(x)$, we have

$$P(100) = R(100) - C(100) = 3000 - 2150 = 850 \text{ dollars}$$

while

$$P(120) = R(120) - C(120) = 3120 - 2790 = 330 \text{ dollars.}$$

Thus profits decline by \$520 on an increase in production from 100 to 120 cameras per week. $\qquad \square$

Social Utility of Consumption

Figure 7.2 shows the graph of a typical demand function giving the selling price p of an item as a function $D(x)$ of the number x of items that the public is willing to consume (that is, purchase) in a given period of time. The shaded region lying between the graph of the

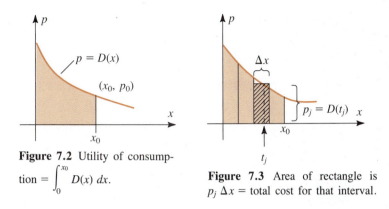

Figure 7.2 Utility of consumption $= \displaystyle\int_0^{x_0} D(x)\, dx.$

Figure 7.3 Area of rectangle is $p_j\, \Delta x =$ total cost for that interval.

demand function and the x-axis for x between 0 and x_0 is called the **social utility of consumption** at level x_0. Since utility refers to the value that we attach to an item, this phrase may be interpreted as the aggregate value placed by the public on its consumption of x_0 items per unit time.

This interpretation is illustrated in Figure 7.3. If we subdivide the interval $[0, x_0]$ into small subintervals of length Δx, the area of the rectangle that approximates the shaded part of the region over any subinterval has the form $p_j\, \Delta x = D(t_j)\, \Delta x$, the price per item at consumption level $x = t_j$ multiplied by Δx, the number of items associated with that subinterval. Since

(price per item) times (number of items) = total cost

the area of the approximating rectangle is the total cost to the public of the purchases corresponding to that subinterval. Summing these individual approximations and taking the limit as the number of subintervals becomes infinitely large gives

$$\text{Utility of consumption} = \int_0^{x_0} D(x)\, dx. \qquad (7)$$
(Total value)

Example 3 The price that a producer of a particular software package must charge in order to sell x packages per month is determined to be $p = D(x) = \dfrac{500}{1 + 0.02x}$ dollars. Find the utility of consumption associated with the purchase and sale of 200 packages per month.

Solution: According to equation (7), the total value to the public of these 200 packages is

$$\int_0^{200} \frac{500}{1 + 0.02x}\, dx = \frac{500}{0.02} \ln(1 + 0.02x)\Big]_0^{200}$$

$$= 25{,}000 \ln 5$$

$$\approx 40{,}236 \text{ dollars.}$$

(See Figure 7.4.) □

Figure 7.4 Utility of consumption in Example 3.

Consumers' Surplus

Of course, not every software package in Example 3 will be sold for the highest price its purchaser is willing to pay. As is the case in "open" economies, the software packages will all be sold at the same price, $p = D(200) = 100$ dollars. Thus the producer will receive not the total value held by the public, \$40,236, but the actual total revenues $(200)(100) = \$20,000$ from the sale of 200 packages at \$100 each. This difference is explained by the fact that of the 200 buyers willing to purchase the package at the price of \$100, some (fewer) would have been willing to pay \$150 for it, and some (even fewer) would have been willing to pay \$250, and so on.

The difference between the total value of a number of items (utility of consumption) and the actual total price that consumers have to pay for these items is called the *consumers' surplus*. (See Figure 7.5.) Since the total paid by consumers for x_0 items, each priced at p_0 dollars, is $x_0 p_0$, consumers' surplus is given by the following equation:

$$\left.\begin{array}{c}\text{Consumers' surplus at}\\ \text{consumption level } x_0\end{array}\right\} = \int_0^{x_0} D(x)\,dx - x_0 p_0, \qquad p_0 = D(x_0). \qquad (8)$$

(See Figure 7.6.)

Figure 7.5 Consumers' surplus. (Refer to Example 3.)

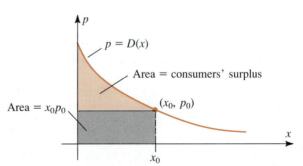

Figure 7.6 Consumers' surplus at consumption level x_0 is

$$\int_0^{x_0} D(x)\,dx - x_0 p_0.$$

**Suppliers'
Surplus**

The notion of suppliers' surplus is similar to that of consumers' surplus. Let $p = S(x)$ be a supply function giving the price p that must be offered in order that suppliers will provide x units of a particular item in a given period of time. (See Figure 7.7.) If the point (x_0, p_0) is on the graph of this function, suppliers will together provide a total of x_0 items to sell at price p_0. But if the price had been less, some (but not all) of these suppliers would still have been willing to produce items for sale. These suppliers benefit from the higher price received by all sellers. Figure 7.7 illustrates that the magnitude of this benefit is the following:

$$\left.\begin{array}{l} \text{Suppliers' surplus at} \\ \text{supply level } x_0 \end{array}\right\} = x_0 p_0 - \int_0^{x_0} S(x)\, dx, \qquad p_0 = S(x_0). \tag{9}$$

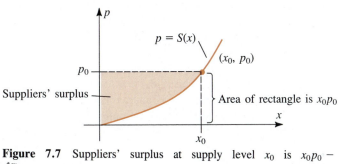

Figure 7.7 Suppliers' surplus at supply level x_0 is $x_0 p_0 - \int_0^{x_0} S(x)\, dx$.

Example 4 Figure 7.8 shows a demand curve $D(x) = \dfrac{2500}{x + 50}$ and a supply curve $S(x) = 0.01x^2$ for a single product. With respect to the equilibrium point $(x_0, p_0) = (50, 25)$, we have from equation (8)

$$\text{Consumers' surplus} = \int_0^{50} \frac{2500}{x + 50}\, dx - (50)(25)$$

$$= 2500 \ln(x + 50)]_0^{50} - 1250$$

$$= 2500 \ln 2 - 1250$$

$$\approx 483 \text{ dollars}$$

while from equation (9) we have

$$\text{Suppliers' surplus} = (50)(25) - \int_0^{50} 0.01x^2\, dx$$

$$= 1250 - \frac{x^3}{300}\bigg]_0^{50}$$

$$= 1250 - \frac{1250}{3}$$

$$\approx 833 \text{ dollars.} \qquad \square$$

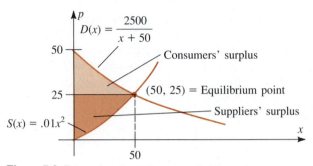

Figure 7.8 Demand and supply curves in Example 4.

Revenue Streams

Certain types of assets, such as rental properties, timber stands, or annuities, produce income for their owners over time. If such a "revenue stream" is assumed to be a continuous function of time, the definite integral can be used to calculate the value of the revenue stream.

Suppose that revenue flows continuously from an asset at a rate of $R(t)$ dollars per year after t years. As Figure 7.9 illustrates, in any interval of time of length Δt the amount of revenue that flows from the asset is approximated by a product of the form $R(t_j)\,\Delta t$, the rate per unit time (at a "test" point) multiplied by the length of time. The total revenue flowing between times $t = a$ and $t = b$ is approximated by a sum of the form $\sum\limits_{j=1}^{n} R(t_j)\,\Delta t$, which in the limit as $n \to \infty$ gives the following integral.

$$\int_a^b R(t)\,dt \qquad\qquad (10)$$

is the (nominal) value of a revenue stream flowing at the rate of $R(t)$ dollars per year from time $t = a$ to time $t = b$. (See Figure 7.10.)

Figure 7.9 Revenue flowing in any time interval is approximated as $R(t_j)\,\Delta t$.

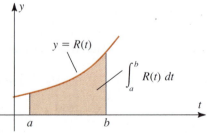

Figure 7.10 Total value of revenue stream from time a to time b is $\int_a^b R(t)\,dt$.

Example 5 The revenue produced by a particular annuity is expected to be $R(t) = 1000\sqrt{t + 4}$ dollars per year for 5 years. Find the total revenue to be produced by the annuity during that time.

Solution: Using expression (10), we find the value of the revenue stream to be

$$\int_0^5 1000\sqrt{t + 4} \, dt = \frac{2(1000)}{3} (t + 4)^{3/2}\Big]_0^5$$

$$= \frac{2000}{3} (9^{3/2} - 4^{3/2})$$

$$= \$12,666.67. \qquad \square$$

Recall that the *present value* of the amount $R(t_j) \, \Delta t$ of a revenue stream that will become available after t_j years is, according to equation (10) of Section 4.5, $P_0(t_j) = e^{-rt_j}R(t_j) \, \Delta t$, assuming a nominal interest rate of r percent compounded continuously. The limit as $n \to \infty$ of the approximating sum $\sum_{j=1}^{n} R(t_j)e^{-rt_j} \, \Delta t$ is therefore the present value of the revenue stream.

$$\int_a^b R(t)e^{-rt} \, dt \qquad (11)$$

is the *present value* of the revenue stream flowing at the rate $R(t)$ dollars per year from time $t = a$ to time $t = b$.

For most functions R more advanced techniques will be required in order for the integral in expression (11) to be evaluated. However, our last example shows how to calculate the present value of a constant revenue stream.

Example 6 Find the present value of a constant revenue stream of $R(t) = R$ dollars per year flowing between $t = a$ and $t = b$ years assuming a nominal interest rate of r percent.

Solution: Using expression (11), we find the value of the revenue stream to be

$$\int_a^b Re^{-rt} \, dt = -\frac{R}{r} e^{-rt}\Big]_a^b$$

$$= -\frac{R}{r}(e^{-rb} - e^{-ra})$$

$$= \frac{R}{r}(e^{-ra} - e^{-rb}). \qquad \square$$

Exercise Set 5.7

In Exercises 1–5 a demand function $p = D(x)$ is given. Find the consumers' surplus at consumption level x_0.

1. $D(x) = 20 - 0.5x$, $x_0 = 10$

2. $D(x) = 40 - 0.02x^2$, $x_0 = 10$

3. $D(x) = 100(40 - x^2)$, $x_0 = 5$

4. $D(x) = \dfrac{500}{x + 10}$, $x_0 = 15$

5. $D(x) = \dfrac{40}{\sqrt{2x + 1}}$, $x_0 = 12$

In Exercises 6–10 a supply function $p = S(x)$ is given. Find the suppliers' surplus at supply level x_0.

6. $S(x) = 5 + \sqrt{x}$, $x_0 = 16$

7. $S(x) = \dfrac{1}{100}x^2$, $x_0 = 6$

8. $S(x) = x\sqrt{9 + x^2}$, $x_0 = 4$

9. $S(x) = e^{0.2x} - 1$, $x_0 = 20$

10. $S(x) = xe^{0.01x^2}$, $x_0 = 10$

11. A manufacturer's marginal cost at production level x is $MC(x) = 200 + 0.4x$ dollars per item. Find the increase in total costs corresponding to an increase in production from $x = 40$ to $x = 50$ items per week.

12. A manufacturer of vacuum cleaners has a marginal cost per vacuum cleaner of $MC(x) = 60 + \dfrac{40}{x + 10}$ dollars per cleaner at production level x per week. Find the increase in total costs resulting from an increase in production from $x = 20$ to $x = 40$ vacuum cleaners per week.

13. The total cost of producing 30 items per week for the manufacturer in Exercise 11 is $C(30) = 7000$ dollars. Find its total cost in producing 40 items per week.

14. The manufacturer in Exercise 12 experiences fixed costs of production of $C(0) = 5000$ dollars per week. Find $C(20)$, its total cost of producing $x = 20$ items per week.

15. The manufacturer in Exercise 11 finds that it has a marginal revenue function $MR(x) = 400 - 0.5x$ dollars at sales level x items per week. Find the increase in revenue resulting from an increase in sales level from $x = 30$ to $x = 50$ items per week.

16. The manufacturer in Exercise 12 has a marginal revenue function $MR(x) = \dfrac{100}{1 + 0.02x}$ at production level x vacuum cleaners per week. Find its total weekly revenues at sales level $x = 30$ cleaners per week.

17. What is the capital formation (total nominal value) of a revenue flow of $A(t) = 3t + 2000$ dollars per year for 3 years?

18. What is the nominal value of a revenue stream flowing at a rate of $A(t) = 100\sqrt{t + 1}$ dollars per year for 8 years?

19. Find the total nominal value of a revenue stream flowing at a rate of $A(t) = 1000e^{-0.2t}$ between years $t_1 = 5$ and $t_2 = 10$.

20. What is the present value of a revenue stream flowing at the constant rate of $A(t) = 2000$ dollars per year for 5 years if the discount rate is assumed to be $r = 0.08$?

21. What is the present value of a cash flow of 5000 dollars per year for 10 years if the discount rate is assumed to be $r = 0.10$?

22. An annuity pays $5000\sqrt{t + 1}$ dollars per year for 8 years. What is the total nominal value of the annuity?

5.8 OTHER APPLICATIONS OF THE DEFINITE INTEGRAL

Many manufactured objects are produced by shaping a rotating piece of stock. For example, in using a lathe to produce a wooden table leg, a craftsman presses a chisel against a rapidly rotating block of wood (Figure 8.1). Similarly, a potter works a ball of clay into a vase by using a potter's wheel, which allows the clay to be rotated at a uniform speed about a central axis (Figure 8.2).

Figure 8.1 Table leg produced on a lathe by pressing a chisel against a rotating piece of wood stock.

Figure 8.2 Pottery produced by shaping clay rotating on a wheel.

Such objects are called *solids of revolution*. Although they are actually three-dimensional objects in space, their volumes can often be calculated as the definite integral of a function of just one variable.

The problem of calculating the volume of such solids is idealized mathematically as follows. Let f be a continuous nonnegative function for $a \leq x \leq b$. Let R denote the region bounded by the graph of f, the x-axis, and the lines $x = a$ and $x = b$ (Figure 8.3). As the region R rotates about the x-axis (Figure 8.4), it sweeps out a **solid of revolution, S.** Just as for the lathe and pottery wheel illustrations, the cross-sections for S taken perpendicular to the x-axis will be circles of radius $r = f(x)$. This is because the cross-section taken at location x is described by rotating about the x-axis the line segment from $(x, 0)$ to $(x, f(x))$.

To find a formula for the volume of S, we begin by developing an approximation to the solid S. We do this by dividing the interval $[a, b]$ into n equal subintervals of length $\Delta x = \dfrac{b - a}{n}$, and with endpoints $a = x_0, x_1, \ldots, x_n = b$. We arbitrarily select one "test number" t_j in each interval $[x_{j-1}, x_j]$ and approximate the value of the function f throughout the interval $[x_{j-1}, x_j]$ by the constant value $f(t_j)$.

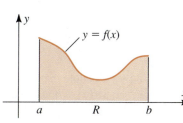

Figure 8.3 Region to be rotated.

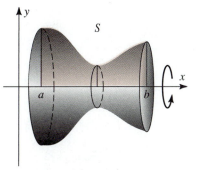

Figure 8.4 Solid obtained by rotating R about the x-axis.

In our approximation, corresponding to each interval $[x_{j-1}, x_j]$, we shall be rotating a rectangle of height $y = f(t_j)$ and width Δx about the x-axis. This will generate a disc, of radius $r_j = f(t_j)$ and thickness Δx. The volume of this disc is therefore

$$\Delta V_j = \pi r_j^2 \, \Delta x = \pi[f(t_j)]^2 \, \Delta x$$

(See Figure 8.5.)

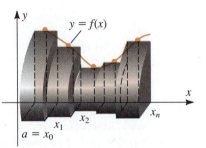

Figure 8.5 Rectangle of radius $r_j = f(t_j)$ generates disc of volume $V_j = \pi f^2(t_j) \, \Delta x$.

Figure 8.6 One quarter of the approximation to the volume of revolution S obtained by assuming f constant on subintervals.

Summing the volumes of these individual discs from 1 to n gives the volume of our approximating solid as

$$\sum_{j=1}^{n} \Delta V_j = \sum_{j=1}^{n} \pi f^2(t_j) \, \Delta x.$$

(See Figure 8.6.) Next, we argue that as $n \to \infty$ and the size Δx of each individual disc becomes small, the volume of our approximating solid should approach the volume of S. That is, we want to *define* the volume V of S by the equation

$$V = \lim_{n \to \infty} \sum_{j=1}^{n} \Delta V_j = \lim_{n \to \infty} \sum_{j=1}^{n} \pi[f(t_j)]^2 \, \Delta x.$$

Since the sum on the right is an approximating sum for the integral of the function $\pi[f(x)]^2$, we have arrived at the following definition:

> Let f be continuous for $a \le x \le b$ and let R denote the region bounded by the graph of f, the x-axis, and the lines $x = a$ and $x = b$. The volume of the solid obtained by rotating R about the x-axis is
>
> $$V = \int_a^b \pi[f(x)]^2 \, dx.$$

(1)

Example 1 Verify that equation (1) produces the formula $V = \dfrac{1}{3}\pi r^2 h$ for the volume of a right circular cone.

Strategy

Label variables.

Use a coordinate system to view cone as solid of revolution.

Find an equation for the line bounding the cross-section from above.

Apply (1).

Solution

Let S be the cone with radius r and height h. We impose a coordinate system on S as illustrated in Figure 8.7. We can then view the cone as the solid obtained by rotating the triangle with vertices $(0, 0)$, $(h, 0)$, and (h, r) about the x-axis. Since the equation for the hypotenuse is $f(x) = \dfrac{r}{h} x$, we apply equation (1) to find that

$$V = \int_0^h \pi \left[\frac{r}{h} x\right]^2 dx = \frac{\pi r^2}{3h^2} x^3 \bigg]_0^h = \frac{1}{3}\pi r^2 h$$

which is the desired formula. ☐

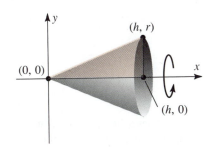

Figure 8.7 Cone as a volume of revolution.

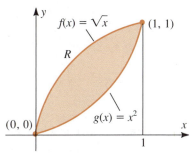

Figure 8.8 Region to be rotated about x-axis.

Example 2 Find the volume of the solid obtained by rotating the region bounded by the graphs of $f(x) = \sqrt{x}$ and $g(x) = x^2$ about the x-axis.

Strategy

Find points where the graphs cross.

Draw solid as difference of two solids of revolution.

Apply (1) to each.

Solution

The two graphs cross at $(0, 0)$ and $(1, 1)$ since the equation $\sqrt{x} = x^2$ implies $x = x^4$ or $x(1 - x^3) = 0$. Since $\sqrt{x} > x^2$ for $0 < x < 1$, the region is bounded above by the graph of $f(x) = \sqrt{x}$ and below by the graph of $g(x) = x^2$. (See Figure 8.8.) As Figure 8.9 indicates, we may view the resulting solid as the solid obtained by rotation of $f(x) = \sqrt{x}$ from which the solid obtained by rotation of $g(x) = x^2$ is removed. The calculation for volume, by equation (1), is therefore

$$V = \int_0^1 \pi(\sqrt{x})^2 dx - \int_0^1 \pi(x^2)^2 dx$$

$$= \frac{\pi}{2} x^2 \bigg]_0^1 - \frac{\pi}{5} x^5 \bigg]_0^1$$

$$= \frac{3\pi}{10}.$$

☐

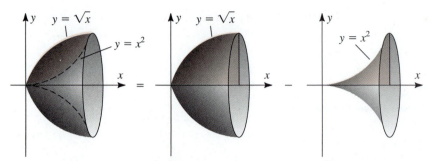

Figure 8.9 Volume obtained by expressing area between curves $\sqrt{x}$ and x^2 as the difference of volumes corresponding to upper curve $\sqrt{x}$ and lower curve x^2.

NOTE: It is important to notice that the answer we obtained, the difference between the two volumes, is given by

$$\int_a^b \pi([f(x)]^2 - [g(x)]^2) \, dx$$

but *not* by

$$\int_a^b \pi(f(x) - g(x))^2 \, dx.$$

The Average Value of a Function

Often a single number is sought that describes the "typical" or "average" value of a function f on an interval $[a, b]$. For example, consider the problem of determining average daily temperature, for the purpose of calculating energy consumption in an office building. By noting the two daily temperature functions g and h in Figures 8.10 and 8.11, we can see that such an average should not be computed simply from the difference between the maximum and minimum temperatures for the day in question. Indeed, intuition suggests that heating costs would be greater on the day whose temperature function h is given

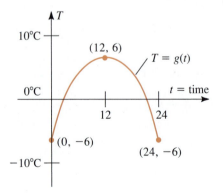

Figure 8.10 Temperature
$g(t) = 6 - \frac{1}{12}(t - 12)^2$.

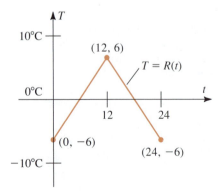

Figure 8.11 Temperature
$$h(t) = \begin{cases} t - 6, & 0 \le t \le 12 \\ 18 - t, & 12 \le t \le 24. \end{cases}$$

in Figure 8.11 since the temperatures $h(t)$ are lower than the temperatures $g(t)$, except at times $t = 0$, $t = 12$, and $t = 24$ hours. Notice, however, that on both days the maximum temperatures (6°C) and minimum temperatures (−6°C) are the same.

These remarks suggest that a measurement of average temperature for a temperature function f, for $a \leq t \leq b$, should reflect not only the difference between maximum and minimum temperatures but also some measurement of how long the temperature function remained at various temperature levels in between. One way to do this is to divide the interval $[a, b]$ into n subintervals of equal length $\Delta t = \dfrac{b - a}{n}$ at points $a = t_0 < t_1 < t_2 < \cdots < t_n = b$. If we select one "test" time s_j in each of the subintervals $[t_{j-1}, t_j]$, the n numbers $f(s_1), f(s_2), \ldots, f(s_n)$ represent temperature readings taken from among equally spaced time intervals. By averaging these values, we arrive at a number

$$\overline{f_n} = \frac{f(s_1) + f(s_2) + \cdots + f(s_n)}{n}$$

reflecting an approximate value for f in each subinterval $[t_{j-1}, t_j]$ of the time interval $[a, b]$. We therefore argue that by letting $n \to \infty$, we should obtain an average f, reflecting *each* individual function value $f(t)$, t in $[a, b]$. That is, we take

$$\overline{f} = \lim_{n \to \infty} \overline{f_n} = \lim_{n \to \infty} \frac{1}{n} \sum_{j=1}^{n} f(s_j). \tag{2}$$

Since $\Delta t = \dfrac{b - a}{n}$, we can write the sum in equation (2) as an approximating sum as follows:

$$\frac{1}{n} \sum_{j=1}^{n} f(s_j) = \sum_{j=1}^{n} f(s_j) \cdot \frac{1}{n}$$

$$= \left(\frac{b - a}{b - a} \right) \sum_{j=1}^{n} f(s_j) \cdot \frac{1}{n}$$

$$= \frac{1}{b - a} \sum_{j=1}^{n} f(s_j) \left(\frac{b - a}{n} \right)$$

$$= \frac{1}{b - a} \sum_{j=1}^{n} f(s_j) \, \Delta t.$$

If f is continuous for t in $[a, b]$, the limit as $n \to \infty$ of this approximating sum is a definite integral:

$$\overline{f} = \lim_{n \to \infty} \frac{1}{b - a} \sum_{j=1}^{n} f(s_j) \, \Delta t = \frac{1}{b - a} \int_{a}^{b} f(t) \, dt.$$

Thus if $f(x)$ is a continuous function for x in $[a, b]$, *we define* the *average value A of* $f(x)$ *on* $[a, b]$ *by*

$$A = \bar{f} = \frac{1}{b - a} \int_a^b f(x) \, dx. \tag{3}$$

When $f(x)$ is nonnegative, we may interpret $\bar{f}$ as the height of a rectangle whose base has length $b - a$ and whose area is $\int_a^b f(x) \, dx$. (See Figure 8.12.)

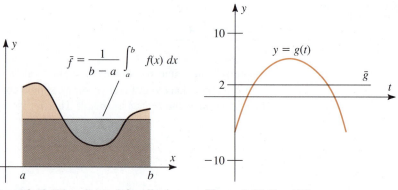

Figure 8.12 When $f(x) \geq 0$ for all x in $[a, b]$, $\bar{f}$ is the height of the rectangle with base $(b - a)$ and area $\int_a^b f(x) \, dx$.

Figure 8.13 $\bar{g} = 2°\text{C}$.

Example 3 Find the average temperature for the temperature functions g and h in Figures 8.10 and 8.11.

Solution: Applying equation (3) to $g(t) = 6 - \dfrac{1}{12}(t - 12)^2$, we obtain

$$\bar{g} = \frac{1}{24 - 0} \int_0^{24} \left[6 - \frac{1}{12}(t - 12)^2 \right] dt$$

$$= \frac{1}{24} \left[6t - \frac{1}{36}(t - 12)^3 \right]_0^{24}$$

$$= \frac{1}{24} \{144 - 2(48)\}$$

$$= 2°\text{C}.$$

(Figure 8.13.)

For the temperature function $h(t) = \begin{cases} t - 6, & 0 \le t \le 12 \\ 18 - t, & 12 \le t \le 24 \end{cases}$, we obtain

$$\bar{h} = \frac{1}{24 - 0} \left\{ \int_0^{12} (t - 6)\, dt + \int_{12}^{24} (18 - t)\, dt \right\}$$

$$= \frac{1}{24} \left\{ \frac{1}{2} t^2 - 6t \Big]_0^{12} + 18t - \frac{1}{2} t^2 \Big]_{12}^{24} \right\}$$

$$= \frac{1}{24} \{(72 - 72) + (144 - 144)\}$$

$$= 0°\text{C}. \qquad \square$$

Example 4 Suppose that the amount of heating oil required to heat a house for one day during the winter months is the product of the difference between 20°C and the average daily temperature multiplied by 0.6 gallons. Find the levels of oil consumption associated with each of the temperature functions in Figures 8.10 and 8.11.

Solution: For temperature function g the oil consumption is

$$C_g = (20°\text{C} - \bar{g}) \cdot (0.6 \text{ gal per } °\text{C})$$
$$= (20 - 2)(0.6) = 10.8 \text{ gal.}$$

For the temperature function h the oil consumption is

$$C_h = (20°\text{C} - \bar{h}) \cdot (0.6 \text{ gal per } °\text{C})$$
$$= (20 - 0)(0.6) = 12 \text{ gal.}$$

Thus our remark about the cost of heating was correct: The function h does lead to a greater heating cost than does g. $\qquad \square$

Exercise Set 5.8

In Exercises 1–5 find the volume of the solid obtained by revolving the region bounded by the graph of the given function and the x-axis, for $a \le x \le b$, about the x-axis.

1. $f(x) = 2x + 1, \quad 1 \le x \le 4$

2. $f(x) = \sqrt{4x - 1}, \quad 1 \le x \le 5$

3. $f(x) = |x - 1|, \quad 0 \le x \le 3$

4. $f(x) = \sqrt{4 - x^2}, \quad 1 \le x \le 2$

5. $f(x) = \dfrac{\sqrt{x + 1}}{x^{3/2}}, \quad 1 \le x \le 2$

In Exercises 6–10 find the volume of the solid obtained by revolving about the x-axis the region bounded by the given curves.

6. $f(x) = x^2, \quad g(x) = x^3$

7. $f(x) = \dfrac{1}{4} x^2, \quad g(x) = x$

8. $f(x) = \dfrac{1}{x}, \quad g(x) = \sqrt{x}, \quad 1 \le x \le 4$

9. $x + y = 4, \quad y = 0, \quad 0 \le x \le 4$

10. $x = y^2, \quad x = 4$

In each of Exercises 11–15 find the average value of the given function on the given interval.

11. $f(x) = 2x - 1, \quad x \text{ in } [-2, 2]$

12. $f(x) = x^2 - 7, \quad x \text{ in } [-1, 3]$

13. $f(x) = \sqrt{x + 2}, \quad x$ in $[-2, 2]$

14. $f(x) = \dfrac{1 - x}{1 + x}, \quad x$ in $[0, 1]$

15. $f(x) = \dfrac{x^3 + 1}{x + 1}, \quad x$ in $[0, 2]$

16. Find the average value of the function $f(x) = 3x + 1$ on the interval $[1, 3]$ by geometry.

17. Find the average value of the function $f(x) = \sqrt{9 - x^2}$ on the interval $[-3, 3]$ by geometry. (*Hint:* The graph of $f(x) = \sqrt{9 - x^2}$ is a semicircle.)

18. Find the average value of the total cost function $C(x) = 10 + 40x + 0.3x^2$ for $0 \le x \le 20$.

19. Find the average value of the total revenue function $R(x) = 120x - 4x^2$ for $0 \le x \le 10$.

20. Find the average value of the traffic flow function

$$q(p) = \dfrac{p}{1 + p^2} \text{ for } 0 \le p \le 4.$$

21. For the total cost function $C(x)$ we have defined average cost to be the quotient $c(x) = \dfrac{C(x)}{x}$.
Explain why average cost is the average value of the marginal cost function on the interval $[0, x]$ when $C(0) = 0$.

22. Find the average value of the electric power consumption function

$$E(t) = 4 - 2\left(\dfrac{t - 13}{7}\right)^4$$

for $6 \le t \le 20$. (See Exercise 24, Section 3.4.)

23. Let r be a positive constant. Find the volume of the solid obtained by revolving the region bounded by the graph of $f(x) = \sqrt{r^2 - x^2}$ and the x-axis about the x-axis. Do you recognize the formula for the volume that you have obtained? (See endpapers.)

SUMMARY OUTLINE OF CHAPTER 5

■ The function F is an **antiderivative** for the function f if F is differentiable and $F'(x) = f(x)$. (Page 250)
The most general form of the antiderivative of f is then

$$\int f(x) \, dx = F(x) + C.$$

■ **Power Rule:** $\displaystyle\int x^n \, dx = \dfrac{1}{n + 1} x^{n+1} + C, \qquad n \ne -1.$ (Page 252)

■ **Other Integration (Antidifferentiation) Rules** (Page 253)

$$\int cf(x) \, dx = c \int f(x) \, dx$$

$$\int [f(x) + g(x)] \, dx = \int f(x) \, dx + \int g(x) \, dx$$

$$\int \dfrac{1}{x} \, dx = \ln |x| + C, \qquad x \ne 0$$

$$\int e^x \, dx = e^x + C$$

■ *Economic Applications of the Integral* (Page 258)

Total cost $= \int MC(x) \, dx = C(x) + A$ (A constant)

Total revenue $= \int MC(x) \, dx = R(x) + B$ (B constant)

■ The **differential** for the function u is (Page 264)

$du = u'(x) \, dx.$

■ Integration by substitution: If $F' = f$, (Page 265)

$$\int f(u(x))u'(x) \, dx = \int f(u) \, du = F(u) + C$$

$$= F(u(x)) + C.$$

■ The **area** of the region R bounded by the graph of the continuous positive function f and the (Page 275)
x-axis between $x = a$ and $x = b$ is the limit of the approximating sum

$$\text{Area of } R = \lim_{n \to \infty} \sum_{j=1}^{n} f(t_j) \, \Delta x.$$

■ The **definite integral** of the continuous function f from $x = a$ to $x = b$ is the limit (Page 275)

$$\int_a^b f(x) \, dx = \lim_{n \to \infty} \sum_{j=1}^{n} f(t_j) \, \Delta x.$$

■ When $f(x) \geq 0$ for all x in $[a, b]$, the area of the region R described above is given by the (Page 276)
definite integral

$$\text{Area of } R = \int_a^b f(x) \, dx.$$

■ *Theorem (Fundamental Theorem of Calculus):* If f is continuous on $[a, b]$, then (Page 278)

$$\int_a^b f(x) \, dx = F(b) - F(a)$$

where $F' = f$ on $[a, b]$.

■ The **area** of the region R bounded above by the graph of the continuous function f and (Page 285)
below by the graph of the continuous function g between $x = a$ and $x = b$, is

$$\text{Area of } R = \int_a^b [f(x) - g(x)] \, dx.$$

■ The **consumers' surplus** associated with the demand function $p = D(x)$ at consumption level (Page 298)
x_0 is

$$\text{Consumers' surplus} = \int_0^{x_0} D(x) \, dx - x_0 p_0, \qquad p_0 = D(x_0).$$

■ The **suppliers' surplus** associated with the supply function $p = S(x)$ at supply level x_0 is (Page 299)

$$\text{Suppliers' surplus} = x_0 p_0 - \int_0^{x_0} S(x) \, dx, \qquad p_0 = S(x_0).$$

■ The **nominal value** of a revenue stream flowing at the rate of $R(t)$ dollars per year from time $t = a$ to $t = b$ is (Page 300)

$$\int_a^b R(t)\, dt.$$

■ The **present value** of the above revenue stream subject to a discount factor of r percent per annum is (Page 301)

$$\int_a^b R(t)e^{-rt}\, dt.$$

■ The **volume** of the solid obtained by revolving about the x-axis the region bounded by the graph of the continuous function f and the x-axis between $x = a$ and $x = b$ is (Page 304)

$$V = \int_a^b \pi[f(x)]^2\, dx.$$

■ The **average value** of the continuous function f on the interval $[a, b]$ is (Page 308)

$$A = \frac{1}{b - a}\int_a^b f(x)\, dx.$$

REVIEW EXERCISES—CHAPTER 5

In Exercises 1–20 find the antiderivative.

1. $\displaystyle\int (6x^2 - 2x + 1)\, dx$

2. $\displaystyle\int (x^2 - 6x)^2\, dx$

3. $\displaystyle\int x\sqrt{9x^4}\, dx$

4. $\displaystyle\int (3\sqrt{x} + 3/\sqrt{x})\, dx$

5. $\displaystyle\int (t + \sqrt[3]{t})^2\, dt$

6. $\displaystyle\int t\sqrt{9 - t^2}\, dt$

7. $\displaystyle\int \frac{x^3 - 7x^2 + 6}{x}\, dx$

8. $\displaystyle\int \frac{x^3 + x^2 - x + 2}{x + 2}\, dx$

9. $\displaystyle\int (2x - 1)(2x + 1)\, dx$

10. $\displaystyle\int \frac{x}{4x^4 + 4x^2 + 1}\, dx$

11. $\displaystyle\int \frac{x}{1 - x^2}\, dx$

12. $\displaystyle\int \frac{x + 1}{4x + 2x^2}\, dx$

13. $\displaystyle\int \sqrt{e^x}\, dx$

14. $\displaystyle\int (e^x + 1)^2\, dx$

15. $\displaystyle\int \frac{1}{x\sqrt{\ln x}}\, dx$

16. $\displaystyle\int \frac{e^x - e^{-x}}{e^x + e^{-x}}\, dx$

17. $\displaystyle\int \frac{x^3 - 1}{x + 1}\, dx$

18. $\displaystyle\int \frac{e^{\sqrt{x}}}{\sqrt{x}}\, dx$

19. $\displaystyle\int e^{2x}(1 - e^{2x})^3\, dx$

20. $\displaystyle\int \frac{2x + 3x^2}{x^3 + x^2 - 7}\, dx$

21. A particle moves along a line with a velocity $v(t) = 2t - (t + 1)^{-2}$.
  a. Find $s(t)$, its position at time t, if $s(0) = 0$. (Recall, $v(t) = s'(t)$.)
  b. Find $a(t)$, its acceleration at time t.

22. The population of a town is growing according to the population function $P(t) = \dfrac{200{,}000}{20 + 40e^{-0.2t}}$. Find the horizontal asymptote.

23. A company's marginal cost of production at production level x units per day is $MC(x) = 120 + 6x$. Its fixed costs are \$500 per day.
  a. Find total daily costs, $C(x)$.
  b. Find $C(20)$.

24. Refer to Exercise 23. The company's marginal revenue at sales level x units per day is $MR(x) = 250$ dollars.
  a. Find total daily revenue, $R(x)$.
  b. Find $R(20)$.

25. Find the daily profit function P for the company in Exercises 23 and 24. Is it profitable at production and sales level $x = 20$ units per day?

26. A population of rabbits is growing at a rate $\dfrac{dP}{dt} = 8e^{0.5t}$ rabbits per month t months after time $t = t_0$. If $P(t_0) = 16$, find the population function P.

27. A company purchases a computer for \$100,000 and estimates that the value $V(t)$ of the computer will decrease over time at the rate $\dfrac{dV}{dt} = -\dfrac{80,000}{(t + 1)^2}$ dollars per year. Find
   a. $V(t)$, the value of the computer after t years.
   b. $V(3)$.
   c. $\displaystyle\lim_{t\to\infty} V(t)$.

28. An investment is growing at a rate of $\dfrac{500 \cdot e^{\sqrt{t}}}{\sqrt{t}}$ dollars per year. Find the value of the investment after 4 years if its initial value was \$1000.

29. The marginal cost of producing a certain item is $MC(x) = 70 + 2x$ at production level x items per month. If the total cost of producing ten items per month is $C(10) = 1000$ dollars
   a. Find the total monthly cost function C.
   b. Find the fixed monthly costs, $C(0)$.

30. A manufacturer experiences a marginal cost of $MC(x) = 40 + 2x$ and a marginal revenue of $MR(x) = 120$ in the production and sale of x radios per week. If the manufacturer's profit from the production and sale of $x = 20$ items per week is $P(20) = 1050$, find $C(0)$, the manufacturer's fixed weekly costs.

In Exercises 31–50 use the Fundamental Theorem of Calculus to find the value of the definite integral.

31. $\displaystyle\int_0^3 (4x - 7)\, dx$

32. $\displaystyle\int_0^2 \frac{1}{x + 3}\, dx$

33. $\displaystyle\int_{-1}^1 (x^2 - 6)^2\, dx$

34. $\displaystyle\int_{-3}^{-1} 6\, dx$

35. $\displaystyle\int_1^4 \sqrt{x + 3}\, dx$

36. $\displaystyle\int_0^1 3e^{2x}\, dx$

37. $\displaystyle\int_0^3 \frac{x}{x + 3}\, dx$

38. $\displaystyle\int_{-1}^1 (x^2 - 3x + 9)\, dx$

39. $\displaystyle\int_0^1 \frac{x + 2}{x - 2}\, dx$

40. $\displaystyle\int_0^2 x^2 e^{2x^3}\, dx$

41. $\displaystyle\int_{-1}^3 (x - 1)(x + 5)\, dx$

42. $\displaystyle\int_4^9 (\sqrt{x} - 1)(\sqrt{x} + 1)\, dx$

43. $\displaystyle\int_4^8 |x - 4|\, dx$

44. $\displaystyle\int_0^4 x(\sqrt{x} + x^{2/3})\, dx$

45. $\displaystyle\int_{-8}^{-1} (x^{1/3} - x^{5/3})\, dx$

46. $\displaystyle\int_0^{\sqrt{8}} \frac{x}{\sqrt{x^2 + 1}}\, dx$

47. $\displaystyle\int_1^8 \frac{\sqrt[3]{x}}{5 + x^{4/3}}\, dx$

48. $\displaystyle\int_0^1 (\sqrt{x} + 5)(x^{1/3} + x)\, dx$

49. $\displaystyle\int_1^8 \frac{x^{2/3} + 3x^{5/2}}{x}\, dx$

50. $\displaystyle\int_0^{\ln 2} (e^x + e^{-x})^2\, dx$

51. Use geometry to find $\displaystyle\int_{-4}^4 \sqrt{16 - x^2}\, dx$.

52. Use geometry to find $\displaystyle\int_{-2}^0 \sqrt{4 - x^2}\, dx$.

53. Find the area of the region bounded by the graph of $f(x) = 9 - \sqrt{x}$ and the x-axis between $x = 0$ and $x = 9$.

54. Find the area of the region bounded by the graph of $f(x) = 10 - 2x$ and the coordinate axes.

55. Find the area of the region bounded by the graphs of $f(x) = 3 + x^{2/3}$ and $g(x) = 3x^2 + 1$.

56. Find the area of the region bounded by the graphs of $f(x) = e^{-x}$ and $g(x) = e^x$ for $0 \le x \le 1$.

57. Find the area of the region bounded by the graphs of $f(x) = \sqrt[3]{x}, \quad g(x) = -x \quad \text{and} \quad h(x) = \dfrac{2}{3}x - \dfrac{10}{3}$.

58. A manufacturer of microwave ovens has a total cost of $C(40) = 6000$ dollars in producing 40 units per week. If the manufacturer's marginal cost function is $MC(x) = 120$ dollars per unit.
   a. Find the total cost of producing $x = 50$ units per week.
   b. Find the manufacturer's total cost function $C(x)$.

59. A company that manufactures lawnmowers has a marginal cost of $MC(x) = 110 + \dfrac{50}{x + 10}$ dollars per lawnmower at production level x lawnmowers per week. Find the increase in total costs resulting from an increase in production from $x = 40$ to $x = 50$ lawn-mowers per week.

60. A company that manufactures refrigerators finds that it has a marginal revenue function of $MR(x) = 600 - 2\sqrt{x}$ dollars at sales level x refrigerators per week. Find the increase in revenues resulting from an increase in sales from $x = 36$ to $x = 49$ refrigerators per week.

61. Find the consumers' surplus for the demand function $D(x) = 40(50 - x^2)$ at consumption level $x_0 = 6$.

62. Find the consumers' surplus for the demand function
$$D(x) = \frac{40\sqrt{x}}{1 + x^{3/2}}$$ at consumption level $x_0 = 16$.

63. Find the suppliers' surplus for the supply function
$$S(x) = \frac{1}{\sqrt{x} + 1} e^{\sqrt{x+1}}$$ at supply level $x_0 = 15$.

64. What is the total nominal value of a revenue flow of $A(t) = 6\sqrt{t} + 3000$ dollars per year for 4 years?

65. Find the nominal value of a revenue stream flowing at a rate of $A(t) = 100e^{0.2t}$ dollars per year for 10 years.

66. Find the present value of a revenue stream flowing at the constant rate of $A(t) = 5000$ dollars per year for 10 years if the discount rate is assumed to be $r = 0.08$.

67. An annuity pays $10{,}000\, t\sqrt{1 + t^2}$ dollars per year for 5 years. What is the total nominal value of the annuity?

68. Find the volume of the solid obtained by revolving the region bounded by the graphs of $f(x) = x^{2/3}$ and $g(x) = x$ about the x-axis.

69. Find the average value of the function $f(x) = \dfrac{x}{x^2 + 7}$ on the interval $[0, 3]$.

70. Find the average value of the function $g(x) = (1 + \sqrt{x})(1 - \sqrt{x})$ on the interval $[1, 4]$.

6

Chapter **6** ***Trigonometric Functions***

Certain models, used to study "periodic" fluctuations of quantities in economics and science, involve functions of angles. In order to develop such functions, we need to introduce a new way of measuring angles.

6.1 RADIAN MEASURE

Until now you have probably used degrees as the only unit of measurement for angles. We say that a right angle measures 90 degrees (90°), that an about-face is a 180° turn, and that a complete circle (rather than just an arc) has 360°.

However, the degree is not the unit of choice for angles in calculus. In defining the **trigonometric functions,** which involve angle measurement, we need to exploit the basic relationship between the circumference and radius of a circle, $C = 2\pi r$, which leads to the definition of the **radian** as a unit for measuring angles.

Figure 1.1 shows that in studying an angle θ associated with a right triangle, we can

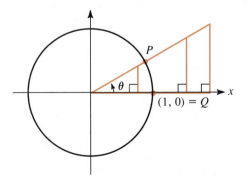

Figure 1.1 Various right triangles with angle θ, each of which is identified with the point P.

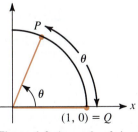

Figure 1.2 An angle of size θ radians determines an arc of length θ between P and $Q = (1, 0)$ on the unit circle.

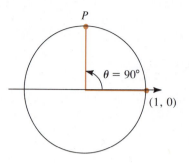

Figure 1.3 $90° = \dfrac{\pi}{2}$ radians.

always position the triangle so that the vertex corresponding to θ lies at the origin, one of the legs adjacent to the right angle lies along the x-axis, and the other such leg is parallel to the y-axis. We now draw a unit circle (a circle of radius $r = 1$) centered at the origin. The hypotenuse of the triangle then determines a line that intersects the unit circle at P.

The point of Figure 1.1 is that the angle θ is identified by the point P and the fixed point $Q = (1, 0)$ on the unit circle; the size of the triangle doesn't matter.

Figure 1.2 shows how we use the unit circle to define the size of the angle θ. The size of the angle θ in *radians* is equal to the length of the arc of the unit circle that is traversed by a point moving counterclockwise along the circle from point $Q = (1, 0)$ to point P. When the length of the arc from P to Q is one unit (equal to the radius of the circle), then $\theta = 1$ radian. Since the circumference of the unit circle is $2\pi(1) = 2\pi$ units, we say that **2π radians equal 360 degrees.** Thus

(a) The radian measure of an angle of size $90°$ is

$$\frac{1}{4}(2\pi) = \frac{\pi}{2} \text{ radians}$$

since $90°$ is one-fourth of $360°$ (Figure 1.3).

(b) The radian measure of an angle of size $180°$ is

$$\frac{1}{2}(2\pi) = \pi \text{ radians}$$

since $180°$ is one-half of $360°$ (Figure 1.4).

(c) The radian measure of an angle size $30°$ is

$$\frac{1}{12}(2\pi) = \frac{\pi}{6} \text{ radians}$$

since $30°$ is $\dfrac{30}{360} = \dfrac{1}{12}$ of $360°$ (Figure 1.5).

Each of statements (a)–(c) follows from the more general relationship between an angle whose measurement is θ_d in degrees and θ_r in radians:

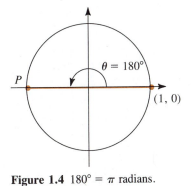

Figure 1.4 $180° = \pi$ radians.

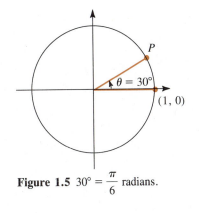

Figure 1.5 $30° = \dfrac{\pi}{6}$ radians.

$$\frac{\theta_d}{360} = \frac{\theta_r}{2\pi}. \tag{1}$$

Equation (1) gives the formulas for converting from degree to radian measure, and vice versa:

$$\theta_r = \frac{2\pi}{360}\theta_d = \frac{\pi}{180}\theta_d \tag{2}$$

$$\theta_d = \frac{180}{\pi}\theta_r. \tag{3}$$

Henceforth, unless an angle measurement is explicitly stated to be in units of degrees, we shall assume that it is in radians.

Example 1

(a) An angle of size $60°$ has radian measure

$$\theta_r = \frac{\pi}{180}(60) = \frac{\pi}{3}. \qquad \text{(Figure 1.6)}$$

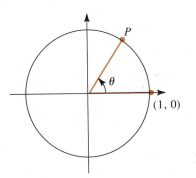

Figure 1.6 $60° = \dfrac{\pi}{3}$ radians.

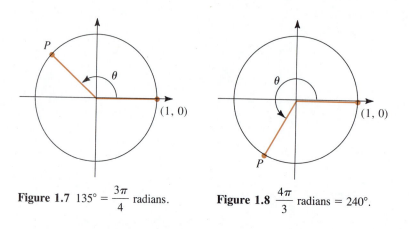

Figure 1.7 $135° = \dfrac{3\pi}{4}$ radians.   **Figure 1.8** $\dfrac{4\pi}{3}$ radians $= 240°$.

(b) An angle of size $135°$ has radian measure

$$\theta_r = \frac{\pi}{180}(135) = \frac{3\pi}{4}. \qquad \text{(Figure 1.7)}$$

(c) An angle of size $\dfrac{4\pi}{3}$ has degree measure

$$\theta_d = \frac{180}{\pi}\left(\frac{4\pi}{3}\right) = 240°. \qquad \text{(Figure 1.8)} \qquad \square$$

Table 1.1 shows degree and radian measures for various angles θ with $0 \le \theta \le 2\pi$.

Table 1.1

Angle in degrees	0	30	45	60	90	120	135	150	180
Angle in radians	0	$\dfrac{\pi}{6}$	$\dfrac{\pi}{4}$	$\dfrac{\pi}{3}$	$\dfrac{\pi}{2}$	$\dfrac{2\pi}{3}$	$\dfrac{3\pi}{4}$	$\dfrac{5\pi}{6}$	π

Angle in degrees	210	225	240	270	300	315	330	360
Angle in radians	$\dfrac{7\pi}{6}$	$\dfrac{5\pi}{4}$	$\dfrac{4\pi}{3}$	$\dfrac{3\pi}{2}$	$\dfrac{5\pi}{3}$	$\dfrac{7\pi}{4}$	$\dfrac{11\pi}{6}$	2π

Extending Radian Measure

In many applications the angle measure θ is used to measure how far a lever or other device has turned, and measurements greater than $360°$ or 2π radians are required. For example, the instruction, "turn the set screw three complete turns outward" tells an auto mechanic to turn a carburetor adjustment screw through $3 \times 360° = 1080°$, or $3 \times 2\pi = 6\pi$ radians.

For $t > 2\pi$ the meaning of the radian measure t is the distance traveled counterclockwise around the unit circle by the point P, from the fixed point $Q = (1, 0)$, as the radius

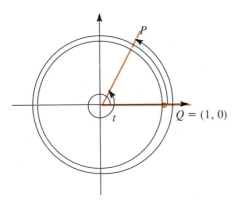

Figure 1.9 t radians corresponds to $\dfrac{t}{2\pi}$ revolutions.

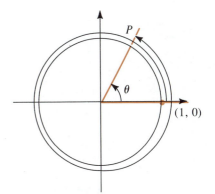

Figure 1.10 The principal angle θ associated with t radians satisfies $t = \theta + 2n\pi$ with $0 \le \theta < 2\pi$.

OP turns through $\dfrac{t}{2\pi}$ revolutions. (See Figure 1.9.) Thus

$$t = 4\pi \quad \text{corresponds to} \quad \frac{4\pi}{2\pi} = 2 \text{ revolutions counterclockwise}$$

and

$$t = \frac{5\pi}{2} \quad \text{corresponds to} \quad \frac{5\pi/2}{2\pi} = \frac{5}{4} \text{ revolutions counterclockwise.}$$

With this interpretation we can identify any positive number t with an angle θ that satisfies the equation

$$t = \theta + 2n\pi \quad \text{for } 0 \le \theta < 2\pi \text{ and } n = 0, 1, 2, \ldots \ . \tag{4}$$

In this case the number θ is called the **principal angle** associated with the number t. (See Figure 1.10.)

Example 2

(a) $t = 7\pi$ radians corresponds to $\dfrac{7\pi}{2\pi} = \dfrac{7}{2}$ revolutions counterclockwise. The principal angle θ associated with $t = 7\pi$ radians is $\theta = \pi$, since equation (4) takes the form

$$7\pi = \pi + 3(2\pi) \qquad (n = 3).$$

(b) $t = \dfrac{9\pi}{2}$ radians corresponds to the principal angle $\theta = \dfrac{\pi}{2}$ since we can write

$$\frac{9\pi}{2} = \frac{\pi}{2} + 2(2\pi) \qquad (n = 2).$$

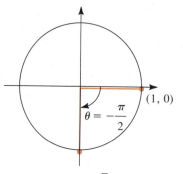

Figure 1.11 $\theta = -\dfrac{\pi}{2}$.

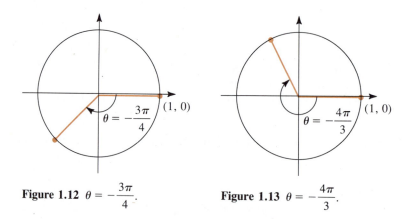

Figure 1.12 $\theta = -\dfrac{3\pi}{4}$.

Figure 1.13 $\theta = -\dfrac{4\pi}{3}$.

Finally, we can extend radian measure to negative numbers simply by letting the point P move in the *clockwise* direction along the unit circle, beginning at $Q = (1, 0)$. Figures 1.11–1.13 show several locations for P corresponding to negative radian measures.

□

Negative values of radian measure are important in simple applications having to do with the motion of revolving parts, where clockwise motion needs to be distinguished from counterclockwise motion. However, the most important application of (signed) radian measure is that it allows us to associate each real number with a radian measure t, as illustrated by the number line in Figure 1.14.

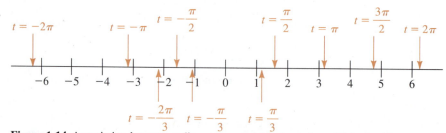

Figure 1.14 Association between radian measure and real numbers determined by constant $\pi = 3.14159 \ldots$.

In the remaining sections of this chapter we shall define and work with the *trigonometric* functions. While you may have worked previously with trigonometric functions of *angles,* this association between radian measure and real numbers will allow us to define the trigonometric functions for numbers other than just those in the interval $[0, 2\pi]$.

Example 3 Using equation (1), we find that

(a) A revolution of 480° counterclockwise corresponds to a radian measure of $t = \dfrac{8\pi}{3}$, since

$$\frac{480}{360} = \frac{t}{2\pi} \quad \text{gives} \quad t = 2\pi\left(\frac{480}{360}\right) = \frac{8\pi}{3}.$$

(The sign of t is positive since the motion is counterclockwise.)

(b) A revolution of 270° clockwise corresponds to a radian measure of $t = -\dfrac{3\pi}{2}$ since

$$\frac{-270}{360} = \frac{t}{2\pi} \quad \text{gives} \quad t = 2\pi\left(\frac{-270}{360}\right) = -\frac{3\pi}{2}.$$

(The sign of t is negative since the motion is clockwise.)

(c) A revolution of 810° clockwise corresponds to a radian measure of $t = -\dfrac{9\pi}{2}$ since

$$-\frac{810}{360} = \frac{t}{2\pi} \quad \text{gives} \quad t = 2\pi\left(\frac{-810}{360}\right) = -\frac{9\pi}{2}.$$

(The sign of t is negative since the motion is clockwise.) □

Exercise Set 6.1

1. Find a radian measure equivalent to the degree measure.
   a. 90°   b. 45°   c. −135°
   d. 30°   e. 60°   f. −150°
   g. 180°   h. 210°

2. Find a radian measure equivalent to the degree measure.
   a. 240°   b. 225°   c. −270°
   d. 300°   e. 330°   f. −315°
   g. 115°   h. 235°

3. Find a degree measure equivalent to the radian measure.
   a. $\dfrac{\pi}{4}$ b. $\dfrac{3\pi}{2}$ c. $-\dfrac{\pi}{12}$
   d. $\dfrac{7\pi}{6}$ e. $\dfrac{7\pi}{8}$ f. $-\dfrac{5\pi}{6}$
   g. $\dfrac{11\pi}{6}$ h. $-\dfrac{3\pi}{4}$

4. Find a radian measure equivalent to a counterclockwise rotation through the given number of degrees.
   a. 720°   b. 480°   c. 750°
   d. 1440°   e. 450°   f. 390°
   g. 540°   h. 690°

5. Find a radian measure equivalent to a clockwise rotation through the given number of degrees.
   a. 45°   b. 270°   c. 30°
   d. 150°   e. 390°   f. 135°
   g. 330°   h. 540°

6. Turning a screw three and one third complete turns counterclockwise corresponds to what radian measure?

7. Determine the radian measure corresponding to the advance of the hour hand on a clock from midnight to each of the following times.

    a. 6 a.m.   **b.** noon

    c. 3 p.m.   **d.** 8 p.m.

In Exercises 8–13 determine the radian measure for the angle described by the figure.

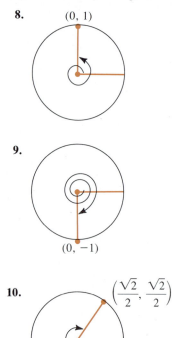

8. $(0, 1)$

9. $(0, -1)$

10. $\left(\dfrac{\sqrt{2}}{2}, \dfrac{\sqrt{2}}{2}\right)$

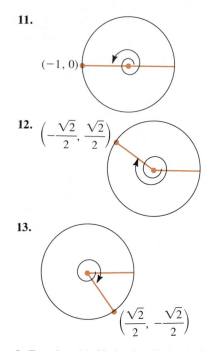

11. $(-1, 0)$

12. $\left(-\dfrac{\sqrt{2}}{2}, \dfrac{\sqrt{2}}{2}\right)$

13. $\left(\dfrac{\sqrt{2}}{2}, -\dfrac{\sqrt{2}}{2}\right)$

In Exercises 14–20 sketch a figure similar to those in Exercises 8–13, describing the given angle.

14. $-\dfrac{5\pi}{4}$ **15.** 11π **16.** $-\dfrac{16\pi}{3}$

17. $\dfrac{13\pi}{4}$ **18.** $\dfrac{9\pi}{4}$ **19.** $-\dfrac{13\pi}{2}$

20. $\dfrac{10\pi}{3}$ **21.** $-\dfrac{7\pi}{4}$ **22.** $\dfrac{14\pi}{3}$

23. -3π **24.** $-\dfrac{17\pi}{6}$

6.2 THE SINE AND COSINE FUNCTIONS

    In attempting to model economic or scientific phenomena, we sometimes encounter the need for functions whose values repeat themselves at regular intervals. For example, Figure 2.1 shows a typical graph of the average daily temperature for a New England city over a period of 3 years, which consists of three arcs that are nearly congruent. Similarly, Figure 2.2 shows a typical "weekly load curve," representing the electric power supplied by an urban power plant. Note that the first five arcs, corresponding to the weekdays Monday through Friday, are nearly identical.

    Figures 2.1 and 2.2 suggest the need to develop **periodic** functions, that is, functions that repeat their values at regularly spaced intervals. None of the functions we have studied up to this point has this property, except the constant functions. However, by

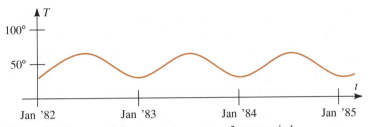

Figure 2.1 Average daily temperature over a 3-year period.

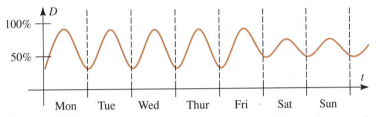

Figure 2.2 Demand for electric power as a percentage of power plant capacity over a 1-week period.

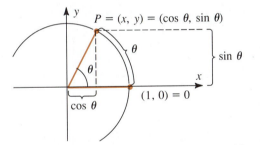

Figure 2.3 If P is a point on the unit circle with coordinated $P = (x, y)$, then

$$x = \cos \theta$$
$$y = \sin \theta.$$

using the notion of radian measure, we may define two such functions, the **sine** and **cosine** functions.

Figure 2.3 shows how these functions are defined for angles θ with $0 \le \theta < 2\pi$. If P is the point on the unit circle lying θ units along the circle from the point $Q = (1, 0)$ in the counterclockwise direction, we define the numbers $\sin \theta$ and $\cos \theta$ (read ''sine of θ'' and ''cosine of θ'') by

$$\sin \theta = y\text{-coordinate of } P \qquad\qquad (1)$$

and

$$\cos \theta = x\text{-coordinate of } P. \qquad\qquad (2)$$

(See Figure 2.3.)

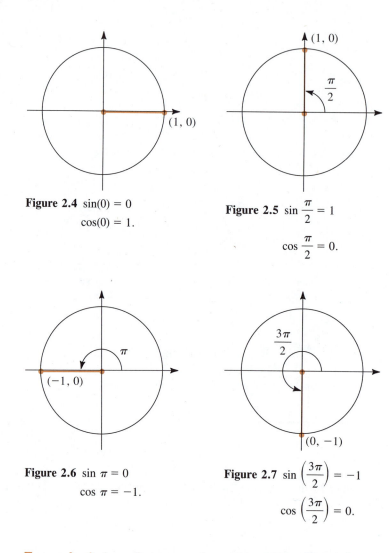

Figure 2.4 $\sin(0) = 0$
$\cos(0) = 1$.

Figure 2.5 $\sin \dfrac{\pi}{2} = 1$

$\cos \dfrac{\pi}{2} = 0$.

Figure 2.6 $\sin \pi = 0$
$\cos \pi = -1$.

Figure 2.7 $\sin \left(\dfrac{3\pi}{2} \right) = -1$

$\cos \left(\dfrac{3\pi}{2} \right) = 0$.

Example 1 According to equations (1) and (2), we have

$$\sin(0) = 0 \quad \text{and} \quad \cos(0) = 1 \qquad \text{(Figure 2.4)}$$

$$\sin \frac{\pi}{2} = 1 \quad \text{and} \quad \cos \frac{\pi}{2} = 0 \qquad \text{(Figure 2.5)}$$

$$\sin \pi = 0 \quad \text{and} \quad \cos \pi = -1 \qquad \text{(Figure 2.6)}$$

$$\sin \frac{3\pi}{2} = -1 \quad \text{and} \quad \cos \frac{3\pi}{2} = 0 \qquad \text{(Figure 2.7)}.$$

In general, values of $\sin \theta$ and $\cos \theta$ are difficult to compute. Because, however, we know the ratios of the lengths of the sides of 30°–60°–90° and 45°–45°–90° triangles, we can calculate sines and cosines of these angles easily. Figure 2.8 shows the lengths of the legs of these triangles when the hypotenuse has length 1. □

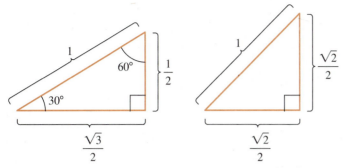

Figure 2.8 Length of sides in 30°–60°–90° and 45°–45°–90° triangles.

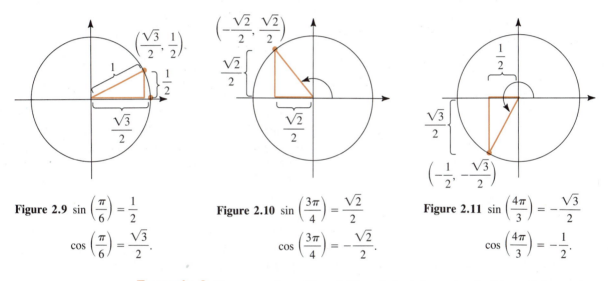

Figure 2.9 $\sin\left(\dfrac{\pi}{6}\right) = \dfrac{1}{2}$

$\cos\left(\dfrac{\pi}{6}\right) = \dfrac{\sqrt{3}}{2}$.

Figure 2.10 $\sin\left(\dfrac{3\pi}{4}\right) = \dfrac{\sqrt{2}}{2}$

$\cos\left(\dfrac{3\pi}{4}\right) = -\dfrac{\sqrt{2}}{2}$.

Figure 2.11 $\sin\left(\dfrac{4\pi}{3}\right) = -\dfrac{\sqrt{3}}{2}$

$\cos\left(\dfrac{4\pi}{3}\right) = -\dfrac{1}{2}$.

Example 2 From equations (1) and (2) and the information in Figure 2.8 we may conclude that

(a) $\sin\left(\dfrac{\pi}{6}\right) = \dfrac{1}{2}$ and $\cos\left(\dfrac{\pi}{6}\right) = \dfrac{\sqrt{3}}{2}$ (Figure 2.9)

(b) $\sin\left(\dfrac{3\pi}{4}\right) = \dfrac{\sqrt{2}}{2}$ and $\cos\left(\dfrac{3\pi}{4}\right) = -\dfrac{\sqrt{2}}{2}$ (Figure 2.10)

(c) $\sin\left(\dfrac{4\pi}{3}\right) = -\dfrac{\sqrt{3}}{2}$ and $\cos\left(\dfrac{4\pi}{3}\right) = -\dfrac{1}{2}$ (Figure 2.11).   □

Table 2.1 gives values of $\sin\theta$ and $\cos\theta$ for angles θ that are multiples of $\pi/6$ and $\pi/4$ in the interval $[0, 2\pi)$. Each may be calculated from equations (1) and (2) and the information in Figure 2.8. Values of the sine and cosine functions for angles other than those in Table 2.1 may be found in Table 3 in Appendix II or may be found using most calculators. In either case it is important to note whether the table or calculator is set up to accept angles measured in degrees or in radians.

Table 2.1

θ	0	$\dfrac{\pi}{6}$	$\dfrac{\pi}{4}$	$\dfrac{\pi}{3}$	$\dfrac{\pi}{2}$	$\dfrac{2\pi}{3}$	$\dfrac{3\pi}{4}$	$\dfrac{5\pi}{6}$
$\sin\theta$	0	$\dfrac{1}{2}$	$\dfrac{\sqrt{2}}{2}$	$\dfrac{\sqrt{3}}{2}$	1	$\dfrac{\sqrt{3}}{2}$	$\dfrac{\sqrt{2}}{2}$	$\dfrac{1}{2}$
$\cos\theta$	1	$\dfrac{\sqrt{3}}{2}$	$\dfrac{\sqrt{2}}{2}$	$\dfrac{1}{2}$	0	$-\dfrac{1}{2}$	$-\dfrac{\sqrt{2}}{2}$	$-\dfrac{\sqrt{3}}{2}$

θ	π	$\dfrac{7\pi}{6}$	$\dfrac{5\pi}{4}$	$\dfrac{4\pi}{3}$	$\dfrac{3\pi}{2}$	$\dfrac{5\pi}{3}$	$\dfrac{7\pi}{4}$	$\dfrac{11\pi}{6}$
$\sin\theta$	0	$-\dfrac{1}{2}$	$-\dfrac{\sqrt{2}}{2}$	$-\dfrac{\sqrt{3}}{2}$	-1	$-\dfrac{\sqrt{3}}{2}$	$-\dfrac{\sqrt{2}}{2}$	$-\dfrac{1}{2}$
$\cos\theta$	-1	$-\dfrac{\sqrt{3}}{2}$	$-\dfrac{\sqrt{2}}{2}$	$-\dfrac{1}{2}$	0	$\dfrac{1}{2}$	$\dfrac{\sqrt{2}}{2}$	$\dfrac{\sqrt{3}}{2}$

Equations (1) and (2) define the values $\sin\theta$ and $\cos\theta$ for angles θ with $0 \le \theta < 2\pi$. We may therefore refer to the *functions* $f(\theta) = \sin\theta$ and $g(\theta) = \cos\theta$ with domains $[0, 2\pi)$. Each is referred to as a *trigonometric* function. (Four additional trigonometric functions will be defined in Section 6.5.)

Extending the Domains of sin θ, cos θ

Just as radian measure can be defined for all real numbers t, we may extend the domains of the functions $\sin\theta$ and $\cos\theta$ to the interval $(-\infty, \infty)$ by use of the notion of the principal angle θ associated with the number t. Recall that this is the angle θ for which

$$t = \theta + 2n\pi, \qquad 0 \le \theta < 2\pi \tag{3}$$

for some integer n.

Using equation (3), we define the functions $\sin t$ and $\cos t$ for all t in $(-\infty, \infty)$ by

$$\sin t = \sin\theta, \quad \text{whenever} \quad t = \theta + 2n\pi, \qquad 0 \le \theta < 2\pi$$
$$\cos t = \cos\theta, \quad \text{whenever} \quad t = \theta + 2n\pi, \qquad 0 \le \theta < 2\pi.$$

That is, $\sin t$ is defined to be equal to $\sin\theta$ where θ is the principal angle associated with the number t, and similarly for $\cos t$. In other words, we have the identities

$$\sin\theta = \sin(\theta + 2n\pi), \qquad n = \pm1, \pm2, \ldots \tag{4a}$$
$$\cos\theta = \cos(\theta + 2n\pi), \qquad n = \pm1, \pm2, \ldots. \tag{4b}$$

Example 3 The entries in Table 2.1, together with identities (4a) and (4b), give

$$\sin(3\pi) = \sin(\pi + 2\pi) = \sin\pi = 0$$
$$\cos(7\pi) = \cos(\pi + 3 \cdot 2\pi) = \cos\pi = -1$$

$$\sin\left(\frac{9\pi}{2}\right) = \sin\left(\frac{\pi}{2} + 2 \cdot 2\pi\right) = \sin\left(\frac{\pi}{2}\right) = 1$$

and

$$\cos\left(-\frac{11\pi}{4}\right) = \cos\left(\frac{5\pi}{4} - 2 \cdot 2\pi\right) = \cos\left(\frac{5\pi}{4}\right) = -\frac{\sqrt{2}}{2}. \qquad \square$$

Identities (4a) and (4b) say that the values of the sine and cosine functions are the same at any number located a multiple of 2π radians away from the number θ as they are at the number θ. For this reason we say that these functions are *periodic* with *period* $T = 2\pi$. This periodicity appears in the graphs of $\sin t$ and $\cos t$ in Figures 2.12 and 2.13. Notice that, for both functions, the maximum value is 1 and the minimum value is -1. That is, $|\sin t| \le 1$ and $|\cos t| \le 1$ for all numbers t, or, in other words, the ranges of the extended sine and cosine functions are $[-1, 1]$.

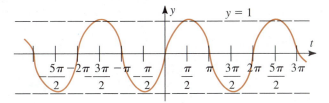

Figure 2.12 Graph of $y = \sin t$.

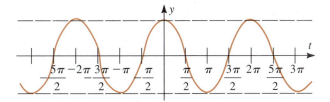

Figure 2.13 Graph of $y = \cos t$.

REMARK: We have used the letters t and θ to represent the independent variable of the trigonometric functions. As in all function notation, though, there is nothing special about the letters used. In fact, in the next section we shall most often use x instead of t.

Example 4 Sketch the graphs of the functions

(a) $f(t) = 3 \cos t$

(b) $g(t) = \sin 4t$.

Solution: Although we must wait to apply most of the graphing techniques of Chapter 3 until the derivatives of $\sin t$ and $\cos t$ are determined in Section 6.3, we can obtain graphs for these two functions by comparing them directly with $\sin t$ and $\cos t$.

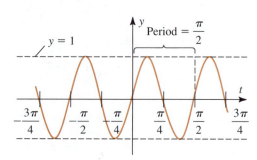

Figure 2.14 Graph of $f(t) = 3\cos t$. **Figure 2.15** Graph of $g(t) = \sin 4t$.

(a) The graph of $f(t) = 3\cos t$ is the same as the graph of $\cos t$ except that the y-coordinate of every point on the graph of $f(t) = 3\cos t$ is three times the y-coordinate of the corresponding point on the graph of $y = \cos t$. We say that the graph of $f(t) = 3\cos t$ has *amplitude* 3 since $|f(t)| \leq 3$ for all t. The graph of $y = \cos t$ has amplitude 1 as we have already noted. See Figure 2.14.

(b) The graph of $g(t) = \sin 4t$ has amplitude 1, since $|\sin 4t| \leq 1$ for all t. But the zeros of $g(t) = \sin 4t$ are spaced closer together than those of $\sin t$. Indeed, since

$$\sin t = 0 \qquad \text{for } t = 0, \pm\pi, \pm 2\pi, \ldots, \pm n\pi, \ldots$$

we have

$$\sin 4t = 0 \qquad \text{for } 4t = 0, \pm\pi, \pm 2\pi, \ldots, \pm n\pi, \ldots.$$

That is, $\sin 4t = 0$ when

$$t = 0, \pm\frac{\pi}{4}, \pm\frac{2\pi}{4}, \ldots, \pm\frac{n\pi}{4}, \ldots.$$

Thus while the function $\sin t$ has period $T = 2\pi$, the graph of $g(t) = \sin 4t$ has period

$$T = \frac{2\pi}{4} = \frac{\pi}{2}.$$

(See Figure 2.15.) □

Right Triangle Interpretation of sin θ, cos θ

For angles θ with $0 < \theta < \pi/2$ the trigonometric functions $\sin \theta$ and $\cos \theta$ may be defined as ratios of sides of a right triangle with base angle θ. If the legs of the triangle are labeled as in Figure 2.16, then

$$\sin \theta = \frac{\text{opp}}{\text{hyp}} \quad \text{and} \quad \cos \theta = \frac{\text{adj}}{\text{hyp}} \tag{5}$$

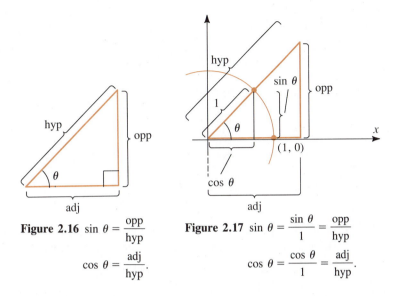

Figure 2.16 $\sin \theta = \dfrac{\text{opp}}{\text{hyp}}$

$\cos \theta = \dfrac{\text{adj}}{\text{hyp}}.$

Figure 2.17 $\sin \theta = \dfrac{\sin \theta}{1} = \dfrac{\text{opp}}{\text{hyp}}$

$\cos \theta = \dfrac{\cos \theta}{1} = \dfrac{\text{adj}}{\text{hyp}}.$

where adj, opp, and hyp are the lengths of the legs adjacent to and opposite the angle θ, and the hypotenuse, respectively. Figure 2.17 illustrates why the definitions in line (4) give the same values for $\sin \theta$ and $\cos \theta$ as our original definitions in equations (1) and (2): The triangle in Figure 2.16 is similar to a triangle inscribed in the unit circle, and the ratios of corresponding sides give

$$\frac{\sin \theta}{1} = \frac{\text{opp}}{\text{hyp}} = \sin \theta, \qquad \frac{\cos \theta}{1} = \frac{\text{adj}}{\text{hyp}} = \cos \theta.$$

Example 5 Find the lengths of the legs of the right triangle in Figure 2.18.

Strategy

We know that hyp = 10 and we need to find opp, so use

$$\frac{\text{opp}}{\text{hyp}} = \sin \theta.$$

Solution

The length of the leg opposite the angle of size $\pi/3$ is labeled y. Since the length of the hypotenuse is $h = 10$, we have from equation (5) that

$$\frac{y}{10} = \sin\left(\frac{\pi}{3}\right) = \frac{\sqrt{3}}{2}.$$

Figure 2.18

Solve for y.

Here we use

$$\frac{\text{adj}}{\text{hyp}} = \cos \theta.$$

Thus

$$y = 10\left(\frac{\sqrt{3}}{2}\right) = \frac{10\sqrt{3}}{2} \approx 8.66$$

Similarly, we have

$$\frac{x}{10} = \cos\left(\frac{\pi}{3}\right) = \frac{1}{2}$$

so

$$x = 10\left(\frac{1}{2}\right) = 5. \qquad \square$$

Example 6 A person walks 2 miles on a treadmill that is inclined 15° from the horizontal. Through what equivalent vertical distance has the person risen? (See Figure 2.19.)

Figure 2.19

Solution: Walking 2 miles on this treadmill is equivalent to climbing a hypotenuse of length 2 miles on a triangle whose base angle is $15° = \dfrac{\pi}{12}$. If the vertical side of this triangle has length y, we have

$$\frac{y}{2} = \sin \frac{\pi}{12}$$

so the desired vertical distance is

$$y = 2 \sin \frac{\pi}{12}$$

$$= 2(0.2588)$$

$$= 0.5176 \text{ miles.}$$

(The value of $\sin \dfrac{\pi}{12}$ may be obtained from Table 3 in Appendix II or from a calculator.)

$\square$

Identities Involving sin θ, cos θ

Various identities involving $\sin \theta$ and $\cos \theta$ are established in courses on trigonometry. For example, the Pythagorean Theorem applied to the smaller of the right triangles in Figure 2.17 gives the identity

$$\boxed{\sin^2 \theta + \cos^2 \theta = 1}$$

which holds for all numbers θ. Other such identities include the following:

$$\sin(\theta \pm \phi) = \sin \theta \cos \phi \pm \cos \theta \sin \phi \qquad \cos 2\theta = \cos^2 \theta - \sin^2 \theta$$

$$\cos(\theta \pm \phi) = \cos \theta \cos \phi \mp \sin \theta \sin \phi \qquad \sin \theta = \cos\left(\frac{\pi}{2} - \theta\right)$$

$$\sin^2 \theta = \frac{1}{2}(1 - \cos 2\theta) \qquad\qquad \cos \theta = \sin\left(\frac{\pi}{2} - \theta\right)$$

$$\cos^2 \theta = \frac{1}{2}(1 + \cos 2\theta) \qquad\qquad \sin(-\theta) = -\sin \theta$$

$$\sin 2\theta = 2 \sin \theta \cos \theta \qquad\qquad \cos(-\theta) = \cos \theta.$$

Exercise Set 6.2

In Exercises 1–6 state the (a) sine and (b) cosine of the angle θ contained in the given triangle.

In Exercises 7–10 find $\sin t$ and $\cos t$ for the angle t described by the figure.

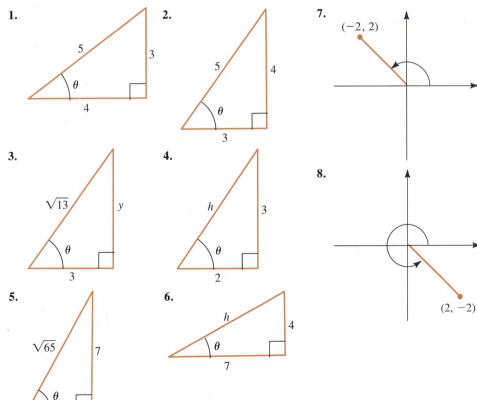

1.

2.

3.

4.

5.

6.

7.

$(-2, 2)$

8.

$(2, -2)$

9.

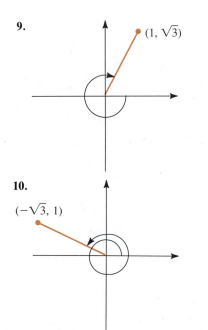

$(1, \sqrt{3})$

10.

$(-\sqrt{3}, 1)$

In Exercises 11–18 find the value of the sine or cosine by first finding the principal angle associated with the given angle.

11. $\sin\left(-\dfrac{\pi}{4}\right)$

12. $\cos(3\pi)$

13. $\cos\left(\dfrac{7\pi}{2}\right)$

14. $\sin\left(\dfrac{11\pi}{3}\right)$

15. $\sin\left(-\dfrac{5\pi}{3}\right)$

16. $\cos\left(-\dfrac{3\pi}{2}\right)$

17. $\cos\left(\dfrac{11\pi}{4}\right)$

18. $\sin\left(-\dfrac{9\pi}{4}\right)$

In Exercises 19–24 sketch the graph of the given function.

19. $f(t) = 2\sin t$

20. $f(t) = -\sin t$

21. $f(t) = \sin 2t$

22. $f(t) = 3\sin 2t$

23. $f(t) = 4\cos(-t)$

24. $f(t) = -\cos 2t$

In Exercises 25–28 find the lengths x and y of the legs of the given triangles.

25.

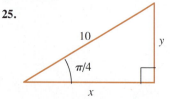

10

y

$\pi/4$

x

26.

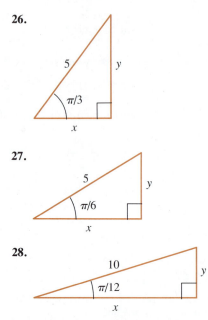

5

y

$\pi/3$

x

27.

5

y

$\pi/6$

x

28.

10

y

$\pi/12$

x

In Exercises 29–34 find the stated value of $\sin t$ or $\cos t$ from Table 3 in Appendix II or by use of a calculator.

29. $\sin 15°$

30. $\cos 70°$

31. $\cos \dfrac{\pi}{8}$

32. $\sin 140°$

33. $\sin \dfrac{2\pi}{5}$

34. $\cos \dfrac{7\pi}{5}$

35. A person walks 1 mile on a treadmill that is inclined $\dfrac{\pi}{12}$ radians from the horizontal. Through what equivalent distance will this person have risen?

36. The temperature on a particular day in St. Louis was determined to be $T(t) = 60 - 15\cos\left(\dfrac{\pi t}{12}\right)$ degrees Fahrenheit at a time t hours after midnight. Find the temperature at
   a. 6 a.m.   **b.** noon   **c.** 4 p.m.

37. The number of hours of daylight in a particular North American city is $D(t) = 12 + 3\sin\left(\dfrac{\pi t}{6} - \dfrac{5\pi}{12}\right)$ hours where t represents the number of months after January 1. Find the number of hours of daylight on
   a. March 15 $\left(t = \dfrac{5}{2}\right)$ **b.** June 15 $\left(t = \dfrac{11}{2}\right)$
   c. December 15 $\left(t = \dfrac{23}{2}\right)$

Figure 2.20 Ferris wheel

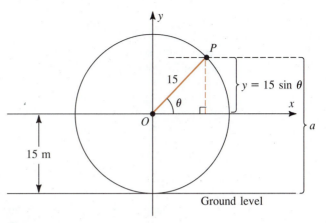

Figure 2.21 Idealized Ferris wheel

38. The height of a particular seat on a Ferris wheel is

$$h = 15 + 15 \sin \theta \text{ meters}$$

where θ is the angle illustrated in Figures 2.20 and 2.21. Find the height of the seat above ground level when θ equals

a. 0 radians **b.** $\dfrac{2\pi}{3}$ radians

c. $\dfrac{3\pi}{2}$ radians

39. Researchers interested in modelling the rate at which animals (including humans) grow know that growth is not uniform. Periods of rapid growth often occur between periods of very slow growth. One model for such types of growth is given by the function

$$h(t) = t + \sin\left(\dfrac{\pi t}{4}\right) + B$$

where B is a constant, t is years since birth, and $h(t)$ is human height. (See Figure 2.22.) According to this model, find

a. $h(0)$ **b.** $h(4)$ **c.** $h(6)$

40. During a certain period of time closing prices on a particular stock exchange were approximately

$$P(t) = 10t + 4 \sin\left(\dfrac{\pi t}{2}\right) + 500$$

where P is price in dollars and t is time in months. Find
a. $P(0)$ **b.** $P(12)$ **c.** $P(15)$

41. Explain why the following formula for the area of the triangle in Figure 2.23 is valid:

$$A = \dfrac{1}{2}xh \sin \theta.$$

Figure 2.23

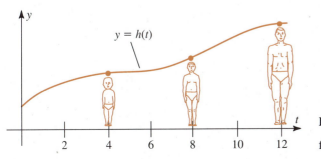

$y = h(t)$

Figure 2.22 The model $h(t) = t + \sin\left(\dfrac{\pi t}{4}\right) + B$ for human growth.

6.3 DERIVATIVES OF SIN X AND COS X

At the end of this section we shall show that the derivatives of the functions sin x and cos x are the following:

$$\frac{d}{dx} \sin x = \cos x \tag{1}$$

$$\frac{d}{dx} \cos x = -\sin x. \tag{2}$$

If u is a differentiable function of x, we may apply the Chain Rule together with equations (1) and (2) to conclude that

$$\frac{d}{dx} \sin u = \cos u \cdot \frac{du}{dx} \tag{3}$$

$$\frac{d}{dx} \cos u = -\sin u \cdot \frac{du}{dx}. \tag{4}$$

Figure 3.1 illustrates the geometric significance of equation (1). The derivative $f'(x_0)$ can be interpreted as the slope of the line tangent to the graph of f at the point where

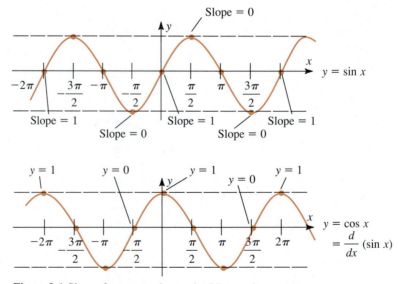

Figure 3.1 Slope of tangent to the graph of $f(x) = \sin x$ at $x = x_0$ if $f'(x_0) = \cos x_0$.

$x = x_0$. As the figure shows, the tangent to the graph of $f(x) = \sin x$ has

slope $m = \cos(0) = 1$ when $x = 0$

slope $m = \cos\left(\dfrac{\pi}{2}\right) = 0$ when $x = \dfrac{\pi}{2}$

slope $m = \cos \pi = -1$ when $x = \pi$

etc. Also, since the function $f(x) = \sin x$ is periodic with period 2π, the "slope function" $f'(x)$ should also be periodic, with period $T = 2\pi$. This is precisely the behavior of its derivative, $f'(x) = \cos x$. You can verify by a similar analysis that properties of the derivative $g'(x) = -\sin x$ are those of the slope function for the function $g(x) = \cos x$.

Example 1 For $f(x) = \sin 4x$, find $f'(x)$.

Solution: Here $f(x) = \sin u$ with $u = 4x$. According to equation (3), we have

$$f'(x) = (\cos 4x) \cdot \frac{d}{dx}(4x)$$

$$= (\cos 4x)(4)$$

$$= 4 \cos 4x. \qquad \square$$

Example 2 Find $\dfrac{d}{dx} \sin x^3$.

Solution: Again applying equation (3), with $u = x^3$, we have

$$\frac{d}{dx} \sin x^3 = (\cos x^3) \cdot \frac{d}{dx}(x^3)$$

$$= (\cos x^3)(3x^2)$$

$$= 3x^2 \cos x^3. \qquad \square$$

Example 3 For $f(x) = x \cos \sqrt{x}$ find $f'(x)$.

Solution: Since f is a product, with factors x and $\cos \sqrt{x}$, we apply the Product Rule together with equation (4):

$$f'(x) = \left[\frac{d}{dx}(x)\right]\cos \sqrt{x} + x\left[\frac{d}{dx} \cos \sqrt{x}\right]$$

$$= (1)\cos \sqrt{x} + x\left[\left(-\sin \sqrt{x}\right)\left(\frac{d}{dx}\sqrt{x}\right)\right] \qquad \text{(Apply Chain Rule)}$$

$$= \cos \sqrt{x} + x(-\sin \sqrt{x})\left(\frac{1}{2\sqrt{x}}\right)$$

$$= \cos \sqrt{x} - \frac{1}{2}\sqrt{x} \sin \sqrt{x}. \qquad \square$$

Example 4 Find $\dfrac{d}{dx}(x + \sin x)^4$.

Solution: The function to be differentiated has the form $f(x) = [u(x)]^4$ with $u(x) = x + \sin x$. Since the derivative of such a function is $f'(x) = 4[u(x)]^3 \cdot u'(x)$, we first apply the Chain Rule to obtain $u'(x) = 1 + \cos x$ and then evaluate $f'(x)$:

$$\frac{d}{dx}(x + \sin x)^4 = 4(x + \sin x)^3(1 + \cos x). \qquad \square$$

Example 5 Find $f'(x)$ if $f(x) = \dfrac{x + \sin x}{x^2}$, $x \neq 0$.

Solution: Applying the Quotient Rule, we obtain

$$f'(x) = \frac{x^2\left[\dfrac{d}{dx}(x + \sin x)\right] - (x + \sin x)\left[\dfrac{d}{dx}x^2\right]}{(x^2)^2}$$

$$= \frac{x^2(1 + \cos x) - (x + \sin x)(2x)}{x^4}$$

$$= \frac{x^2(\cos x - 1) - 2x \sin x}{x^4}$$

$$= \frac{x(\cos x - 1) - 2 \sin x}{x^3}. \qquad \square$$

Example 6 Find the maximum and minimum values of the function $f(x) = \sqrt{2 + \sin x}$ for x in the interval $[0, 2\pi]$.

Strategy

Solution

This problem is worked in the same way as the optimization problems in Chapter 3. The only difference is that the function involves the sine, which we can now differentiate.

Find f'.

For $f(x) = \sqrt{2 + \sin x} = (2 + \sin x)^{1/2}$ we have

$$f'(x) = \frac{1}{2}(2 + \sin x)^{-1/2} \cdot \cos x$$

$$= \frac{\cos x}{2\sqrt{2 + \sin x}}.$$

Set $f'(x) = 0$ to find critical numbers.

Thus $f'(x)$ equals zero when $\cos x = 0$, which occurs at the numbers

$$x = \frac{\pi}{2} \quad \text{and} \quad x = \frac{3\pi}{2}$$

in $[0, 2\pi]$. Since $f'(x)$ is defined for all x in $[0, 2\pi]$, the only critical numbers for f are $x = \dfrac{\pi}{2}$ and $\dfrac{3\pi}{2}$. Checking the value $f(x)$ at each critical number and endpoint gives

Calculate $f(x)$ for each critical number and endpoint.

$$f(0) = \sqrt{2 + \sin(0)} = \sqrt{2 + 0}$$
$$= \sqrt{2} \qquad \text{(endpoint)}$$

$$f\!\left(\frac{\pi}{2}\right) = \sqrt{2 + \sin\frac{\pi}{2}} = \sqrt{2 + 1}$$
$$= \sqrt{3} \qquad \text{(critical number)}$$

$$f\!\left(\frac{3\pi}{2}\right) = \sqrt{2 + \sin\frac{3\pi}{2}} = \sqrt{2 + (-1)}$$
$$= 1 \qquad \text{(critical number)}$$

$$f(2\pi) = \sqrt{2 + \sin 2\pi} = \sqrt{2 + 0}$$
$$= \sqrt{2} \qquad \text{(endpoint).}$$

Select maximum and minimum from among these values.

The maximum value of f on $[0, 2\pi]$ is therefore $f\!\left(\dfrac{\pi}{2}\right) = \sqrt{3}$, and the minimum is $f\!\left(\dfrac{3\pi}{2}\right) = 1$. The graph of f appears in Figure 3.2. □

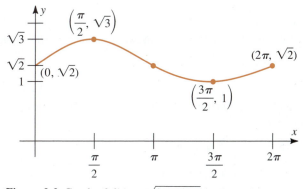

Figure 3.2 Graph of $f(x) = \sqrt{2 + \sin x}$ in Example 6.

Example 7 A Ferris wheel 30 meters high turns at the rate of 3 revolutions per minute. (See Figure 3.3.) Find

(a) a function describing the rate at which the altitude of a passenger changes, and
(b) the positions at which this rate is the greatest.

Solution: Figure 3.4 represents the passenger's location as a point P on the idealized Ferris wheel. If we superimpose an xy-coordinate system with its origin O at the center of the Ferris wheel, then P is a point on a circle of radius $r = 15$ meters. If P has coordinates $P = (x, y)$ and if θ is the angle between the ray OP and the positive x-axis, then the

Figure 3.3 Ferris wheel.

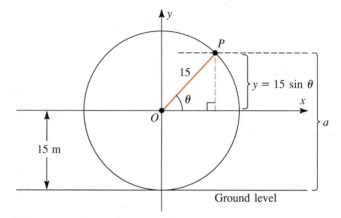

Figure 3.4 Idealized Ferris wheel.

definition of the sine function tells us that

$$\sin \theta = \frac{y}{15}$$

so

$$y = 15 \sin \theta.$$

Now since the x-axis lies 15 meters above ground level, the *altitude h* of the point P is

$$h = 15 + y$$
$$= 15 + 15 \sin \theta \text{ meters.} \qquad (5)$$

Since the angle θ is a function of t (time), equation (5) allows us to express the altitude of P as a function of t:

$$h = 15 + 15 \sin \theta(t) \text{ meters.}$$

Thus the *rate* at which altitude changes with respect to time is the derivative $h'(t)$:

$$\frac{dh}{dt} = \frac{d}{dt}[15 + 15 \sin \theta(t)]$$
$$= 15 \cos \theta(t) \cdot \frac{d\theta}{dt} \text{ meters per minute} \qquad (6)$$

according to equation (3). Since we are given that $\dfrac{d\theta}{dt} = 3$ revolutions per minute, and θ in line (5) is measured in *radians,* we must convert revolutions to radians (1 revolution $= 2\pi$ radians). We obtain

$$\frac{d\theta}{dt} = \left(3\frac{\text{rev}}{\text{min}}\right)\left(2\pi\frac{\text{rad}}{\text{rev}}\right) = 6\pi \text{ rad/min.}$$

Substituting this number for $\dfrac{d\theta}{dt}$ in equation (6) gives the desired rate as

$$\frac{dh}{dt} = 15 \cos \theta(t) \cdot 6\pi$$

$$= 90\pi \cos \theta(t) \text{ meters per minute.} \tag{7}$$

To find the positions at which this rate is the greatest, we set the derivative of $\dfrac{dh}{dt}$ (which is the *second* derivative of *h*) equal to zero and solve for θ. Since

$$\frac{d^2h}{dt^2} = \frac{d}{dt}[90\pi \cos \theta(t)]$$

$$= -90\pi \sin \theta(t) \cdot \frac{d\theta}{dt}$$

$$= -90\pi \sin \theta(t)(6\pi)$$

$$= -540\pi^2 \sin \theta(t)$$

we will have $\dfrac{d^2h}{dt^2} = 0$ when $\sin \theta = 0$, which occurs when $\theta = 0$ and when $\theta = \pi$.

These critical numbers correspond to what you have already experienced from riding a Ferris wheel. The point at which you are rising most rapidly corresponds to $\theta = 0$ (that is, halfway up), and the point at which you are falling most rapidly corresponds to $\theta = \pi$ (halfway down). (See Figures 3.5 and 3.6.) □

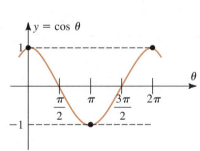

Figure 3.5 $\cos \theta$ achieves its maximum value at $\theta = 0$ and its minimum value at $\theta = \pi$.

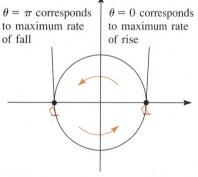

Figure 3.6 Maximum rates of rise and fall occur at the halfway points.

Deriving Formulas (1) and (2)

To establish formula (1) for the derivative of the function $f(x) = \sin x$, we need to make use of the following limit:

$$\lim_{\theta \to 0} \frac{\sin \theta}{\theta} = 1. \tag{8}$$

A geometric argument for this limit can be given. However, you may observe this limit, experimentally quite easily. Just select a sequence of numbers θ that approach zero (from either direction) and calculate the ratio $\dfrac{\sin \theta}{\theta}$ for each, using a calculator or a computer. You will obtain results similar to those in Table 3.1.

Table 3.1 Values of $\dfrac{\sin \theta}{\theta}$ as $\theta \to 0$ (θ is in radians)

θ	$\sin \theta$	$\dfrac{\sin \theta}{\theta}$	θ	$\sin \theta$	$\dfrac{\sin \theta}{\theta}$
.5	.479426	.958851	$-.5$	$-.479426$	.958851
.1	.099833	.998334	$-.1$	$-.099833$	.998334
.05	.049979	.999583	$-.05$	$-.049979$	.999583
.01	.010000	.999983	$-.01$	$-.010000$	.999983
.005	.005000	.999996	$-.005$	$-.005000$	.999996
.001	.001000	.999999	$-.001$	$-.001000$	.999999

In addition to the limit in line (8) we need to make use of the limit

$$\lim_{\theta \to 0} \left(\frac{\cos \theta - 1}{\theta} \right) = 0 \tag{9}$$

which is explained in Exercise 40. Finally, we need to use the formula from trigonometry for the sine of the sum of two angles:

$$\sin(\theta + \phi) = \sin \theta \cos \phi + \cos \theta \sin \phi. \tag{10}$$

We are now ready to use the basic definition of the derivative to find $f'(x)$ for $f(x) = \sin x$:

$$f'(x) = \lim_{h \to 0} \frac{f(x + h) - f(x)}{h} \qquad \text{(definition of } f'(x)\text{)}$$

$$= \lim_{h \to 0} \frac{\sin(x + h) - \sin x}{h} \qquad (f(x) = \sin x)$$

$$= \lim_{h \to 0} \frac{[\sin x \cos h + \cos x \sin h] - \sin x}{h} \qquad \text{(formula (10))}$$

$$= \lim_{h \to 0} \left\{ \sin x \left(\frac{\cos h - 1}{h} \right) + \cos x \left(\frac{\sin h}{h} \right) \right\}$$

$$= \sin x \left[\lim_{h \to 0} \left(\frac{\cos h - 1}{h} \right) \right] + \cos x \left[\lim_{h \to 0} \left(\frac{\sin h}{h} \right) \right]$$

$$= \sin x \cdot (0) + \cos x \cdot (1) \qquad \text{(equations (8), (9))}$$

$$= \cos x.$$

This establishes formula (1), that $\dfrac{d}{dx} \sin x = \cos x$.

To establish formula (2), that $\dfrac{d}{dx} \cos x = -\sin x$, we use formula (3) and the identities

$$\sin\left(\frac{\pi}{2} - x\right) = \cos x; \qquad \cos\left(\frac{\pi}{2} - x\right) = \sin x. \tag{11}$$

We obtain

$$\frac{d}{dx} \cos x = \frac{d}{dx} \sin\left(\frac{\pi}{2} - x\right) \qquad \text{(equations (11))}$$

$$= \cos\left(\frac{\pi}{2} - x\right) \cdot \frac{d}{dx}\left(\frac{\pi}{2} - x\right) \qquad \text{(formula (3))}$$

$$= (-1)\cos\left(\frac{\pi}{2} - x\right)$$

$$= -\sin x. \qquad \text{(equations (11))}.$$

Exercise Set 6.3

In Exercises 1–20 find the derivative of the given function.

1. $f(x) = \sin 3x$

2. $f(x) = \cos 2x$

3. $y = x \sin x$

4. $f(x) = \sin\left(\dfrac{\pi}{2} - x\right)$

5. $f(x) = x^3 \cos 2x$

6. $f(\theta) = e^{\sin \theta}$

7. $y = \sin x \cos x$

8. $f(x) = \sqrt{x} + \cos x$

9. $f(x) = \sin\sqrt{1 + x^2}$

10. $f(x) = \dfrac{\sin x}{x}$

11. $y = \cos(\ln t)$

12. $y = \dfrac{1 - \sin x}{1 + \cos x}$

13. $f(x) = \sin^2 x + \cos^2 x$

14. $f(x) = x \cos\sqrt{1 - x^2}$

15. $f(t) = \ln(t + \sin t)$

16. $y = (\cos \theta)e^{\cos \theta}$

17. $y = \dfrac{1 + \sin x}{3 + x^2}$

18. $f(x) = \sin(x^3 + \sqrt{x})$

19. $f(x) = x \sin\left(\dfrac{1}{x}\right)$

20. $y = \sin^3 x \cos^5 x$

21. Find an equation for the line tangent to the graph of $y = \sin 2x$ at the point $\left(\dfrac{\pi}{6}, \dfrac{\sqrt{3}}{2}\right)$.

22. Find an equation for the line tangent to the graph of $y = e^{\cos t}$ at the point where $t = \pi/2$.

23. A particle moves along a line so that after t seconds its position is $s(t) = \dfrac{\sin 2t}{3 + \cos^2 t}$. Find its velocity at time $t = \dfrac{\pi}{4}$. (Units for distance are feet.)

24. A particle moves along a line so that at time t its position is $s(t) = 6 \sin 3t$. (Units are meters for distance, seconds for time.)
   a. At which times does it change direction?
   b. What is its maximum velocity?

25. Why do the functions $f(x) = \sin^2 x$ and $g(x) = 1 - \cos^2 x$ have the same derivatives?

26. Find $f''(x)$ for $f(x) = \sqrt{x} \sin x$.

27. Find $f''(t)$ for $f(t) = \sin \ln t$.

In Exercises 28–31 use implicit differentiation to find $\dfrac{dy}{dx}$.

28. $x = \sin y$

29. $\cos xy = x$

30. $\sin(x + y) + \cos y = 1$

31. $\sin y = x \cos y$

In Exercises 32–35 find the maximum and minimum values of the given function on the given interval.

32. $f(x) = \sin 2x, \quad x$ in $\left[0, \dfrac{\pi}{2}\right]$

33. $f(x) = \sin x + \cos x, \quad x$ in $[0, 2\pi]$

34. $f(x) = \sin x - \cos x, \quad x$ in $\left[-\dfrac{\pi}{2}, \dfrac{\pi}{2}\right]$

35. $f(x) = x - \sin x, \quad x$ in $\left[\dfrac{\pi}{2}, \dfrac{3\pi}{2}\right]$

In Exercises 36–41 use information provided by the first and second derivatives of the given function to sketch its graph.

36. $f(x) = \sin 3x$

37. $f(x) = \cos\left(x - \dfrac{\pi}{2}\right)$

38. $f(x) = x \sin x$

39. $f(x) = x + \sin x$

40. $f(x) = \sin x + \cos x$

41. $f(x) = \dfrac{\sin x}{2 + \cos x}$, $-\pi \le x \le \pi$

42. A water trough is to be constructed from three metal sheets each 1 meter wide and 6 meters long plus end panels in the shape of trapezoids. (See Figure 3.7.) Find the angle θ at which the long panels should be joined so as to provide a trough of maximum volume. (See Figure 3.8.)

Figure 3.7

Figure 3.8

43. Researchers interested in modelling the rate at which animals grow have used the model

$$y = t + \sin\left(\dfrac{\pi t}{4}\right) + B$$

where y is height in centimeters, B is birth height, and t is in units of months from birth. Find the maximum and minimum values of the rate of growth, $\dfrac{dy}{dt}$, and the times during the first year at which they occur.

44. The temperature on a particular day in St. Louis was determined to be $T(t) = 60 - 15 \cos\left(\dfrac{\pi t}{12}\right)$ degrees Fahrenheit t hours after midnight.
   a. What was the maximum temperature, and when did it occur?
   b. What was the minimum temperature and when did it occur?
   c. When was the temperature increasing most rapidly?

45. The number of hours of daylight in a particular North American city is $D(t) = 12 + 3 \sin\left(\dfrac{\pi t}{6} - \dfrac{5\pi}{12}\right)$, where t represents months after January 1. Find the time t at which
   a. the days are the longest.
   b. the days are the shortest.

46. Explain the steps in the following demonstration that $\lim\limits_{\theta \to 0} \dfrac{\cos \theta - 1}{\theta} = 0$:

$$\lim_{\theta \to 0} \dfrac{\cos \theta - 1}{\theta} = \lim_{\theta \to 0}\left[\left(\dfrac{\cos \theta - 1}{\theta}\right)\left(\dfrac{\cos \theta + 1}{\cos \theta + 1}\right)\right]$$

$$= \lim_{\theta \to 0} \dfrac{\cos^2 \theta - 1}{\theta(1 + \cos \theta)}$$

$$= \lim_{\theta \to 0} \dfrac{-\sin^2 \theta}{\theta(1 + \cos \theta)}$$

$$= \lim_{\theta \to 0}\left(\dfrac{\sin \theta}{\theta}\right)\left(\dfrac{-\sin \theta}{1 + \cos \theta}\right)$$

$$= (1)\left(\dfrac{0}{1 + 1}\right)$$

$$= 0.$$

6.4 INTEGRALS OF SINE AND COSINE

The differentiation formulas $\dfrac{d}{dx} \sin x = \cos x$ and $\dfrac{d}{dx} \cos x = -\sin x$ lead immediately to the following antiderivatives of the sine and cosine functions.

$$\int \cos x \, dx = \sin x + C \tag{1}$$

$$\int \sin x \, dx = -\cos x + C. \tag{2}$$

These formulas, together with the method of substitutions, allow us to evaluate a variety of integrals involving the sine and cosine functions.

Example 1 Find $\displaystyle\int 3 \cos 3x \, dx$.

Solution: If we let $u = 3x$, then $du = 3 \, dx$, and we obtain

$$\int 3 \cos 3x \, dx = \int \underbrace{\cos 3x}_{\cos u} \cdot \underbrace{3 \, dx}_{du} = \int \cos u \, du$$

$$= \sin u + C \qquad \text{(by equation (1))}$$

$$= \sin 3x + C.$$

This result is easy to verify: $\dfrac{d}{dx}(\sin 3x + C) = \cos 3x \cdot \dfrac{d}{dx}(3x) = 3 \cos 3x.$ □

Example 2 Find $\displaystyle\int \sin 5x \, dx$.

Solution: This integral is similar to that of Example 1. If we let

$$u = 5x, \quad \text{then} \quad du = 5 \, dx.$$

Thus $dx = \dfrac{1}{5} \, du$. Using these substitutions and equation (2), we obtain

$$\int \sin 5x \, dx = \int \sin u \cdot \frac{1}{5} \, du = \frac{1}{5} \int \sin u \, du$$

$$= \frac{1}{5}(-\cos u) + C$$

$$= -\frac{1}{5} \cos 5x + C.$$ □

Example 3 Find $\displaystyle\int_0^{\pi/4} \cos 2x \, dx$.

Solution: By the method of Example 1 we find that the antiderivative for $\cos 2x$ is

$$\int \cos 2x \, dx = \frac{1}{2} \sin 2x + C.$$

The definite integral may therefore be evaluated using this antiderivative and the Fundamental Theorem:

$$\int_0^{\pi/4} \cos 2x \, dx = \frac{1}{2} \sin 2x \Big|_0^{\pi/4}$$

$$= \frac{1}{2} \sin \frac{\pi}{2} - \frac{1}{2} \sin(0)$$

$$= \frac{1}{2}(1) - \frac{1}{2}(0)$$

$$= \frac{1}{2}.$$ □

Example 4 Find $\displaystyle\int x^2 \sin x^3 \, dx$.

Strategy

Make the substitution

$u = x^3$.

Determine du and then solve for the remaining factors of the integrand.

Make substitution.

Apply equation (2).

Solution

If we let

$$u = x^3, \quad \text{then} \quad du = 3x^2 \, dx.$$

Solving the second equation for the factor $x^2 \, dx$ that appears in the integrand gives $x^2 \, dx = \dfrac{1}{3} \, du$. Making these substitutions, we then obtain

$$\int x^2 \sin x^3 \, dx = \int (\sin x^3) x^2 \, dx$$

$$= \int \sin u \cdot \frac{1}{3} \, du$$

$$= \frac{1}{3} \int \sin u \, du$$

$$= -\frac{1}{3} \cos u + C$$

Substitute back in terms of x.
$$= -\frac{1}{3} \cos x^3 + C. \qquad \Box$$

Example 5 Find $\displaystyle\int_0^{\pi/2} \frac{\sin x \cos x}{\sqrt{1 + \sin^2 x}}\, dx.$

Strategy

Begin by finding antiderivative for
$\dfrac{\sin x \cos x}{\sqrt{1 + \sin^2 x}}$.

Try the substitution

$u = 1 + \sin^2 x$

since this expression is underneath the radical sign.

Determine du.

Make the substitution.

Find the antiderivative using the power rule. Substitute back in terms of x.

Evaluate definite integral using the Fundamental Theorem of Calculus.

Solution

Since this is a definite integral, we first find the antiderivative

$$\int \frac{\sin x \cos x}{\sqrt{1 + \sin^2 x}}.$$

To do so, we begin by making a substitution for the expression under the radical sign. We let

$$u = 1 + \sin^2 x.$$

Then

$$du = \left[\frac{d}{dx}(1 + \sin^2 x)\right] dx = 2 \sin x \cos x\, dx.$$

Thus the factor $\sin x \cos x\, dx$ in the integrand is just $\dfrac{1}{2}\, du$, and we obtain

$$\int \frac{\sin x \cos x\, dx}{\sqrt{1 + \sin^2 x}} = \int \frac{\frac{1}{2}\, du}{\sqrt{u}}$$

$$= \int \frac{1}{2} u^{-1/2}\, du$$

$$= u^{1/2} + C.$$

$$= \sqrt{1 + \sin^2 x} + C$$

The definite integral may therefore be evaluated as follows:

$$\int_0^{\pi/2} \frac{\sin x \cos x\, dx}{\sqrt{1 + \sin^2 x}} = \sqrt{1 + \sin^2 x}\;\Big|_0^{\pi/2}$$

$$= \sqrt{1 + 1^2} - \sqrt{1 + 0^2}$$

$$= \sqrt{2} - 1. \qquad \Box$$

Applications of Integrals Involving sin x, cos x

Applications involving integrals of the sine and cosine functions are handled just as the applications of the definite integral that were discussed in Chapter 5. The following two examples are typical. Others appear in the exercise set.

Example 6 A revenue stream (interest from a bond, or rent from a building, for example) flows to the owner of an asset at a rate of

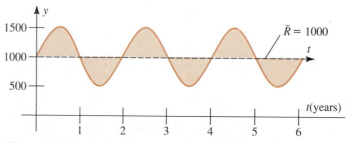

Figure 4.1 Revenue function $R(t) = 1000 + 500 \sin \pi t$ showing deviations above and below average value $\bar{R} = 1000$ on interval $[0, 6]$.

$$R(t) = 1000 + 500 \sin \pi t \qquad \text{(See Figure 4.1.)}$$

dollars per year t years after the asset is purchased. Find the total amount produced by the revenue stream during the first 6 years after the asset is purchased.

Solution: The total T produced by the revenue stream is given by the formula $T = \int_0^6 R(t)\, dt$. (See equation (10), Section 5.7.) Thus

$$T = \int_0^6 (1000 + 500 \sin \pi t)\, dt$$

$$= 1000t - \frac{500}{\pi} \cos \pi t \Big|_0^6$$

$$= \left(6000 - \frac{500}{\pi} \cos 6\pi\right) - \left(0 - \frac{500}{\pi} \cos(0)\right)$$

$$= \left(6000 - \frac{500}{\pi}\right) - \left(-\frac{500}{\pi}\right)$$

$$= 6000 \text{ dollars.} \qquad \qquad \square$$

REMARK: In this case the total revenue is the same as that produced by the *constant* revenue stream $R(t) \equiv 1000$ over the period $t = 0$ to $t = 6$. That is because the additional revenue flowing during the intervals $(0, 1)$, $(2, 3)$, and $(4, 5)$ is exactly offset by the shortfall that occurs in the intervals $(1, 2)$, $(3, 4)$, and $(5, 6)$. This is another way of saying that the *average value* of the revenue function $R(t) = 1000 + 500 \sin \pi t$ over the interval $[0, 6]$ is $\bar{R} = 1000$. (See Exercise 30.)

Example 7 Find the area of the region bounded by the graphs of $f(x) = \sin x$ and $g(x) = \cos x$ for $0 \le x \le \pi$. (See Figure 4.2.)

Strategy
Determine where the two graphs intersect.

Solution
In calculating the area of a region bounded by two curves, we must know which curve is on top. To determine this we first determine where

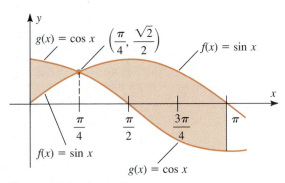

Figure 4.2 Region in Example 7.

the curves intersect by solving the equation $f(x) = g(x)$. That is, we must find the numbers x for which

$$\sin x = \cos x, \qquad 0 \le x \le \pi.$$

Checking the values of $\sin x$ and $\cos x$ in Table 2.1, Section 6.2, we find that $x = \pi/4$ is the only such number. Since

Determine which curve is on top in each interval.

$$\cos x \ge \sin x \qquad \text{for } x \text{ in } [0, \pi/4] \text{ and}$$
$$\sin x \ge \cos x \qquad \text{for } x \text{ in } [\pi/4, \pi]$$

the area of the region is (see (4), Section 5.6)

Calculate area using formula

$$A = \int_a^b [f(x) - g(x)] \, dx$$

when $f(x) \ge g(x)$ for all x in $[a, b]$.

$$A = \int_0^{\pi/4} [\cos x - \sin x] \, dx + \int_{\pi/4}^{\pi} [\sin x - \cos x] \, dx$$

$$= (\sin x + \cos x) \Big|_0^{\pi/4} + (-\cos x - \sin x) \Big|_{\pi/4}^{\pi}$$

$$= \left[\left(\frac{\sqrt{2}}{2} + \frac{\sqrt{2}}{2} \right) - (0 + 1) \right] + \left[(-(-1) - 0) - \left(-\frac{\sqrt{2}}{2} - \frac{\sqrt{2}}{2} \right) \right]$$

$$= 2\sqrt{2}. \qquad \qquad \Box$$

Exercise Set 6.4

In Exercises 1–22 evaluate the given integral.

1. $\displaystyle\int \cos 6x \, dx$

2. $\displaystyle\int \sin 3x \, dx$

3. $\displaystyle\int_0^{\pi} 3 \sin(\pi - x) \, dx$

4. $\displaystyle\int_0^{\sqrt{\pi/2}} x \cos x^2 \, dx$

5. $\displaystyle\int x^2 \sin(1 + x^3) \, dx$

6. $\displaystyle\int \sin^2 x \cos x \, dx$

7. $\displaystyle\int t \cos(\pi - t^2) \, dt$

8. $\displaystyle\int \frac{\cos \sqrt{t}}{\sqrt{t}} \, dt$

9. $\displaystyle\int_0^{\pi} \cos^3 t \sin t \, dt$

10. $\displaystyle\int \frac{\sin x}{\cos^2 x} \, dx$

11. $\displaystyle\int \cos x \sqrt{1 - \sin x} \, dx$

12. $\displaystyle\int_0^1 \sin x \sqrt{\cos x} \, dx$

13. $\displaystyle\int (x + 1)\sin(x^2 + 2x) \, dx$

14. $\displaystyle\int_0^1 e^x \sin(e^x) \, dx$

15. $\displaystyle\int \frac{\cos(\ln x)}{x}\,dx$

16. $\displaystyle\int \frac{\cos x}{\sin x}\,dx$

17. $\displaystyle\int_0^{\pi/2} (\sin t)e^{\cos t}\,dt$

18. $\displaystyle\int \sin t\,(\cos t)e^{\sin^2 t}\,dt$

19. $\displaystyle\int \sin^2 x\,dx$ $\left(\text{\textit{Hint:} Use the identity } \sin^2 x = \frac{1}{2} - \frac{1}{2}\cos 2x.\right)$

20. $\displaystyle\int \cos^2 x\,dx$ $\left(\text{\textit{Hint:} Use the identity } \cos^2 x = \frac{1}{2} + \frac{1}{2}\cos 2x.\right)$

21. $\displaystyle\int x \sin^2 x^2 \cos x^2\,dx$

22. $\displaystyle\int \frac{\sin \sqrt{x}\,\cos \sqrt{x}}{\sqrt{x}}\,dx$

23. Find the area of the region bounded by the graph of $f(x) = \sin \pi x$ and the x-axis between $x = 0$ and $x = 1$.

24. Find the total amount of money produced by a revenue stream yielding $R(t) = 1000 + 400 \cos 2\pi t$ dollars per year between $t = 0$ and $t = 4$ years.

25. Find the average value of the function $f(t) = \cos \pi t$ on the interval $[0, 1]$.

26. Find the area of the region bounded by the graphs of $f(x) = x$ and $g(x) = \sin x$ for $0 \le x \le \pi$.

27. Find the area of the region bounded by the graphs of $f(x) = \cos\left(\frac{1}{2}x\right)$ and $g(x) = \frac{1}{2}x - \frac{\pi}{2}$ for $0 \le x \le \pi$.

28. Find the volume of the solid obtained by revolving about the x-axis the region bounded above by the graph of $y = \sin x$ and below by the x-axis for $0 \le x \le \pi$. (See Exercise 19.)

29. Find the volume of the solid obtained by revolving about the x-axis the region bounded above by the graph of $y = \cos x$ and below by the x-axis for $0 \le x \le \pi/2$. (See Exercise 20.)

30. Find the average value of the revenue function $R(x) = 1000 + 500 \sin \pi t$ on the interval $[0, 6]$.

6.5 OTHER TRIGONOMETRIC FUNCTIONS

For a given right triangle containing the angle θ the numbers $\sin \theta$ and $\cos \theta$ are only two of six possible ratios of lengths of sides. In this section we describe how each of the remaining four ratios defines a trigonometric function, and how these functions are related to the sine and cosine functions.

Figure 5.1 shows a rocket that has been launched vertically x yards from a camera that is inclined at an angle θ from the horizontal. If we let y represent the altitude at which the rocket will be sighted by the camera, we have described a right triangle with angle θ for which the side opposite the angle θ has length y and the side adjacent to θ has length x. The **tangent** of this angle θ, denoted by $\tan \theta$, is the ratio

$$\tan \theta = \frac{y}{x}. \qquad \left(\frac{\text{opposite}}{\text{adjacent}}\right) \tag{1}$$

Figures 5.2–5.4 show familiar 30°–60°–90° and 45°–45°–90° right triangles and the facts that

$$\tan \frac{\pi}{6} = \frac{1}{\sqrt{3}}, \qquad \tan \frac{\pi}{3} = \sqrt{3}, \qquad \text{and} \qquad \tan \frac{\pi}{4} = 1$$

each of which follows from equation (1).

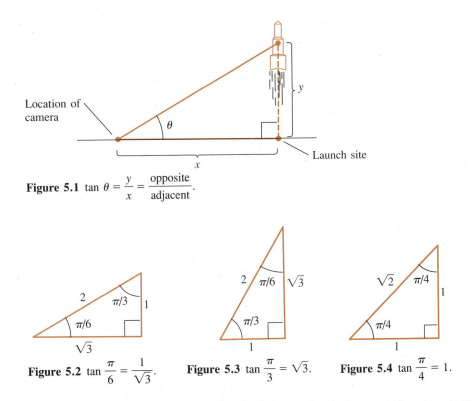

Figure 5.1 $\tan \theta = \dfrac{y}{x} = \dfrac{\text{opposite}}{\text{adjacent}}$.

Figure 5.2 $\tan \dfrac{\pi}{6} = \dfrac{1}{\sqrt{3}}$.

Figure 5.3 $\tan \dfrac{\pi}{3} = \sqrt{3}$.

Figure 5.4 $\tan \dfrac{\pi}{4} = 1$.

Values of $\tan \theta$ for other angles appear in the table of trigonometric functions in Table 3 of Appendix II. The tangent function is useful when two of the numbers x, y, and $\tan \theta$ in equation (1) are known and we desire to calculate the third.

Example 1 An architect wishes to design a ramp inclined at 30° leading from the ground level to a second-story door in a parking garage. How far from the building must the ramp begin if the bottom of the door is 12 feet above ground level? (See Figure 5.5.)

Figure 5.5 $\tan \dfrac{\pi}{6} = \dfrac{12}{x}$.

Solution: From Figure 5.5 we can see that the ramp, the ground level, and the building wall form a right triangle with base angle $= 30° = \dfrac{\pi}{6}$, with opposite side of length 12 feet, and with adjacent side of length x feet. From equation (1) we have that

$$\tan \frac{\pi}{6} = \frac{12}{x}. \tag{2}$$

Since $\tan \dfrac{\pi}{6} = \dfrac{1}{\sqrt{3}}$ (Figure 5.2) equation (2) becomes

$$\frac{1}{\sqrt{3}} = \frac{12}{x}.$$

Thus $x = 12\sqrt{3}$ feet ≈ 20.8 feet. □

Other Trigonometric Functions

When θ is an angle of a right triangle with opposite side of length y, adjacent side of length x, and hypotenuse of length h, three other trigonometric functions of θ may be defined: the **cotangent** of θ, written cot θ, the **secant** of θ, written sec θ, and the **cosecant** of θ, written csc θ, as follows:

$$\cot \theta = \frac{x}{y} \left(\frac{\text{adjacent}}{\text{opposite}} \right)$$

$$\sec \theta = \frac{h}{x} \left(\frac{\text{hypotenuse}}{\text{adjacent}} \right)$$

$$\csc \theta = \frac{h}{y} \left(\frac{\text{hypotenuse}}{\text{opposite}} \right)$$

Figure 5.6

(See Figure 5.6.)

Figure 5.7

Example 2 For the right triangle in Figure 5.7 we have

$$\tan \theta = \frac{3}{4} \qquad \sec \theta = \frac{5}{4}$$

$$\cot \theta = \frac{4}{3} \qquad \csc \theta = \frac{5}{3}. \qquad □$$

Since the sine and cosine of the angle θ in Figure 5.6 may be expressed as $\sin \theta = \dfrac{y}{h}$ and $\cos \theta = \dfrac{x}{h}$, we may express the other four trigonometric functions in terms of $\sin \theta$ and $\cos \theta$ as follows:

$$\tan \theta = \frac{\sin \theta}{\cos \theta} \qquad \left(\text{since } \frac{y/h}{x/h} = \frac{y}{x} \right) \tag{3}$$

$$\cot \theta = \frac{\cos \theta}{\sin \theta} \quad \left(\text{since } \frac{x/h}{y/h} = \frac{x}{y}\right) \tag{4}$$

$$\sec \theta = \frac{1}{\cos \theta} \quad \left(\text{since } \frac{1}{x/h} = \frac{h}{x}\right) \tag{5}$$

and

$$\csc \theta = \frac{1}{\sin \theta} \quad \left(\text{since } \frac{1}{y/h} = \frac{h}{y}\right). \tag{6}$$

The advantage of using equations (3)–(6) to define these four trigonometric functions is that the right-hand sides of these equations are defined for all numbers θ for which the denominators are not zero. This allows us to use periodicity to define the function $f(x) = \tan x$ for all numbers x in $(-\infty, \infty)$ for which $\cos x \neq 0$ $\left(\text{that is, for all } x \text{ except } x = \pm\frac{\pi}{2}, \pm\frac{3\pi}{2}, \ldots\right)$, and similarly for the functions $\cot x$, $\sec x$ and $\csc x$.

Graphs of these four functions appear in Figures 5.8–5.11.

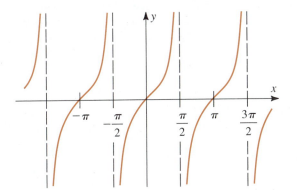

Figure 5.8 Graph of $y = \tan x$.

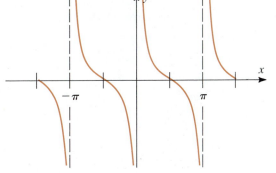

Figure 5.9 Graph of $y = \cot x$.

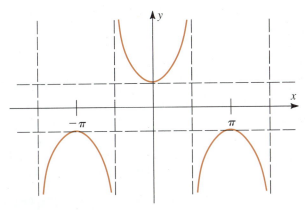

Figure 5.10 Graph of $y = \sec x$.

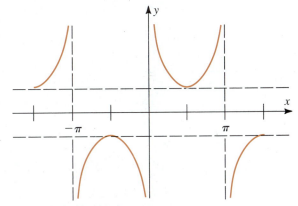

Figure 5.11 Graph of $y = \csc x$.

Example 3 According to equations (3)–(6) we have

(a) $\tan\left(\dfrac{7\pi}{4}\right) = \dfrac{\sin\left(\dfrac{7\pi}{4}\right)}{\cos\left(\dfrac{7\pi}{4}\right)} = \dfrac{\left(-\dfrac{\sqrt{2}}{2}\right)}{\left(\dfrac{\sqrt{2}}{2}\right)} = -1$

(b) $\cot\left(-\dfrac{\pi}{2}\right) = \dfrac{\cos\left(-\dfrac{\pi}{2}\right)}{\sin\left(-\dfrac{\pi}{2}\right)} = \dfrac{0}{-1} = 0$

(c) $\sec(5\pi) = \dfrac{1}{\cos(5\pi)} = \dfrac{1}{-1} = -1$

(d) $\csc\left(-\dfrac{3\pi}{4}\right) = \dfrac{1}{\sin\left(-\dfrac{3\pi}{4}\right)} = \dfrac{1}{\left(-\dfrac{\sqrt{2}}{2}\right)} = -\dfrac{2}{\sqrt{2}}.$ □

Derivatives

Since the four functions $\tan x$, $\cot x$, $\sec x$, and $\csc x$ are defined as quotients involving $\sin x$ and $\cos x$, their derivatives can be found by using the quotient rule. For example, the derivative of the function $y = \tan x$ is

$$\frac{d}{dx} \tan x = \frac{d}{dx}\left(\frac{\sin x}{\cos x}\right)$$

$$= \frac{\cos x\left(\dfrac{d}{dx}\sin x\right) - \sin x\left(\dfrac{d}{dx}\cos x\right)}{\cos^2 x}$$

$$= \frac{\cos x(\cos x) - \sin x(-\sin x)}{\cos^2 x}$$

$$= \frac{\cos^2 x + \sin^2 x}{\cos^2 x}$$

$$= \frac{1}{\cos^2 x} \qquad (\text{since } \cos^2 x + \sin^2 x = 1)$$

$$= \sec^2 x \qquad \left(\text{since } \frac{1}{\cos x} = \sec x,\ \cos x \neq 0\right).$$

That is

$$\frac{d}{dx} \tan x = \sec^2 x. \tag{7}$$

In Exercise 42 you are asked to use the same technique to establish the following additional differentiation formulas.

$$\frac{d}{dx} \cot x = -\csc^2 x \tag{8}$$

$$\frac{d}{dx} \sec x = \sec x \tan x \tag{9}$$

$$\frac{d}{dx} \csc x = -\csc x \cot x. \tag{10}$$

Example 4 Find $f'(x)$ for $f(x) = x \sec x$.

Strategy

Recognize f as a product $g(x)h(x)$ with $g(x) = x$ and $h(x) = \sec x$. Apply product rule and formula (9).

Solution

By the product rule and formula (9) we have

$$f'(x) = \left(\frac{d}{dx} x\right)\sec x + x\left(\frac{d}{dx} \sec x\right)$$

$$= (1)\sec x + x(\sec x \tan x)$$

$$= \sec x(1 + x \tan x). \qquad \square$$

Example 5 Find $f'(x)$ for $f(x) = \ln \tan x$.

Strategy

Recognize f as $f(x) = \ln u$ with $u = \tan x$.

Apply Chain Rule

and formula (7).

Simplify result by expressing $\tan x$ and $\sec x$ in terms of $\sin x$ and $\cos x$.

Solution

Since $f(x) = \ln \tan x$ is a composite function, the Chain Rule and equation (7) give

$$f'(x) = \frac{1}{\tan x} \cdot \frac{d}{dx} \tan x$$

$$= \left(\frac{1}{\tan x}\right)\sec^2 x$$

$$= \left(\frac{\cos x}{\sin x}\right)\left(\frac{1}{\cos x}\right)^2$$

$$= \left(\frac{1}{\sin x}\right)\left(\frac{1}{\cos x}\right)$$

$$= \sec x \cdot \csc x. \qquad \square$$

REMARK: It is often easiest to simplify a complicated product of trigonometric functions by first expressing all of them in terms of the sine and cosine, as was done in Example 5.

Integrals

Differentiation formulas (7)–(10) immediately give the following integration formulas.

$$\int \sec^2 x \, dx = \tan x + C \tag{11}$$

$$\int \csc^2 x \, dx = -\cot x + C \tag{12}$$

$$\int \sec x \tan x \, dx = \sec x + C \tag{13}$$

$$\int \csc x \cot x \, dx = -\csc x + C. \tag{14}$$

In addition, the following integration formulas can be verified by differentiation.

$$\int \tan x \, dx = \ln |\sec x| + C \tag{15}$$

$$\int \cot x \, dx = \ln |\sin x| + C \tag{16}$$

$$\int \sec x \, dx = \ln |\sec x + \tan x| + C \tag{17}$$

$$\int \csc x \, dx = \ln |\csc x - \cot x| + C. \tag{18}$$

Example 6 Verify integration formula (15):

$$\int \tan x \, dx = \ln |\sec x| + C.$$

Solution: Because of the absolute value sign, we must examine the two cases $\sec x > 0$ and $\sec x < 0$ separately.

(a) If $\sec x > 0$, then $|\sec x| = \sec x$, and

$$\frac{d}{dx} \ln \sec x = \frac{1}{\sec x} \cdot \frac{d}{dx} \sec x$$

$$= \frac{1}{\sec x} (\sec x \tan x)$$

$$= \tan x.$$

(b) If $\sec x < 0$, then $|\sec x| = -\sec x > 0$, and

$$\frac{d}{dx} \ln |\sec x| = \frac{d}{dx} \ln(-\sec x)$$

$$= \frac{1}{(-\sec x)} \frac{d}{dx}(-\sec x)$$

$$= \frac{1}{(-\sec x)}(-\sec x \tan x)$$

$$= \tan x.$$

Thus in either case formula (15) holds. □

Example 7 Find $\displaystyle\int_0^{\pi/4} \sec x \, dx$.

Solution: Using formula (17), we find that

$$\int_0^{\pi/4} \sec x \, dx = \ln \left| \sec x + \tan x \right|_0^{\pi/4}$$

$$= \ln \left| \sec\left(\frac{\pi}{4}\right) + \tan\left(\frac{\pi}{4}\right) \right| - \ln \left| \sec(0) + \tan(0) \right|$$

$$= \ln|\sqrt{2} + 1| - \ln |1 + 0|$$

$$= \ln(\sqrt{2} + 1).$$ □

REMARK: When working with the four trigonometric functions $\tan x$, $\cot x$, $\sec x$, and $\csc x$, we should note that these functions are not defined for all x, as are the functions $\sin x$ and $\cos x$. It is particularly important in calculations involving definite integrals of these functions to ensure that the integrands are defined for all x involved in the integration. For example, $\displaystyle\int_0^{\pi} \tan x \, dx$ is not defined, since $\tan x$ is undefined for $x = \dfrac{\pi}{2}$, which lies in the interval $[0, \pi]$.

Exercise Set 6.5

In Exercises 1–4 find (a) $\tan \theta$, (b) $\cot \theta$, (c) $\sec \theta$, and (d) $\csc \theta$ for the given triangle and angle.

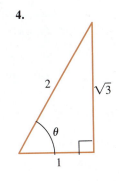

1.   **2.**   **3.**   **4.**

5. Figure 5.12 shows two points A and B on opposite sides of a straight river bed. A third point C is located 50 m from point A as indicated in Figure 5.12. A is directly opposite point B. What is the distance between points A and B if $\theta = \dfrac{\pi}{6}$?

Figure 5.12

6. What is the answer to the question in Exercise 5 if $\theta = \dfrac{\pi}{3}$?

7. A child stands 30 meters from a point directly below the location of an airplane. If the angle between ground level and the child's line of sight to the airplane is 30°, what is the altitude of the airplane? (Neglect the child's height.)

In Exercises 8–23 find the derivative of the given function.

8. $y = \tan \pi x$

9. $f(x) = x \sec x$

10. $f(x) = \sec(x + \pi/6)$

11. $y = \cot 6x$

12. $y = e^x \sec x$

13. $f(x) = \sec x \cdot \tan x$

14. $f(x) = \ln \sec x$

15. $y = \dfrac{1}{\sec x}$

16. $f(x) = \sqrt{1 + \sec^2 x}$

17. $f(x) = \cot^2 3x$

18. $y = x^3 \csc(1 - x)$

19. $y = \tan x^3$

20. $f(x) = \dfrac{\tan(x + \pi)}{x}$

21. $f(x) = \sec \sqrt{1 + x^2}$

22. $y = e^{\sec \pi x}$

23. $f(x) = \cot \ln \sqrt{x}$

In Exercises 24–35 find the integral.

24. $\displaystyle\int \cot 2x \, dx$

25. $\displaystyle\int \sec^2 4x \, dx$

26. $\displaystyle\int_0^{\pi/4} 3 \sec x \tan x \, dx$

27. $\displaystyle\int x \sec^2 x^2 \, dx$

28. $\displaystyle\int \tan \pi x \, dx$

29. $\displaystyle\int_0^{\pi/4} \sec(\pi + x) \, dx$

30. $\displaystyle\int \dfrac{\sec^2 \sqrt{x}}{\sqrt{x}} \, dx$

31. $\displaystyle\int (x - \csc^2 x) \, dx$

32. $\displaystyle\int_0^{\pi/\sqrt{2}} x \tan(\pi - x^2) \, dx$

33. $\displaystyle\int \csc(\pi x)\cot(\pi x) \, dx$

34. $\displaystyle\int \dfrac{\sec^2 \sqrt{2x + 1}}{\sqrt{2x + 1}} \, dx$

35. $\displaystyle\int (\csc^2 \pi x - \cot \pi x) \, dx$

36. Find the area of the region bounded by the graph of $y = \tan x$ and the x-axis for $0 \le x \le \pi/4$.

37. Find the area of the region bounded by the graphs of $y = \tan x$ and $y = \dfrac{4}{\pi} x$. $\left(\textit{Hint: } \text{Superimpose the graph of } y = \dfrac{4}{\pi} x \text{ on the graph of } y = \tan x \text{ to identify the region.}\right)$

38. Find the volume of the solid obtained by revolving about the x-axis the region bounded above by the graph of $y = \sec x$ and below by the x-axis for $-\pi/4 \le x \le \pi/4$.

39. Find the volume of the solid obtained by revolving about the x-axis the region bounded above by the graph of $y = \csc x$ and below by the x-axis for $\dfrac{\pi}{4} \le x \le \dfrac{3\pi}{4}$.

40. Find the average value of the function $y = \sec x$ on the interval $[0, \pi/4]$.

41. What are the vertical asymptotes for the graph of $y = \tan x$? Why?

42. Verify differentiation formulas (8)–(10).

43. Verify integration formulas (16)–(18).

SUMMARY OUTLINE OF CHAPTER 6

■ Angles are measured in **radians** and **degrees:** (Page 316)

2π radians $= 360°$.

■ If $P = (x, y)$ is a point on the unit circle corresponding to an angle θ (see Figure 6.1), then (Page 323)
(B constant)

$\sin \theta = y$ and $\cos \theta = x$.

Figure 6.1

■ If opp, adj, and hyp are the lengths of the sides of a right triangle containing the angle θ, as in (Page 328)
Figure 6.2, then

$$\sin \theta = \frac{\text{opp}}{\text{hyp}} \quad \text{and} \quad \cos \theta = \frac{\text{adj}}{\text{hyp}}.$$

Figure 6.2

■ Identities involving $\sin \theta$ and $\cos \theta$: (Page 331)

$$\left\{ \begin{array}{l} \sin(\theta + 2n\pi) = \sin \theta \\ \cos(\theta + 2n\pi) = \cos \theta \end{array} \right\}$$ Extending definitions of $\sin \theta$, $\cos \theta$ to all numbers θ.

$$\sin^2 \theta + \cos^2 \theta = 1$$

$$\sin^2 \theta = \frac{1}{2}(1 - \cos 2\theta)$$

$$\cos^2 \theta = \frac{1}{2}(1 + \cos 2\theta).$$

■ Derivatives of sine and cosine functions: (Page 334)

$$\frac{d}{dx} \sin u = \cos u \cdot \frac{du}{dx}$$

$$\frac{d}{dx} \cos u = -\sin u \cdot \frac{du}{dx}.$$

■ Integrals of sine and cosine functions: (Page 343)

$$\int \sin u \, du = -\cos u + C$$

$$\int \cos u \, du = \sin u + C.$$

■ Four additional trigonometric functions are defined as follows: (Page 350)

$$\tan \theta = \frac{\sin \theta}{\cos \theta} = \frac{\text{opp}}{\text{adj}} \qquad \sec \theta = \frac{1}{\cos \theta} = \frac{\text{hyp}}{\text{adj}}$$

$$\cot \theta = \frac{\cos \theta}{\sin \theta} = \frac{\text{adj}}{\text{opp}} \qquad \csc \theta = \frac{1}{\sin \theta} = \frac{\text{hyp}}{\text{opp}}.$$

(See Figure 6.2.)

■ Derivatives of these four functions are (Page 352)

$$\frac{d}{dx} \tan x = \sec^2 x \qquad \frac{d}{dx} \sec x = \sec x \tan x$$

$$\frac{d}{dx} \cot x = -\csc^2 x \qquad \frac{d}{dx} \csc x = -\cot x.$$

■ Integrals involving these four functions are (Page 353)

$$\int \sec^2 x \, dx = \tan x + C$$

$$\int \sec x \tan x \, dx = \sec x + C$$

$$\int \csc^2 x \, dx = -\cot x + C$$

$$\int \csc x \cot x \, dx = -\csc x + C$$

$$\int \tan x \, dx = \ln |\sec x| + C$$

$$\int \cot x \, dx = \ln |\sin x| + C$$

$$\int \sec x \, dx = \ln |\sec x + \tan x| + C$$

$$\int \csc x \, dx = \ln |\csc x - \cot x| + C.$$

REVIEW EXERCISES—CHAPTER 6

1. Convert the following degree measures to radian measures.

  a. $60°$ **b.** $35°$ **c.** $120°$
  d. $210°$ **e.** $10°$ **f.** $75°$

2. Convert the following radian measures to degree measures.

  a. $\dfrac{\pi}{6}$ **b.** $\dfrac{3\pi}{4}$ **c.** $\dfrac{7\pi}{4}$

  d. $\dfrac{3\pi}{2}$ **e.** $\dfrac{7\pi}{6}$ **f.** $\dfrac{11\pi}{12}$

3. Find the principal angle equivalent, in radians, for each of the following.

  a. $-\dfrac{\pi}{2}$ **b.** $\dfrac{7\pi}{2}$ **c.** $\dfrac{11\pi}{4}$

  d. $-\dfrac{5\pi}{3}$ **e.** -7π **f.** 9π

4. Find the following values of the sine and cosine functions.

  a. $\sin \dfrac{5\pi}{6}$ **b.** $\cos\left(-\dfrac{3\pi}{2}\right)$ **c.** $\sin \dfrac{9\pi}{2}$

  d. $\cos \dfrac{11\pi}{3}$ **e.** $\sin\left(-\dfrac{5\pi}{3}\right)$ **f.** $\sin(8\pi)$

  g. $\cos \dfrac{7\pi}{2}$ **h.** $\cos 3\pi$ **i.** $\sin\left(-\dfrac{11\pi}{6}\right)$

In Exercises 5–20 find the derivative.

5. $y = \sin(\pi - x)$ **6.** $f(x) = \pi \cos(3x)$

7. $f(x) = \tan 3x$ **8.** $y = \sin \sqrt{x}$

9. $y = \sin(\cos x)$ **10.** $f(x) = \dfrac{x}{\cos x}$

11. $y = \ln \tan x$ **12.** $y = \sqrt{1 + \cos x}$

13. $f(x) = e^{\sec \pi x}$ **14.** $f(x) = \dfrac{\cos x}{\tan x}$

15. $y = (x + \sec x)^3$ **16.** $y = (\cos x - \cot x)^2$

17. $f(x) = \dfrac{e^{\cos x}}{1 + \sin x}$ **18.** $f(x) = \tan e^x$

19. $y = \ln \sqrt{4 + \cos x}$ **20.** $f(x) = \sec x - \tan 2x$

In Exercises 21–36 find the integral.

21. $\displaystyle \int \sin 4x \, dx$ **22.** $\displaystyle \int \pi \cos(\pi + x) \, dx$

23. $\displaystyle \int \sec \pi x \tan \pi x \, dx$ **24.** $\displaystyle \int \sec^2 x \tan x \, dx$

25. $\displaystyle \int \sqrt{\tan x} \sec^2 x \, dx$ **26.** $\displaystyle \int \dfrac{\cos x}{\pi + \sin x} \, dx$

27. $\displaystyle \int \sec^2(3x)e^{\tan 3x} \, dx$ **28.** $\displaystyle \int \dfrac{\sec^2 \sqrt{x}}{\sqrt{x}} \, dx$

29. $\displaystyle \int \cos^4 x \sin x \, dx$ **30.** $\displaystyle \int \sin x \sqrt{1 + \cos x} \, dx$

31. $\displaystyle\int_0^{\pi/4} \cos\left(x + \frac{\pi}{2}\right) dx$ **32.** $\displaystyle\int_{-\pi/4}^{\pi/4} \sec^2 x \, dx$

33. $\displaystyle\int_0^{\pi/4} \sec x \tan x e^{\sec x} \, dx$ **34.** $\displaystyle\int_{\pi/4}^{\pi/2} x \csc^2 x^2 \, dx$

35. $\displaystyle\int_0^1 \sin \pi x \cos \pi x \, dx$ **36.** $\displaystyle\int_0^{\pi/4} \frac{\sec^2 x}{1 + \tan x} \, dx$

37. A predator–prey model is one consisting of two species, say, x's and y's (foxes and rabbits, for example), where the x's prey on the y's for food. A typical solution to a predator–prey model involves functions $x(t)$ and $y(t)$, representing the size of the two populations as functions of time, of the form

$$x(t) = 10 + 2 \sin\left(t + \frac{\pi}{2}\right)$$

$$y(t) = 20 + 5 \cos\left(t + \frac{\pi}{2}\right).$$

   a. Graph these two functions on the same pair of axes.
   b. Find the maximum and minimum values of $x(t)$.
   c. Find the maximum and minimum values of $y(t)$.

38. A mass connected to a spring oscillates back and forth according to the function

$$x(t) = 10 \cos\left(\pi t - \frac{\pi}{4}\right)$$

where $x(t)$ is the location of the mass at time t relative to its rest position.

   a. What is the location of the mass at time $t = 0$?
   b. What is the velocity of the mass at time t? (*Hint:* Recall $v(t) = x'(t)$.)
   c. What is the farthest the mass travels from its rest position?

39. Find the slope of the line tangent to the graph of $y = \tan x$ at the point $(\pi/4, 1)$.

40. Find an equation for the line tangent to the graph of $y = \sec x$ at the point $(\pi/3, 2)$.

41. Find the area of the region bounded by the graph of $y = x \sec x^2$ and the x-axis for $0 \le x \le \sqrt{\dfrac{\pi}{3}}$.

42. Find the area of the region bounded by the graph of $y = \sec x \cdot \tan x$ and the x-axis for $-\dfrac{\pi}{3} \le x \le \dfrac{\pi}{3}$.

43. Find the average value of the function $f(x) = \csc x$ for $\dfrac{\pi}{6} \le x \le \dfrac{\pi}{3}$.

44. Find the volume of the solid obtained by revolving about the x-axis the region bounded by the graph of $y = \sqrt{1 + \sin x}$ and the x-axis for $0 \le x \le \pi$.

7 Techniques of Integration

Chapter

7.1 REVIEW OF METHOD OF SUBSTITUTIONS

Much of the calculus that we have discussed involves evaluating definite integrals using the Fundamental Theorem of Calculus

$$\int_a^b f(x)\, dx = F(b) - F(a) \tag{1}$$

where $F' = f$. Both the process of evaluating the definite integral in equation (1) and finding an antiderivative (or indefinite integral) F for f are referred to as **integration**.

Up until this point all antiderivatives F have been found by recognizing the integrand f as the derivative of some known function or by using a simple substitution. The purpose of this chapter is to develop additional techniques for evaluating more general types of integrals. Before presenting these additional techniques, we summarize those integration rules we have already developed and review briefly the method of substitutions.

Constants and Powers

1. $\displaystyle\int a\, dx = ax + C$

2. $\displaystyle\int x^n\, dx = \frac{x^{n+1}}{n+1} + C, \qquad n \neq -1$

3. $\displaystyle\int \frac{1}{x}\, dx = \ln|x| + C, \qquad x \neq 0$

Exponential Functions

4. $\displaystyle\int e^x\, dx = e^x + C$

5. $\displaystyle\int a^x\, dx = \frac{1}{\ln a} a^x + C, \qquad a \neq 1,\, a > 0$

Trigonometric Functions

6. $\displaystyle\int \sin x \, dx = -\cos x + C$

7. $\displaystyle\int \cos x \, dx = \sin x + C$

8. $\displaystyle\int \sec^2 x \, dx = \tan x + C$

9. $\displaystyle\int \csc^2 x \, dx = -\cot x + C$

10. $\displaystyle\int \sec x \cdot \tan x \, dx = \sec x + C$

11. $\displaystyle\int \csc x \cdot \cot x \, dx = -\csc x + C$

12. $\displaystyle\int \tan x \, dx = \ln |\sec x| + C$

13. $\displaystyle\int \cot x \, dx = \ln |\sin x| + C$

14. $\displaystyle\int \sec x \, dx = \ln |\sec x + \tan x| + C$

15. $\displaystyle\int \csc x \, dx = \ln |\csc x - \cot x| + C$

The exercise set of this section offers additional practice in the use of these integration formulas together with the technique of **substitutions**. Recall that an integral of the form

$$\int f(g(x)) \cdot g'(x) \, dx \tag{2}$$

may be evaluated by means of the substitution

$$u = g(x); \qquad du = g'(x) \, dx.$$

We obtain

$$\int f(g(x)) \, g'(x) \, dx = \int f(u) \, du.$$

Thus the integral in line (2) can be evaluated if an antiderivative F for f can be found.

Example 1 Find $\displaystyle\int \cos x \cdot e^{\sin x}\, dx$.

Strategy

The key is to notice that $\cos x$ is the derivative of the exponent $\sin x$. Thus try the substitution

$u = \sin x$.

Solution

We make the substitution

$$u = \sin x; \qquad du = \cos x\, dx.$$

We obtain

$$\int \cos x \cdot e^{\sin x}\, dx = \int \underbrace{e^{\sin x}}_{e^u}\ \underbrace{\cos x\, dx}_{du}$$

Apply formula 4 from the list of integration formulas.

$$= \int e^u\, du$$

Substitute back in terms of x.

$$= e^u + C$$
$$= e^{\sin x} + C.$$ □

Example 2 Find $\displaystyle\int_0^1 \frac{x}{\sqrt{x^2 + 4}}\, dx$.

Strategy

To use the Fundamental Theorem, first find an antiderivative of the integrand.
Since the derivative of $x^2 + 4$ is $2x$, we try the substitution

$u = x^2 + 4$

$du = 2x\, dx.$

Apply formula 2 with $n = -\dfrac{1}{2}$.

Solution

First, we find $\displaystyle\int \frac{x}{\sqrt{x^2 + 4}}\, dx$.

We let $u = x^2 + 4$. Then $du = 2x\, dx$, so $x\, dx = \dfrac{1}{2}\, du$. We obtain

$$\int \frac{x}{\sqrt{x^2 + 4}}\, dx = \int \underbrace{(x^2 + 4)^{-1/2}}_{u^{-1/2}} \cdot \underbrace{x\, dx}_{\frac{1}{2}\, du}$$

$$= \int u^{-1/2} \cdot \frac{1}{2}\, du$$

$$= \frac{1}{2}\int u^{-1/2}\, du$$

$$= \frac{1}{2}(2u^{1/2}) + C$$

Substitute back in terms of x.

$$= \sqrt{x^2 + 4} + C.$$

Then the definite integral is

Use this antiderivative to compute the definite integral from (1).

$$\int_0^1 \frac{x}{\sqrt{x^2 + 4}}\, dx = \sqrt{x^2 + 4}\ \Big]_0^1$$

$$= \sqrt{5} - 2.$$ □

Example 3 Find $\int \dfrac{x^2 - 3}{x^3 - 9x} \, dx.$

Solution: We try the substitution

$$u = x^3 - 9x; \qquad du = (3x^2 - 9) \, dx.$$

Then $(x^2 - 3) \, dx = \dfrac{1}{3}(3x^2 - 9) \, dx = \dfrac{1}{3} \, du,$ and we obtain

$$\int \frac{x^2 - 3}{x^3 - 9x} \, dx = \int \frac{\dfrac{1}{3} \, du}{u}$$

$$= \frac{1}{3} \int \frac{du}{u}$$

$$= \frac{1}{3} \ln |u| + C \qquad \text{(by formula 3)}$$

$$= \frac{1}{3} \ln |x^3 - 9x| + C. \qquad \qquad \square$$

Exercise Set 7.1

In Exercises 1–32 evaluate the integral using a substitution where necessary.

1. $\displaystyle\int \cos x \sqrt{4 + \sin x} \, dx$

2. $\displaystyle\int_0^2 x\sqrt{1 + x^2} \, dx$

3. $\displaystyle\int_3^4 (x^2 - 3)\sqrt{x^3 - 9x} \, dx$

4. $\displaystyle\int \frac{x}{1 + x} \, dx$

5. $\displaystyle\int_0^1 xe^{x^2} \, dx$

6. $\displaystyle\int_0^1 \sin \pi x \, dx$

7. $\displaystyle\int \frac{x^4}{1 + x^5} \, dx$

8. $\displaystyle\int \frac{x^3}{\sqrt{1 - x^4}} \, dx$

9. $\displaystyle\int \frac{e^x}{1 + e^x} \, dx$

10. $\displaystyle\int \tan x \cdot \sec^2 x \, dx$

11. $\displaystyle\int_0^{\pi/4} \tan^3 x \cdot \sec^2 x \, dx$

12. $\displaystyle\int \frac{x + 2}{x^2 + 4x + 7} \, dx$

13. $\displaystyle\int (x + 3)(x^2 + 6x + 4)^3 \, dx$

14. $\displaystyle\int_0^1 \frac{\cos \pi x}{4 + \sin \pi x} \, dx$

15. $\displaystyle\int \sec x \tan x \cdot e^{\sec x} \, dx$

16. $\displaystyle\int x^{1/3}\sqrt{1 + x^{4/3}} \, dx$

17. $\displaystyle\int_1^4 \frac{e^{\sqrt{x}}}{\sqrt{x}} \, dx$

18. $\displaystyle\int \frac{\sin x \cos x}{1 + \sin^2 x} \, dx$

19. $\displaystyle\int \sin^3 x \cos x \, dx$

20. $\displaystyle\int_0^{\sqrt{\pi/2}} x \tan x^2 \, dx$

21. $\displaystyle\int \frac{dx}{\cos^2 x}$

22. $\displaystyle\int \frac{\sec^2 x \, dx}{\sqrt{1 + 2 \tan x}}$

23. $\displaystyle\int_1^2 \frac{dx}{x^2 + 2x + 1}$

24. $\displaystyle\int_0^3 \frac{2^{\sqrt{x+1}}}{\sqrt{x + 1}} \, dx$

25. $\displaystyle\int e^x \tan(1 + e^x) \, dx$

26. $\displaystyle\int x \sec(\pi + x^2) \, dx$

27. $\displaystyle\int \frac{\cos x \, dx}{\sqrt{1 - \sin^2 x}}$

28. $\displaystyle\int (e^x + 7)^2 \, dx$

29. $\displaystyle\int_e^{e^2} \frac{1}{x \ln x} \, dx$

30. $\displaystyle\int \frac{(1 + 2\sqrt{x})^3}{\sqrt{x}} \, dx$

31. $\displaystyle\int \frac{1}{\sqrt{x}(1 + \sqrt{x})} \, dx$

32. $\displaystyle\int_1^2 \frac{x^2 - 3x - 4}{x - 4} \, dx$

33. Find the area of the region bounded above by the graph of $f(x) = x \sin x^2$ and below by the x-axis between $x = 0$ and $x = \dfrac{\sqrt{\pi}}{2}$.

34. Find the volume of the solid obtained by revolving about the x-axis the region bounded by the graph of $y = \sin x$ and the x-axis between $x = 0$ and $x = \pi$.

(*Hint:* Use the identity $\sin^2 x = \dfrac{1}{2} - \dfrac{1}{2} \cos 2x$.)

35. A supplier of microwave ovens determines that the sell-

ing price p at which consumers will purchase x ovens per week is given by the demand function $p = D(x) = \dfrac{20{,}000}{20 + 5x}$. If the supplier wishes to sell ovens at a rate of $x = 16$ per week, the corresponding selling price must, therefore, be $p_0 = D(x_0) = \dfrac{20{,}000}{20 + 5 \cdot 16} = 200$ dollars per oven. Find the consumers' surplus $\displaystyle\int_0^{16} D(x)\, dx - (16)(200)$, corresponding to this equilibrium point $(x_0, p_0) = (16, 200)$.

7.2 INTEGRATION BY PARTS

The Product Rule tells us how to differentiate a product of two functions. **Integration by parts** is a method that is sometimes helpful in finding an antiderivative for the product of two functions. It says

$$\int f(x)g'(x)\, dx = f(x)g(x) - \int f'(x)g(x)\, dx. \tag{1}$$

Equation (1) is called the *integration by parts formula*. It allows us to replace the integral of a product of two functions, f and $g' = \dfrac{dg}{dx}$, by the *difference* of

(a) the product fg (involving an *integral* of the original function g'), and

(b) the integral $\displaystyle\int f'g\, dx$ (involving the *derivative* of the original function f, and an *integral* of the original function g').

The advantage in using the integration by parts formula is that the integral on the right-hand side of equation (1) is often easier to evaluate than the integral on the left. To use equation (1), we must be able to envision an integrand as the product of two functions, one called f, the other g'.

Example 1 Find $\displaystyle\int xe^x\, dx$.

Solution: If we choose to let

$$f(x) = x \quad \text{and} \quad g'(x) = e^x$$

it follows that

$$f'(x) = 1 \quad \text{and} \quad g(x) = e^x$$

since $g(x) = e^x$ is an antiderivative (integral) for $g'(x) = e^x$. The integration by parts formula then gives

$$\int \underbrace{x}_{f(x)} \underbrace{e^x}_{g'(x)} dx = \underbrace{x}_{f(x)} \underbrace{e^x}_{g(x)} - \int \underbrace{(1)}_{f'(x)} \underbrace{e^x}_{g(x)} dx$$

$$= xe^x - \int e^x dx$$

$$= xe^x - e^x + C.$$

Thus

$$\int xe^x dx = xe^x - e^x + C. \qquad \square$$

Example 2 Find $\displaystyle\int x\sqrt{x + 5}\ dx$.

Strategy

Use integration by parts. Choose

$f(x) = x$

since $f'(x) = 1$ is simpler than $f(x)$.

Solution

We choose

$$f(x) = x \quad \text{and} \quad g'(x) = \sqrt{x + 5} = (x + 5)^{1/2}.$$

Then

$$f'(x) = 1 \quad \text{and} \quad g(x) = \frac{2}{3}(x + 5)^{3/2}.$$

By the integration by parts formula

Apply formula (1).

$$\int \underbrace{x}_{f} \underbrace{(x + 5)^{1/2}}_{g'} dx = \underbrace{x}_{f} \cdot \underbrace{\frac{2}{3}(x + 5)^{3/2}}_{g} - \int \underbrace{1}_{f'} \cdot \underbrace{\frac{2}{3}(x + 5)^{3/2}}_{g} dx$$

$$= \frac{2}{3}x(x + 5)^{3/2} - \frac{4}{15}(x + 5)^{5/2} + C. \qquad \square$$

REMARK 1: In applying the integration by parts formula, we often find more than one choice for which function to take as f and which for g'. In general, we try to choose these functions in such a way that the integral on the right-hand side of equation (1) is "simpler" than the one on the left. For example, in Example 1 we did not make the choice

$$f(x) = e^x, \qquad g'(x) = x$$

since we would have then had

$$f'(x) = e^x, \qquad g(x) = \frac{1}{2}x^2$$

and equation (1) would have given

$$\int xe^x \, dx = \frac{1}{2}x^2e^x - \int \frac{1}{2}x^2e^x \, dx. \tag{2}$$

Although equation (2) is not incorrect, it is not helpful since we have no hint as to how to evaluate the integral on its right-hand side.

REMARK 2: There is no need to include an arbitrary constant C when finding an antiderivative g for g' when using equation (1) since this equation holds for *any* antiderivative of g'. The simplest choice is therefore to let the arbitrary constant C equal zero.

We can use the notation of differentials to write equation (1) in a simple form. Recall, if we let $u = f(x)$, we define the differential du by the equation $du = f'(x) \, dx$. Thus if we let

$$u = f(x) \quad \text{and} \quad dv = g'(x) \, dx$$

then

$$du = f'(x) \, dx \quad \text{and} \quad v = g(x)$$

and equation (1) becomes

$$\int u \, dv = uv - \int v \, du. \tag{3}$$

Example 3 Find $\int x \cos x \, dx$.

Strategy

Choose $u = x$, since the differential du is simply $du = dx$.

Then $dv = \cos x \, dx$, so $v = \sin x$.

Apply equation (3).

Solution

We let

$$u = x \quad \text{and} \quad dv = \cos x \, dx.$$

Then

$$du = dx \quad \text{and} \quad v = \sin x.$$

Equation (3) then gives

$$\int \underbrace{x}_{u} \underbrace{\cos x \, dx}_{dv} = \underbrace{x}_{u} \underbrace{\sin x}_{v} - \int \underbrace{\sin x}_{v} \underbrace{dx}_{du}$$

$$= x \sin x - (-\cos x) + C$$

$$= x \sin x + \cos x + C. \qquad \square$$

Often more than one application of integration by parts is required, as the following example shows.

Example 4 Find $\int x^2 e^{x+3} \, dx$.

Strategy

Take $u = x^2$ since differentiation reduces the exponent by one.

Solution

We take

$$u = x^2, \qquad dv = e^{x+3} \, dx.$$

Then

Find du and v.

$$du = 2x \, dx, \qquad v = e^{x+3}$$

and integration by parts gives

Apply equation (3).

$$\int x^2 e^{x+3} \, dx = x^2 e^{x+3} - \int 2x e^{x+3} \, dx$$

$$= x^2 e^{x+3} - 2 \int x e^{x+3} \, dx. \qquad (4)$$

The integral on the right requires a second application of integration by parts. In

$$\int x e^{x+3} \, dx$$

In $\int x e^{x+3} \, dx$ again take $u = x$ since differentiation will give $du = dx$.

we take

$$u = x, \qquad dv = e^{x+3} \, dx.$$

Then

Find du and v.

$$du = dx, \qquad v = e^{x+3}$$

and integration by parts gives

Apply equation (3) again.

$$\int x e^{x+3} \, dx = x e^{x+3} - \int e^{x+3} \, dx$$

$$= x e^{x+3} - e^{x+3} + C_1. \qquad (5)$$

Combine equations (4) and (5).

Combining equations (4) and (5), we obtain

$$\int x^2 e^{x+3} \, dx = x^2 e^{x+3} - 2[x e^{x+3} - e^{x+3} + C_1]$$

$$= (x^2 - 2x + 2)e^{x+3} + C$$

where $C = -2C_1$. (Since C_1 is an arbitrary constant, so is $C = -2C_1$, but in a simpler form.) □

Sometimes integration by parts can be used to evaluate an integral of the form $\int f(x) \, dx$ by taking $u = f(x)$ and $dv = dx$, as in the following example.

Example 5 Use integration by parts to show that

$$\int \ln x \, dx = x \ln x - x + C, \qquad x > 0.$$

Solution: Since the only factor of the integrand is $f(x) = \ln x$, we take

$$u = \ln x, \qquad dv = dx.$$

Then

$$du = \frac{1}{x} \, dx, \qquad v = x$$

and equation (3) gives

$$\int \ln x \, dx = x \ln x - \int x \left(\frac{1}{x} \right) dx$$

$$= x \ln x - \int 1 \, dx$$

$$= x \ln x - x + C$$

as claimed. $\square$

Example 6 Find $\int e^x \cos x \, dx$.

Solution: This time there is no clue as to which function to take as u. We arbitrarily choose

$$u = e^x, \qquad dv = \cos x \, dx.$$

Then

$$du = e^x, \qquad v = \sin x$$

and the integration by parts formula gives

$$\int e^x \cos x \, dx = e^x \sin x - \int e^x \sin x \, dx. \qquad (6)$$

Now the integral on the right-hand side of equation (6) seems no simpler than the integral on the left. However, let's try another application of the integration by parts formula in the integral $\int e^x \sin x \, dx$. Again taking $u = e^x$, we have

$$u = e^x, \qquad dv = \sin x \, dx$$

and

$$du = e^x \, dx, \qquad v = -\cos x.$$

Thus

$$\int e^x \sin x \, dx = -e^x \cos x - \int (-\cos x) e^x \, dx$$

$$= -e^x \cos x + \int e^x \cos x \, dx. \qquad (7)$$

Now note that the desired integral has reappeared on the right-hand side of equation (7)! Combining equations (6) and (7) gives

$$\int e^x \cos x \, dx = e^x \sin x - \left\{ -e^x \cos x + \int e^x \cos x \, dx \right\}$$

$$= e^x(\sin x + \cos x) - \int e^x \cos x \, dx.$$

We may now add $\int e^x \cos x \, dx$ to both sides to obtain

$$2 \int e^x \cos x \, dx = e^x(\sin x + \cos x)$$

so

$$\int e^x \cos x \, dx = \frac{1}{2} e^x(\sin x + \cos x) + C. \qquad \Box$$

Our last example illustrates how the integration by parts formula can be applied to definite integrals.

Example 7 Find $\displaystyle\int_1^e \sqrt{x} \ln x \, dx$.

Solution: We first find $\displaystyle\int \sqrt{x} \ln x \, dx$ by taking

$$u = \ln x, \qquad dv = \sqrt{x} \, dx = x^{1/2} \, dx.$$

Then

$$du = \frac{1}{x} \, dx, \qquad v = \frac{2}{3} x^{3/2}$$

so integration by parts gives

$$\int \sqrt{x} \ln x \, dx = \frac{2}{3} x^{3/2} \ln x - \int \frac{2}{3} x^{3/2} \left(\frac{1}{x} \right) dx$$

$$= \frac{2}{3} x^{3/2} \ln x - \frac{2}{3} \int x^{1/2} \, dx$$

$$= \frac{2}{3} x^{3/2} \ln x - \frac{4}{9} x^{3/2} + C.$$

Then by the Fundamental Theorem we have

$$\int_1^e \sqrt{x} \ln x \, dx = \left(\frac{2}{3} x^{3/2} \ln x - \frac{4}{9} x^{3/2} \right) \Bigg]_1^e$$

$$= \left(\frac{2}{3}e^{3/2}\ln e - \frac{4}{9}e^{3/2}\right) - \left(\frac{2}{3}\cdot 1^{3/2}\ln(1) - \frac{4}{9}\cdot 1^{3/2}\right)$$

$$= \frac{2}{3}e^{3/2} - \frac{4}{9}e^{3/2} + \frac{4}{9}$$

$$= \frac{2}{9}e^{3/2} + \frac{4}{9}. \qquad\qquad \square$$

Justifying the Integration by Parts Formula

The Product Rule, applied to the product fg, gives

$$\frac{d}{dx}[f(x)g(x)] = f'(x)g(x) + f(x)g'(x).$$

Solving for the last term on the right, we obtain

$$f(x)g'(x) = \frac{d}{dx}[f(x)g(x)] - f'(x)g(x). \qquad\qquad (8)$$

Now the meaning of equation (8) is that the functions on either side of the equation are the same. Thus they must have the same indefinite integrals. That is

$$\int f(x)g'(x)\, dx = \int \left\{\frac{d}{dx}[f(x)g(x)]\right\} dx - \int f'(x)g(x)\, dx.$$

Since

$$\int \left\{\frac{d}{dx}[f(x)g(x)]\right\} dx = f(x)g(x) + C \qquad\qquad (9)$$

we can write*

$$\int f(x)g'(x)\, dx = f(x)g(x) - \int f'(x)g(x)\, dx$$

as claimed, since the constant C in equation (9) may be combined with the constant of integration in $\int f'(x)g(x)\, dx$.

Exercise Set 7.2

In Exercises 1–22 evaluate the given integral.

1. $\displaystyle\int xe^{-x}\, dx$

2. $\displaystyle\int (x-4)e^x\, dx$

3. $\displaystyle\int x\ln x\, dx$

4. $\displaystyle\int_1^e x^2\ln x\, dx$

---

*Here we use the fact that one antiderivative of the derivative of u is the function u itself. That is, $\displaystyle\int u'(x)\, dx = u(x) + C.$

5. $\int_0^3 x\sqrt{x+1}\ dx$

6. $\int \dfrac{x}{\sqrt{x+4}}\ dx$

7. $\int x \sin x\ dx$

8. $\int x^2 \cos x\ dx$

9. $\int x(2x+1)^5\ dx$

10. $\int x \ln x^2\ dx$

11. $\int x^2 e^x\ dx$

12. $\int x^3 \ln x\ dx$

13. $\int_0^{\pi/4} x \sec^2 x\ dx$

14. $\int_0^{\pi/4} x \sec x \tan x\ dx$

15. $\int_0^1 x^2 e^{-x}\ dx$

16. $\int_1^e (\ln x)^2\ dx$

17. $\int \dfrac{x^3}{\sqrt{1+x^2}}\ dx$

18. $\int x^3\sqrt{1+x^2}\ dx$

19. $\int e^{2x} \cos x\ dx$

20. $\int \sin(\ln x)\ dx$

21. $\int x^3 e^{-x^2}\ dx$

22. $\int e^{\sqrt{x}}\ dx$

23. Find the area of the region bounded by the graphs of $y = \ln(1+x)$, $y = x$, and $x = 1$.

24. Find the volume of the solid obtained by revolving about the x-axis the region bounded by the graph of $y = \ln x$ and the x-axis for $1 \le x \le e$.

25. The present value of a revenue stream flowing at the rate of $R(t) = 100t$ dollars per year for the next 5 years, assuming a nominal rate of interest of $r = 10\%$ compounded continuously, is given by the integral

$$\int_0^5 R(t)e^{-rt}\ dx = \int_0^5 100te^{-0.10t}\ dt.$$

Find the value of this integral.

26. Find the value of a revenue stream flowing at the rate of $R(t) = 1000t$ dollars per year for the next 10 years, assuming a nominal rate of interest of $r = 5\%$ compounded continuously. (See Exercise 25.)

27. Find the average value of the function $f(x) = \ln x$ for x in the interval $[1, e]$.

7.3 THE USE OF INTEGRAL TABLES

For many functions the techniques of this chapter will not succeed in finding an antiderivative. For some of these functions the reason for this failure is that more sophisticated techniques are required. For others, antiderivatives simply do not exist, at least in the form we have come to expect in this text.

Fortunately, a great many antiderivatives have been discovered and tabulated. Hundreds appear in various standard books of mathematical tables. A sample of 34 of these formulas appears on the endpapers of this book. Our purpose here is to demonstrate how these formulas are used.

The integral formulas 1–34 are written in terms of the variable u. The key to using them is to remember the notation of a *substitution:*

$$\int f'(g(x))g'(x)\ dx = \int f'(u)\ du = f(u) + C$$

under the substitution $u = g(x)$, $du = g'(x)\ dx$.

Example 1 Integral formula 24 states that

$$\int \ln u\ du = u \ln u - u + C.$$

This answers the question, "What is the antiderivative of the natural logarithm function?" With $u = x$ we have $du = dx$, and the formula simply states that

$$\int \ln x \, dx = x \ln x - x + C.$$ □

Example 2 To evaluate the integral $\displaystyle\int \frac{dx}{x(3 + 5x)}$, we use formula 10:

$$\int \frac{du}{u(a + bu)} = -\frac{1}{a} \ln \left| \frac{a + bu}{u} \right| + C$$

with $u = x$, $a = 3$, and $b = 5$. We obtain

$$\int \frac{dx}{x(3 + 5x)} = -\frac{1}{3} \ln \left| \frac{3 + 5x}{x} \right| + C.$$ □

Example 3 To evaluate the integral $\displaystyle\int \ln (4x) \, dx$, we again use formula 24:

$$\int \ln u \, du = u \ln u - u + C.$$

However, since our integral is not quite of this form, we need first to make the substitution

$$u = 4x, \qquad du = 4 \, dx.$$

Thus $dx = \dfrac{1}{4} \, du$, and using formula 24, we obtain

$$\begin{aligned}
\int \ln(4x) \, dx &= \int (\ln u)\left(\frac{1}{4}\right) du \\
&= \frac{1}{4} \int \ln u \, du \\
&= \frac{1}{4}[u \ln u - u + C] \\
&= \frac{1}{4}(4x) \ln (4x) - \frac{1}{4}(4x) + \frac{1}{4}C \\
&= x \ln (4x) - x + C.
\end{aligned}$$

(We write C rather than $C/4$ in the last line since C represents only an unknown constant. If a specific value of C were known we would have written $C/4$ rather than C.) □

Example 3 illustrates how we must often execute a substitution before a given integral matches one in our list of formulas. Our final example requires this step also.

Example 4 Evaluate the integral $\int x^3\sqrt{9 + 4x^2}\, dx$.

Strategy

Choose a candidate from the list of integral formulas.

Determine what substitution is required.

Find du and dx.

Apply substitution.

Apply formula 11.

"Substitute back" x^2 in place of u, and write C rather than $C/2$.

Solution

Inspecting our list of formulas suggests that formula 11 might prove to be a match for our integral:

$$\int u\sqrt{a + bu}\, du = -\frac{2(2a - 3bu)(a + bu)^{3/2}}{15b^2} + C.$$

If this were the case we would need to apply the substitution

$$u = x^2, \qquad du = 2x\, dx.$$

Thus $x\, dx = \frac{1}{2}\, du$ and, using formula 11 with $a = 9$ and $b = 4$, we obtain

$$\int x^3\sqrt{9 + 4x^2}\, dx = \int x^2\sqrt{9 + 4x^2}\, x\, dx$$

$$= \int u\sqrt{9 + 4u}\left(\frac{1}{2}\right)du$$

$$= \frac{1}{2}\int u\sqrt{9 + 4u}\, du$$

$$= \frac{1}{2}\left[-\frac{2(2 \cdot 9 - 3 \cdot 4 \cdot u)(9 + 4u)^{3/2}}{15 \cdot 4^2} + C\right]$$

$$= -\frac{2(18 - 12x^2)(9 + 4x^2)^{3/2}}{2 \cdot 15 \cdot 16} + C$$

$$= -\frac{(3 - 2x^2)(9 + 4x^2)^{3/2}}{40} + C. \qquad \square$$

Exercise Set 7.3

In each of Exercises 1–22 use the integral tables on the endpapers to evaluate the given integral.

1. $\int (4 + 3x)^5\, dx$

2. $\int \frac{dx}{(3 - 2x)^2}$

3. $\int \frac{x\, dx}{(5 + x)^2}$

4. $\int \frac{dx}{x(6 + 3x)}$

5. $\int x\sqrt{7 + 2x}\, dx$

6. $\int x\sqrt{1 + x}\, dx$

7. $\int \frac{x\, dx}{\sqrt{3 + 2x}}$

8. $\int \frac{dx}{9 - x^2}$

9. $\int \frac{dx}{\sqrt{16 + x^2}}$

10. $\int xe^{6x}\, dx$

11. $\int \frac{x\, dx}{(5 + x^2)^2}$

12. $\int \frac{x\, dx}{x^2(3 + 5x^2)}$

13. $\int \frac{dx}{9 - 16x^2}$

14. $\int x\sqrt{x^4 + 1}\, dx$

15. $\int \frac{dx}{x\sqrt{x^4 + 9}}$

16. $\int x^3 e^{x^2}\, dx$

17. $\int \frac{4x}{1 + e^{x^2}}\, dx$

18. $\int \frac{x\, dx}{7 - 2e^{x^2}}$

19. $\int \frac{dx}{\sqrt{x}(1 + e^{\sqrt{x}})}$

20. $\int \frac{dx}{x\ln x^2}$

21. $\int x\sin^3(x^2)\, dx$

22. $\int e^x \cos^3(e^x)\, dx$

7.4 IMPROPER INTEGRALS

Up to this point the definite integral $\int_a^b f(x)\,dx$ has been defined only when the function f is continuous on the closed bounded interval $[a, b]$. There are certain situations in which we wish to extend this definition to the cases (i) where one or both of the limits of integration is infinite, or (ii) where the integrand f has a vertical asymptote (hence, a discontinuity) at some number in $[a, b]$. These are the types of situations in which we define **improper integrals**.

Here is a typical example of how an improper integral of type (i) occurs.

Example 1 (Revenue streams) A large stand of timber may be harvested annually to produce $10,000 worth of hardwood per year. Thus over a 10-year period the woodlot will yield a total of $10 \times (\$10,000) = \$100,000$ worth of timber.

We may think of the value of this timber as a *revenue stream* that produces $10,000 per year. Although this revenue stream produces a nominal value of $100,000 over a 10-year period, the *present value* of this 10-year stream is considerably less than $100,000, since the revenue is received over time, and the opportunity to invest it at current interest rates is lost until the money is actually received. Formula (11) in Section 5.7 gives the *present value* of such a revenue stream as

$$P_0 = \int_0^{10} 10,000e^{-rt}\,dt$$

where r is the nominal interest rate, under continuous compounding, that is available to investors (bank depositors, for example). If, for example, we expect a nominal rate of $r = .10 = 10\%$ to prevail over this period, then the present value of this revenue stream is

$$
\begin{aligned}
P_0 &= \int_0^{10} 10,000e^{-.10t}\,dt \\
&= -\frac{10,000}{0.10}e^{-.10t}\Bigg]_0^{10} \\
&= 100,000\left(1 - \frac{1}{e}\right) \\
&\approx 63,212 \text{ dollars.}
\end{aligned}
$$

Figure 4.1 illustrates this notion. The nominal value of the revenue stream ($100,000) is represented by the area enclosed by the rectangles, and the present value, discounted by the factor e^{-rt}, is represented by the area of the shaded region.

Now suppose we ask for the value of this revenue stream ''in perpetuity.'' That is, assuming that the timber stand will continue to produce revenue of $10,000 per year forever, what is the present value of the revenue stream it represents? (That is, what is a fair price at which to sell the woodlot from the standpoint of revenue?) This is sometimes called the *capital value* of a revenue-producing asset.

Although the time period over which the woodlot will produce revenue is infinite, the present value of this revenue stream is not infinite. To see this, note that the present value

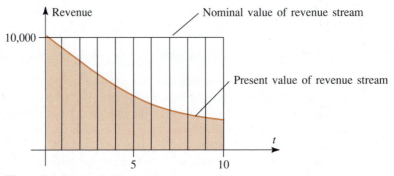

Figure 4.1 Nominal value versus present value.

of this revenue stream up to time T years is

$$P_0 = \int_0^T 10{,}000e^{-.10t}\, dt \qquad \text{(formula (11), Section 5.7)}$$

$$= -\frac{10{,}000}{0.10}e^{-.10t}\Bigg]_0^T$$

$$= 100{,}000\left(1 - \frac{1}{e^{.10T}}\right).$$

Thus in the limit as T approaches infinity, we obtain the present value

$$P_0 = \lim_{T\to\infty}\left[100{,}000\left(1 - \frac{1}{e^{.10T}}\right)\right]$$

$$= 100{,}000 \text{ dollars.}$$

This is the present value of a revenue stream paying \$10,000 per year forever, assuming the availability of continuous compounding of interest at a nominal rate of 10%.

Figure 4.2 gives the geometric interpretation of this result. As T approaches infinity,

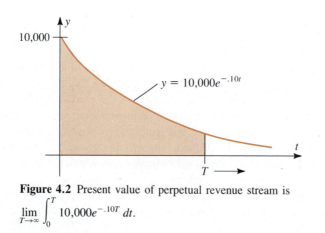

Figure 4.2 Present value of perpetual revenue stream is
$\lim\limits_{T\to\infty}\int_0^T 10{,}000e^{-.10T}\, dt.$

(that is, as the right-hand boundary of the shaded region moves to the right without bound), the **area** of the region lying under the present-value curve approaches 100,000. □

Type I Improper Integrals

The preceding example gives one reason why we define the improper integral of the type $\int_{a}^{\infty} f(x)\, dx$ as the limit as $t \to \infty$ of the definite integral $\int_{a}^{t} f(x)\, dx$.

DEFINITION 1

If f is continuous on the interval $[a, \infty)$, we define the **improper integral** $\int_{a}^{\infty} f(x)\, dx$ as the limit

$$\int_{a}^{\infty} f(x)\, dx = \lim_{t \to \infty} \int_{a}^{t} f(x)\, dx. \tag{1}$$

When the limit in equation (1) exists, we say that the improper integral $\int_{a}^{\infty} f(x)\, dx$ **converges**. When this limit does not exist, we say that the improper integral **diverges**.

Figure 4.3 shows the geometric interpretation of the improper integral $\int_{a}^{\infty} f(x)\, dx$ for the case in which $f(x) \geq 0$ for $x \geq a$. If the function f can take negative values for some numbers x, then the improper integral is still defined by (1), but it cannot be interpreted easily as an area.

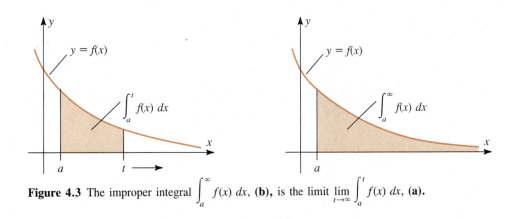

Figure 4.3 The improper integral $\int_{a}^{\infty} f(x)\, dx$, **(b)**, is the limit $\lim\limits_{t \to \infty} \int_{a}^{t} f(x)\, dx$, **(a).**

Example 2 Evaluate the improper integral $\int_1^\infty e^{-x}\,dx$.

Strategy

Evaluate the integral

$$\int_1^t e^{-x}\,dx.$$

Solution

Since an antiderivative for e^{-x} is $-e^{-x}$, we have

$$\int_1^t e^{-x}\,dx = -e^{-x}\Big]_1^t = -e^{-t} - (-e^{-1})$$

$$= \frac{1}{e} - e^{-t}.$$

Thus

Find the limit as $t \to \infty$ of this result.

$$\int_1^\infty e^{-x} = \lim_{t\to\infty}\int_1^t e^{-x}\,dx$$

$$= \lim_{t\to\infty}\left(\frac{1}{e} - e^{-t}\right)$$

$$= \frac{1}{e}$$

$$\text{since } \lim_{t\to\infty} e^{-t} = \lim_{t\to\infty}\left(\frac{1}{e^t}\right) = 0. \qquad \square$$

Example 3 The improper integral

$$\int_2^\infty \frac{1}{\sqrt{x}}\,dx$$

diverges. To see this, note that

$$\int_2^\infty \frac{1}{\sqrt{x}}\,dx = \lim_{t\to\infty}\int_2^t x^{-1/2}\,dx$$

$$= \lim_{t\to\infty} 2\sqrt{x}\Big]_2^t$$

$$= \lim_{t\to\infty}[2\sqrt{t} - 2\sqrt{2}]$$

$$= +\infty. \qquad \square$$

Improper integrals of Type I may also involve an infinite *lower* limit of integration. Figure 4.4 illustrates the definition of an improper integral of the form $\int_{-\infty}^b f(x)\,dx$:

$$\int_{-\infty}^b f(x)\,dx = \lim_{t\to -\infty}\int_t^b f(x)\,dx. \tag{2}$$

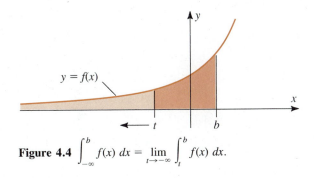

Figure 4.4 $\displaystyle\int_{-\infty}^{b} f(x)\, dx = \lim_{t \to -\infty} \int_{t}^{b} f(x)\, dx.$

As before, when the limit on the right-hand side of equation (2) exists, we say that the improper integral converges. Otherwise it is said to diverge.

Example 4 Evaluate the improper integral $\displaystyle\int_{-\infty}^{1} e^{2x}\, dx.$

Solution: Using equation 2, we have

$$\int_{-\infty}^{1} e^{2x}\, dx = \lim_{t \to -\infty} \int_{t}^{1} e^{2x}\, dx$$

$$= \lim_{t \to -\infty} \frac{1}{2} e^{2x}\Big]_{t}^{1}$$

$$= \lim_{t \to -\infty} \left(\frac{1}{2} e^{2} - \frac{1}{2} e^{2t} \right)$$

$$= \frac{1}{2} e^{2} - \lim_{t \to -\infty} \frac{1}{2} e^{2t}$$

$$= \frac{1}{2} e^{2}$$

since $\displaystyle\lim_{t \to -\infty} \frac{1}{2} e^{2t} = 0.$ $\square$

REMARK: Improper integrals of the form $\displaystyle\int_{-\infty}^{\infty} f(x)\, dx$ may also be defined. We say that the improper integral $\displaystyle\int_{-\infty}^{\infty} f(x)\, dx$ converges if, for any constant a, *both* the improper integrals $\displaystyle\int_{-\infty}^{a} f(x)\, dx$ and $\displaystyle\int_{a}^{\infty} f(x)\, dx$ converge. In this case the value of the integral is

$$\int_{-\infty}^{\infty} f(x)\, dx = \int_{-\infty}^{a} f(x)\, dx + \int_{a}^{\infty} f(x)\, dx.$$

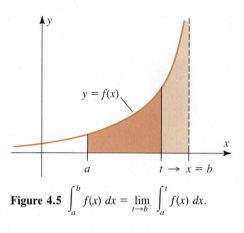

Figure 4.5 $\displaystyle\int_a^b f(x)\ dx = \lim_{t \to b^-} \int_a^t f(x)\ dx.$

**Type II
Improper
Integrals**

A second type of improper integral occurs when the integrand in a definite integral becomes infinite at one or both limits of integration. For example, suppose that the function f is defined and continuous on the interval $[a, b)$, but that $\lim\limits_{x \to b^-} f(x)$ fails to exist (Figure 4.5). In this case the definite integral $\displaystyle\int_a^b f(x)\ dx$ as defined in Chapter 5 does not exist. However, we define the associated *improper integral* by using a left-hand limit as follows.

DEFINITION 2

If f is continuous on $[a, b)$ but $\lim\limits_{x \to b^-} f(x)$ fails to exist, we define the *improper integral* $\displaystyle\int_a^b f(x)\ dx$ as the limit

$$\int_a^b f(x)\ dx = \lim_{t \to b^-} \int_a^t f(x)\ dx \tag{3}$$

provided that this limit exists.

As with improper integrals of Type I, we say that the improper integral in equation (3) converges if the limit on the right-hand side of equation (3) exists. Figure 4.5 illustrates that when $f(x) \geq 0$ for $a < x < b$, we may interpret the value of the improper integral in equation (3) as the area of the region bounded by the graph of f and the x-axis for $a \leq x < b$, which may be finite even though this region has infinite extent in the y-direction.

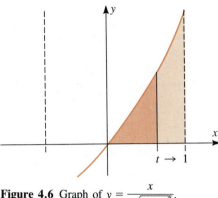

Figure 4.6 Graph of $y = \dfrac{x}{\sqrt{1 - x^2}}$.

Example 5 Determine whether the improper integral

$$\int_0^1 \frac{x}{\sqrt{1 - x^2}}\, dx$$

converges or diverges.

Strategy

Determine where the integrand fails to exist.

Apply Definition 2.
Use a substitution $u = 1 - x^2$,
$du = -2x\, dx$ to obtain

$$\int \frac{x}{\sqrt{1 - x^2}}\, dx = \int \frac{-(1/2)\, du}{\sqrt{u}}$$

$$= -\sqrt{u} + C.$$

Solution

The integrand becomes infinite at $x = 1$ since $\sqrt{1 - (1)^2} = 0$. Thus

$$\int_0^1 \frac{x}{\sqrt{1 - x^2}} = \lim_{t \to 1^-} \int_0^t \frac{x}{\sqrt{1 - x^2}}\, dx$$

$$= \lim_{t \to 1^-} \left\{ -\sqrt{1 - x^2} \Big]_0^t \right\}$$

$$= \lim_{t \to 1^-} \{1 - \sqrt{1 - t^2}\}$$

$$= 1.$$

The integral converges to 1. (See Figure 4.6.) □

When an integrand f becomes infinite at the *left* endpoint of an interval $[a, b]$, the improper integral is defined by analogy with equation (3) using the right-hand limit a. That is

$$\int_a^b f(x)\, dx = \lim_{t \to a^+} \int_t^b f(x)\, dx \tag{4}$$

if f is continuous on $(a, b]$ but $\lim_{x \to a^+} f(x)$ fails to exist.

Example 6 The improper integral $\displaystyle\int_{-2}^2 (x + 2)^{-3/2}\, dx$ fails to exist. To see this, use equation (4) to find that

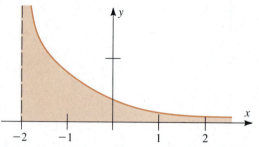

Figure 4.7 Graph of $y = (x + 2)^{-3/2}$. Area of shaded region is infinite.

$$\int_{-2}^{2} (x + 2)^{-3/2} \, dx = \lim_{t \to -2^+} \int_{t}^{2} (x + 2)^{-3/2} \, dx$$

$$= \lim_{t \to -2^+} \left\{ -2(x + 2)^{-1/2} \Big]_{t}^{2} \right\}$$

$$= \lim_{t \to -2^+} \left\{ \frac{-2}{\sqrt{4}} - \frac{-2}{\sqrt{t + 2}} \right\}$$

$$= +\infty$$

since $\lim\limits_{t \to -2^+} \dfrac{2}{\sqrt{t + 2}} = +\infty$. (See Figure 4.7.) $\qquad\square$

Exercise Set 7.4

In Exercises 1–14 determine whether the improper integral of Type I converges. Evaluate those that converge.

1. $\displaystyle\int_{2}^{\infty} \frac{1}{x^2} \, dx$

2. $\displaystyle\int_{1}^{\infty} \frac{1}{x^3} \, dx$

3. $\displaystyle\int_{0}^{\infty} e^{-x} \, dx$

4. $\displaystyle\int_{1}^{\infty} x e^{x^2} \, dx$

5. $\displaystyle\int_{1}^{\infty} \frac{1}{\sqrt{x}} \, dx$

6. $\displaystyle\int_{0}^{\infty} e^{\pi x} \, dx$

7. $\displaystyle\int_{0}^{\infty} \frac{x}{\sqrt{4 + x^2}} \, dx$

8. $\displaystyle\int_{1}^{\infty} \frac{e^{-\sqrt{x}}}{\sqrt{x}} \, dx$

9. $\displaystyle\int_{0}^{\infty} \frac{1}{x + 2} \, dx$

10. $\displaystyle\int_{0}^{\infty} x\sqrt{1 + x^2} \, dx$

11. $\displaystyle\int_{-\infty}^{-2} \frac{1}{x^2} \, dx$

12. $\displaystyle\int_{-\infty}^{0} e^{x} \, dx$

13. $\displaystyle\int_{-\infty}^{0} \frac{1}{\sqrt{1 - x}} \, dx$

14. $\displaystyle\int_{0}^{\infty} \frac{x}{1 + x^2} \, dx$

In Exercises 15–22 determine whether the integral of Type II converges. Evaluate those that converge.

15. $\displaystyle\int_{0}^{1} \frac{1}{\sqrt{1 - x}} \, dx$

16. $\displaystyle\int_{0}^{1} \frac{1}{x} \, dx$

17. $\displaystyle\int_{0}^{1} \frac{1}{x^2} \, dx$

18. $\displaystyle\int_{1}^{3} \frac{x}{\sqrt{9 - x^2}} \, dx$

19. $\displaystyle\int_{1}^{e} \frac{1}{x \ln x} \, dx$

20. $\displaystyle\int_{0}^{1} (1 - x)^{-2/3} \, dx$

21. $\displaystyle\int_{1}^{3} \frac{1}{x^2 - 2x + 1} \, dx$

22. $\displaystyle\int_{0}^{2} \frac{1}{(x - 2)^2} \, dx$

23. Find the present value of a perpetual revenue stream paying $100 per year. Assume that the prevailing rate of interest, compounded continuously, will remain constant at the nominal rate $r = 10\%$.

24. Find the present value of the revenue stream in Exercise 23 if the prevailing rate of interest is assumed to remain constant at $r = 5\%$.

25. A business expects to generate profits at a rate of 1 million dollars per year indefinitely. Assuming a constant prevailing rate of interest of $r = 10\%$
   (a) Find the present value of its anticipated profits over the next 10 years.
   (b) Find the present value of all anticipated future profits.

26. What is a fair price to pay today for a stand of timber that will yield $1000 worth of lumber per year indefinitely, assuming that operating costs are negligible and that prevailing interest rates remain constant at $r = 0.05$?

27. Find the volume of the solid obtained by revolving

about the x-axis the region bounded by the graph of $y = e^{-x}$ and the y-axis for $0 \le x < \infty$.

28. For what set of numbers p does the improper integral
$$\int_1^\infty x^p \, dx \text{ converge?}$$

29. For what set of numbers p does the integral $\int_0^1 \frac{1}{x^p} \, dx$ converge?

30. The Ajax Corporation wishes to raise money by selling perpetual bonds that would pay the owners of the bonds $1000 per year per bond indefinitely. What is the value of one of these bonds to a potential investor who believes that the prevailing rate of interest will remain constant at
   a. 5% per year?
   b. 10% per year?
   c. 15% per year?

7.5 RULES FOR APPROXIMATING INTEGRALS

Many functions, such as $\sqrt{x^2 + 1}$ or e^{x^2}, do not have antiderivatives that are easily found. For such functions the Fundamental Theorem of Calculus is not usable, and we can only *approximate* definite integrals involving these functions.

One way to approximate the definite integral

$$\int_a^b f(x) \, dx$$

is by using approximating sums of the form

$$\int_a^b f(x) \, dx \approx \sum_{j=1}^n f(t_j) \, \Delta x, \quad \text{where} \quad \Delta x = \frac{b - a}{n}$$

and where t_j is an arbitrary number in $[x_{j-1}, x_j]$ \hfill (1)

as described in Chapter 5 (and as implemented in Program 4 in Appendix I). In this section we present two additional methods for approximating $\int_a^b f(x) \, dx$.

The Trapezoidal Rule

The sum in (1) involves approximating "slices" of the region R bounded by the graph of f with rectangles. Figure 5.1 illustrates the fact that an approximating rectangle often does not fit this graph very well. There may be large regions under the graph of f that are not included in the rectangle, and vice versa.

We can try to improve on these "rectangular" estimates by drawing a line from

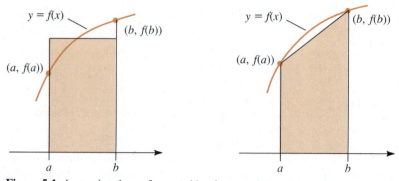

Figure 5.1 Approximations of trapezoids often provide a "better fit" than do approximations by rectangles.

$(a, f(a))$ to $(b, f(b))$ in Figure 5.1 and using the area of the resulting trapezoid in the same way that we previously used approximating rectangles. In doing so, we will make use of the formula from geometry for the area A of the trapezoid with altitude* h and bases of length b and B:

$$A = \frac{1}{2}(B + b)h. \tag{2}$$

To use this idea in approximating the area of the region R bounded by the graph of f and the x-axis for $a \le x \le b$, we divide the interval $[a, b]$ into n subintervals of equal length $\Delta x = \dfrac{b - a}{n}$, and we denote the endpoints of these subintervals by the numbers

$$a = x_0 < x_1 < x_2 < \cdots < x_n = b.$$

But instead of using the area of an approximating rectangle in each subinterval, we use the approximation

$$\int_{x_{j-1}}^{x_j} f(x)\, dx \approx \frac{1}{2}[f(x_{j-1}) + f(x_j)]\, \Delta x \tag{3}$$

which, according to equation (2), is the area of the trapezoid with bases $B = f(x_{j-1})$ and $b = f(x_j)$, and altitude $h = \Delta x$. (See Figures 5.2 and 5.3.)

Summing the approximations in line (3), one for each of the n subintervals, gives the following approximation:

$$\int_a^b f(x)\, dx \approx \frac{\Delta x}{2}\{[f(x_0) + f(x_1)] + [f(x_1) + f(x_2)] + [f(x_2) + f(x_3)]$$

$$+ \cdots + [f(x_{n-2}) + f(x_{n-1})] + [f(x_{n-1}) + f(x_n)]\}.$$

---

*Here we use "altitude" to mean the distance between the parallel sides of the trapezoid; don't confuse it with "height." See Figure 5.3.

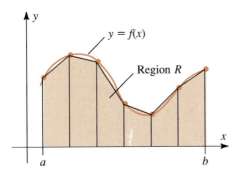

Figure 5.2 Approximating the region R by trapezoids.

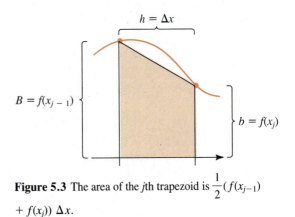

Figure 5.3 The area of the jth trapezoid is $\frac{1}{2}(f(x_{j-1}) + f(x_j)) \, \Delta x$.

Since each function value $f(x_j)$, except $f(x_0)$ and $f(x_n)$, appears *twice* in the above sum (why?), we can restate this approximation as follows.

Trapezoidal Rule: Let f be continuous on $[a, b]$, let $\Delta x = \dfrac{b-a}{n}$, and let

$$x_0 = a, \quad x_1 = a + \Delta x, \quad x_2 = a + 2\,\Delta x, \quad \ldots, \quad x_n = a + n\,\Delta x = b.$$

Then

$$\int_a^b f(x)\, dx \approx \frac{b-a}{2n}\{f(x_0) + 2f(x_1) + 2f(x_2) + \cdots + 2f(x_{n-1}) + f(x_n)\}.$$

(4)

Note that while the motivation for stating the Trapezoidal Rule was based on area considerations, the statement of the Trapezoidal Rule does not require that $f(x) \geq 0$ for all x in $[a, b]$. That is, the Trapezoidal Rule may be used to approximate the definite integral of a continuous function f whether or not that integral corresponds to the area of a region bounded by the graph of f and the x-axis.

Example 1 Approximate $\displaystyle\int_1^4 \frac{1}{x}\, dx$ using the Trapezoidal Rule with $n = 6$.

Solution: With $a = 1$, $b = 4$, and $n = 6$ we have $\Delta x = \dfrac{4-1}{6} = \dfrac{1}{2}$, so

$$x_0 = 1, \quad x_1 = \frac{3}{2}, \quad x_2 = 2, \quad x_3 = \frac{5}{2}, \quad x_4 = 3, \quad x_5 = \frac{7}{2}, \quad x_6 = 4.$$

(See Figure 5.4.)

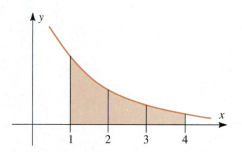

Figure 5.4 Approximations in Example 1.

The Trapezoidal Rule thus gives the approximation

$$\int_1^4 \frac{1}{x}\,dx \approx \frac{3}{12}\left\{1 + \frac{4}{3} + 1 + \frac{4}{5} + \frac{2}{3} + \frac{4}{7} + \frac{1}{4}\right\} = \frac{2361}{1680} = 1.4054.$$ □

REMARK 1: From the geometry of Figure 5.4 we can see that the approximation in Example 1 is actually *larger* than the actual value of the integral. This is because the graph of f is concave up on $[1, 4]$.

REMARK 2: Notice that if we evaluate the definite integral in Example 1 using the Fundamental Theorem, we find that

$$\int_1^4 \frac{1}{x}\,dx = \ln x \Big]_1^4 \;=\; \ln 4 - \ln 1$$

$$= \ln 4 \qquad (\ln 1 = 0).$$

Combining this with the result of Example 1 gives the approximation

$$\ln 4 \approx 1.4054.$$

More generally, this shows how we can use the Trapezoidal Rule to approximate values of the natural logarithm function as

$$\ln x = \int_1^x \frac{1}{t}\,dt, \qquad x > 0.$$

This is an important observation because it shows one way in which values of a transcendental function can be approximated without resorting to a table of values or a calculating device.

ERROR IN TRAPEZOIDAL APPROXIMATIONS: In more advanced courses it is proved that the error in using the Trapezoidal Rule to approximate the integral $\int_a^b f(x)\,dx$ with n subdivisions has the following bound:

$$|\text{Error}| \leq \frac{(b - a)^3 M}{12n^2} \tag{5}$$

where M is the maximum value of $f''(x)$ on the interval $[a, b]$.

Example 2 Find a bound on the error for the approximation in Example 1.

Solution: For $f(x) = \dfrac{1}{x}$, $f''(x) = \dfrac{2}{x^3}$, which has a maximum value of $M = 2$ on the interval $[1, 4]$. Thus by inequality (5) we have

$$|\text{Error}| \le \frac{3^3}{12(6^2)}(2) = \frac{1}{8} = 0.125. \qquad \square$$

Program 5 in Appendix I is a BASIC program implementing the Trapezoidal Rule to approximate the integral $\displaystyle\int_a^b \frac{1}{x}\,dx$. Running this program on a computer, we obtained the results in Table 5.1. The student with access to a computer is encouraged to experiment with modifications of this program and to compare its results with integrals previously calculated using the Fundamental Theorem.

Table 5.1

$[A, B]$	n	S
$[1, 4]$	5	1.41348
$[1, 4]$	20	1.38805
$[1, 4]$	100	1.38636
$[1, 4]$	250	1.38631
$[1, 20]$	5	3.76955
$[1, 20]$	20	3.06566
$[1, 20]$	100	2.99872
$[1, 20]$	250	2.99621

Simpson's Rule   This approximation procedure uses parabolic arcs, rather than line segments, to approximate portions of the graph of $y = f(x)$. It uses the fact that if x_0, x_1, and x_2 are three values of x with

$$x_1 - x_0 = x_2 - x_1 = \Delta x$$

then the definite integral of the quadratic function passing through the three points (x_0, y_0), (x_1, y_1), and (x_2, y_2) equals

$$\frac{\Delta x}{3}[y_0 + 4y_1 + y_2]. \qquad (6)$$

(You're asked to complete the proof of this in Exercise 37.) We apply this result to obtain Simpson's Rule. As before, we divide $[a, b]$ into n subintervals, but now we must require that n be an *even* integer. We then approximate the integral

$$\int_{x_{2j-2}}^{x_{2j}} f(x)\,dx$$

over each *pair* of subintervals by expression (6). Thus we are approximating the actual value of the integral of $f(x)$ over $[x_{2j-2}, x_{2j}]$ by the integral of the approximating quadratic function. We obtain the approximation

$$\int_a^b f(x)\, dx \approx \frac{\Delta x}{3}\{[\,f(x_0) + 4f(x_1) + f(x_2)] + [\,f(x_2) + 4f(x_3) + f(x_4)]$$

$$+ [\,f(x_4) + 4f(x_5) + f(x_6)] + \cdots$$
$$+ [\,f(x_{n-4}) + 4f(x_{n-3}) + f(x_{n-2})]$$
$$+ [\,f(x_{n-2}) + 4f(x_{n-1}) + f(x_n)]\}.$$

Again, notice that each $f(x_{2j})$ is counted twice, except for $f(x_0)$ and $f(x_n)$, for the same reason as in the Trapezoidal Rule. We can state this approximation rule as follows.

Simpson's Rule: Let n be an even integer and let $\Delta x = \dfrac{b-a}{n}$. Let

$$x_0 = a, \quad x_1 = a + \Delta x, \quad x_2 = a + 2\,\Delta x, \quad \ldots, \quad x_n = a + n\,\Delta x = b.$$

Then

$$\int_a^b f(x)\, dx \approx \frac{b-a}{3n}\{f(x_0) + 4f(x_1) + 2f(x_2) + 4f(x_3)$$

$$+2f(x_4) + \cdots + 2f(x_{n-2}) + 4f(x_{n-1}) + f(x_n)\}.$$

(See Figure 5.5.)

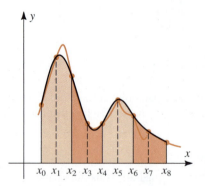

$x_0\ x_1\ x_2\ x_3\ x_4\ x_5\ x_6\ x_7\ x_8$ **Figure 5.5** Approximation by Simpson's Rule.

Example 3 Use Simpson's Rule with $n = 6$ to approximate $\displaystyle\int_1^4 \frac{1}{x}\, dx$ (that is, ln 4).

Solution: With $x_0, x_1, \ldots, x_6$ as in Example 1 we obtain from the statement of Simpson's Rule the approximation

$$\int_1^4 \frac{1}{x}\, dx = \frac{4-1}{3\cdot 6}\left\{1 + 4\left(\frac{2}{3}\right) + 2\left(\frac{1}{2}\right) + 4\left(\frac{2}{5}\right) + 2\left(\frac{1}{3}\right) + 4\left(\frac{2}{7}\right) + \frac{1}{4}\right\}$$

$$= \frac{3497}{2520} = 1.3877 \qquad \qquad \Box$$

ERROR IN SIMPSON'S RULE APPROXIMATIONS: For the approximation provided by Simpson's Rule it is proved in more advanced courses that

$$|\text{Error}| \leq \frac{(b-a)^5 M}{180n^4} \tag{7}$$

where M is the maximum value of $|f^{(4)}(x)|$, the fourth derivative, on the interval $[a, b]$.

Example 4 Find how large n must be in order that the approximation of the integral in Example 3 by Simpson's Rule will be accurate to within 0.005.

Solution: For $f(x) = \dfrac{1}{x}$, $f^{(4)}(x) = 24x^{-5}$. The maximum value of this fourth derivative on $[1, 4]$ is $M = 24$. According to inequality (7), we must find n sufficiently large that

$$\frac{(4-1)^5}{180n^4}(24) \leq 0.005.$$

Solving this inequality for n, we find that

$$n^4 \geq \frac{(24)(243)}{(0.005)(180)} = 6480.$$

Here $n = 10$ will suffice since $10^4 = 10{,}000 > 6480$. □

Program 6 in Appendix I is a BASIC program that implements Simpson's Rule to approximate the integral $\displaystyle\int_a^b \frac{1}{x}\,dx$. Table 5.2 was obtained by use of this program.

Table 5.2

$[A, B]$	n	S
$[1, 4]$	6	1.38770
$[1, 4]$	20	1.38631
$[1, 4]$	100	1.38629
$[1, 4]$	250	1.38629
$[1, 20]$	6	3.21696
$[1, 20]$	20	3.00677
$[1, 20]$	100	2.99577
$[1, 20]$	250	2.99573

In conclusion, several observations should be made about the two approximation procedures presented in this section.

(1) While the Trapezoidal Rule approximates curves by line segments, Simpson's Rule fits second-degree curves to the given curve. Thus we would expect Simpson's Rule to be more accurate for the same value of n.
(2) Since the error formula for the Trapezoidal Rule involves f'', the Trapezoidal Rule gives exact information for first-degree polynomials, for which $f''(x) \equiv 0$.

(3) Since, however, the error formula for Simpson's Rule involves the fourth derivative of f, Simpson's Rule gives exact results for polynomials of degree three or less, for which $f^{(4)}(x) \equiv 0$.

Exercise Set 7.5

In Exercises 1–10 use the Trapezoidal Rule with the given value of n to approximate the integral.

1. $\displaystyle\int_0^4 (x^3 - 7x + 4)\, dx, \quad n = 4$

2. $\displaystyle\int_0^2 \frac{1}{1 + x^2}\, dx, \quad n = 4$

3. $\displaystyle\int_0^\pi \sin x\, dx, \quad n = 4$

4. $\displaystyle\int_0^\pi \sin x\, dx, \quad n = 6$

5. $\displaystyle\int_0^4 \frac{1}{1 + x^2}\, dx, \quad n = 8$

6. $\displaystyle\int_0^2 \sqrt{1 + x^2}\, dx, \quad n = 4$

7. $\displaystyle\int_0^4 \cos\left(\frac{\pi x}{4}\right) dx, \quad n = 4$

8. $\displaystyle\int_0^2 \frac{x}{1 + x^3}\, dx, \quad n = 4$

9. $\displaystyle\int_2^4 \frac{1}{1 - x^2}\, dx, \quad n = 8$

10. $\displaystyle\int_{-4}^4 x \sin\left(\frac{\pi x}{4}\right) dx, \quad n = 8$

In Exercises 11–20 use Simpson's Rule to approximate the integral.

11. In Exercise 1

12. In Exercise 2

13. In Exercise 3

14. In Exercise 4

15. In Exercise 5

16. In Exercise 6

17. In Exercise 7

18. In Exercise 8

19. In Exercise 9

20. In Exercise 10

(Calculator) In Exercises 21–26 use the Trapezoidal Rule with $n = 6$ to approximate the given integral.

21. $\displaystyle\int_0^{\pi/4} \frac{1}{1 + x^2}\, dx$

22. $\displaystyle\int_0^3 \sqrt{1 + x^2}\, dx$

23. $\displaystyle\int_1^3 \sqrt{x^2 + 2}\, dx$

24. $\displaystyle\int_0^1 \sin x^2\, dx$

25. $\displaystyle\int_0^2 x \sin\left(\frac{\pi x}{2}\right) dx$

26. $\displaystyle\int_0^{\pi/4} \tan \sqrt{x}\, dx$

(Calculator) In Exercises 27–32 use Simpson's Rule with $n = 6$ to approximate the integral.

27. In Exercise 21

28. In Exercise 22

29. In Exercise 23

30. In Exercise 24

31. In Exercise 25

32. In Exercise 26

33. (Computer) Modify Program 5 in Appendix I to approximate the integral in Exercise 21 with $n = 5, 20, 100,$ and 250.

34. (Computer) Modify Program 6 in Appendix I to approximate the integral in Exercise 21 with $n = 6, 20, 100,$ and 250.

35. (Calculator) Use the Trapezoidal Rule with $n = 6$ to approximate

a. $\displaystyle\int_0^3 \sin \sqrt{x}\, dx.$ **b.** $\displaystyle\int_0^{\pi/3} \frac{1}{\sqrt{\cos x}}\, dx.$

36. (Calculator) Repeat Exercise 35 using Simpson's Rule.

37. Verify Equation (6) by completing the following steps. (Refer to Figure 5.6.)

a. Let $x_0 = -\Delta x$, $x_1 = 0$, and $x_2 = \Delta x$ as in Figure 5.6.

b. Let the parabola through (x_0, y_0), (x_1, y_1), and (x_2, y_2) be denoted as $y = ax^2 + bx + c$.

c. Use the resulting values

$y_0 = a(\Delta x)^2 - b(\Delta x) + c$
$y_1 = c,$ and
$y_2 = a(\Delta x)^2 + b(\Delta x) + c$

to show that

$$\int_{-\Delta x}^{\Delta x} (ax^2 + bx + c)\, dx = \frac{\Delta x}{3}[y_0 + 4y_1 + y_2].$$

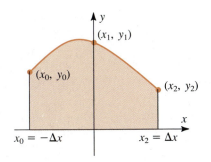

Figure 5.6

38. a. Use the Trapezoidal Rule with $n = 8$ to approximate ln 3. (See Example 1.)

b. What is the bound on the error in the approximation in part (a) according to equation (5)?

c. Use a calculator or Table 2 in Appendix II to compare the actual value of ln 3 with your approximation from part (a). What is the difference? How does this compare with the estimate of the errors in part (b)?

39. a. Use Simpson's Rule with $n = 8$ to approximate ln 5.

b. What is the bound on the error in the approximation in part (a) according to equation (7)?

c. Find the actual value of ln 5 from Table 2 in Appendix II or by using a calculator. What is the difference between this actual value and your approximation in part (a)? How does this compare with the estimate of the error in part (b)?

SUMMARY OUTLINE OF CHAPTER 7

■ **Method of Substitutions** (Page 362)

$$\int f(g(x))g'(x)\, dx = \int f(u)\, du = F(u) + C$$

$$= F(u(x)) + C$$

where $f = F'$ and $u = g(x)$.

■ **Integration by Parts** (Page 365)

$$\int u\, dv = uv - \int v\, du$$

or

$$\int f(x)g'(x)\, dx = f(x)g(x) - \int f'(x)g(x)\, dx.$$

■ **Improper Integrals** (Page 377)

$$\int_a^\infty f(x)\, dx = \lim_{t \to \infty} \int_a^t f(x)\, dx \qquad \text{(Type I)}$$

$$\int_{-\infty}^b f(x)\, dx = \lim_{t \to -\infty} \int_t^b f(x)\, dx \qquad \text{(Type I)}$$

$$\int_a^b f(x)\, dx = \lim_{t \to b^-} \int_a^t f(x)\, dx \qquad \text{(Type II)}$$

$$\int_a^b f(x)\, dx = \lim_{t \to a^+} \int_t^b f(x)\, dx \qquad \text{(Type II)}$$

■ *Trapezoidal Rule* (Page 385)

$$\int_a^b f(x)\, dx \approx \frac{b - a}{2n}\{f(x_0) + 2f(x_1) + 2f(x_2) + \cdots + 2f(x_{n-1}) + f(x_n)\}$$ (1)

■ *Simpson's Rule* (Page 388)

$$\int_a^b f(x)\, dx \approx \frac{b - a}{3n}\{f(x_0) + 4f(x_1) + 2f(x_2) + 4f(x_3) + \cdots + 2f(x_{n-2}) + 4f(x_{n-1}) + f(x_n)\}$$ (2)

In (1) and (2) $a = x_0 < x_1 < x_2 < \cdots < x_n = b$ with $x_j = a + j\,\Delta x = a + j\left(\dfrac{b - a}{n}\right)$, $j = 1, 2, \ldots, n$.

In (2) n must be even.

REVIEW EXERCISES—CHAPTER 7

In Exercises 1–30 evaluate the integral.

1. $\displaystyle\int \sqrt{x + 1}\, dx$

2. $\displaystyle\int x \cos x^2\, dx$

3. $\displaystyle\int \frac{x}{1 + x^2}\, dx$

4. $\displaystyle\int \frac{e^{\sqrt{x+1}}}{\sqrt{x + 1}}\, dx$

5. $\displaystyle\int_0^2 (x - 2)(x + 3)\, dx$

6. $\displaystyle\int \frac{x}{\sqrt{1 - x^2}}\, dx$

7. $\displaystyle\int_0^{\pi/6} \sin^3 x \cos x\, dx$

8. $\displaystyle\int e^{2x} \ln e^{2x}\, dx$

9. $\displaystyle\int x\sqrt{x^2 + 9}\, dx$

10. $\displaystyle\int_0^1 xe^{3x}\, dx$

11. $\displaystyle\int_0^2 \frac{dx}{(x - 3)^2}$

12. $\displaystyle\int xe^{-3x}\, dx$

13. $\displaystyle\int x \ln x\, dx$

14. $\displaystyle\int x^2 \sin x\, dx$

15. $\displaystyle\int_0^3 \frac{1}{\sqrt[3]{x + 1}}\, dx$

16. $\displaystyle\int_{-2}^{-1} x(x + 3)^6\, dx$

17. $\displaystyle\int_0^{1/4} \sec^2 \pi x\, dx$

18. $\displaystyle\int xe^{-x}\, dx$

19. $\displaystyle\int x^2 \ln x\, dx$

20. $\displaystyle\int_1^\infty \frac{dx}{\sqrt[3]{x + 2}}$

21. $\displaystyle\int x \csc^2 x\, dx$

22. $\displaystyle\int \sin(\ln x)\, dx$

23. $\displaystyle\int_{-\infty}^0 e^x\, dx$

24. $\displaystyle\int \frac{x\, dx}{\sqrt{9 + x^2}}$

25. $\displaystyle\int_{e^2}^\infty \frac{dx}{x \ln x}$

26. $\displaystyle\int_2^\infty \frac{x}{5 + x^2}\, dx$

27. $\displaystyle\int_1^4 \frac{x}{\sqrt{16 - x^2}}\, dx$

28. $\displaystyle\int (x + 2) \ln x\, dx$

29. $\displaystyle\int \frac{2x + 4}{x^2 + 4x + 3}\, dx$

30. $\displaystyle\int \tan 3x\, dx$

Chapter 8   *Multivariable Calculus*

8.1 FUNCTIONS OF SEVERAL VARIABLES

Up to this point all functions that we have encountered have shared the property of having only a single independent variable. That is, they have all had the form $y = f(x)$. But a function may have more than one independent variable. Here are two examples of functions of the form $z = f(x, y)$. That is, functions with *two* independent variables.

(i) A manufacturing company produces two products, bicycles and roller skates. Its fixed costs of production are $1200 per week. Its variable costs of production are $40 for each bicycle produced and $15 for each pair of roller skates. Its total weekly costs in producing x bicycles and y pairs of roller skates are therefore

$$C(x, y) = 1200 + 40x + 15y.$$

For example, in producing $x = 20$ bicycles and $y = 30$ pairs of roller skates per week the manufacturer experiences total costs of

$$C(20, 30) = 1200 + 40(20) + 15(30)$$
$$= 2450 \text{ dollars.}$$

(ii) The Cobb-Douglas production function

$$f(K, L) = \lambda K^\alpha L^{1 - \alpha}$$

is a model used by economists to study the relationship between levels of labor, L, and capital goods, K, supplied in a manufacturing process and the resulting level of production, f. Here λ and α are constants, with $0 < \alpha < 1$, and L and K are the two independent variables. For example, in the Cobb-Douglas model

$$f(K, L) = 100K^{1/4}L^{3/4}$$

the inputs of $K = 256$ units of capital and $L = 16$ units of labor result in

$$f(256, 16) = 100(256)^{1/4}(16)^{3/4}$$
$$= 100(4)(8)$$
$$= 3200$$

units of production.

Functions can be specified with more than two independent variables. For example, the amount of interest I that accrues in a savings account from a single initial deposit of P dollars for t years under continuous compounding of interest at rate r is actually a function of the three independent variables P, r, and t:

$$I(P, r, t) = P(e^{rt} - 1).$$

More generally, we would write a function of the n independent variables x_1, x_2, . . . , x_n in the form

$$y = f(x_1, x_2, \ldots, x_n).$$

You will encounter many examples of functions of several variables in this chapter. While the number of independent variables may change from example to example, what will not change is the essential property that a *unique* value of the function will be determined for each choice of the independent variables.

Example 1 For the function $f(x, y) = x^2 + 4xy - 2y^3$ we have

$$f(2, 3) = 2^2 + 4(2)(3) - 2(3)^3 = 4 + 24 - 54 = -26$$
$$f(4, 0) = 4^2 + 4(4)(0) - 2(0)^3 = 16 + 0 - 0 = 16$$

and

$$f(-3, 1) = (-3)^2 + 4(-3)(1) - 2(1)^3 = 9 - 12 - 2 = -5. \qquad \square$$

Example 2 A manufacturer of automobile tires produces three different types— regular, snow, and racing tires. If the regular tires sell for \$60 each, the snow tires for \$50 each, and the racing tires for \$100 each, find a function giving the manufacturer's total receipts from the sale of x regular tires, y snow tires, and z racing tires.

Solution: Since the receipts from the sale of any tire type is the price per tire times the number of tires sold, the total receipts are

$$R(x, y, z) = 60x + 50y + 100z.$$

For example, receipts from the sale of ten tires of each type would be

$$R(10, 10, 10) = 60(10) + 50(10) + 100(10)$$
$$= 2100 \text{ dollars.} \qquad \square$$

Three-dimensional Coordinate Systems

Graphing a function of two variables of the form $z = f(x, y)$ requires a **three-dimensional coordinate system** in which to plot the three variables x, y, and z. Figure 1.1 illustrates the convention for coordinatizing three-dimensional space. The usual xy-plane is positioned horizontally, with the z-axis intersecting the xy-plane at the common location

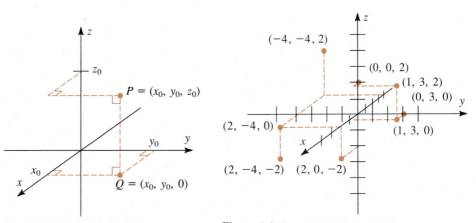

Figure 1.1 Coordinatizing xyz space.

Figure 1.2 Coordinates of several points in space.

of the zeros of each of the three axes. (This point is called the **origin**.) Points of the form (x_0, y_0) in the xy-plane are assigned the coordinates $(x_0, y_0, 0)$ in space, and the point (x_0, y_0, z_0) is located $|z_0|$ units above or below the point $(x_0, y_0, 0)$, depending on the sign of z_0. Several particular points are labeled in Figure 1.2.

Example 3 Plot the points $P = (2, 1, -3)$ and $Q = (-2, -2, 3)$.

Solution: These points are plotted in Figure 1.3. □

Graphs of Functions of Two Variables

We may use a three-dimensional coordinate system to sketch the graph of a function of two variables. Figure 1.4 shows the usual convention for sketching the graph of a function of the form $z = f(x, y)$. For each pair of numbers (x_0, y_0) for which $f(x_0, y_0)$ is defined, the point (x_0, y_0, z_0) is plotted, with $z_0 = f(x_0, y_0)$. Thus the value $z_0 = f(x_0, y_0)$ is repre-

Figure 1.3 Plotting the points P and Q in Example 3.

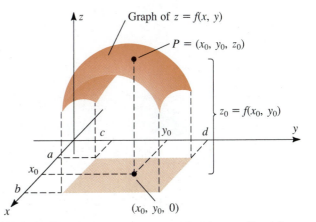

Figure 1.4 The graph of a continuous function $z = f(x, y)$ is a surface in space.

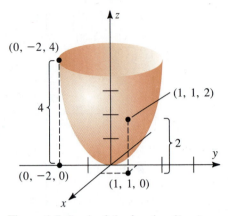

Figure 1.5 Graph of the function $f(x, y) = x^2 + y^2$.

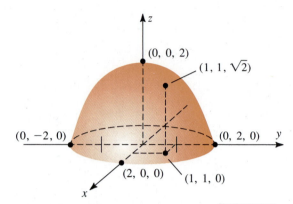

Figure 1.6 Graph of the function $z = \sqrt{4 - x^2 - y^2}$.

sented as the distance of the point (x_0, y_0, z_0) above or below the point $(x_0, y_0, 0)$, depending on the sign of $f(x_0, y_0)$.

For example, Figure 1.5 shows the graph of the function $f(x, y) = x^2 + y^2$, including several particular points, while Figure 1.6 shows the graph of the "hemisphere" $z = \sqrt{4 - x^2 - y^2}$.

Level Curves

Another way to describe the graph of a function of two variables is by the use of *level curves*. Figures 1.7 and 1.8 show the basic idea: For various numbers c we plot the points (x, y) whose coordinates satisfy the equation $c = f(x, y)$. For a given number c the set of all points so obtained constitutes a **level curve.** This level curve may be interpreted as the intersection of the horizontal plane $z = c$ with the graph of the equation $z = f(x, y)$. Although the graph of the function $z = f(x, y)$ is sketched in a three-dimensional coordinate system, its level curves are sketched in the coordinate plane.

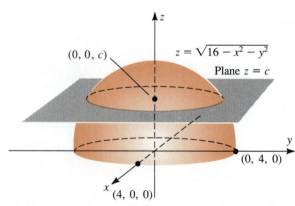

Figure 1.7 Level curves lie at intersection of horizontal plane with graph of $z = f(x, y)$.

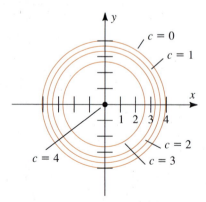

Figure 1.8 Level curves for the function $f(x, y) = \sqrt{16 - x^2 - y^2}$.

Figure 1.9 A picture of a mountainous region.

Figure 1.10 A topographical map of the region.

One common application of level curves is in contour maps (see Figures 1.9 and 1.10). Other applications are in the construction of maps showing regions of equal temperature (isothermal maps) or maps of equal atmospheric pressure (isobaric maps).

Domains

Unless specified otherwise, the **domain** of a function $z = f(x, y)$ of two variables is the set of all pairs (x, y) for which $f(x, y)$ is defined. Similarly, the domain of a function $y = f(x_1, x_2, \ldots, x_n)$ of n variables is the set of all n-tuples $(x_1, x_2, \ldots, x_n)$ for which $f(x_1, x_2, \ldots, x_n)$ is defined. The following example shows two typical situations in which the domain of a function $z = f(x, y)$ fails to include all points in the plane.

Example 4 Find the domain of the function

(a) $f(x, y) = \dfrac{6}{2 - x - y}$, (b) $g(x, y) = \sqrt{4 - x^2 - y^2}$

Solution:

(a) The function f is defined for all pairs (x, y) for which the denominator $2 - x - y$ is not zero. Thus we must exclude all pairs (x, y) for which

$$2 - x - y = 0 \quad \text{or} \quad y = 2 - x.$$

Thus the domain of f includes all points (x, y) except those points lying on the line $y = 2 - x$.

(b) The function g is defined only when the quantity $4 - x^2 - y^2$ underneath the radical sign is nonnegative. That is, when $4 - x^2 - y^2 \geq 0$, or $x^2 + y^2 \leq 4$. Thus the domain is the set of all (x, y) for which $\sqrt{x^2 + y^2} \leq 2$. These are the points on and inside the circle with center at the origin and radius $r = 2$ as illustrated in Figure 1.6.

□

Continuity

The graph of a *continuous* function of two variables will be a surface in space without any "holes" or "tears," just as the graph of a continuous function $y = f(x)$ of one variable is a curve with the same property. A more precise characterization of continuity is beyond our needs here since almost all functions that we shall encounter in this chapter will be continuous.

Exercise Set 8.1

In each of Exercises 1–8 find the indicated values of the given function.

1. $f(x, y) = 3x + 7y$
   a. $f(2, 3)$ b. $f(0, -6)$ c. $f(-1, 5)$

2. $f(x, y) = 4xy^2 - 3x^2y + 6$
   a. $f(0, 0)$ b. $f(1, -2)$ c. $f(3, 5)$

3. $f(x, y) = \sqrt{xy} - \dfrac{1}{xy}$
   a. $f(1, 1)$ b. $f(-2, -2)$ c. $f(1, 9)$

4. $f(x, y) = xe^y - ye^x$
   a. $f(0, 2)$ b. $f(-2, \ln 2)$ c. $f(4, 1)$

5. $f(x, y, z) = x^2 - 2xy + xz^3$
   a. $f(1, 2, -1)$ b. $f(3, -3, 0)$ c. $f(-1, 1, 4)$

6. $f(x, y, z) = xye^z + \ln(x + y) + 3$
   a. $f(2, 1, 0)$ b. $f(2, 2, -1)$ c. $f(0, 1, 3)$

7. $f(x, y, z) = \sqrt{x^2 + y^2 + z^2}$
   a. $f(0, 1, 0)$ b. $f(-2, 1, 2)$ c. $f(3, -3, \sqrt{7})$

8. $f(x, y, z) = x\sqrt{y} + \sqrt[3]{xz} - y^{2/3}z^{-1/3}$
   a. $f(1, 1, 0)$ b. $f(2, 1, 4)$ c. $f(0, 1, -8)$

9. Plot the following points on a three-dimensional coordinate system.
   a. $(1, 1, 2)$ b. $(0, 0, 3)$
   c. $(-2, 2, 4)$ d. $(1, -3, -2)$

10. Plot the following points on a three-dimensional coordinate system.
    a. $(2, -3, -2)$ b. $(-2, -2, -2)$
    c. $(-3, 3, -2)$ d. $(2, 1, -3)$

In each of Exercises 11–16 state the domain of the given function.

11. $f(x, y) = 3xy^2 - 5x^3y + 7$

12. $f(x, y) = \sqrt{x^2 + y^2}$

13. $f(x, y) = \dfrac{10x^2}{y - x}$

14. $f(x, y) = \ln xy$

15. $f(x, y) = 4x^{2/3}y^{-3/4}$

16. $f(x, y, z) = \sqrt{16 - (x^2 + y^2 + z^2)}$

17. A merchant sells two types of brooms, plastic and fiber. The plastic brooms sell for \$10 each, while the selling price of the fiber brooms is \$15. Write a function $z = R(x, y)$ giving the merchant's weekly revenues from the sale of x plastic and y fiber brooms.

18. The merchant in Exercise 17 assigns \$20 per week of its fixed costs to the sale of brooms. If the plastic brooms cost the merchant \$4 each and the fiber brooms cost the merchant \$8 each, find the function $z = C(x, y)$ giving the merchant's total weekly cost from the sale of x plastic and y fiber brooms.

19. Find the weekly profit function $z = P(x, y)$ for the merchant in Exercises 17 and 18.

20. The amount of \$500 is invested in a savings account paying r percent interest compounded continuously.
    a. Find the function $z = P(r, t)$ giving the amount on deposit after t years.
    b. Find $P(0.03, 5)$.
    c. Find $P(0.10, 7)$.

21. Write a function of two variables $z = P_0(r, T)$ giving the *present* value, discounted under the assumption of continuous compounding of interest at the nominal rate of r percent per year, of an investment worth \$10,000 in T years.

22. For the Cobb-Douglas production function
    $$f(K, L) = 100K^{1/4}L^{3/4}$$
    a. Find the production level $f(81, 16)$ resulting from $K = 81$ units of capital and $L = 16$ units of labor.
    b. Find the production level $f(16, 81)$.

23. A monopolist sells its product in two distinct markets, one domestic and one foreign. In the domestic market the demand function giving the relationship between selling price p and the number x of items the monopolist can expect to sell per month is given by the equation
    $$p = 200 - 2x$$
    while in the foreign market the relationship between price, q, and sales level, y, is
    $$q = 150 - 3y.$$
    The manufacturer's monthly cost of producing x items for the domestic market and y items for the foreign market is
    $$C(x, y) = 500 + 4x + 5y.$$

a. Find the monthly revenue function

$$R(x, y) = xp + yq$$

in terms of x and y.

b. Find the monthly profit function

$$z = P(x, y) = R(x, y) - C(x, y).$$

24. A medical researcher determines that the effect $E(x, t)$ of a dosage of x units of a certain hypertension drug on a patient t hours after it is administered is given by the function

$$E(x, t) = 20x^{3/2}e^{-0.05t}.$$

a. Find $E(16, 20)$.   **b.** Find $E(64, 10)$.

25. Show that in the Cobb-Douglas production function $f(K, L) = \lambda K^{\alpha}L^{1-\alpha}$, with $0 < \alpha < 1$ and λ a constant, doubling the size of both inputs K and L will result in a doubling of the output, $f(K, L)$. Economists refer to this property by saying that such models exhibit *constant returns to scale*.

8.2 PARTIAL DIFFERENTIATION

For a function f of a single variable, the derivative f' measures the rate at which the values $f(x)$ change as the independent variable x changes. For functions $y = f(x_1, x_2, \ldots, x_n)$ of more than one variable we may also ask for the rate at which $f(x_1, x_2, \ldots, x_n)$ changes as the independent variables change, but the question is much more complex. Which of the independent variables are to change, and how? And just what does the answer mean?

Fortunately, a great deal can be learned in such situations from the **partial derivatives** of the function $y = f(x_1, x_2, \ldots, x_n)$. A partial derivative of a function of several variables is just the rate at which the values of the function change as *one* of the independent variables changes and all others are held constant. Thus one partial derivative is obtained for each independent variable.

DEFINITION 1

For the function $z = f(x, y)$ and the point (x_0, y_0) in the domain of f

(a) the partial derivative with respect to x at (x_0, y_0) is the limit

$$\frac{\partial f}{\partial x}(x_0, y_0) = \lim_{h \to 0} \frac{f(x_0 + h, y_0) - f(x_0, y_0)}{h} \tag{1}$$

(b) the partial derivative with respect to y at (x_0, y_0) is the limit

$$\frac{\partial f}{\partial y}(x_0, y_0) = \lim_{h \to 0} \frac{f(x_0, y_0 + h) - f(x_0, y_0)}{h}. \tag{2}$$

The partial derivative $\dfrac{\partial f}{\partial x}(x, y)$ is defined whenever the point (x, y) is in the domain of the function f and the limit in equation (1) exists. Its value at (x, y) is the *rate* at which the values $f(x, y)$ change *as x varies but y is held constant*. Thus the partial derivative $\dfrac{\partial f}{\partial x}(x, y)$ may be calculated by applying appropriate differentiation rules for functions of x alone, treating x as the independent variable and y as a constant.

Similarly, the partial derivative $\dfrac{\partial f}{\partial y}(x, y)$ may be found by treating y as the single independent variable and x as a constant.

Example 1 For the function $f(x, y) = x^3 + 4x^2y^3 + y^2$

(a) $\dfrac{\partial f}{\partial x}(x, y) = 3x^2 + 8xy^3$

since when we treat y as a constant, we have

$$\frac{d}{dx}(x^3) = 3x^2; \qquad \frac{d}{dx}(4x^2y^3) = 8xy^3; \qquad \frac{d}{dx}(y^2) = 0.$$

(b) $\dfrac{\partial f}{\partial y}(x, y) = 12x^2y^2 + 2y$

since when we treat x as a constant, we have

$$\frac{d}{dy}(x^3) = 0; \qquad \frac{d}{dy}(4x^2y^3) = 12x^2y^2; \qquad \frac{d}{dy}(y^2) = 2y.$$

(c) $\dfrac{\partial f}{\partial x}(1, 3) = 3(1)^2 + 8(1)(3)^3 = 3 + 8(27) = 219.$

(d) $\dfrac{\partial f}{\partial y}(2, -4) = 12(2)^2(-4)^2 + 2(-4) = 12(4)(16) - 8 = 760.$ □

Example 2 A small factory produces two types of machine parts, bearings and grease seals. The plant manager determines that the total daily cost of production of x hundred bearings and y hundred seals is

$$C(x, y) = 400 + 20x + 8y - 4\sqrt{xy}.$$

For the daily production schedule $x = 10$ and $y = 40$ total daily cost is

$$C(10, 40) = 400 + 20(10) + 8(40) - 4\sqrt{(10)(40)}$$
$$= 840 \text{ dollars.}$$

The partial derivative

$$\frac{\partial C}{\partial x}(x, y) = 20 - \frac{2y}{\sqrt{xy}}$$

gives the rate at which total cost changes with respect to change in x, the number of (hundred) bearings produced. The *rate*

$$\frac{\partial C}{\partial x}(10, 40) = 20 - \frac{2(40)}{\sqrt{(10)(40)}} = 16$$

means that an increase in bearing production from $x = 10$ to $x = 11$, while seal production remains fixed at $y = 40$, will result in an increase in total cost of approximately \$16. Similarly, since

$$\frac{\partial C}{\partial y} = 8 - \frac{2x}{\sqrt{xy}}$$

we have

$$\frac{\partial C}{\partial y}(10, 40) = 8 - \frac{2(10)}{\sqrt{(10)(40)}} = 7$$

which means that an increase in seal production from $y = 40$ to $y = 41$ hundred units, while bearing production remains fixed at $x = 10$, will result in an increase in total cost of approximately \$7. $\qquad\square$

Notation for Partial Derivatives

There are several commonly used symbols for partial derivatives. For the function $z = f(x, y)$ of two variables

(i) the symbols z_x, $\dfrac{\partial z}{\partial x}$, f_x, and $\dfrac{\partial f}{\partial x}$ all mean $\dfrac{\partial f}{\partial x}(x, y)$;

(ii) the symbols z_y, $\dfrac{\partial z}{\partial y}$, f_y, and $\dfrac{\partial f}{\partial y}$ all mean $\dfrac{\partial f}{\partial y}(x, y)$;

(iii) the expression ``$\dfrac{\partial}{\partial x}$'' means, ``the partial derivative with respect to x of.'' Thus

$$\frac{\partial}{\partial x} f(x, y) \quad \text{means} \quad \frac{\partial f}{\partial x}(x, y)$$

$$\frac{\partial}{\partial x}(x^2 y^2) = 2xy^2, \quad \text{and}$$

$$\frac{\partial}{\partial x}(y \ln x + x^3 - \sqrt{y}) = y\left(\frac{1}{x}\right) + 3x^2.$$

(iv) the expression ``$\dfrac{\partial}{\partial y}$'' means, ``the partial derivative with respect to y of.'' Thus

$$\frac{\partial}{\partial y} f(x, y) \quad \text{means} \quad \frac{\partial f}{\partial y}(x, y)$$

$$\frac{\partial}{\partial y}(x^2 y^2) = 2x^2 y, \quad \text{and}$$

$$\frac{\partial}{\partial y}(y \ln x + x^3 - \sqrt{y}) = \ln x - \frac{1}{2\sqrt{y}}.$$

Example 3 For the function $f(x, y) = x^2 e^{y^3} + \sqrt{2x + 3y}$ find (a) $\dfrac{\partial f}{\partial x}$ and (b) $\dfrac{\partial f}{\partial y}$.

Strategy

(a) Treat y, and any terms involving y alone, as constants.

First term has form

$$\frac{d}{dx}(x^2C) = 2xC$$

with $C = e^{y^3}$.

Second term has form

$$\frac{d}{dx}(2x + C)^{1/2}$$

$$= \frac{1}{2}(2x + C)^{-1/2}(2)$$

with $C = 3y$.

(b) Treat x, and any terms involving x alone, as constants.

First term has form

$$\frac{d}{dy}(Ce^{y^3}) = 3y^2Ce^{y^3}$$

with $C = x^2$.

Second term has form

$$\frac{d}{dy}(C + 3y)^{1/2}$$

$$= \frac{1}{2}(C + 3y)^{-1/2}(3)$$

with $C = 2x$.

Solution

(a) Differentiating with respect to x, we find

$$\frac{\partial f}{\partial x} = \frac{\partial}{\partial x}(x^2e^{y^3} + \sqrt{2x + 3y})$$

$$= \frac{\partial}{\partial x}(x^2e^{y^3}) + \frac{\partial}{\partial x}[(2x + 3y)^{1/2}]$$

$$= \left[\frac{d}{dx}(x^2)\right]e^{y^3} + \frac{1}{2}(2x + 3y)^{-1/2}\left[\frac{\partial}{\partial x}(2x + 3y)\right]$$

$$= [2x]e^{y^3} + \frac{1}{2}(2x + 3y)^{-1/2}(2)$$

$$= 2xe^{y^3} + (2x + 3y)^{-1/2}.$$

(b) Differentiating with respect to y, we find

$$\frac{\partial f}{\partial y} = \frac{\partial}{\partial y}(x^2e^{y^3} + \sqrt{2x + 3y})$$

$$= \frac{\partial}{\partial y}(x^2e^{y^3}) + \frac{\partial}{\partial y}[(2x + 3y)^{1/2}]$$

$$= x^2\left[\frac{d}{dy}(e^{y^3})\right] + \frac{1}{2}(2x + 3y)^{-1/2}\left[\frac{\partial}{\partial y}(2x + 3y)\right]$$

$$= x^2[e^{y^3}(3y^2)] + \frac{1}{2}(2x + 3y)^{-1/2}(3)$$

$$= 3x^2y^2e^{y^3} + \frac{3}{2}(2x + 3y)^{-1/2}.$$

□

Example 4 For the Cobb-Douglas production function

$$f(K, L) = 20K^{1/4}L^{3/4}$$

(a) Find the rate at which production changes with respect to capital K, called the *marginal productivity of capital*, when $K = 16$ units and $L = 81$ units.

(b) Find the rate at which production changes with respect to labor L, called the *marginal productivity of labor*, when $K = 16$ units and $L = 81$ units.

Solution

(a) The marginal productivity of capital is the partial derivative

$$\frac{\partial f}{\partial K} = \frac{\partial}{\partial K}(20K^{1/4}L^{3/4})$$

$$= 20\left[\frac{d}{dK}(K^{1/4})\right]L^{3/4}$$

$$= 20\left[\frac{1}{4}K^{-3/4}\right]L^{3/4}$$

$$= 5K^{-3/4}L^{3/4}.$$

The desired rate is therefore

$$\frac{\partial f}{\partial K}(16, 81) = 5(16^{-3/4})(81)^{3/4}$$

$$= \frac{5(81)^{3/4}}{16^{3/4}}$$

$$= \frac{5 \cdot 27}{8}$$

$$= \frac{135}{8}.$$

This means that an increase in capital from level $K = 16$ to level $K = 17$ units will result in an increase of approximately 135/8 units of production.

(b) The marginal productivity of labor is

$$\frac{\partial f}{\partial L} = \frac{\partial}{\partial L}(20K^{1/4}L^{3/4})$$

$$= 20K^{1/4}\left[\frac{d}{dL}(L^{3/4})\right]$$

$$= 20K^{1/4}\left[\frac{3}{4}L^{-1/4}\right]$$

$$= 15K^{1/4}L^{-1/4}$$

The desired rate is therefore

$$\frac{\partial f}{\partial L}(16, 81) = 15(16^{1/4})(81^{-1/4})$$

$$= \frac{15(16^{1/4})}{81^{1/4}}$$

$$= \frac{15(2)}{3}$$

$$= 10 \text{ units.}$$

This means that an increase in labor from $L = 81$ to $L = 82$ units will result in an increase of approximately 10 units of production. □

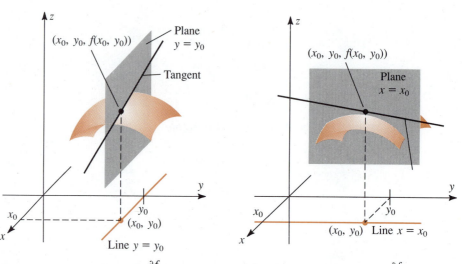

Figure 2.1 Partial derivative $\dfrac{\partial f}{\partial x}(x_0, y_0)$ is the slope of the line tangent to the trace of $z = f(x, y)$ in the plane $y = y_0$.

Figure 2.2 Partial derivative $\dfrac{\partial f}{\partial y}(x_0, y_0)$ is the slope of the line tangent to the trace of $z = f(x, y)$ in the plane $x = x_0$.

A Geometric Interpretation of Partial Derivatives

Figures 2.1 and 2.2 illustrate the geometric interpretations of the partial derivatives in Definition 1. To interpret $\dfrac{\partial f}{\partial x}(x_0, y_0)$, we note that the set of all points (x, y, z) in space, with $y = y_0$ fixed, is a plane. The intersection of this plane with the graph of $z = f(x, y)$ is a curve in the plane $y = y_0$, which is called the **trace** of $z = f(x, y)$ in the plane $y = y_0$. This curve may be viewed as the graph of the function of one variable

$$h(x) = f(x, y_0) \qquad (y_0 \text{ fixed}).$$

Since

$$h'(x_0) = \frac{\partial f}{\partial x}(x_0, y_0)$$

the partial derivative $\dfrac{\partial f}{\partial x}(x_0, y_0)$ gives the slope of the line tangent to this trace at the point $P_0 = (x_0, y_0, z_0)$ with $z_0 = f(x_0, y_0)$. That is, $\dfrac{\partial f}{\partial x}(x_0, y_0)$ *is the slope of the graph of $z = f(x, y)$ in the direction of the x-axis at the point P_0.* Similarly, the partial derivative $\dfrac{\partial f}{\partial y}(x_0, y_0)$ gives the slope of the graph of $z = f(x, y)$ in the direction of the y-axis at the point P_0.

Example 5 A company produces two types of tools, hammers and screwdrivers. The company's profit from the sale of x hundred hammers and y hundred screwdrivers per week is

$$P(x, y) = 20x - x^2 + 40y - y^2.$$

Find the slopes of the graph of $z = P(x, y)$ in the directions of the x and y axes at the point $(10, 20, P(10, 20)) = (10, 20, 500)$.

Solution: The partial derivative of P with respect to x is

$$\frac{\partial P}{\partial x} = \frac{\partial}{\partial x}(20x - x^2 + 40y - y^2)$$
$$= 20 - 2x$$

so the slope of the graph of $z = P(x, y)$ at the point $(10, 20, 500)$ in the direction of the x-axis is

$$\frac{\partial P}{\partial x}(10, 20) = 20 - 2(10) = 0.$$

Similarly, since

$$\frac{\partial P}{\partial y} = \frac{\partial}{\partial y}(20x - x^2 + 40y - y^2)$$
$$= 40 - 2y$$

the slope of the graph of $z = P(x, y)$ at the point $(10, 20, 500)$ in the direction of the y-axis is

$$\frac{\partial P}{\partial y}(10, 20) = 40 - 2(20) = 0.$$

Figure 2.3 shows the significance of the fact that both of these slopes are zero at the point $Q = (10, 20, 500)$. In fact, Q is the high point on the graph of $z = P(x, y)$. We shall exploit this idea further to find maximum and minimum values of functions of two variables in Section 8.3. □

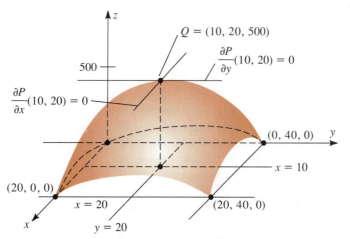

Figure 2.3 Graph of profit function $z = P(x, y)$ in Example 5. At high point both partial derivatives are zero.

Functions of More Than Two Variables

Partial derivatives are defined for functions of more than two variables by the same idea used in Definition 1: Hold all variables constant except one, and differentiate the resulting function of a single variable as before. For example, if $w = f(x, y, z)$ is a function of the three independent variables x, y, and z, the three partial derivatives are defined as follows:

$$\frac{\partial f}{\partial x}(x, y, z) = \lim_{h \to 0} \frac{f(x + h, y, z) - f(x, y, z)}{h} \tag{3}$$

$$\frac{\partial f}{\partial y}(x, y, z) = \lim_{h \to 0} \frac{f(x, y + h, z) - f(x, y, z)}{h} \tag{4}$$

$$\frac{\partial f}{\partial z}(x, y, z) = \lim_{h \to 0} \frac{f(x, y, z + h) - f(x, y, z)}{h} \tag{5}$$

Unfortunately, we lose our geometric interpretation in the presence of more than two independent variables. However, we do have the basic interpretation of the derivative as a rate of change. For example, the partial derivative $\frac{\partial f}{\partial x}$ in equation (3) is the rate at which the values $f(x, y, z)$ change with respect to change in x. Also, we make use of subscript notation for partial derivatives for functions of more than two variables just as in the two-variable case. That is, if $w = f(x, y, z)$

$$\frac{\partial f}{\partial z}(x, y, z) = \frac{\partial}{\partial z} f(x, y, z) = f_z(x, y, z) = w_z.$$

Example 6 Let $f(x, y, z) = \sqrt{x} e^{y/z}$, $z \neq 0$, $x \geq 0$. Then

$$\frac{\partial f}{\partial x}(x, y, z) = \left\{ \frac{d}{dx} \sqrt{x} \right\} e^{y/z} = \frac{1}{2\sqrt{x}} \cdot e^{y/z}$$

$$\frac{\partial f}{\partial y}(x, y, z) = \sqrt{x} \left\{ \frac{\partial}{\partial y} (e^{y/z}) \right\} = \sqrt{x} e^{y/z} \cdot \frac{1}{z} = \frac{\sqrt{x}}{z} e^{y/z}$$

$$\frac{\partial f}{\partial z}(x, y, z) = \sqrt{x} \left\{ \frac{\partial}{\partial z} (e^{y/z}) \right\} = \sqrt{x} e^{y/z} \left(\frac{-y}{z^2} \right) = \frac{-y\sqrt{x}}{z^2} e^{y/z}. \qquad \square$$

Higher Order Partial Derivatives

Repeated applications of partial differentiation lead to **higher order partial derivatives.** There is nothing terribly complicated about this concept, except that we must be very careful about notation since we encounter **mixed partial derivatives,** in which one differentiation is performed with respect to a particular variable, followed by another differentiation with respect to a different variable.

We shall use the following notation:

$$\frac{\partial^2 f}{\partial x^2}(x, y) = \frac{\partial^2}{\partial x^2} f(x, y) \qquad \text{means} \qquad \frac{\partial}{\partial x} \left(\frac{\partial f}{\partial x}(x, y) \right).$$

$$\frac{\partial^2 f}{\partial y \partial x}(x, y) = \frac{\partial^2}{\partial y \partial x} f(x, y) \qquad \text{means} \qquad \frac{\partial}{\partial y} \left(\frac{\partial f}{\partial x}(x, y) \right).$$

$$\frac{\partial^2 f}{\partial x \partial y}(x, y) = \frac{\partial^2}{\partial x \partial y} f(x, y) \qquad \text{means} \qquad \frac{\partial}{\partial x}\left(\frac{\partial f}{\partial y}(x, y)\right).$$

$$\frac{\partial^2 f}{\partial y^2}(x, y) = \frac{\partial^2}{\partial y^2} f(x, y) \qquad \text{means} \qquad \frac{\partial}{\partial y}\left(\frac{\partial f}{\partial y}(x, y)\right).$$

Note that the order in which the differentiations are performed is indicated by reading the "denominator" of the derivative notation from *right* to *left*. Similar definitions hold for third and higher order partial derivatives.

Example 7 For the function $f(x, y) = x^2 y^3 + e^{4x} \ln y$

$$\frac{\partial^2 f}{\partial x^2}(x, y) = \frac{\partial}{\partial x}\left(\frac{\partial f}{\partial x}(x, y)\right) = \frac{\partial}{\partial x}(2xy^3 + 4e^{4x} \ln y) = 2y^3 + 16e^{4x} \ln y$$

$$\frac{\partial^2 f}{\partial y \partial x}(x, y) = \frac{\partial}{\partial y}\left(\frac{\partial f}{\partial x}(x, y)\right) = \frac{\partial}{\partial y}(2xy^3 + 4e^{4x} \ln y) = 6xy^2 + \frac{4}{y}e^{4x}$$

$$\frac{\partial^2 f}{\partial x \partial y}(x, y) = \frac{\partial}{\partial x}\left(\frac{\partial f}{\partial y}(x, y)\right) = \frac{\partial}{\partial x}\left(3x^2 y^2 + \frac{1}{y}e^{4x}\right) = 6xy^2 + \frac{4}{y}e^{4x}$$

$$\frac{\partial^2 f}{\partial y^2}(x, y) = \frac{\partial}{\partial y}\left(\frac{\partial f}{\partial y}(x, y)\right) = \frac{\partial}{\partial y}\left(3x^2 y^2 + \frac{1}{y}e^{4x}\right) = 6x^2 y - \frac{1}{y^2}e^{4x}$$

$$\frac{\partial^3 f}{\partial y^3}(x, y) = \frac{\partial}{\partial y}\left(\frac{\partial^2 f}{\partial y^2}(x, y)\right) = \frac{\partial}{\partial y}\left(\frac{\partial^2 f}{\partial y^2}\right) = \frac{\partial}{\partial y}\left(6x^2 y - \frac{1}{y^2}e^{4x}\right)$$

$$= 6x^2 + \frac{2}{y^3}e^{4x}. \qquad \qquad \square$$

Example 8 For the function $f(x, y) = 4x^2 y - 6xy^3$ show that

$$\frac{\partial^2 f}{\partial x \partial y}(x, y) = \frac{\partial^2 f}{\partial y \partial x}(x, y).$$

Solution: We have

$$\frac{\partial f}{\partial y}(x, y) = \frac{\partial}{\partial y}(4x^2 y - 6xy^3) = 4x^2 - 18xy^2$$

so

$$\frac{\partial^2 f}{\partial x \partial y}(x, y) = \frac{\partial}{\partial x}(4x^2 - 18xy^2) = 8x - 18y^2.$$

Also, since

$$\frac{\partial f}{\partial x}(x, y) = \frac{\partial}{\partial x}(4x^2 y - 6xy^3) = 8xy - 6y^3$$

we have

$$\frac{\partial^2 f}{\partial y \partial x} = \frac{\partial}{\partial y}(8xy - 6y^3) = 8x - 18y^2.$$

Thus

$$\frac{\partial^2 f}{\partial x \partial y}(x, y) = 8x - 18y^2 = \frac{\partial^2 f}{\partial y \partial x}(x, y)$$

as claimed. ☐

Equality of Mixed Partials

The fact that the mixed partial derivatives $\dfrac{\partial^2 f}{\partial x \partial y}$ and $\dfrac{\partial^2 f}{\partial y \partial x}$ are the same for the functions in both Example 7 and Example 8 is not a coincidence. Although this is not the case for all functions of two variables, it is true when the function $z = f(x, y)$ and various of its partial derivatives are continuous, as the following theorem states. Its proof is beyond the scope of this text.

THEOREM 1

If the function $z = f(x, y)$ and the partial derivatives

$$\frac{\partial f}{\partial x}, \quad \frac{\partial f}{\partial y}, \quad \frac{\partial^2 f}{\partial x \partial y}, \quad \text{and} \quad \frac{\partial^2 f}{\partial y \partial x}$$

are all continuous near the point (x_0, y_0), then

$$\frac{\partial^2 f}{\partial x \partial y}(x_0, y_0) = \frac{\partial^2 f}{\partial y \partial x}(x_0, y_0).$$

Exercise Set 8.2

In Exercises 1–14 find (a) $\dfrac{\partial f}{\partial x}(x, y)$ and (b) $\dfrac{\partial f}{\partial y}(x, y)$.

1. $f(x, y) = 3x - 6y + 5$

2. $f(x, y) = x^2 - 2xy + y^3$

3. $f(x, y) = 4xy^2 - 3x^3y + y^5$

4. $f(x, y) = \dfrac{y}{x}$

5. $f(x, y) = \sqrt{x + y}$

6. $f(x, y) = x\sqrt{y} - y\sqrt{x}$

7. $f(x, y) = \dfrac{x + y}{x - y}$

8. $f(x, y) = e^{x^2 - 2y}$

9. $f(x, y) = \ln \sqrt{x^2 + y^2}$

10. $f(x, y) = xe^{\sqrt{y}}$

11. $f(x, y) = x^{2/3}y^{-1/3} + \sqrt{\dfrac{y}{x}}$

12. $f(x, y) = \dfrac{x - e^y}{y + e^x}$

13. $f(x, y) = \sqrt{x + 3e^y}$

14. $f(x, y) = \ln\left(\dfrac{x + y}{3 + x}\right)$

In Exercises 15–20 find z_x and z_y.

15. $z = \dfrac{x}{y^2} - \dfrac{y}{x^2}$

16. $z = 4 - x^{2/3}\sqrt{y + 1}$

17. $z = ye^{\sqrt{x} - 1}$

18. $z = \ln(x + ye^x)$

19. $z = \sqrt{xy + e^{xy}}$

20. $z = \dfrac{x^{5/2} + y^{2/3}}{y - x}$

In Exercises 21–28 find the indicated partial derivative at the given point.

21. For $f(x, y) = 2x(y - 7)$ find $\dfrac{\partial f}{\partial x}(3, 5)$.

22. For $f(x, y) = \dfrac{x}{y + x}$ find $\dfrac{\partial f}{\partial y}(2, -3)$.

23. For $f(x, y) = \sqrt{y^2 + 2x}$ find $\dfrac{\partial f}{\partial y}(4, 1)$.

24. For $f(x, y) = ye^{-2x} + x \ln y$ find $\dfrac{\partial f}{\partial y}(1, 4)$.

25. For $f(x, y, z) = xy^2 + xz^3 - xz$ find $\dfrac{\partial f}{\partial z}(1, 3, 5)$.

26. For $f(x, y, z) = xe^{yz}$ find $\dfrac{\partial f}{\partial y}(2, 0, 3)$.

27. For $f(x, y, z) = xe^{yz}$ find $\dfrac{\partial f}{\partial z}(2, 0, 3)$.

28. For $f(x, y, z) = \sqrt{xyz}$ find $\dfrac{\partial f}{\partial y}(2, 3, 6)$.

29. For $f(x, y) = x^2 - 2xy + y^2$ find
   a. $\dfrac{\partial^2 f}{\partial x^2}$ **b.** $\dfrac{\partial^2 f}{\partial x \partial y}$ **c.** $\dfrac{\partial^2 f}{\partial y \partial x}$ **d.** $\dfrac{\partial^2 f}{\partial y^2}$

30. For $f(x, y) = xe^{y-x}$ find
   a. $\dfrac{\partial^2 f}{\partial x^2}$ **b.** $\dfrac{\partial^2 f}{\partial y \partial x}$ **c.** $\dfrac{\partial^2 f}{\partial x \partial y}$ **d.** $\dfrac{\partial^2 f}{\partial y^2}$

31. For $f(x, y) = \ln (x^2 + y^2)$ find
   a. $\dfrac{\partial^2 f}{\partial x^2}$ **b.** $\dfrac{\partial^2 f}{\partial y \partial x}$ **c.** $\dfrac{\partial^2 f}{\partial x \partial y}$ **d.** $\dfrac{\partial^2 f}{\partial y^2}$

32. For the Cobb-Douglas production function
$$f(K, L) = 40K^{1/3}L^{2/3}$$
   a. Find the marginal productivity of capital, $\dfrac{\partial f}{\partial K}$, at capital level $K = 27$ units and labor level $L = 8$ units.
   b. Find the marginal productivity of labor, $\dfrac{\partial f}{\partial L}$, at capital level $K = 27$ units and labor level $L = 8$ units.

33. Recalling that the partial derivative $\dfrac{\partial f}{\partial K}$ is the *rate* at which the value of the function $z = f(K, L)$ will change with respect to change in K, use the result of Exercise 32a to estimate the change in production resulting from an increase in capital level from $K = 27$ units to $K = 28$ units if the labor level remains constant at $L = 8$ units.

34. A manufacturer of two different types of drugs determines that its monthly revenues from the sale of x units of drug A and y units of drug B is
$$R(x, y) = 40x^2 + 80y^2 - 100x - 200y - 20\sqrt{xy} \text{ dollars.}$$
   a. Find $\dfrac{\partial R}{\partial x}(4, 9)$, the rate at which revenues will increase with respect to increase in sales of drug A, at sales level $x = 4$ units and $y = 9$ units.
   b. $\dfrac{\partial R}{\partial y}(4, 9)$, the rate at which revenues will increase with respect to increase in sales of drug B at sales level $x = 4$ units and $y = 9$ units.

35. Suppose that a person's level of satisfaction resulting from the consumption of x slices of pizza and y glasses of soda in a certain period of time is given by the *utility function* $u(x, y) = 4x^{2/3} + 6y^{3/2} - xy^2$.
   a. Find the marginal utility $\dfrac{\partial u}{\partial x}(8, 4)$.
   b. Find the marginal utility $\dfrac{\partial u}{\partial y}(2, 4)$.

   How would you interpret these results?

36. The amount of interest earned by a deposit of $P_0 = 100$ dollars in a savings account paying a nominal interest rate of r percent per year compounded continuously for t years is given by the function of two variables $P(r, t) = 100(e^{rt} - 1)$.
   a. Find $\dfrac{\partial P}{\partial t}(0.10, 5)$
   b. Use the answer to part **(a)** to estimate the increase in earned interest if the time period is increased from $t = 5$ to $t = 6$ years and the interest rate remains fixed at $r = 10$ percent.

37. For the production function $f(K, L) = 100\sqrt{KL}$
   a. Find $\dfrac{\partial f}{\partial K}(3, 27)$, the marginal productivity of capital when $K = 3$ and $L = 27$.
   b. Find $\dfrac{\partial f}{\partial L}(3, 27)$, the corresponding marginal productivity of labor.

38. An automobile manufacturer determines that the demand for a certain model of automobile is given by a function $z = f(x, y)$ where x is the list price of the automobile and y is the amount of money spent on advertising the automobile per unit time.

a. What would you expect the *sign* of the quantity $\dfrac{\partial f}{\partial x}(x, y)$ to be? Why?

b. What would you expect the sign of $\dfrac{\partial f}{\partial y}(x, y)$ to be? Why?

39. A grocer sells two types of eggs, white and brown. The grocer determines that when the white eggs sell for x dollars per dozen and the brown eggs sell for y dollars per dozen the weekly demand for white eggs is

$$W(x, y) = 150 - 30x^2 + 20y$$

and the weekly demand for brown eggs is

$$B(x, y) = 200 + 40x - 30y^2.$$

a. Show that $\dfrac{\partial W}{\partial x}(1, 1) < 0$ but $\dfrac{\partial W}{\partial y}(1, 1) > 0$, and explain these results.

b. Show that $\dfrac{\partial B}{\partial x}(1, 1) > 0$ but $\dfrac{\partial B}{\partial y}(1, 1) < 0$, and explain these results.

40. Consider a market occupied by two products, A and B (perhaps among others). Let p_A be the selling price of product A, let p_B be the selling price of product B, let $D_A(p_A, p_B)$ be the demand for product A (number purchased per unit time), and let $D_B(p_A, p_B)$ be the demand for product B at prices p_A and p_B.

a. Explain why we would always expect to have

$$\frac{\partial D_A}{\partial p_A} \le 0 \quad \text{and} \quad \frac{\partial D_B}{\partial p_B} \le 0.$$

b. If $\dfrac{\partial D_A}{\partial p_B} > 0$ and $\dfrac{\partial D_B}{\partial p_A} > 0$ the products are called *competitive* since an increase in the price of one causes demand for the other to increase. The white and brown eggs in Exercise 39 are one such example. Find others.

c. If $\dfrac{\partial D_A}{\partial p_B} < 0$ and $\dfrac{\partial D_B}{\partial p_A} < 0$ the products are called *complementary*. In this case an increase in the price of either causes a decrease in the demand for the other. Give examples of complementary products.

8.3 RELATIVE MAXIMA AND MINIMA

Just as for functions of a single variable, we wish to know how to find maximum and minimum values of functions of several variables. Our discussion of how to do so will be largely confined to finding *relative extrema* for functions of two variables.

DEFINITION 2

The number $z_0 = f(x_0, y_0)$ is called a **relative maximum** for the function $z = f(x, y)$ if there exists a circle C with center (x_0, y_0) and radius $r \ne 0$ so that, for all points inside C, $f(x, y)$ is defined and

$$f(x_0, y_0) \ge f(x, y).$$

The number $f(x_0, y_0)$ is a **relative minimum** for $z = f(x, y)$ if, for all points (x, y) inside such a circle C, $f(x, y)$ is defined and

$$f(x_0, y_0) \le f(x, y).$$

A **relative extremum** refers to either a relative maximum or a relative minimum.

In other words, a relative maximum $f(x_0, y_0)$ is the largest value of the function $z = f(x, y)$ for all points (x, y) "near" (x_0, y_0), and similarly for relative minima.

Example 1 Figure 3.1 shows a portion of the graph of the function $f(x, y) = 2x + 4y - x^2 - y^2 - 1$. This function has a relative maximum value at the point $(1, 2)$ of

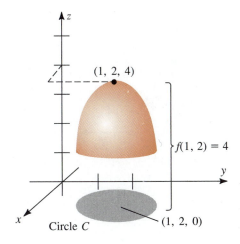

Figure 3.1 $f(1, 2) = 4$ is relative maximum for the function $f(x, y) = 2x + 4y - x^2 - y^2 - 1$.

$z = f(1, 2) = 4$, as you can observe by calculating values of $f(x, y)$ near this point. Note also that both partial derivatives $\dfrac{\partial f}{\partial x}$ and $\dfrac{\partial f}{\partial y}$ are zero at this point. Indeed, since

$$\frac{\partial f}{\partial x}(x, y) = 2 - 2x; \qquad \frac{\partial f}{\partial y}(x, y) = 4 - 2y$$

we have $\dfrac{\partial f}{\partial x}(1, 2) = 2 - 2(1) = 0$ and $\dfrac{\partial f}{\partial y}(1, 2) = 4 - 2(2) = 0$. $\square$

Example 2 Figure 3.2 shows a portion of the graph of $f(x, y) = e^{x^2 + y^2}$. This function has a relative minimum value at $(0, 0)$ of $f(0, 0) = e^0 = 1$, where the exponent $x^2 + y^2$ has its minimum value. Again, note that both partial derivatives are zero at this point:

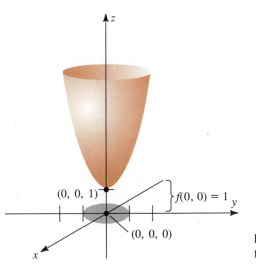

Figure 3.2 $f(0, 0) = 1$ is relative minimum for $f(x, y) = e^{x^2 + y^2}$.

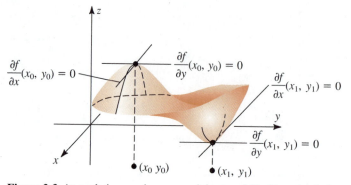

Figure 3.3 At a relative maximum or minimum of f both partial derivatives, if they exist, must equal zero.

$$\frac{\partial f}{\partial x}(x, y) = 2xe^{x^2+y^2} \quad \text{so} \quad \frac{\partial f}{\partial x}(0, 0) = 0$$

$$\frac{\partial f}{\partial y}(x, y) = 2ye^{x^2+y^2} \quad \text{so} \quad \frac{\partial f}{\partial y}(0, 0) = 0.$$

Figure 3.3 illustrates why both partial derivatives, if they exist, must equal zero at a point (x_0, y_0) corresponding to a relative extremum:

(i) If $f(x_0, y_0)$ is a relative maximum and $\dfrac{\partial f}{\partial x}(x, y)$ exists for all (x, y) near (x_0, y_0), then by holding y fixed at $y = y_0$ we obtain a differentiable function of the single variable x, $g(x) = f(x, y_0)$. Since $f(x_0, y_0)$ is a relative maximum for f, $g(x_0)$ must be a relative maximum for g. Thus we must have

$$g'(x_0) = \frac{\partial f}{\partial x}(x_0, y_0) = 0.$$

(ii) Similarly, if $f(x_0, y_0)$ is a relative maximum, we may fix x at $x = x_0$ and obtain the function $h(y) = f(x_0, y)$, which has a maximum at $y = y_0$. Thus

$$h'(y_0) = \frac{\partial f}{\partial y}(x_0, y_0) = 0.$$

Statements (i) and (ii) apply to relative minima as well. Geometrically, these conclusions simply state that if $f(x_0, y_0)$ is a relative extremum, the slopes of the tangents to the graph of $z = f(x, y)$ parallel to the x and y axes at the point $(x_0, y_0, f(x_0, y_0))$, if they exist, must equal zero. $\square$

The following theorem summarizes these observations.

THEOREM 2   If the number $z = f(x_0, y_0)$ is a relative maximum or minimum for the function $z = f(x, y)$, and if both partial derivatives $\frac{\partial f}{\partial x}(x_0, y_0)$ and $\frac{\partial f}{\partial y}(x_0, y_0)$ exist, then we must have

$$\frac{\partial f}{\partial x}(x_0, y_0) = \frac{\partial f}{\partial y}(x_0, y_0) = 0.$$

Theorem 2 suggests a very straightforward procedure for finding points where extrema may occur: Find those points at which both partial derivatives simultaneously equal zero. As for functions of a single variable, we shall refer to such points as **critical points.** (Extrema can also occur at points where one or both partial derivatives fail to exist, but we shall not consider such functions here. See Exercise 26 for one such example.)

Before stating how to determine whether a critical point corresponds to a relative maximum or relative minimum, if either, we should note that a critical point need not correspond to an extremum at all. Figure 3.4 shows the graph of the function $f(x, y) = y^2 - x^2$. Since

$$\frac{\partial f}{\partial x}(x, y) = \frac{\partial}{\partial x}(y^2 - x^2) = -2x$$

and

$$\frac{\partial f}{\partial y}(x, y) = \frac{\partial}{\partial y}(y^2 - x^2) = 2y$$

it is easy to see that $\frac{\partial f}{\partial x}(0, 0) = \frac{\partial f}{\partial y}(0, 0) = 0$, so the point $(0, 0)$ is a critical point. But it is just as easy to see that the value $f(0, 0) = 0$ is neither a relative maximum nor a relative minimum. With y fixed at $y = 0$ the point $(0, 0, 0)$ corresponds to a relative maximum for the function $g(x) = f(x, 0) = -x^2$, but with x fixed at $x = 0$ the point $(0, 0, 0)$ corresponds to a relative minimum for the function $h(y) = f(0, y) = y^2$. (See Figure 3.4.) Such points are called **saddle points.**

The following theorem is a sort of "Second Derivative Test" for functions of two

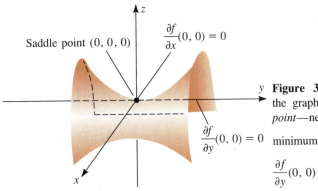

Figure 3.4 The point $(0, 0, 0)$ on the graph of $z = y^2 - x^2$ is a *saddle point*—neither a relative maximum or minimum, even though $\frac{\partial f}{\partial x}(0, 0) = \frac{\partial f}{\partial y}(0, 0) = 0$.

Saddle point $(0, 0, 0)$

$\frac{\partial f}{\partial x}(0, 0) = 0$

$\frac{\partial f}{\partial y}(0, 0) = 0$

variables. We shall use it to classify critical points as relative maxima, relative minima, or saddle points. Its proof is omitted.

THEOREM 3
Second Derivative Test

Suppose that all second order partial derivatives of $z = f(x, y)$ are continuous in a circle with center (x_0, y_0) and that $\dfrac{\partial f}{\partial x}(x_0, y_0) = \dfrac{\partial f}{\partial y}(x_0, y_0) = 0$. Let

$$A = \frac{\partial^2 f}{\partial x^2}(x_0, y_0), \qquad B = \frac{\partial^2 f}{\partial y \partial x}(x_0, y_0), \qquad C = \frac{\partial^2 f}{\partial y^2}(x_0, y_0)$$

and

$$D = B^2 - AC.$$

Then

(i) If $D < 0$ and $A < 0$, $f(x_0, y_0)$ is a relative maximum.
(ii) If $D < 0$ and $A > 0$, $f(x_0, y_0)$ is a relative minimum.
(iii) If $D > 0$, $(x_0, y_0, f(x_0, y_0))$ is a saddle point.
(iv) If $D = 0$, no conclusions may be drawn.

Example 3 We can use the Second Derivative Test to verify that the critical point $(1, 2)$ for the function

$$f(x, y) = 2x + 4y - x^2 - y^2 - 1$$

of Example 1 indeed corresponds to a relative maximum. Since

$$\frac{\partial f}{\partial x}(x, y) = 2 - 2x \quad \text{and} \quad \frac{\partial f}{\partial y}(x, y) = 4 - 2y$$

we have

$$\frac{\partial^2 f}{\partial x^2} = -2; \qquad \frac{\partial^2 f}{\partial y \partial x} = 0; \qquad \frac{\partial^2 f}{\partial y^2} = -2.$$

Thus $A = -2, B = 0, C = -2$, and $D = 0^2 - (-2)(-2) = -4$. Since $D < 0$ and $A < 0$, the critical point $(1, 2)$ corresponds to a relative maximum by part (i) of Theorem 3.

□

Example 4 Find and classify all extreme points for the function $f(x, y) = 4x - 2y - x^2 - 2y^2 + 2xy - 10 = 0$.

Strategy

Set $\dfrac{\partial f}{\partial x} = 0$.

Solution

We begin by finding the critical point(s). Setting

$$\frac{\partial f}{\partial x} = 4 - 2x + 2y = 0$$

gives the equation

$$-2x + 2y = -4. \tag{1}$$

Similarly, setting

Set $\dfrac{\partial f}{\partial y} = 0.$

$$\frac{\partial f}{\partial y} = -2 - 4y + 2x = 0$$

gives the equation

$$2x - 4y = 2. \tag{2}$$

Solve the resulting pair of equations by substitution (or by addition) to find the critical point(s).

We must therefore find the simultaneous solution of the pair of equations (1) and (2):

$$-2x + 2y = -4 \tag{3}$$

$$2x - 4y = 2. \tag{4}$$

Solving the first equation for x gives

$$-2x = -4 - 2y$$

or

$$x = 2 + y. \tag{5}$$

Substituting this expression for x into equation (4) gives

$$2(2 + y) - 4y = 2$$

or

$$4 - 2y = 2.$$

Thus

$$-2y = -2.$$

so

$$y = 1.$$

Substituting this value of y back into equation (5) then gives

$$x = 2 + (1) = 3.$$

Calculate the second order partials and determine A, B, C, and D for each critical point.

The single critical point is therefore $(3, 1)$. To test this critical point, we calculate

$$A = \frac{\partial^2 f}{\partial x^2}(3, 1) = -2$$

$$B = \frac{\partial^2 f}{\partial y \partial x}(3, 1) = 2$$

$$C = \frac{\partial^2 f}{\partial y^2}(3, 1) = -4$$

and

$$D = B^2 - AC = (2)^2 - (-2)(-4) = -4.$$

Apply the Second Derivative Test.

Since $D < 0$ and $A < 0$ the critical point $(3, 1)$ corresponds to the *relative maximum* value

$$f(3, 1) = 4(3) - 2(1) - 3^2 - 2(1^2) + 2(3)(1) - 10$$
$$= -5. \qquad \square$$

Example 5 Find and classify all relative extrema for the function

$$f(x, y) = x^4 + y^4 - 4xy.$$

Strategy

Set $\dfrac{\partial f}{\partial x} = 0$ and solve.

Solution

To find the critical points we begin by setting

$$\frac{\partial f}{\partial x} = 4x^3 - 4y = 0$$

which gives the equation

$$y = x^3. \tag{6}$$

Set $\dfrac{\partial f}{\partial y} = 0$ and solve.

Similarly, setting $\dfrac{\partial f}{\partial y}$ equal to zero gives

$$\frac{\partial f}{\partial y} = 4y^3 - 4x = 0$$

or

$$x = y^3. \tag{7}$$

Solve the two resulting equations by substitution, obtaining a single equation for y and then finding the corresponding values for x.

Since both equations (6) and (7) must hold at a critical point, we substitute $x = y^3$ from equation (7) into equation (6) to obtain

$$y = x^3 = (y^3)^3 = y^9$$

or

$$y = y^9. \tag{8}$$

Each pair (x_0, y_0) gives a critical point for $f(x, y)$.

Thus either $y = 0$ or, after dividing both sides of equation (8) by y, $y^8 = 1$. This last equation has solutions $y = -1$ and $y = 1$. Using equation (7), we then find that

$$\begin{array}{lll}
\text{if} & y = 0, & x = 0^3 = 0 \\
\text{if} & y = -1, & x = (-1)^3 = -1 \\
\text{if} & y = 1, & x = (1)^3 = 1.
\end{array}$$

We therefore have three critical points to check: $(0, 0)$, $(-1, -1)$, and $(1, 1)$.

Calculate all second order partials.

The second partials of $f(x, y)$ are

$$\frac{\partial^2 f}{\partial x^2} = \frac{\partial}{\partial x}(4x^3 - 4y) = 12x^2$$

$$\frac{\partial^2 f}{\partial y \partial x} = \frac{\partial}{\partial y}(4x^3 - 4y) = -4$$

$$\frac{\partial^2 f}{\partial y^2} = \frac{\partial}{\partial y}(4y^3 - 4x) = 12y^2.$$

At each critical point substitute the coordinates to obtain A, B, C, and D and apply Theorem 3.

Thus at the critical point $(0, 0)$ we have

$$A = 12(0)^2 = 0, \qquad B = -4, \qquad C = 12(0)^2 = 0$$

and

$$D = (-4)^2 - (0)(0) = 16 > 0.$$

Thus $(0, 0, 0)$ is a saddle point.
  At the critical point $(-1, -1)$ we have

$$A = 12(-1)^2 = 12, \qquad B = -4, \qquad C = 12(-1)^2 = 12$$

and

$$D = (-4)^2 - (12)(12) = -128 < 0.$$

Thus since $A > 0$, $f(-1, -1) = -2$ is a relative minimum.
  Finally, at the critical point $(1, 1)$ we have

$$A = 12(1)^2, \qquad B = -4, \qquad C = 12(1)^2 = 12$$

and

$$D = (-4)^2 - (12)(12) = -128 < 0.$$

Since, again, $A > 0$, $f(1, 1) = -2$ is a relative minimum. □

Applications of Relative Extrema

The following examples are typical of applied problems in which the desired maximum or minimum value of a function occurs at a relative extremum. Although there are other ways in which absolute extrema can occur (such as at points where partial derivatives fail to exist, or at "boundary points" of the domain of the function), we shall not consider such problems in this text.

Example 6 A company produces electronic typewriters and word processors. It sells the electronic typewriters for $100 each and the word processors for $300 each. The company has determined that its weekly total cost in producing x electronic typewriters and y word processors is given by the joint cost function

$$C(x, y) = 2000 + 50x + 80y + x^2 + 2y^2.$$

Find the numbers x and y of machines that the company should manufacture and sell weekly in order to maximize profits.

Solution: Since the typewriters sell for $100 each and the word processors for $300 each, the weekly revenue from the sale of x typewriters and y word processors is

$$R(x, y) = 100x + 300y.$$

The weekly profit function to be maximized is therefore

$$\begin{aligned} P(x, y) &= R(x, y) - C(x, y) \\ &= (100x + 300y) - (2000 + 50x + 80y + x^2 + 2y^2) \\ &= 50x + 220y - x^2 - 2y^2 - 2000. \end{aligned}$$

Setting the two partial derivatives equal to zero gives

$$\frac{\partial P}{\partial x} = 50 - 2x = 0$$

so

$$x = 25$$

and

$$\frac{\partial P}{\partial y} = 220 - 4y = 0$$

so

$$y = 55.$$

To test the single critical point $(25, 55)$, we note that

$$A = \frac{\partial^2 P}{\partial x^2} = -2; \qquad B = \frac{\partial^2 P}{\partial y \partial x} = 0; \qquad C = \frac{\partial^2 P}{\partial y^2} = -4$$

and $D = B^2 - AC = 0 - (-2)(-4) = -8$. Since $D < 0$ and $A < 0$, the critical point $(25, 55)$ corresponds to a relative maximum. The production schedule for maximum profit is therefore $x = 25$ typewriters and $y = 55$ word processors, which yields a profit of

$$P = (25, 55) = 50(25) + 220(55) - 25^2 - 2(55)^2 - 2000$$
$$= 4675 \text{ dollars.} \qquad \square$$

Example 7 A grocer sells two types of eggs, white and brown, which compete with each other for sales depending on how they are priced. The grocer has determined that when the white eggs sell for x dollars per dozen and the brown eggs for y dollars per dozen, the daily sales of the white eggs will be

$$W(x, y) = 30 - 15x + 3y$$

dozen, while the daily sales of the brown eggs will be

$$B(x, y) = 20 - 12y + 2x$$

dozen. Find the prices x and y for which the grocer's daily revenue from the sale of eggs will be a maximum.

Solution: The grocer's daily revenue from selling white eggs at price x and brown eggs at price y will be

$$R(x, y) = xW(x, y) + yB(x, y)$$
$$= x(30 - 15x + 3y) + y(20 - 12y + 2x)$$
$$= 30x - 15x^2 + 5xy + 20y - 12y^2.$$

To find the critical point(s), we set the partial derivatives equal to zero:

$$\frac{\partial R}{\partial x} = 30 - 30x + 5y = 0$$

gives

$$30x - 5y = 30$$

or

$$6x - y = 6 \tag{9}$$

and

$$\frac{\partial R}{\partial y} = 5x + 20 - 24y = 0$$

gives

$$5x - 24y = -20. \tag{10}$$

To find the simultaneous solution of equations (9) and (10), we first solve (9) for y, obtaining

$$y = 6x - 6 \tag{11}$$

which we use to substitute for y in equation (10) to obtain

$$5x - 24(6x - 6) = -20$$

which gives

$$-139x + 144 = -20$$

or

$$x = \frac{164}{139} \approx 1.18.$$

From equation (11) we then obtain

$$y = 6\left(\frac{164}{139}\right) - 6 = \frac{150}{139} \approx 1.08.$$

To test the critical point $\left(\frac{164}{139}, \frac{150}{139}\right)$, we calculate

$$A = \frac{\partial^2 R}{\partial x^2} = -30, \qquad B = \frac{\partial^2 R}{\partial y \partial x} = 5, \qquad C = \frac{\partial^2 R}{\partial y^2} = -24$$

Figure 3.5 Container in Example 8.

and $D = B^2 - AC = 25 - (-30)(-24) = -695$. Since $D < 0$ and $A < 0$, the critical point corresponds to a relative maximum by Theorem 3. The grocer will therefore maximize egg revenues by pricing the white eggs at $1.18 per dozen and the brown eggs at $1.08 per dozen. □

Example 8 A container company wishes to design an open rectangular container with a capacity of 144 cubic feet. If the material for the bottom costs $4 per square foot and the material for the sides costs $3 per square foot, find the dimensions for which the cost of materials will be a minimum.

Solution: Let the container have length x feet, width y feet, and height z feet, as in Figure 3.5. Since the area of the bottom is xy square feet, the cost of the bottom will be $4xy$. Similarly, the cost of the front and back panels will be $3xz$ each, and the cost of the end panels will be $3yz$ each. The total cost function to be minimized is therefore

$$C(x, y, z) = 4xy + 2(3xz) + 2(3yz)$$
$$= 4xy + 6xz + 6yz.$$

Although this is a function of three variables, there is an *auxiliary* equation that we can use to substitute for one of the variables in terms of the other two. It is simply the statement that volume equals 144 cubic feet. That is

$$xyz = 144 \quad \text{or} \quad z = \frac{144}{xy}.$$

Substituting this expression for z into the cost function gives the function of two variables:

$$C(x, y) = 4xy + 6x\left(\frac{144}{xy}\right) + 6y\left(\frac{144}{xy}\right)$$
$$= 4xy + \frac{864}{y} + \frac{864}{x}.$$

We now proceed to minimize C as before. The equation

$$\frac{\partial C}{\partial x}(x, y) = 4y - \frac{864}{x^2} = 0$$

gives

$$y = \frac{216}{x^2} \tag{12}$$

and the equation

$$\frac{\partial C}{\partial y}(x, y) = 4x - \frac{864}{y^2} = 0$$

gives, using equation (12)

$$x = \frac{216}{y^2} = \frac{216}{\left(\frac{216}{x^2}\right)^2} = \frac{x^4}{216}.$$

Thus

$$x^3 = 216 \quad \text{so} \quad x = \sqrt[3]{216} = 6.$$

With $x = 6$ equation (12) gives $y = \frac{216}{6^2} = 6$. We leave it to you to verify that the single critical point $(6, 6)$ actually yields a relative minimum. The dimensions for minimum cost are therefore $x = 6$ ft, $y = 6$ ft, and $z = \frac{144}{6^2} = 4$ ft. □

Functions of More Than Two Variables

The discussion of this section has focused on finding relative extrema for functions of two variables. For functions of more than two variables Theorem 2 generalizes directly. For example, if the function $w = f(x, y, z)$ of three variables has a relative extremum at the point (x_0, y_0, z_0) and each of its three partial derivatives exists at this point, then we must have

$$\frac{\partial f}{\partial x}(x_0, y_0, z_0) = \frac{\partial f}{\partial y}(x_0, y_0, z_0) = \frac{\partial f}{\partial z}(x_0, y_0, z_0) = 0.$$

However, the Second Derivative Test does not generalize quite so easily. For functions of more than two variables the classification of critical points is considerably more difficult and will not be pursued here.

Exercise Set 8.3

In each of Exercises 1–16 find all critical points for f and determine whether each corresponds to a relative maximum, a relative minimum, or a saddle point.

1. $f(x, y) = x^2 + y^2 + 4y + 4$

2. $f(x, y) = 4x - x^2 - y^2 + 6$

3. $f(x, y) = x^2 + y^2 + 4x - 2y + 11$

4. $f(x, y) = 2x - 6y - x^2 - 2y^2 + 10$

5. $f(x, y) = x^2 - y^2 + 6x + 4y + 5$

6. $f(x, y) = 7 - 2x + 2y - x^2 - y^2$

7. $f(x, y) = xy + 9$

8. $f(x, y) = x^2 + y^4 - 2x - 4y^2 + 5$

9. $f(x, y) = 5x^2 + y^2 - 10x - 6y + 15$

10. $f(x, y) = x^2 + y^3 - 3y$

11. $f(x, y) = x^3 - y^3$

12. $f(x, y) = e^{x^2 - 2x + y^2 + 4}$

13. $f(x, y) = x^3 + y^3 + 4xy$

14. $f(x, y) = 2x^2 - y^2 + 3x + y - xy + 10$

15. $f(x, y) = x^2 + y^2 + x - 2y + xy + 5$

16. $f(x, y) = x^2 + 3y^2 - 2x + 3y + 2xy - 6$

17. A sewing machine manufacturer sells its machines in two markets, foreign and domestic. It determines that the profit resulting from the sale of x machines in the domestic market and y machines in the foreign market per week is

$$P(x, y) = 30x + 90y - 0.5x^2 - 2y^2 - xy.$$

Find the number of machines that should be sold in the domestic and foreign markets per week in order to maximize profits.

18. For the tool company in Example 5, Section 8.2, with profit equation

$$P(x, y) = 20x - x^2 + 40y - y^2$$

find the point (x, y) for which profit is a maximum.

19. A company determines that the productivity of its manufacturing plant resulting from weekly expenditures of x thousand dollars in labor and y thousand dollars in equipment is

$$P(x, y) = 40x + 80y - 2x^2 - 10y^2 - 4xy.$$

Find the point (x, y) for which productivity is a maximum.

20. A company determines that if x units of labor and y units of capital equipment are used to meet its weekly production schedule, the total cost of operating the factory is

$$C(x, y)$$
$$= 10,000 + 10x^2 + 15y^2 - 100x - 200y + 10xy.$$

Find the mix (x, y) of labor and equipment that minimizes costs.

21. The prices p and q of two items are related to the number of items, x and y, which will be sold during a fixed period of time by the demand equations

$$p = 20 - x; \qquad q = 46 - \frac{5}{2}y.$$

The total cost of producing x items of the first kind and y items of the second kind during this period is

$$C(x, y) = 100 + 4x + 2y + xy.$$

Find the number of items, x and y, for which profit will be a maximum.

22. A grocer sells two brands of coffee, A and B. Brand A is a premium coffee that costs the grocer $2 per pound. Brand B, a generic brand, costs only $1 per pound. The grocer determines that if he charges x dollars per pound for Brand A and y dollars per pound for Brand B, he can sell

$$D_A(x, y) = 30 - 5x + y$$

pounds of brand A and

$$D_B(x, y) = 40 - 4y + x$$

pounds of brand B per week.
a. Find the grocer's weekly revenue function $R(x, y) = xD_A(x, y) + yD_B(x, y)$ from the sale of coffee.
b. Assuming the grocer's only cost in selling coffee is the price paid per pound, find the grocer's weekly profit function $P = R - C$.
c. Find the values of x and y for which P is a maximum.

23. A rectangular box, with a top, is to hold 16 cubic meters. Find the dimensions that produce the least expensive box if the material for the side walls is half as expensive as the material for the top and bottom.

24. Find the dimensions of the closed rectangular box of volume $V = 8000$ cm^3 and of minimum surface area.

25. Find the dimensions of the rectangular package of largest volume that can be mailed under the restrictions that length plus girth cannot exceed 84 inches. (Girth is the perimeter of the cross-section taken perpendicular to the length.)

26. Figure 3.6 shows the graph of the function $f(x, y) = \sqrt{x^2 + y^2}$.
a. Explain why the function $f(x, y)$ has a relative minimum value of $f(0, 0) = 0$ at $(0, 0)$.
b. Show that neither $\dfrac{\partial f}{\partial x}(0, 0)$ nor $\dfrac{\partial f}{\partial y}(0, 0)$ is defined.

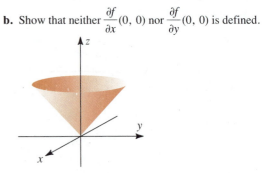

Figure 3.6 Graph of $f(x, y) = \sqrt{x^2 + y^2}$.

8.4 OPTIMIZATION PROBLEMS WITH CONSTRAINTS

Frequently, especially in business applications, it is necessary to find the maximum or minimum value of a function of two variables subject to a **constraint,** which is an additional condition that must be satisfied by the independent variables. The presence of such a constraint can change both the optimum (maximum or minimum) value of the function as well as the points at which the optimum occurs.

For example, suppose that a company determines that the profit resulting from the production and sale of x standard and y deluxe dishwashers per day is given by the profit function

$$P(x, y) = 40x + 20y - x^2 - y^2 \text{ dollars.} \tag{1}$$

In order to find the production schedule (x, y) for which profit is a maximum, we would apply the technique of Section 8.3. Setting the partial derivative $\dfrac{\partial P}{\partial x}$ equal to zero, we obtain

$$\frac{\partial P}{\partial x}(x, y) = 40 - 2x = 0$$

which gives $2x = 40$, or $x = 20$. Setting $\dfrac{\partial P}{\partial y} = 0$, we obtain

$$\frac{\partial P}{\partial y} = 20 - 2y = 0$$

which gives $2y = 20$, or $y = 10$. Thus the only critical point is $(x, y) = (20, 10)$. You can verify, using the Second Derivative Test, that this critical point corresponds to a relative maximum. Thus profit will be a maximum when the company manufactures 20 standard and 10 deluxe dishwashers per day. The maximum profit is $P(20, 10) = 500$ dollars per day.

But now suppose that the production capacity of the company's manufacturing plant is limited to a total of 20 dishwashers per day. This is an example of a *constraint,* which we can represent by the equation

$$x + y = 20. \tag{2}$$

Equation (2) simply states that the number of standard and deluxe dishwashers produced per day must equal 20. Clearly, this constraint prohibits the production schedule $(x, y) = (20, 10)$.

Finding the production schedule (x, y) for which the profit function P is a maximum subject to the constraint given by equation (2) is an example of an *optimization problem with a constraint,* which we can state as follows.

Example 1 Find the maximum value of the profit function

$$P(x, y) = 40x + 20y - x^2 - y^2 \tag{3}$$

subject to the constraint

$$x + y = 20. \tag{4}$$

Solution: Solving equation (4) for y gives

$$y = 20 - x. \tag{5}$$

Substituting this expression for y in equation (3) then gives the profit function P as a function of the single variable x:

$$\begin{aligned} P(x) &= 40x + 20(20 - x) - x^2 - (20 - x)^2 \\ &= 40x + 400 - 20x - x^2 - (400 - 40x + x^2) \\ &= 60x - 2x^2. \end{aligned}$$

To find the maximum value of P, we set

$$P'(x) = 60 - 4x = 0$$

and conclude that $4x = 60$, or $x = 15$. Since $P''(15) = -4 < 0$, the Second Derivative Test verifies that this critical number corresponds to a relative maximum value for P. Equation (5) then gives $y = 20 - 15 = 5$. The maximum profit therefore occurs when $x = 15$ standard and $y = 5$ deluxe dishwashers are manufactured per day. It is

$$\begin{aligned} P(15, 5) &= 40(15) + 20(5) - (15)^2 - 5^2 \\ &= 450 \text{ dollars.} \end{aligned}$$

□

The Geometry of Constraints

Figure 4.1 illustrates the effect of a constraint on the problem of finding the maximum or minimum value of a function. In this case the problem is to find the maximum value of the function

$$f(x, y) = 16 - x^2 - y^2 \tag{6}$$

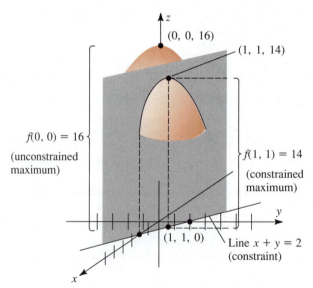

Figure 4.1 Maximum value of $f(x, y) = 16 - x^2 - y^2$, subject to the constraint $x + y = 2$, is $f(1, 1) = 14$.

subject to the constraint

$$x + y = 2. \tag{7}$$

If the constraint were not present, the maximum value of f would be $f(0, 0) = 16$ since the highest point on the graph of $z = f(x, y)$ is $(0, 0, 16)$, which lies directly above the point $(0, 0, 0)$ in the xy-plane.

However, since the constraint restricts the domain of f to the points (x, y) lying on the line $x + y = 2$, the constrained maximum corresponds to the highest point lying on the curve C, which is the part of the graph of f lying above the line $x + y = 2$.

To find this maximum, we proceed as before to solve the constraint equation (7) for y, obtaining $y = 2 - x$, and substitute this expression for y into equation (6). This gives the function whose graph corresponds to the curve C:

$$\begin{aligned} f(x) &= 16 - x^2 - (2 - x)^2 \\ &= 16 - x^2 - (x^2 - 4x + 4) \\ &= 12 + 4x - 2x^2. \end{aligned}$$

The maximum value of this function is found by setting $f'(x) = 0$. Since

$$f'(x) = 4 - 4x$$

the equation $f'(x) = 0$ gives $4x - 4 = 0$, or $x = 1$. Equation (7) then gives $y = 1$. Thus the constrained maximum is $f(1, 1) = 16 - 1^2 - 1^2 = 14$, which corresponds to the ''high'' point on the curve C, namely, $(1, 1, 14)$.

The Method of Lagrange Multipliers

Up to this point we have approached the problem of finding the maximum or minimum value of $z = f(x, y)$ subject to a constraint by solving the constraint equation for one of the independent variables and substituting this result into the expression for $f(x, y)$. This reduces f to a function of a single variable to which the techniques of Chapter 3 can be applied.

However, this method is not always ideal, since the constraint equation may be difficult to solve, or the resulting function of a single variable may be difficult to work with.

The mathematician Joseph L. Lagrange (1736–1813) discovered a different method for solving such problems, which is often superior to the substitution method. In order to describe Lagrange's method, we need to think of the constraint equation in the form

$$g(x, y) = 0. \qquad \text{(constraint)}$$

This can always be done by simply moving all nonzero terms to one side of the equation. For example, the constraint equation $x + y = 20$ in Example 1 can be written

$$x + y - 20 = 0$$

by subtracting 20 from both sides. Thus $g(x, y) = x + y - 20$ in this example. The following theorem, whose proof is beyond the scope of this text, is the basis for Lagrange's method.

THEOREM 4 **Method of Lagrange**	Let f and g have continuous partial derivatives. If a relative maximum or minimum value of the function f subject to the constraint

$$g(x, y) = 0$$

occurs at the point (x, y), then there is a number λ for which the point (x, y, λ) is a critical point for the function $L = f + \lambda g$.

The variable λ that appears in Theorem 4 is called the **Lagrange multiplier.** Since the critical points of the function $L = f + \lambda g$ are found by setting all partial derivatives equal to zero, the **method of Lagrange multipliers** is:

(i) Form the function $L = f + \lambda g$, with values

$$L(x, y, \lambda) = f(x, y) + \lambda g(x, y).$$

(ii) Set all partial derivatives of L equal to zero, obtaining the equations

$$\frac{\partial L}{\partial x} = 0, \quad \text{or} \quad \frac{\partial f}{\partial x} + \lambda \frac{\partial g}{\partial x} = 0 \tag{L1}$$

$$\frac{\partial L}{\partial y} = 0, \quad \text{or} \quad \frac{\partial f}{\partial y} + \lambda \frac{\partial g}{\partial y} = 0 \tag{L2}$$

$$\frac{\partial L}{\partial \lambda} = 0, \quad \text{or} \quad g(x, y) = 0. \tag{L3}$$

(iii) For the simultaneous solutions (x, y, λ) of equations (L1)–(L3) inspect the values $f(x, y)$ to obtain the desired maximum or minimum.

Example 2 Find the maximum value of the utility function

$$u(x, y) = xy$$

subject to the constraint that

$$5x + 2y = 20.$$

(This problem would arise, for example, if the utility of consumption of x pizzas and y sodas is xy, the cost of pizzas is \$5 each, the cost of sodas is \$2 each, and the total amount available for snacks is \$20.)

Strategy Identify the constraint function $g(x, y)$.	*Solution* Here we can write the constraint as $g(x, y) = 0$ with

$$g(x, y) = 5x + 2y - 20.$$

Form the function $L(x, y, \lambda)$ $= f(x, y) + \lambda g(x, y)$.

The Lagrange function is therefore

$$L(x, y, \lambda) = xy + \lambda(5x + 2y - 20).$$

Set the partials equal to zero.

Setting the partial derivatives equal to zero gives the equations

$$\frac{\partial L}{\partial x} = 0$$

$$y + 5\lambda = 0 \tag{8}$$

$$\frac{\partial L}{\partial y} = 0$$

$$x + 2\lambda = 0 \tag{9}$$

$$\frac{\partial L}{\partial \lambda} = 0$$

$$5x + 2y - 20 = 0. \tag{10}$$

Solve the first equation for y, the second for x, and insert the results in the third equation.

Solve the resulting equation for λ. Then find x, y.

Substituting the values $y = -5\lambda$ and $x = -2\lambda$ from equations (8) and (9) into equation (10) gives

$$5(-2\lambda) + 2(-5\lambda) - 20 = 0$$

or

$$-20\lambda = 20.$$

Thus $\lambda = -1$, so $x = -2(-1) = 2$ and $y = -5(-1) = 5$. We leave it to you to verify that the value

$$u(2, 5) = 2 \cdot 5 = 10$$

is the maximum value of utility rather than the minimum. □

Example 3 Use the method of Lagrange multipliers to find the maximum and minimum values of the function

$$f(x, y) = xy$$

subject to the constraint that $x^2 + y^2 = 8$.

Solution: The constraint is $g(x, y) = 0$ with $g(x, y) = x^2 + y^2 - 8$, so the Lagrange function is

$$L(x, y, \lambda) = xy + \lambda(x^2 + y^2 - 8).$$

Setting the three partial derivatives equal to zero gives

$$\left(\frac{\partial L}{\partial x} = 0\right) \qquad y + 2\lambda x = 0, \quad \text{or} \quad y = -2\lambda x \tag{11}$$

$$\left(\frac{\partial L}{\partial y} = 0\right) \qquad x + 2\lambda y = 0, \quad \text{or} \quad x = -2\lambda y \tag{12}$$

$$\left(\frac{\partial L}{\partial \lambda} = 0\right) \qquad x^2 + y^2 - 8 = 0. \tag{13}$$

Multiplying both sides of equation (11) by y gives

$$y^2 = -2\lambda xy$$

and multiplying both sides of equation (12) by x gives

$$x^2 = -2\lambda xy.$$

It then follows that $y^2 = x^2$, so equation (13) becomes

$$2x^2 - 8 = 0, \quad \text{or} \quad x^2 = 4.$$

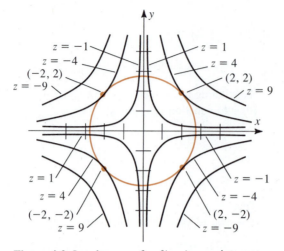

Figure 4.2 Level curves for $f(x, y) = xy$ intersect constraint curve $x^2 + y^2 = 8$ at critical points for Lagrange function $L(x, y, \lambda) = f(x, y) + \lambda g(x, y)$.

Thus $x = \pm 2$, and since $y^2 = x^2$, $y = \pm x$. The four critical points to be checked are therefore $(-2, -2)$, $(-2, 2)$, $(2, -2)$, and $(2, 2)$. We find that

$$f(-2, -2) = (-2)(-2) = 4 \qquad \text{(constrained maximum)}$$
$$f(-2, 2) = (-2)(2) = -4 \qquad \text{(constrained minimum)}$$
$$f(2, -2) = (2)(-2) = -4 \qquad \text{(constrained minimum)}$$
$$f(2, 2) = (2)(2) = 4. \qquad \text{(constrained maximum)}$$

The maximum value of f subject to the constraint is therefore 4, and the minimum is -4.

Figure 4.2 illustrates the geometry of this example. The constraint describes the circle with center at the origin and radius $\sqrt{8}$, which intersects the level curves for the graph of $f(x, y) = xy$ at the critical points. □

Constraints often arise in business and economics applications due to limitations on budget, manufacturing capacity, or other economic factors. The following is one such example.

Example 4 A manufacturing company determines that the level of production resulting from the use of x units of labor and y units of capital per hour to produce a certain product is given by the Cobb-Douglas production function

$$P(x, y) = 100x^{1/4}y^{3/4}.$$

Each unit of labor costs $200 and each unit of capital costs $100. If a total of $800 worth of labor and capital is to be used per hour, find the amounts of labor and capital for which production will be a maximum.

Solution: Since labor costs $200 per unit, the cost of x units of labor is $200x$. Similarly, the cost of y units of capital is $100y$. The *budget* constraint is therefore

$$200x + 100y = 800.$$

The problem is thus to maximize

$$P(x, y) = 100x^{1/4}y^{3/4}$$

subject to the constraint

$$g(x, y) = 200x + 100y - 800 = 0.$$

The Lagrange function is

$$L(x, y, \lambda) = 100x^{1/4}y^{3/4} + \lambda(200x + 100y - 800).$$

Setting the three partial derivatives equal to zero gives the equations

$$\left(\frac{\partial L}{\partial x} = 0\right) \qquad 25x^{-3/4}y^{3/4} + 200\lambda = 0 \tag{14}$$

$$\left(\frac{\partial L}{\partial y} = 0\right) \qquad 75x^{1/4}y^{-1/4} + 100\lambda = 0 \tag{15}$$

$$\left(\frac{\partial L}{\partial \lambda} = 0\right) \qquad 200x + 100y - 800 = 0. \tag{16}$$

Equation (14) gives

$$\lambda = \frac{-25}{200}x^{-3/4}y^{3/4} = -\frac{1}{8}x^{-3/4}y^{3/4}$$

and equation (15) gives

$$\lambda = \frac{-75}{100}x^{1/4}y^{-1/4} = -\frac{3}{4}x^{1/4}y^{-1/4}.$$

Thus

$$-\frac{1}{8}x^{-3/4}y^{3/4} = -\frac{3}{4}x^{1/4}y^{-1/4}. \tag{17}$$

Multiplying both sides of equation (17) by $-8x^{3/4}y^{1/4}$ gives

$$y = 6x. \tag{18}$$

Substituting this expression for y into equation (16) then gives

$$200x + 100(6x) - 800 = 0$$

or

$$800x = 800.$$

Thus $x = 1$ and, by equation (18), $y = 6$.

Since it is clear that smaller values of production can be obtained (say, $P(4, 0) = 0$ with $x = 4$ and $y = 0$), the value $P(1, 6) = 100(1)^{1/4}(6)^{3/4} = 100(6)^{3/4}$ must be the maximum production level attainable within this budget constraint. □

Functions of Three Variables

The method of Lagrange multipliers also applies to the problem of finding the maximum or minimum value of a function of three variables $w = f(x, y, z)$ subject to a constraint of the form $g(x, y, z) = 0$. The Lagrange function is again $L = f + \lambda g$. That is

$$L(x, y, z, \lambda) = f(x, y, z) + \lambda g(x, y, z)$$

and the method is again to set all partial derivatives of L equal to zero and solve the resulting set of (four) equations. To illustrate this technique, we shall rework Example 8 from Section 8.3.

Example 5 A container company wishes to design an open rectangular container with capacity 144 cubic feet. If the material for the bottom costs $4 per square foot and the material for the sides costs $3 per square foot, find the dimensions for which the cost of materials will be a minimum.

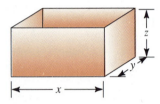

Figure 4.3 Container in Example 5.

Solution: If we label the dimensions of the container x, y, and z as in Figure 4.3, the area of the front and back panels is xz each, so their total cost is $2(3)xz = 6xz$ dollars. Similarly, the cost of the two end panels is $2(3)yz = 6yz$ dollars, and the cost of the bottom is $4xy$ dollars. The total cost of the material is therefore

$$C(x, y, z) = 4xy + 6xz + 6yz$$

which we must minimize subject to the volume constraint

$$V(x, y, z) = xyz - 144 = 0.$$

The Lagrange function is therefore

$$L(x, y, z, \lambda) = 4xy + 6xz + 6yz + \lambda(xyz - 144).$$

Setting the four partial derivatives equal to zero gives the equations

$$\left(\frac{\partial L}{\partial x} = 0\right) \qquad 4y + 6z + \lambda yz = 0 \tag{19}$$

$$\left(\frac{\partial L}{\partial y} = 0\right) \qquad 4x + 6z + \lambda xz = 0 \tag{20}$$

$$\left(\frac{\partial L}{\partial z} = 0\right) \qquad 6x + 6y + \lambda xy = 0 \tag{21}$$

$$\left(\frac{\partial L}{\partial \lambda} = 0\right) \qquad xyz - 144 = 0. \tag{22}$$

This time it is helpful to begin by multiplying both sides of equation (19) by x, both sides of equation (20) by y and both sides of equation (21) by z and then solving each for the form λxyz. We obtain

$$4xy + 6xz = -\lambda xyz \tag{23}$$
$$4xy + 6yz = -\lambda xyz \tag{24}$$
$$6xz + 6yz = -\lambda xyz. \tag{25}$$

Equating the left-hand sides of equations (23) and (24) then gives

$$4xy + 6xz = 4xy + 6yz$$

so

$$6xz = 6yz.$$

Dividing both sides by $6z$ (which cannot be zero since z is a dimension of a box) then gives

$$x = y. \tag{26}$$

Similarly, equating the left-hand sides of equations (24) and (25) gives

$$4xy + 6yz = 6xz + 6yz$$

so

$$4xy = 6xz$$

and

$$z = \frac{4xy}{6x} = \frac{2}{3}y. \tag{27}$$

From equations (26) and (27) we now have $y = x$ and $z = \frac{2}{3}y = \frac{2}{3}x$. Returning to equation (22), we find that

$$x(x)\left(\frac{2}{3}x\right) - 144 = 0$$

or

$$\frac{2}{3}x^3 = 144.$$

Thus

$$x^3 = \frac{3}{2}(144) = 216$$

so

$$x = \sqrt[3]{216} = 6$$
$$y = 6$$

and

$$z = \frac{2}{3}(6) = 4$$

as found in Section 8.3.

□

Exercise Set 8.4

In Exercises 1–10 find the minimum and maximum values of the function f subject to the given constraint.

1. $f(x, y) = xy$ subject to $x + 2y = 4$.

2. $f(x, y) = x^2 + 2y^2$ subject to $x - y + 3 = 0$.

3. $f(x, y) = x^2 - 8x + y^2 + 4y - 6$ subject to $2x = y - 5$.

4. $f(x, y) = y - x$ subject to $x^2 + y^2 = 2$.

5. $f(x, y) = xy$ subject to $x^2 + y^2 = 32$.

6. $f(x, y) = x^2 + y$ subject to $x^2 + y^2 = 9$.

7. $f(x, y) = x^3 - y^3$ subject to $x - y = 2$.

8. $f(x, y, z) = x + 2y - z$ subject to $x^2 + y^2 + z^2 = 24$.

9. $f(x, y, z) = xyz$ subject to $x^2 + y^2 + z^2 = 12$.

10. $f(x, y, z) = x + y + z$ subject to $x^2 + y^2 + z^2 = 12$.

11. Find the production levels x and y for which the production function

$$P(x, y) = 60x^{1/4}y^{3/4}$$

is a maximum when subjected to the constraint $20x + 10y = 80$.

12. Find the critical point for the Lagrange function

$$L(x, y, \lambda) = 8x^2 + 2y^2 - 4xy + \lambda(4x + 2y - 20).$$

Does this point correspond to a maximum or a minimum for the function $f(x, y) = 8x^2 + 2y^2 - 4xy$ subject to the constraint $4x + 2y = 20$?

13. A manufacturer of television sets produces both color and black and white sets. The weekly profit from the production and sale of x color and y black and white sets is

$$P(x, y) = 200x + 100y - 4x^2 - 2y^2.$$

In order to produce one color set 20 units of labor are required, while only 10 units of labor are required to produce a black and white set. If 600 units of labor are available per week, how many of each type of set should the manufacturer produce in order to maximize profit?

14. Find the maximum value of the production function

$$P(x, y) = 10x + 25y - 5xy$$

subject to the budget constraint $4x + 2y = 40$.

15. Find the maximum value of the profit function

$$P(x, y) = 20x + 40y - x^2 - y^2$$

subject to the capacity constraint that $x + y = 40$.

16. A rectangular box, with a top, is to hold 16 cubic meters. Find the dimensions that produce the least expensive box if the material for the side walls is half as expensive as the material for the top and the bottom.

17. Find the dimensions of the rectangular package of largest volume that can be mailed under the restrictions that length plus girth cannot exceed 84 inches. (Girth is the perimeter of the cross-section taken perpendicular to the length.)

18. A shipper wishes to design an open rectangular container. The material for the sides costs $4 per square foot and the material for the bottom costs $5 per square foot. If the total cost of materials is not to exceed $960, find the dimensions of the container of maximum volume.

19. A cylindrical can, with a top and a bottom, is to be manufactured using 100 cm² of tin, ignoring waste. What dimensions produce the can of maximum volume? (See Figure 4.4.)

20. A builder wishes to design a rectangular house containing V cubic meters of heated space, so as to minimize heating costs. One wall of the building is to face south.

Figure 4.4

The annual heating costs are estimated to be $4 per square meter of floor space, $3 per square meter for all exterior walls not facing south, and $2 per square meter for exterior wall space facing south. What dimensions will produce the most energy-efficient building? (See Figure 4.5.)

Figure 4.5

21. Find the dimensions for the cylindrical jar with a volume of 2 liters, which has minimum exterior surface area. Assume that the jar has a lid.

22. Recall that for the production function $z = P(x, y)$, where x represents units of labor and y represents units of capital used in production per unit time, the partial derivative $\dfrac{\partial P}{\partial x}$ is called the *marginal productivity of labor* and the partial derivative $\dfrac{\partial P}{\partial y}$ is called the *marginal productivity of capital*. For the production function $P(x, y) = 100x^{1/4}y^{3/4}$ in Example 4

a. Find $\dfrac{\partial P}{\partial x}(1, 6)$, the marginal productivity of labor at the optimum production level.

b. Find $\dfrac{\partial P}{\partial y}(1, 6)$, the marginal productivity of capital at the optimum production level.

c. Recall that the unit cost of labor in this example is 200, while the unit cost of capital is 100. Show that the ratio $\dfrac{200}{100} = 2$ also equals the ratio $\dfrac{\dfrac{\partial P}{\partial x}(1, 6)}{\dfrac{\partial P}{\partial y}(1, 6)}$ of the marginal productivities of labor and capital.

d. Show that the same conclusion holds for the production function in Exercise 11.

8.5 THE METHOD OF LEAST SQUARES

Students often ask where the functions of the calculus come from. For example, when a text states that, "the profit resulting from the sale of x toasters will be $P(x) = 30x - x^2$," how was the function P determined? Obviously, many (indeed, most) textbook functions are simply made up. But in real applications researchers often spend a great deal of time collecting data about how two or more variables are related before actually deciding what function gives the desired relationship. And often the resulting function is not a perfect match to the data, but rather, a "best fit," according to some agreed-upon meaning of this term.

One of the simplest functions that can describe the relationship between two variables is the linear function $y = mx + b$. The purpose of this section is to describe how we can determine from a set of data (x_1, y_1), (x_2, y_2), . . . , (x_n, y_n) the "best" straight line "fitting" this set of data. The technique we shall describe is called the **Method of Least Squares.**

For example, suppose that a calculus teacher wishes to know the relationship, if any, between students' scores on a mathematics achievement test given at the beginning of the course and their scores on the final examination. The teacher administers the achievement test to six students at the beginning of the course, records the scores, $x_1, x_2, \ldots, x_6$ and then waits until the end of the course and records the scores of the same students' final exams $y_1, y_2, \ldots, y_6$. The data are given in Table 5.1.

Table 5.1. Scores for six students on an achievement test (x_j) and a final examination in calculus (y_j)

j	1	2	3	4	5	6
x_j	6	4	8	7	3	9
y_j	7	4	9	5	5	8

The meaning of the entries in Table 5.1 is that the first student scored $x_1 = 6$ on the achievement test and $y_1 = 7$ on the final examination, the second student scored $x_2 = y_2 = 4$ on each, and so on.

Figure 5.1 shows the result of plotting the six data points (x_1, y_1), (x_2, y_2), $\ldots$, (x_6, y_6) in the xy-plane. While it is clear that these six points do not all lie on a common line, it does seem that the points lie generally grouped about some line, so we might ask, "What straight line best describes the relationship between x and y, that is, 'best fits' these data?"

Figure 5.2 shows the line ℓ that we wish to determine—the line that strikes the "happy medium" in a sense that we now make precise.

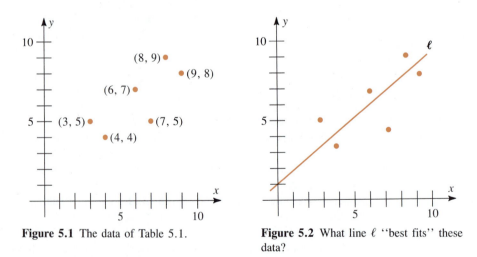

Figure 5.1 The data of Table 5.1.

Figure 5.2 What line ℓ "best fits" these data?

The Principle of Least Squares

Figure 5.3 shows a line ℓ with equation $y = mx + b$ and several data points (x_1, y_1), (x_2, y_2), $\ldots$, (x_n, y_n). For each data point (x_j, y_j) we define the **error** E_j to be the (vertical) distance between the data point (x_j, y_j) and the point on ℓ with the same x-coor-

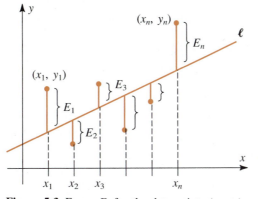

Figure 5.3 Errors E_j for the data points (x_1, y_1), (x_2, y_2), . . . , (x_n, y_n) and the line ℓ.

dinate. Since the y-coordinate of the point on ℓ with x-coordinate x_j is $(mx_j + b)$, the error associated with the data point (x_j, y_j) is

$$E_j = y_j - (mx_j + b)$$
$$= y_j - mx_j - b.$$

One approach to defining the "best-fitting" line ℓ would be to find the line for which the sum of these errors, $E_1 + E_2 + \cdots + E_n$, is a minimum. The difficulty in this is that since, in general, some data points lie above ℓ and some lie below ℓ, the errors $E_1, E_2, \ldots , E_n$ are of differing sign, and some of the error will therefore "cancel out." To prevent this, we work instead with the sum of the *squares* of the errors.

PRINCIPLE OF LEAST SQUARES: The line ℓ with equation $y = mx + b$ that best fits the data points (x_1, y_1), (x_2, y_2), . . . , (x_n, y_n) is the line for which the sum of the squares of the errors

$$E_1^2 + E_2^2 + \cdots + E_n^2 = \sum_{j=1}^{n} E_j^2 = \sum_{j=1}^{n} (y_j - mx_j - b)^2$$

is a minimum.

Since all of the numbers $x_1, x_2, \ldots , x_n$ and $y_1, y_2, \ldots , y_n$ are known, the Principle of Least Squares simply requires that we find the minimum value of the function of two variables

$$f(m, b) = \sum_{j=1}^{n} [y_j - mx_j - b]^2$$

$$= (y_1 - mx_1 - b)^2 + (y_2 - mx_2 - b)^2 + \cdots + (y_n - mx_n - b)^2.$$

Once we have found the numbers m and b for which this function is a minimum, we shall know the equation $y = mx + b$ for the best-fitting straight line ℓ, called the *regression line*.

At the end of this section we shall show how partial differentiation can be used to find m and b. The result is the following.

THEOREM 5
Method of Least Squares

The equation of the straight line ℓ that best fits the data points (x_1, y_1), $(x_2, y_2), \ldots, (x_n, y_n)$ according to the Principle of Least Squares has equation

$$y = mx + b$$

where

$$m = \frac{n \sum_{j=1}^{n} x_j y_j - \left(\sum_{j=1}^{n} x_j\right)\left(\sum_{j=1}^{n} y_j\right)}{n \sum_{j=1}^{n} x_j^2 - \left(\sum_{j=1}^{n} x_j\right)^2}. \tag{1}$$

and

$$b = \frac{\left(\sum_{j=1}^{n} x_j^2\right)\left(\sum_{j=1}^{n} y_j\right) - \left(\sum_{j=1}^{n} x_j\right)\left(\sum_{j=1}^{n} x_j y_j\right)}{n \sum_{j=1}^{n} x_j^2 - \left(\sum_{j=1}^{n} x_j\right)^2}. \tag{2}$$

In the statement of Theorem 5

$$\sum_{j=1}^{n} x_j \qquad \text{means} \quad x_1 + x_2 + \cdots + x_n,$$

$$\sum_{j=1}^{n} x_j^2 \qquad \text{means} \quad x_1^2 + x_2^2 + \cdots + x_n^2,$$

$$\left(\sum_{j=1}^{n} x_j\right)^2 \qquad \text{means} \quad (x_1 + x_2 + \cdots + x_n)^2,$$

and so on.

Although the equations for m and b may seem a bit overwhelming, they are really only tedious at worst, although you must be very careful in calculating the various expressions. The advantage of stating them in this general form is that they can be used for any number of data points.

Example 1 Find the equation for the least squares regression line for the data in the example given at the beginning of this section.

Solution: In order to apply formulas (1) and (2), we complete the following table for the data in Table 5.1.

Table 5.2. Entries in formulas (1) and (2) for the data in Table 5.1

j	x_j	y_j	x_j^2	$x_j y_j$
1	6	7	36	42
2	4	4	16	16
3	8	9	64	72
4	7	5	49	35
5	3	5	9	15
6	9	8	81	72
Σ	37	38	255	252

From the bottom row of Table 5.2 we have the totals

$$\sum_{j=1}^{6} x_j = 37; \qquad \sum_{j=1}^{n} y_j = 38; \qquad \sum_{j=1}^{n} x_j^2 = 255; \qquad \sum_{j=1}^{n} x_j y_j = 252.$$

Using these totals in formulas (1) and (2), with $n = 6$, gives

$$m = \frac{6(252) - (37)(38)}{6(255) - (37)^2} = \frac{106}{161} \approx 0.66$$

and

$$b = \frac{(255)(38) - (37)(252)}{6(255) - (37)^2} = \frac{366}{161} \approx 2.27.$$

The least squares regression line therefore has equation $y = \left(\dfrac{106}{161}\right)x + \left(\dfrac{366}{161}\right)$, which can be approximated by

$$y = 0.66x + 2.27. \qquad \qquad \square$$

Regression Lines as Predictors

One of the most important uses of regression lines is as predictors. Once we have determined the least squares line $y = mx + b$ from a set of data, we can use this equation to predict the value of y that will result from the selection of an additional value of x. Of course, the result is only a prediction, since the regression line does not fit the data precisely.

Example 2   Use the regression line obtained in Example 1 to predict the final exam score for a calculus student who scores a 5 on the achievement test.

Solution: With $x = 5$ the least squares regression line $y = 0.66x + 2.27$ predicts a final exam score of

$$y = 0.66(5) + 2.27$$
$$= 5.57.$$ □

Example 3 Table 5.3 shows the United States' Federal Funds discount rate for five months of 1982 beginning in June.

(a) Find a least squares regression line for these data.
(b) Use this line to predict from the data what the discount rate would have been in December 1982.

Table 5.3. United States' Federal Funds discount rate for five months of 1982

Month	June	July	August	September	October
Rate	14%	13%	10%	10%	9%

Solution: When data correspond to time, it is often convenient to let the independent variable represent time after some starting date. We therefore let x represent the number of months after June 1982, and we let y denote the discount rate. We may then restate the data in Table 5.3 as follows for the $n = 5$ months in question:

Table 5.4

n	x_j	y_j	x_j^2	$x_j y_j$
1	0	14	0	0
2	1	13	1	13
3	2	10	4	20
4	3	10	9	30
5	4	9	16	36
Σ	10	56	30	99

From the bottom row of the table we can see that

$$\sum_{j=1}^{5} x_j = 10; \quad \sum_{j=1}^{5} y_j = 56; \quad \sum_{j=1}^{5} x_j^2 = 30; \quad \sum_{j=1}^{5} x_j y_j = 99.$$

Using these sums and $n = 5$ in formulas (1) and (2) gives

$$m = \frac{5(99) - (10)(56)}{5(30) - (10)^2} = \frac{-65}{50} = -\frac{13}{10}$$

and

$$b = \frac{(30)(56) - (10)(99)}{5(30) - (10)^2} = \frac{690}{50} = \frac{69}{5}.$$

Thus

(a) The least squares regression line is

$$y = -\frac{13}{10}x + \frac{69}{5}.$$

(b) Since December 1982 is $x = 6$ months after June 1982, the predicted discount rate for December 1982 is

$$y = -\frac{13}{10}(6) + \frac{69}{5}$$

$$= 6 \text{ percent.}$$

(In fact, the actual rate was just under 9%, which illustrates the difficulty in using straight lines to predict something as volatile as interest rates.)   □

Proof of Theorem 5:   To find the relative minimum value of the sum of squares function

$$f(m, b) = \sum_{j=1}^{n} (y_j - mx_j - b)^2$$

we first set $\dfrac{\partial f}{\partial m}$ equal to zero, obtaining

$$0 = \frac{\partial f}{\partial m} = \sum_{j=1}^{n} 2(y_j - mx_j - b)(-x_j)$$

$$= -2 \sum_{j=1}^{n} x_j y_j + 2m \sum_{j=1}^{n} x_j^2 + 2b \sum_{j=1}^{n} x_j.$$

Thus

$$m \sum_{j=1}^{n} x_j^2 + b \sum_{j=1}^{n} x_j = \sum_{j=1}^{n} x_j y_j. \tag{3}$$

We next set $\dfrac{\partial f}{\partial b}$ equal to zero, obtaining

$$0 = \frac{\partial f}{\partial b} = \sum_{j=1}^{n} 2(y_j - mx_j - b)(-1)$$

$$= -2 \sum_{j=1}^{n} y_j + 2m \sum_{j=1}^{n} x_j + 2nb. \qquad \left(\sum_{j=1}^{n} b = nb. \right)$$

Thus

$$b = \frac{1}{n}\left(\sum_{j=1}^{n} y_j - m \sum_{j=1}^{n} x_j\right).$$ (4)

Substituting the expression for b in equation (4) into equation (3) then gives

$$m \sum_{j=1}^{n} x_j^2 + \frac{1}{n}\sum_{j=1}^{n} x_j\left(\sum_{j=1}^{n} y_j - m \sum_{j=1}^{n} x_j\right) = \sum_{j=1}^{n} x_j y_j$$

so

$$m\left[\sum_{j=1}^{n} x_j^2 - \frac{1}{n}\left(\sum x_j\right)^2\right] = \sum_{j=1}^{n} x_j y_j - \frac{1}{n}\left(\sum_{j=1}^{n} x_j\right)\left(\sum_{j=1}^{n} y_j\right)$$

or

$$m = \frac{n \sum_{j=1}^{n} x_j y_j - \left(\sum_{j=1}^{n} x_j\right)\left(\sum_{j=1}^{n} y_j\right)}{n \sum_{j=1}^{n} x_j^2 - \left(\sum_{j=1}^{n} x_j\right)^2}.$$

We leave it to you to verify that substituting the expression for m into equation (4) gives

$$b = \frac{\left(\sum_{j=1}^{n} x_j^2\right)\left(\sum_{j=1}^{n} y_j\right) - \left(\sum_{j=1}^{n} x_j\right)\left(\sum_{j=1}^{n} x_j y_j\right)}{n \sum_{j=1}^{n} x_j^2 - \left(\sum_{j=1}^{n} x_j\right)^2}.$$

Verifying that this critical point actually yields the relative minimum is beyond the scope of this text. ∎

Exercise Set 8.5

In each of Exercises 1–5 find the equation for the least squares regression line $y = mx + b$ corresponding to the given data.

1.

x	4	3	5	5
y	9	6	7	8

2.

x	12	6	10	18	3	5
y	4	14	9	2	16	16

3.

x	52	46	69	54	61	48
y	74	66	94	91	84	80

4.

x	6	8	9	10	12	6	11
y	2	5	5	7	9	4	8

5.

x	0	1	1	2	2	3	3	3	4	4
y	1	0	2	1	2	2	1	3	2	3

6. Personal savings, as a percent of disposable income, is shown for 3 years, 1981–1983, in Table 5.5.
   a. Find the least squares regression line for these data.
   b. Use this line to predict the personal savings rate for 1985.

Table 5.5

Year	1981	1982	1983
Rate	5.6	6.2	5.3

7. A manufacturer of railroad cars received orders for new cars according to Table 5.6 for the years 1983–1988. (Orders are in thousands.)
   a. Find the least squares regression line.
   b. Use this line to predict the number of orders the company would have received in 1989.

Table 5.6

Year	1983	1984	1985	1986	1987	1988
Orders	10	14	8	4	2	3

8. A study produced the data of Table 5.7 showing the number of deaths per 100,000 male automobile drivers versus the age of these drivers.
   a. Find the least squares regression line giving the number of deaths as a function of age.
   b. How many deaths does this model predict for males age 65?

Table 5.7

Age	25	35	45	55
Deaths per 100,000	52	30	26	21

9. Demand for leaded gasoline in the United States for the years 1977–1983 is given approximately by the entries in Table 5.8.
   a. Find the least squares regression line giving the demand for leaded gasoline as a function of time.
   b. What demand does this model predict for the year 1986?

Table 5.8

Year	1977	1978	1979	1980	1981	1982	1983
Demand in billions of gallons	78	75	65	54	51	50	43

10. The percentage of United States adults aged 25–29 who had graduated from college is shown for each of 5 years in Table 5.9.
   a. Find the least squares regression line giving this percentage as a function of time.
   b. What percentage does this model predict for 1990?

Table 5.9

Year	1940	1950	1960	1970	1980
Percentage	6	7	11	16	23

8.6 DOUBLE INTEGRALS

Figure 6.1 recalls the definition of the definite integral of a continuous function of a single variable as the limit of an approximating sum. When $f(x) \geq 0$ the terms in the approximating sum can be interpreted as areas of rectangles that approximate a region in the xy-plane bounded by the graph of $y = f(x)$.

Figure 6.2 illustrates how the notion of a definite integral is generalized for continuous functions of two variables. The integral is defined over a region R, rather than an interval, which is "partitioned" by a rectangular grid into small rectangles of area $\Delta A = \Delta x \, \Delta y$. In each such rectangle lying within R a "test point" (s_j, t_j) is chosen. The products $f(s_j, t_j) \, \Delta x \, \Delta y$, one for each rectangle lying in R, are then summed, giving an approxi-

mating sum $\sum_{j=1}^{n} f(s_j, t_j) \, \Delta x \, \Delta y$. The limit of this approximating sum, as n becomes infi-

nitely large (and the dimensions Δx and Δy of the small rectangles approach zero), is called the **double integral** of the function $z = f(x, y)$ over the region R:

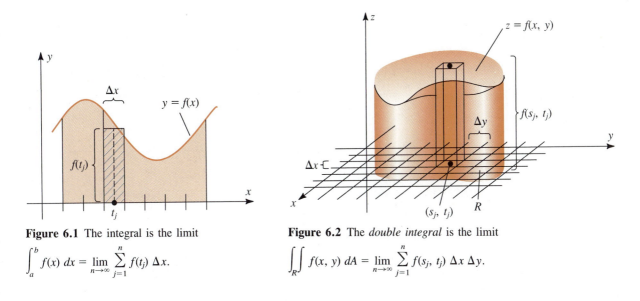

Figure 6.1 The integral is the limit

$$\int_a^b f(x)\,dx = \lim_{n\to\infty} \sum_{j=1}^n f(t_j)\,\Delta x.$$

Figure 6.2 The *double integral* is the limit

$$\iint_R f(x,\,y)\,dA = \lim_{n\to\infty} \sum_{j=1}^n f(s_j,\,t_j)\,\Delta x\,\Delta y.$$

$$\iint_R f(x,\,y)\,dA = \lim_{n\to\infty} \sum_{j=1}^n f(s_j,\,t_j)\,\Delta x\,\Delta y. \tag{1}$$

Rather than concern ourselves with the technical description of just what the limit in equation (1) means, we shall rely on your intuitive understanding of approximating sums from the single variable case and simply state how several types of double integrals can be evaluated. Before doing so, however, we note the relationship between double integrals and volumes of certain solids, the analog of the relationship between definite integrals and areas of certain regions.

The Double Integral as Volume

When $f(x,\,y) \geq 0$ for all $(x,\,y)$ in R, each of the terms $f(s_j,\,t_j)\,\Delta x\,\Delta y$ in the approximating sum in equation (1) may be interpreted as the volume of a rectangular prism whose base has area $\Delta x\,\Delta y$ and whose height is $f(s_j,\,t_j)$. As $n \to \infty$ these prisms more nearly fill the volume of the solid lying between the graph of $z = f(x,\,y)$ and the xy-plane. We therefore define the volume of this solid to be the double integral in equation (1).

If $z = f(x,\,y)$ is continuous and $f(x,\,y) \geq 0$ for all $(x,\,y)$ in R, the volume of the solid bounded above by the graph of $z = f(x,\,y)$ and below by the region R is

$$V = \iint_R f(x,\,y)\,dA. \tag{2}$$

(See Figure 6.3.)

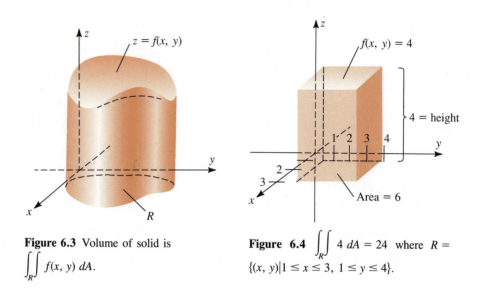

Figure 6.3 Volume of solid is
$$\iint_R f(x, y) \, dA.$$

Figure 6.4 $\displaystyle\iint_R 4 \, dA = 24$ where $R = \{(x, y) \,|\, 1 \le x \le 3, \ 1 \le y \le 4\}$.

Example 1 Figure 6.4 shows the portion of the graph of the constant function $f(x, y) = 4$ above the rectangle $R = \{(x, y) \,|\, 1 \le x \le 3, \ 1 \le y \le 4\}$. Since the area of R is $2 \cdot 3 = 6$, the volume of the rectangular prism lying between the graph of $z = f(x, y)$ and R is $4 \cdot 6 = 24$. Thus by equation (2)

$$\iint_R f(x, y) \, dA = 24.$$ □

Evaluating Double Integrals

The following theorem, which we state without proof, shows how double integrals over certain types of regions may be evaluated.

THEOREM 6

Let $z = f(x, y)$ be a continuous function on the region R.

(i) If there exist continuous functions $x = g_1(y)$ and $x = g_2(y)$ for which

$$R = \{(x, y) \,|\, g_1(y) \le x \le g_2(y), \ c \le y \le d\}$$

then

$$\iint_R f(x, y) \, dA = \int_c^d \left\{ \int_{g_1(y)}^{g_2(y)} f(x, y) \, dx \right\} dy. \tag{3}$$

(See Figure 6.5.)

(ii) If there exist continuous functions $y = h_1(x)$ and $y = h_2(x)$ for which

$$R = \{(x, y) \,|\, a \le x \le b, \ h_1(x) \le y \le h_2(x)\}$$

then

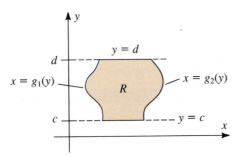

Figure 6.5 A region of the form $R = \{(x, y)|g_1(y) \le x \le g_2(y), c \le y \le d\}$.

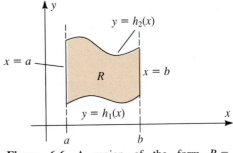

Figure 6.6 A region of the form $R = \{(x, y)|a \le x \le b, h_1(x) \le y \le h_2(x)\}$.

$$\iint_R f(x, y)\, dA = \int_a^b \left\{ \int_{h_1(x)}^{h_2(x)} f(x, y)\, dy \right\} dx. \tag{4}$$

(See Figure 6.6.)

The integrals on the right-hand sides of equations (3) and (4) are called **iterated** integrals because they are evaluated in two steps, "inside out":

Step 1: The inside integral is evaluated by treating one independent variable as a constant. The result is a function of the other independent variable.

Step 2: The resulting function is then integrated over the limits indicated by the "outside integral sign."

The order of the differentials dx and dy indicates which type of integration is performed first.

Example 2 Evaluate the iterated integral

$$\int_0^4 \left\{ \int_1^{\sqrt{y}} (x + y)\, dx \right\} dy.$$

Solution: We begin by evaluating the integral $\int_1^{\sqrt{y}} (x + y)\, dx$ inside the braces as a function of x alone, treating y as a constant. Since the antiderivative of the function $f(x, y) = x + y$ *with respect to x alone* is

$$\int (x + y)\, dx = \frac{1}{2}x^2 + xy + C$$

the "inside" integral is

$$\int_1^{\sqrt{y}} (x + y)\, dx = \frac{1}{2}x^2 + xy \bigg]_1^{\sqrt{y}}$$

$$= \left[\frac{1}{2}(\sqrt{y})^2 + (\sqrt{y})y\right] - \left[\frac{1}{2}(1)^2 + (1)y\right]$$

$$= y^{3/2} - \frac{1}{2}y - \frac{1}{2}.$$

This is the function of y alone that we must then integrate from $y = 0$ to $y = 4$. The complete solution is

$$\int_0^4 \left\{ \int_1^{\sqrt{y}} (x + y)\, dx \right\} dy$$

$$= \int_0^4 \left\{ \frac{1}{2}x^2 + xy \right]_1^{\sqrt{y}} \right\} dy$$

$$= \int_0^4 \left\{ \left[\frac{1}{2}(\sqrt{y})^2 + (\sqrt{y})y\right] - \left[\frac{1}{2}(1)^2 + (1)y\right] \right\} dy$$

$$= \int_0^4 \left(y^{3/2} - \frac{1}{2}y - \frac{1}{2} \right) dy$$

$$= \frac{2}{5}y^{5/2} - \frac{1}{4}y^2 - \frac{1}{2}y \Big]_0^4$$

$$= \left[\frac{2}{5}(4)^{5/2} - \frac{1}{4}(4)^2 - \frac{1}{2}(4)\right] - \left[\frac{2}{5}(0) - \frac{1}{4}(0) - \frac{1}{2}(0)\right]$$

$$= \frac{2}{5}(32) - \frac{1}{4}(16) - \frac{1}{2}(4)$$

$$= \frac{34}{5}. \qquad \square$$

Example 3 Evaluate $\int_2^4 \left\{ \int_1^3 (2x + y)\, dy \right\} dx.$

Strategy

First, find an antiderivative for $(2x + y)$ *with respect to* y, and evaluate between $y = 1$ and $y = 3$, carrying x along as a constant.

Then integrate the resulting function of x over the outside limits from $x = 2$ to $x = 4$.

Solution

$$\int_2^4 \left\{ \int_1^3 (2x + y)\, dy \right\} dx$$

$$= \int_2^4 \left\{ 2xy + \frac{1}{2}y^2 \right]_1^3 \right\} dx$$

$$= \int_2^4 \left\{ \left[2x(3) + \frac{1}{2}(3)^2\right] - \left[2x(1) + \frac{1}{2}(1)^2\right] \right\} dx$$

$$= \int_2^4 (4x + 4)\, dx$$

$$= 2x^2 + 4x \Big]_2^4$$

$$= (2 \cdot 4^2 + 4 \cdot 4) - (2 \cdot 2^2 + 4 \cdot 2)$$

$$= 32. \qquad \square$$

It is important to compare the solutions to Examples 2 and 3. In Example 2 the "inside" differential was dx, so the first integration was with respect to x. In Example 3 the inside differential was dy, so the first integration was with respect to dy.

Notation for Iterated Integrals

We usually omit the braces in writing iterated integrals. Thus

$$\int_c^d \int_{g_1(y)}^{g_2(y)} f(x, y) \, dx \, dy \quad \text{means} \quad \int_c^d \left\{ \int_{g_1(y)}^{g_2(y)} f(x, y) \, dx \right\} dy$$

and

$$\int_a^b \int_{h_1(x)}^{h_2(x)} f(x, y) \, dy \, dx \quad \text{means} \quad \int_a^b \left\{ \int_{h_1(x)}^{h_2(x)} f(x, y) \, dy \right\} dx$$

In using this notation, however, it is important to follow the order of the differentials and integrate "inside out." Using this notation, we can write the results of Examples 2 and 3 as

$$\int_0^4 \int_1^{\sqrt{y}} (x + y) \, dx \, dy = \frac{34}{5}$$

and

$$\int_2^4 \int_1^3 (2x + y) \, dy \, dx = 32.$$

From Double Integrals to Iterated Integrals

In using Theorem 6 to evaluate a double integral of the form $\iint_R f(x, y) \, dA$, we must first express the region R using inequalities. The type of description obtained then determines how $\iint_R f(x, y) \, dA$ may be expressed as an iterated integral.

Example 4 Evaluate $\iint_R \dfrac{y}{x} \, dA$ where R is the region in Figure 6.7.

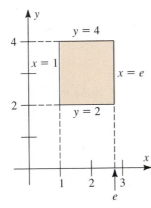

Figure 6.7 The region R in Example 4.

Solution: The region R is a rectangle described by the inequalities

$$R: \quad 1 \le x \le e, \quad 2 \le y \le 4.$$

Since both variables are bounded by constants, we could begin by integrating with respect to either one. If we choose to integrate first with respect to x, using equation (3), we have

$$\iint_R \frac{y}{x} \, dA = \int_2^4 \int_1^e \frac{y}{x} \, dx \, dy$$

$$= \int_2^4 \left\{ y \ln x \, \bigg]_1^e \right\} dy$$

$$= \int_2^4 (y \ln e - y \ln 1) \, dy$$

$$= \int_2^4 y \, dy \qquad (\ln e = 1; \ln 1 = 0)$$

$$= \frac{1}{2} y^2 \bigg]_2^4$$

$$= 6.$$

If we had chosen to integrate first with respect to y, using equation (4), we would have obtained,

$$\iint_R \frac{y}{x} \, dA = \int_1^e \int_2^4 \frac{y}{x} \, dy \, dx$$

$$= \int_1^e \left\{ \frac{y^2}{2x} \bigg]_2^4 \right\} dx$$

$$= \int_1^e \left\{ \frac{4^2}{2x} - \frac{2^2}{2x} \right\} dx$$

$$= \int_1^e \frac{6}{x} \, dx$$

$$= 6 \ln x \,]_1^e$$

$$= 6.$$

For double integrals taken over rectangles either order of integration is possible, and both yield the same result. □

Example 5 Evaluate $\iint_R 2xy \, dA$ where R is the region in Figure 6.8.

Solution: Here R is described by the inequalities

$$R: \quad -2 \le x \le 2, \quad 0 \le y \le \sqrt{4 - x^2}.$$

Since the inequality for y involves the function $h(x) = \sqrt{4 - x^2}$, which is not constant, we must use equation (4), integrating first with respect to y. We obtain

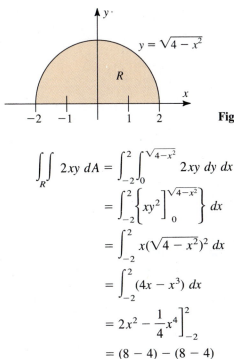

Figure 6.8 The region R in Example 5.

$$\iint_R 2xy \, dA = \int_{-2}^{2} \int_{0}^{\sqrt{4-x^2}} 2xy \, dy \, dx$$

$$= \int_{-2}^{2} \left\{ xy^2 \right]_{0}^{\sqrt{4-x^2}} \right\} dx$$

$$= \int_{-2}^{2} x(\sqrt{4-x^2})^2 \, dx$$

$$= \int_{-2}^{2} (4x - x^3) \, dx$$

$$= 2x^2 - \frac{1}{4}x^4 \Big]_{-2}^{2}$$

$$= (8 - 4) - (8 - 4)$$

$$= 0. \qquad \qquad \square$$

REMARK: It is not unusual for the value of a double integral to be zero, or even a negative number. It is only in the case $f(x, y) \geq 0$ that the double integral corresponds to a volume. In Example 5 the function $f(x, y) = 2xy$ is positive in the first quadrant but negative in the second quadrant. Thus "canceling" occurs, and the answer is not the volume of a solid, but simply the value of the double integral.

Example 6 Find the volume of the solid lying below the graph of $f(x, y) = 8 - x - y$ and above the triangular region R in the xy-plane with vertices $(0, 0, 0)$, $(2, 0, 0)$, and $(0, 4, 0)$. (See Figure 6.9.)

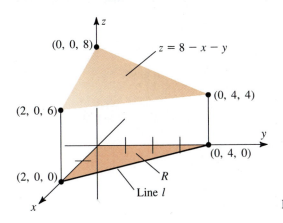

Figure 6.9 Solid in Example 6.

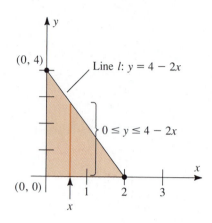

Figure 6.10 For any point (x, y) in R with $0 \le x \le 2$ we have $0 \le y \le 4 - 2x$.

Strategy

Find an equation for the hypotenuse of R.

Using this equation, write the inequalities in x and y describing the region R.

Use the definition of volume and equation (4) to write volume as an iterated integral.

Integrate first with respect to y.

Then integrate with respect to x.

Result is volume.

Solution

The region R is shown in Figure 6.10. The slope of the line ℓ through the points $(2, 0)$ and $(0, 4)$ is

$$m = \frac{4 - 0}{0 - 2} = -2$$

so the line ℓ has equation

$$y - 4 = -2(x - 0), \quad \text{or}$$
$$y = -2x + 4.$$

Thus the region R is described by the inequalities

$$0 \le x \le 2, \quad \text{and} \quad 0 \le y \le 4 - 2x.$$

Using equations (2) and (4), we find that

$$\text{Volume} = \iint_R (8 - x - y) \, dA$$

$$= \int_0^2 \int_0^{4-2x} (8 - x - y) \, dy \, dx$$

$$= \int_0^2 \left\{ 8y - xy - \frac{1}{2}y^2 \Big]_0^{(4-2x)} \right\} dx$$

$$= \int_0^2 \left[8(4 - 2x) - x(4 - 2x) - \frac{1}{2}(4 - 2x)^2 \right] dx$$

$$= \int_0^2 (24 - 12x) \, dx$$

$$= 24x - 6x^2 \Big]_0^2$$

$$= (48 - 24)$$

$$= 24.$$

We leave it as an exercise for you to show that this volume could also have been calculated as

$$\text{Volume} = \int_0^4 \int_0^{2-1/2\,y} (8 - x - y)\, dx\, dy.$$ □

Interchanging the Order of Integration

There are occasions on which we need to interchange the order of integration in a given double integral. This usually occurs when an antiderivative cannot be found with respect to the "inside" variable. For example, in the double integral

$$\int_0^1 \int_{y^2}^1 y e^{x^2}\, dx\, dy \tag{5}$$

we are in deep trouble, since we cannot find a (formal) antiderivative for the integrand $y e^{x^2}$ with respect to x.

However, such integrals can often be evaluated by reversing the order of integration. In particular, the antiderivative of $y e^{x^2}$ with respect to y is easily seen to be the function $\frac{1}{2} y^2 e^{x^2}$. But if we are going to reverse the order in which the antidifferentiations are performed, we must also determine correct limits of integration corresponding to this new order of integration. We do so as follows:

To reverse the order of integration in the iterated integral

$$\int_c^d \int_{h_1(y)}^{h_2(y)} f(x, y)\, dx\, dy$$

(i) Identify the region Q for which the iterated integral can be written as the double integral

$$\int_c^d \int_{h_1(y)}^{h_2(y)} f(x, y)\, dx\, dy = \int_Q \int f(x, y)\, dA.$$

(ii) Find constants a and b and continuous functions g_1 and g_2, so that the region Q can be expressed as

$$Q = \{(x, y)\,|\,a \le x \le b,\ g_1(x) \le y \le g_2(x)\}.$$

(iii) Rewrite the iterated integral as

$$\int_c^d \int_{h_1(y)}^{h_2(y)} f(x, y)\, dx\, dy = \int_Q \int f(x, y)\, dA$$

$$= \int_a^b \int_{g_1(x)}^{g_2(x)} f(x, y)\, dy\, dx.$$

Although the procedure is stated for reversing the order of integration from $dx\, dy$ to $dy\, dx$, the procedure for changing from order $dy\, dx$ to order $dx\, dy$ is analogous. Also, there is no guarantee that the resulting integral can be more easily evaluated than the original one.

Example 7 Use the procedure for reversing order of integration to evaluate the iterated integral

$$\int_0^1 \int_{y^2}^1 ye^{x^2} \, dx \, dy.$$

Solution: From the given limits of integration the region Q is described by the inequalities

$$y^2 \le x \le 1, \qquad 0 \le y \le 1.$$

That is, Q is the region bounded between the graphs of $x = y^2$ and $x = 1$ for $0 \le y \le 1$. From Figure 6.11 (which must always be sketched when applying this method), we can see that Q is regular and can also be described by the inequalities

$$0 \le x \le 1, \qquad 0 \le y \le \sqrt{x}.$$

(See Figure 6.12.)

Beginning with the given integral, we therefore reverse the order of integration as follows:

$$\int_0^1 \int_{y^2}^1 ye^{x^2} \, dx \, dy = \iint_Q ye^{x^2} \, dA = \int_0^1 \int_0^{\sqrt{x}} ye^{x^2} \, dy \, dx$$

$$= \int_0^1 \left\{ \frac{1}{2} y^2 e^{x^2} \Big]_{y=0}^{y=\sqrt{x}} \right\} dx$$

$$= \int_0^1 \frac{1}{2} xe^{x^2} \, dx$$

$$= \frac{1}{4} e^{x^2} \Big]_0^1$$

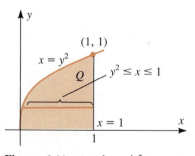

Figure 6.11 $Q = \{(x, y)|y^2 \le x \le 1, 0 \le y \le 1\}$.

Figure 6.12 $Q = \{(x, y)|0 \le x \le 1, 0 \le y \le \sqrt{x}\}$.

$$= \frac{1}{4}(e - 1)$$

$$\approx 0.43 \qquad \qquad \qquad \square$$

REMARK: It is important to note that we cannot simply interchange limits of integration when we interchange the order of integration. There is no alternative to sketching the region Q and working out the new limits of integration from knowledge of the boundary of Q.

Example 8 Evaluate the iterated integral

$$\int_0^1 \int_0^{\sqrt{1-x}} xy^2 \, dy \, dx$$

by first reversing the order of integration.

Solution: The given limits of integration are

$$0 \le x \le 1, \qquad 0 \le y \le \sqrt{1 - x}$$

which describe the region Q bounded above by the graph of $y = \sqrt{1 - x}$ and below by the x-axis for $0 \le x \le 1$ (see Figure 6.13). By solving the equation $y = \sqrt{1 - x}$ for x, we find that this region may also be described by the inequalities

$$0 \le x \le 1 - y^2, \qquad 0 \le y \le 1.$$

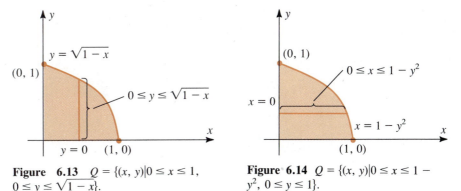

Figure 6.13 $Q = \{(x, y)|0 \le x \le 1, 0 \le y \le \sqrt{1 - x}\}.$

Figure 6.14 $Q = \{(x, y)|0 \le x \le 1 - y^2, 0 \le y \le 1\}.$

(See Figure 6.14.) We may therefore evaluate the iterated integral as

$$\int_0^1 \int_0^{\sqrt{1-x}} xy^2 \, dy \, dx = \int_Q xy^2 \, dA = \int_0^1 \int_0^{1-y^2} xy^2 \, dx \, dy$$

$$= \int_0^1 \left\{ \frac{1}{2} x^2 y \Big]_{x=0}^{x=1-y^2} \right\} dy$$

$$= \int_0^1 \frac{1}{2}(1 - y^2)^2 y^2 \, dy$$

$$= \int_0^1 \left(\frac{1}{2} y^2 - y^4 + \frac{1}{2} y^6 \right) dy$$

$$= \frac{1}{6} y^3 - \frac{1}{5} y^5 + \frac{1}{14} y^7 \bigg]_0^1$$

$$= \frac{4}{105}. \qquad \square$$

REMARK: If we attempt to evaluate the integral of Example 8 without first interchanging the order of integration, the following intermediate step will appear:

$$\int_0^1 \frac{1}{3} x(1 - x)^{3/2} dx.$$

Evaluation of this integral requires the technique of integration by parts, which involves somewhat more work than does the method of the example. As you work more problems, you will learn to recognize which order of integration is likely to be easier.

Exercise Set 8.6

In Exercises 1–10 evaluate the iterated integral.

1. $\displaystyle\int_0^2 \int_0^1 (x - y) \, dx \, dy$
2. $\displaystyle\int_0^1 \int_1^3 (2x + 4y) \, dy \, dx$

3. $\displaystyle\int_{-1}^1 \int_0^2 xy \, dy \, dx$
4. $\displaystyle\int_{-2}^1 \int_0^1 x\sqrt{y} \, dy \, dx$

5. $\displaystyle\int_0^2 \int_0^x (x - 2y) \, dy \, dx$
6. $\displaystyle\int_{-1}^2 \int_0^x y\sqrt{x^3 + 1} \, dy \, dx$

7. $\displaystyle\int_0^1 \int_0^1 ye^{x-y^2} \, dy \, dx$
8. $\displaystyle\int_{-1}^0 \int_{-1}^{y+1} (xy - x) \, dx \, dy$

9. $\displaystyle\int_0^1 \int_{-x}^{\sqrt{x}} \frac{y}{1 + x} \, dy \, dx$
10. $\displaystyle\int_1^2 \int_0^{y^3} e^{x/y} \, dx \, dy$

In Exercises 11–16 evaluate $\displaystyle\iint_R f(x, y) \, dA$

11. $\displaystyle\iint_R (x + y^2) \, dA, \qquad R = \{(x, y) \mid 0 \leq x \leq 1, \, 0 \leq y \leq 2\}$

12. $\displaystyle\iint_R (x^2 + y^2) \, dA, \qquad R = \{(x, y) \mid 0 \leq x \leq a, \, 0 \leq y \leq b\}$

13. $\displaystyle\iint_R \frac{xy}{\sqrt{x^2 + y^2}} \, dA, \qquad R = \{(x, y) \mid 1 \leq x \leq 2, \, 1 \leq y \leq 2\}$

14. $\displaystyle\iint_R \frac{x^2}{1+y}\,dA,$ $\quad R = \{(x, y) \mid 0 \le x \le 1,\ 0 \le y \le e^x - 1\}$

15. $\displaystyle\iint_R xy\,dA,$ $\quad R = \{(x, y) \mid y \le x \le \sqrt{y},\ 0 \le y \le 1\}$

16. $\displaystyle\iint_R x\,dA,$ $\quad R = \{(x, y) \mid 0 \le x \le \sqrt{1 - y^2},\ 0 \le y \le 1\}$

In Exercises 17–20 sketch the region Q determined by the limits of integration, interchange the order of integration, and evaluate the given integral, where possible.

17. $\displaystyle\int_{-1}^{1}\int_{0}^{x+1} (x + y)\,dy\,dx$

18. $\displaystyle\int_{0}^{1}\int_{x^2}^{1} xe^{y^2}\,dy\,dx$

19. $\displaystyle\int_{0}^{1}\int_{0}^{y} xy^2\,dx\,dy$

20. $\displaystyle\int_{-2}^{0}\int_{x^2}^{4} xe^{y^2}\,dy\,dx$

In each of Exercises 21–24 use a double integral to find the volume of the solid indicated by the figure.

21.

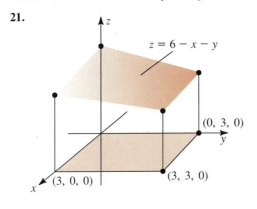

$z = 6 - x - y$

(0, 3, 0)

(3, 0, 0)

(3, 3, 0)

22.

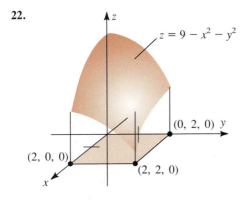

$z = 9 - x^2 - y^2$

(2, 0, 0)

(0, 2, 0)

(2, 2, 0)

23.

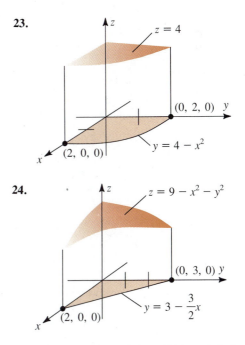

$z = 4$

(0, 2, 0)

(2, 0, 0)

$y = 4 - x^2$

24.

$z = 9 - x^2 - y^2$

(0, 3, 0)

(2, 0, 0)

$y = 3 - \dfrac{3}{2}x$

25. Find the volume of the solid bounded above by the graph of $f(x) = 16 - 4x - 2y$ and below by the rectangle $R = \{(x, y) \mid 0 \le x \le 2,\ 0 \le y \le 1\}$.

26. Find the volume of the solid bounded above by the graph of $f(x, y) = \dfrac{e^{\sqrt{x}}}{\sqrt{xy}}$ and below by the rectangle $R = \{(x, y) \mid 1 \le x \le 4,\ 1 \le y \le 9\}$.

27. Find the volume of the solid bounded above by the graph of $f(x, y) = 4 + x + y$ and below by the region enclosed by the triangle with vertices $(0, 0, 0)$, $(2, 0, 0)$, and $(0, 3, 0)$.

28. Find the area of the region R bounded by the graphs of $y = 4 - x^2$ and $y = x + 2$ by finding the volume of the solid bounded above by the graph of $f(x, y) \equiv 1$ and below by the region R.

29. Use the method of Exercise 28 to find the area of the region bounded by the graphs of $y = x^2$ and $y = x^3$.

SUMMARY OUTLINE OF CHAPTER 8

■ The **partial derivatives,** $\dfrac{\partial f}{\partial x}$ and $\dfrac{\partial f}{\partial y}$, for the function $z = f(x, y)$ of two variables are (Page 399)

$$\frac{\partial f}{\partial x}(x, y) = \lim_{h \to 0} \frac{f(x + h, y) - f(x, y)}{h}$$

$$\frac{\partial f}{\partial y}(x, y) = \lim_{h \to 0} \frac{f(x, y + h) - f(x, y)}{h}.$$

■ Other **notation** for partial derivatives of $z = f(x, y)$ is (Page 401)

$$\frac{\partial f}{\partial x}(x, y) = \frac{\partial f}{\partial x} = f_x = z_x, \text{ etc.}$$

■ The **second order** partial derivatives for $z = f(x, y)$ are (Page 406)

$$\frac{\partial^2 f}{\partial x^2}(x, y) = \frac{\partial}{\partial x}\left(\frac{\partial f}{\partial x}(x, y)\right); \qquad \frac{\partial^2 f}{\partial y \partial x}(x, y) = \frac{\partial}{\partial y}\left(\frac{\partial f}{\partial x}(x, y)\right)$$

$$\frac{\partial^2 f}{\partial x \partial y}(x, y) = \frac{\partial}{\partial x}\left(\frac{\partial f}{\partial y}(x, y)\right); \qquad \frac{\partial^2 f}{\partial y^2}(x, y) = \frac{\partial}{\partial y}\left(\frac{\partial f}{\partial y}(x, y)\right).$$

■ **Theorem:** If $f(x_0, y_0)$ is a relative extremum for $z = f(x, y)$, and if $\dfrac{\partial f}{\partial x}(x_0, y_0)$ and $\dfrac{\partial f}{\partial y}(x_0, y_0)$ (Page 412)

exist, then $\dfrac{\partial f}{\partial x}(x_0, y_0) = \dfrac{\partial f}{\partial y}(x_0, y_0) = 0$.

■ **Theorem:** If all second order partial derivatives of $z = f(x, y)$ are continuous, if $\dfrac{\partial f}{\partial x}(x_0, y_0) =$ (Page 414)

$\dfrac{\partial f}{\partial y}(x_0, y_0) = 0$, and if

$$A = \frac{\partial^2 f}{\partial x^2}(x_0, y_0), \quad B = \frac{\partial^2 f}{\partial y \partial x}(x_0, y_0)$$

$$C = \frac{\partial^2 f}{\partial y^2}(x_0, y_0)$$

and $D = B^2 - AC$, then

(i) $D < 0$ and $A < 0$ imply $f(x_0, y_0)$ is a relative maximum.
(ii) $D < 0$ and $A > 0$ imply $f(x_0, y_0)$ is a relative minimum.
(iii) $D > 0$ implies $(x_0, y_0, f(x_0, y_0))$ is a saddle point.
(iv) $D = 0$ yields no conclusion.

■ **Theorem:** If f and g have continuous partials, the relative maxima and minima of f subject to (Page 426)
the **constraint**

$g(x, y) = 0$

occur at critical points for the **Lagrange** function $L = f + \lambda g$.

■ **Theorem:** The straight line that best fits the data points (x_1, y_1), (x_2, y_2), . . . , (x_n, y_n) in the   (Page 436)
sense of "least squares" has equation $y = mx + b$ where

$$m = \frac{n \sum\limits_{j=1}^{n} x_j y_j - \left(\sum\limits_{j=1}^{n} x_j\right)\left(\sum\limits_{j=1}^{n} y_j\right)}{n \sum\limits_{j=1}^{n} x_j^2 - \left(\sum\limits_{j=1}^{n} x_j\right)^2}.$$

and

$$b = \frac{\left(\sum\limits_{j=1}^{n} x_j^2\right)\left(\sum\limits_{j=1}^{n} y_j\right) - \left(\sum\limits_{j=1}^{n} x_j\right)\left(\sum\limits_{j=1}^{n} x_j y_j\right)}{n \sum\limits_{j=1}^{n} x_j^2 - \left(\sum\limits_{j=1}^{n} x_j\right)^2}.$$

■ The **double integral** of the continuous function $z = f(x, y)$ over the region R is   (Page 443)

$$\iint\limits_{R} f(x, y)\, dA = \lim_{n \to \infty} \sum_{j=1}^{n} f(s_j, t_j)\, \Delta x\, \Delta y$$

where one test point (s_j, t_j) lies in each of the rectangles of area $\Delta x\, \Delta y$, which together approximate
the region R.

■ If $f(x, y) \geq 0$, the **volume** of the solid bounded above by the graph of $z = f(x, y)$ and below   (Page 443)
by the region R in the xy-plane is

$$V = \iint\limits_{R} f(x, y)\, dA.$$

■ **Theorem:** If $z = f(x, y)$ is continuous on R, then   (Page 444)

(i) $\displaystyle \iint\limits_{R} f(x, y)\, dA = \int_{c}^{d} \int_{g_1(y)}^{g_2(y)} f(x, y)\, dx\, dy$

   if $R = \{(x, y) \,|\, g_1(y) \leq x \leq g_2(y),\ c \leq y \leq d\}$

(ii) $\displaystyle \iint\limits_{R} f(x, y)\, dA = \int_{a}^{b} \int_{h_1(x)}^{h_2(x)} f(x, y)\, dy\, dx$

   if $R = \{(x, y) \,|\, a \leq x \leq b,\ h_1(x) \leq y \leq h_2(x)\}$

where the functions g_1, g_2, h_1, and h_2 are assumed continuous.

REVIEW EXERCISES—CHAPTER 8

1. State the domain of the function $f(x, y) = \sqrt{25 - x^2 - y^2}$.

2. For the function $f(x, y) = x^3 - y \ln(3 + x)$ find
   a. $f(0, 2)$ **b.** $f(1, -2)$

3. For the function $f(x, y) = \dfrac{y - x}{y + x}$ find

   a. $f(2, -1)$ **b.** $f(1, -2)$

4. Plot the following points in a three-dimensional coordinate space.

   a. $(2, 3, 1)$ **b.** $(1, -3, -2)$ **c.** $(-2, -2, 3)$

5. State the points (x, y) at which the function $f(x, y) = \dfrac{1 + x^2}{1 - xy}$ is discontinuous.

In each of Exercises 6–15 find $\dfrac{\partial f}{\partial x}$ and $\dfrac{\partial f}{\partial y}$.

6. $f(x, y) = x^4 + 2xy - 3y^2$ **7.** $f(x, y) = (x - y)^3 + \ln xy$

8. $f(x, y) = (4 - xy)^5 + \sqrt{xy}$ **9.** $f(x, y) = \dfrac{xy}{x + y}$

10. $f(x, y) = xye^{x+y}$ **11.** $f(x, y) = x^{2/3}y^{1/3} - \dfrac{x^{3/4}}{y^{1/4}}$

12. $f(x, y) = \ln\sqrt{x^2 + y^2}$ **13.** $f(x, y) = xy\sqrt{y^2 - x^2}$

14. $f(x, y) = \dfrac{e^{xy}}{1 + xy}$ **15.** $f(x, y) = (xy^2 - x^2y)^{2/3}$

In Exercises 16 and 17 find all three first order partial derivatives

16. $f(x, y, z) = \dfrac{xyz}{x + y + z}$

17. $f(x, y, z) = \ln \sqrt{x^2 + 4y^2 + 2z^2}$

In Exercises 18 and 19 find the second order partial derivatives $\dfrac{\partial^2 f}{\partial x^2}$, $\dfrac{\partial^2 f}{\partial x \partial y}$, and $\dfrac{\partial^2 f}{\partial y^2}$.

18. $f(x, y) = ye^{x+y}$ **19.** $f(x, y) = \sqrt{y^2 - x^2}$

In Exercises 20–26 find and classify all relative extrema.

20. $f(x, y) = 4y^2 - 2x^2$

21. $f(x, y) = 3x^2 + xy - 6y^2$

22. $f(x, y) = x^2y + xy^2 + 4x + 4y$

23. $f(x, y) = e^{x^2 - 4xy}$

24. $f(x, y) = \ln(1 + x^2 + y^2)$

25. $f(x, y) = e^{1 + x^2 - y^2}$

26. $f(x, y) = 6x^2 - 2x - 3xy + y^2 + 5y + 5$

In Exercises 27–32 evaluate the iterated integral.

27. $\displaystyle\int_{-1}^{1}\int_{0}^{2} (2x + 3y)\, dx\, dy$ **28.** $\displaystyle\int_{0}^{3}\int_{0}^{1} \dfrac{xy}{\sqrt{x^2 + y^2}}\, dx\, dy$

29. $\displaystyle\int_{0}^{1}\int_{0}^{x} xy\sqrt{x^2 + y^2}\, dy\, dx$ **30.** $\displaystyle\int_{0}^{1}\int_{0}^{y} y^4 e^{xy^2}\, dx\, dy$

31. $\displaystyle\int_{0}^{1}\int_{0}^{x} xy(7 + y^2)\, dy\, dx$ **32.** $\displaystyle\int_{-1}^{1}\int_{-y}^{y} (xy^3 + x^3y)\, dx\, dy$

33. A monopolist can sell x items per week in its domestic market at price p where

$$p = 80 - 2x$$

and it can sell y items per week in its foreign market at price q where

$$q = 120 - 4y.$$

The monopolist's total weekly cost in producing x items for the domestic market and y items for the foreign market is

$$C(x, y) = 1000 + 20x + 20y.$$

   a. Find the weekly revenue function $R(x, y) = xp + yq$.

   b. Find the weekly profit function $P = R - C$.

   c. Find the weekly production level (x, y) for which revenue is a maximum.

   d. Find the weekly production level (x, y) for which profit is a maximum.

34. For the monopolist in Exercise 33 find the production schedule (x, y) for which profit is a maximum subject to the constraint that a maximum of 20 items per week can be produced for sale.

35. A company determines that the productivity of a manufacturing plant is given by the function

$$P(x, y) = 200\sqrt{xy}$$

where x is units of capital and y is units of labor scheduled per unit time.

   a. Find the production level corresponding to $x = 4$ units of capital and $y = 16$ units of labor.

   b. Using $\dfrac{\partial P}{\partial x}$, approximate the increase in production resulting from an increase from $x = 4$ to $x = 5$ units of capital, if labor is held constant at $y = 16$ units.

36. An equipment-leasing firm uses the depreciation formula $V(r, t) = 10,000e^{-(0.05 + r)t}$ to find the value of a \$10,000 tractor after t years when r is the prevailing rate of interest.

a. Find $V(0.10, 2)$, the value of the tractor after 2 years if interest rates remain at 10%.

b. Estimate, using your answer to part **(a)**, the amount by which the value of the tractor will decrease between year 2 and year 3 if interest rates remain constant at 10%.

37. A house cleaning firm determines that its weekly profit from cleaning x houses and y apartments is

$$P(x, y) = 25x + 5y - x^2 - xy \text{ dollars.}$$

What weekly schedule (x, y) should it follow in order to maximize profits?

38. A closed rectangular box is to contain 1000 cm³. If the material for the top and bottom costs 2 cents/cm² and the material for the sides costs 3 cents/cm², what are the dimensions for which cost is a minimum?

39. Find the maximum value of the function $f(x, y, z) = x + 2y - 3z + 1$ subject to the constraint that $x^2 + y^2 + z^2 = 14$.

40. Find the rectangular solid of maximum volume that can be inscribed in the ellipsoid $\dfrac{x^2}{4} + \dfrac{y^2}{9} + \dfrac{z^2}{4} = 1$. That is, find the maximum value for the function $V(x, y, z) = (2x)(2y)(2z) = 8xyz$ subject to the constraint that $\dfrac{x^2}{4} + \dfrac{y^2}{9} + \dfrac{z^2}{4} = 1$. (See Figure 8.1.)

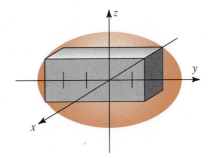

Figure 8.1

9 Chapter Infinite Series

9.1 INFINITE SEQUENCES

Up to this point we have worked almost exclusively with continuous functions, even though many of the functions that arise in business and economic applications are defined only for nonnegative integers. In doing so, we have taken the view that such discontinuous functions can be approximated by continuous functions to which the techniques of the differential and integral calculus can be applied.

For example, in saying that the selling price p and the daily demand x for bicycles are related by the equation $p = 100 - 2x$, we referred to the differentiable function

$$p(x) = 100 - 2x, \qquad x \geq 0$$

defined for all real numbers $x \geq 0$, although this relationship really involves the function

$$p(n) = 100 - 2n, \qquad n = 0, 1, 2, 3, \ldots$$

whose domain is the nonnegative integers, because demand refers to the number of whole bicycles that will be purchased per day.

In some instances it is more appropriate to use techniques that are designed specifically for functions whose domains are subsets of the integers, rather than assuming these functions to be defined for all real numbers. The purpose of this chapter is to develop these techniques, primarily the theory of *infinite sequences* and *infinite series*. Although the initial motivation for these topics will come from functions defined only for integers, we shall see later in the chapter that the theory we develop for infinite series (in particular, *Taylor series*) can be used to analyze certain types of continuous functions.

Infinite Sequences

An **infinite sequence** is simply an infinite list of numbers, such as

$$\{1, 2, 3, 4, \ldots\} \tag{1}$$

or

$$\left\{1, \frac{1}{2}, \frac{1}{3}, \frac{1}{4}, \frac{1}{5}, \ldots\right\} \tag{2}$$

or

$$\{1, e, e^2, e^3, e^4, \ldots\}. \tag{3}$$

The order in which the numbers are listed is important, and we must always indicate which is the first number in a sequence, which is the second, etc. We often make this ordering explicit by using the positive integers, as subscripts, to **index** the terms of a sequence by writing

$$\{a_1, a_2, a_3, \ldots\} \quad \text{or} \quad \{a_n\}, \qquad \text{where } n = 1, 2, 3, \ldots. \tag{4}$$

We say that equation (4) is the **general form** of an infinite sequence, and that a_n (without the braces) is the **general term**.

Example 1

(a) The general term $a_n = n$ generates the sequence $\{1, 2, 3, 4, \ldots\}$ where $a_1 = 1$, $a_2 = 2$, $a_3 = 3$, $a_4 = 4$, . . . , etc.

(b) The general term $a_n = \dfrac{1}{n}$ gives the sequence $\left\{1, \dfrac{1}{2}, \dfrac{1}{3}, \dfrac{1}{4}, \dfrac{1}{5}, \ldots\right\}$ where $a_1 = 1$, $a_2 = \dfrac{1}{2}$, $a_3 = \dfrac{1}{3}$, $a_4 = \dfrac{1}{4}$, ...

(c) The general term $a_n = e^{n-1}$ gives the sequence $\{1, e, e^2, e^3, e^4, \ldots\}$ where $a_1 = e^0$, $a_2 = e^1$, $a_3 = e^2$, $a_4 = e^3$, ...

(d) Using the general terms identified above, we can abbreviate each of these sequences using equation (4) as follows:

$$\{1, 2, 3, 4, \ldots\} = \{n\}, \qquad n = 1, 2, \ldots$$

$$\left\{1, \frac{1}{2}, \frac{1}{3}, \frac{1}{4}, \ldots\right\} = \left\{\frac{1}{n}\right\}, \qquad n = 1, 2, \ldots$$

$$\{1, e, e^2, e^3, \ldots\} = \{e^{n-1}\}, \qquad n = 1, 2, \ldots. \qquad \square$$

Example 2 Write out the first few terms of the sequence $\{a_n\}$ whose general term is

(a) $a_n = \dfrac{n}{n + 1}$

(b) $a_n = \sin\left(\dfrac{n\pi}{2}\right).$

Solution:

(a) Since the terms of a sequence are obtained from the general term a_n by setting $n = 1$, 2, 3, . . . , we find that

$$a_1 = \frac{1}{1 + 1} = \frac{1}{2} \qquad a_3 = \frac{3}{3 + 1} = \frac{3}{4}$$

$$a_2 = \frac{2}{2 + 1} = \frac{2}{3} \qquad a_4 = \frac{4}{4 + 1} = \frac{4}{5}$$

etc., so

$$\left\{\frac{n}{n+1}\right\} = \left\{\frac{1}{2}, \frac{2}{3}, \frac{3}{4}, \frac{4}{5}, \ldots\right\}.$$

(b) Substituting $n = 1, 2, 3, \ldots$ in $a_n = \sin\left(\frac{n\pi}{2}\right)$ gives

$$a_1 = \sin\left(\frac{1 \cdot \pi}{2}\right) = \sin\left(\frac{\pi}{2}\right) = 1$$

$$a_2 = \sin\left(\frac{2\pi}{2}\right) = \sin(\pi) = 0$$

$$a_3 = \sin\left(\frac{3\pi}{2}\right) = -1$$

$$a_4 = \sin\left(\frac{4\pi}{2}\right) = \sin(2\pi) = 0$$

$$a_5 = \sin\left(\frac{5\pi}{2}\right) = 1$$

etc. Thus

$$\left\{\sin\frac{n\pi}{2}\right\} = \{1, 0, -1, 0, 1, 0, -1, 0, \ldots\}. \qquad \square$$

REMARK: Note in part (b) of Example 2 that we include those terms of the sequence that are zeros, and that the terms of a sequence can be negative numbers as well as positive numbers.

Example 3 If $100 is deposited in a savings account paying 10% interest compounded annually, the amounts $P(n)$ on deposit at the end of $n = 1, 2, 3, \ldots$ years can be thought of as the terms of the sequence whose general term is $a_n = P(n)$:

$$\{a_n\} = \{P(n)\} = \{\$100(1.10)^n\}, \qquad n = 1, 2, \ldots$$
$$= \{\$100(1.1), \$100(1.1)^2, \$100(1.1)^3, \ldots\}$$
$$= \{\$110, \$121, \$133.10, \ldots\}. \qquad \square$$

The following formal definition captures our notion that a sequence is a list of numbers indexed by the positive integers.

DEFINITION 1

An **infinite sequence** is a function whose domain is the positive integers. That is, the sequence $\{a_n\}$ is the set of values of a function f for which

$$f(n) = a_n, \qquad n = 1, 2, 3, \ldots.$$

Thinking of sequences as functions allows us to apply the techniques previously used with functions to sequences. For example, we can add two sequences to form a new sequence by adding the corresponding terms:

$$\{a_n\} + \{b_n\} = \{a_n + b_n\}, \qquad n = 1, 2, \ldots.$$

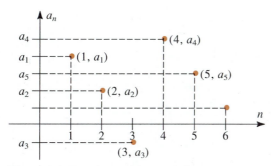

Figure 1.1 Graphing a sequence $\{a_n\}$.

And we can multiply a sequence by a constant by multiplying each term by that constant:

$$c\{a_n\} = \{ca_n\}, \qquad n = 1, 2, \ldots .$$

Finally, we can graph the sequence $\{a_n\}$ by plotting the points $(n, f(n)) = (n, a_n)$ for $n = 1, 2, 3, \ldots$ (but *not* connecting these points). See Figure 1.1.

Example 4 For the sequences

$$\{a_n\} = \{2n\} = \{2, 4, 6, 8, \ldots\} \quad \text{and}$$
$$\{b_n\} = \{n + 5\} = \{6, 7, 8, 9, \ldots\}$$

(a) the sequence $\{c_n\} = \{a_n + b_n\}$ is

$$\{c_n\} = \{2 + 6, 4 + 7, 6 + 8, 8 + 9, \ldots\}$$
$$= \{8, 11, 14, 17, \ldots\}$$

for which the general term is

$$c_n = a_n + b_n = (2n) + (n + 5) = 3n + 5.$$

(b) The sequence $\{d_n\} = 6\{a_n\} = \{6a_n\}$ is

$$\{d_n\} = \{6 \cdot 2, 6 \cdot 4, 6 \cdot 6, 6 \cdot 8, \ldots\}$$
$$= \{12, 24, 36, 48, \ldots\}$$

with general term

$$d_n = 6 \cdot a_n = 6(2n) = 12n. \qquad \square$$

Example 5 Graph the sequences whose general terms are

(a) $a_n = 2n + 1$

(b) $a_n = 2 + \dfrac{(-1)^n}{n}$.

Solution: The sequences are

(a) $\{a_n\} = \{2n + 1\} = \{3, 5, 7, 9, \ldots\}$
(See Figure 1.2.) $\qquad \square$

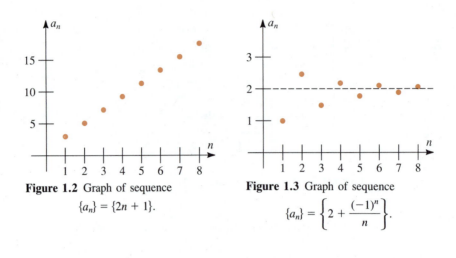

Figure 1.2 Graph of sequence
$\{a_n\} = \{2n + 1\}$.

Figure 1.3 Graph of sequence
$\{a_n\} = \left\{2 + \dfrac{(-1)^n}{n}\right\}$.

(b) $\{a_n\} = \left\{2 + \dfrac{(-1)^n}{n}\right\} = \left\{2 - 1,\ 2 + \dfrac{1}{2},\ 2 - \dfrac{1}{3},\ 2 + \dfrac{1}{4},\ \ldots\right\}$

$= \left\{1,\ \dfrac{5}{2},\ \dfrac{5}{3},\ \dfrac{9}{4},\ \ldots\right\}.$

(See Figure 1.3.) $\qquad\qquad\qquad\qquad\qquad\qquad\qquad\qquad\qquad$ □

REMARK: Recall that $(-1)^n = 1$ when n is even and $(-1)^n = -1$ when n is odd. This is how we signify a term whose sign *alternates* with increasing n.

Limits of Sequences

There is an important difference between the sequence $\{2n + 1\}$ in Figure 1.2 and the sequence $\left\{2 + \dfrac{(-1)^n}{n}\right\}$ in Figure 1.3. In the first sequence the terms increase at a constant rate, not approaching any particular number. In the language of Chapter 2 we would say that

$$\lim_{n \to \infty} \{2n + 1\} = +\infty$$

since the terms of this sequence increase without bound as $n \to \infty$.

However, the terms of the sequence $2 + \dfrac{(-1)^n}{n}$ "approach" the number $L = 2$ as $n \to \infty$ since the term $\dfrac{(-1)^n}{n}$ approaches zero as $n \to \infty$. Thus we write

$$\lim_{n \to \infty} \left\{2 + \dfrac{(-1)^n}{n}\right\} = 2$$

and we say that "the limit of the sequence $\left\{2 + \dfrac{(-1)^n}{n}\right\}$ is the number 2." Thus the limit of an infinite sequence is just the infinite limit "$\lim\limits_{x\to\infty} f(x)$," which we discussed in Chapter 2, except that the general term (function) $f(n) = a_n$ is defined only for positive integers.

DEFINITION 2

$L = \lim\limits_{n\to\infty} a_n$ means that the terms of the sequence $\{a_n\}$ approach the unique number L as n increases without bound.

Example 6 $\lim\limits_{n\to\infty}\left(\dfrac{3n + 5}{n}\right) = \lim\limits_{n\to\infty}\left(3 + \dfrac{5}{n}\right) = 3$ since $\dfrac{5}{n}$ approaches zero as $n\to\infty$. □

Example 7 $\lim\limits_{n\to\infty} \sin\dfrac{\pi}{n} = \sin(0) = 0$ since $\dfrac{\pi}{n}$ approaches zero as $n\to\infty$. □

Example 8 $\lim\limits_{n\to\infty} \sin\left(\dfrac{n\pi}{2}\right)$ does not exist since

$$\left\{\sin\left(\dfrac{n\pi}{2}\right)\right\} = \left\{\sin\dfrac{\pi}{2}, \sin\pi, \sin\dfrac{3\pi}{2}, \sin 2\pi, \ldots\right\}$$
$$= \{1, 0, -1, 0, 1, 0, -1, \ldots\}$$

and these terms do not approach any *single* (that is, *unique*) number as $n\to\infty$. Rather, they oscillate back and forth among the three numbers 1, 0, and -1. □

Since sequences are functions, we may use the same algebra in evaluating limits of sequences that we use in evaluating infinite limits of the form $\lim\limits_{x\to\infty} f(x)$, including the following **properties of limits of sequences**:

$$\lim\limits_{n\to\infty}\{a_n + b_n\} = \lim\limits_{n\to\infty}\{a_n\} + \lim\limits_{n\to\infty}\{b_n\} \tag{5}$$

$$\lim\limits_{n\to\infty}\{ca_n\} = c\,\lim\limits_{n\to\infty}\{a_n\} \tag{6}$$

$$\lim\limits_{n\to\infty}\{a_n b_n\} = \left(\lim\limits_{n\to\infty}\{a_n\}\right)\left(\lim\limits_{n\to\infty}\{b_n\}\right) \tag{7}$$

$$\lim\limits_{n\to\infty}\left\{\dfrac{a_n}{b_n}\right\} = \dfrac{\lim\limits_{n\to\infty}\{a_n\}}{\lim\limits_{n\to\infty}\{b_n\}}, \text{ provided } \lim\limits_{n\to\infty}\{b_n\} \neq 0. \tag{8}$$

In writing equations (5)–(8), we assume that each of the limits on the right-hand sides of the equations exists.

Example 9 Find $\lim\limits_{n\to\infty} \dfrac{6n^3 + 5n^2 + 7}{4n^3 - 2n + 2}$.

Divide all terms by n^3 (the highest power of n present).

Solution

$$\lim_{n\to\infty} \frac{6n^3 + 5n^2 + 7}{4n^3 - 2n + 2} = \lim_{n\to\infty} \left(\frac{6n^3 + 5n^2 + 7}{4n^3 - 2n + 2} \right) \left(\frac{\dfrac{1}{n^3}}{\dfrac{1}{n^3}} \right)$$

$$= \lim_{n\to\infty} \frac{6 + \dfrac{5}{n} + \dfrac{7}{n^3}}{4 - \dfrac{2}{n^2} + \dfrac{2}{n^3}}$$

$$= \frac{6 + 0 + 0}{4 - 0 + 0}$$

Use properties (5)–(7) to take the limit of each individual term, using the facts that

$$\lim_{n\to\infty} \frac{1}{n} = 0, \quad \lim_{n\to\infty} \frac{1}{n^2} = 0$$

and $\lim_{n\to\infty} \dfrac{1}{n^3} = 0.$

$$= \frac{3}{2}. \qquad \square$$

Example 10 Find $\lim_{n\to\infty} [\ln(n + 4) - \dfrac{1}{2} \ln(n)]$.

Strategy
Use properties of natural logarithms that

$r \ln a = \ln a^r$ and

$\ln a - \ln b = \ln \dfrac{a}{b}.$

Divide each term in numerator by $\sqrt{n}$.

Solution

$$\lim_{n\to\infty} \left[\ln(n + 4) - \frac{1}{2} \ln n \right] = \lim_{n\to\infty} [\ln(n + 4) - \ln(n^{1/2})]$$

$$= \lim_{n\to\infty} \ln \frac{n + 4}{\sqrt{n}}$$

$$= \lim_{n\to\infty} \ln \left(\sqrt{n} + \frac{4}{\sqrt{n}} \right)$$

$$= +\infty. \qquad \square$$

When $\lim_{n\to\infty} a_n = L$ exists, we say that the sequence $\{a_n\}$ **converges** to L. When the number $L = \lim_{n\to\infty} a_n$ does not exist, we say that the sequence $\{a_n\}$ **diverges**. Thus the sequence in Example 9 converges to the number $L = \dfrac{3}{2}$, while the sequence in Example 10 diverges.

Example 11 Find $\lim_{n\to\infty} (-1)^n \left(\dfrac{n + 1}{n} \right)$.

Strategy

Because of the factor of $(-1)^n$, examine even and odd terms separately.

Solution

For even integers n we have $(-1)^n = 1$, so

$$\lim_{n \to \infty} (-1)^n \left(\frac{n + 1}{n} \right) = \lim_{n \to \infty} (1) \left(1 + \frac{1}{n} \right)$$

Divide through by n.

$$= 1$$

(even integers only).

However, for odd integers n

$$(-1)^n = -1$$

so

Determine whether odd terms approach the same limit as even terms.

$$\lim_{n \to \infty} (-1)^n \left(\frac{n + 1}{n} \right) = \lim_{n \to \infty} (-1) \left(1 + \frac{1}{n} \right)$$

$$= -1$$

(odd integers only).

This shows that the terms do not approach a *single* number L as $n \to \infty$, so the limit $\lim_{n \to \infty} (-1)^n \left(\frac{n + 1}{n} \right)$ does not exist. Thus the sequence $\left\{ (-1)^n \left(\frac{n + 1}{n} \right) \right\}$ diverges. (See Figure 1.4.) □

The last two examples concern limits of two important sequences that we have studied previously, those for compound interest and approximating sums.

Example 12 (Continuous compounding of interest). In Chapter 4 we developed the following formula for the amount of money $P_n(1)$ on deposit one year after placing an

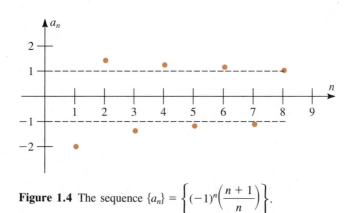

Figure 1.4 The sequence $\{a_n\} = \left\{ (-1)^n \left(\frac{n + 1}{n} \right) \right\}$.

initial amount P_0 in a savings account paying fractional interest rate r compounded n times per year:

$$P_n(1) = \left(1 + \frac{r}{n}\right)^n P_0.$$

This expression may be viewed as the general term of the sequence $\{a_n\} = \{P_n(1)\}$. Letting $n \to \infty$, we obtain the limit

$$P(1) = \lim_{n \to \infty}\left(1 + \frac{r}{n}\right)^n P_0$$

$$= e^r P_0 \tag{9}$$

where we have used the fact that $\lim_{n \to \infty}\left(1 + \frac{r}{n}\right)^n = e^r$ as developed in Chapter 4. Equation (9) is the formula for *continuous* compounding of interest for the period of one year. For continuous compounding of interest over a period of t years we use the limit

$$P(t) = \lim_{n \to \infty}\left(1 + \frac{r}{n}\right)^{nt} P_0$$

$$= \left[\lim_{n \to \infty}\left(1 + \frac{r}{n}\right)^n\right]^t P_0$$

$$= [e^r]^t P_0$$

$$= e^{rt} P_0$$

also developed in Chapter 4. $\qquad\qquad\qquad\qquad\qquad\qquad\qquad\qquad\qquad\qquad\qquad\square$

Example 13 (Approximating sums) Each of the procedures for approximating the definite integral $\int_a^b f(x)\,dx$ that we have described (approximating sums, Trapezoidal Rule, Simpson's Rule) is really a procedure for generating the n th term in a sequence of approximations, where n is the number of subintervals into which the interval $[a, b]$ is divided. Figure 1.5 shows one term of the sequence $\{S_n\}$ of approximating sums for the

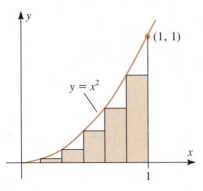

Figure 1.5 The definite integral $\int_0^1 x^2\,dx$ is the limit of a sequence of approximating sums

$$\int_0^1 x^2\,dx = \lim_{n \to \infty} S_n = \lim_{n \to \infty} \sum_{j=1}^n t_j^2\,\Delta x.$$

function $f(x) = x^2$ on the interval $[0,1]$, using the minimum value of $f(x) = x^2$ in each subinterval. In this case we know that

$$\int_0^1 x^2 \, dx = \lim_{n \to \infty} \sum_{j=1}^n \left(\frac{j-1}{n}\right)^2 \left(\frac{1}{n}\right)$$

$$= \frac{1}{3}$$

as demonstrated in Chapter 5.

□

Exercise Set 9.1

In each of Exercises 1–10 write out the first five terms of the sequence whose general term is given.

1. $\{2n + 3\}$

2. $\left\{\dfrac{n}{2n + 1}\right\}$

3. $\left\{\dfrac{2n - 1}{n + 5}\right\}$

4. $\left\{\dfrac{(-1)^n}{1 + n^2}\right\}$

5. $\left\{\dfrac{(-1)^n}{e^n}\right\}$

6. $\left\{\dfrac{20n}{1 + n^2}\right\}$

7. $\left\{\dfrac{(n - 1)(n + 1)}{2n^2 + 2n + 2}\right\}$

8. $\{\ln(n + 2)\}$

9. $\left\{(-1)^{n+1} \sin\left(\dfrac{n\pi}{4}\right)\right\}$

10. $\left\{\dfrac{1 + n}{5 + n^3}\right\}$

In each of Exercises 11–30 determine whether the sequence converges or diverges. For those that converge find the limits.

11. $\left\{\dfrac{n}{3n + 6}\right\}$

12. $\left\{\dfrac{2n - 1}{n + 3}\right\}$

13. $\{\sqrt{5}\}$

14. $\left\{\dfrac{n - 1}{n^3 + 3}\right\}$

15. $\left\{\dfrac{e^n + 3}{e^n}\right\}$

16. $\left\{\dfrac{9 - n^2}{2n^2 + 1}\right\}$

17. $\left\{\dfrac{n^2 - 2n + 1}{3 + n^2}\right\}$

18. $\left\{\dfrac{n^2 + 3n + 5}{n^5 + 2}\right\}$

19. $\{\ln(n + 1) - \ln(n)\}$

20. $\{e^{(n-1)/n}\}$

21. $\{\sin n\pi\}$

22. $\left\{\dfrac{(-1)^n n^2}{9 - n^2}\right\}$

23. $\left\{\cos \dfrac{n - 1}{n^2}\right\}$

24. $\left\{\dfrac{e^n - e^{-n}}{e^n + e^{-n}}\right\}$

25. $\left\{\dfrac{1}{n} - \dfrac{1}{n + 1}\right\}$

26. $\left\{\dfrac{2^n}{3^n + 4^n}\right\}$

27. $\left\{\left(1 + \dfrac{1}{n}\right)^n\right\}$

28. $\left\{\left(1 - \dfrac{1}{n}\right)^n\right\}$

29. $\left\{\dfrac{\sqrt{n}}{1 + \sqrt{n}}\right\}$

30. $\left\{\ln\left(\dfrac{n^2 + 6}{9 + n^2}\right)\right\}$

31. A biological population grows in number so that after n years it has $N(n) = \dfrac{A}{B + Ce^{-n}}$ members. Find $\lim\limits_{n \to \infty} N(n)$, its long-term size.

32. Find $\lim\limits_{n \to \infty} N(n)$ for the growth model $N(n) = \dfrac{A}{B + Ce^n}$.

33. Let the terms of the sequence $\{a_n\}$ be the amount on deposit n years after an initial deposit of $1000 is made in a savings account paying 8% interest compounded annually. Find
   a. a_1
   b. a_2
   c. a_n

34. Let the terms of the sequence $\{a_n\}$ be the present value of an investment that will be worth $1000 in n years, assuming continuous compounding of interest at the nominal rate $r = .10$. Find
   a. a_1
   b. a_2
   c. a_n

35. Does the sequence in Exercise 33 converge? Why or why not?

36. What is the limit of the sequence in Exercise 34? What does this mean?

37. The *Fibonacci numbers* are the sequence

$$\{1, 1, 2, 3, 5, 8, 13, 21, 34, \ldots\}$$

which is defined by the following rules:

a. $a_1 = 1$
b. $a_2 = 1$

c. $a_{n+2} = a_n + a_{n+1}, \quad n = 1, 2, 3, \ldots$

This sequence is sometimes described as the solution to the following problem. Assume that we begin with a single pair of rabbits (male and female), that this pair and each resulting pair of rabbits become fertile one month after birth, that each pair of fertile rabbits gives birth to one new pair of rabbits each month, and that no rabbit dies. Let a_n be the number of pairs of rabbits present n months after this experiment begins. Explain why the Fibonacci sequence describes the sequence $\{a_n\}$.

9.2 INFINITE SERIES

In Chapter 8 we introduced the concept of an improper integral of the form $\displaystyle\int_a^\infty f(x)\, dx$.

One reason for doing this was to extend certain applications of the definite integral (such as the calculation of area) to intervals extending infinitely far to the right. For these same reasons we sometimes need to use a similar concept that applies to functions that are defined only for integers. This is the concept of *infinite series*. The following example shows one situation to which this concept applies.

Example 1 An investor has the opportunity to place $100,000 in an investment account that pays 20% interest compounded annually. The investor plans to withdraw one-half of the total amount on deposit annually, just after the interest is credited to the account.

(a) Find $S(1)$, the sum that the investor will receive at the end of the first year.
(b) Find $S(n)$, the sum that the investor will receive *(total)* over the first n years.
(c) Find $\displaystyle\lim_{n\to\infty} S(n)$, the sum of all payments due to the investor and his heirs in perpetuity.

(Work in nominal dollars, rather than convert to present values.)

Solution

(a) After the first annual payment the total in the account will be

$$T(1) = (1.2)(\$100{,}000) = \$120{,}000.$$

The investor will withdraw half this amount

$$W(1) = (0.5)(1.2)(\$100{,}000) = \$60{,}000$$

leaving the amount

$$R(1) = (0.5)(1.2)(\$100{,}000) = \$60{,}000$$

remaining in the account. Thus $S(1) = W(1) = \$60{,}000$.

(b) After the second annual interest payment the account will contain

$$T(2) = (1.2)R(1) = (0.5)(1.2)^2(\$100{,}000) = \$72{,}000$$

of which the amount withdrawn will be

$W(2) = (0.5)T(2) = (0.5)^2(1.2)^2(\$100,000) = \$36,000.$

The sum of the withdrawals during the first 2 years will therefore be

$$
\begin{aligned}
S(2) &= W(1) + W(2) \\
&= (0.5)(1.2)(\$100,000) + (0.5)^2(1.2)^2(\$100,000) \\
&= \left[\frac{3}{5} + \left(\frac{3}{5} \right)^2 \right] (\$100,000) \\
&= \$96,000.
\end{aligned}
$$

Continuing in this way, we find that the total of all withdrawals during the first n years will be

$$
\begin{aligned}
S(n) &= \left[\frac{3}{5} + \left(\frac{3}{5} \right)^2 + \left(\frac{3}{5} \right)^3 + \cdots + \left(\frac{3}{5} \right)^n \right] (\$100,000) \\
&= \frac{3}{5} \left[1 + \frac{3}{5} + \left(\frac{3}{5} \right)^2 + \cdots + \left(\frac{3}{5} \right)^{n-1} \right] (\$100,000). \tag{1}
\end{aligned}
$$

To simplify this expression, we use the formula

$$
1 + x + x^2 + \cdots + x^k = \frac{1 - x^{k+1}}{1 - x} \tag{2}
$$

which we shall verify later in this section. With $k = n - 1$ and $x = \dfrac{3}{5}$ we find that

$$
\begin{aligned}
1 + \frac{3}{5} + \left(\frac{3}{5} \right)^2 + \cdots + \left(\frac{3}{5} \right)^{n-1} &= \frac{1 - \left(\dfrac{3}{5} \right)^n}{1 - \dfrac{3}{5}} \\
&= \frac{5}{2} \left[1 - \left(\frac{3}{5} \right)^n \right]. \tag{3}
\end{aligned}
$$

Substituting the right-hand side of equation (3) into equation (1) in place of the expression in brackets gives

$$
\begin{aligned}
S(n) &= \frac{3}{5} \left(\frac{5}{2} \right) \left[1 - \left(\frac{3}{5} \right)^n \right] (\$100,000) \\
&= \$150,000 \left[1 - \left(\frac{3}{5} \right)^n \right] \tag{4}
\end{aligned}
$$

as the total of all withdrawals over the first n years. (See Figure 2.1.)

(c) The sum of all annual withdrawals, in perpetuity (meaning forever), is just the limit as $n \to \infty$ of the expression for $S(n)$ in equation (4):

$$
\begin{aligned}
S = \lim_{n \to \infty} S(n) &= \lim_{n \to \infty} \$150,000 \left[1 - \left(\frac{3}{5} \right)^n \right] \\
&= \$150,000[1 - 0] \\
&= \$150,000
\end{aligned}
$$

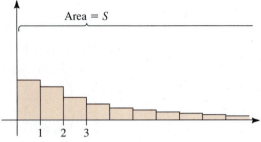

Figure 2.1 $S(n)$ is the sum of the annual withdrawals

$$S(n) = W(1) + W(2) + \cdots + W(n).$$

Figure 2.2 $S = \lim\limits_{n \to \infty} S(n)$ is the sum of all annual withdrawals in perpetuity.

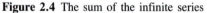

Figure 2.3 The improper integral

$$\int_0^\infty f(x)\,dx$$

is the *limit* of the partial integrals

$$\int_0^t f(x)\,dx:$$

$$\int_0^\infty f(x)\,dx = \lim_{t \to \infty} \int_0^t f(x)\,dx.$$

Figure 2.4 The sum of the infinite series

$$\sum_{k=1}^\infty a_k = a_1 + a_2 + a_3 + \cdots$$

is the *limit* of the sequence of partial sums

$$\sum_{k=1}^\infty a_k = \lim_{n \to \infty} S_n = \lim_{n \to \infty} \sum_{k=1}^n a_k.$$

since $\lim\limits_{n \to \infty} \left(\dfrac{3}{5}\right)^n = 0$ (a fact that you can observe experimentally by using a calculator, for example).

Figure 2.2 shows that this limit S corresponds to the area associated with the infinitely many rectangles with widths one and heights $W(1)$, $W(2)$, $W(3)$, ... $\square$

The point of Example 1 is that we can think of summing infinitely many function values $f(1), f(2), f(3), \ldots$ in the same way that we evaluate the improper integral $\int_a^\infty f(x)\,dx$ of a continuous function: by first summing finitely many terms, and then evaluating the limit at infinity of the resulting expression. (See Figures 2.3 and 2.4.) This is the concept of *infinite series*, which we define as follows.

DEFINITION 3

An **infinite series** is an expression of the form

$$\sum_{k=1}^{\infty} a_k = a_1 + a_2 + a_3 + \cdots.$$

For such infinite series the **partial sums** $S_1, S_2, S_3, \ldots$ are defined by

$$S_1 = a_1$$
$$S_2 = a_1 + a_2$$
$$S_3 = a_1 + a_2 + a_3$$
$$\vdots$$

$$S_n = a_1 + a_2 + a_3 + \cdots + a_n = \sum_{k=1}^{n} a_k, \text{ etc.}$$

The infinite series $\sum_{k=1}^{\infty} a_k$ is said to **converge** to the *sum S* if

$$S = \lim_{n \to \infty} S_n = \lim_{n \to \infty} \sum_{k=1}^{n} a_k \tag{5}$$

exists. (That is, the sum of an infinite series is the limit at infinity of the sequence of its partial sums.) If the limit in equation (5) does not exist, the infinite series is said to **diverge**.

Example 2 The total of the annual withdrawals in Example 1 is an infinite series:

$$W(1) + W(2) + W(3) + \cdots$$

$$= \left(\frac{3}{5}\right)(100,000) + \left(\frac{3}{5}\right)^2(100,000) + \left(\frac{3}{5}\right)^3(100,000) + \cdots$$

$$= \sum_{k=1}^{\infty} \left(\frac{3}{5}\right)^k (100,000).$$

In that example the partial sums for the series were found to be

$$S_n = S(n) = \left[\frac{3}{5} + \left(\frac{3}{5}\right)^2 + \cdots + \left(\frac{3}{5}\right)^n\right](100,000)$$

$$= 150,000\left[1 - \left(\frac{3}{5}\right)^n\right]$$

and the sum of the series is

$$S = \lim_{n \to \infty} S_n = 150,000.$$

That is

$$\sum_{k=1}^{\infty} \left(\frac{3}{5}\right)^k (100{,}000) = 150{,}000.$$

Thus this series *converges*. (This is an example of a *geometric* series, which we shall discuss later in this chapter.) □

Example 3 The infinite series

$$\sum_{k=1}^{\infty} (-1)^k = -1 + 1 + (-1) + 1 + (-1) + \cdots$$

is an example of a divergent series. To see this, simply note the sequence of partial sums:

$$
\begin{aligned}
S_1 &= -1 \\
S_2 &= -1 + 1 = S_1 + 1 = 0 \\
S_3 &= -1 + 1 + (-1) = S_2 - 1 = -1 \\
S_4 &= -1 + 1 + (-1) + 1 = S_3 + 1 = 0 \\
&\qquad\vdots \\
S_{2n} &= S_{2n-1} + 1 = -1 + 1 = 0 \\
S_{2n+1} &= S_{2n} - 1 = 0 - 1 = -1 \\
&\qquad\vdots
\end{aligned}
$$

Thus the sequence of partial sums is

$$\{S_n\} = \{-1, 0, -1, 0, -1, 0, -1, \ldots\}$$

which oscillates between -1 and 0, so $\lim_{n\to\infty} S_n$ does not exist. This series, therefore,

diverges. □

Geometric Series

A **geometric series** is an infinite series of the form

$$1 + x + x^2 + \cdots + x^n + \cdots = \sum_{k=0}^{\infty} x^k. \tag{6}$$

That is, a geometric series is one involving positive integral powers of some number x. Note that the first term of this series is $1 = x^0$, which is why the index k begins at $k = 0$. The partial sums of a geometric series may be found using formula (2) given earlier:

$$1 + x + x^2 + \cdots + x^n = \frac{1 - x^{n+1}}{1 - x}, \qquad x \neq 1. \tag{7}$$

To verify equation (7), we note that

$$
\begin{aligned}
(1 - x)&(1 + x + x^2 + \cdots + x^n) \\
&= 1(1 + x + x^2 + \cdots + x^n) - x(1 + x + x^2 + \cdots + x^n)
\end{aligned}
$$

$$= (1 + x + x^2 + \cdots + x^n) - (x + x^2 + x^3 + \cdots + x^{n+1})$$
$$= 1 - x^{n+1}$$

since all of the powers of x between x and x^n are added in the first parentheses and subtracted in the second parentheses. Dividing the left-hand and right-hand sides of this equation by $(1 - x)$ then gives formula (7).

Using formula (7), we find that the partial sums S_n of the geometric series in equation (6) can be written

$$S_n = 1 + x + x^2 + \cdots + x^n \qquad\qquad (8)$$
$$= \frac{1 - x^{n+1}}{1 - x}, \qquad x \neq 1.$$

To determine whether the sequence $\{S_n\}$ converges, we need to know something about $\lim\limits_{n\to\infty} x^{n+1}$. This limit depends on the number x. A fact that we shall not prove, but that you can observe using a calculator, is that $\lim\limits_{n\to\infty} x^{n+1}$ (a) equals zero if $|x| < 1$, (b) equals 1 if $x = 1$, and (c) fails to exist if $x = -1$ (by Example 3) or if $|x| > 1$.

Since the case $x = 1$ is already ruled out in equation (8), the limit $\lim\limits_{n\to\infty} S_n$ exists only when $|x| < 1$. In this case we have

$$\lim_{n\to\infty} S_n = \lim_{n\to\infty} \frac{1 - x^{n+1}}{1 - x} = \frac{1 - 0}{1 - x} = \frac{1}{1 - x}$$

since $\lim\limits_{n\to\infty} x^{n+1} = 0$. The behavior of geometric series is therefore the following.

THEOREM 1

The geometric series

$$\sum_{k=0}^{\infty} x^k = 1 + x + x^2 + x^3 + \cdots$$

(a) converges to $\dfrac{1}{1 - x}$ if $|x| < 1$

(b) diverges if $|x| \geq 1$.

We may summarize Theorem 1 by saying that

$$\sum_{k=0}^{\infty} x^k = \frac{1}{1 - x} \qquad \text{if } |x| < 1. \qquad\qquad (9)$$

The following examples illustrate how equation (9) may be used.

Example 4 The series

$$1 + \frac{1}{2} + \frac{1}{4} + \frac{1}{8} + \cdots + \frac{1}{2^k} + \cdots$$

is a geometric series whose general term is $a_k = \dfrac{1}{2^k} = \left(\dfrac{1}{2}\right)^k$. It therefore has the form of the geometric series in equation (9) with $x = \dfrac{1}{2}$. Thus since $|x| = \left|\dfrac{1}{2}\right| = \dfrac{1}{2} < 1$, the series converges, with sum

$$\sum_{k=0}^{\infty} \left(\frac{1}{2}\right)^k = \frac{1}{1 - \dfrac{1}{2}} = \frac{1}{\dfrac{1}{2}} = 2. \qquad \Box$$

Example 5 The series

$$1 + \frac{2}{3} + \frac{4}{9} + \frac{8}{27} + \cdots + \frac{2^k}{3^k} + \cdots$$

has the form of equation (9) with $x = \dfrac{2}{3}$, since $x^k = \dfrac{2^k}{3^k} = \left(\dfrac{2}{3}\right)^k$. Thus since $|x| = \left|\dfrac{2}{3}\right| = \dfrac{2}{3} < 1$, the series converges, with sum

$$\sum_{k=0}^{\infty} \frac{2^k}{3^k} = \sum_{k=0}^{\infty} \left(\frac{2}{3}\right)^k = \frac{1}{1 - \dfrac{2}{3}} = \frac{1}{\dfrac{1}{3}} = 3. \qquad \Box$$

Example 6 The series

$$1 + 2 + 4 + 8 + \cdots + 2^k + \cdots$$

is a geometric series with general term $a_k = x^k = 2^k$. Since $x = 2$, we have $|x| = |2| = 2 > 1$, so this geometric series diverges. $\qquad \Box$

Example 7 The series

$$\frac{1}{2} + \frac{1}{4} + \frac{1}{8} + \cdots + \frac{1}{2^k} + \cdots$$

has the form of the geometric series in equation (9) with $x^k = \dfrac{1}{2^k} = \left(\dfrac{1}{2}\right)^k$, *so* $x = \dfrac{1}{2}$. *But the first term* $x^0 = \left(\dfrac{1}{2}\right)^0 = 1$ *is missing.* Since $|x| = \left|\dfrac{1}{2}\right| = \dfrac{1}{2} < 1$, the series converges, but we must subtract the missing term 1 from the formula for the sum of this series:

$$\frac{1}{2} + \frac{1}{4} + \frac{1}{8} + \cdots + \frac{1}{2^k} = \sum_{k=1}^{\infty} \left(\frac{1}{2}\right)^k$$

$$= \sum_{k=0}^{\infty} \left(\frac{1}{2}\right)^k - \left(\frac{1}{2}\right)^0$$

Note the change in the starting index.

$$= \frac{1}{1 - \frac{1}{2}} - 1$$

$$= 2 - 1$$

$$= 1.$$

$\square$

REMARK: Example 7 raises an important point about infinite series. If the series $\sum_{k=1}^{\infty} a_k$ converges, then adding or subtracting a *finite* number of terms will not change the fact that the series converges, *but it will change the value of the sum* (by whatever amount was added or subtracted). Thus, for example, since

$$1 + \frac{1}{3} + \left(\frac{1}{3}\right)^2 + \cdots + \left(\frac{1}{3}\right)^k + \cdots = \sum_{k=0}^{\infty} \left(\frac{1}{3}\right)^k = \frac{1}{1 - \frac{1}{3}} = \frac{3}{2}$$

we have

$$\frac{1}{3} + \left(\frac{1}{3}\right)^2 + \cdots + \left(\frac{1}{3}\right)^k + \cdots$$

$$= \sum_{k=1}^{\infty} \left(\frac{1}{3}\right)^k = \sum_{k=0}^{\infty} \left(\frac{1}{3}\right)^k - 1 = \frac{3}{2} - 1 = \frac{1}{2}$$

$$\left(\frac{1}{3}\right)^2 + \left(\frac{1}{3}\right)^3 + \cdots + \left(\frac{1}{3}\right)^k + \cdots$$

Note change in starting index from $k = 2$ to $k = 0$.

$$= \sum_{k=2}^{\infty} \left(\frac{1}{3}\right)^k = \sum_{k=0}^{\infty} \left(\frac{1}{3}\right)^k - 1 - \frac{1}{3} = \frac{3}{2} - 1 - \frac{1}{3} = \frac{1}{6}$$

etc. When working with the formula for the sum of a *geometric* series, be very careful to note the value of the index corresponding to the first term. If it is not $k = 0$, you must adjust the formula for the sum accordingly.

Properties of Infinite Series

We may perform simple algebra on convergent infinite series just as we do for finite sums. Specifically, if $\sum a_k$ and $\sum b_k$ are convergent infinite series and c is a constant

$$\sum c a_k = c \sum a_k \tag{10}$$

and

$$\sum (a_k + b_k) = \sum a_k + \sum b_k. \tag{11}$$

That is, multiplying each term of a convergent series by a constant c multiplies the sum by this same constant, and the sum of series formed by pairwise addition of the terms of two convergent series is found by adding the individual sums.

REMARK 1: In writing equations (10) and (11), we have not indicated initial values for the indices since these equations are true independent of whether the series begins at $k = 0$, $k = 1$, etc. (However, the series in equation (11) must all begin at the same value of the index.)

REMARK 2: This sort of algebra is valid only if all the series involved are convergent. Adding a convergent series termwise to a divergent series just gives another divergent series.

Example 8 Find the sum of the series

$$\sum_{k=0}^{\infty} \frac{3 \cdot 2^k + 3^k}{5^k} = 4 + \frac{9}{5} + \frac{21}{25} + \frac{51}{125} + \cdots.$$

Solution: We use formula (9) together with equations (10) and (11) as follows:

$$\sum_{k=0}^{\infty} \frac{3 \cdot 2^k + 3^k}{5^k} = \sum_{k=0}^{\infty} \left(\frac{3 \cdot 2^k}{5^k} + \frac{3^k}{5^k} \right)$$

$$= 3 \sum_{k=0}^{\infty} \frac{2^k}{5^k} + \sum_{k=0}^{\infty} \frac{3^k}{5^k}$$

$$= 3 \sum_{k=0}^{\infty} \left(\frac{2}{5} \right)^k + \sum_{k=0}^{\infty} \left(\frac{3}{5} \right)^k$$

$$= 3 \left(\frac{1}{1 - \frac{2}{5}} \right) + \left(\frac{1}{1 - \frac{3}{5}} \right)$$

$$= 3 \left(\frac{5}{3} \right) + \frac{5}{2}$$

$$= \frac{15}{2}.$$

$\square$

Exercise Set 9.2

In Exercises 1–8 write out the first four terms of the infinite series.

1. $\displaystyle\sum_{k=1}^{\infty} (2k + 3)$

2. $\displaystyle\sum_{k=1}^{\infty} \sqrt{1 + k^2}$

3. $\displaystyle\sum_{k=0}^{\infty} \frac{1}{6^k}$

4. $\displaystyle\sum_{k=0}^{\infty} \frac{1 + 3^k}{5^k}$

5. $\displaystyle\sum_{k=1}^{\infty} \frac{(-1)^k}{1 + k^2}$

6. $\displaystyle\sum_{k=0}^{\infty} \frac{2^k}{5^k}$

7. $\displaystyle\sum_{k=1}^{\infty} \frac{\cos(\pi k)}{1 + k}$

8. $\displaystyle\sum_{k=1}^{\infty} k^k$

In Exercises 9–22 determine whether the series converges. If it does, find its sum.

9. $\displaystyle\sum_{k=0}^{\infty} \frac{1}{4^k}$

10. $\displaystyle\sum_{k=0}^{\infty} \frac{3^k}{2^k}$

11. $\displaystyle\sum_{k=0}^{\infty} \frac{3^k}{7^k}$

12. $\displaystyle\sum_{k=0}^{\infty} \left(\frac{1}{\sqrt{2}}\right)^k$

13. $\displaystyle\sum_{k=1}^{\infty} \frac{1}{3^k}$

14. $\displaystyle\sum_{k=2}^{\infty} \frac{1}{2^k}$

15. $\displaystyle\sum_{k=1}^{\infty} \frac{7^k}{5^k}$

16. $\displaystyle\sum_{k=2}^{\infty} \frac{9^k}{7^k}$

17. $\displaystyle\sum_{k=0}^{\infty} \frac{1 + 3^k}{4^k}$

18. $\displaystyle\sum_{k=0}^{\infty} \frac{3 \cdot 2^k}{3^k}$

19. $\displaystyle\sum_{k=1}^{\infty} \frac{2^k - 3^k}{4^k}$

20. $\displaystyle\sum_{k=0}^{\infty} \frac{2^k + 5^k}{4^k}$

21. $\displaystyle\sum_{k=1}^{\infty} \frac{5 \cdot 3^k - 3 \cdot 5^k}{7^k}$

22. $\displaystyle\sum_{k=0}^{\infty} \frac{7 \cdot 2^{k+1} - 3 \cdot 4^{k+1}}{5^{k+1}}$

23. The repeating decimal $0.3333\overline{3}$ may be written as a geometric series as follows:

$$0.3333\overline{3} = \frac{3}{10} + \frac{3}{100} + \frac{3}{1000} + \cdots$$

$$= 3\left(\frac{1}{10}\right) + 3\left(\frac{1}{10}\right)^2 + 3\left(\frac{1}{10}\right)^3 + \cdots$$

$$= 3 \sum_{k=1}^{\infty} \left(\frac{1}{10}\right)^k$$

Show that the sum of this series is $\dfrac{1}{3}$.

24. Write the repeating decimal $0.2222\overline{2}$ as a geometric series. (See Exercise 23.) What is the sum of this series in fraction form?

25. A life insurance agent is to be paid a commission according to the following scheme: $1000 in year 1, half

the year 1 amount in year 2, half the year 2 amount in year 3, etc.

a. What is the total of all amounts received by the agent in the first 5 years?

b. What is the total of all amounts to be received by the agent in perpetuity?

26. When dropped from a height h, a ball rebounds to height $\dfrac{2}{3}h$.

a. Write an infinite series describing the total distance traveled by the ball.

b. What is the sum of this series?

27. In a certain economic model, every dollar received by an individual or corporation will result in an expenditure of 75 cents (that is, a transfer to another individual or corporation). What is the total of all expenditures including and resulting from an initial expenditure of $1 million dollars, according to this model?

28. A marketing firm determines that for every 100 high school students who are told a piece of information (positive or negative) about a particular college, 60 of these students will each pass the information on to one additional student. In total, how many students will learn about such a piece of information if 500 are told initially?

29. Assume that the present value of an amount $P(n)$ to be received after n years is $P_0 = e^{-rn}P(n)$. Consider an annuity that pays the owner $P(n) = \$1000$ per year in perpetuity.

a. Show that the present value of the sum of the first n payments is

$$S(n) = \sum_{k=1}^{n} 1000e^{-rk}.$$

b. Find an expression for $S(n)$ using equation (7).

c. Show that the present value of all payments to be received in perpetuity is

$$S = \sum_{k=1}^{\infty} 1000e^{-rk}.$$

d. Explain why the series in part (c) is a convergent geometric series.

e. Find S.

30. Explain why the geometric series $\displaystyle\sum_{k=0}^{\infty} x^k$ diverges for $x = 1$.

31. A paper carrier agrees to be paid $20 for the first week's delivery of a daily paper and, for each week thereafter, to be paid 75% of the previous week's pay. If this scheme continues in perpetuity, how much will the paper carrier be paid?

32. Explain why the following formulas are true:

a. $\displaystyle\sum_{k=1}^{\infty} x^k = \frac{x}{1-x}$, $\quad |x| < 1$.

b. $\displaystyle\sum_{k=2}^{\infty} x^k = \frac{x^2}{1-x}$, $\quad |x| < 1$.

9.3 TESTS FOR CONVERGENCE

We began our discussion of infinite series with geometric series because we can actually calculate the sum of a convergent geometric series. For most other infinite series we shall only be able to determine whether or not the series converges but not be able to find its sum. However, we shall see later that even this is useful information. In this section we present three criteria for determining whether an infinite series converges.

The Integral Test

Figure 3.1 shows the (positive) terms $a_1, a_2, a_3, \ldots$ of an infinite series $\displaystyle\sum_{k=1}^{\infty} a_k$ represented as areas of rectangles. Each rectangle has width equal to one, and the heights are just the terms of the series $\displaystyle\sum_{k=1}^{\infty} a_k$. Also shown is the graph of a continuous decreasing function f which has the property that $f(k) = a_k$, $k = 1, 2, \ldots$. That is, the graph of f passes through the upper right-hand corner of each rectangle.

Figure 3.1 suggests that since the function f is decreasing, the area of the first n rectangles $(a_1 + a_2 + \cdots + a_n)$ is less than the area of the region bounded by the graph of f and the x-axis between $x = 0$ and $x = n$. That is

$$S_n = \sum_{k=1}^{n} a_k < \int_0^n f(x)\, dx. \tag{1}$$

Now as $n \to \infty$ the integral on the right-hand side of inequality (1) becomes the improper integral $\displaystyle\int_0^{\infty} f(x)\, dx$. Since the left-hand side of inequality (1) is a partial sum for the series

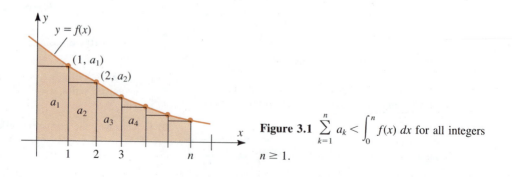

Figure 3.1 $\displaystyle\sum_{k=1}^{n} a_k < \int_0^n f(x)\, dx$ for all integers $n \geq 1$.

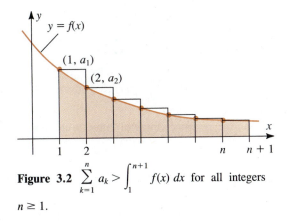

Figure 3.2 $\displaystyle\sum_{k=1}^{n} a_k > \int_{1}^{n+1} f(x)\, dx$ for all integers

$n \geq 1$.

$\displaystyle\sum_{k=1}^{\infty} a_k$, it follows that *if the improper integral* $\displaystyle\int_{0}^{\infty} f(x)\, dx$ *converges, the infinite series*

$\displaystyle\sum_{k=1}^{\infty} a_k$ *also converges.* That is because the partial sums for $\displaystyle\sum_{k=1}^{\infty} a_k$ can never exceed the

number $\displaystyle\int_{0}^{\infty} f(x)\, dx$. (The details of a careful proof of this statement require more theory of

infinite sequences than we have developed, but the essence of the argument is illustrated in Figure 3.1.)

In Figure 3.2 we represent the same terms $a_1, a_2, \ldots$ and function f, except that for each k the rectangle with area a_k extends to the right of the point (k, a_k) rather than to the left. Since f is decreasing, this means that the area of the first n rectangles is *greater* than the area of the region bounded by the graph of f and the x-axis for $1 \leq x \leq n + 1$. Thus

$$\int_{1}^{n+1} f(x)\, dx < \sum_{k=1}^{n} a_k = S_n. \tag{2}$$

From inequality (2) we draw a different conclusion: Since the a_k's are positive, if the

improper integral $\displaystyle\int_{1}^{\infty} f(x)\, dx$ diverges, the "partial integrals" on the left-hand side of

inequality (2) become infinitely large as $n \to \infty$. This means that the infinite series $\displaystyle\sum_{k=1}^{\infty} a_k$

must also diverge, since the partial sums $\displaystyle\sum_{k=1}^{n} a_k$ become infinitely large as $n \to \infty$. That is,

if the improper integral $\displaystyle\int_{1}^{\infty} f(x)\, dx$ *diverges, so must the infinite series* $\displaystyle\sum_{k=1}^{\infty} a_k$.

Combining these two statements gives us the statement of the Integral Test.

THEOREM 2 **Integral Test**	Let each of the terms of the infinite series $\sum_{k=1}^{\infty} a_k$ be positive, and let f be a continuous decreasing function with $f(k) = a_k$ for $k = 1, 2, \ldots$. Then

$$\sum_{k=1}^{\infty} a_k \text{ converges if and only if } \int_{1}^{\infty} f(x)\, dx \text{ converges.}$$

That is, the series and the improper integral either both converge or both diverge.

To use the integral test on the series $\sum_{k=1}^{\infty} a_k$, we must first ensure that all terms of the series are positive. We then find the required decreasing function f (usually by simply replacing k in the general term for $\sum_{k=1}^{\infty} a_k$ with x) and check to see whether the improper integral $\int_{1}^{\infty} f(x)\, dx$ converges.

Example 1 The series $\sum_{k=1}^{\infty} \dfrac{1}{k} = 1 + \dfrac{1}{2} + \dfrac{1}{3} + \dfrac{1}{4} + \cdots$ is called the **harmonic series.** Determine whether the harmonic series converges.

Solution: The series has general term $a_k = \dfrac{1}{k}$, so we use the function $f(x) = \dfrac{1}{x}$. Since f is continuous and decreasing on $[1, \infty)$, the Integral Test applies. The improper integral is

$$\int_{1}^{\infty} \frac{1}{x}\, dx = \lim_{t \to \infty} \int_{1}^{t} \frac{1}{x}\, dx$$

$$= \lim_{t \to \infty} \ln x \Big]_{1}^{t}$$

$$= \lim_{t \to \infty} (\ln t - \ln 1)$$

$$= +\infty.$$

(Here we use the fact that $\lim_{t \to \infty} \ln t = +\infty$, which we established in Chapter 4.) Thus the improper integral diverges and, according to the Integral Test, so does the harmonic series. $\square$

The harmonic series

$$\sum_{k=1}^{\infty} \frac{1}{k} = 1 + \frac{1}{2} + \frac{1}{3} + \frac{1}{4} + \cdots$$

diverges.

Example 2 Determine whether the series $\displaystyle\sum_{k=1}^{\infty} \frac{1}{(k+1)^2} = \frac{1}{4} + \frac{1}{9} + \frac{1}{16} + \cdots$ converges.

Solution: The function corresponding to the general term $a_k = \dfrac{1}{(k+1)^2}$ is $f(x) = \dfrac{1}{(x+1)^2}$, which is continuous and decreasing on $[1, \infty)$. Since

$$\int_1^{\infty} \frac{1}{(x+1)^2}\,dx = \lim_{t\to\infty} \int_1^t (x+1)^{-2}\,dx$$

$$= \lim_{t\to\infty} -(x+1)^{-1}\Big]_1^t$$

$$= \lim_{t\to\infty} \left(\frac{1}{2} - \frac{1}{t+1}\right)$$

$$= \frac{1}{2}$$

the improper integral converges. Thus, by the Integral Test, the series $\displaystyle\sum_{k=1}^{\infty} \frac{1}{(k+1)^2}$ also converges. □

REMARK: Note, however, that we do not claim that the *sum* of the series is $\dfrac{1}{2}$. That would be incorrect. The series and the improper integral have the same convergence properties, but they do not have the same values, in general. (Can you explain why this is so, using Figures 3.1 and 3.2?)

Example 3 A **p-series** is a series of the form

$$\sum_{k=1}^{\infty} \frac{1}{k^p} = 1 + \frac{1}{2^p} + \frac{1}{3^p} + \frac{1}{4^p} + \cdots, \qquad p > 0.$$

Determine for which values of p a p-series converges.

Strategy

Identify an appropriate function f.

Verify that f satisfies the requirements of the Integral Test.

Solution

We use the function $f(x) = \dfrac{1}{x^p} = x^{-p}$ where p is constant. Since

$$f'(x) = -px^{-p-1} = \frac{-p}{x^{p+1}}$$

is negative for all $x > 0$, the function f is decreasing on $[1, \infty)$. Thus the Integral Test applies.

Set up the improper integral and evaluate it, carrying p along as a constant.

Now

$$\int_1^\infty x^{-p}\, dx = \lim_{t \to \infty} \int_1^t x^{-p}\, dx$$

$$= \lim_{t \to \infty} \left. \frac{x^{-p+1}}{1-p} \right]_1^t, \qquad p \neq 1$$

$$= \left\{ \lim_{t \to \infty} \frac{t^{1-p}}{1-p} \right\} - \frac{1}{1-p}.$$

Determine the values of p for which the improper integral converges.
Apply the Integral Test.

This limit will exist only if the exponent for t is negative; that is, only if $1 - p < 0$, which gives $p > 1$.

Thus the improper integral (hence, the series) converges if $p > 1$ and diverges otherwise. (Note that the case $p = 1$, excluded above, yields the harmonic series, which diverges.) This result is sufficiently important that we summarize it below. □

The p-series

$$\sum_{k=1}^\infty \frac{1}{k^p} = 1 + \frac{1}{2^p} + \frac{1}{3^p} + \frac{1}{4^p} + \cdots, \qquad p > 0$$

converges only if $p > 1$.

As we remarked in Section 9.2, adding or subtracting a finite number of terms doesn't change the convergence properties of an infinite series. If the starting index is not 1, then we simply compare the series to the improper integral with the appropriate lower limit; for example

$$\sum_{k=3}^\infty a_k \text{ converges if and only if } \int_3^\infty f(x)\, dx \text{ converges.}$$

For this reason, when we are discussing only convergence rather than the actual value of the series, we often omit the index and write just $\Sigma\, a_k$.

The Comparison Test

Like the Integral Test, the Comparison Test is motivated by area considerations. However, instead of comparing a series with an improper integral, we compare it with another series.

THEOREM 3
Comparison Test

Let $\Sigma\, a_k$ and $\Sigma\, b_k$ be a series with positive terms. If

$$0 < a_k \leq b_k$$

for each $k = 1, 2, 3, \ldots$, then:

(a) If $\Sigma\, b_k$ converges, so does $\Sigma\, a_k$.
(b) If $\Sigma\, a_k$ diverges, so does $\Sigma\, b_k$.

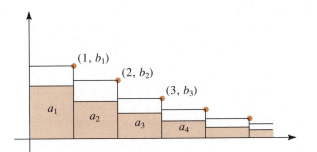

Figure 3.3 Comparison Test.

In other words, a positive series in which each term is *smaller* than the corresponding term of a *convergent* series must also converge. And a series in which each term is *larger* than the corresponding term of a positive *divergent* series must also diverge.

Figure 3.3 conveys the idea underlying the proof of Theorem 3, which we omit. If $\Sigma\, b_k$ converges, the area of the rectangles corresponding to the terms $b_1, b_2, \ldots$ is finite. Since the rectangles corresponding to the terms $a_1, a_2, \ldots$ lie inside those corresponding to the terms $b_1, b_2, \ldots$, the area of these rectangles must also be finite, so $\Sigma\, a_k$ converges. Conversely, if $\Sigma\, a_k$ diverges, the area enclosed by the rectangles corresponding to $a_1, a_2, \ldots$ is infinite. Thus the (larger) area associated with $b_1, b_2, \ldots$, must be infinite, so $\Sigma\, b_k$ diverges.

The following examples illustrate how the Comparison Test is used.

Example 4 The series $\displaystyle\sum \frac{1}{1 + k^3}$ converges, by comparison with the convergent p-series $\displaystyle\sum \frac{1}{k^3}$ since

$$\frac{1}{1 + k^3} < \frac{1}{k^3}. \qquad \square$$

Example 5 The series $\displaystyle\sum \frac{1 + \ln k}{k}$ diverges by comparison with the (divergent) harmonic series $\displaystyle\sum \frac{1}{k}$ since

$$\frac{1 + \ln k}{k} > \frac{1}{k} \qquad \text{if } k > 1. \qquad \square$$

Example 6 To determine whether the series $\displaystyle\sum \frac{\sqrt{k} - 1}{k^2 + 2}$ converges we make the comparison.

$$\frac{\sqrt{k} - 1}{k^2 + 2} < \frac{\sqrt{k}}{k^2 + 2} < \frac{\sqrt{k}}{k^2} = \frac{1}{k^{3/2}}, \qquad k \geq 1.$$

Since $\sum \dfrac{1}{k^{3/2}}$ is a p-series with $p = \dfrac{3}{2} > 1$, it converges. Thus $\sum \dfrac{\sqrt{k} - 1}{k^2 + 2}$ also con-

verges, by the Comparison Test. □

REMARK 1: Successful application of the Comparison Test to the series $\sum a_k$ depends on finding another series $\sum b_k$ with which you can make a useful comparison. If you suspect that $\sum a_k$ converges, you should look for a *convergent series* $\sum b_k$ with $a_k \leq b_k$. But if you suspect $\sum a_k$ diverges, look for a divergent series $\sum b_k$ with $a_k \geq b_k$.

REMARK 2: It is important to remember that the Integral Test and the Comparison Test apply only to *series with positive terms*. If a series contains negative terms, these tests do not apply.

The Limit Comparison Test

The following variation on the basic Comparison Test is often easier to apply.

THEOREM 4
Limit Comparison Test

Let $\sum a_k$ and $\sum b_k$ be series with $a_k > 0$, $b_k > 0$ for all $k = 1, 2, \ldots$. If the limit

$$\rho = \lim_{k \to \infty} \frac{a_k}{b_k}$$

exists, and $\rho \neq 0$, then either both series converge or both series diverge.

The Limit Comparison Test states that if the ratio of the general terms of two series tends to a positive limit, then the two series have the same convergence property—either both converge or both diverge. This is most useful in determining the convergence of series whose general terms are rational functions or quotients of functions involving rational exponents.

Before proving this theorem, we consider several examples of its use.

Example 7 The series $\displaystyle\sum_{k=1}^{\infty} \frac{1}{2k + 1}$ diverges, as can be shown by Theorem 4. Compar-

ing this series with the harmonic series $\displaystyle\sum b_k = \sum_{k=1}^{\infty} \frac{1}{k}$ we find that

$$\rho = \lim_{k \to \infty} \left(\frac{\dfrac{1}{2k + 1}}{\dfrac{1}{k}} \right) = \lim_{k \to \infty} \frac{k}{2k + 1} = \frac{1}{2} > 0.$$

Since the limit $\rho > 0$ exists, and since the harmonic series diverges, so must the given series. □

Example 8 To apply the Limit Comparison Test to the series

$$\sum_{k=1}^{\infty} \frac{\sqrt{k} + 2}{k^2 + k + 1},$$

we observe that the general term is a quotient containing a highest exponent of 1/2 in its numerator and a highest exponent of 2 in its denominator. This suggests a comparison with the series whose general term is $b_k = \dfrac{k^{1/2}}{k^2} = \dfrac{1}{k^{3/2}}$. We therefore take $\Sigma\, b_k$ to be the

p-series $\sum_{k=1}^{\infty} \dfrac{1}{k^{3/2}}$. We obtain

$$\rho = \lim_{k \to \infty} \left(\frac{\dfrac{\sqrt{k} + 2}{k^2 + k + 1}}{\dfrac{1}{k^{3/2}}} \right) = \lim_{k \to \infty} \left(\frac{k^2 + 2k^{3/2}}{k^2 + k + 1} \right) = 1 > 0.$$

We may now conclude that since the p-series $\sum_{k=1}^{\infty} \dfrac{1}{k^{3/2}}$ converges, so does the series

$$\sum_{k=1}^{\infty} \frac{\sqrt{k} + 2}{k^2 + k + 1}. \qquad \qquad \square$$

Example 9 Determine whether the series $\displaystyle\sum_{k=1}^{\infty} \sin\left(\frac{\pi}{k}\right)$ converges.

Solution: Let's systematically try each of the techniques available thus far:

(a) The Integral Test is not appropriate, since the improper integral $\displaystyle\int_{1}^{\infty} \sin\left(\frac{\pi}{x}\right) dx$ cannot easily be evaluated.

(b) The basic Comparison Test depends on our ability to choose a valid comparison series. We do have the comparison

$$0 < \sin\left(\frac{\pi}{k}\right) < \frac{\pi}{k}, \qquad k = 1, 2, 3, \ldots$$

but the Comparison Test gives no information since the series $\displaystyle\sum \frac{\pi}{k} = \pi \sum \frac{1}{k}$ diverges.

However, the Limit Comparison Test may be applied by recalling that we have previously established the limit

$$\rho = \lim_{k \to \infty} \frac{\sin\left(\dfrac{\pi}{k}\right)}{\left(\dfrac{\pi}{k}\right)} = \lim_{x \to 0^+} \frac{\sin x}{x} = 1.$$

Thus by Theorem 4, the series $\sum_{k=1}^{\infty} \sin\left(\dfrac{\pi}{k}\right)$ has the same convergence property as does the

series $\sum_{k=1}^{\infty} \dfrac{\pi}{k}$. That is, $\sum_{k=1}^{\infty} \sin\left(\dfrac{\pi}{k}\right)$ diverges. □

Proof of the Limit Comparison Test: As we did for the (basic) Comparison Test, we present only a sketch of the proof in order to focus attention on the basic idea.

If we assume that $\rho = \lim\limits_{k \to \infty} \dfrac{a_k}{b_k} > 0$, then there must exist positive numbers L and M and a positive integer K, so that

$$0 < L < \frac{a_k}{b_k} < M$$

for all $k > K$. Multiplying through by $b_k > 0$ gives the statements

$$a_k < Mb_k \tag{3}$$

and

$$a_k > Lb_k. \tag{4}$$

Since a_k, b_k, L, and M are positive numbers, we may use the basic Comparison Test to conclude from inequality (3) that $\Sigma\, a_k$ converges if $\Sigma\, b_k$ converges and, from inequality (4), that $\Sigma\, a_k$ diverges if $\Sigma\, b_k$ diverges. These two conclusions establish the statement of the Limit Comparison Test.

A NECESSARY CONDITION: The following is a condition that must be satisfied by any convergent series.

THEOREM 5 **(Necessary** **Condition)**	If the infinite series $\Sigma\, a_k$ converges, then $\lim\limits_{k \to \infty} a_k = 0$.

Theorem 5 says that in order for $\Sigma\, a_k$ to converge, the size of the general term a_k must approach zero as $k \to \infty$. But although necessary, this condition is not strong enough to guarantee that a series converges, as the following example shows.

Example 10 The harmonic series $\sum_{k=1}^{\infty} \dfrac{1}{k}$ diverges. (See Example 1.) However, the harmonic series does satisfy the criterion of Theorem 3, since

$$\lim_{k \to \infty} a_k = \lim_{k \to \infty} \frac{1}{k} = 0.$$
□

What, then, is the use of Theorem 5? The answer is that Theorem 5 can identify a divergent series when $\lim\limits_{k \to \infty} a_k \neq 0$. This will save you the work of applying the Integral

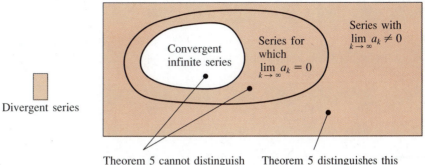

Theorem 5 cannot distinguish   Theorem 5 distinguishes this
between these two cases.   case from those with $\lim\limits_{k \to \infty} a_k = 0$.

Figure 3.4 Diagram on the applicability of Theorem 5.

Test or the Comparison Test for these series.

Example 11 Determine whether the series $\sum\limits_{k=1}^{\infty} \dfrac{k}{5+k}$ converges.

Solution: Checking the limit as $k \to \infty$ of the general term, we find that

$$\lim_{k \to \infty} a_k = \lim_{k \to \infty} \frac{k}{k+5}$$

$$= \lim_{k \to \infty} \left(\frac{k}{k+5}\right)\left(\frac{\frac{1}{k}}{\frac{1}{k}}\right)$$

$$= \lim_{k \to \infty} \frac{1}{1+\dfrac{5}{k}}$$

$$= 1.$$

Since this limit is not zero, the series cannot converge, according to Theorem 5. □

Figure 3.4 summarizes how Theorem 5 is used to detect divergent series.

Exercise Set 9.3

In Exercises 1–8 use the Integral Test to determine whether the series converges or diverges.

1. $\sum \dfrac{1}{k+2}$

2. $\sum \dfrac{1}{3k+5}$

3. $\sum \dfrac{k}{(k^2+1)^{3/2}}$

4. $\sum \dfrac{k}{k^2+1}$

5. $\sum \dfrac{1}{k \ln k}$

6. $\sum k e^{-k^2}$

7. $\sum k^2 e^{-k^3}$ **8.** $\sum ke^k$

25. $\sum \dfrac{\cos^2 \pi k}{k^2}$ **26.** $\sum \dfrac{k+1}{k}$

In Exercises 9–16 use the Comparison Test to determine whether the series converges or diverges.

27. $\sum \dfrac{e^{\sqrt{k}}}{\sqrt{k}}$ **28.** $\sum \cos\left(\dfrac{\pi k}{4}\right)$

9. $\sum \dfrac{3}{k}$ **10.** $\sum \dfrac{1}{1 + k^2}$

29. $\sum \dfrac{2k^2 - 2}{k^4 + 3k}$ **30.** $\sum \dfrac{k^2 - 2}{k^2 + 2}$

11. $\sum \dfrac{1}{2 + k^{3/2}}$ **12.** $\sum \dfrac{\sqrt{k}}{1 + k^3}$

31. $\sum \dfrac{k}{\sqrt{1 + k^2}}$ **32.** $\sum \dfrac{4}{1 + k^2}$

13. $\sum \dfrac{5 + k}{k^2}$ **14.** $\sum \dfrac{\sqrt{k}}{1 + k}$

33. $\sum \dfrac{1}{\sqrt{4k(k+1)}}$ **34.** $\sum \dfrac{2k}{\sqrt{k^3 - 1}}$

15. $\sum \dfrac{k}{1 + k^3}$ **16.** $\sum \dfrac{3}{\sqrt{k} + 2}$

35. $\sum \dfrac{2^k}{5 + 3^{k+1}}$ **36.** $\sum \dfrac{2}{(k + 3)^{3/2}}$

In Exercises 17–20 use the Limit Comparison Test to determine whether the given series converge or diverge.

37. $\sum \dfrac{\ln k}{k}$ **38.** $\sum \dfrac{\ln k}{1 + \ln k}$

17. $\sum \dfrac{k + 3}{2k^2 + 1}$ **18.** $\sum \dfrac{k^2 - 4}{k^3 + k + 5}$

39. $\sum \dfrac{1 + k}{1 + k^2}$ **40.** $\sum \dfrac{k^2}{k^4 + 2k + 1}$

19. $\sum \dfrac{k + \sqrt{k}}{k + k^3}$ **20.** $\sum \dfrac{2k + 2}{\sqrt{k^3 + 2}}$

41. $\sum \dfrac{2^k}{k + 5}$ **42.** $\sum \dfrac{2k^2}{\sqrt{k^3 + 3}}$

In Exercises 21–48 determine whether the series converges or diverges and state which test applies.

43. $\sum \dfrac{k}{(k + 2)2^k}$ **44.** $\sum \dfrac{\sin(\pi k)}{\sqrt{1 + k}}$

21. $\sum \dfrac{k^2 + 5}{k^2}$ **22.** $\sum \dfrac{3^k}{3 + 5^k}$

45. $\sum \dfrac{2k + 2}{\sqrt{k^3 + 2}}$ **46.** $\sum \dfrac{\ln k}{k^3}$

23. $\sum \dfrac{k}{4 + k^3}$ **24.** $\sum \dfrac{1}{\sqrt{k} + 1}$

47. $\sum \dfrac{k}{2 + \ln k}$ **49.** $\sum \dfrac{k + 3}{(k + 2)2^k}$

9.4 THE RATIO TEST AND ABSOLUTE CONVERGENCE

The Ratio Test is another test for convergence of series with positive terms. Rather than comparing the series with an integral or another series, the ratio test uses the limit of the ratios of successive terms of the series itself.

THEOREM 6

Let $\Sigma\, a_k$ be a series with positive terms. Assume that the limit

$$\rho = \lim_{k \to \infty} \frac{a_{k+1}}{a_k}$$

exists. Then:

   (i) If $0 \le \rho < 1$ the series $\Sigma\, a_k$ converges.
  (ii) If $\rho > 1$ the series $\Sigma\, a_k$ diverges.
 (iii) If $\rho = 1$ there is no conclusion.

To use the Ratio Test on a series with positive terms we do the following:

(a) Identify the general term a_k, involving the index k.
(b) Find the expression a_{k+1} by replacing k by $k + 1$ in the general term a_k.
(c) Form the ratio $\dfrac{a_{k+1}}{a_k}$ and simplify this expression as much as possible.
(d) Find the limit $\rho = \lim\limits_{k \to \infty} \dfrac{a_{k+1}}{a_k}$.
(e) Apply the Ratio Test, according to the number ρ.

In applying the Ratio Test, we must remember that the outcome $\rho = 1$ leads to no conclusion. In this case the series may either converge or diverge, and you must try another test.

Example 1 For the series $\displaystyle\sum_{k=1}^{\infty} \frac{k}{2^k} = \frac{1}{2} + \frac{2}{2^2} + \frac{3}{2^3} + \cdots$ the general term is

$$a_k = \frac{k}{2^k}$$

so

$$a_{k+1} = \frac{k + 1}{2^{k+1}}.$$

The limit ρ is therefore

$$\rho = \lim_{k \to \infty} \frac{a_{k+1}}{a_k} = \lim_{k \to \infty} \frac{\left(\dfrac{k + 1}{2^{k+1}}\right)}{\left(\dfrac{k}{2^k}\right)}$$

$$= \lim_{k \to \infty} \left(\frac{k + 1}{2^{k+1}}\right)\left(\frac{2^k}{k}\right) \qquad \text{(multiply by reciprocal of denominator)}$$

$$= \lim_{k \to \infty} \frac{1}{2}\left(\frac{k+1}{k}\right) \qquad \text{(rearrange, and cancel } 2^k \text{ from numerator and denominator)} \quad .$$

$$= \lim_{k \to \infty} \frac{1}{2}\left(1 + \frac{1}{k}\right)$$

$$= \frac{1}{2} \qquad \left(\text{since } \lim_{k \to \infty} \frac{1}{k} = 0\right).$$

Thus since $\frac{1}{2} < 1$, the Ratio Test guarantees that the series converges. □

Example 2 For the geometric series $\Sigma\, x^k$ the general term is

$$a_k = x^k$$

so

$$a_{k+1} = x^{k+1}.$$

The ratio ρ is therefore

$$\rho = \lim_{k \to \infty} \frac{a_{k+1}}{a_k} = \lim_{k \to \infty} \frac{x^{k+1}}{x^k} = \lim_{k \to \infty} x = x.$$

Thus the Ratio Test confirms some of what we already know about the convergence of the geometric series when $x \geq 0$: $\Sigma\, x^k$ converges when $x = \rho < 1$ and diverges when $x = \rho > 1$. (In case $\rho = x = 1$ the Ratio Test is inconclusive, but we already know that the series $\Sigma\, 1^k = 1 + 1 + 1 + \cdots$ diverges.) □

The Ratio Test is often useful in dealing with series involving factorial expressions. Recall that $k!$ (k factorial) means the product of all integers from 1 to k inclusive:

$$k! = k(k-1)(k-2) \cdot \cdots \cdot 2 \cdot 1.$$

Thus $3! = 3 \cdot 2 \cdot 1 = 6$, $5! = 5 \cdot 4 \cdot 3 \cdot 2 \cdot 1 = 120$, etc.

Because of their common factors, ratios of factorials can often be simplified greatly. For example

$$\frac{(k+1)!}{k!} = \frac{(k+1)(k)(k-1) \cdot \cdots \cdot 2 \cdot 1}{k(k-1) \cdot \cdots \cdot 2 \cdot 1} = k + 1$$

$$\frac{(k+2)!}{k!} = \frac{(k+2)(k+1)(k)(k-1) \cdot \cdots \cdot 2 \cdot 1}{k(k-1) \cdot \cdots \cdot 2 \cdot 1} = (k+2)(k+1)$$

etc. The next two examples involve such factorials.

Example 3 Determine whether the series $\displaystyle\sum_{k=1}^{\infty} \frac{k+2}{k!}$ converges.

Strategy

Identify a_k.

Solution

For this series the general term is

$$a_k = \frac{k+2}{k!}$$

Find a_{k+1} by replacing each k with $(k + 1)$ and simplify.

so

$$a_{k+1} = \frac{(k + 1) + 2}{(k + 1)!} = \frac{k + 3}{(k + 1)!}.$$

The ratio limit is therefore

Form ratio and simplify, using the fact that

$$\frac{k!}{(k + 1)!} = \frac{k!}{(k + 1)k!}$$

$$= \frac{1}{k + 1}.$$

$$\rho = \lim_{k \to \infty} \frac{a_{k+1}}{a_k} = \lim_{k \to \infty} \frac{\left(\dfrac{k + 3}{(k + 1)!} \right)}{\left(\dfrac{k + 2}{k!} \right)}$$

$$= \lim_{k \to \infty} \frac{k + 3}{(k + 1)!} \cdot \frac{k!}{k + 2}$$

$$= \lim_{k \to \infty} \left(\frac{k + 3}{k + 2} \right) \left[\frac{k!}{(k + 1)!} \right]$$

$$= \lim_{k \to \infty} \left(\frac{k + 3}{k + 2} \right) \left(\frac{1}{k + 1} \right)$$

Evaluate limit using fact that

$$\lim_{k \to \infty} \frac{k + 3}{k + 2} = 1.$$

$$= 1 \cdot 0$$

$$= 0.$$

Apply conclusions of Ratio Test.

Thus $\rho = 0 < 1$, so the series converges, according to the Ratio Test.

$\square$

Example 4 Determine whether the series $\sum \dfrac{k^k}{k!}$ converges.

Strategy

Identify a_k.
Find a_{k+1} by replacing each k by $(k+1)$ in a_k.

Solution

Here the general term is

$$a_k = \frac{k^k}{k!}.$$

so

$$a_{k+1} = \frac{(k + 1)^{k+1}}{(k + 1)!}$$

and the ratio limit is

Form the ratio $\dfrac{a_{k+1}}{a_k}$ and simplify.

$$\rho = \lim_{k \to \infty} \frac{\dfrac{(k + 1)^{k+1}}{(k + 1)!}}{\dfrac{k^k}{k!}}$$

$$= \lim_{k \to \infty} \frac{(k + 1)^{k+1}}{(k + 1)!} \cdot \frac{k!}{k^k}$$

$$= \lim_{k \to \infty} \frac{k!}{(k + 1)!} \cdot \frac{(k + 1)^{k+1}}{k^k}$$

Use fact that

$$\frac{k!}{(k+1)!} = \frac{1}{k+1}.$$

Use fact that

$$\lim_{k \to \infty} \left(1 + \frac{1}{k}\right)^k = e$$

from Chapter 4 to evaluate limit.

Apply conclusions of Ratio Test.

$$= \lim_{k \to \infty} \frac{1}{k+1} \cdot \frac{(k+1)^{k+1}}{k^k}$$

$$= \lim_{k \to \infty} 1 \cdot \left(\frac{k+1}{k}\right)^k$$

$$= \lim_{k \to \infty} \left(1 + \frac{1}{k}\right)^k$$

$$= e \approx 2.71828 \ldots .$$

Since $\rho = e > 1$ the series diverges. $\quad\square$

Example 5 If we attempt to use the Ratio Test on the series $\sum \dfrac{k+1}{k^2}$, we find that the ratio limit is

$$\rho = \lim_{k \to \infty} \frac{a_{k+1}}{a_k} = \lim_{k \to \infty} \frac{\dfrac{[(k+1)+1]}{(k+1)^2}}{\dfrac{k+1}{k^2}}$$

$$= \lim_{k \to \infty} \frac{k+2}{(k+1)^2} \cdot \frac{k^2}{k+1}$$

$$= \lim_{k \to \infty} \frac{k^2(k+2)}{(k+1)^3}$$

$$= \lim_{k \to \infty} \frac{k^3 + 2k^2}{k^3 + 3k^2 + 3k + 1}$$

$$= 1.$$

Thus since $\rho = 1$, the Ratio Test is inconclusive. However, we can compare the general term $a_k = \dfrac{k+1}{k^2}$ with the general term $b_k = \dfrac{1}{k}$ of the harmonic series $\sum \dfrac{1}{k}$ as follows.

$$a_k = \frac{k+1}{k^2} > \frac{k}{k^2} = \frac{1}{k} = b_k.$$

Since the harmonic series diverges, the Comparison Test determines that the series $\sum \dfrac{k+1}{k^2}$ also diverges, as its general term is larger than that of the harmonic series. $\quad\square$

Series with Negative Terms

When an infinite series $\sum a_k$ contains negative terms, we cannot apply the integral, comparison, or ratio tests directly, since they are valid only for series with positive terms. There are, however, two types of series with negative terms for which there are straightforward tests for convergence.

Alternating Series

An **alternating series** is a series such as

$$\sum_{k=0}^{\infty} (-1)^k a_k = a_0 - a_1 + a_2 - a_3 + a_4 - \cdots$$

where $a_k > 0$ for all k. That is, the terms of an alternating series alternate signs. Examples of alternating series are:

(i) $\displaystyle\sum_{k=1}^{\infty} \frac{(-1)^k}{k} = -1 + \frac{1}{2} - \frac{1}{3} + \frac{1}{4} - \frac{1}{5} + \cdots$ (here k begins at 1)

(ii) $\displaystyle\sum_{k=0}^{\infty} \frac{(-1)^k}{1 + k^2} = 1 - \frac{1}{2} + \frac{1}{5} - \frac{1}{10} + \cdots$ (here k begins at 0)

(iii) $\displaystyle\sum_{k=0}^{\infty} \frac{\cos k\pi}{\sqrt{k + 1}} = 1 - \frac{1}{\sqrt{2}} + \frac{1}{\sqrt{3}} - \frac{1}{\sqrt{4}} + \cdots$ (here k begins at 0).

The following theorem gives a very simple criterion for identifying convergent alternating series.

THEOREM 7
Alternating Series Test

Let $a_k > 0$ for all k. The alternating series $\Sigma(-1)^k a_k$ converges if both of the following hold:

(i) $a_k > a_{k+1}$ for all k larger than some fixed integer K.
(ii) $\displaystyle\lim_{k \to \infty} a_k = 0$.

Condition (i) says that the absolute values of the terms eventually must be decreasing, and condition (ii) says that, in the limit, these terms must decrease to zero. Note that each of the series in Examples (i)–(iii) above meet both criteria, so each of these is a convergent alternating series.

Example 6 Determine whether the alternating harmonic series

$$\sum_{k=1}^{\infty} \frac{(-1)^{k+1}}{k} = 1 - \frac{1}{2} + \frac{1}{3} - \frac{1}{4} + \cdots$$

converges.

Solution: This series is an alternating series of the form $\Sigma(-1)^{k+1} a_k$ with $a_k = \dfrac{1}{k}$. Since

$$a_k = \frac{1}{k} > \frac{1}{k + 1} = a_{k+1}$$

the terms of a_k are decreasing. Moreover

$$\lim_{k \to \infty} a_k = \lim_{k \to \infty} \frac{1}{k} = 0.$$

Thus both requirements of the alternating series test are satisfied, so the series converges. □

Example 7 Does the series $\displaystyle\sum_{k=0}^{\infty} \frac{(-1)^k k^2}{4 + k^2}$ converge?

Solution: This series has the form $\sum (-1)^k a_k$ with $a_k = \dfrac{k^2}{1 + k^2}$. Since

$$\lim_{k \to \infty} a_k = \lim_{k \to \infty} \frac{k^2}{1 + k^2}$$

$$= \lim_{k \to \infty} \left(\frac{k^2}{1 + k^2} \right) \left(\frac{\dfrac{1}{k^2}}{\dfrac{1}{k^2}} \right)$$

$$= \lim_{k \to \infty} \frac{1}{\dfrac{1}{k^2} + 1}$$

$$= 1$$

the second condition of the alternating series test fails. This series diverges because the necessary condition of Theorem 5 fails. $\qquad\square$

Figure 4.1 For the alternating series $\sum (-1)^k a_k$, the partial sums oscillate back and forth on the number line.

Figure 4.1 illustrates the reason why an infinite series that satisfies the requirements of Theorem 7 must converge. Because each term of the series is of an opposite sign from the one preceding it, the partial sums for the series oscillate back and forth on the number line. Because the absolute values of the terms decrease, the distance between successive partial sums decreases. Finally, since the size of the terms of the series approaches zero, the distance between successive partial sums approaches zero in the limit. This means that the sequence of partial sums converges.

Absolute Convergence

When an infinite series $\sum a_k$ contains negative terms but is not an alternating series, we may test the series for *absolute convergence*, which is defined as follows.

DEFINITION 4

The series $\sum a_k$ is said to **converge absolutely** if the series of absolute values $\sum |a_k|$ converges.

In other words, the series $\Sigma\, a_k$ converges *absolutely* if, by changing the signs of all negative terms, we obtain a convergent series.

Example 8 The series $\displaystyle\sum_{k=1}^{\infty} \frac{(-1)^{k+1}}{k^2} = 1 - \frac{1}{4} + \frac{1}{9} - \frac{1}{16} + \cdots$ converges absolutely since in this case

$$\sum |a_k| = \sum \left| \frac{(-1)^{k+1}}{k^2} \right| = \sum \frac{1}{k^2}$$

which is a convergent p-series $(p = 2)$. □

Example 9 The series $\displaystyle\sum_{k=1}^{\infty} (-1)^{k+1} = 1 - 1 + 1 - 1 + 1 - \cdots$ does *not* converge absolutely since

$$\sum \left| (-1)^{k+1} \right| = \sum 1 = 1 + 1 + 1 + \cdots$$

which is a divergent series. □

It is important to note the relationship between absolute convergence (that is, convergence of the series $\Sigma\, |a_k|$) and convergence of the series $\Sigma\, a_k$ itself. The following theorem states that the series $\Sigma\, a_k$ converges if it is absolutely convergent.

THEOREM 8   If the series $\Sigma\, |a_k|$ converges, so does the series $\Sigma\, a_k$. That is, every absolutely convergent series converges.

Theorem 8 is simple to apply. If the series $\Sigma\, a_k$ contains negative terms, we test the positive series $\Sigma\, |a_k|$ for convergence. If $\Sigma\, |a_k|$ converges, so does $\Sigma\, a_k$. However, if $\Sigma\, |a_k|$ diverges, we can draw no conclusion about $\Sigma\, a_k$.

Example 10 To test the series $\displaystyle\sum_{k=0}^{\infty} \frac{(-1)^k k^2}{(k+1)!}$ for convergence, we apply the ratio test to the series

$$\sum_{k=0}^{\infty} \left| \frac{(-1)^k k^2}{(k+1)!} \right| = \sum_{k=0}^{\infty} \frac{k^2}{(k+1)!}.$$

We obtain the ratio limit

$$\rho = \lim_{k \to \infty} \frac{|a_{k+1}|}{|a_k|} = \lim_{k \to \infty} \frac{\dfrac{(k+1)^2}{[(k+1)+1]!}}{\dfrac{k^2}{(k+1)!}}$$

$$= \lim_{k \to \infty} \frac{(k+1)^2}{(k+2)!} \cdot \frac{(k+1)!}{k^2}$$

$$= \lim_{k \to \infty} \frac{(k+1)!}{(k+2)!} \cdot \frac{(k+1)^2}{k^2}$$

$$= \lim_{k \to \infty} \frac{1}{k+2}\left(\frac{k+1}{k}\right)^2$$

$$= \lim_{k \to \infty} \frac{1}{k+2}\left(1 + \frac{1}{k}\right)^2$$

$$= 0.$$

Thus since $\rho = 0 < 1$, the series $\displaystyle\sum_{k=0}^{\infty} \left|\frac{(-1)^k k^2}{(k+1)!}\right|$ converges, so the original series

$\displaystyle\sum_{k=0}^{\infty} \frac{(-1)^k k^2}{(k+1)!}$ converges absolutely. □

Example 11 Determine whether the series

$$\sum_{k=0}^{\infty} \frac{\cos \pi k}{2^k} = 1 - \frac{1}{2} + \frac{1}{4} - \frac{1}{8} + \cdots$$

converges.

Solution: Testing for absolute convergence, we find that

$$\sum_{k=0}^{\infty} \left|\frac{\cos \pi k}{2^k}\right| = \sum_{k=0}^{\infty} \frac{|\cos \pi k|}{2^k} = \sum_{k=0}^{\infty} \frac{1}{2^k} = \sum_{k=0}^{\infty} \left(\frac{1}{2}\right)^k .$$

which converges since it is a geometric series. The series $\displaystyle\sum_{k=0}^{\infty} \frac{\cos \pi k}{2^k}$ therefore converges

absolutely. (The Alternating Series Test also can be used to show that this series converges.) □

If a series does not converge absolutely, it still may converge. (That is, $\Sigma\, a_k$ may converge even though $\Sigma\, |a_k|$ diverges.) Such series are said to be *conditionally* convergent. Our last example is one such series.

Example 12 For the series $\displaystyle\sum_{k=1}^{\infty} \frac{(-1)^{k+1}}{k} = 1 - \frac{1}{2} + \frac{1}{3} - \frac{1}{4} + \frac{1}{5} - \cdots$ the series of absolute values is just the harmonic series

$$\sum_{k=1}^{\infty} \left| \frac{(-1)^{k+1}}{k} \right| = \sum_{k=1}^{\infty} \frac{1}{k} = 1 + \frac{1}{2} + \frac{1}{3} + \frac{1}{4} + \cdots$$

which we have shown to diverge. Thus the series $\displaystyle\sum_{k=1}^{\infty} \frac{(-1)^{k+1}}{k}$ does not converge absolutely. But this series does converge conditionally, as the Alternating Series Test showed in Example 6. □

Exercise Set 9.4

In Exercises 1–12 use the Ratio Test, where applicable, to determine whether the given series converges or diverges.

1. $\displaystyle\sum \frac{2^k}{k+2}$

2. $\displaystyle\sum \frac{2^k}{k!}$

3. $\displaystyle\sum k^2 e^{-k}$

4. $\displaystyle\sum \frac{k!}{2^k}$

5. $\displaystyle\sum \frac{\ln k}{e^k}$

6. $\displaystyle\sum \frac{(3k)!}{(k!)^2}$

7. $\displaystyle\sum \frac{k+2}{1+k^2}$

8. $\displaystyle\sum \frac{\sqrt{k}}{k!}$

9. $\displaystyle\sum \frac{k!}{ke^k}$

10. $\displaystyle\sum \frac{k^2 e^k}{k!}$

11. $\displaystyle\sum \frac{e^{3k}}{k!}$

12. $\displaystyle\sum \frac{1}{(k+1)!}$

In Exercises 13–18 determine whether the alternating series converges.

13. $\displaystyle\sum_{k=0}^{\infty} \frac{(-1)^k}{k+2} = \frac{1}{2} - \frac{1}{3} + \frac{1}{4} - \frac{1}{5} + \cdots$

14. $\displaystyle\sum_{k=1}^{\infty} \frac{(-1)^{k+1}}{k^2} = 1 - \frac{1}{4} + \frac{1}{9} - \frac{1}{16} + \cdots$

15. $\displaystyle\sum_{k=1}^{\infty} \frac{\cos \pi k}{k} = -1 + \frac{1}{2} - \frac{1}{3} + \frac{1}{4} - \cdots$

16. $\displaystyle\sum_{k=1}^{\infty} \frac{(-1)^k k}{k+5} = -\frac{1}{5} + \frac{2}{6} - \frac{3}{7} + \frac{4}{8} - \cdots$

17. $\displaystyle\sum_{k=1}^{\infty} \frac{(-1)^k k^2}{\sqrt{k+1}} = -\frac{1}{\sqrt{2}} + \frac{4}{\sqrt{3}} - \frac{9}{\sqrt{4}} + \frac{16}{\sqrt{5}} - \cdots$

18. $\displaystyle\sum_{k=1}^{\infty} \frac{(-1)^{k+1} k}{\sqrt{k+1}} = \frac{1}{\sqrt{2}} - \frac{2}{\sqrt{3}} + \frac{3}{\sqrt{4}} - \frac{4}{\sqrt{5}} + \cdots$

In Exercises 19–30 determine whether the series converges absolutely. If the series does not converge absolutely, determine whether it converges conditionally or diverges.

19. $\displaystyle\sum_{k=1}^{\infty} \frac{(-1)^k}{k^2}$

20. $\displaystyle\sum_{k=1}^{\infty} \frac{(-1)^k}{2k+1}$

21. $\displaystyle\sum_{k=1}^{\infty} \frac{(-1)^k \cdot k}{(k+1)!}$

22. $\displaystyle\sum_{k=1}^{\infty} \frac{(-1)^k \sqrt{k}}{k!}$

23. $\displaystyle\sum_{k=1}^{\infty} \frac{(-1)^k}{1+\sqrt{k}}$

24. $\displaystyle\sum_{k=1}^{\infty} \frac{\cos \pi k}{k}$

25. $\displaystyle\sum_{k=1}^{\infty} \frac{1}{k} \sin \frac{k\pi}{2}$

26. $\displaystyle\sum_{k=1}^{\infty} \frac{(-1)^k k^3}{2^k}$

27. $\displaystyle\sum_{k=1}^{\infty} \frac{(-1)^k \sqrt{k}}{1 + k^2}$

28. $\displaystyle\sum_{k=1}^{\infty} \frac{(-1)^k}{1 + k^2}$

29. $\displaystyle\sum_{k=1}^{\infty} \frac{(-1)^{k+1} k!}{5^k}$

30. $\displaystyle\sum_{k=2}^{\infty} \frac{(-1)^k \cdot k}{\ln \sqrt{k}}$

31. Let $\Sigma(-1)^k a_k$ be an alternating series with $a_k > a_{k+1}$ for all k. Suppose that $\displaystyle\lim_{k \to \infty} a_k = L$ with $L \neq 0$ (thus violating condition ii of the Alternating Series Test). Explain why this series cannot converge.

32. Complete the following steps in this intuitive explanation of why the Ratio Test is true.

a. Assume that $\rho < 1$, where $\rho = \displaystyle\lim_{k \to \infty} \frac{a_{k+1}}{a_k}$.

b. Let $\alpha = \dfrac{\rho + 1}{2}$. Explain why $\alpha < 1$.

c. Since $\rho = \displaystyle\lim_{k \to \infty} \frac{a_{k+1}}{a_k}$, explain why we must have

$$\frac{a_{k+1}}{a_k} < \alpha \text{ for all } k \text{ equal to or larger than some}$$

fixed integer, say, N.

d. Since $\dfrac{a_{N+1}}{a_N} < \alpha$ show that

$$a_{N+1} < \alpha a_N$$
$$a_{N+2} < \alpha a_{N+1} < \alpha^2 a_N$$
$$a_{N+3} < \alpha a_{N+2} < \alpha^2 a_{N+1} < \alpha^3 a_N$$

and, in general

$$a_{N+p} < \alpha^p a_N .$$

e. Explain why the series $\Sigma \, \alpha^k$ is a geometric series that converges.

f. Use the results of parts (d) and (e), together with the Comparison Test, to explain why the series $\Sigma \, a_k$ must converge when $\rho < 1$.

g. Find an analogous argument to show why the series $\Sigma \, a_k$ must diverge when $\rho > 1$.

9.5 TAYLOR POLYNOMIALS

We have seen many applications of the transcendental functions e^x, $\ln x$, and the trigonometric functions. As important as these functions are, however, at this point we have no clues as to how to calculate values of these functions other than resorting to a calculator, computer, or table of values. Even though such "aids" are readily available, our study of calculus would be incomplete if we did not raise the question of how the tables are constructed or how the calculator is programmed.

Taylor polynomials can be used to approximate values of transcendental functions to any desired degree of accuracy. We shall present the rudiments of Taylor polynomials here and discuss their important relationship to infinite Taylor series in the next two sections.

Approximating the Function $f(x) = e^x$

Figure 5.1 shows the graph of the function $f(x) = e^x$. It is easy to see that the value of this function at $x = 0$ is $f(0) = e^0 = 1$. But how might we approximate values of $f(x) = e^x$ for x near zero?

Figure 5.2 shows one approach—simply to approximate the values $f(x) = e^x$ by the constant value $P_0(x) = f(0) = 1$. That is, we could use the value $f(0) = e^0 = 1$ to approximate values of $f(x)$ for x near 0. It is easy to see that, except for x's very close to 0, this is not a very good approximation.

Figure 5.3 suggests a better way to approximate values of $f(x) = e^x$ near $a = 0$, that is, by using the first degree polynomial

$$P_1(x) = f(0) + f'(0)x$$
$$= 1 + x. \qquad \text{(For } f(x) = e^x, f'(0) = e^0 = 1.\text{)}$$

(1)

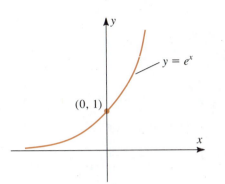

Figure 5.1 The function $f(x) = e^x$ has value $f(0) = 1$ at $x = 0$.

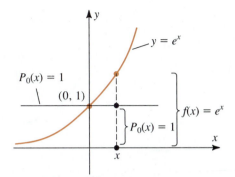

Figure 5.2 Approximating values $f(x) = e^x$ by the constant function $P_0(x) = f(0) = 1$.

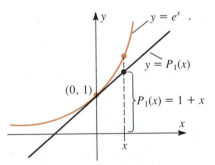

Figure 5.3 Approximating values of $f(x) = e^x$ by the first degree polynomial $P_1(x) = 1 + x$.

This function has the same value as $f(x) = e^x$ at $a = 0$, but it also has the same first derivative as f at $x = 0$. To see this, simply note that

$$P_1'(x) = f'(0) = 1$$

for all x, so $P_1'(0) = f'(0)$. In a sense, the graph of P_1 is ''shaped more like'' the graph of f than is the graph of P_0.

Figure 5.4 shows yet a better way to approximate $f(x)$ near $x = 0$: Use the second degree polynomial

$$P_2(x) = f(0) + f'(0)x + \frac{f''(0)}{2}x^2$$

$$= 1 + x + \frac{1}{2}x^2. \qquad \text{(For } f(x) = e^x, f''(0) = e^0 = 1.)$$

(2)

The polynomial $P_2(x)$ has the following properties:

(i) $P_2(0) = f(0)$.
(ii) Since $P_2'(x) = f'(0) + f''(0)x$,

$$P_2'(0) = f'(0).$$

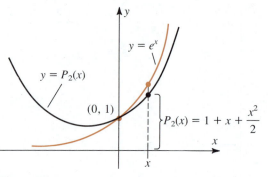

Figure 5.4 Approximating values of $f(x) = e^x$ by the second degree polynomial $P_2(x) = 1 + x + \dfrac{x^2}{2}$.

(iii) Since $P''_2(x) = f''(0)$,

$$P''_2(0) = f''(0).$$

That is, $P_2(x)$ and each of its first two derivatives have the same values as $f(x) = e^x$ and its first two derivatives at $x = 0$. The graph of P_2 is even more like the graph of f than its two predecessors.

Table 5.1 illustrates our claim that the values $P_0(x)$, $P_1(x)$, and $P_2(x)$ provide increasingly better approximations to the values $f(x) = e^x$ for x near zero. (Note that the closer x is chosen to zero, the better the approximation is.)

Table 5.1. Values of $f(x) = e^x$ and approximating polynomials $P_0(x)$, $P_1(x)$, $P_2(x)$ to four-decimal-place accuracy for various numbers x

x	$P_0(x)$	$P_1(x)$	$P_2(x)$	$f(x) = e^x$
-1.0	1.0	0.0	0.5000	0.3679
-0.5	1.0	0.5	0.6250	0.6065
-0.2	1.0	0.8	0.8200	0.8187
-0.1	1.0	0.9	0.9050	0.9048
0	1.0	1.0	1.0	1.0
0.1	1.0	1.1	1.1050	1.052
0.2	1.0	1.2	1.2200	1.2214
0.5	1.0	1.5	1.6250	1.6487
1.0	1.0	2.0	2.5000	2.7183

As you might suspect, we can provide even better approximations to $f(x) = e^x$ by using a third degree polynomial, with the requirements that it and each of its first three derivatives have the same values as $f(x) = e^x$ and its first three derivatives at $x = 0$. We leave it to you to verify that the polynomial

$$P_3(x) = f(0) + f'(0)x + \frac{f''(0)}{2}x^2 + \frac{f'''(0)}{3!}x^3$$

$$= 1 + x + \frac{1}{2}x^2 + \frac{1}{6}x^3$$

(3)

satisfies these requirements and to calculate the values $P_3(x)$ for the x's listed in Table 5.1.

The advantage in using these polynomials to approximate values of the function $f(x) = e^x$ is that values of polynomials are easy to compute.

Example 1 Use the polynomial P_3 in equation (3) to approximate the value $\sqrt[3]{e} = e^{1/3}$.

Solution: With $x = \dfrac{1}{3}$ we have, from equation (3), that

$$P_3\left(\frac{1}{3}\right) = 1 + \frac{1}{3} + \frac{1}{2}\left(\frac{1}{3}\right)^2 + \frac{1}{6}\left(\frac{1}{3}\right)^3$$

$$= 1 + \frac{1}{3} + \frac{1}{18} + \frac{1}{162}$$

$$= \frac{162 + 54 + 9 + 1}{162}$$

$$= \frac{113}{81} \approx 1.3951.$$

The value of $\sqrt[3]{e}$, to four-decimal-place accuracy, is $\sqrt[3]{e} \approx 1.3956$. □

Taylor Polynomials

The preceding discussion on approximating values of the exponential function with polynomials can be extended to apply to a wide variety of functions. The idea, however, remains the same: To approximate values of the function f near $x = 0$ with a polynomial of degree n, use a polynomial whose value at $x = 0$ and those of its first n derivatives are the same as the value of f and those of its first n derivatives at $x = 0$.

Assuming that $f(0)$ and each of the derivatives $f'(0)$, $f''(0)$, . . . , $f^{(n)}(0)$ exist, there will be only one such polynomial, which we refer to as the nth *Taylor polynomial*.

DEFINITION 5

If the function f and each of its first n derivatives are defined for $x = 0$, the **nth Taylor polynomial** expanded about $x = 0$ is the function

$$P_n(x) = f(0) + f'(0)x + \frac{f''(0)}{2}x^2 + \frac{f'''(0)}{3!}x^3 + \cdots + \frac{f^{(n)}(0)}{n!}x^n.$$

Thus

$$P_1(x) = f(0) + f'(0)x$$

$$P_2(x) = f(0) + f'(0)x + \frac{f''(0)}{2}x^2$$

$$P_3(x) = f(0) + f'(0)x + \frac{f''(0)}{2}x^2 + \frac{f'''(0)}{6}x^3$$

etc.

Example 2 Find the first three Taylor polynomials for the function $f(x) = \ln(1 + x)$ expanded about $x = 0$.

Solution: First, we need to know the first three derivatives of $f(x) = \ln(1 + x)$. They are

$$f'(x) = \frac{1}{1 + x} = (1 + x)^{-1}$$

$$f''(x) = -(1 + x)^{-2}$$

$$f'''(x) = 2(1 + x)^{-3}$$

Thus $f(0) = \ln(1 + 0) = 0$, $f'(0) = 1$, $f''(0) = -1$, and $f'''(0) = 2$. Using these values in Definition 5 gives

$$P_1(x) = 0 + 1 \cdot x = x$$

$$P_2(x) = 0 + 1 \cdot x + \frac{(-1)}{2}x^2 = x - \frac{1}{2}x^2$$

$$P_3(x) = 0 + 1 \cdot x + \frac{(-1)}{2}x^2 + \frac{2}{6}x^3 = x - \frac{1}{2}x^2 + \frac{1}{3}x^3.$$

We leave it to you to verify that for $f(x) = \ln(1 + x)$

$$P_4(x) = x - \frac{1}{2}x^2 + \frac{1}{3}x^3 - \frac{1}{4}x^4$$

and

$$P_5(x) = x - \frac{1}{2}x^2 + \frac{1}{3}x^3 - \frac{1}{4}x^4 + \frac{1}{5}x^5$$

as the pattern for $P_1(x)$, $P_2(x)$, and $P_3(x)$ suggests. □

   Table 5.2 shows values of each of the Taylor polynomials $P_1(x)$, . . . , $P_5(x)$ for $f(x) = \ln(1 + x)$ in Example 2 for various choices of x near zero, as well as the values of $\ln(1 + x)$ correct to four-decimal-place accuracy. Again, note that the accuracy of these approximations increases as the degree of the polynomial increases, and that the approximations are better for x's nearer zero.

Table 5.2. Values of Taylor polynomials for $f(x) = \ln(1 + x)$ for x near zero, to four-decimal-place accuracy.

x	$P_1(x)$	$P_2(x)$	$P_3(x)$	$P_4(x)$	$P_5(x)$	$\ln(1 + x)$
$-.4$	-0.4	-0.4800	$-.5013$	$-.5077$	-0.5098	-0.5108
$-.3$	-0.3	-0.3450	$-.3540$	$-.3560$	-0.3565	-0.3567
$-.2$	-0.2	-0.2200	$-.2227$	$-.2231$	-0.2231	-0.2231
$-.1$	-0.1	-0.1050	$-.1053$	$-.1054$	-0.1054	-0.1054
0	0.0	0.0	0.0	0.0	0.0	0.0
0.1	1.0	0.0950	0.0953	0.0953	0.0953	0.0953
0.2	0.2	0.1800	0.1827	0.1823	0.1823	0.1823
0.3	0.3	0.2550	0.2640	0.2620	0.2625	0.2624
0.4	0.4	0.3200	0.3413	0.3349	0.3370	0.3365

The next example shows how the entries of Table 5.2 are computed.

Example 3 Use the results of Example 2 to approximate $\ln\left(\dfrac{3}{2}\right)$.

Solution: With $x = \dfrac{1}{2}$ so that $\ln(1 + x) = \ln\left(1 + \dfrac{1}{2}\right) = \ln\left(\dfrac{3}{2}\right)$ we obtain the approximations

$$P_1\left(\frac{1}{2}\right) = \frac{1}{2} = 0.5000$$

$$P_2\left(\frac{1}{2}\right) = \frac{1}{2} - \frac{1}{2}\left(\frac{1}{2}\right)^2 = \frac{1}{2} - \frac{1}{8} = \frac{3}{8} = 0.3750$$

$$P_3\left(\frac{1}{2}\right) = \frac{1}{2} - \frac{1}{2}\left(\frac{1}{2}\right)^2 + \frac{1}{3}\left(\frac{1}{2}\right)^3 = \frac{1}{2} - \frac{1}{8} + \frac{1}{24} = \frac{5}{12} \approx 0.4167$$

$$P_4\left(\frac{1}{2}\right) = \frac{1}{2} - \frac{1}{2}\left(\frac{1}{2}\right)^2 + \frac{1}{3}\left(\frac{1}{2}\right)^3 - \frac{1}{4}\left(\frac{1}{2}\right)^4 = \frac{77}{192} \approx 0.4010$$

$$P_5\left(\frac{1}{2}\right) = \frac{1}{2} - \frac{1}{2}\left(\frac{1}{2}\right)^2 + \frac{1}{3}\left(\frac{1}{2}\right)^3 - \frac{1}{4}\left(\frac{1}{2}\right)^4 + \frac{1}{5}\left(\frac{1}{2}\right)^5 = \frac{391}{960} \approx 0.4073.$$

The value of $\ln\left(\dfrac{3}{2}\right)$, approximated to four-decimal-place accuracy, is $\ln\left(\dfrac{3}{2}\right) \approx 0.4055$.

□

Example 4 Find the first five Taylor polynomials for the function $f(x) = \sin x$ expanded about $x = 0$.

Solution: The values of $f(x) = \sin x$ and its first five derivatives at $x = 0$ are as follows:

$$f(x) = \sin x \qquad f(0) = \sin(0) = 0$$
$$f'(x) = \cos x \qquad f'(0) = \cos(0) = 1$$
$$f''(x) = -\sin x \qquad f''(0) = -\sin(0) = 0$$
$$f'''(x) = -\cos x \qquad f'''(0) = -\cos(0) = -1$$
$$f^{(4)}(x) = \sin x \qquad f^{(4)}(0) = \sin(0) = 0$$
$$f^{(5)}(x) = \cos x \qquad f^{(5)}(0) = \cos(0) = 1.$$

Using these values in Definition 5 gives the Taylor polynomials

$$P_1(x) = 0 + 1 \cdot x = x$$

$$P_2(x) = 0 + 1 \cdot x + \frac{0}{2}x^2 = x$$

$$P_3(x) = 0 + 1 \cdot x + \frac{0}{2}x^2 + \frac{(-1)}{3!}x^3 = x - \frac{x^3}{3!}$$

$$P_4(x) = 0 + 1 \cdot x + \frac{0}{2}x^2 + \frac{(-1)}{3!}x^3 + \frac{0}{4!}x^4 = x - \frac{x^3}{3!}$$

$$P_5(x) = 0 + 1 \cdot x + \frac{0}{2}x^2 + \frac{(-1)}{3!}x^3 + \frac{0}{4!}x^4 + \frac{1}{5!}x^5 = x - \frac{x^3}{3!} + \frac{x^5}{5!}.$$

Notice that

$$P_1(x) = P_2(x) = x$$

$$P_3(x) = P_4(x) = x - \frac{x^3}{3!} \qquad \text{(See Figure 5.5.)}$$

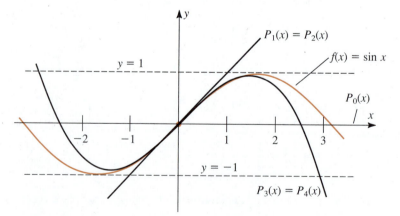

Figure 5.5 Five Taylor polynomials for $f(x) = \sin x$ expanded about $x = 0$. Note that $P_1(x)$ and $P_2(x)$ coincide, as do $P_3(x)$ and $P_4(x)$.

and

$$P_5(x) = x - \frac{x^3}{3!} + \frac{x^5}{5!}.$$ □

This example shows that an nth Taylor polynomial may actually have a degree less than n.

Example 5 Use the Taylor polynomial P_3 found in Example 4 to approximate $\sin(0.2)$ (where 0.2 is in radian measure).

Solution: With $x = 0.2$ we have

$$\sin(0.2) \approx P_3(0.2) = 0.2 - \frac{(0.2)^3}{3!}$$

$$= 0.2 - \frac{0.008}{6}$$

$$= 0.2 - 0.0013$$

$$= 0.1987.$$

The value of $\sin(0.2)$ to four-decimal-place accuracy is $\sin(0.2) \approx 0.1987$. □

Expanding About $x = a$

There is no reason why the values $f(x)$ should be approximated only for x near zero. We may, for example, wish to approximate values of the square root function $f(x) = \sqrt{x}$ for x near 4. Since we know the value $f(4) = \sqrt{4} = 2$, in this case it makes sense to base our approximation scheme at $x = 4$, rather than at $x = 0$. This is done as follows.

DEFINITION 6

If the function f and each of its first n derivatives are defined at $x = a$, the **nth Taylor polynomial expanded about $x = a$ is**

$$P_n(x) = f(a) + f'(a)(x - a) + \frac{f''(a)}{2}(x - a)^2 + \frac{f'''(a)}{3!}(x - a)^3$$

$$+ \cdots + \frac{f^{(n)}(a)}{n!}(x - a)^n.$$

Note that the nth Taylor polynomial expanded about $x = a$ differs from the nth Taylor polynomial expanded about $x = 0$ in that (i) f and its derivatives are evaluated at $x = a$ rather than at $x = 0$ and (ii) the polynomial involves powers of $(x - a)$ rather than powers of x.

Example 6 Show that at $x = a$, values of the third Taylor polynomial P_3, and each of its first three derivatives are the same as the values of f and its first three derivatives.

Solution: Since

$$P_3(x) = f(a) + f'(a)(x - a) + \frac{f''(a)}{2}(x - a)^2 + \frac{f'''(a)}{3!}(x - a)^3$$

we have $P_3(a) = f(a)$. Next, note that

$$P_3'(x) = f'(a) + f''(a)(x - a) + \frac{f'''(a)}{2}(x - a)^2 \quad \left(\text{since } \frac{3}{3!} = \frac{3}{6} = \frac{1}{2}\right)$$

so

$$P_3'(a) = f'(a).$$

Continuing by differentiating P_3, we obtain

$$P_3''(x) = f''(a) + f'''(a)(x - a)$$

so

$$P_3''(a) = f''(a).$$

Finally, since $P_3'''(x) = f'''(a)$, we have $P_3'''(a) = f'''(a)$. □

The point of Example 6 is that the nth Taylor polynomial and its first n derivatives have the same value at $x = a$ as the function f and its first n derivatives, regardless of whether or not a is zero.

Example 7 Find the first three Taylor polynomials for the function $f(x) = \sqrt{x}$ expanded about $a = 4$.

Solution: The values of f and its first three derivatives at $a = 4$ are found as follows:

$$f(x) = \sqrt{x} = x^{1/2} \qquad f(4) = 4^{1/2} = 2$$

$$f'(x) = \frac{1}{2}x^{-1/2} \qquad f'(4) = \frac{1}{2} \cdot 4^{-1/2} = \frac{1}{2} \cdot \frac{1}{2} = \frac{1}{4}$$

$$f''(x) = -\frac{1}{4}x^{-3/2} \qquad f''(4) = -\frac{1}{4} \cdot 4^{-3/2} = -\frac{1}{4} \cdot \frac{1}{8} = -\frac{1}{32}$$

$$f'''(x) = \frac{3}{8}x^{-5/2} \qquad f'''(4) = \frac{3}{8} \cdot 4^{-5/2} = \frac{3}{8} \cdot \frac{1}{32} = \frac{3}{256}.$$

With these values Definition 6 gives

$$P_1(x) = 2 + \frac{1}{4}(x - 4)$$

$$P_2(x) = 2 + \frac{1}{4}(x - 4) - \frac{1}{64}(x - 4)^2$$

$$P_3(x) = 2 + \frac{1}{4}(x - 4) - \frac{1}{64}(x - 4)^2 + \frac{1}{512}(x - 4)^3. \qquad □$$

Table 5.3 shows the result of using these Taylor polynomials to approximate values of $f(x) = \sqrt{x}$ near $x = 4$.

Table 5.3

x	$P_1(x)$	$P_2(x)$	$P_3(x)$	$f(x) = \sqrt{x}$
3.5	1.875	1.8711	1.8708	1.8708
3.8	1.95	1.9494	1.9494	1.9494
3.9	1.975	1.9748	1.9748	1.9748
4.0	2.0	2.0	2.0	2.0
4.1	2.025	2.0248	2.0248	2.0248
4.2	2.05	2.0494	2.0494	2.0494
4.5	2.125	2.1211	2.1213	2.1213

Summary

In this section we have demonstrated that values of transcendental functions (or of any differentiable function, for that matter) may be approximated by Taylor polynomials, and that the accuracy of these approximations seems to increase with the degree of the polynomial. Taylor's theorem, which we discuss in Section 9.6, enables us to estimate the accuracy of these approximations. Using this theorem, and the preceding ideas about infinite series, we shall show in Section 9.7 that Taylor polynomials can be viewed, in general, as partial sums of infinite *Taylor series*, which converge to the function values of interest for appropriate choices of the independent variable x.

Exercise Set 9.5

In Exercises 1–8 find the nth Taylor polynomial for the given function expanded about $x = 0$.

1. $f(x) = e^{-x}, \quad n = 3$

2. $f(x) = \cos x, \quad n = 2$

3. $f(x) = \cos x, \quad n = 4$

4. $f(x) = \ln(2 + x), \quad n = 4$

5. $f(x) = \dfrac{1}{1 + x}, \quad n = 2$

6. $f(x) = \tan x, \quad n = 2$

7. $f(x) = \sqrt{x + 2}, \quad n = 2$

8. $f(x) = \sqrt{1 + x}, \quad n = 2$

In Exercises 9–18 find the nth Taylor polynomial expanded about $x = a$.

9. $f(x) = \sin x, \quad a = \pi/4, \quad n = 2$

10. $f(x) = \cos x, \quad a = \pi/3, \quad n = 2$

11. $f(x) = e^x, \quad a = 1, \quad n = 4$

12. $f(x) = \ln x, \quad a = 1, \quad n = 4$

13. $f(x) = \sqrt{x}, \quad a = 9, \quad n = 3$

14. $f(x) = \sqrt{1 + x}, \quad a = 3, \quad n = 3$

15. $f(x) = e^{x^2}, \quad a = 0, \quad n = 2$

16. $f(x) = \sin x, \quad a = \pi/2, \quad n = 4$

17. $f(x) = x \cos x, \quad a = \pi/4, \quad n = 2$

18. $f(x) = \dfrac{1}{1 + x}, \quad a = 1, \quad n = 2$

19. Use the third Taylor polynomial for e^{-x} expanded about $x = 0$ to approximate $\dfrac{1}{e} = e^{-1}$. (See Exercise 1.)

20. Use the second Taylor polynomial for $f(x) = \cos x$ expanded about $x = 0$ to approximate $\cos \dfrac{\pi}{12}$. (See Exercise 2.)

21. Use the third Taylor polynomial for $f(x) = \sqrt{x}$ expanded about $x = 9$ to approximate $\sqrt{10}$. (See Exercise 13.)

22. Let $P_n(x)$ be the nth Taylor polynomial for $f(x) = e^x$ expanded about $x = 0$. Show that $P'_n(x) = P_{n-1}(x)$.

23. Let $P_n(x)$ be the nth Taylor polynomial for $f(x) = e^{-x}$ expanded about $x = 0$. Show that $P'_n(x) = -P_{n-1}(x)$.

24. Let $P_5(x)$ be the fifth Taylor polynomial for $f(x) = \sin x$ expanded about $x = 0$, and let $Q_5(x)$ be the fifth Taylor polynomial for $g(x) = \cos x$ expanded about $x = 0$. Show that $P'_5(x) = Q_5(x)$.

25. Let $Q_4(x)$ be the fourth Taylor polynomial for $g(x) = \cos x$, and let $P_4(x)$ be the fourth Taylor polynomial for $f(x) = \sin x$, both expanded about $x = 0$. Show that $Q'_4(x) = -P_4(x)$.

9.6 TAYLOR'S THEOREM

If f is a differentiable function of x, we have seen in Section 9.5 how to approximate values $f(x)$ by values of the nth Taylor polynomial $P_n(x)$. If we write

$$f(x) = P_n(x) + R_n(x) \qquad (1)$$

then the term $R_n(x) = f(x) - P_n(x)$ represents the error in this approximation. (See Figure 6.1.) The term $R_n(x)$ is referred to as the *remainder* in a Taylor approximation. Taylor's Theorem provides an estimate on the size of the remainder $R_n(x)$ in equation (1).

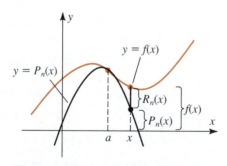

Figure 6.1 $R_n(x) = f(x) - P_n(x)$ is the *remainder* in a Taylor approximation.

THEOREM 9	Let n be a positive integer. If the function f and its derivatives f', f'', ... , $f^{(n)}$ are

THEOREM 9
(Taylor's Theorem)

Let n be a positive integer. If the function f and its derivatives f', f'', ... , $f^{(n)}$ are continuous on the interval $[c, d]$ and if the derivative $f^{(n+1)}(t)$ exists for each $t \in (c, d)$, then for any numbers x and a in $[c, d]$

$$f(x) = f(a) + f'(a)(x - a) + \frac{f''(a)}{2}(x - a)^2 \tag{2}$$

$$+ \cdots + \frac{f^{(n)}(a)}{n!}(x - a)^n + R_n(x)$$

with

$$|R_n(x)| \leq \frac{M}{(n + 1)!}|x - a|^{n+1} \tag{3}$$

where M is any number so that

$$|f^{(n+1)}(t)| \leq M \tag{4}$$

for all t between a and x.

Since equation (2) is just the restatement of equation (1) displaying the terms of the Taylor polynomial $P_n(x)$ expanded about a, inequality (3) says that the error in approximating the function value $f(x)$ by the value $P_n(x)$ of the nth Taylor polynomial expanded about a is no greater than $\frac{M}{(n + 1)!}|x - a|^{n+1}$, where M is the maximum value of $|f^{(n+1)}(t)|$ for t between the numbers a and x. The proof of Taylor's theorem is given at the end of this section.

Example 1 Use Taylor's Theorem to estimate the accuracy in approximating values of the function $f(x) = e^x$ for x in the interval $[0, 1]$ using the third Taylor polynomial $P_3(x)$ expanded about $a = 0$.

Solution: For the function $f(x) = e^x$ all derivatives are $f^{(n)}(x) = e^x$, so the hypotheses of Taylor's Theorem are satisfied. Since $f(x) = e^x$ is an increasing function for all x, the maximum value of the derivative $f^{(4)}(x) = e^x$ on the interval $[0, 1)$ is $f^{(4)}(1) = e^1 = e$. If we allow the assumption that $e \leq 3$, an upper bound M for the value of this derivative on $[0, 1]$ is $M = 3$. Thus in place of inequality (4) we have

$$|f^{(4)}(t)| \leq 3 \quad \text{for} \quad 0 \leq t \leq 1.$$

In Section 9.5 we have already shown that the third Taylor polynomial expanded about $a = 0$ is

$$P_3(x) = 1 + x + \frac{x^2}{2} + \frac{x^3}{6}.$$

According to Taylor's theorem, the approximation

$$e^x \approx 1 + x + \frac{x^2}{2} + \frac{x^3}{6} \qquad (5)$$

for x in the interval $[0, 1]$ is in error by no more than

$$|R_3(x)| \leq \frac{3}{(3+1)!}|x - 0|^{3+1} \qquad (n = 3)$$

$$= \frac{3}{4!}|x|^4$$

$$= \frac{|x|^4}{8}.$$

Thus, for example

(a) With $x = 1$ the approximation (5) is

$$e \approx 1 + 1 + \frac{1}{2} + \frac{1}{6} = \frac{8}{3}$$

which is accurate to within $|R_3(1)| \leq \frac{1}{8}$.

(b) With $x = \frac{1}{3}$ the approximation (5) is

$$\sqrt[3]{e} = e^{1/3} \approx 1 + \frac{1}{3} + \frac{(1/3)^2}{2} + \frac{(1/3)^3}{6} = \frac{113}{81}$$

which is accurate to within $\left|R_3\left(\frac{1}{3}\right)\right| \leq \frac{1}{8}\left(\frac{1}{3}\right)^4 = \frac{1}{648}$. □

Example 2 Use the Taylor polynomial $P_3(x)$ expanded about $a = 1$ to approximate $\ln(1.5)$ and estimate the accuracy of this approximation.

Strategy
Calculate the derivatives f', f'', and f''' and their values at $a = 1$.

Solution
We have

$$f(x) = \ln x; \qquad f(1) = 0$$

$$f'(x) = \frac{1}{x}; \qquad f'(1) = 1$$

$$f''(x) = -\frac{1}{x^2}; \qquad f''(1) = -1$$

$$f'''(x) = \frac{2}{x^3}; \qquad f'''(1) = 2$$

Form the Taylor polynomial $P_3(x)$ and use its value at $x = 1.5$ to approximate $\ln(1.5)$.

The third Taylor polynomial expanded about $a = 1$ is therefore

$$P_3(x) = (x - 1) - \frac{1}{2}(x - 1)^2 + \frac{2}{3!}(x - 1)^3$$

and with $x = 1.5$ the approximation is

$$\ln(1.5) \approx P_3(1.5) = (0.5) - \frac{1}{2}(0.5)^2 + \frac{1}{3}(0.5)^3$$

$$= 0.416\overline{6}. \tag{6}$$

Find the bound M on $|f^{(n+1)}(t)|$ for t in the interval $|a, x|$.

To estimate the accuracy, we note that the absolute value of the fourth derivative is

$$|f^{(4)}(x)| = \left|\frac{-6}{x^4}\right| = \frac{6}{x^4}$$

which is a decreasing function on the interval $[a, x] = [1, 1.5]$. Its maximum value on this interval is therefore

$$M = |f^{(4)}(1)| = \frac{6}{1^4} = 6.$$

Apply Taylor's Theorem to estimate the error.

Inequality (3) then gives the estimate

$$|R_3(1.5)| \leq \frac{6}{(3 + 1)!}(1.5 - 1)^{(3+1)}$$

$$= \frac{1}{64}$$

which shows that the approximation in line (6) is accurate to within $\frac{1}{64} = 0.015625$. □

Example 3 Suppose we wish to use the approximation

$$\sin x \approx x \tag{7}$$

for angles x with $|x| < \frac{\pi}{45}(=4°)$. What is the maximum possible error in this approximation?

Solution: The approximation in line (7) can be written

$$\sin x \approx P_2(x) \tag{8}$$

where $P_2(x)$ is the second Taylor polynomial for the function $f(x) = \sin x$ expanded about $a = 0$. (See Example 4, Section 9.5. Note that $g(x) = x$ is also the first Taylor polynomial $P_1(x)$ for $\sin x$, but taking $n = 2$ is preferable since it allows us to verify better accuracy for the approximation.)

Thus $n = 2$ in approximation (7). Since the third derivative of $f(x) = \sin x$ is $f^{(3)}(x) = -\cos x$, and since $|\cos x| \leq 1$ for all x, we have

$$|f^{(3)}(x)| = |-\cos x| \leq 1$$

for all x. We may therefore use $M = 1$ in inequality (3). With $a = 0$, $n = 2$, and $|x - a| = |x - 0| = |x| < \dfrac{\pi}{45}$, Taylor's Theorem gives the bound

$$
\begin{aligned}
|R_3(x)| &\le \frac{1}{(2 + 1)!} |x - 0|^3 \\
&= \frac{1}{6} |x|^3 \\
&< \frac{1}{6}\left(\frac{\pi}{45}\right)^3 \\
&\approx 0.00006
\end{aligned}
$$

for the accuracy in approximation (8). $\qquad\qquad$ □

Proving Taylor's Theorem

The proof that follows uses two properties of definite integrals of continuous functions. The proofs of these properties are sketched in Exercises 13 and 14.

PROPERTY 1: If $f(x) \le g(x)$ for $a \le x \le b$, then

$$
\int_a^b f(x)\, dx \le \int_a^b g(x)\, dx.
$$

PROPERTY 2: $\left| \displaystyle\int_a^b f(x)\, dx \right| \le \displaystyle\int_a^b |f(x)|\, dx.$

Proof of Taylor's Theorem: We assume in this proof that $a < x$. The case $x < a$ follows in the same way.

The Taylor polynomial

$$
P_n(x) = f(a) + f'(a)(x - a) + \frac{f''(a)}{2}(x - a)^2 + \cdots + \frac{f^{(n)}(a)}{n!}(x - a)^n
$$

was defined so that

$$
f(a) = P_n(a); \qquad f'(a) = P_n'(a); \qquad \ldots\ ; \qquad f^{(n)}(a) = P_n^{(n)}(a). \tag{9}
$$

Also, since $P_n(x)$ is a polynomial of degree n

$$
P_n^{(n+1)}(x) = 0 \qquad \text{for all } x. \tag{10}
$$

Now the remainder is defined by the equation

$$
R_n(x) = f(x) - P_n(x).
$$

Differentiating both sides of this equation and using equations (9), we conclude that

$$
R_n(a) = R_n'(a) = R_n''(a) = \cdots = R_n^{(n)}(a) = 0 \tag{11}
$$

and, using equation (10), we have

$$
R^{(n+1)}(t) = f^{(n+1)}(t) \quad \text{for} \quad a < t < x. \tag{12}
$$

We now proceed to integrate both sides of equation (12). Using equation (11), we obtain

$$\int_a^x f^{(n+1)}(t)\, dt = \int_a^x R_n^{(n+1)}(t)\, dt = R_n^{(n)}(t)\Big|_a^x = R_n^{(n)}(x). \tag{13}$$

Using the bound $|f^{(n+1)}(t)| \leq M$, equation (13), and properties (1) and (2), we conclude that

$$\left| R_n^{(n)}(x) \right| = \left| \int_a^x f^{(n+1)}(t)\, dt \right| \qquad \text{(by equation (13))} \tag{14}$$

$$\leq \int_a^x \left| f^{(n+1)}(t) \right|\, dt \qquad \text{(by property 2)}$$

$$\leq \int_a^x M\, dt \qquad \text{(by the bound on } |f^{(n+1)}|)$$

$$= M(x - a), \qquad a \leq x \leq d.$$

Then since

$$\int_a^x R_n^{(n)}(t)\, dt = R_n^{(n-1)}(t)\Big|_a^x = R_n^{(n-1)}(x) - R_n^{(n-1)}(a) = R_n^{n-1}(x)$$

it follows from inequality (14) and properties (1) and (2) that

$$\left| R_n^{n-1}(x) \right| = \left| \int_a^x R_n^{(n)}(t)\, dt \right|$$

$$\leq \int_a^x \left| R_n^{(n)}(t) \right|\, dt \qquad \text{(by property 2)}$$

$$\leq \int_a^x M(t - a)\, dt \qquad \text{(by equation (14))}$$

$$= \frac{M}{2}(x - a)^2, \qquad a \leq x \leq d.$$

Continuing by integrating in this way, we obtain the inequalities

$$\left| R_n^{(n-k)}(x) \right| \leq \frac{M}{(k+1)!}(x - a)^{k+1}, \qquad k = 2, 3, \ldots, n. \tag{15}$$

The desired inequality

$$\left| R_n(x) \right| \leq \frac{M}{(n+1)!}(x - a)^{n+1}$$

is a special case of inequality (15) with $k = n$. ∎

Exercise Set 9.6

In Exercises 1–8 use Taylor's theorem to find an approximation to the given function value using a Taylor polynomial of degree n expanded about $x = a$ and estimate the accuracy of the approximation.

1. $\ln(1.4)$, $a = 1$, $n = 3$

2. $\cos 36°$, $a = \dfrac{\pi}{4}$, $n = 2$

3. $\sqrt{3.91}$, $a = 4$, $n = 2$

4. $e^{0.2}$, $a = 0$, $n = 3$

5. $\sin 48°$, $a = \dfrac{\pi}{4}$, $n = 4$

6. $\sqrt[3]{10}$, $a = 8$, $n = 2$

7. $\tan \dfrac{\pi}{12}$, $a = 0$, $n = 3$

8. $\sqrt{50}$, $a = 49$, $n = 3$

9. Estimate the accuracy of the approximation

$$e^x \approx 1 + x + \frac{x^2}{2}$$

for x in the interval $[0, 1]$.

10. Find a bound on the error associated with the approximation

$$\sin x \approx x - \frac{x^3}{3!} \qquad \text{for } |x| < \frac{\pi}{4}.$$

11. How large must the integer n be chosen so that the approximation

$$\sin x \approx x - \frac{x^3}{31} + \ldots \quad \text{with } n \text{ finite.}$$

is accurate to within 0.001 for angles x satisfying the inequality $|x| < 0.2$ (in radians)? (*Hint:* Since $|f^{(n)}(x)| \le 1$ for all derivatives of $f(x) = \sin x$, Taylor's Theorem gives $|R_n(x)| \le \dfrac{1}{(n+1)!}|x|^{n+1}$. You must therefore find n sufficiently large that

$$\frac{1}{(n+1)!}(0.2)^{(n+1)} \le 0.001.)$$

12. How large must n be chosen so that the approximation

$$\cos x \approx 1 - \frac{x^2}{2} + \ldots \quad \text{with } n \text{ finite.}$$

is accurate to within 0.0001 for angles x satisfying the inequality $|x| < 0.2$ (in radians)? (See Exercise 11.)

13. Let f and g be continuous on $[a, b]$ with $f(x) \le g(x)$ for all $a \le x \le b$. Prove that

$$\int_a^b f(x)\, dx \le \int_a^b g(x)\, dx$$

as follows.
a. Define h by $h(x) = g(x) - f(x)$.
b. Verify that $h(x) \ge 0$ for $a \le x \le b$.
c. Write $g(x) = f(x) + h(x)$ and show that

$$\int_a^b g(x)\, dx = \int_a^b f(x)\, dx + \int_a^b h(x)\, dx.$$

d. Explain why $\displaystyle\int_a^b h(x)\, dx \ge 0$.
e. Conclude from parts (c) and (d) that

$$\int_a^b f(x)\, dx \le \int_a^b g(x)\, dx.$$

14. Let f be continuous on $[a, b]$. Prove that

$$\left| \int_a^b f(x)\, dx \right| \le \int_a^b |f(x)\, dx|\, dx$$

as follows.
a. Explain why $-|f(x)| \le f(x) \le |f(x)|$, $a \le x \le b$.
b. Conclude from part (a) and Exercise 13 that

$$-\int_a^b |f(x)|\, dx \le \int_a^b f(x)\, dx \le \int_a^b |f(x)|\, dx$$

c. Conclude from part (b) that

$$\left| \int_a^b f(x)\, dx \right| \le \int_a^b |f(x)|\, dx.$$

9.7 TAYLOR SERIES

In Sections 9.5 and 9.6 we approximated values of the function f for x near zero using the nth Taylor polynomial

$$P_n(x) = f(0) + f'(0) + \frac{f''(0)}{2}x^2 + \cdots + \frac{f^{(n)}(0)}{n!}x^n. \tag{1}$$

Using the conventions $0! = 1$, $1! = 1$, and $f^{(0)} = f$, we can express the Taylor polynomial in equation (1) in the form

$$P_n(x) = \sum_{k=0}^{n} \frac{f^{(k)}(0)}{k!} x^k. \tag{2}$$

The Taylor polynomial $P_n(x)$ in equation (2) is defined only when f and its first n derivatives are defined for $x = 0$. Several examples in Section 9.5 suggested that the values $P_n(x)$ become increasingly good approximations to $f(x)$ as n increases.

When the function f and *all* its derivatives $f^{(n)}(x)$, $n = 1, 2, 3, \ldots$ are defined for $x = 0$, we may extend the notion of Taylor polynomials to (infinite) **Taylor series** by taking the limit as $n \to \infty$ of the sum in equations (1) and (2). That is, we view the Taylor polynomial $P_n(x)$ as a partial sum of the Taylor series (expanded about $x = 0$):

$$\sum_{k=0}^{\infty} \frac{f^{(k)}}{k!} x^k = f(0) + f'(0)x + \frac{f''(0)}{2} x^2 + \cdots + \frac{f^{(k)}(0)}{k!} x^k + \cdots. \tag{3}$$

Taylor series are of interest because under certain conditions on the function f the series in equation (3) converges to the value $f(x)$. Thus we have a method for expressing values of the function f as sums of certain infinite series. Before saying more about the general situation, we look at the particular case of the function $f(x) = e^x$.

Example 1 Find the Taylor series expanded about $x = 0$ for the function $f(x) = e^x$.

Solution: For the function $f(x) = e^x$ all derivatives are also e^x: $f'(x) = e^x$, $f''(x) = e^x$, etc. Thus $f(0) = e^0 = 1$, $f'(0) = e^0 = 1$, and $f^{(k)}(0) = e^0 = 1$ for all $k = 2, 3, \ldots$. Equation (3) becomes

$$\sum_{k=0}^{\infty} \frac{1}{k!} x^k = 1 + x + \frac{x^2}{2} + \frac{x^3}{3!} + \cdots + \frac{x^k}{k!} + \cdots. \tag{4}$$

You should recognize the sum of the first $n + 1$ terms in this series as the nth Taylor polynomial for $f(x) = e^x$ expanded about $x = 0$ as discussed in Section 9.5. □

Example 2 For which numbers x does the Taylor series in Example 1 converge?

Solution: For any particular number x the right-hand side of equation (4) is just an infinite series of constants, whose general term is $a_k = \dfrac{x^k}{k!}$. We may therefore test this series for absolute convergence using the Ratio Test. We find that

$$\rho = \lim_{k \to \infty} \frac{|a_{k+1}|}{|a_k|} = \lim_{k \to \infty} \frac{\left| \dfrac{x^{k+1}}{(k+1)!} \right|}{\left| \dfrac{x^k}{k!} \right|}$$

$$= \lim_{k \to \infty} \frac{|x|^{k+1}}{(k+1)!} \cdot \frac{k!}{|x|^k}$$

$$= \lim_{k \to \infty} \frac{k!}{(k+1)!} \cdot \frac{|x|^{k+1}}{|x|^k}$$

$$= \lim_{k \to \infty} \left(\frac{1}{k+1}\right) |x|$$

$$= 0 \cdot |x|$$

$$= 0.$$

(Remember, x doesn't depend on k in this calculation, which is why $\lim_{k \to \infty} \left(\frac{1}{k+1}\right) |x| = 0 \cdot |x| = 0$.) Thus the ratio limit is $\rho = 0$ *for any number* x, so the Taylor series $\Sigma \dfrac{x^k}{k!}$ converges absolutely *for all* x. □

REMARK 1: There is a stronger statement to be made about the Taylor series for e^x in Example 2. Not only does it converge for all x, but *it converges to the function value* e^x for all x. (See Exercise 21.) Thus we can write

$$e^x = \sum_{k=0}^{\infty} \frac{x^k}{k!}$$

$$= 1 + x + \frac{x^2}{2} + \cdots + \frac{x^k}{k!} + \cdots$$

for all x. In particular, setting $x = 1$ gives an infinite series that converges to the number e:

$$e = 1 + 1 + \frac{1}{2} + \frac{1}{3!} + \cdots + \frac{1}{k!} + \cdots$$

$$= \sum_{k=0}^{\infty} \frac{1}{k!}.$$

Similarly, setting $x = -1$ gives an alternating series that converges to $e^{-1} = \dfrac{1}{e}$:

$$\frac{1}{e} = 1 - 1 + \frac{1}{2} - \frac{1}{3!} + \frac{1}{4!} - \cdots + \frac{(-1)^k}{k!} + \cdots$$

$$= \sum_{k=0}^{\infty} \frac{(-1)^k}{k!}.$$

There are some exotic examples of functions for which the corresponding Taylor series converge to numbers other than the function values, but this does not occur for any of the functions considered in this text. For a function f discussed in this chapter, if the Taylor series for f converges at x, then it converges to the value $f(x)$.

REMARK 2: A Taylor series may converge only for numbers x within a certain interval. When this is the case, we have to check for convergence separately at each endpoint—the series may converge at both endpoints, only at one, or at neither, as shown in the following examples.

Example 3 Find a Taylor series expanded about $x = 0$ for the function $f(x) = \ln(1 + x)$ and determine the numbers x for which the series converges.

Solution: We calculate the terms of the Taylor series just as we calculated the Taylor polynomial for $f(x) = \ln(1 + x)$ in Example 2, Section 9.5. We have

$$f(x) = \ln(1 + x) \qquad\qquad f(0) = \ln(1 + 0) = 0$$

$$f'(x) = \frac{1}{1 + x} = (1 + x)^{-1} \qquad f'(0) = (1 + 0)^{-1} = 1$$

$$f''(x) = -1(1 + x)^{-2} \qquad f''(0) = -(1 + 0)^{-2} = -1$$

$$f'''(x) = 2(1 + x)^{-3} \qquad f'''(0) = 2(1 + 0)^{-3} = 2$$

$$f^{(4)}(x) = -6(1 + x)^{-4} = -3!(1 + x)^{-4} \qquad f^{(4)}(0) = -3!(1 + 0)^{-4} = -3!$$

$$\vdots \qquad\qquad\qquad\qquad \vdots$$

$$f^{(k)}(x) = (-1)^{k+1}(k - 1)!(1 + x)^{-k} \qquad f^{(k)}(0) = (-1)^{k+1}(k - 1)!(1 + 0)^{-k}$$
$$= (-1)^{k+1}(k - 1)!$$

The Taylor series for $f(x) = \ln(1 + x)$ is therefore

$$\sum_{k=1}^{\infty} \frac{(-1)^{k+1}(k - 1)!}{k!} x^k = \sum_{k=1}^{\infty} \frac{(-1)^{k+1}}{k} x^k$$

$$= x - \frac{x^2}{2} + \frac{x^3}{3} - \frac{x^4}{4} + \cdots + \frac{(-1)^{k+1}}{k} x^k + \cdots. \tag{5}$$

Using the general term $a_k = \dfrac{(-1)^{k+1}}{k} x^k$, we test the series for absolute convergence using the Ratio Test. We obtain

$$\rho = \lim_{k \to \infty} \frac{|a_{k+1}|}{|a_k|} = \lim_{k \to \infty} \frac{\left| \dfrac{(-1)^{k+2}}{k + 1} x^{k+1} \right|}{\left| \dfrac{(-1)^{k+1}}{k} x^k \right|}$$

$$= \lim_{k \to \infty} \left(\frac{k}{k + 1} \right) \frac{|x|^{k+1}}{|x|^k}$$

$$= \lim_{k \to \infty} \left(\frac{k}{k + 1} \right) |x|$$

$$= |x|$$

since $\lim\limits_{k \to \infty} \dfrac{k}{k + 1} = 1$. Thus the Taylor series will converge absolutely when $\rho = |x| < 1$. We leave it as an exercise for you to show that this Taylor series diverges for any x with $|x| > 1$. (See Exercises 16 and 17.)

This leaves the two cases $x = 1$ and $x = -1$ (corresponding to $|x| = 1$) to be checked individually.

Case $x = 1$: When $x = 1$, the Taylor series in equation (5) becomes the series

$$\sum_{k=1}^{\infty} \frac{(-1)^{k+1}}{k}(1)^k = 1 - \frac{1}{2} + \frac{1}{3} - \frac{1}{4} + \frac{1}{5} + \cdots.$$

This is an alternating series that satisfies the requirements of the Alternating Series Test. It therefore converges.

Case $x = -1$: When $x = -1$ the Taylor series becomes

$$\sum_{k=1}^{\infty} \frac{(-1)^{k+1}}{k}(-1)^k = (-1) - \frac{(-1)^2}{2} + \frac{(-1)^3}{3} - \frac{(-1)^4}{4} + \cdots$$

$$= -\left(1 + \frac{1}{2} + \frac{1}{3} + \frac{1}{4} + \cdots\right).$$

This is the negative of the harmonic series, which diverges.

We have therefore determined that the Taylor series for $f(x) = \ln(1 + x)$ in equation (5):

(a) converges for $|x| < 1$
(b) diverges for $|x| > 1$
(c) converges for $x = 1$
(d) diverges for $x = -1$.

That is, the Taylor series converges only for those x in the interval $(-1, 1]$. □

Example 4 We have already determined in Theorem 1 of Section 9.2 that for $|x| < 1$ the function $f(x) = \dfrac{1}{1 - x}$ may be expressed as the infinite geometric series

$$\frac{1}{1 - x} = \sum_{k=0}^{\infty} x^k = 1 + x + x^2 + x^3 + \cdots + x^k + \cdots. \tag{6}$$

We leave it as an exercise for you to show that the terms of this series have the form $a_k = \dfrac{f^{(k)}(0)}{k!}x^k$. (See Exercise 18.) Thus equation (6) is a Taylor series for the function $f(x) = \dfrac{1}{1 - x}$, which converges for x in the interval $(-1, 1)$. □

Example 5 We leave it as additional exercises for you to show that the following are the Taylor series associated with the functions $f(x) = \sin x$ and $g(x) = \cos x$, and that both series converge for all values of x:

$$\sin x = \sum_{k=0}^{\infty} \frac{(-1)^k}{(2k + 1)!}x^{2k+1} = x - \frac{x^3}{3!} + \frac{x^5}{5!} - \frac{x^7}{7!} + \cdots \tag{7}$$

$$\cos x = \sum_{k=0}^{\infty} \frac{(-1)^k}{(2k)!} x^{2k} = 1 - \frac{x^2}{2!} + \frac{x^4}{4!} - \frac{x^6}{6!} + \cdots. \tag{8}$$

(See Exercises 19 and 20.) □

More General Taylor Series

Up to this point all examples of Taylor series have been expanded about $x = 0$. (Taylor series expanded about $a = 0$ are called **Maclaurin** series.) But, just as Taylor polynomials can be expanded about an arbitrary point $x = a$, so can Taylor series, provided that $f(a)$ and all derivatives $f^{(n)}(a)$ exist. The general form for **a Taylor series expanded about $x = a$** is

$$\sum_{k=0}^{\infty} \frac{f^{(k)}(a)}{k!}(x - a)^k = f(a) + f'(a)(x - a) + \frac{f''(a)}{2}(x - a)^2$$

$$+ \frac{f'''(a)}{3!}(x - a)^3 + \cdots + \frac{f^{(k)}(a)}{k!}(x - a)^k + \cdots. \tag{9}$$

Example 6 The Taylor series for the function $f(x) = \sin x$ expanded about $a = \dfrac{\pi}{4}$ is found as follows:

$$f(x) = \sin x \qquad f\left(\frac{\pi}{4}\right) = \frac{\sqrt{2}}{2}$$

$$f'(x) = \cos x \qquad f'\left(\frac{\pi}{4}\right) = \frac{\sqrt{2}}{2}$$

$$f''(x) = -\sin x \qquad f''\left(\frac{\pi}{4}\right) = -\frac{\sqrt{2}}{2}$$

$$f'''(x) = -\cos x \qquad f'''\left(\frac{\pi}{4}\right) = -\frac{\sqrt{2}}{2}$$

$$f^{(4)}(x) = \sin x \qquad f^{(4)}\left(\frac{\pi}{4}\right) = \frac{\sqrt{2}}{2}$$

$$\text{etc.} \qquad\qquad \text{etc.}$$

Using equation (9), we obtain

$$\sin x = \frac{\sqrt{2}}{2} + \frac{\sqrt{2}}{2}\left(x - \frac{\pi}{4}\right) - \frac{\sqrt{2}}{2}\frac{\left(x - \frac{\pi}{4}\right)^2}{2} - \frac{\sqrt{2}}{2}\frac{\left(x - \frac{\pi}{4}\right)^3}{3!} + \frac{\sqrt{2}}{2}\frac{\left(x - \frac{\pi}{4}\right)^4}{4!} + \cdots.$$

To find the interval of convergence for this series, we note that $|a_k| = \left| \dfrac{\sqrt{2}}{2} \dfrac{\left(x - \frac{\pi}{4}\right)^k}{k!} \right|.$

Applying the Ratio Test gives

$$\rho = \lim_{k \to \infty} \frac{\left| \dfrac{\sqrt{2}}{2} \dfrac{\left(x - \dfrac{\pi}{4}\right)^{k+1}}{(k+1)!} \right|}{\left| \dfrac{\sqrt{2}}{2} \dfrac{\left(x - \dfrac{\pi}{4}\right)^{k}}{k!} \right|} = \lim_{k \to \infty} \frac{k!}{(k+1)!} \left| x - \frac{\pi}{4} \right|$$

$$= \lim_{k \to \infty} \left(\frac{1}{k+1} \right) \left| x - \frac{\pi}{4} \right|$$

$$= 0.$$

Thus the Ratio Test shows that this Taylor series converges for all x. □

Properties of Taylor Series

Taylor series (and the more general *power series,* which we shall not discuss here) have several important properties that account for their usefulness. To keep the discussion simple, we shall discuss these properties only in general terms.

INTERVAL OF CONVERGENCE: The Taylor series in equation (9) will always converge when $x = a$, since all terms except possibly $f(a)$ equal zero. This may be the only number x for which a Taylor series converges. If not, the set of all x for which the series converges will be either

(a) all real numbers x, or
(b) an interval, whose midpoint is a, which may or may not include one or both of its endpoints. This interval is called the **interval of convergence.**

ALGEBRA OF SERIES: Within the interval of convergence we can multiply a series by a constant, term by term, and the resulting series will converge to that same multiple of the corresponding function. That is, for x in the interval of convergence and any real number c

$$cf(x) = c \sum_{k=0}^{\infty} \frac{f^{(k)}(a)}{k!}(x - a)^k = \sum_{k=0}^{\infty} c\frac{f^{(k)}(a)}{k!}(x - a)^k.$$

Moreover, when two Taylor series share a common interval of convergence, they may be added term by term, and the resulting series will converge to the sum of the corresponding functions for each x in that interval. Thus within a common interval of convergence the algebra of Taylor series is the same as that of polynomial functions.

DIFFERENTIATION OF TAYLOR SERIES: Within its interval of convergence (except possibly at endpoints), a Taylor series for f may be differentiated term by term, just as if it were an "infinitely long polynomial," and the resulting series will converge to the series for $f'(x)$. For example, we have

$$\frac{d}{dx}e^x = \frac{d}{dx}\left[1 + x + \frac{x^2}{2!} + \frac{x^3}{3!} + \frac{x^4}{4!} + \cdots\right]$$

$$= \frac{d}{dx}(1) + \frac{d}{dx}(x) + \frac{d}{dx}\left(\frac{x^2}{2}\right) + \frac{d}{dx}\left(\frac{x^3}{3!}\right) + \frac{d}{dx}\left(\frac{x^4}{4!}\right) + \cdots$$

$$= 0 + 1 + \frac{2x}{2} + \frac{3x^2}{3!} + \frac{4x^3}{4!} + \cdots$$

$$= 1 + x + \frac{x^2}{2} + \frac{x^3}{3!} + \cdots$$

$$= e^x$$

and

$$\frac{d}{dx}\sin x = \frac{d}{dx}\left[x - \frac{x^3}{3!} + \frac{x^5}{5!} - \frac{x^7}{7!} + \cdots\right]$$

$$= \frac{d}{dx}(x) - \frac{d}{dx}\left(\frac{x^3}{3!}\right) + \frac{d}{dx}\left(\frac{x^5}{5!}\right) - \frac{d}{dx}\left(\frac{x^7}{7!}\right) + \cdots$$

$$= 1 - \frac{3x^2}{3!} + \frac{5x^4}{5!} - \frac{7x^6}{7!} + \cdots$$

$$= 1 - \frac{x^2}{2} + \frac{x^4}{4!} - \frac{x^6}{6!} + \cdots$$

$$= \cos x.$$

Example 7 The Taylor series for $f(x) = \dfrac{1}{1-x}$ is

$$\frac{1}{1-x} = 1 + x + x^2 + x^3 + \cdots + x^k + \cdots$$

which converges for all x with $|x| < 1$. Since the derivative of f is $f'(x) = \dfrac{1}{(1-x)^2}$, this derivative has Taylor series

$$\frac{1}{(1-x)^2} = \frac{d}{dx}(1) + \frac{d}{dx}(x) + \frac{d}{dx}(x^2) + \frac{d}{dx}(x^3) + \cdots + \frac{d}{dx}(x^k)$$

$$= 1 + 2x + 3x^2 + \cdots + kx^{k-1} + \cdots$$

which converges for all x with $|x| < 1$. □

CONCLUSION: We have seen that Taylor polynomials are useful in approximating values of functions that are themselves difficult to compute. The question of whether successive approximations by Taylor polynomials actually converge to the desired function value may be recast as the question of whether the corresponding Taylor series converges to the function value at the designated value of x. This may be determined by using the Ratio Test to identify the width of the interval of convergence for the given function. This leaves only the question of the convergence of the Taylor series at the endpoints (if any) of the interval of convergence. This may be determined by using the tests for convergence of series of constants.

Exercise Set 9.7

In Exercises 1–10 find the Taylor series for the given function expanded about $x = a$.

1. $f(x) = e^{2x}, \quad a = 0$

2. $f(x) = e^{-x}, \quad a = 0$

3. $f(x) = \dfrac{1}{1+x}, \quad a = 0$

4. $f(x) = \sqrt{1+x}, \quad a = 0$

5. $f(x) = \ln x, \quad a = 1$

6. $f(x) = \dfrac{1}{x}, \quad a = 1$

7. $f(x) = \sin x, \quad a = \dfrac{\pi}{2}$

8. $f(x) = \cos x, \quad a = \dfrac{\pi}{4}$

9. $f(x) = \sin x, \quad a = \dfrac{\pi}{3}$

10. $f(x) = x \sin x, \quad a = 0$

11. Determine the interval of convergence for the Taylor series in Exercise 1.

12. Determine the interval of convergence for the Taylor series in Exercise 3.

13. Determine the interval of convergence for the Taylor series in Exercise 5.

14. Find a Taylor series for $f(x) = \dfrac{1}{(1+x)^2}$ expanded about $a = 0$ by differentiating the Taylor series for $g(x) = \dfrac{1}{1+x}$. (See Exercise 3.)

15. Find a Taylor series for $f(x) = \cos x$ expanded about $a = \dfrac{\pi}{2}$ by differentiating the Taylor series for $g(x) = \sin x$. (See Exercise 7.)

16. Explain why the Taylor series

$$\sum_{k=1}^{\infty} \frac{(-1)^{k+1}}{k} x^k = 1 - x + \frac{x^2}{2} - \frac{x^3}{3} + \frac{x^4}{4} - \cdots$$

in equation (5) diverges for $x > 1$. [*Hint:* Use the Alternating Series Test and the fact that $\lim\limits_{k \to \infty} \dfrac{x^k}{k} = +\infty$ if $x > 1$.]

17. Explain why the Taylor series in Exercise 16 diverges for $x < -1$ by comparing it with the harmonic series.

18. Verify that the terms of the geometric series

$$\frac{1}{1-x} = \sum_{k=0}^{\infty} x^k = 1 + x + x^2 + x^3 + \cdots + x^k + \cdots$$

have the form $a_k = \dfrac{f^{(k)}(0)}{k!} x^k$ for $f(x) = \dfrac{1}{1-x}$.

19. Show that the Taylor series for $f(x) = \sin x$ expanded about $x = 0$ is that given by equation (7).

20. Show that the Taylor series for $f(x) = \cos x$ expanded about $x = 0$ is that given by equation (8).

21. Show that the Taylor Series for the function $f(x) = e^x$ expanded about $a = 0$ converges *to the function f* as follows:

a. Show that for any fixed number x, the derivative $f^{(n+1)}$ satisfies the inequality

$$\left| f^{(n+1)}(t) \right| \le e^x, \qquad -\infty < t \le x.$$

b. Let $M = e^x$. Use Taylor's Theorem to show that

$$|R_n(t)| \le \frac{M}{(n+1)!} |x|^{(n+1)}, \qquad t \le x$$

where $R_n(t) = f(t) - P_n(t)$.

c. Show that $\lim\limits_{n \to \infty} \dfrac{|x|^{(n+1)}}{(n+1)!} = 0$ by noting that if $N \ge x$, we can write

$$\frac{|x|^{n+1}}{(n+1)!} = \left(|x| \cdot \frac{|x|}{2} \cdot \frac{|x|}{3} \cdot \ldots \cdot \frac{|x|}{N} \right)$$

$$\cdot \left(\frac{|x|}{N+1} \cdot \ldots \cdot \frac{|x|}{(n+1)} \right).$$

Argue that, as $n \to \infty$, the factor in the first set of parentheses remains fixed while the factor in the second set of parentheses must approach zero.

d. Conclude from parts (b) and (c) that

$$\lim_{n \to \infty} |R_n(t)| = 0, \qquad t \le x.$$

e. Conclude from part (d) that

$$\lim_{n \to \infty} P_n(t) = \sum_{k=1}^{\infty} \frac{t^k}{k!} = e^t, \qquad t \le x.$$

f. Since x was arbitrary in part (a), conclude from part (e) that the Taylor series for e^x converges to e^x for all x.

9.8 NEWTON'S METHOD

In solving optimization problems earlier in this book, we needed to be able to find the critical numbers that arise as solutions of the equation $f'(x) = 0$, that is, to find the *zeros* of the derivative f'. This is only one of many circumstances in mathematics in which we need to find the zeros of a particular function. While the techniques of algebra will enable us to find the zeros of certain limited classes of functions (namely, polynomials of degree 4 or less), there is no general method that succeeds in finding the zeros of all, or even most, functions.

Over three centuries ago the famous mathematician and physicist Isaac Newton (1642–1727) discovered a method for *approximating* zeros of functions that can be described in terms of Taylor polynomials. Instead of using the values of a function and its derivatives at a particular number to approximate values of the function at nearby numbers, this method uses an initial approximation for the zero of the function (a number, say, x_1) and values of the function and its derivative at this initial approximation to find a better approximation for the zero, say, x_2. The method is then repeated to generate a **sequence** of approximations $\{x_1, x_2, x_3, x_4, \ldots\}$, which under certain conditions converges to the desired zero of the function.

We may explain the method as follows. Suppose f is a differentiable function, and we wish to approximate a zero of f that is known to lie between the numbers a and b. We make an initial guess for the zero by choosing any number in the interval $[a, b]$, say, x_1. Then, using the function value $f(x_1)$ and the value $f'(x_1)$ of the derivative f' at x_1, we form the first Taylor polynomial

$$P_1(x) = f(x_1) + f'(x_1)(x - x_1) \tag{1}$$

for f at x_1. Figure 8.1 reminds you that the graph of this Taylor polynomial is the line tangent to the graph of f at the point $(x_1, f(x_1))$. Although we do not know the zero c of the function f, we can use equation (1) to find a zero x_2 for the Taylor polynomial P_1 by solving the equation $P(x_2) = 0$. That is, we set

$$P(x_2) = f(x_1) + f'(x_1)(x_2 - x_1) = 0$$

and solve to obtain

$$x_2 = x_1 - \frac{f(x_1)}{f'(x_1)}. \tag{2}$$

Newton's famous observation is that, for many functions, this root is a better approxima-

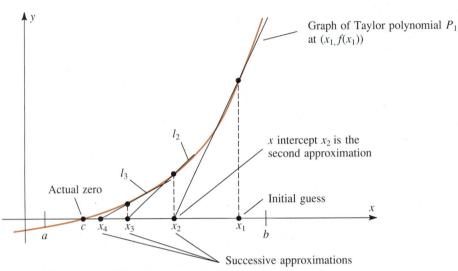

Figure 8.1 Newton's Method for approximating zero of $y = f(x)$.

tion for the root c of f than is the initial approximation x_1. When this is the case, we may repeat the procedure by forming the first Taylor polynomial $P_1(x)$ for f at x_2 and then finding the zero x_3 of this polynomial, which yet will be a better approximation to the zero c of f. Continuing in this way, we generate a sequence of approximations $\{x_n\}$, each of which is obtained from the former via the equation

$$x_{n+1} = x_n - \frac{f(x_n)}{f'(x_n)} \tag{3}$$

Equation (3) is called **the formula for the approximation of a root of the differentiable function f by Newton's Method.**

Example 1 Use Newton's Method to approximate the zero of the function $f(x) = x^3 - 10$ lying between $x = 0$ and $x = 4$.

Strategy

First, verify that a zero exists in [0, 4].

Solution

Before applying the method, we should verify that the given function indeed has a zero between the given numbers. Since $f(0) = -10$ and $f(4) = 54$ are of opposite sign and f is continuous, $f(x)$ must equal zero for some $x \in (0, 4)$.

Make an initial guess for the first approximation.

We begin by making an initial guess for c, say, $x_1 = 3$. Then, from equation (3), with $n = 1$, we obtain the second approximation

Apply (3), using

$f'(x) = 3x^2$.

$$x_2 = 3 - \frac{f(3)}{f'(3)}$$

$$= 3 - \frac{(3^3 - 10)}{3(3)^2}$$

$$= 3 - \frac{17}{27} \approx 2.37.$$

Apply (3) again,

The next approximation, x_3, is obtained from equation (3) by using $x_2 = 2.37$ and $n = 2$:

and so on.

$$x_3 = 2.37 - \frac{f(2.37)}{f'(2.37)}$$

$$= 2.37 - \frac{(2.37)^3 - 10}{3(2.37)^2}$$

$$\approx 2.17$$

etc.

□

Although the formula for Newton's Method is simple to state, the hand calculations quickly become tedious and the chance for error grows rapidly. However, this is precisely the sort of problem that is easy to implement on a hand calculator or on a computer. (A BASIC program for implementing Newton's Method is listed in Appendix I.) Table 8.1 contains the results obtained by using a computer to continue the calculations of this example through $n = 4$ iterations.

Table 8.1

n	x_n	$f(x_n)$	$f'(x_n)$	x_{n+1}
1	3.	17.	27.	2.370370
2	2.370370	3.318295	16.855967	2.173509
3	2.173509	0.267958	14.172419	2.154602
4	2.154602	0.002324	13.926924	2.154435

The advantage of listing partial calculations in a table such as this is that each approximation to the zero, x_n (column 2), can be compared directly to the function value (column 3) at each step. The results obtained in the second and fifth column of Table 8.1 suggest that $x_4 = 2.154602$ is a good approximation to the desired zero.

The next example shows that Newton's Method can be used to approximate roots regardless of where the particular root lies.

Example 2 Use Newton's Method to approximate $\sqrt[3]{a}$.

Solution: Finding the number x for which $\sqrt[3]{a} = x$ is equivalent to solving the equation $a = x^3$, which in turn is equivalent to finding a zero for the function $f(x) = x^3 - a$. This is the problem treated in Example 1 with $a = 10$. □

Example 3 Use Newton's Method to find the point in the right half plane where the graphs of $f(x) = 2x^3$ and $g(x) = 5x + 1$ intersect (see Figure 8.2).

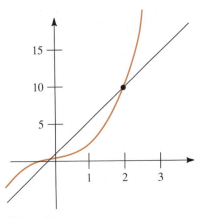

Figure 8.2

Solution: Finding a number x for which

$$2x^3 = 5x + 1$$

is equivalent to finding a zero of the function $h(x) = 2x^3 - (5x + 1)$.

For this function the approximation scheme (3) becomes

$$x_{n+1} = x_n - \frac{2x_n^3 - (5x_n + 1)}{6x_n^2 - 5}.$$

Since $h(0) = -1$ and $h(2) = 5$, we know that a zero must lie within the interval $[0, 2]$. Using a first approximation $x_1 = 2$, we obtain the information contained in Table 8.2 for four iterations of Newton's Method.

Table 8.2

n	x_n	$f(x_n)$	$f'(x_n)$	x_{n+1}
1	2.0	5.0	19.0	1.736842
2	1.736842	0.794576	13.099723	1.676186
3	1.676186	0.037894	11.857599	1.672990
4	1.672990	0.000103	11.793380	1.672982

The desired point of intersection is approximately $(1.672982, 9.364910)$.   □

When the approximations produced by Newton's Method approach the desired zero, we say that the method **converges** to that zero. Unfortunately, depending on the function and the initial approximation, Newton's Method may not converge to the desired zero. We comment on some of the reasons for this here, and pursue some of the details in the Exercise Set.

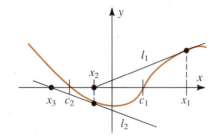

Figure 8.3 Initial approximation x_1 leads to zero c_2 where zero c_1 was desired.

1. The function may have more than one zero. Depending on the initial approximation x_1, the method may converge to a zero other than the one desired (see Figure 8.3).
2. An approximation x_k may be obtained for which $f'(x_k) = 0$. In this case the denominator in the expression for x_{k+1} is zero, so the method fails.
3. The slope of the graph of f may be such that the approximations simply do not converge, but instead move away from the desired zero (see Figure 8.4 and Exercise 15), or simply oscillate between two or more distinct approximations.

Figure 8.4 Iterates move away from zero rather than converging.

Newton's Method is not a theorem, but rather, a procedure that produces useful approximations in a large number of cases and that, unfortunately, fails in other cases. The fact that the method does not converge for all functions is not a criticism of Newton's idea, but rather, an invitation to you to pursue a deeper study of the geometric and analytic properties of functions that allow techniques such as Newton's Method to succeed. The relevant courses in which to pursue these issues are those on advanced calculus, numerical analysis, and modern analysis.

Exercise Set 9.8

In Exercises 1–8 use Newton's Method to approximate the zero of the given function lying between the given values $x = a$ and $x = b$. In each case first verify that such a zero indeed exists. If a calculator or computer is not available to you, only two or three iterations should be performed. If a calculator or computer is available, a table such as Table 8.1 should be constructed.

1. $f(x) = x^2 - \dfrac{7}{2}x + \dfrac{3}{4}$, $\quad a = 0$, $\quad b = 2$

2. $f(x) = 1 - x - x^2$, $\quad a = -2$, $\quad b = -1$

3. $f(x) = 1 - x - x^2$, $\quad a = 0$, $\quad b = 1$

4. $f(x) = x^3 + x - 3$, $\quad a = 1$, $\quad b = 2$

5. $f(x) = \sqrt{x + 3} - x$, $\quad a = 1$, $\quad b = 3$

6. $f(x) = x^3 + x^2 + 3$, $\quad a = -3$, $\quad b = -1$

7. $f(x) = x^4 - 5$, $\quad a = -2$, $\quad b = -1$

8. $f(x) = 5 - x^4$, $\quad a = 1$, $\quad b = 2$

9. Use the results of Exercise 5 to find a point of intersection between the graphs of $f(x) = \sqrt{x + 3}$ and $g(x) = x$.

10. Use Newton's Method to approximate
    a. $\sqrt{40}$ b. $\sqrt[3]{49}$
    c. $\sqrt[5]{18}$ d. $\sqrt{37}$

11. What results from applying Newton's Method if the (lucky) first guess is precisely the desired zero?

12. Use a calculator or computer to find an approximation to the point where the graphs of $f(x) = 2x$ and $g(x) = \tan x$ intersect for $0 < x \le \pi/2$.

13. Attempt to use Newton's Method to find the zero of $f(x) = (x - 2)^{1/3}$ with initial guess $x_1 = 3$. What happens? Why?

14. The graphs of the functions f and g in Example 3 cross at two other points. Find them.

SUMMARY OUTLINE OF CHAPTER 9

■ An **infinite sequence** is a function whose domain is the positive integers.   (Page 462)

■ L is the **limit** of the sequence $\{a_n\}$, written $L = \lim\limits_{n \to \infty} a_n$ if the terms a_n approach the unique   (Page 465)
number L as n increases without bound.

■ Properties of limits of sequences: (Page 465)

   (i) $\lim\limits_{n \to \infty} \{a_n + b_n\} = \lim\limits_{n \to \infty} \{a_n\} + \lim\limits_{n \to \infty} \{b_n\}$

   (ii) $\lim\limits_{n \to \infty} \{c a_n\} = c \cdot \lim\limits_{n \to \infty} \{a_n\}$

   (iii) $\lim\limits_{n \to \infty} \{a_n b_n\} = \left(\lim\limits_{n \to \infty} \{a_n\} \right) \left(\lim\limits_{n \to \infty} \{b_n\} \right)$

   (iv) $\lim\limits_{n \to \infty} \left\{ \dfrac{a_n}{b_n} \right\} = \dfrac{\lim\limits_{n \to \infty} \{a_n\}}{\lim\limits_{n \to \infty} \{b_n\}}$, $\quad$ provided $\lim\limits_{n \to \infty} \{b_n\} \ne 0$,

      assuming that the limits on the right-hand sides of the equations exist.

■ An **infinite series** is an expression of the form   (Page 473)

$$\sum_{k=1}^{\infty} a_k = a_1 + a_2 + a_3 + \cdots.$$

■ The **partial sum** S_n of an infinite series $\sum\limits_{k=1}^{\infty} a_k$ is   (Page 473)

$$S_n = \sum_{k=1}^{n} a_k, \quad n = 1, 2, \ldots.$$

■ The infinite series Σa_k **converges** if the limit (Page 473)

$$S = \lim_{n \to \infty} S_n = \lim_{n \to \infty} \sum_{k=1}^{n} a_k$$

exists.

■ *Theorem:* The **geometric series** (Page 475)

$$\sum_{k=0}^{\infty} x^k = 1 + x + x^2 + x^3 + \cdots$$

 (i) converges to the sum $\dfrac{1}{1-x} = \displaystyle\sum_{k=0}^{\infty} x^k$ if $|x| < 1$

(ii) diverges if $|x| \geq 1$.

■ *Theorem (Integral Test):* Let $a_k \geq 0$ for all k and let f be a continuous decreasing function (Page 482) with $f(k) = a_k$ for $k = 1, 2, \ldots$. Then

$$\sum a_k \text{ converges if and only if } \int_1^\infty f(x)\ dx \text{ converges.}$$

■ The **harmonic** series (Page 482)

$$\sum_{k=1}^{\infty} \frac{1}{k} = 1 + \frac{1}{2} + \frac{1}{3} + \frac{1}{4} + \cdots$$

diverges.

■ The **p-series** $\displaystyle\sum_{k=1}^{\infty} \frac{1}{k^p} = 1 + \frac{1}{2^p} + \frac{1}{3^p} + \frac{1}{4^p} + \cdots$, $p > 0$ converges only if $p > 1$. (Page 484)

■ *Theorem (Comparison Test):* Let Σa_k and Σb_k be series with (Page 484)

$$0 < a_k \leq b_k$$

for each $k = 1, 2, 3, \ldots$. Then

(a) If $\Sigma\, b_k$ converges, so does $\Sigma\, a_k$.
(b) If $\Sigma\, a_k$ diverges, so does $\Sigma\, b_k$.

■ *Theorem (Limit Comparison Test):* Let $\Sigma\, a_k$ and $\Sigma\, b_k$ be series with positive terms. If the (Page 486) limit $\rho = \displaystyle\lim_{k \to \infty} \frac{a_k}{b_k}$ exists and is nonzero, then either both series converge or both series diverge.

■ *Theorem (Necessary Condition):* If the series Σa_k converges, then $\displaystyle\lim_{k \to \infty} a_k = 0$. (Page 488)

■ *Theorem (Ratio Test):* Let $\Sigma\, a_k$ be a series with positive terms and let ρ be the limit (Page 491)

$$\rho = \lim_{k \to \infty} \frac{a_{k+1}}{a_k}.$$

Then

 (i) If $0 \leq \rho < 1$ the series $\Sigma\, a_k$ converges.

(ii) If $\rho > 1$ the series $\Sigma \, a_k$ diverges.

(iii) If $\rho = 1$ there is no conclusion.

■ **Theorem (Alternating Series Test):** Let $a_k > 0$ for all k. The alternating series $\sum(-1)^k a_k$ converges if and only if both (Page 495)

(i) $a_k > a_{k+1}$ for all k with $k \geq K$ (K fixed)

(ii) $\lim\limits_{k \to \infty} a_k = 0$.

■ The series $\Sigma \, a_k$ is said to **converge absolutely** if $\Sigma \, |a_k|$ converges. (Page 496)

■ **Theorem:** If the series $\Sigma \, |a_k|$ converges, so does the series $\Sigma \, a_k$. (Page 497)

■ The **nth Taylor polynomial** for f expanded about $x = a$ is (Page 507)

$$P_n(x) = \sum_{k=0}^{n} \frac{f^{(k)}(a)}{k!}(x-a)^k$$

$$= f(a) + f'(a)(x-a) + \frac{f''(a)}{2}(x-a)^2$$

$$+ \cdots + \frac{f^{(n)}(a)}{n!}(x-a)^n.$$

■ **Taylor's Theorem** states that if $R_n(x) = f(x) - P_n(x)$, then under appropriate conditions (Page 511)

$$|R_n(x)| \leq \frac{M}{(n+1)!}|x-a|^{n+1}$$

where the constant M is determined by the derivative $f^{(n+1)}$.

■ The **Taylor series** for f expanded about $x = a$ is (Page 517)

$$\sum_{k=0}^{\infty} \frac{f^{(k)}(a)}{k!}(x-a)^k = f(a) + f'(a)(x-a)$$

$$+ \frac{f''(a)}{2}(x-a)^2$$

$$+ \frac{f'''(a)}{3!}(x-a)^3 + \cdots.$$

■ **Newton's Method** for approximating a zero of a differentiable function f uses the sequence of approximations (Page 526)

$$x_{n+1} = x_n - \frac{f(x_n)}{f'(x_n)}$$

REVIEW EXERCISES—CHAPTER 9

In Exercises 1–8 determine whether the given sequence converges. If it does, find its limit.

1. $\left\{\sin \dfrac{n\pi}{2}\right\}$

2. $\left\{\dfrac{1}{\sqrt{n+1}}\right\}$

3. $\{\ln \sqrt{n} - \ln n\}$

4. $\left\{\left(1 + \dfrac{1}{n}\right)^{2n}\right\}$

5. $\left\{\dfrac{n^2 + 3n - 5}{4 + 2n^2}\right\}$

6. $\left\{\dfrac{4 + \sqrt{n}}{n+3}\right\}$

7. $\left\{ \dfrac{3^n + 4^n}{5^n} \right\}$

8. $\{\ln n - \ln(n + 1)\}$

In Exercises 9–12 find the sum of the series.

9. $\displaystyle\sum_{k=0}^{\infty} \dfrac{1}{3^k}$

10. $\displaystyle\sum_{k=0}^{\infty} \dfrac{1 + 2^k}{5^k}$

11. $\displaystyle\sum_{k=1}^{\infty} \dfrac{2^k}{3^k}$

12. $\displaystyle\sum_{k=1}^{\infty} \dfrac{7}{4^k}$

In Exercises 13–30 determine whether the series converges or diverges, and state which test you used.

13. $\displaystyle\sum_{k=1}^{\infty} \dfrac{1}{k^2}$

14. $\displaystyle\sum_{k=1}^{\infty} \dfrac{1}{k(k + 1)}$

15. $\displaystyle\sum_{k=1}^{\infty} \dfrac{3^k}{k^2}$

16. $\displaystyle\sum_{k=1}^{\infty} \dfrac{3^{2k}}{k^3}$

17. $\displaystyle\sum_{k=1}^{\infty} \dfrac{k^2}{k!}$

18. $\displaystyle\sum_{k=0}^{\infty} \dfrac{k}{k + 3}$

19. $\displaystyle\sum_{k=0}^{\infty} \dfrac{k - 1}{k + 1}$

20. $\displaystyle\sum_{k=0}^{\infty} \dfrac{3}{k + 2}$

21. $\displaystyle\sum_{k=1}^{\infty} \dfrac{1}{\sqrt{k + 1}}$

22. $\displaystyle\sum_{k=1}^{\infty} \dfrac{1}{k \ln k}$

23. $\displaystyle\sum_{k=1}^{\infty} \dfrac{1}{k^2 + 1}$

24. $\displaystyle\sum_{k=1}^{\infty} \dfrac{(-1)^k}{2k + 1}$

25. $\displaystyle\sum_{k=0}^{\infty} \dfrac{(-1)^k k}{\sqrt{k + 1}}$

26. $\displaystyle\sum_{k=0}^{\infty} \dfrac{\cos \pi k}{k + 1}$

27. $\displaystyle\sum_{k=1}^{\infty} \dfrac{k \cdot 2^k}{k!}$

28. $\displaystyle\sum_{k=1}^{\infty} (-1)^{k+1} \dfrac{\sqrt{k}}{k + 2}$

29. $\displaystyle\sum_{k=0}^{\infty} \dfrac{1}{(k + 2)^{3/2}}$

30. $\displaystyle\sum_{k=1}^{\infty} \dfrac{(-1)^k k}{(k + 1)!}$

In Exercises 31–38 find the nth Taylor polynomial for the function f expanded about $x = a$.

31. $f(x) = e^{-3x}, \quad a = 0, \quad n = 3$

32. $f(x) = \cos x, \quad a = \pi/6, \quad n = 3$

33. $f(x) = \tan x, \quad a = 0, \quad n = 2$

34. $f(x) = e^{x^2}, \quad a = 0, \quad n = 3$

35. $f(x) = \ln(1 + x^2), \quad a = 0, \quad n = 2$

36. $f(x) = \sin x, \quad a = \pi/6, \quad n = 3$

37. $f(x) = \sqrt{x}, \quad a = 9, \quad n = 3$

38. $f(x) = \sec x, \quad a = \pi/4, \quad n = 3$

39. Find a Taylor series for the function $f(x) = xe^x$ expanded about $x = 0$.

40. For which values of x does the Taylor series in Exercise 39 converge?

41. Find a Taylor series for the function $f(x) = \dfrac{x}{1 - x}$ expanded about $a = 0$.

42. For which numbers x does the Taylor series in Exercise 41 converge?

43. Find the Taylor series for the polynomial $f(x) = x^3 + 2x^2 - x + 3$
   a. expanded about $x = 0$.
   b. expanded about $x = 1$.

44. a. Express the repeating decimal $0.7\overline{777}\ldots$ as a geometric series.
   b. Use the formula for the sum of a geometric series to find the rational number $\dfrac{p}{q}$ to which the series in part a converges.

45. An annuity is to pay \$500 at the end of year 1, one third of this amount at the end of year 2, and, at the end of each succeeding year, one third of the amount paid in the preceding year. What is the total amount paid by the annuity in perpetuity?

46. Use Newton's Method to approximate the solution of $x^2 - x - 1 = 0$ lying in the interval $[0, 2]$.

47. Use Newton's Method to approximate the solution of $\cot x = x$ lying in the interval $(0, \pi/2]$.

48. Use Newton's Method to approximate $\sqrt[3]{175}$.

49. Use Newton's Method to approximate the maximum value of the function $f(x) = x \sin x$ for $x \in [0, 3]$. (*Hint:* What condition holds for f' at a maximum value for f?)

Chapter 10

Differential Equations

10.1 INTRODUCTION TO DIFFERENTIAL EQUATIONS

In Chapter 3 we introduced the notion of the derivative f' of the function f as the *rate of change* of f. Applications involving this rate of change have included the following:

 (i) The derivative $MC(x) = C'(x)$ is the rate at which total cost C changes with respect to change in the quantity x.
 (ii) The derivative $P'(t)$ gives the rate at which the size P of a population changes with respect to change in time t.
(iii) The derivative $v(t) = s'(t)$ gives the velocity of an object that is moving along a line and that is located at a distance $s(t)$ from the origin at time t. That is, $v(t)$ is the rate of change of position with respect to change in time.

When we attempt to develop mathematical models describing biological, economic, or physical processes, we often begin by describing the relationship between a certain function of interest (such as population size) and a growth rate (such as population growth) associated with that function. This leads to an equation such as

$$P'(t) = kP(t) \qquad\qquad (1)$$

which we referred to as the *Law of Natural Growth* for the size of a population $P(t)$ in Section 4.5. That is, we obtain an equation involving one or more derivatives of a function (as well as, perhaps, the function itself and possibly even other functions). Such equations are called **differential equations.**

Thus a differential equation is an equation involving one or more derivatives of an unknown function. A **solution** of a differential equation is a function that is differentiable as many times as the equation requires, and that satisfies the equation when substituted for the unknown and its derivatives.

Example 1 In Chapter 4 we saw that a solution of the differential equation $P'(t) = kP(t)$ in equation (1) has the form $P(t) = Ce^{kt}$ where C is an arbitrary constant. To verify that this function satisfies equation (1), we calculate

$$P'(t) = \frac{d}{dt}(Ce^{kt})$$

$$= C \cdot ke^{kt}$$

$$= k(Ce^{kt}).$$

Substituting this expression for $P'(t)$ in equation (1) then gives the identity

$$k(Ce^{kt}) = k(Ce^{kt})$$

so the differential equation is "solved." It can be proved that *every* function that satisfies equation (1) has the form $P(t) = Ce^{kt}$. ☐

Example 2 The function $y = \sin x$ satisfies the differential equation $\frac{d^2y}{dx^2} + y = 0$. To show this, we compute the second derivative:

$$\frac{dy}{dx} = \cos x \quad \text{so} \quad \frac{d^2y}{dx^2} = -\sin x.$$

Thus substitution in the differential equation gives

$$\frac{d^2y}{dx^2} + y = (-\sin x) + \sin x$$

$$= 0$$

as required. Thus $y = \sin x$ is a solution of the differential equation. ☐

Order

The **order** of a differential equation is the order of the highest derivative in the equation. Thus equation (1) is a **first order** differential equation, while the equation in Example 2 is of **second order.** We shall not consider differential equations of order higher than two in this text.

Solutions of Differential Equations

If a differential equation has any solutions at all, then it generally has infinitely many solutions, for the same reason that a given function has infinitely many antiderivatives. For example, consider the simple differential equation

$$\frac{dy}{dx} = 2x + 1. \tag{2}$$

For this differential equation it's easy to compute the antiderivative on each side, and we get

$$\int \left(\frac{dy}{dx}\right) dx = \int (2x + 1)\, dx$$

$$y = x^2 + x + C. \tag{3}$$

The constant of integration C can take any one of an infinite number of real values. For each value of C we obtain a different equation (2) called a *particular solution*. The form (3), which expresses all of the particular solutions in one compact expression, is called the *general solution* of (2). Two particular solutions of (2) are

$$y_1 = x^2 + x + 7 \quad \text{and} \quad y_2 = x^2 + x + \pi.$$

You can verify by differentiating both of these functions, and the general solution (3), that all of them indeed satisfy the differential equation (2).

In general, we expect to find one arbitrary constant in the general solution of a first order differential equation and two arbitrary constants in the general solution of a second order differential equation.

Example 3 In Example 2 we showed that the function $y_1 = \sin x$ is a particular solution of the differential equation

$$\frac{d^2y}{dx^2} + y = 0.$$

Another particular solution is

$$y_2 = \cos x.$$

To see this, note that $\dfrac{dy_2}{dx} = -\sin x$ and $\dfrac{d^2y_2}{dx^2} = -\cos x$, so

$$\frac{d^2y_2}{dx^2} + y_2 = -\cos x + \cos x = 0$$

as required.

In fact, the general solution of this differential equation is the following:

$$\begin{aligned} y &= C_1 y_1 + C_2 y_2 \\ &= C_1 \sin x + C_2 \cos x \end{aligned} \tag{4}$$

where C_1 and C_2 are arbitrary (and independent) constants. To see that this function is a solution of the differential equation, we note that

$$\frac{dy}{dx} = C_1 \cos x - C_2 \sin x$$

and

$$\frac{d^2y}{dx^2} = -C_1 \sin x - C_2 \cos x$$

so

$$\frac{d^2y}{dx^2} + y = (-C_1 \sin x - C_2 \cos x) + (C_1 \sin x + C_2 \cos x)$$

$$= 0$$

as required. It is a fact, which we shall not prove, that any particular solution of this differential equation has the form of the function y in equation (4), with specific values of C_1 and C_2. For instance, the particular solution $y_1 = \sin x$ has $C_1 = 1$ and $C_2 = 0$. □

Example 4 A particular solution of the differential equation

$$\frac{d^2y}{dx^2} + 2\frac{dy}{dx} - 3y = 0$$

is $y_1 = e^{-3x}$. This is because

$$\frac{dy_1}{dx} = -3e^{-3x} \quad \text{and} \quad \frac{d^2y_1}{dx^2} = 9e^{-3x}$$

so

$$\frac{d^2y_1}{dx^2} + 2\frac{dy_1}{dx} - 3y_1 = 9e^{-3x} + 2(-3e^{-3x}) - 3e^{-3x}$$

$$= (9 - 6 - 3)e^{-3x}$$

$$= 0.$$

A second solution is $y_2 = e^x$. We leave this for you to verify, as well as the fact that the general solution

$$y = C_1 e^{-3x} + C_2 e^x$$

satisfies the differential equation. □

Initial Value Problems

Once the general solution of a differential equation is known, a particular solution may be determined if information about the value of a solution, and possibly of its derivative(s), at a particular number x is given. For example, the general solution of the differential equation

$$f'(x) = 3f(x) \tag{5}$$

is $f(x) = Ce^{3x}$. If we are given the additional information that

$$f(0) = 5 \tag{6}$$

we can substitute $x = 0$ in the general solution and use equation (6) to conclude that

$$f(0) = Ce^{3 \cdot 0} = Ce^0 = C = 5$$

so the particular solution of equation (5) satisfying the *initial value* $f(0) = 5$ in equation (6) is

$$f(x) = 5e^{3x}.$$

An **initial value** is simply the value of a solution, or of one of its derivatives, at a particular number x (not necessarily zero). For a first order differential equation one initial value (a value of the solution) is required to determine a particular solution from the general solution. For a second order differential equation we must specify two initial values (the value of the solution and the value of its derivative, at the *same* number x) in order to determine a particular solution. A differential equation, together with enough initial values to determine a unique particular solution, is called an **initial value problem.**

Example 5 Solve the initial value problem

$$\frac{dy}{dt} = 3t^2 + 2 \tag{7}$$

$$y(0) = 5.$$

Solution: Since the right-hand side of equation (7) is a function of the independent variable t alone, we find the general solution by integration:

$$\int \left(\frac{dy}{dt} \right) dt = \int (3t^2 + 2) \, dt$$

$$y(t) = t^3 + 2t + C.$$

We determine the constant C by setting $t = 0$ in the general solution and substituting the initial value $y(0) = 5$:

$$y(0) = 0^3 + 2 \cdot 0 + C = C = 5.$$

Thus $C = 5$, and the solution of the initial value problem is

$$y = t^3 + 2t + 5. \qquad \Box$$

Example 6 A manufacturer of dehumidifiers determines that its marginal cost of producing units at production level x units per week is $MC(x) = 70 + 0.2x$ dollars per unit. Find its total cost $C(x)$ in producing x units per week if its fixed cost of production is $C(0) = \$2000$ per week.

Solution: Since marginal cost is the derivative $MC(x) = C'(x)$ of total cost, we must solve the initial value problem:

$$C'(x) \;=\; 70 + 0.2x \tag{8}$$

$$C(0) \;=\; 2000. \tag{9}$$

The general solution C of equation (8) is found by integration:

$$\int C'(x) \, dx = \int (70 + 0.2x) \, dx$$

$$C(x) = 70x + 0.1x^2 + K$$

where K is a constant. To determine the constant K, we set $x = 0$ in the expression for $C(x)$ and use equation (9):

$$C(0) = 70 \cdot 0 + (0.1) \cdot 0^2 + K = 2000$$

so $K = 2000$. The solution is therefore

$$C(x) = 70x + 0.1x^2 + 2000 \text{ dollars per week.}$$ □

Graphical Solutions

We can sometimes get a notion of the behavior of the solutions of a first order differential equation by remembering that the derivative f' gives the slope of the line tangent to the graph of f. If the form of the differential equation allows us to determine values of the derivative f' corresponding to particular points in the xy-plane, we can sketch in small tangent lines with slopes f' at various points. The resulting sketch, called a **direction field**, helps us develop an idea of how graphs of various solutions must behave.

Example 7 Figure 1.1 shows tangents sketched onto the plane as determined by the differential equation

$$\frac{dy}{dx} = \frac{1}{2}y. \tag{10}$$

Slopes of these tangents are obtained by selecting various numbers y and noting that the corresponding slope is always half of the selected number y. $\Big($Note that the slopes do not depend on x, since x does not appear in the expression for $\dfrac{dy}{dx}$.$\Big)$ Figure 1.2 shows several solution curves $y = Ce^{x/2}$ associated with the differential equation (10), for various values of C. □

Figure 1.1 Direction field for differential equation

$$\frac{dy}{dx} = \frac{1}{2}y.$$

Figure 1.2 Solution curves $y = Ce^{x/2}$ through direction field in Figure 1.1.

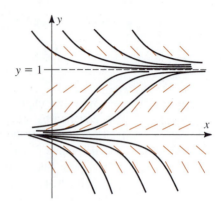

Figure 1.3 Direction field for differential equation

$$\frac{dy}{dx} = y(1 - y).$$

Figure 1.4 Solution curves associated with direction field in Figure 1.3.

Example 8 Figures 1.3 and 1.4 show the direction field and several corresponding solution curves for the logistic differential equation

$$\frac{dy}{dx} = y(1 - y).$$

Note that the derivative equals zero for $y = 0$ and $y = 1$, and that the sign of the derivative $\frac{dy}{dx}$ is:

   (i) negative for $y > 1$, since $1 - y < 0$.
  (ii) positive for $0 < y < 1$, since $1 - y > 0$.
 (iii) negative for $y < 0$, since $1 - y > 0$.

We shall show in Section 10.2 how the technique of separation of variables can be used to find explicit solutions of this differential equation. ☐

Exercise Set 10.1

In Exercises 1–10 verify that the given general solution satisfies the differential equation.

1. $\dfrac{dy}{dx} = 6y;$ $y = Ce^{6x}$

2. $\dfrac{dy}{dx} = \dfrac{x}{y};$ $y = \sqrt{x^2 + C}$

3. $\dfrac{d^2y}{dx^2} + 4y = 0;$ $y = A \sin 2x + B \cos 2x$

4. $f''(t) - 9f(t) = 0;$ $f(t) = Ae^{3t} + Be^{-3t}$

5. $\dfrac{dy}{dt} = 2 - 4y;$ $y = \dfrac{1}{2} + Ce^{-4t}$

6. $f'(t) = 2 - 0.02f(t);$ $f(t) = 100 + Ce^{-0.02t}$

7. $\dfrac{dy}{dx} = 10(22 - y);$ $y = 22 - Ce^{-10x}$

8. $\dfrac{d^2y}{dx^2} + 5\dfrac{dy}{dx} + 6y = 0;$ $y = Ae^{-2x} + Be^{-3x}$

9. $f''(t) + 4f'(t) + 4f(t) = 0;$ $f(t) = Ae^{-2t} + Bte^{-2t}$

10. $\dfrac{dy}{dt} + 2ty = t;$ $y = Ce^{-t^2} + \dfrac{1}{2}$

In Exercises 11–16 find the general solution of the given differential equation by integration.

11. $\dfrac{dy}{dx} = 9 - x$

12. $f'(t) = e^{2t} + 2$

13. $\dfrac{dy}{dt} = te^t$

14. $\dfrac{dy}{dx} = \sec^2 x$

15. $\dfrac{d^2y}{dt^2} = \dfrac{t + 1}{\sqrt{t}}, \quad t > 0$

16. $\dfrac{d^2y}{dt^2} = \cos t$

In Exercises 17–20 solve the initial value problem.

17. $\dfrac{dy}{dx} = x\sqrt{x^2 + 1}$

$y(0) = 1$

18. $\dfrac{dy}{dx} = \dfrac{x}{y}$

$y(0) = 1$

(See Exercise 2.)

19. $\dfrac{dy}{dt} = 2 - 4y$

$y(0) = 3$

(See Exercise 5.)

20. $\dfrac{dy}{dt} = \sec^2 t$

$y(0) = \pi$

In Exercises 21–24 sketch a direction field for the differential equation and several solution curves.

21. $\dfrac{dy}{dx} = -y$

22. $\dfrac{dy}{dx} = y(4 - y)$

23. $\dfrac{dy}{dx} = y(y - 3)$

24. $f'(t) = 1 - f(t)$

25. The value of a bottle of a certain rare wine is increasing at a rate of $\sqrt{t + 1}$ dollars per year when the wine is t years old. The wine sold for \$5 per bottle when new.
   a. Write an initial value problem that determines the value $V(t)$ of the bottle of wine after t years.
   b. Find the solution of this initial value problem.

26. A manufacturer of a single product determines that its marginal cost at production level x units per day is $MC(x) = 150 + 2x$ dollars and that its fixed cost of production is \$400 per day.
   a. Write an initial value problem that determines the manufacturer's total cost $C(x)$ of producing x units per day.
   b. Find the solution of this initial value problem.

27. An individual's marginal utility $MU(x)$ at the consumption level of x units per day (of, say, candy) is given by the function $MU(x) = 6 - 2x$. Let $U(x)$ be the individual's utility function so that $MU(x) = U'(x)$. Write an initial value problem describing the individual's utility function for this commodity under the assumption that $U(0) = 0$.

28. Find a differential equation satisfied by the family of functions
   a. $f(t) = Ce^{6t}$
   b. $f(t) = A \sin 3t + B \cos 3t$

10.2 APPLICATIONS TO POPULATION GROWTH

The Law of Natural Growth (Section 4.5) describes situations in which the rate of change of a quantity is proportional to its current size. There are other possible relationships between rate of change and size. We consider two such types of growth models in this section, both of which lead to solutions involving exponential functions.

Limited Growth   Figure 2.1 shows what happens to the temperature of a cool drink when it is removed from a refrigerator and placed in a warm room. Note that the temperature rises rapidly at first, but increases more slowly as it approaches room temperature. This phenomenon is an

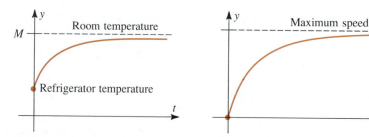

Figure 2.1 Rise of temperature $y(t)$ as a function of time for a cool drink placed in a warm room.

Figure 2.2 "Learning curve" for typing speed t days after beginning instruction.

example of Newton's law of cooling, which states that the rate at which the temperature of the drink will change is proportional to the *difference* between its current temperature and that of its surrounding environment. In mathematical terms this law can be stated

$$\frac{dy}{dt} = k(M - y) \tag{1}$$

where $y(t)$ is the temperature of the drink at time t, M is the temperature in the room, and k is a constant.

A second example of such models is the "learning curve" in Figure 2.2. In this case the function y represents an individual's typing speed t days after beginning a first course on typing. Here speed increases most rapidly in the beginning, then slows as the individual's natural maximum speed is approached. Curves such as those in Figures 2.1 and 2.2 are referred to as *diffusion* curves by sociologists because they model the way certain types of information or propaganda spread throughout a population—rapidly at first, then more slowly as nearly everyone comes to know the information.

The solution of the differential equation

$$\frac{dy}{dt} = k(M - y)$$

has the form

$$y(t) = M(1 - Ce^{-kt}) \tag{2}$$

where C is a constant. (You are asked to verify this in Exercise 13.) When each of the constants M, C, and k is positive, the derivative

$$\frac{dy}{dt} = kMCe^{-kt}$$

is positive for all t. Thus the solution y is increasing for all t. Moreover, the second derivative

$$\frac{d^2y}{dt^2} = -k^2MCe^{-kt}$$

is negative for all t, so the graph of y is concave down for all t. Finally, since $\lim\limits_{t \to \infty} e^{-kt} =$

$\lim\limits_{t \to \infty} \dfrac{1}{e^{kt}} = 0$, we have

$$\lim_{t \to \infty} y(t) = \lim_{t \to \infty} M(1 - Ce^{-kt})$$
$$= M(1 - C \cdot 0)$$
$$= M.$$

Thus the values of the solution y approach the number M as t becomes large. (See Figure 2.3.) For this reason M is sometimes called the **carrying capacity** of the system that is being modelled by equation (1), which is called the **Law of Limited Growth.**

Example 1 Suppose that the percentage of the population that knows the result of an election t hours after the result is announced is

$$P(t) = 100(1 - Ce^{-kt}). \tag{3}$$

If 40% of the population knows the result 2 hours after it is announced, when will 80% of the population know it?

Strategy

Use the initial value $P_0 = 0$ to solve for C.

Solution

Since no one knows the result until it has been announced, we have the initial value $P(0) = 0$. Inserting this data in equation (3) gives

$$0 = P(0) = 100(1 - Ce^{-k \cdot 0})$$
$$= 100(1 - C).$$

Thus $1 - C = 0$, so $C = 1$, and

$$P(t) = 100(1 - e^{-kt}). \tag{4}$$

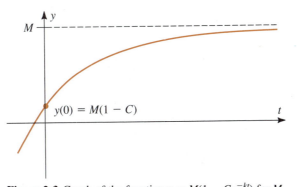

Figure 2.3 Graph of the function $y = M(1 - Ce^{-kt})$ for M, C, and k positive. This function is increasing and concave down for all t.

Use data $P(2) = 40$ to obtain an equation involving the constant k.

Setting $t = 2$ in equation (4) and using the additional data $P(2) = 40$ then gives

$$40 = 100(1 - e^{-2k}).$$

Thus

$$1 - e^{-2k} = \frac{40}{100} = 0.4$$

so

$$e^{-2k} = 1 - 0.4 = 0.6$$

Take natural logs of both sides to solve for k.

and

$$-2k = \ln(0.6).$$

Thus

$$k = -\frac{\ln(0.6)}{2}$$

Write the explicit form of the function $P(t)$.

and the function $P(t)$ is

$$P(t) = 100(1 - e^{(\ln 0.6/2)t}).$$

Set up the equation $P(T) = 80$ and solve for T.

The condition that 80% of the population has heard the result after T hours is $P(T) = 80$, or

$$80 = 100(1 - e^{(\ln 0.6/2)/T})$$

which gives

$$1 - e^{(\ln 0.6/2)/T} = \frac{80}{100}$$

or

$$e^{(\ln 0.6/2)/T} = 1 - 0.8 = 0.2.$$

Take natural logs of both sides again.

Thus

$$\left(\frac{\ln 0.6}{2}\right)T = \ln(0.2)$$

so

$$T = \frac{2[\ln(0.2)]}{\ln(0.6)}$$

$$\approx 6.3 \text{ hours.} \qquad \square$$

Logistic Growth

The logistic growth model resembles the model for the Law of Natural Growth in that the rate of change depends on the current size of the quantity. But it also borrows from the model for limited growth the existence of a maximum possible size M for the quantity or population. (*M* is called the *carrying capacity* of the system, as before.) The logistic growth model assumes that

$$\{\text{Rate of change}\} \propto \{\text{Current size}\} \cdot \{\text{Unutilized capacity for growth}\}. \qquad (5)$$

In order to translate this law into a mathematical equation involving the size of a population $P(t)$ at time t, we note that for unutilized capacity for growth to exist we must have $P < M$, and that the fraction of full capacity available for additional growth will be $\left(\dfrac{M - P}{M}\right)$. We therefore express the logistic growth law in equation (5) as

$$\frac{dP}{dt} = kP\left(\frac{M - P}{M}\right)$$

where k is again called the growth constant. The solution of this differential equation is

$$P(t) = \frac{M}{1 + Ce^{-kt}}$$

where C is an arbitrary constant that is determined by an initial value for the solution.

The graph of a typical solution $y = P(t)$ is shown in Figure 2.4. Note that P is increasing for all $t > 0$, but that its graph is concave up on $(0, a)$ and concave down on (a, ∞), with an inflection point at $(a, P(a))$.

Logistic growth typically occurs when exponential growth would have resulted had the environment not been limited in size or resources. Fisheries provide a typical example, and many studies have been done that have resulted in logistic growth curves. If a pond is stocked with a small number of fish, the growth in the fish population is nearly exponential for awhile, since the limits of the pond's size are not noticeable to a small fish population. But as the size of the population increases the competition for finite food resources and the negative effects of crowding exert a downward pressure on the population's growth rate. As in the case of limited growth, we find that

$$\lim_{t \to \infty} P(t) = \lim_{t \to \infty} \frac{M}{1 + Ce^{-kt}} = \frac{M}{1 + 0} = M$$

which says that the population size approaches the carrying capacity M in the long run.

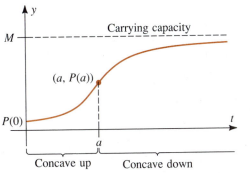

Figure 2.4 Graph of typical logistic curve

$$P(t) = \frac{M}{1 + Ce^{-kt}}.$$

Example 2 A breeding pond is initially stocked with 10 fish. After 5 weeks the pond contains 30 fish. If the growth in the fish population is logistic and the maximum number of fish that the pond can support is 600, find the number of fish that will be present in the pond after 20 weeks.

Solution: Assuming logistic growth, the fish population has size

$$N(t) = \frac{M}{1 + Ce^{-kt}}$$

where t is time in weeks. We are given that the carrying capacity of the pond is $M = 600$ fish. Using this constant and the initial value $P(0) = 10$ gives

$$10 = \frac{600}{1 + Ce^{-k \cdot 0}} = \frac{600}{1 + C}.$$

We can now solve for C:

$$\frac{600}{1 + C} = 10 \quad \text{gives} \quad 1 + C = \frac{600}{10} = 60$$

so $C = 59$. We can therefore write the function P as

$$P(t) = \frac{600}{1 + 59e^{-kt}}.$$

Inserting the data $P(5) = 30$ gives an equation in k alone:

$$30 = \frac{600}{1 + 59e^{-5k}}$$

so

$$1 + 59e^{-5k} = \frac{600}{30} = 20.$$

Thus

$$59e^{-5k} = 19$$

so

$$e^{-5k} = \frac{19}{59}$$

and

$$k = -\frac{1}{5} \ln \frac{19}{59}.$$

Thus

$$P(t) = \frac{600}{1 + 59e^{((1/5)\ln(19/59))t}}$$

so the size of the population after 20 weeks is

$$P(20) = \frac{600}{1 + 59e^{((1/5)\ln(19/59))20}}$$

$$= \frac{600}{1 + 59e^{4\ln(19/59)}}$$

$$= \frac{600}{1 + 59 \cdot (19/59)^4}$$

$$\approx 367 \text{ fish.}$$

□

Exercise Set 10.2

1. In Example 1 approximately what percentage of the population knows the result of the election 5 hours after it is announced?

2. In Example 2 how many fish will be present in the pond after ten weeks?

3. In Example 2 when will the pond contain 300 fish?

4. When word processing equipment was made available to staff in a certain corporation, it was found that t months after its introduction $W(t) = 140(1 - e^{-0.5t})$ employees used the equipment at least once per day.
   a. How many employees used word processing equipment after six months?
   b. What is the number of employees that can eventually be expected to use the equipment?

5. A marketing firm predicts that for a newly introduced personal computer the number of units that will be sold during the first month is

$$N(x) = 10,000(1 - e^{-0.002x})$$

where x is the number of dollars spent on introductory advertising. For what level of advertising x will sales increase most rapidly as a function of x?

6. An epidemiologist predicts that t days after the outbreak of winter colds

$$N(t) = \frac{500}{1 + 10e^{-kt}}$$

residents of a certain town will have caught a cold.
   a. If 100 of the residents have caught a cold after ten days, what is k?
   b. When will 200 of the residents have caught a cold?

7. The number of subscribers to cable television t months after its introduction in a certain city is expected to be

$$N(t) = \frac{80,000}{1 + 200e^{-0.2t}}.$$

   a. How many subscribers will there be after six months?
   b. How many subscribers will ultimately exist?

8. Find the inflection point for the graph of the function

$$f(x) = \frac{1}{1 + e^{-t}}.$$

9. In a certain town t days after the outbreak of influenza

$$N(t) = \frac{400}{1 + 100e^{-t}}$$

individuals have the flu. When is the rate at which the influenza is spreading the greatest?

10. A rumor spreads through an office complex so that t hours after the rumor has begun

$$N(t) = \frac{50}{1 + 49e^{-t}}$$

people have heard the rumor. How many people have heard it after $t = 4$ hours?

11. The learning curve $N(t) = 30(1 - e^{-0.05t})$ models the number of vocabulary words that a student in an elementary French course can master in t minutes of study in a single study session.
   a. How many vocabulary words can a student master in 20 minutes?
   b. When will a student have mastered 20 words?

12. A bottle of soda at temperature 6°C is removed from a refrigerator and placed in a room at temperature 22°C. If, after 10 minutes, the temperature of the soda has risen to 14°C, find its temperature after 20 minutes according to Newton's law of cooling.

13. Verify that the function $y = M(1 - Ce^{-kt})$ satisfies the differential equation $\dfrac{dy}{dt} = k(M - y)$.

14. Verify that the function $P(t) = \dfrac{M}{1 + Ce^{-kt}}$ satisfies the logistic differential equation $\dfrac{dP}{dt} = kP\left(\dfrac{M - P}{M}\right)$.

10.3 SEPARATION OF VARIABLES

There are several techniques, in addition to simple integration, that can be used to solve first order differential equations. Of these, **separation of variables** is the simplest. This technique applies to differential equations that have the form

$$\frac{dy}{dx} = \frac{f(x)}{g(y)}, \qquad g(y) \neq 0 \tag{1}$$

which we refer to as *separable* differential equations. The reason for this terminology is that we may multiply both sides of equation (1) by $g(y)$, thus obtaining the equation

$$g(y) \cdot \frac{dy}{dx} = f(x). \tag{2}$$

In doing so, we say that we have separated the variables in equation (1) since in equation (2) the left-hand side involves only the solution y and its derivative (assuming these exist) while the right-hand side is a function of x alone. Here is why this is helpful, in general:

(i) If we can find an antiderivative G for the function g, the left-hand side of equation (2) can be written as

$$\frac{d}{dx}G(y) = G'(y) \cdot \frac{dy}{dx} = g(y) \cdot \frac{dy}{dx}. \tag{3}$$

(ii) If we can find an antiderivative F for f, we can write the right-hand side of equation (2) as

$$\frac{d}{dx}F(x) = f(x). \tag{4}$$

(iii) Using equations (3) and (4), we can then rewrite equation (2) as

$$\frac{d}{dx}G(y) = \frac{d}{dx}F(x). \tag{5}$$

It then follows that

$$G(y) = F(x) + C \tag{6}$$

since both F and G have the same derivatives with respect to x (Theorem 1, Chapter 5).

REMARK 1: Equation (6) represents the general solution of equation (1) since the derivative $\frac{dy}{dx}$ has been eliminated. However, this may be an equation that involves y implicitly rather than as an explicit function of x.

REMARK 2: The differential notation $dy = \left(\frac{dy}{dx}\right) dx$ is often used to abbreviate the preceding discussion by saying, beginning with equation (1), that

$$\frac{dy}{dx} = \frac{f(x)}{g(y)} \quad \text{implies that} \quad g(y) \, dy = f(x) \, dx$$

so

$$\int g(y) \, dy = \int f(x) \, dx$$

and we obtain the solution by "integrating both sides."

Example 1 Use the technique of separation of variables to find the general solution of the differential equation

$$\frac{dy}{dx} = \frac{x}{y}, \quad y \neq 0.$$

Strategy

Separate the variables.

Solution

Multiplying both sides by y gives

$$y \cdot \frac{dy}{dx} = x$$

so

Find antiderivatives of both sides.

$$\int y \cdot \left(\frac{dy}{dx}\right) dx = \int x \, dx.$$

Thus

$$\frac{1}{2}y^2 + C_1 = \frac{1}{2}x^2 + C_2$$

Solve for y^2, combining arbitrary constants.

so

$$y^2 + 2C_1 = x^2 + 2C_2$$

or

$$y^2 = x^2 + C$$

where $C = 2C_2 - 2C_1$. Graphs of this equation for various values of C appear in Figure 3.1.

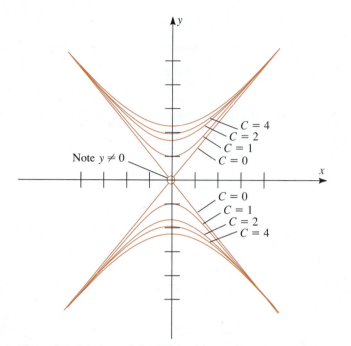

Figure 3.1 Solutions of the differential equation

$$\frac{dy}{dx} = \frac{x}{y}.$$

Check result by differentiation.

To check this result, differentiate implicitly both sides of the equation $y^2 = x^2 + C$ to obtain

$$2y \cdot \frac{dy}{dx} = 2x$$

or

$$\frac{dy}{dx} = \frac{x}{y}.$$

Each solution consists of one branch above the x-axis and one branch below it. □

Example 2 Find the solution of the initial value problem

$$\frac{dy}{dx} = 2xy^2, \qquad y \neq 0$$

$$y(0) = \frac{1}{2}.$$

Solution: This time we separate variables by dividing both sides of the differential equation by y^2. This gives

$$y^{-2} \cdot \frac{dy}{dx} = 2x$$

so

$$\int y^{-2} \left(\frac{dy}{dx} \right) dx = \int 2x \, dx$$

and therefore

$$-y^{-1} + C_1 = x^2 + C_2.$$

Thus

$$-\frac{1}{y} = x^2 + C, \qquad \text{where } C = C_2 - C_1$$

so

$$y = \frac{-1}{x^2 + C}$$

is the general solution of the differential equation.

The initial condition $y(0) = \dfrac{1}{2}$ means that when $x = 0$, we must have

$$\frac{1}{2} = \frac{-1}{0^2 + C} = \frac{-1}{C}$$

so $C = -2$. The solution of the initial value problem is therefore

$$y = \frac{-1}{x^2 - 2} = \frac{1}{2 - x^2}. \qquad \qquad \square$$

Example 3 The equation that we have referred to as the Law of Natural Growth

$$\frac{dP}{dt} = kP$$

is a separable differential equation. Dividing both sides by the nonzero function P gives

$$\frac{1}{P} \cdot \frac{dP}{dt} = k.$$

Integrating with respect to t then gives

$$\int \frac{1}{P} \left(\frac{dP}{dt} \right) dt = \int k \, dt$$

so

$$\ln |P| + C_1 = kt + C_2$$

or

$$\ln P = kt + C_3, \qquad C_3 = C_2 - C_1, \qquad P > 0.$$

Taking exponentials of both sides then gives

$$e^{\ln P} = e^{kt+C_3} = e^{C_3} \cdot e^{kt}.$$

Writing $C = e^{C_3}$ (since e to a constant power is just another constant) and noting that $e^{\ln P} = P$ gives

$$P = Ce^{kt}$$

as claimed in Chapter 4. Note that, as required, P is positive for all t.   □

Example 4 In Section 2 we developed a logistic model for growth in a limited environment:

$$\frac{dP}{dt} = kP\left(\frac{K - P}{K}\right) \tag{7}$$

where K is a constant representing the carrying capacity or maximum sustainable population size for the environment and the population is of size $P(t)$ at time t.

We may separate the variables in equation (7) by dividing both sides by the nonzero factor $P(K - P)$, obtaining

$$\frac{1}{P(K - P)} \cdot \frac{dP}{dt} = \frac{k}{K}$$

so

$$\int \frac{1}{P(K - P)} \cdot \left(\frac{dP}{dt}\right) dt = \int \frac{k}{K}\, dt. \tag{8}$$

We leave it as an exercise for you to verify that the integrand on the left is

$$\frac{1}{P(K - P)} = \frac{1}{K} \cdot \frac{1}{P} + \frac{1}{K} \cdot \frac{1}{(K - P)}.$$

The integral on the left-hand side of equation (8) is therefore

$$
\begin{aligned}
\int \frac{1}{P(K - P)} \left(\frac{dP}{dt}\right) dt &= \int \frac{1}{P(K - P)}\, dP \\
&= \frac{1}{K} \int \left(\frac{1}{P} + \frac{1}{K - P}\right) dP \\
&= \frac{1}{K}[\ln P - \ln(K - P)] + C_1, \qquad 0 < P < K \\
&= \frac{1}{K} \ln\left(\frac{P}{K - P}\right) + C_1.
\end{aligned}
$$

Since the integral on the right-hand side of equation (8) is $\dfrac{k}{K}t + C_2$, we have

$$\frac{1}{K}\ln\left(\frac{P}{K-P}\right) + C_1 = \frac{k}{K}t + C_2.$$

Thus

$$\ln\left(\frac{P}{K-P}\right) = kt + C, \qquad \text{where } C = K(C_2 - C_1)$$

so

$$\frac{P}{K-P} = e^{kt+C}.$$

Solving for P, we obtain

$$Pe^{-kt-C} = K - P$$

so

$$P(1 + Me^{-kt}) = K, \qquad \text{where } M = e^{-C}$$

and

$$P(t) = \frac{K}{1 + Me^{-kt}}. \tag{9}$$

The graph of this solution, for $0 < P(0) < K$, appears in Figure 3.2. Note that $\displaystyle\lim_{t\to\infty} P(t) = K$, so the carrying capacity of the environment is a horizontal asymptote of the population curve. □

Example 5 Sociologists model the movement of information (such as a rumor) through a population by assuming that the rate at which information spreads is proportional to the product of the proportion of the population that already knows the information and the proportion of the population that does not yet know the information. If we let $P(t)$ be the proportion of the population knowing the information at time t, then this model corresponds to the differential equation

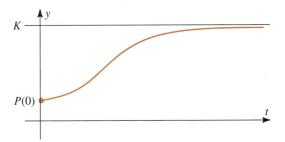

Figure 3.2 Solution of logistic growth equation in Example 4.

$$\frac{dP}{dt} = kP(1 - P). \tag{10}$$

Now equation (10) is just the equation (7) for logistic growth with $K = 1$. The solution is therefore

$$P(t) = \frac{1}{1 + Me^{-kt}}$$

as given by equation (9) with $K = 1$. $\square$

Generalizing the Law of Natural Growth

The differential equation $\frac{dy}{dt} = ky$ for natural growth can be generalized in several ways. One such generalization is to differential equations of the form

$$\frac{dy}{dt} = ky - b \tag{11}$$

where b is a constant. This model applies, for example, to a fishery when the size of the fish population is small relative to the carrying capacity, and fish are harvested at the constant rate of b fish per unit time.

Letting $a = \frac{b}{k}$, we can write equation (11) as

$$\frac{dy}{dt} = k(y - a), \qquad a \text{ constant.} \tag{12}$$

The interpretation of equation (12) is now that the rate of change of y is proportional to the *difference* between the population size y and the constant a.

The following example shows how the method of separation of variables may be applied to equation (12).

Example 6 Find the general solution of the differential equation

$$\frac{dy}{dt} = k(y - a), \qquad y - a \neq 0. \tag{13}$$

Solution: Dividing both sides by $y - a$ gives

$$\frac{1}{y - a} \cdot \frac{dy}{dt} = k$$

so

$$\int \frac{1}{y - a} \left(\frac{dy}{dt} \right) dt = \int k \, dt.$$

Integrating both sides then gives

$$\ln |y - a| + C_1 = kt + C_2$$

so

$$\ln |y - a| = kt + C_3, \qquad C_3 = C_2 - C_1.$$

Taking exponentials of both sides gives

$$\begin{aligned}
|y - a| &= e^{kt + C_3} \\
&= e^{C_3} \cdot e^{kt} \\
&= C_4 e^{kt}, \qquad C_4 = e^{C_3}.
\end{aligned}$$

so

$$\begin{aligned}
y - a &= \pm C_4 e^{kt} \\
&= C e^{kt}, \qquad C = \pm C_4.
\end{aligned}$$

Thus

$$y = a + C e^{kt}$$

where the arbitrary constant C (which may be either positive or negative) is determined by an initial value. $\square$

Since we shall use the result of Example 6 in our last three examples, we restate it as follows.

> The solution of the differential equation
>
> $$\frac{dy}{dt} = k(y - a) \tag{14}$$
>
> is
>
> $$y = a + C e^{kt}. \tag{15}$$

Example 7 Find the solution of the initial value problem

$$\frac{dy}{dt} = 2 - y \tag{16}$$

$$y(0) = 1. \tag{17}$$

Solution: To compare equation (16) with equation (14), we rewrite it in the form

$$\frac{dy}{dt} = -(y - 2).$$

Thus equation (16) has the form of equation (14) with $k = -1$ and $a = 2$. The solution of the differential equation, according to equation (15), is

$$y = 2 + C e^{-kt}.$$

Setting $t = 0$ and applying the initial value $y(0) = 1$ from equation (17) gives

Figure 3.3 Solution of differential equation

$$\frac{dy}{dt} = 2 - y \text{ with } y(0) = 1$$

is $y = 2 - e^{-t}$.

Figure 3.4 Diagram for "mixing" problems showing inputs, thorough mixing, and output.

$$1 = y(0) = 2 + Ce^0 = 2 + C$$

so $C = 1 - 2 = -1$. The solution of the initial value problem is therefore

$$y = 2 - e^{-t}.$$

(See Figure 3.3.) □

Mixing Problems

Differential equations of the form of equation (14) arise in problems involving the mixing of two or more liquids. The typical situation is depicted in Figure 3.4. The liquids are entering a container at a specified rate, the contents of the container are being kept well mixed, and the resulting mixture is being removed at a specified rate.

Example 8 A 500-liter tank in which chocolate milk is being mixed contains 460 liters of milk and 40 liters of chocolate syrup initially. Syrup and milk are then added to the tank at the rate of 2 liters per minute of syrup and 8 liters per minute of milk. Simultaneously, the mixture is withdrawn at the rate of 10 liters per minute. Assuming perfect mixing of the milk and syrup, find the function y giving the amount of syrup $y(t)$ in the tank at time t.

Solution: If y represents the amount of syrup in the tank, then $\dfrac{dy}{dt}$ is the rate of change of y, which is the rate at which syrup enters the tank (2 liters per minute) *minus* the rate at which syrup leaves the tank. The rate at which syrup leaves the tank depends on how much syrup is in the mixture; it is

$$\frac{y \text{ liters of syrup}}{500 \text{ liters of mixture}} \times (10 \text{ liters of mixture per minute withdrawn})$$

$$= \frac{y}{50} = 0.02y \text{ liters of syrup per minute.}$$

Thus

$$\frac{dy}{dt} = 2 - 0.02y$$

is the differential equation describing this process. This equation can be written in the form of equation (14) as

$$\frac{dy}{dt} = -0.02(y - 100),$$

so $k = -0.02$ and $a = 100$ in equation (14). The solution is, according to equation (15)

$$y = 100 + Ce^{-0.02t}. \tag{18}$$

Since 40 liters of syrup are present initially, we have the initial value $y(0) = 40$. Setting $t = 0$ in equation (18) and using this initial value gives

$$40 = 100 + Ce^0 = 100 + C$$

so $C = -60$. The desired solution is therefore

$$y = (100 - 60e^{-0.02t}) \text{ liters.}$$

The graph of this solution appears in Figure 3.5. Values of the solution for various times t appear in Table 3.1. □

Table 3.1. Calculated values $y(t)$ in Figure 3.4

t minutes	0	10	20	30	40	50	60	100	150	200
$y = 100 - 60e^{-0.02t}$ liters	40.0	50.9	59.8	67.1	73.1	77.9	81.9	91.9	97.0	98.9

REMARK: Figure 3.4 illustrates an important property of solutions of differential equations of the form $\frac{dy}{dt} = k(y - a)$, namely, when $k < 0$, the solution approaches a horizontal asymptote as $t \to \infty$. For the solution y of Example 8 we have

$$\lim_{t \to \infty} y = \lim_{t \to \infty} (100 - 60e^{-0.02t}) = 100$$

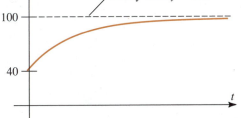

Figure 3.5 Solution of the mixing problem in Example 8.

as Figure 3.4 suggests. The asymptote $y = 100$ is called the *steady state* part of the solution, while the term $60e^{-0.02t}$ is referred to as the *transient* part of the solution. The result of Example 8 is precisely what intuition suggests—regardless of the initial concentration of syrup, the steady state concentration will be $\dfrac{100}{500} = 0.2$, since 2 liters of syrup enter for every 8 liters of milk.

Example 9 Newton's law of cooling states that the rate at which an object gains or loses heat is proportional to the difference in temperature between the object and its surroundings. A bottle of soda at temperature 6°C is removed from a refrigerator and placed in a room at temperature 22°C. If after 10 minutes the temperature of the soda is 14°C, find its temperature $y(t)$ as a function of t.

Solution: According to Newton's law, the rate $\dfrac{dy}{dt}$ at which the temperature of the soda changes is proportional to the quantity $(y - 22)$. That is

$$\frac{dy}{dt} = k(y - 22)$$

for some constant k. This equation has the form of equation (14) with $a = 22$. The general solution is therefore

$$y = 22 + Ce^{kt} \tag{19}$$

and our task is to determine the constants C and k. Since we are given the initial value $y(0) = 6°$, we may set $t = 0$ in equation (19) and conclude that

$$6 = y(0) = 22 + Ce^0 = 22 + C$$

so $C = -16$, and the solution has the form

$$y = 22 - 16e^{kt}. \tag{20}$$

To determine k, we use the additional information that $y(10) = 14°$ by setting $t = 10$ in equation (20):

$$14 = y(10) = 22 - 16e^{10k}$$

or

$$16e^{10k} = 8.$$

Thus

$$e^{10k} = \frac{1}{2}$$

so

$$\ln(e^{10k}) = 10k = \ln\left(\frac{1}{2}\right) = -\ln 2$$

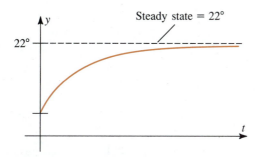

Figure 3.6 Graph of temperature function $y(t)$ in Example 9.

and

$$k = -\frac{\ln 2}{10}.$$

The temperature $y(t)$ of the soda at time t is therefore

$$y(t) = 22 - 16e^{(-\ln 2/10)t}.$$

The graph of this solution appears in Figure 3.6, and values $y(t)$ for various numbers t are given in Table 3.2. Again, note that the solution contains a steady state part ($y = 22°$), which $y(t)$ approaches as $t \to \infty$.   □

Table 3.2. Values of the temperature function y in Example 9

t	0	4	8	12	16	20	24	28	32	36
$y(t)$	6.0	9.87	12.81	15.04	16.72	18.00	18.97	19.70	20.26	20.68

Exercise Set 10.3

In Exercises 1–10 use the technique of separation of variables to find the general solution of the differential equation.

1. $\dfrac{dy}{dx} = \dfrac{y}{x}, \quad x \neq 0$

2. $\dfrac{dy}{dx} = \dfrac{x}{y^2}, \quad y \neq 0$

3. $\dfrac{dy}{dx} = -4xy^2$

4. $\dfrac{dy}{dx} = \dfrac{\sqrt{x}}{\sqrt{y}}, \quad y > 0$

5. $\dfrac{dy}{dx} = e^{x-y}$

6. $\dfrac{dy}{dx} = \dfrac{y \ln x}{x}$

7. $e^y \dfrac{dy}{dt} - t^2 = 0$

8. $\dfrac{dy}{dt} + y \sin t = 0$

9. $\dfrac{dy}{dx} = (1 + x)(1 + y)$

10. $2t\dfrac{dy}{dt} = \dfrac{t^2 + 1}{y^2}$

In Exercises 11–20 find a solution of the initial value problem.

11. $y' = 5y, \quad y(1) = 4e^5$

12. $yy' = x + 2, \quad y(0) = 3$

13. $\dfrac{dy}{dx} = xy, \quad y(2) = 1$

14. $\dfrac{dy}{dx} = -\dfrac{y}{x^2}, \quad y(1) = 2$

15. $\dfrac{dy}{dx} = y \cos x, \quad y(0) = \pi$

16. $y' = \sqrt{y}, \quad y(0) = 1$

17. $y' = 2(y - 1), \quad y(0) = 1$

18. $\dfrac{dy}{dt} = 5(10 - y), \quad y(0) = 5$

19. $y' - 3y = 6, \quad y(0) = 2$

20. $\dfrac{dy}{dt} + 2y = 2, \quad y(0) = 4$

21. Find a solution of the logistic differential equation $\dfrac{dP}{dt} = 2P\left(\dfrac{100 - P}{100}\right)$. What is the carrying capacity? (See Example 4.)

22. Let $P(t)$ be a solution of the differential equation $\dfrac{dP}{dt} = P(1 - P)$ with $P(0) > 0$. What is $\lim\limits_{t \to \infty} P(t)$? (See Example 4.)

23. Radioactive uranium-238 decays according to the differential equation $\dfrac{dU}{dt} = kU$ with $k < 0$. The half-life (the time required for the uranium to decay to half the original amount, as explained in Example 2 of Section 4.5) is known to be 4.5 billion years. What is k?

24. The population of a city is growing according to the Law of Natural Growth. If its population in 1970 was 0.2 million and its population in 1980 was 0.5 million, what will its population be in 1990?

25. A tank contains 200 liters of brine. The initial concentration of salt is 0.5 kilogram per liter of brine. Fresh water is then added at a rate of 3 liters per minute and the brine is drawn off from the bottom of the tank at the same rate. Assume perfect mixing.
  a. Find a differential equation for the number of kilograms of salt, $p(t)$, in the solution at time t.
  b. How much salt is present after 20 minutes?
  c. What is the steady state concentration of salt?

26. A bottle of water at temperature 26°C is placed in a room at temperature 12°C. If the temperature of the water after 30 minutes is 20°C, find its temperature after 2 hours.

27. A person initially places $100 in a savings account that pays 10% per year compounded continuously. If the person makes additional deposits of $100 per year continuously through each year

a. Find a differential equation for $P(t)$, the amount on deposit after t years.
b. Find the amount on deposit after 7 years.

28. Show that the graph of the solution to the initial value problem

$$\frac{dy}{dt} = y(1 - y), \qquad y(0) = \frac{1}{4}$$

has an inflection point where $y = \dfrac{1}{2}$. (*Hint:* Find an equation for $\dfrac{d^2y}{dt^2}$ by differentiating both sides of the differential equation for y with respect to t. When is $\dfrac{d^2y}{dt^2}$ equal to zero?)

29. According to the sociologists' model of Example 5, if 100 people in a city of 100,000 know a rumor on Monday, and 500 know it on Wednesday, how many will know it on Saturday?

30. In the chemical reaction called inversion of raw sugar the inversion rate is proportional to the amount of raw sugar remaining. If 100 kilograms of raw sugar is reduced to 75 kilograms in 6 hours, how long will it be until
  a. half the raw sugar has been inverted?
  b. 90% of the raw sugar has been inverted?

31. A pan of boiling water is removed from a burner and cools to 80°C in 3 minutes. Find its temperature after 10 minutes, according to Newton's law of cooling, if the surrounding temperature is 30°C.

32. A room containing 1000 cu ft of air is initially free of pollutants. Air containing 100 parts of pollutants per cu ft is pumped into the room at a rate of 50 cu ft per minute. The air in the room is kept perfectly mixed, and air is removed from the room at the same rate of 50 cu ft per minute. Find the function $P(t)$ giving the number of parts of pollutants per cubic foot in the room air after t minutes.

33. A tank contains a mixture of 500 gallons of salt brine (salt and water) initially containing 2 pounds of salt per gallon. A second mixture, containing 6 pounds of salt per gallon, is then pumped in at a rate of 10 gallons per minute. The resulting mixture is drawn off at the same rate. Assuming perfect mixing, find the function that describes the concentration of salt (pounds per gallon) in the mixture in the tank t minutes after mixing begins.

34. Influenza spreads through a university community at a rate proportional to the product of the proportion of those already infected and the proportion of those not yet infected. Assuming that those infected remain infected, find the proportion $P(t)$ infected after t days if initially 10% were infected and after 3 days 30% were infected.

35. An annuity is initially set up containing $10,000 on which interest is compounded continuously at a rate of 10% per year. The owner of the annuity withdraws $2000 per year in a continuous manner. When does the value of the annuity reach zero?

36. Certain biological populations grow in size according to the Gompertz growth equation

$$\frac{dy}{dt} = -ky \ln y$$

where k is a positive constant. Find the general solution of this differential equation.

37. The Weber-Fechner law in psychology is a model for the rate of change of the reaction y to a stimulus of strength s that is given by the differential equation

$$\frac{dy}{ds} = k\frac{y}{s}$$

where k is a positive constant. Find the general solution of this differential equation.

10.4 SECOND ORDER EQUATIONS

By adding a term involving the second derivative to the differential equation $y' = ky$ for natural growth, we obtain one type of *second order* differential equation,*

$$\frac{d^2y}{dt^2} + a\,\frac{dy}{dt} + by = 0 \tag{1}$$

where a and b are constants. From our experience with first order equations we might suspect that equation (1) has a solution of the form $y = e^{rt}$ where r is a constant. If this were true, the derivative of this proposed solution would be

$$\frac{dy}{dt} = re^{rt}, \quad \text{and} \quad \frac{d^2y}{dt^2} = r^2e^{rt}.$$

Substituting into equation (1) would therefore give

$$r^2e^{rt} + a(re^{rt}) + b(e^{rt}) = 0$$

or

$$(r^2 + ar + b)e^{rt} = 0. \tag{2}$$

Since values of the exponential function $y = e^{rt}$ are always nonzero, the only way in which equation (2) can be true is if the *characteristic polynomial* $r^2 + ar + b$ equals zero. That is, r must be a (real) *root* of the quadratic polynomial $r^2 + ar + b$.

---

*More precisely, equation (1) is referred to as a second order, constant-coefficient, linear, homogeneous equation. The term homogeneous refers to the fact that the right-hand side is zero. This is but one type of second order equation. There are many others, but we shall not pursue them here.

> The function $y = e^{rt}$ is a solution of the differential equation
>
> $$\frac{d^2y}{dt^2} + a\frac{dy}{dt} + by = 0$$
>
> only if the constant r is a real root of the characteristic polynomial
>
> $$r^2 + ar + b.$$

Thus we expect to find two, one, or zero solutions of the differential equation (1) of the form $y = e^{rt}$, depending on the number of real roots r of the associated characteristic polynomial.

Example 1 For the differential equation

$$\frac{d^2y}{dt^2} - \frac{dy}{dt} - 2y = 0$$

the constant coefficients are $a = -1$ and $b = -2$, so the characteristic polynomial is

$$r^2 - r - 2 = (r - 2)(r + 1).$$

This characteristic polynomial has real roots $r_1 = 2$ and $r_2 = -1$. Two solutions of the differential equation are therefore

$$y_1 = e^{r_1 t} = e^{2t} \quad \text{and} \quad y_2 = e^{r_2 t} = e^{-t}$$

as you can verify. ☐

When the characteristic polynomial $r^2 + ar + b$ does not have distinct real roots, one of two remaining possibilities will occur. Without proof, we shall simply state these possible outcomes and provide an example of each.

> If the characteristic polynomial $r^2 + ar + b$ of the differential equation
>
> $$\frac{d^2y}{dt^2} + a\left(\frac{dy}{dt}\right) + by = 0 \tag{3}$$
>
> has the *single* repeated real root r, then both of the functions
>
> $$y_1 = e^{rt} \quad \text{and} \quad y_2 = te^{rt}$$
>
> are solutions of the differential equation (3).

Example 2 The differential equation

$$\frac{d^2y}{dt^2} + 4\left(\frac{dy}{dt}\right) + 4y = 0$$

has the characteristic polynomial

$$r^2 + 4r + 4 = (r + 2)^2$$

which has the single repeated real root $r = -2$. According to the preceding statement, the differential equation has the two solutions

$$y_1 = e^{-2t} \quad \text{and} \quad y_2 = te^{-2t}$$

which you can verify by direct substitution.   □

If the characteristic polynomial $r^2 + ar + b$ of the differential equation

$$\frac{d^2y}{dt^2} + a\left(\frac{dy}{dt}\right) + by = 0 \qquad (4)$$

has no real roots, then it has two distinct complex (imaginary) roots $r_1 = \alpha + i\beta$ and $r_2 = \alpha - i\beta$ where α and β are real numbers with $\beta \neq 0$, and i is the complex number defined by the equation $i^2 = -1$. In this case the differential equation (4) will have the two solutions

$$y_1 = e^{\alpha t} \sin \beta t \quad \text{and} \quad y_2 = e^{\alpha t} \cos \beta t. \qquad (5)$$

Example 3 The differential equation

$$y'' + 2y' + 2y = 0 \qquad (6)$$

has characteristic polynomial $r^2 + 2r + 2$ with roots, according to the quadratic formula

$$r = \frac{-2 \pm \sqrt{2^2 - 4(1)(2)}}{2(1)}$$

$$= \frac{-2 \pm \sqrt{-4}}{2}$$

$$= \frac{-2 \pm (\sqrt{4})(\sqrt{-1})}{2}$$

$$= -1 \pm i, \qquad i^2 = -1.$$

Thus the characteristic polynomial has roots $r = \alpha \pm \beta i$ with $\alpha = -1$ and $\beta = 1$. According to equation (5), differential equation (6) has solutions

$$y_1 = e^{-t} \sin t \quad \text{and} \quad y_2 = e^{-t} \cos t$$

as you can verify by direct substitution.   □

Example 4 Differential equations of the form

$$y'' + by = 0 \qquad (7)$$

have characteristic polynomial $r^2 + b = 0$. Since this gives $r^2 = -b$, in the case when $b > 0$, we obtain the complex roots $r = \pm\sqrt{b}i$. Thus, with $\alpha = 0$ and $\beta = \sqrt{b}$, equation (5) says that differential equation (7) has solutions

$$y_1 = \sin \sqrt{b}t \quad \text{and} \quad y_2 = \cos \sqrt{b}t.$$

We shall return to this type of equation in studying oscillatory motion later in this section.   □

The General Solution of Equation (1)

If y_1 and y_2 are any two solutions of equation (1), then so is any *linear combination* of these solutions of the form

$$y(t) = Ay_1(t) + By_2(t) \tag{8}$$

where A and B are constants. (See Exercise 28.) If the functions y_1 and y_2 in equation (8) are not multiples of each other, the function $y(t)$ is called the *general solution* of equation (1) since it can be shown that any solution can be written in this form for an appropriate choice of the constants A and B.

Example 5

(i) The general solution of the differential equation $\dfrac{d^2y}{dt^2} - \dfrac{dy}{dt} - 2y = 0$ in Example 1 is

$$y = Ae^{2t} + Be^{-t}.$$

(ii) The general solution of the differential equation $\dfrac{d^2y}{dt^2} + 4\dfrac{dy}{dt} + 4y$ in Example 2 is

$$y = Ae^{-2t} + Bte^{-2t}.$$

(iii) The general solution of the differential equation

$$y'' + 2y' + 2y = 0$$

in Example 3 is

$$y = Ae^{-t} \sin t + Be^{-t} \cos t.$$

(iv) The general solution of the differential equation

$$y'' + by = 0, \qquad b > 0$$

in Example 4 is

$$y = A \sin \sqrt{b}t + B \cos \sqrt{b}t. \qquad \square$$

Initial Conditions

In general, two initial conditions are required to determine a unique solution of a second order differential equation, one for the solution y and one for its derivative $\dfrac{dy}{dt}$, *specified at the same number* t_0. Since the general solution of equation (1) has the form given by equation (8), two initial conditions will produce two linear equations for the constants A and B, from which these numbers may be determined.

Example 6 Find the unique solution of the initial value problem consisting of the differential equation

$$y'' - 2y' - 3y = 0 \tag{9}$$

and the initial conditions

$$y(0) = 0$$
$$y'(0) = 4.$$

Strategy

First, find the general solution by

a. finding the characteristic polynomial

b. identifying its roots r_1 and r_2

c. combining the solutions $e^{r_1 t}$ and $e^{r_2 t}$ as in equation (8).

Solution

The characteristic polynomial for equation (9) is

$$r^2 - 2r - 3 = (r - 3)(r + 1)$$

which has roots $r_1 = 3$ and $r_2 = -1$. The general solution of equation (9) is therefore

$$y(t) = Ae^{3t} + Be^{-t} \tag{10}$$

so the derivative y' has the form

$$y'(t) = 3Ae^{3t} - Be^{-t}. \tag{11}$$

Next, substitute initial conditions into general solution y, and its derivative y', to obtain two equations in A and B.

Substituting the initial condition $y(0) = 0$ into equation (10) gives the equation

$$A + B = 0. \tag{12}$$

Substituting the initial condition $y'(0) = 4$ into equation (11) gives the equation

$$3A - B = 4. \tag{13}$$

Solve the two equations for A and B by substitution or by elimination.

Substituting $B = -A$ from equation (12) into equation (13) gives

$$3A - (-A) = 4$$

or

$$4A = 4.$$

Thus $A = 1$ and $B = -A = -1$.

Find particular solution by substituting for A and B in general solution.

Equation (10) now becomes

$$y(t) = e^{3t} - e^{-t}$$

the desired solution. □

Modelling Oscillatory Motion

A version of equation (1) that occurs frequently in applications is

$$\frac{d^2y}{dt^2} + by = 0, \qquad b > 0. \tag{14}$$

As stated in Example 4, the periodic function $y_1 = \sin \sqrt{b}t$ is easily seen to be a solution of equation (14), since

$$\frac{dy_1}{dt} = \sqrt{b} \cos \sqrt{b}t \quad \text{and} \quad \frac{d^2y_1}{dt^2} = -b \sin \sqrt{b}t$$

so

$$\frac{d^2y_1}{dt^2} + by_1 = -b \sin \sqrt{b}t + b(\sin \sqrt{b}t)$$

$$= 0$$

as required. Similarly, the periodic function $y_2 = \cos \sqrt{b}t$ is also a solution of equation (14), as you can verify. Thus the general solution of a differential equation of the form of equation (14) must be a combination of the "oscillatory" functions $\sin \sqrt{b}t$ and $\cos \sqrt{b}t$.

> The differential equation
> $$\frac{d^2y}{dt^2} + by = 0, \qquad b > 0 \tag{15}$$
> has general solution
> $$y = A \sin \sqrt{b}t + B \cos \sqrt{b}t. \tag{16}$$

Example 7 Find the general solution of the differential equation
$$y'' + 9y = 0.$$

Solution: Since the differential equation has the form of equation (15), with $b = 9$ the general solution is
$$y = A \sin 3t + B \cos 3t. \qquad \square$$

Because solutions of equation (14) are made up of sine and cosine functions, versions of this equation occur in models for the motion of vibrating physical systems involving springs, pendulums, etc., and applications of this type may be found in texts oriented toward engineering and the physical sciences. A different application involving oscillatory phenomena occurs in a biological setting.

Example 8 (predator–prey system) Assume that a population of predators (foxes, say) depends on a population of prey (rabbits, in this case) for its food supply. More specifically, we assume that:

 (i) The rate of growth of the predator population depends on the size of the prey population (i.e., more rabbits means more foxes).
(ii) The rate of decline of the prey population depends on the size of the predator population (i.e., more foxes means *fewer* rabbits).

Figure 4.1 suggests that these assumptions lead to oscillating sizes for both populations.

In order to treat this situation mathematically, we shall make several simplifying assumptions. First, we shall denote the rabbit population by $x(t)$ and the fox population by $y(t)$, where t represents time, and we shall assume that both functions are at least twice differentiable for all t. Then we translate assumptions (i) and (ii) into the specific relations

(i') $\dfrac{dy}{dt} = a(x - \bar{x}), \quad a > 0, \quad \bar{x} > 0$

(ii') $\dfrac{dx}{dt} = -b(y - \bar{y}), \quad b > 0, \quad \bar{y} > 0.$

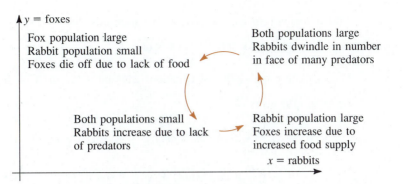

Figure 4.1 A predator–prey system.

Here $\bar{x}$ and $\bar{y}$ represent certain fixed "equilibrium" sizes for the two populations. Since $a > 0$, assumption (i') states that the fox population will increase when the rabbit population is larger than $\bar{x}$ and decrease when the rabbit population falls below $\bar{x}$. Similarly, hypothesis (ii') states that the rabbit population will decline when the fox population exceeds $\bar{y}$ and grow otherwise.

To simplify these equations further, we change to the variable X and Y via the substitutions

$$X = x - \bar{x} \quad \text{and} \quad Y = y - \bar{y}. \tag{17}$$

Then $\dfrac{dX}{dt} = \dfrac{dx}{dt}, \dfrac{dY}{dt} = \dfrac{dy}{dt}$, and our system of equations becomes

$$\frac{dY}{dt} = aX \tag{18}$$

$$\frac{dX}{dt} = -bY. \tag{19}$$

We can now convert this "system" of two first order equations to a second order equation as follows. First, we differentiate both sides of equation (19) with respect to t to obtain

$$\frac{d^2X}{dt^2} = -b\frac{dY}{dt}.$$

Next, we substitute for $\dfrac{dY}{dt}$ in this equation using equation (18) to obtain

$$\frac{d^2X}{dt^2} = -abX$$

or

$$\frac{d^2X}{dt^2} + abX = 0. \tag{20}$$

Now equation (20) has the form of equation (15), so its general solution is

$$X(t) = A \sin(\sqrt{ab}\,t) + B \cos(\sqrt{ab}\,t).$$ (21)

Differentiating both sides of equation (21) and using the fact that $Y = -\dfrac{1}{b}\dfrac{dX}{dt}$ from equation (19) gives

$$Y(t) = -\frac{1}{b}[\sqrt{ab}\,A \cos(\sqrt{ab}\,t) - \sqrt{ab}\,B \sin(\sqrt{ab}\,t)]$$

$$= -\sqrt{\frac{a}{b}}\,A \cos(\sqrt{ab}\,t) + \sqrt{\frac{a}{b}}\,B \sin(\sqrt{ab}\,t).$$

Reversing the substitutions made in equation (17), we have the following conclusion.

The predator–prey system

$$\frac{dy}{dt} = a(x - \bar{x}), \quad a > 0, \quad \bar{x} > 0$$ (22)

$$\frac{dx}{dt} = -b(y - \bar{y}), \quad b > 0, \quad \bar{y} > 0$$ (23)

has the general solution

$$x(t) = \bar{x} + A \sin(\sqrt{ab}\,t) + B \cos(\sqrt{ab}\,t)$$ (24)

$$y(t) = \bar{y} - \sqrt{\frac{a}{b}}[A \cos(\sqrt{ab}\,t) - B \sin(\sqrt{ab}\,t)].$$ (25)

The form of the solutions in equations (24) and (25) supports the intuitive suggestion that the two population sizes will oscillate about their equilibrium sizes. However, this conclusion depends heavily on the specific assumptions that we made about the nature of the original equations, and various other models, with different outcomes, are possible.

□

The final example is a particular case of our model.

Example 9 Find the solutions for the predator–prey system

$$\frac{dy}{dt} = 2(x - 20)$$

$$\frac{dx}{dt} = -(y - 10)$$

if at time $t = 0$ there are $x(0) = 25$ rabbits and $y(0) = 12$ foxes present.

Solution: This system is a predator–prey system as in equations (22) and (23) with

$\bar{x} = 20$, $\bar{y} = 10$, $a = 2$, and $b = 1$. Its general solution, given by equations (22) and (23), is

$$x(t) = 20 + A \sin \sqrt{2}t + B \cos \sqrt{2}t \tag{26}$$

$$y(t) = 10 - \sqrt{2}[A \cos \sqrt{2}t - B \sin \sqrt{2}t]. \tag{27}$$

Substituting $t = 0$ in equation (26) and using the initial condition $x(0) = 25$ gives

$$25 = 20 + A \sin(0) + B \cos(0)$$
$$= 20 + B$$

so $B = 5$. Substituting $t = 0$ in equation (27) and using the initial condition $y(0) = 12$ gives the equation

$$12 = 10 - \sqrt{2}[A \cos(0) - B \sin(0)]$$
$$= 10 - A\sqrt{2}$$

so $A = \dfrac{10 - 12}{\sqrt{2}} = -\sqrt{2}$. Thus

$$x(t) = 20 - \sqrt{2} \sin \sqrt{2}t + 5 \cos \sqrt{2}t$$
$$y(t) = 10 + 2 \cos \sqrt{2}t + 5\sqrt{2} \sin \sqrt{2}t.$$

These solutions are graphed on the same axes in Figure 4.2. □

Figure 4.2 Solutions of a predator–prey system.

Exercise Set 10.4

In Exercises 1–6 find the general solution of the differential equation in the form $y = Ae^{r_1 t} + Be^{r_2 t}$ by finding two distinct real roots of the characteristic polynomial.

1. $\dfrac{d^2 y}{dt^2} + 5 \dfrac{dy}{dt} + 6y = 0$ **2.** $y'' + 3y' - 4y = 0$

3. $y'' - y = 0$

4. $\dfrac{d^2 y}{dt^2} - \dfrac{dy}{dt} - 6y = 0$

5. $\dfrac{d^2 y}{dt^2} + 3 \dfrac{dy}{dt} - 10y = 0$ **6.** $y'' - 4y = 0$

In Exercises 7–10 find the general solution of the differential equation in the form $y = Ae^{rt} + Bte^{rt}$ by finding the single root r of the characteristic polynomial.

7. $y'' + 2y' + y = 0$

8. $y'' - 4y' + 4y = 0$

9. $\dfrac{d^2y}{dt^2} + 6\dfrac{dy}{dt} + 9y = 0$

10. $y'' - 2y' + y = 0$

In Exercises 11–18 find the general solution of the differential equation in the form $y = A \sin \sqrt{b}t + B \cos \sqrt{b}t$

11. $\dfrac{d^2y}{dt^2} + 4y = 0$

12. $y'' + 16y = 0$

13. $y'' = -5y$

14. $\dfrac{d^2y}{dt^2} = -25y$

15. $y'' - 4y' + 13y = 0$

16. $y'' - 6y' + 25y = 0$

17. $y'' + y' + \dfrac{y}{2} = 0$

18. $y'' + y' + y = 0$

In Exercises 19–24 find the solution of the initial value problem

19. $y'' - y' - 6y = 0$
$y(0) = 0$
$y'(0) = 5$

20. $y'' - 5y' + 6y = 0$
$y(0) = 2$
$y'(0) = 0$

21. $y'' + 2y' + y = 0$
$y(0) = 3$
$y'(0) = 1$

22. $y'' - 4y' + 4y = 0$
$y(0) = 2$
$y'(0) = 5$

23. $y'' + 9y = 0$
$y(0) = 3$
$y'(0) = -3$

24. $y'' + y = 0$
$y(0) = 0$
$y'(0) = 2$

In Exercises 25–26 find the general solution by first converting the system of two first order equations into a second order equation using the method of Example 8.

25. $\dfrac{dy}{dt} = 4x$

$\dfrac{dx}{dt} = -y$

26. $\dfrac{dy}{dt} = 2(x - 3)$

$\dfrac{dx}{dt} = 8(y + 5)$

27. Find the solution of the predator–prey system in Example 9 if initially $x(0) = 100$ rabbits and $y(0) = 40$ foxes are present.

28. Find the solution of the predator–prey system

$$\dfrac{dy}{dt} = x - 25$$

$$\dfrac{dx}{dt} = -2(y - 16)$$

subject to the initial conditions $x(0) = y(0) = 20$.

29. What is the solution of the predator–prey system in Exercise 28 under the initial conditions $x(0) = 25$ and $y(0) = 16$?

30. Discuss the nature of the solutions of the predator–prey system

$$\dfrac{dy}{dt} = ax \qquad a > 0$$

$$\dfrac{dx}{dt} = -by \qquad b > 0$$

versus the nature of the solutions of this same system when $b < 0$. Would it make sense to refer to this as a predator–prey system when $b < 0$?

31. Show that the function $y = te^{-2t}$ is a solution of the differential equation $y'' + 4y' + 4y = 0$ in Example 2.

32. Show that if y_1 and y_2 are solutions of the second order differential equation $y'' + ay' + by = 0$, then so is the linear combination $y = Ay_1 + By_2$. (*Hint:* Find $\dfrac{dy}{dt}$ and $\dfrac{d^2y}{dt^2}$, insert in the differential equation, and use the fact that y_1 and y_2 are solutions.)

10.5 QUALITATIVE THEORY OF DIFFERENTIAL EQUATIONS

Unfortunately, it is not always a simple matter to find a solution for a given differential equation, just as it is often not possible to find an explicit antiderivative for a given function. Often, however, we are able to infer **properties** of the (possibly unknown) solutions of a differential equation from the differential equation itself. The techniques for doing so are part of what is called **qualitative theory** for differential equations. This term refers to properties of solutions of a differential equation.

We shall discuss qualitative theory only for equations of the form

$$y' = f(y) \tag{1}$$

where f is a differentiable function of y. Such equations are called **autonomous** because they do not involve an independent variable that in differential equations is usually taken to represent time. Thus *autonomous* means time-independent.

The general idea behind the techniques presented here is that equation (1) describes the behavior of the **derivative** $y' = \dfrac{dy}{dt}$ of the solution y in terms of the current value $y(t)$ of the solution. That is, equation (1) tells us something about the **slopes** of the solution curves at each value of a solution y. Geometrically, this means that equation (1) specifies the slopes of solution curves along horizontal lines of the ty-plane. Thus we can make inferences about the shapes of solution curves from equation (1) just as we made inferences about the graphs of **antiderivatives** from the values of a function in Chapter 5.

We shall illustrate some of the basic ideas in qualitative theory by discussing two examples in detail. Then we shall state some generalizations.

Example 1 The differential equation

$$y' = y^2 - 1 \tag{2}$$

has the form $y' = f(y)$ where f is the function

$$f(y) = y^2 - 1 = (y - 1)(y + 1). \tag{3}$$

We make the following observations: □

(i) Equation (2) has the constant solutions $y_1 = -1$ and $y_2 = 1$ because these constants are the zeros for f. (Since the derivative of a constant function is zero, both sides of the differential equation $y' = f(y)$ equal zero when y is a constant function with value equal to a root of f.) We plot the constant solutions y_1 and y_2 on the yt axes in Figure 5.1 for $t \geq 0$.

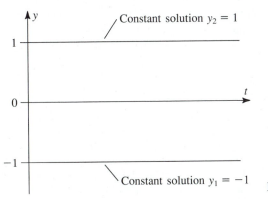

Figure 5.1 Two constant solutions.

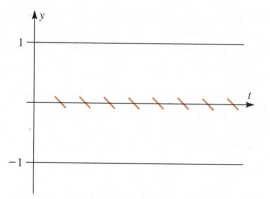

Figure 5.2 Tangents have slope $m = -1$ at $y = 0$.

(ii) On any solution curve at height $y = 0$, the slope must equal $[y(0)]^2 - 1 = 0^2 - 1 = -1$. Thus we sketch "little tangents" along the horizontal line $y = 0$, all of slope $y' = -1$, as illustrated in Figure 5.2.

(iii) Similarly, by choosing various other constant values $y = c$, we find the following slopes y' along the horizontal lines $y = c$:

$y = c$	Slope $= y' = c^2 - 1$
$c = \dfrac{1}{2}$	$y' = -\dfrac{3}{4}$
$c = \dfrac{3}{2}$	$y' = \dfrac{5}{4}$
$c = -\dfrac{1}{2}$	$y' = -\dfrac{3}{4}$
$c = -\dfrac{3}{2}$	$y' = \dfrac{5}{4}$

These tangents are sketched along the horizontal lines $y = c$ in Figure 5.3. We can generalize this observation by noting that equation (2) specifies the *sign* of the derivative $y'(t)$ of any solution in terms of its current value $y(t)$. Using the first derivative test, we may therefore conclude that solutions y must increase or decrease for various ranges of y as follows (and as illustrated in Figure 5.3):

Range of y	Sign of Derivative $y'(t) = y^2(t) - 1$	Behavior of Solution
$y < -1$	positive	increasing
$-1 < y < 1$	negative	decreasing
$y > 1$	positive	increasing

(iv) Knowing that the graph of a solution for equation (2) must have tangents as sketched in Figure 5.3, we use the "direction field" to sketch several possible solution curves as indicated in Figure 5.4.

Figure 5.3 Slopes of solutions of equation (2) for various values of y.

Figure 5.4

(v) To verify the assumptions made about the concavity of the curves sketches in Figure 5.4, we recall that the concavity of the function y is determined by its second derivative y''. Since $y' = f(y) = 1 - y^2$, we have, upon differentiating both sides and using the chain rule, that

$$y'' = \frac{d}{dt}[f(y)] = \frac{d}{dt}(1 - y^2) = \left[\frac{d}{dy}(1 - y^2)\right]\frac{dy}{dt}$$

$$= -2y(1 - y^2)$$

$$= -2y(1 - y)(1 + y)$$

from which we can see that the concavity of solutions is as follows:

Range for y	Concavity of Curves (Sign of $y'' = -2y(1 - y)(1 + y)$)
$y < -1$	negative
$-1 < y < 0$	positive
$0 < y < 1$	negative
$1 < y$	positive

The concavity is thus as labeled in Figure 5.5.

Thus although we have not obtained explicit solutions of equation (2), we have concluded quite a bit about the *nature* of the solutions for $t \geq 0$. ☐

Example 2 For the differential equation

$$y' = \sin y \tag{4}$$

describe the nature of solutions beginning at $t = 0$.

Strategy

Follow the same analysis as in Example 1. Find the zeros of f.

Solution

The equation has the form $y' = f(y)$ with $f(y) = \sin y$. The equation $f(y) = 0$ is therefore $\sin y = 0$, which has solutions

$$y = 0, \ \pm\pi, \ \pm 2\pi, \ \ldots$$

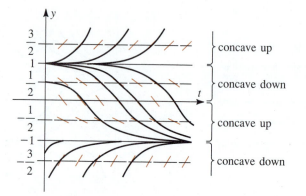

Figure 5.5 Solution curves for equation (2).

Note that each zero of f gives a constant solution of (4).

Use equation (4) and the first derivative test to determine whether solutions y increase or decrease between consecutive zeros of f.

Thus equation (4) has constant solutions

$$y(t) = n\pi, \; n = 0, \; \pm 1, \; \pm 2, \; \ldots$$

as you can verify by direct substitution. (See Figure 5.6.) For values of y in the intervals between these zeros equation (4) determines the behavior of solutions y as follows:

Range for y	Sign of $y'(t)$	Behavior of y
$0 < y < \pi$	positive	increasing
$\pi < y < 2\pi$	negative	decreasing
$2n\pi < y < (2n + 1)\pi$	positive	increasing
$(2n - 1)\pi < y < 2n\pi$	negative	decreasing

Thus we may sketch tangents at constant values of y as in Figure 5.6.

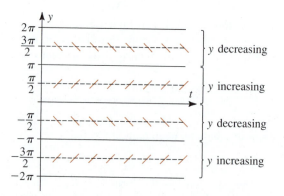

Figure 5.6 Slopes to solution curves for equation (4).

To determine concavity of solution curves, differentiate both sides of equation (4) and apply theory of concavity to the expression for y''. Check the sign of y'' between its zeros, which are

$$t = \pm \frac{n\pi}{2}, \qquad n = 0, \pm 1, \ldots$$

Using the Chain Rule, we find the second derivative y'' by differentiating both sides of equation (4) and substituting for y' from equation (4):

$$y'' = \frac{d}{dt}[y'] = \frac{d}{dt}[\sin y]$$

$$= \frac{d}{dy}[\sin y]\frac{dy}{dt}$$

$$= (\cos y)(\sin y).$$

From the zeros of y'' and its sign on the resulting intervals, we conclude the following about the concavity of solution curves:

Range for y	Sign of $y''(t)$	Concavity
$0 < y < \dfrac{\pi}{2}$	positive	up
$\dfrac{\pi}{2} < y < \pi$	negative	down
$\pi < y < \dfrac{3\pi}{2}$	positive	up
$\dfrac{3\pi}{2} < y < 2\pi$	negative	down

etc.

Use all preceding information to sketch graphs.

From this information we sketch solution curves as in Figure 5.7. □

From the results of Examples 1 and 2 it should not surprise you that solutions of autonomous differential equation of the form

$$y' = f(y) \tag{5}$$

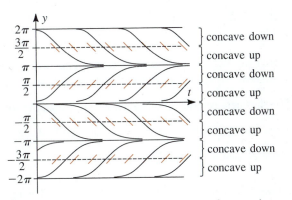

Figure 5.7 Concavity and solution curves for equation (4) for $-2\pi \leq y \leq 2\pi$.

have, for suitably well-behaved functions f, the following properties:

A-I: Each zero y_0 of the function f is a constant solution $y = y_0$ of equation (5). The graphs of these constant solutions slice the ty-plane into horizontal strips.

A-II: Any solution y with initial value $y(0)$ in one of these strips has all values $y(t)$ in the same strip for $t > 0$. That is, solution curves do not cross the horizontal lines that are graphs of constant solutions.

A-III: As t increases, the graph of any nonconstant solution either approaches the graph of a constant solution (as do all solutions in Example 2) or becomes infinitely large in absolute value (as do solutions y in Example 1 with $y(0) > 1$).

We do not claim to have proved statement A-I through A-III. Rather, your familiarity with properties of the first and second derivative, together with the explanations in Examples 1 and 2, should suggest the plausibility of these statements. They are proved in more advanced courses.

Initial Value Problems

The preceding observations are sometimes useful in inferring behavior of the unique solution of an initial value problem of the form

$$y' = f(y)$$
$$y(0) = y_0$$

for small values of $t > 0$. By analyzing the differential equation $y' = f(y)$ as in Examples 1 and 2, and by then noting the location of the initial value $y(0)$, we can infer the behavior of the particular solution of the initial value problem from the behavior of the general solution of the differential equation $y' = f(y)$ in the strip containing the initial value $y(0)$.

Example 3 Describe the short-term behavior of the solution of the initial value problem

$$y' = y^2 - 1 \tag{6}$$

$$y(0) = \frac{1}{2}. \tag{7}$$

Solution: From the behavior of the differential equation $y' = y^2 - 1$ as analyzed in Example 1 we know that it has successive constant solutions $y_1(t) = -1$ and $y_2(t) = 1$. Since $y(0) = \frac{1}{2}$ lies between these constants, we know that the solution of the initial value problem has values $y(t)$ with $-1 < y(t) < 1$ for all $t > 0$, according to property A-II. Moreover, since equation (6) and (7) give that $y'(0) = \left(\frac{1}{2}\right)^2 - 1 = -\frac{3}{4}$, we know that the solution y *decreases* from its initial value $y(0) = \frac{1}{2}$. Thus the solution y must behave initially somewhat like the curve sketched in Figure 5.8. Note that we *cannot* determine,

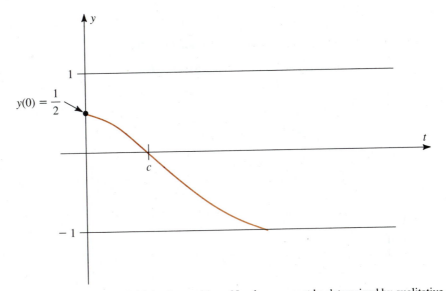

Figure 5.8 Solution to initial value problem. Number c cannot be determined by qualitative theory.

however, the number $t = c$ at which this solution crosses the t-axis. That is a particular piece of **quantitative** information that our **qualitative** theory cannot provide. □

Example 4 Describe the behavior of the solution of the initial value problem

$$y' = 5y(20 - y) \tag{8}$$

$$y(0) = 25. \tag{9}$$

Solution: First, note that this problem may be described as an instance of a **logistic growth model** for a system with carrying capacity $K = 20$ and initial population size $y(0) = 25$, which is *larger* than the carrying capacity. (See Section 10.2.)

We note next that the function $f(y) = 5y(20 - y)$ on the right-hand side of equation (8) has zeros $y_1 = 0$ and $y_2 = 20$. Thus the solution y with initial condition $y(0) = 25$ will have its graph lying entirely *above* the horizontal line $y = 20$.

Next, we note that the tangent to the graph of this solution at $(0, 25)$ has slope

$$\begin{aligned} y'(0) &= 5y(0)[20 - y(0)] \\ &= 5(25)(20 - 25) \\ &= -625 \end{aligned} \tag{10}$$

so the solution decreases to the right of $t = 0$.

Finally, since the second derivative of this solution is

$$y'' = \frac{d}{dt}[y'] = \frac{d}{dt}[5y(20 - y)]$$

$$= \frac{d}{dt}[100y - 5y^2]$$

$$= 100y' - 10yy'$$

we find, using equations (9) and (10), that

$$y''(0) = 100[y'(0)] - 10[y(0)][y'(0)]$$
$$= 100(-625) - 10(25)(-625)$$
$$= 93{,}750$$

which is positive. Thus we may expect the graph of y to be concave up to the right of $t = 0$.

From this information and properties A-I through A-III we infer that the graph of the solution of the initial value problem is similar to that sketched in Figure 5.9. The important *qualitative* conclusion is that the population modeled by equation (8) with initial size $y(0) = 25$ will *decrease* in size toward the carrying capacity $K = 20$. ☐

Figure 5.9

Exercise Set 10.5

In Exercises 1–10 sketch solution curves for the given differential equation from the qualitative information obtained from the equation itself.

1. $y' = y(3 - y)$

2. $y' = y + 2$

3. $y' = y^2$

4. $y' = y(4 + y)$

5. $y' = e^y$

6. $y' = \dfrac{1}{1 + y^2}$

7. $y' = \cos y$

8. $y' = y(1 - y^2)$

9. $y' = y(1 - y)(2 - y)$

10. $y' = \dfrac{1}{1 + y}$

In each of Exercises 11–20 sketch the solution of the initial value problem for $t \geq 0$ as best you can by using qualitative theory.

11. $y' = 4y(10 - y)$
$y(0) = 5$

12. $y' = 4y(10 - y)$
$y(0) = 20$

13. $y' = \sin y$
$y(0) = -\dfrac{\pi}{4}$

14. $y' = \cos y$
$y(0) = \pi$

15. $y' = y(1 - y^2)$
$y(0) = -2$

16. $y' = e^y$
$y(0) = -2$

17. $y' = y^2$
$y(0) = -1$

18. $y' = y^2$
$y(0) = 3$

19. $y' = (y + 3)(y - 2)$
$y(0) = 1$

20. $y' = (y - 1)(y - 4)$
$y(0) = 3$

21. Two solutions, y_1 and y_2, of the differential equation $y' = y(1 - y)$ have initial values $y_1(0) = \dfrac{1}{2}$ and $y_2(0) = \dfrac{3}{4}$. Do values of these solutions become closer together or farther apart as t increases? Why?

22. Same as Exercise 21 except that the initial values are $y_1(0) = \dfrac{1}{2}$ and $y_2(0) = -\dfrac{1}{2}$.

23. For the differential equation $y' = y(y - 1)(y - 2)$, does the solution y with $y(0) = \dfrac{3}{2}$ increase or decrease for $t > 0$?

24. Same as Exercise 23 except that the solution y has initial value $y(0) = 3$.

25. For the differential equation $y' = y(y-1)(y - 2)$ one constant solution $y = c$ has the property that all solutions with initial values $y(0)$ sufficiently close to c have the horizontal line $y = c$ as an asymptote. (We say that all such solutions are **asymptotic** to the solution $y = c$.) Find c.

26. One constant solution $y(t) = c$ of the differential equation $y' = (y + 3)(y - 2)$ has the property that all solutions with initial values sufficiently close to c move *away* from the value constant solution $y(t) = c$ as t increases. Find c.

27. Find the constant solution of the differential equation $y' = 30y(60 - 4y)$ with the property that all nearby solutions are asymptotic to it. (See Exercise 25.)

28. In the logistic growth model $y' = 3y(20 - 3y)$ will a solution with initial value $y(0) = 8$ increase or decrease?

29. In the differential equation $y' = ky(20 - y)$ determine the sign of k if the solution with initial value $y(0) = 10$ is known to decrease for $t > 0$.

30. In the differential equation $y' = ky(1 - y^2)$ determine the sign of k if the solution with initial value $y(0) = -2$ is known to decrease for $t > 0$.

10.6 APPROXIMATING SOLUTIONS OF DIFFERENTIAL EQUATIONS

In this brief chapter we are able only to skim the surface of the body of techniques and principles that constitute the theory of differential equations. However, there remain many differential equations for which simple solutions cannot be found. In such cases, however, we can resort to certain approximation procedures to obtain information about the solution, just as we can use the Trapezoidal Rule or Simpson's Rule to approximate a definite integral when the corresponding antiderivative cannot be found.

To give you an idea of how such approximation procedures work, we discuss one such procedure, called Euler's method, in this section. It is named for the Swiss mathematician Leonhard Euler (1707–1783). Although this method is seldom used in practice because more sophisticated methods have greater accuracy, it is a good first example of how such approximation schemes are constructed. Students interested in studying more advanced methods should consult texts on differential equations or numerical analysis.

Euler's Method   **Euler's method** may be used to approximate the solution to an initial value problem of the form

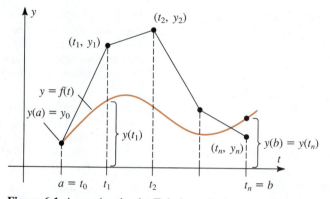

Figure 6.1 Approximation by Euler's method to the solution y of the initial value problem

$$y' = f(t, y), \quad y(a) = y_0.$$

$$\frac{dy}{dt} = f(t, y) \qquad (1)$$

$$y(a) = y_0$$

on some interval $[a, b]$. (We assume that f and its partial derivatives are continuous for all t and y, although less restrictive conditions can be given.)

Figure 6.1 illustrates the basic idea to be pursued. Assume that $y = f(t)$ is a particular solution of the given differential equation. We divide the interval $[a, b]$ into n subintervals of equal length $h = \dfrac{b - a}{n}$, obtaining the $(n + 1)$ endpoints

$$a = t_0 < t_1 < t_2 < \cdots < t_n = b.$$

Then, beginning at the left endpoint $t_0 = a$, we use the given information about the solution y (namely, the initial value $y(a) = y_0$) and its derivative to approximate the value $y(t_1)$, and we call the approximation y_1. (Note, as suggested by Figure 6.1, that, in general, the approximation y_1 will not equal the function value $y(t_1)$, which we do not know.) Next, we use the assumed value y_1 and the number $t_1 = a + h$ to approximate $y(t_2)$, and we call this approximation y_2. Continuing in this way, we finally approximate $y(t_n) = y(b)$ from the approximate value y_{n-1} and the number $t_{n-1} = a + (n - 1)h$. We therefore obtain *approximations* $y_1, y_2, \ldots, y_n$ to the (unknown) values of the solution $y(t_1), y(t_2), \ldots, y(t_n)$ at finitely many numbers t_j in $[a, b]$. The choice of the endpoint b and the number of intervals, n, depend on which values of the solution you wish to approximate and with what accuracy.

Figure 6.2 illustrates how we get started in applying Euler's method. To find the approximation y_1 to the value $y(t_1)$, we follow the line tangent to the graph of y at the point $P_0 = (t_0, y_0)$ until we reach the point $P_1 = (t_1, y_1)$ with t-coordinate t_1. Since the slope of this line is $y'(t_0) = f(t_0, y_0)$, it follows that

$$\frac{y_1 - y_0}{t_1 - t_0} = f(t_0, y_0)$$

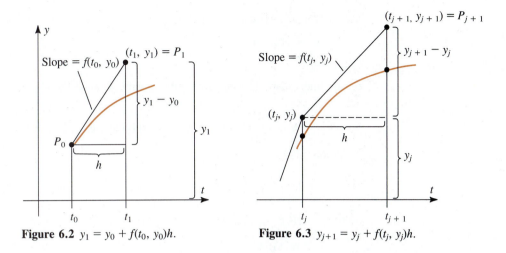

Figure 6.2 $y_1 = y_0 + f(t_0, y_0)h.$ **Figure 6.3** $y_{j+1} = y_j + f(t_j, y_j)h.$

and since $h = t_1 - t_0$

$$y_1 = y_0 + f(t_0, y_0)h. \tag{2}$$

Once we have obtained the approximation y_1 from equation (2), we *pretend* that y_1 is the actual value of the solution y at $t = t_1$. Since we don't know the actual value $y(t_1)$, the approximation y_1 is the "next best thing." This "pretending" is necessary in order that we may carry out the same procedure to obtain the approximation y_2 for $y(t_2)$. Starting with the point (t_1, y_1) and the slope $f(t_1, y_1)$,* we obtain the y-coordinate of the point $P_2 = (t_2, y_2)$ from the equation

$$\frac{y_2 - y_1}{t_2 - t_1} = f(t_1, y_1).$$

That is, since $h = t_2 - t_1$

$$y_2 = y_1 + f(t_1, y_1)h.$$

Continuing in this way, we obtain each succeeding approximation y_{j+1} from the preceding y_j and t_j by the equation

$$\boxed{y_{j+1} = y_j + f(t_j, y_j)h_j \qquad j = 0, 1, \ldots, n - 1} \tag{3}$$

which is the summary statement of Euler's method. (See Figure 6.3.)

The following two examples show how a first order differential equation may be put in the form of equation (1), and how the solution of an associated initial value problem may be approximated using Euler's method.

---

*The pretending occurs here in two places. Since we do not actually know $y(t_j)$, we do not know the actual slope $f(t_j, y(t_j))$, which we approximate by $f(t_j, y_j)$. Neither do we know the point of tangency, $(t_j, y(t_j))$, so we use (t_j, y_j) instead.

Example 1 Use Euler's method to approximate the solution of the initial value problem

$$y' = y(4 - y) \tag{4}$$
$$y(0) = 1$$

on the interval $[0, 1]$ using four subintervals of equal length.

Solution: With $[0, 1]$ divided into four subintervals of length $h = \dfrac{1-0}{4} = \dfrac{1}{4}$ we have endpoints

$$t_0 = 0, \qquad t_1 = \frac{1}{4}, \qquad t_2 = \frac{1}{2}, \qquad t_3 = \frac{3}{4}, \qquad \text{and} \qquad t_4 = 1.$$

With $y_0 = y(0) = 1$ and $f(t, y) = y(4 - y)$ equation (3) gives

$$y_1 = y_0 + y_0(4 - y_0)h$$
$$= 1 + 1(4 - 1)\left(\frac{1}{4}\right)$$
$$= 1 + \frac{3}{4}$$
$$= 1.75$$
$$y_2 = y_1 + y_1(4 - y_1)h$$
$$= 1.75 + 1.75(4 - 1.75)(0.25)$$
$$\cong 2.7344$$

etc. Table 6.1 shows the approximations y_1, y_2, y_3, y_4 together with the actual values $y(t_1)$, $y(t_2), y(t_3), y(t_4)$ of the solution $y = \dfrac{4}{1 + 3e^{-4t}}$ of the initial value problem to four-decimal place accuracy. (This solution is obtained by the method of Section 10.3.) □

Table 6.1. Approximate and actual values of the solution to the initial value problem in Example 1 ($h = 0.25$)

t_j	0	0.25	0.50	0.75	1.0
y_1	1	1.75	2.7344	3.5995	3.9599
$y(t_j)$	1	1.9015	2.8449	3.4802	3.7917

Example 2 Approximate the solution to the initial value problem in Example 1 on the interval $[0, 1]$ using $n = 10$ subintervals of equal size.

Solution: This time we have $h = \dfrac{1-0}{10} = \dfrac{1}{10}$, so Euler's equation becomes

$$y_{j+1} = y_j + y_j(4 - y_j)(0.10), \qquad j = 0, 1, 2, \ldots, 9.$$

Table 6.2 shows the resulting approximations, together with the actual values of the solution, to four decimal place accuracy. (Note the improvement in accuracy compared with Example 1.) □

Table 6.2. Values of the approximations and actual solution to the initial value problem in Example 2 ($h = 0.10$)

t_j	0	0.1	0.2	0.3	0.4	0.5	0.6	0.7	0.8	0.9	1.0
y_j	1	1.3	1.651	2.0388	2.4387	2.8194	3.1523	3.4195	3.6180	3.7562	3.8478
$y(t_j)$	1	1.3285	1.7036	2.1013	2.4911	2.8449	3.1443	3.3829	3.5642	3.6970	3.7917

Example 3 Use Euler's method to approximate the solution of the initial value problem

$$y' + 2y = e^t \tag{5}$$
$$y(0) = 1$$

on the interval $[0, 1]$.

Solution: We begin by moving all terms in equation (5) except the derivative y' to the right-hand side, obtaining

$$\frac{dy}{dt} = e^t - 2y.$$

Thus $f(t, y) = e^t - 2y$ in equation (3) for Euler's method. We obtain two approximations.

(a) Using $n = 4$ subdivisions we have $h = \dfrac{1 - 0}{4} = 0.25$ and equation (3) becomes

$$y_{j+1} = y_j + (e^{t_j} - 2y_j)(0.25), \qquad j = 0, 1, 2, 3.$$

Thus since $y_0 = y(0) = 1$ is given

$$\begin{aligned}
y_1 &= y_0 + (e^{t_0} - 2 \cdot y_0)(0.25) \\
&= 1 + (e^0 - 2 \cdot 1)(0.25) \\
&= 1 - 0.25 \\
&= 0.75
\end{aligned}$$

$$\begin{aligned}
y_2 &= y_1 + (e^{t_1} - 2y_1)(0.25) \\
&= 0.75 + [e^{0.25} - 2(0.75)](0.25) \\
&= 0.6960
\end{aligned}$$

etc. Table 6.3 shows the same approximations $y_1, y_2, y_3,$ and y_4 to four-decimal-place accuracy, as well as the actual values $y(t_1), y(t_2), y(t_3),$ and $y(t_4)$ for the actual solution $y = \dfrac{1}{3}(2e^{-2t} + e^t)$.

Table 6.3. Approximate and actual values for the initial value problem in Example 3 ($h = 0.25$)

t_j	1	0.25	0.5	0.75	1.0
y_j	1	0.75	0.6960	0.7602	0.9093
$y(t_j)$	1	0.8324	0.7948	0.8544	0.9963

(b) The same solution, approximated using ten subintervals of $[0, 1]$ of length $h = \dfrac{1 - 0}{10} = 0.1$, is obtained using the Euler equation

$$y_{j+1} = y_j + (e^{t_j} - 2y_j)(0.1), \qquad j = 0, 1, \ldots, 9.$$

Table 6.4 shows the results of this approximation. □

Table 6.4. Approximate and actual values of the solution of the initial value problem in Example 3 ($h = 0.1$)

t_j	0	0.1	0.2	0.3	0.4	0.5	0.6	0.7	0.8	0.9	1.0
y_j	1	0.9	0.8305	0.7866	0.7642	0.7606	0.7733	0.8009	0.8421	0.8962	0.9629
$y(t_j)$	1	0.9142	0.8540	0.8158	0.7968	0.7948	0.8082	0.8356	0.8764	0.9301	0.9963

CONCLUSION: Euler's method is one procedure for approximating the solution of an initial value problem involving a first order differential equation. While we have used examples in which an actual solution may be obtained for purposes of comparison, Euler's method may be applied to differential equations whose solutions are not known.

Obviously, there is a fair amount of tedium in applying Euler's method. Calculators and especially computers can reduce this tedium to a minimum. Program 7 in Appendix II is a BASIC Computer program that was used to perform the calculations in Tables 6.1–6.4. You may find it useful in working the exercises.

Several sources of error enter into the calculations that are involved in Euler's method. Since we obtain only approximations y_j instead of the actual value $y(t_j)$ at each step, the slope calculation $\dfrac{dy}{dt} = f(t_j, y_j)$ contains an error due to the difference $|y_j - y(t_j)|$, and the point of assumed tangency, (t_j, y_j) contains an error due to the same factor. Finally, roundoff error in such calculations is inescapable. In general, however, we can say that accuracy decreases as we attempt to approximate values of the solution farther away from the number t_0 where the initial condition is prescribed, and that accuracy increases as the stepsize h decreases. These important issues of accuracy are pursued in more advanced courses.

Exercise Set 10.6

In each of Exercises 1–10 use Euler's method to approximate the value of the solution to the initial value problem at the endpoints of four subintervals of $[0, 1]$ of size $h = 0.25$.

1. $y' = 2y$

  $y(0) = 1$

2. $y' = 9y(1 - y)$

  $y(0) = \dfrac{1}{2}$

3. $y' + 2y = 4$

  $y(0) = 1$

4. $\dfrac{dy}{dx} = \dfrac{y + 1}{x + 1}$

  $y(0) = 0$

5. $\dfrac{dy}{dx} = x(y + 1)$

  $y(0) = 1$

6. $\dfrac{dy}{dx} = 2 - 4y$

  $y(0) = 1$

7. $\dfrac{dy}{dt} + 2ty = t$

  $y(0) = \dfrac{3}{2}$

8. $e^y \cdot \dfrac{dy}{dt} - t^2 = 0$

  $y(0) = 0$

9. $\dfrac{dy}{dt} = 2y - 2t$

  $y(0) = \dfrac{3}{2}$

10. $\dfrac{dy}{dt} - 2y = e^{2t} \sin t$

  $y(0) = 0$

11. Use equations (14) and (15) of Section 10.3 to find the algebraic solution of the initial value problem in Exercise 3. Compute the actual value of the solution at each endpoint and compare it with the value from Euler's method.

12. An investor opens a savings account paying 10% interest compounded continuously with an initial deposit of $1000. The investor makes deposits of $1000 per year in a continuous manner.
  a. Show that the function P, giving the amount on deposit after t years, satisfies the differential equation $P'(t) = 0.10P(t) + 1000$.
  b. Use Euler's method to approximate the values $P(1)$, $P(2)$, $P(3)$, and $P(4)$.

SUMMARY OUTLINE OF CHAPTER 10

■ A **differential equation** is an equation involving one or more derivatives of an unknown function. (Page 534)

■ A **solution** of a differential equation is a function that has as many derivatives as the order of the equation and that satisfies the equation. (Page 534)

■ The **order** of a differential equation is the order of the highest derivative in the equation. (Page 535)

■ An **initial value problem** consists of a differential equation together with enough data to determine a unique solution. (Page 537)

■ A **separable** differential equation has the form (Page 548)

$$\dfrac{dy}{dx} = \dfrac{f(x)}{g(y)}.$$

■ We **separate the variables** in this equation as follows: (Page 549)

$$g(y)\, dy = f(x)\, dx$$

so

$$\int g(y)\left(\dfrac{dy}{dx}\right) dx = \int f(x)\, dx$$

and the solution follows by integration, when possible.

■ The differential equation $\dfrac{dy}{dt} = ky$ for natural growth has general solution $y = Ce^{kt}$.   (Page 535)

■ The differential equation $\dfrac{dy}{dt} = k\left(\dfrac{K - y}{K}\right)$ for logistic growth has general solution   (Page 545)

$y = \dfrac{K}{1 + Me^{-kt}}$, M constant.

■ The differential equation $\dfrac{dy}{dt} = k(y - a)$, which is separable, has general solution   (Page 555)

$y = a + Ce^{kt}$.

■ The differential equation $y'' + ay' + by = 0$ has a solution of the form $y = e^{rt}$ if and only if r (Page 562)
is a real root of the characteristic polynomial $r^2 + ar + b$.

■ The differential equation $y'' + by = 0$, with $b > 0$, has the general solution   (Page 566)
$y = A \sin \sqrt{b}t + B \cos \sqrt{b}t$.

■ The predator–prey system   (Page 568)

$\dfrac{dy}{dt} = a(x - \bar{x})$, $a > 0$

$\dfrac{dx}{dt} = -b(y - \bar{y})$, $b > 0$

has general solution

$x(t) = \bar{x} + A \sin \sqrt{ab}t + B \cos \sqrt{ab}t$

$y(t) = \bar{y} - \sqrt{\dfrac{a}{b}}[A \cos \sqrt{ab}t - B \sin \sqrt{ab}t]$.

Euler's method for approximating solutions of the initial value problem   (Page 581)

$\dfrac{dy}{dt} = f(t, y)$

$y(a) = y_0$

at n equally spaced numbers t_j in the interval $[a, b]$ is

$y_{j+1} = y_j + f(t_j, y_j)h$, $j = 0, 1, 2, \ldots, n - 1$

where

$h = \dfrac{b - a}{n}$ and $t_j = a + jh$, $j = 0, 1, 2, \ldots, n.$

REVIEW EXERCISES—CHAPTER 10

In Exercises 1–24 find the general solution of the differential equation.

1. $\dfrac{dx}{dt} = x - 3$

2. $\dfrac{dy}{dx} = 4 - \sqrt{x}$

3. $\dfrac{dy}{dx} = \dfrac{\sec \sqrt{x}\,\tan \sqrt{x}}{\sqrt{x}}$

4. $\dfrac{d^2y}{dt^2} = \sqrt{1 + t}$

5. $\dfrac{dy}{dt} = 3ty^2$

6. $\dfrac{dy}{dx} = \dfrac{\sqrt{x + 1}}{y}$

7. $\dfrac{dy}{dt} = (1 + t)(2 + y)$

8. $ty' = \dfrac{t^2 + 3}{y}$

9. $y' + 2y = 4$

10. $y' = y(3 - y)$

11. $y' + y \cos t = 0$

12. $y' = 4y \ln y$

13. $(t - 1)y' = t - 1$

14. $\dfrac{dy}{dt} + \dfrac{1}{t^2}y = 0$

15. $\dfrac{dy}{dx} = y(2 - y)$

16. $ty' + 2y = 0$

17. $\dfrac{dy}{dt} = \dfrac{t + 1}{y}$

18. $\dfrac{dy}{dt} = \dfrac{y}{t + 1}$

19. $y'' + 5y' + 6y = 0$

20. $\dfrac{d^2y}{dx^2} - 9y = 0$

21. $\dfrac{d^2y}{dx^2} - 5\dfrac{dy}{dx} - 14y = 0$

22. $y'' + y' - 6y = 0$

23. $y'' + 8y' + 16y = 0$

24. $\dfrac{d^2y}{dt^2} = 3\dfrac{dy}{dt} + 10y$

In Exercises 25–36 find the solution of the initial value problem.

25. $\dfrac{dy}{dx} = \dfrac{x}{\sqrt{1 + x^2}}$

$y(0) = 3$

26. $\dfrac{d^2y}{dt^2} = 1 + t$

$y(0) = 1$

$y'(0) = 0$

27. $y' = 9y$

$y(0) = 3$

28. $\dfrac{dy}{dt} = ty$

$y(1) = 3$

29. $\dfrac{dy}{dx} = y \sin x$

$y(0) = \pi$

30. $y' = 4(1 - y)$

$y(0) = 1$

31. $y' = 10(5 - t)$

$y(0) = 2$

32. $\dfrac{d^2y}{dx^2} = -9y$

$y(0) = 3$

$y'(0) = 9$

33. $\dfrac{d^2y}{dx^2} + 16y = 0$

$y(0) = 5$

$y'(0) = 0$

34. $\dfrac{dy}{dt} = 5y(10 - y)$

$y(0) = 5$

35. $y'' - 6y' + 9y = 0$

$y(0) = 0$

$y'(0) = 6$

36. $y'' - 16y = 0$

$y(0) = 2$

$y'(0) = 8$

In Exercises 37–40 sketch the direction field and several solution curves for the differential equation

37. $\dfrac{dy}{dt} = 2y(4 - y)$

38. $\dfrac{dy}{dx} = xy$

39. $\dfrac{dy}{dt} = y + 3$

40. $\dfrac{dy}{dx} = \dfrac{x}{y}$

41. The value of a certain piece of real estate in San Diego has been increasing at a rate of $V'(t) = \sqrt{36 + t}$ thousand dollars per year since 1975. If its value in 1975 was \$200,000, find its value $V(t)$, t years after 1975.

42. A manufacturer of refrigerators determines that its marginal revenue at production level x refrigerators per week is $60 + 2x - 0.2x^2$ dollars. Use the fact that $R(0) = 0$ to find the manufacturer's total revenue function $R(t)$.

43. Find a differential equation satisfied by the function
 a. $y = Ce^{2t}$
 b. $y = C_1e^t + C_2e^{-t}$

44. A tank contains 500 liters of brine with an initial concentration of salt of 1 kilogram of salt per liter. A second mixture of brine containing 2 kilograms of salt per liter is then added at a rate of 50 liters per minute. The tank is kept well mixed, and brine is drawn off the bottom at the same rate, 50 liters per minute. Find the concentration of salt t minutes after the second mixture begins entering the first.

45. An automobile radiator contains 5 gallons of pure anti-freeze. The owner begins adding freshwater at the rate of 1 gallon per minute, with the engine running to ensure complete mixing, and draining the radiator at the same rate of 1 gallon per minute. Find the concentration $y(t)$ of antifreeze per gallon t minutes after the owner begins this process.

46. An investor places $5000 in a savings account paying 10% interest compounded continuously and pledges to make additional deposits of $2000 per year in a continual manner.

a. Find a differential equation for $P(t)$, the amount on deposit in this account t years after it is opened.

b. Use Euler's method to approximate $P(1)$, $P(2)$, $P(3)$, and $P(4)$.

47. A cold drink is removed from a refrigerator at temperature 40°F and is placed on a sunporch where the surrounding temperature is 80°F. After 5 minutes the temperature of the drink is 50°F. Find its temperature after 10 minutes.

11 Applications to Probability Theory

11.1 DISCRETE PROBABILITY DISTRIBUTIONS

The goal of probability theory is to develop mathematical models for predicting the outcomes of ''experiments'' based on assumptions about the likelihood of certain associated events occurring. Those assumptions are stated in the form of a function called a **probability distribution**. We shall not attempt to develop the basic principles of probability theory that let us determine which function to use for this purpose. Rather, we shall assume that a probability distribution is already at hand, and show how the definite integral is used to calculate probabilities for various events. Before discussing continuous distributions (Section 11.2), we establish the terminology and basic principles for the discrete distributions in this section.

Outcomes, Events, and Probabilities

By an **outcome** of an experiment, we mean one of the simple results that can occur. Part of the description of an experiment consists of specifying what the set of all possible outcomes (called the **sample space** for the experiment) will be.

For example, in the experiment of rolling a single die, the possible outcomes are the numbers that may be observed, 1 through 6. The sample space for this experiment is therefore the set

$$S = \{1, 2, 3, 4, 5, 6\}$$

which contains six elements (outcomes). By comparison, the occurrence ''observe a number less than 4'' is not called an outcome, since it corresponds to more than one simple outcome (namely, 1, 2, and 3). We use the term **event** to refer to a collection of one or more outcomes. Thus an event E is a subset of a sample space S. In the example E: ''observe a number less than 4,'' we have

$$E = \{1, 2, 3\}.$$

We want to distinguish outcomes from events because a single trial of an experiment can result in only one outcome, while it could correspond to more than one event.

Example 1 In the experiment of rolling a single die, let E_1 through E_3 be the events

E_1:   observe a number less than 4
E_2:   observe an even number
E_3:   observe a number that is not 1.

Then each of these events is a subset of the sample space $S = \{1, 2, 3, 4, 5, 6\}$ as follows:

$$E_1 = \{1, 2, 3\}$$
$$E_2 = \{2, 4, 6\}$$
$$E_3 = \{2, 3, 4, 5, 6\}.$$

Now note that each of the outcomes 1 to 6 in S is a member of one or more events, as follows:

Outcome	Event
1	E_1
2	E_1, E_2, E_3
3	E_1, E_3
4	E_2, E_3
5	E_3
6	E_2, E_3

□

Probabilities

Once a sample space S (that is, the set of possible simple outcomes) for an experiment has been identified, we assign a number called a **probability** to each possible outcome in S. In order for our model to be useful in predicting outcomes of actual experiments, the probability associated with a possible outcome should be the **relative frequency** with which that outcome will occur in a large number of repetitions of the experiment. (From a strictly mathematical point of view, however, this requirement is not necessary. Probabilities need only be assigned according to two rules that we shall state later.)

Example 2 In the experiment of a single toss of a fair coin, we would expect to obtain a head roughly half the time and a tail roughly half the time. (Indeed, this is the definition of a "fair" coin.) We therefore assign probabilities to the sample space $S = \{\text{head, tail}\}$ as follows

$$Pr(\text{head}) = \frac{1}{2}$$

$$Pr(\text{tail}) = \frac{1}{2}.$$

□

Example 3 In the experiment of rolling a fair die, each of the outcomes 1 to 6 is equally likely. The relative frequency with which each would be expected to occur "in the long run" is therefore $\frac{1}{6}$. Thus the probability associated with each outcome is $\frac{1}{6}$. That is,

$$Pr(1) = Pr(2) = Pr(3) = Pr(4) = Pr(5) = Pr(6) = \frac{1}{6}.$$

□

Example 4 A box contains two red balls and three white balls. One ball is selected at random. What is the probability that it is (a) red or (b) white?

Solution: The fact that the ball is selected ''at random'' means that each of the five balls is just as likely to be selected as any of the others. Since two of the five balls are red, we would expect that, in many repetitions of this experiment, we would obtain a red ball roughly $\frac{2}{5}$ of the time. Similarly, the relative frequency with which a white ball will be chosen is $\frac{3}{5}$. These are the probabilities associated with the sample space $S = \{$red ball, white ball$\}$:

$$Pr(\text{red ball}) = \frac{2}{5}$$

$$Pr(\text{white ball}) = \frac{3}{5}.$$ □

Rules for Probabilities

Since the probability associated with an outcome is thought of as a *relative frequency* (meaning the number of times this outcome occurs divided by the total number of times the experiment is performed), probabilities must satisfy two rules.

> **Rule 1:** The probability associated with any outcome is a number in the interval [0, 1].
>
> **Rule 2:** The sum of the probabilities associated with the set of all outcomes in a finite sample space is 1.

Rule 2 can be thought of as meaning that ''*something* must happen.''

Probabilities for Events

Once probabilities have been assigned to individual outcomes in a sample space, probabilities for events may be calculated according to the following rule.

> **Rule 3:** The probability associated with an event is the sum of the probabilities associated with the simple outcomes in that event.

Rule 3 is sometimes called the *addition law* for probabilities associated with (disjoint or mutually exclusive) outcomes.

Example 5 For the events in the die rolling experiment of Example 1 we have

$$Pr(E_1) = Pr(1) + Pr(2) + Pr(3) = \frac{1}{6} + \frac{1}{6} + \frac{1}{6} = \frac{1}{2}$$

$$Pr(E_2) = Pr(2) + Pr(4) + Pr(6) = \frac{1}{6} + \frac{1}{6} + \frac{1}{6} = \frac{1}{2}$$

$$Pr(E_3) = Pr(2) + Pr(3) + Pr(4) + Pr(5) + Pr(6)$$

$$= \frac{1}{6} + \frac{1}{6} + \frac{1}{6} + \frac{1}{6} + \frac{1}{6}$$

$$= \frac{5}{6}.$$

□

Example 6 A car rental firm has determined from past experiments that the probabilities of a blowout occurring on the four tires of one of its cars are as follows:

$$Pr(\text{left front}) = 0.3$$
$$Pr(\text{right front}) = 0.25$$
$$Pr(\text{left rear}) = 0.25$$
$$Pr(\text{right rear}) = 0.2.$$

Since the event E_1 of a front-tire blowout contains the first two outcomes listed above, we have

$$P(E_1) = 0.3 + 0.25 = 0.55.$$

Similarly, the event E_2 of a rear-tire blowout contains the last two outcomes, so

$$P(E_2) = 0.25 + 0.2 = 0.45.$$

□

Random Variables and Probability Functions

From this point on we shall restrict our discussion to sample spaces in which the outcomes are represented by numbers, such as the die-rolling experiment of Example 1 where the sample space is $S = \{1, 2, 3, 4, 5, 6\}$. We may therefore use a variable such as X to represent an outcome in the sample space. Such a variable is called a **random variable**.*

When the outcomes in a sample space are represented by values of a random variable, the assignment of probabilities to the outcomes can be thought of as a function for which the domain is a sample space. We refer to this as a **probability function**, written Pr. We use the following notation with probability functions:

$Pr\{X = a\}$ means the probability associated with outcome a,

while

$Pr\{X \text{ in } E\}$ means the probability associated with the event E.

Example 7 Using this notation for the die-rolling experiment in Example 5, we would write

(a) $Pr\{X = 3\} = \dfrac{1}{6}$

(b) $Pr\{X = 5\} = \dfrac{1}{6}$

---

*A more sophisticated approach would be to say that a random variable is a function defined on a sample space. We shall not require this more general notion of the random variable.

(c) $Pr\{X \text{ in } E_1\} = \dfrac{1}{2}$

etc. □

Probability Distributions

A sample space together with a probability function Pr defined on the sample space is called the **probability distribution** for the random variable X. Probability distributions can be specified by giving either a table of all elements of the sample space and their probabilities, or a **probability histogram**, which is just a bar graph showing the probability associated with each outcome.

Example 8 The probability distribution for the random variable X representing the possible outcomes of a single roll of a fair die is given in Table 1.1 and by the probability histogram in Figure 1.1. □

Table 1.1

Outcome X	1	2	3	4	5	6
Probability $Pr\{X\}$	$\dfrac{1}{6}$	$\dfrac{1}{6}$	$\dfrac{1}{6}$	$\dfrac{1}{6}$	$\dfrac{1}{6}$	$\dfrac{1}{6}$

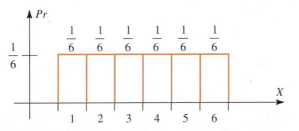

Figure 1.1 Probability distribution for die-rolling experiment.

Example 9 Consider the experiment consisting of two tosses of a fair coin. Let the random variable X represent the number of heads obtained. Find the probability distribution for X.

Solution: The sample space for X consists of the possible outcomes $S = \{0, 1, 2\}$. To determine how to assign probabilities to each of these outcomes, we must determine the relative frequency with which each is expected to occur. In terms of the *order* in which the results of this experiment occur, there are two ways in which $X = 1$ head can occur, while there is only one way that $X = 0$ heads or $X = 2$ heads can occur:

Results, in Order	Value of X
tail, tail	$X = 0$
tail, head	$X = 1$
head, tail	$X = 1$
head, head	$X = 2$

Since each of these four results seems equally likely to occur, we assign each the probability (relative frequency) $\frac{1}{4}$. This means that the outcome $X = 1$ head will occur with relative frequency $\frac{1}{4} + \frac{1}{4} = \frac{1}{2}$, while outcomes $X = 0$ and $X = 2$ will occur with relative frequency $\frac{1}{4}$. These are the probabilities assigned to the three possible values of X, as shown in Table 1.2. Figure 1.2 shows the probability histogram for this distribution. ☐

Table 1.2

Outcome X	0	1	2
Probability $Pr\{X\}$	$\frac{1}{4}$	$\frac{1}{2}$	$\frac{1}{4}$

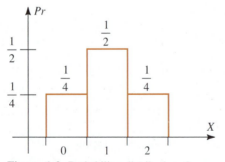

Figure 1.2 Probability distribution for random variable X in Example 9.

Example 10 Table 1.3 shows the possible outcomes for the experiment of rolling two fair dice, where the random variable X represents the sum of the numbers showing on the two dice. The probabilities assigned to the possible outcomes (values of X) are the relative frequencies determined by noting how many of the 36 distinct possible pairs (a, b) are associated with that outcome.

The distribution for the random variable X is shown by the probability histogram in Figure 1.3. ☐

Table 1.3

Outcome X	Pairs (a, b) corresponding to X	Probability
2	(1, 1)	$Pr\{X = 2\} = \dfrac{1}{36}$
3	(1, 2), (2, 1)	$Pr\{X = 3\} = \dfrac{2}{36}$
4	(1, 3), (3, 1), (2, 2)	$Pr\{X = 4\} = \dfrac{3}{36}$
5	(1, 4), (4, 1), (2, 3), (3, 2)	$Pr\{X = 5\} = \dfrac{4}{36}$
6	(1, 5), (5, 1), (2, 4), (4, 2), (3, 3)	$Pr\{X = 6\} = \dfrac{5}{36}$
7	(1, 6), (6, 1), (2, 5), (5, 2), (3, 4), (4, 3)	$Pr\{X = 7\} = \dfrac{6}{36}$
8	(2, 6), (6, 2), (3, 5), (5, 3), (4, 4)	$Pr\{X = 8\} = \dfrac{5}{36}$
9	(3, 6), (6, 3), (4, 5), (5, 4)	$Pr\{X = 9\} = \dfrac{4}{36}$
10	(4, 6), (6, 4), (5, 5)	$Pr\{X = 10\} = \dfrac{3}{36}$
11	(5, 6), (6, 5)	$Pr\{X = 11\} = \dfrac{2}{36}$
12	(6, 6)	$Pr\{X = 12\} = \dfrac{1}{36}$

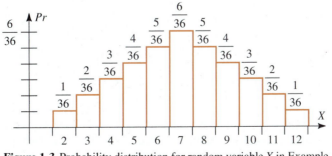

Figure 1.3 Probability distribution for random variable X in Example 10.

Example 11 For the experiment in which the random variable X represents the sum obtained in a single toss of two fair dice (Example 10) we can define many events; probabilities for some of them are

$$Pr\{X = 5\} = \frac{4}{36} = \frac{1}{9}$$

$$Pr\{X = 2, 3, \text{ or } 4\} = \frac{1}{36} + \frac{2}{36} + \frac{3}{36} = \frac{6}{36} = \frac{1}{6}$$

$$Pr\{X \leq 6\} = \frac{1}{36} + \frac{2}{36} + \frac{3}{36} + \frac{4}{36} + \frac{5}{36} = \frac{15}{36}$$

$$Pr\{3 < X \leq 5\} = Pr\{X = 4\} + Pr\{X = 5\} = \frac{3}{36} + \frac{4}{36} = \frac{7}{36}$$

$$Pr\{X > 8\} = Pr\{X = 9\} + Pr\{X = 10\} + Pr\{X = 11\} + Pr\{X = 12\}$$

$$= \frac{4}{36} + \frac{3}{36} + \frac{2}{36} + \frac{1}{36}$$

$$= \frac{5}{18}.$$ □

Expected Value   For a random variable X associated with a finite sample space $\{x_1, x_2, \ldots, x_n\}$ the concept of *expected value* is a generalization of the notion of the average of the numbers $x_1, x_2, \ldots, x_n$.

DEFINITION 1   Let X be a random variable on the sample space $\{x_1, x_2, \ldots, x_n\}$ and let p_j be the probability $p_j = Pr\{X = x_j\}$ for $j = 1, 2, \ldots, n$. The **expected value of X**, written $E(X)$, is the sum

$$E(X) = x_1 p_1 + x_2 p_2 + \cdots + x_n p_n.$$

Notice that when each of the outcomes x_j has equal probability $p_j = \dfrac{1}{n}$, the expected value of X is

$$E(X) = x_1 \left(\frac{1}{n}\right) + x_2 \left(\frac{1}{n}\right) + \cdots + x_n \left(\frac{1}{n}\right)$$

$$= \frac{x_1 + x_2 + \cdots + x_n}{n}$$

the average of the numbers $x_1, x_2, \ldots, x_n$. Thus the expected value $E(X)$ is a sort of "weighted average," where the outcomes with higher probabilities make larger contributions to the average.

Example 12

(a) For the experiment of rolling a single die the expected value of the outcome X is

$$E(X) = 1\left(\frac{1}{6}\right) + 2\left(\frac{1}{6}\right) + 3\left(\frac{1}{6}\right) + 4\left(\frac{1}{6}\right) + 5\left(\frac{1}{6}\right) + 6\left(\frac{1}{6}\right)$$

$$= \frac{1 + 2 + 3 + 4 + 5 + 6}{6}$$

$$= \frac{7}{2}$$

which is just the average of the six (equally likely) outcomes.

(b) For the experiment consisting of the rolling of a pair of dice the expected value of the sum is, according to the probabilities in Table 1.3 and Definition 1

$$E(X) = 2\left(\frac{1}{36}\right) + 3\left(\frac{2}{36}\right) + 4\left(\frac{3}{36}\right) + 5\left(\frac{4}{36}\right) + 6\left(\frac{5}{36}\right)$$

$$+ 7\left(\frac{6}{36}\right) + 8\left(\frac{5}{36}\right) + 9\left(\frac{4}{36}\right) + 10\left(\frac{3}{36}\right)$$

$$+ 11\left(\frac{2}{36}\right) + 12\left(\frac{1}{36}\right)$$

$$= \frac{2 + 6 + 12 + 20 + 30 + 42 + 40 + 36 + 30 + 22 + 12}{36}$$

$$= \frac{252}{36}$$

$$= 7.$$

This should not be surprising, once you have noted the symmetry in Figure 1.3. The expected value simply corresponds to the center of this pyramid-like distribution.

□

REMARK: Note, as in Example 12, part (a), that the expected value of a random variable need not be one of the numbers in the sample space, just as the average of a set of numbers need not be one of the numbers in the set.

The following example shows how the concept of expected value can be used in decision making.

Example 13
A venture capital firm determines, based on past experience, that for each $100 invested in a high technology startup company a return of $400 is experienced 20% of the time, a return of $100 is experienced 40% of the time, and a total loss is experienced 40% of the time. What is the firm's expected return, based on these data?

Solution: If we let the random variable X represent the three outcomes $x_1 = \$400$,

$x_2 = \$100$, and $x_3 = 0$, the probabilities associated with these outcomes are given in Table 1.4.

Table 1.4

x_i	$400	$100	0
p_i	0.20	0.40	0.40

The expected value of X is therefore

$$E(X) = (\$400)(0.2) + (\$100)(0.4) + (0)(0.4)$$
$$= \$80 + \$40 + 0$$
$$= \$120.$$

This means that, on average, the firm can expect a return of $120 on each $100 invested in a high technology startup company. Whether this result leads to a decision to invest depends on whether the firm is satisfied in this rate of return, and over what time period it will occur. □

Variance

The expected value of a random variable is a measure of its central tendency, that is, what values are most likely to occur. In contrast, a random variable's *variance* is a measure of how far apart the possible outcomes are spread, again weighted by their respective probabilities.

DEFINITION 2

Let X be a random variable for the sample space $\{x_1, x_2, \ldots, x_n\}$ and let $p_j = Pr\{X = x_j\}$, for $j = 1, 2, \ldots, n$. Let $E(X)$ be the expected value of X. The **variance** of X, denoted by $Var(X)$, is the number

$$Var(X) = E([X - E(X)]^2) = [x_1 - E(X)]^2 p_1 + [x_2 - E(X)]^2 p_2$$
$$+ \cdots + [x_n - E(X)]^2 p_n.$$

Example 14 For the experiment of rolling two dice (Examples 10 and 12), where the expected value of their sum is $E(X) = 7$, the variance is

$$Var(X) = (2-7)^2\left(\frac{1}{36}\right) + (3-7)^2\left(\frac{2}{36}\right) + (4-7)^2\left(\frac{3}{36}\right) + (5-7)^2\left(\frac{4}{36}\right)$$

$$+ (6-7)^2\left(\frac{5}{36}\right) + (7-7)^2\left(\frac{6}{36}\right) + (8-7)^2\left(\frac{5}{36}\right) + (9-7)^2\left(\frac{4}{36}\right)$$

$$+ (10-7)^2\left(\frac{3}{36}\right) + (11-7)^2\left(\frac{2}{36}\right) + (12-7)^2\left(\frac{1}{36}\right)$$

$$= \frac{25 + 16 \cdot 2 + 9 \cdot 3 + 4 \cdot 4 + 1 \cdot 5 + 1 \cdot 5 + 4 \cdot 4 + 9 \cdot 3 + 16 \cdot 2 + 25}{36}$$

$$= \frac{210}{36} = \frac{35}{6}.$$ □

Example 15 For the probability distribution experienced by the venture capitalist in Example 13

$$\text{Var}(X) = (400 - 120)^2(0.2) + (100 - 120)^2(0.4) + (0 - 120)^2(0.4)$$
$$= (78{,}400)(0.2) + (400)(0.4) + (14{,}400)(0.4)$$
$$= 21{,}600.$$

□

The variance of a random variable is a number that is more difficult to interpret intuitively than is the expected value. But in comparing two random variables with the same expected value, we can, in general, say that the distribution of the random variable with the larger variance is more ''spread out'' than the other.

Standard Deviation

The **standard deviation** of a random variable X is simply the square root of its variance. The standard deviation of X is denoted by the Greek letter σ (sigma). Thus in Example 14 we have

$$\sigma = \sqrt{\text{Var}(X)} = \sqrt{\frac{35}{6}}$$

while in Example 15 we have

$$\sigma = \sqrt{\text{Var}(X)} = \sqrt{21{,}600} \approx 147.$$

Exercise Set 11.1

1. Table 1.5 shows a sample space and the associated probabilities. Find the probability of
  a. the event $E_1 = \{1, 3\}$.
  b. the event $E_2 = \{1, 4, 5\}$.
  c. the event $E_3 = \{2, 3, 5\}$.

Table 1.5

x_i	1	2	3	4	5
p_i	$\dfrac{1}{3}$	$\dfrac{1}{3}$	$\dfrac{1}{6}$	$\dfrac{1}{12}$	$\dfrac{1}{12}$

2. Table 1.6 shows a sample space and the associated probabilities. Find the probability of
  a. the event $E_1 = \{10, 20\}$.
  b. the event $E_2 = \{15, 20, 30\}$.
  c. the event $E_3 = \{10, 15, 30\}$.

Table 1.6

x_i	10	15	20	25	30
p_i	0.1	0.1	0.3	0.2	0.3

3. Draw a probability histogram for the probability distribution in Exercise 1.

4. Draw a probability histogram for the probability distribution in Exercise 2.

5. Refer to Example 10. For the experiment of rolling two fair dice find the probability of
  a. the event of getting a 3 or a 9.
  b. the event of getting a number larger than 7.
  c. the event of not getting a 6.

6. For the experiment consisting of one roll of a fair die
  a. Find the probability of rolling a 5.
  b. Find the probability of rolling a 2, 3, or 4.
  c. Find the probability of not rolling a 3.

7. A poker hand contains three red cards and two black cards. A single card is chosen at random from the hand. Find the probability that the card is
  a. red.
  b. black.

8. A box contains three red and x white balls. When a

single ball is withdrawn at random, the probability that it is red is $\frac{1}{4}$. How many white balls are in the box?

9. A set of billiard balls contains 15 balls, numbered 1 through 15. The balls numbered 1–8 are a solid color. The balls numbered 9–15 are striped. A single ball is selected at random. Find the probability that it is
   a. a solid color.
   b. an even-numbered ball.
   c. a striped ball with an even number.

10. Mary is at her desk from 9:00 a.m. until 1:00 p.m., and John is at his desk from 11:00 a.m. until 6:00 p.m. If a telephone call arrives at a time randomly chosen between 9:00 a.m. and 6:00 p.m., what is the probability that
    a. both John and Mary are at their desks?
    b. John is at his desk?
    c. neither John nor Mary is at his or her desk?

11. A dart strikes the target in Figure 1.4 at random. What is the probability that it hits the bull's eye? (Assume that no dart can miss the target completely.)

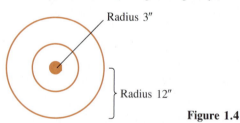

Radius 3″

Radius 12″

Figure 1.4

12. A card is drawn from an ordinary deck of 52 cards. Given that it is a face card, what is the probability that
    a. it is a heart?
    b. it is a queen?
    c. it is not a jack?

13. For the probability distribution in Table 1.7 find
    a. the expected value $E(X)$.
    b. the variance, $\text{Var}(X)$, and the standard deviation σ.

Table 1.7

x_i	0	1	2	3
p_i	0.2	0.2	0.3	0.3

14. For the probability distribution in Table 1.8 find
    a. the expected value $E(X)$.
    b. the variance, $\text{Var}(X)$, and the standard deviation σ.

Table 1.8

x_i	1	2	3	4
p_i	0.1	0.3	0.3	0.3

15. You have the opportunity to buy a lottery ticket that will pay $10 with probability 0.2, $100 with probability 0.1, and nothing with probability 0.7. What is a fair price for the ticket? (That is, what is the expected value of the payoff?)

16. On a rainy day a student can earn $5 doing housework, while on a day when it does not rain the student can earn $40 doing yardwork. If the probability of rain is 30%, what is the student's expected income? (That is, what is the expected value of the student's daily earnings?)

17. An insurance company sells a $100,000 hazard insurance policy on a house. Experience shows that in a year's time the company will have to pay out $100,000 on the policy with probability 0.005, $50,000 on the policy with probability 0.025, and nothing on the policy with probability 0.97. What is the expected value of the company's annual payout on the policy?

11.2 CONTINUOUS PROBABILITY DISTRIBUTIONS

In many situations the outcome of an experiment can be one of infinitely many real numbers. For example

— The depth of water in a 6-inch rain gauge could be any real number in the interval [0, 6].
— The percentage of alcohol in an alcohol/water mixture could be any real number in the interval [0, 100].
— The duration of a telephone call could be any positive real number, that is, any number in the interval (0, ∞).

A random variable X that can equal any number in an interval I (where I can be either of finite or infinite length) is called a **continuous random variable**. Each of the variables described previously is a continuous random variable.

Probability Density Functions

Since a continuous random variable can equal any of the infinitely many numbers in an interval (its sample space), we cannot expect to assign a probability individually to each of the numbers in the sample space as we can do for discrete random variables. Instead, probabilities of the form $Pr\{a \leq X \leq b\}$ are calculated using the concept of a *probability density function*.

DEFINITION 3

A continuous nonnegative function f with the property that

$$Pr\{a \leq X \leq b\} = \int_a^b f(x) \, dx \tag{1}$$

is called a **probability density function** for the random variable X.

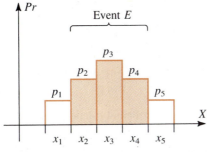

Figure 2.1 In the discrete case probability of event

$E = \{x_2, x_3, x_4\}$ is
$Pr\{x_2 \leq X \leq x_4\} = p_2 + p_3 + p_4.$

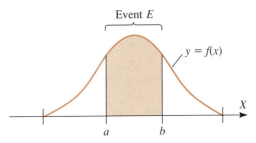

Figure 2.2 In the continuous case probability of event

$E = [a, b]$ is

$$Pr\{a \leq X \leq b\} = \int_a^b f(x) \, dx.$$

Figures 2.1 and 2.2 show how the graph of a probability density function is the analogue of a probability histogram for discrete random variables. In each case the shaded area corresponds to the probability of an event, and the area under the entire graph must be 1. Rather than being too precise about properties of probability density functions at this point, we shall simply show by example how Definition 3 is used, and then consider two frequently occurring continuous probability distributions (the normal and exponential distributions) in Sections 11.3 and 11.4.

Example 1 Figure 2.3 shows the *uniform* probability distribution on the sample space $S = [0, 4]$ with probability density function $f(x) \equiv \dfrac{1}{4}$. According to Definition 3, for this

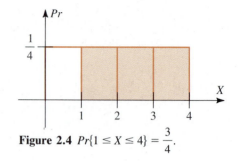

Figure 2.3 Uniform probability distribution on the interval [0, 4].

Figure 2.4 $Pr\{1 \leq X \leq 4\} = \dfrac{3}{4}$.

distribution

(a) $Pr\{0 \leq X \leq 2\} = \displaystyle\int_0^2 \frac{1}{4}\, dx = \left. \frac{x}{4} \right]_0^2$

$$= \frac{1}{2} - 0$$

$$= \frac{1}{2}$$

(b) $Pr\{1 \leq X \leq 4\} = \displaystyle\int_1^4 \frac{1}{4}\, dx = \left. \frac{x}{4} \right]_1^4 = 1 - \frac{1}{4} = \frac{3}{4}$ (see Figure 2.4)

(c) $Pr\{0 \leq X \leq 4\} = \displaystyle\int_0^4 \frac{1}{4}\, dx = \left. \frac{x}{4} \right]_0^4 = 1 - 0 = 1.$ □

Example 2 Figure 2.5 shows the probability density function

$$f(x) = \begin{cases} \dfrac{x}{4}, & 0 \leq x \leq 2 \\[2ex] 1 - \dfrac{x}{4}, & 2 < x \leq 4 \end{cases}$$

on the sample space [0, 4]. In this case we have, for the associated continuous random variable X, that

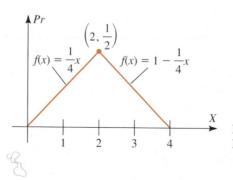

Figure 2.5 Probability density function in Example 2.

(a) $Pr\{0 \le X \le 1\} = \int_0^1 \frac{x}{4}\,dx = \frac{x^2}{8}\Big]_0^1 = \frac{1}{8}$

(b) $Pr\{1 \le X \le 3\} = \int_1^2 \frac{x}{4}\,dx + \int_2^3 \left(1 - \frac{x}{4}\right) dx$

$$= \frac{x^2}{8}\Big]_1^2 + \left(x - \frac{x^2}{8}\right)\Big]_2^3$$

$$= \left(\frac{4}{8} - \frac{1}{8}\right) + \left[\left(3 - \frac{9}{8}\right) - \left(2 - \frac{4}{8}\right)\right]$$

$$= \frac{3}{4} \qquad \text{(see Figure 2.6)}$$

(c) $Pr\{0 \le X \le 4\} = \int_0^2 \frac{1}{4}x\,dx + \int_2^4 \left(1 - \frac{x}{4}\right) dx$

$$= \frac{x^2}{8}\Big]_0^2 + \left(x - \frac{x^2}{8}\right)\Big]_2^4$$

$$= \frac{4}{8} + \left[\left(4 - \frac{16}{8}\right) - \left(2 - \frac{4}{8}\right)\right]$$

$$= 1. \qquad \qquad \Box$$

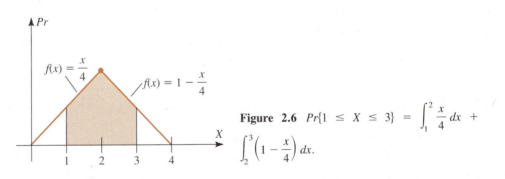

Figure 2.6 $Pr\{1 \le X \le 3\} = \int_1^2 \frac{x}{4}\,dx + \int_2^3 \left(1 - \frac{x}{4}\right) dx.$

Properties of Probability Density Functions

Examples 1 and 2 illustrate two essential properties of the probability density function f. First, because probabilities cannot be negative, the integral in equation (1) must be non-negative for all choices of the interval $[a, b]$. Therefore, we must have $f(x) \ge 0$ for all x in the sample space for the random variable X. Second, since the probability associated with an entire sample space is always one, the integral of f over the entire sample space must equal one (as in parts c of Examples 1 and 2). If the sample space for the continuous random variable X is the interval $[a, b]$, we can write this requirement as

$$\int_a^b f(x)\,dx = 1. \tag{2}$$

If the sample space for X is the entire real number line, this requirement can be written

$$\int_{-\infty}^{\infty} f(x) \, dx = 1 \tag{3}$$

where the integral in equation (3) is improper.

Example 3 Find the number a so that the function $f(x) = ax(6 - x)$ is a probability density function on the sample space $[0, 6]$.

Solution: We must find a so that $f(x) \geq 0$ on $[0, 6]$ and so

$$\int_{0}^{6} ax(6 - x) \, dx = 1.$$

Evaluating this definite integral gives

$$\int_{0}^{6} ax(6 - x) \, dx = a \int_{0}^{6} (6x - x^2) \, dx$$

$$= a\left(3x^2 - \frac{x^3}{3}\right)\Bigg]_{0}^{6}$$

$$= a(108 - 72)$$

$$= 36a.$$

Since we require $\int_{0}^{6} ax(6 - x) \, dx = 36a = 1$, we have $a = \dfrac{1}{36}$. The probability density function $f(x) = \dfrac{x}{36}(6 - x)$, which is indeed nonnegative for $0 \leq x \leq 6$, is illustrated in Figure 2.7. □

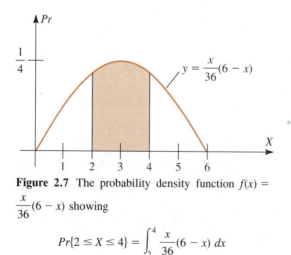

Figure 2.7 The probability density function $f(x) = \dfrac{x}{36}(6 - x)$ showing

$$Pr\{2 \leq X \leq 4\} = \int_{2}^{4} \frac{x}{36}(6 - x) \, dx$$

$$= \frac{13}{27}.$$

Example 4 For the probability density function in Example 3 find $Pr\{2 \le X \le 4\}$.

Solution: According to Definition 3, we have

$$
\begin{aligned}
Pr\{2 \le X \le 4\} &= \int_2^4 \frac{x}{36}(6 - x)\, dx \\
&= \frac{1}{36} \int_2^4 (6x - x^2)\, dx \\
&= \frac{1}{36} \left(3x^2 - \frac{x^3}{3} \right) \Big]_2^4 \\
&= \frac{1}{36} \left[\left(3 \cdot 16 - \frac{64}{3} \right) - \left(3 \cdot 4 - \frac{8}{3} \right) \right] \\
&= \frac{13}{27}. \qquad \text{(See Figure 2.7.)} \qquad \square
\end{aligned}
$$

REMARK: Note that Definition 3 implies that the probability associated with any single number $X = c$ in the sample space for a continuous random variable is zero, since we obtain

$$
Pr\{X = c\} = \int_c^c f(x)\, dx = 0
$$

because the value of a definite integral over an interval consisting of a single number is zero. Since the probability associated with a single number is zero, it does not matter whether endpoints are included when specifying the probability of an event corresponding to an interval. That is

$$
Pr\{a \le X \le b\} = Pr\{a \le X < b\} = Pr\{a < X \le b\} = Pr\{a < X < b\}
$$

for continuous random variables X.

Distribution Functions

A *distribution function*, like a probability density function, determines the probability of an event associated with a random variable. However, the distribution function specifies the cumulative probability associated with all values of the random variable less than or equal to a given number.

DEFINITION 4

A **distribution function** associated with the random variable X is a function F with the property that

$$
F(x) = Pr(X \le x)
$$

for all x in the sample space for X.

For distribution functions there is no requirement that the associated random variable X be continuous or that X have a continuous probability density function. However, when X

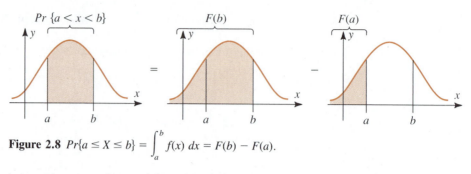

Figure 2.8 $Pr\{a \le X \le b\} = \int_a^b f(x)\ dx = F(b) - F(a)$.

is a continuous random variable with a continuous probability density function f, we have the following relationship between f and the associated distribution function F:

$$\int_a^b f(x)\ dx = Pr\{a \le X \le b\} \qquad \text{(Def. 3)}$$
$$= Pr\{X \le b\} - Pr\{X \le a\} \qquad \text{(see Figure 2.8)}$$
$$= F(b) - F(a).$$

That is

$$\int_a^b f(x)\ dx = F(b) - F(a). \tag{4}$$

Note the resemblance between equation (4) and the Fundamental Theorem of Calculus. The distribution function F gives the *area* $\int_a^b f(x)\ dx$ of the region bounded by the probability density function on an interval $[a, b]$.

Since equation (4) is satisfied by any antiderivative F for f, we may find the distribution function F corresponding to the continuous random variable X by finding the general antiderivative for f and using known information about X to determine the constant of integration. The next example shows how this is done.

Example 5 Find the distribution function for the probability density function

$$f(x) = \frac{1}{4} \quad \text{for} \quad 0 \le x \le 4$$

(the *uniform* probability distribution on $[0, 4]$) in Example 1.

Solution: An antiderivative F for f must be of the form

$$F(x) = \int \frac{1}{4}\ dx = \frac{x}{4} + C. \tag{5}$$

Since the sample space $[0, 4]$ is finite, we know that the probability associated with its left endpoint only is zero. That is, from Definition 4 we know that

$$F(0) = Pr\{X \le 0\} = 0. \tag{6}$$

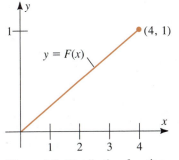

Figure 2.9 Distribution function

$$F(x) = \frac{x}{4} \quad \text{for} \quad f(x) = \frac{1}{4}$$

on interval [0, 4].

Setting $x = 0$ in equation (5) and using equation (6) gives

$$0 = F(0) = \frac{0}{4} + C$$

so $C = 0$. The distribution function is therefore

$$F(x) = \frac{x}{4}. \qquad \text{(See Figure 2.9.)}$$

Notice that $F(4) = 1$. That is, the distribution function has value 1 at the right endpoint of the sample space. $\qquad\square$

Example 6 Find the distribution function F for the probability density function of Example 3:

$$f(x) = \frac{x}{36}(6 - x), \qquad 0 \le x \le 6.$$

Solution: An antiderivative for f is

$$F(x) = \int \frac{x}{36}(6 - x)\, dx = \int \left(\frac{x}{6} - \frac{x^2}{36} \right) dx$$

$$= \frac{x^2}{12} - \frac{x^3}{108} + C.$$

Since 0 is the left endpoint of the sample space [0, 6], we have $F(0) = 0$, so

$$0 = F(0) = \frac{0^2}{12} - \frac{0^3}{108} + C = C.$$

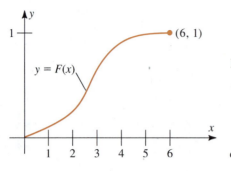

Figure 2.10 Distribution function

$$F(x) = \frac{x^2}{12} - \frac{x^3}{108} \quad \text{for}$$

$$f(x) = \frac{x}{36}(6 - x)$$

on the interval [0, 6].

Thus

$$F(x) = \frac{x^2}{12} - \frac{x^3}{108}.$$

(See Figure 2.10.) Note again that $F(6) = 1$. □

Expected Value

For a continuous random variable X defined on a finite interval $[a, b]$, with continuous probability density function f, the *expected value* of X is given by the definite integral

$$E(X) = \int_a^b xf(x)\, dx. \tag{7}$$

Figure 2.11 shows an approximating sum for the integral in equation (7) and illustrates why equation (7) is the appropriate generalization of the concept of expected value for discrete random variables. In the approximating sum each "test value" t_i of the random variable X is multiplied by the corresponding "probability" $f(t_i)\, \Delta x$.

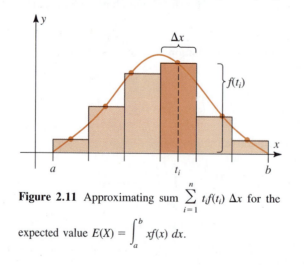

Figure 2.11 Approximating sum $\sum_{i=1}^{n} t_i f(t_i)\, \Delta x$ for the

expected value $E(X) = \int_a^b xf(x)\, dx$.

Example 7 The expected value for the random variable X with uniform probability density function

$$f(x) = \frac{1}{4}, \qquad 0 \le x \le 4$$

in Example 1 is

$$E(X) = \int_0^4 x\left(\frac{1}{4}\right) dx = \int_0^4 \frac{x}{4} \, dx$$

$$= \frac{x^2}{8}\bigg]_0^4$$

$$= 2.$$

This is not surprising, since 2 is the midpoint of the sample space [0, 4], and the probability density is constant across this interval. □

Example 8 Find the expected value of the random variable X with probability density function

$$f(x) = \frac{x}{36}(6 - x)$$

illustrated in Figure 2.7.

Solution: Since this probability density function is a parabola with axis of symmetry $x = 3$, which is also the midpoint of the sample space [0, 6], you might suspect that the symmetry of this density function would produce an expected value of X of $E(X) = 3$. This is indeed the case, since, according to equation (7)

$$E(X) = \int_0^6 x\left[\frac{x}{36}(6 - x)\right] dx$$

$$= \int_0^6 \left(\frac{x^2}{6} - \frac{x^3}{36}\right) dx$$

$$= \left(\frac{x^3}{18} - \frac{x^4}{144}\right)\bigg]_0^6$$

$$= \frac{216}{18} - \frac{1296}{144}$$

$$= 12 - 9$$

$$= 3.$$ □

Variance

The variance associated with the continuous random variable X on the sample space $[a, b]$ is

$$\text{Var}(X) = E([X - E(X)]^2)$$

$$= \int_a^b [x - E(X)]^2 f(x) \, dx \qquad (8)$$

where f is again the probability density function for the random variable X. As for discrete random variables, the variance is the expected value of the square of the deviation of X from its expected value.

Example 9 Find the variance $\mathrm{Var}(X)$ for the random variable X with uniform probability density

$$f(x) = \frac{1}{4}, \qquad 0 \le x \le 4$$

in Example 1.

Solution: Since we found that X has expected value $E(X) = 2$ (Example 7), equation (8) gives

$$\mathrm{Var}(X) = \int_0^4 (x - 2)^2 \cdot \frac{1}{4}\, dx$$

$$= \frac{1}{12}(x - 2)^3 \Big]_0^4$$

$$= \frac{1}{12} 2^3 - \frac{1}{12}(-2)^3$$

$$= \frac{4}{3}.$$

$\square$

Exercise Set 11.2

1. Label each random variable as either discrete or continuous.
   a. the depth of water in a swimming pool
   b. the number of teeth in a person's mouth
   c. a person's weight
   d. the wind speed at sunrise
   e. the temperature at sunrise
   f. the number of hairs on a person's head
   g. your annual income

2. For the probability density function

$$f(x) = \begin{cases} \dfrac{1}{6}, & 0 \le x \le 3 \\[2mm] \dfrac{1}{4}, & 3 < x \le 5 \end{cases}$$

   on the sample space $[0, 5]$ find

   a. $Pr\{0 \le X \le 2\}$

   b. $Pr\{2 \le X \le 4\}$
   c. $Pr\{0 \le X \le 5\}$

3. For the probability density function

$$f(x) = \frac{3}{16}\sqrt{x}, \qquad 0 \le x \le 4$$

   on the sample space $[0, 4]$ find

   a. $Pr\{0 \le X \le 1\}$
   b. $Pr\{2 \le X \le 4\}$
   c. $Pr\{0 \le X \le 4\}$

4. For the probability density function

$$f(x) = \frac{3}{32}x(4 - x), \qquad 0 \le x \le 4$$

   on the sample space $[0, 4]$ find

   a. $Pr\{0 \le X \le 2\}$

b. $Pr\{1 \le X \le 4\}$

c. $Pr\{0 \le X \le 4\}$

5. For the probability density function

$$f(x) = \begin{cases} \dfrac{x}{9}, & 0 \le x \le 3 \\[2mm] \dfrac{2}{3} - \dfrac{1}{9}x, & 3 \le x \le 6 \end{cases}$$

on the sample space $[0, 6]$ find

a. $Pr\{0 \le X \le 2\}$

b. $Pr\{1 \le X \le 5\}$

c. $Pr\{2 \le X \le 6\}$

6. Find the number a so that the function $f(x) = ax(1 - x)$ is a probability density function for the sample space $[0, 1]$.

7. Find the number a so that the function $f(x) = ae^{-x}$ is a probability density function on the interval $[0, \ln 2]$.

8. Find the expected value of the random variable X with uniform probability density function

$$f(x) = \frac{1}{5}, \qquad 0 \le x \le 5$$

on the sample space $[0, 5]$.

9. Find the expected value of the random variable X with probability density function

$$f(x) = \frac{x}{8}, \qquad 0 \le x \le 4$$

on the sample space $[0, 4]$.

10. Find the expected value of the random variable X with the probability density function in Exercise 4.

11. Find the variance of the random variable X in Exercise 8.

12. Find the variance of the random variable X in Exercise 9.

13. Find the variance of the random variable X in Exercise 10.

14. A random variable X has distribution function $F(x) = \dfrac{x^2}{16}$ on the sample space $[0, 4]$. Find

a. $Pr\{0 \le X \le 2\}$

b. $Pr\{1 \le X \le 2\}$

c. the probability density function f.

15. A random variable X has distribution function $F(x) = 3x^2 - 2x^3$ on the sample space $[0, 1]$. Find

a. $Pr\left\{0 \le X \le \dfrac{1}{2}\right\}$

b. $Pr\left\{\dfrac{1}{2} \le X \le 1\right\}$

c. the probability density function f.

16. Find the distribution function for the probability density function in Exercise 3.

17. Find the distribution function for the probability density function in Exercise 4.

18. Find the distribution function for the probability density function in Exercise 8.

19. A biologist determines that the random variable X giving the time at which a sleeping animal awakens during a 4-minute experiment has probability density function

$$f(x) = \frac{5}{4}\left(\frac{1}{1 + x}\right)^2, \qquad 0 \le x \le 4.$$

a. Find the probability that the animal awakens during the first 2 minutes of the experiment.

b. Find the probability that the animal awakens after 1 minute but before the experiment ends.

20. A climatologist believes that the random variable X giving the amount of rainfall expected in a certain region per year has probability density function

$$f(x) = \frac{\pi}{60} \sin\left(\frac{\pi x}{30}\right), \qquad 0 \le x \le 30.$$

Find

a. $Pr\{0 \le X \le 15\}$

b. $Pr\{15 \le X \le 30\}$

21. A research firm determines that the random variable X giving the time, in years, after which a new small business will fail has probability density function

$$f(t) = \frac{k}{(1 + t)^2}, \qquad t \ge 0$$

where k is a constant.

a. Find k by evaluating the improper integral

$$\int_0^\infty \frac{k}{(1 + t)^2}\, dt.$$

b. Find $Pr\{0 \le X \le 9\}$, the probability that a new small business fails within 9 years of its creation.

11.3 THE NORMAL PROBABILITY DISTRIBUTION

One of the most commonly occurring distributions for a continuous random variable X is the *normal* probability distribution. IQ (intelligence quotient), adult height, and adult weight are examples of variables whose distributions are approximately normal. We shall define a normal random variable X in terms of its probability density function.

DEFINITION 5

Let μ and σ be real numbers with $\sigma > 0$. The continuous random variable X with probability density function*

$$f(x) = \frac{1}{\sqrt{2\pi}\sigma} e^{-1/2(x - \mu/\sigma)^2}$$

is called a **normal random variable** with parameters μ and σ. We refer to such a random variable as an $N(\mu, \sigma)$ random variable.

The term *parameter* refers to a fixed constant used in defining a mathematical expression. Definition 5 requires values of two parameters, μ and σ, to determine the normal random variable $N(\mu, \sigma)$. Figure 3.1 shows the graph of the probability density function f for an $N(\mu, \sigma)$ random variable and the significance of the parameters μ and σ:

(a) The vertical line $x = \mu$ is the *axis of symmetry* for the graph of f.

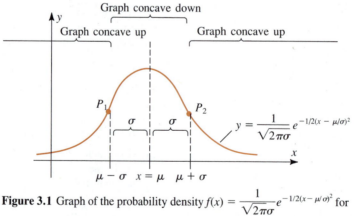

Figure 3.1 Graph of the probability density $f(x) = \dfrac{1}{\sqrt{2\pi}\sigma} e^{-1/2(x - \mu/\sigma)^2}$ for the normal random variable $N(\mu, \sigma)$.

---

*It is a fact, which we shall not prove, that

$$\int_{-\infty}^{\infty} f(x)\, dx = \int_{-\infty}^{\infty} \frac{1}{\sqrt{2\pi}\sigma} e^{-1/2(x - \mu/\sigma)^2}\, dx = 1$$

as required for probability density functions that are defined on the sample space $(-\infty, \infty)$ consisting of the entire real line. Because the sample space is not a finite interval, the integral must be improper.

(b) The graph of $y = f(x)$ has *inflection* points P_1 and P_2 lying σ units to the right and left of the axis of symmetry $x = \mu$.

The Standard Normal Random Variable N(0, 1)

Properties (a) and (b) are most easily demonstrated for the *standard* normal random variable $N(0, 1)$, that is, the normal random variable whose probability density function has parameters $\mu = 0$ and $\sigma = 1$. (See Figure 3.2.)

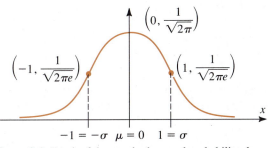

Figure 3.2 Graph of the standard normal probability density function $f(x) = \dfrac{1}{\sqrt{2\pi}} e^{-1/2(x^2)}$.

This probability density function is

$$f(x) = \frac{1}{\sqrt{2\pi}} e^{-1/2(x^2)}.$$

Since x appears to the second power, the value of f at $-x$ is the same as the value of f at x. This is what we mean by saying that the graph of f is symmetric about the line $x = \mu = 0$. (Geometrically, this says that the part of the graph of f lying to the left of the y-axis is the mirror image, in the y-axis, of the part of the graph lying to the right of the y-axis.)

The maximum value of f is found by setting $f'(x) = 0$. Doing so, we obtain

$$f'(x) = \frac{1}{\sqrt{2\pi}} e^{-1/2(x^2)}(-x) = \frac{-x}{\sqrt{2\pi}} e^{-1/2(x^2)}$$

which is zero only when $x = 0$. The point $(0, f(0)) = \left(0, \dfrac{1}{\sqrt{2\pi}}\right)$ therefore corresponds to the relative maximum for f. (Note that this occurs where $x = \mu$.)

To find the inflection points, we set $f''(x) = 0$:

$$f''(x) = \frac{d}{dx}\left(\frac{-x}{\sqrt{2\pi}} e^{-1/2(x^2)}\right)$$

$$= \frac{-1}{\sqrt{2\pi}} e^{-1/2(x^2)} + \frac{-x}{\sqrt{2\pi}} e^{-1/2(x^2)}(-x)$$

$$= (x^2 - 1) \cdot \frac{1}{\sqrt{2\pi}} e^{-1/2(x^2)}$$

so $f''(x) = 0$ when $x^2 = 1$, which occurs when $x = \pm 1 = \pm \sigma$. Since $f''(x) < 0$ for $-1 < x < 1$ and $f''(x) > 0$ for $x < -1$ and for $x > 1$, this shows that Figure 3.2 is correct:

(i) The graph of f is concave down for $-1 < x < 1$, that is, for x within $\sigma = 1$ unit of $\mu = 0$.

(ii) The graph of f is concave up for $|x| > 1$, that is, for x lying more than $\sigma = 1$ unit from $\mu = 0$.

(iii) The points $(1, f(1)) = \left(1, \dfrac{1}{\sqrt{2\pi e}}\right)$

and $(-1, f(-1)) = \left(-1, \dfrac{1}{\sqrt{2\pi e}}\right)$ are inflection points. That is, the inflection points lie $\sigma = 1$ unit from the axis of symmetry $x = \mu = 0$.

We leave it as an exercise for you to verify that conclusions (a) and (b) hold for the general case of an $N(\mu, \sigma)$ normal random variable.

Expected Value and Variance

The significance of the parameters μ and σ for an $N(\mu, \sigma)$ random variable is broader than just the geometric properties described above:

— The *expected value* of the $N(\mu, \sigma)$ normal random variable X is

$$E(X) = \mu.$$

— The *variance* of the $N(\mu, \sigma)$ normal random variable X is

$$\text{Var}(X) = \sigma^2.$$

As for discrete random variables, we refer to the square root of the variance of a continuous random variable X as the *standard deviation* for X. Thus the standard deviation for the $N(\mu, \sigma)$ normal random variable is just the parameter σ. Figures 3.3 and 3.4 illustrate two important rules of thumb for $N(\sigma, \mu)$ random variables X. Regardless of particular values of the parameters σ and μ:

(i) $Pr\{\mu - \sigma \leq X \leq \mu + \sigma\} \approx 0.68$

(ii) $Pr\{\mu - 2\sigma \leq X \leq \mu + 2\sigma\} \approx 0.95$.

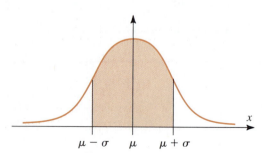

Figure 3.3 $Pr\{\mu - \sigma \leq X \leq \mu + \sigma\}$ is approximately 0.68.

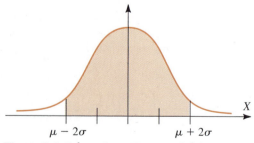

Figure 3.4 $Pr\{\mu - 2\sigma \leq X \leq \mu + 2\sigma\}$ is approximately 0.95.

That is, the probability that a randomly selected value of the $N(\mu, \sigma)$ random variable X will lie within one standard deviation of the expected value μ is roughly 68%. The probability that a randomly selected value of X will lie within two standard deviations of the expected value μ is roughly 95%. The remaining part of this section concerns calculating probabilities associated with normal random variables.

Probabilities for $N(0, 1)$ Variables

It is common practice to denote the standard normal random variable ($\mu = 0$, $\sigma = 1$) by the letter Z. Since the probability density function for this $N(0, 1)$ variable is

$$f(z) = \frac{1}{\sqrt{2\pi}} e^{-z^2/2}$$

the probability $Pr\{Z \le a\}$ is given by the probability distribution function F with values

$$F(a) = Pr\{Z \le a\} = \int_{-\infty}^{a} \frac{1}{\sqrt{2\pi}} e^{-z^2/2} \, dz. \tag{1}$$

(See Figure 3.5.) Values of this distribution function for various values of $a \ge 0$ are given in Table 4, Appendix II. These values have been calculated by evaluating the improper integral in equation (1). The following examples show how Table 4, together with the symmetry of the graph of the probability density function f, is used to calculate probabilities associated with the $N(0, 1)$ variable Z.

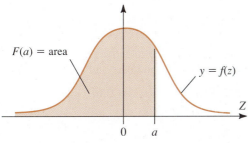

Figure 3.5 For the $N(0, 1)$ variable Z

$$Pr\{Z \le a\} = F(a) = \int_{-\infty}^{a} \frac{1}{\sqrt{2\pi}} e^{-z^2/2} \, dx.$$

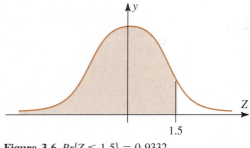

Figure 3.6 $Pr\{Z \le 1.5\} = 0.9332.$

Example 1 For the $N(0, 1)$ variable Z, find $Pr\{Z \le 1.5\}$.

Solution: Since Table 4 gives $Pr\{Z \le a\}$, we find the entry in the body of Table 4 corresponding to the number $a = 1.50$. It is $Pr\{Z \le 1.50\} = 0.9332$. (See Figure 3.6.) □

Example 2 For the $N(0, 1)$ variable Z, find $Pr\{Z \ge 0.85\}$.

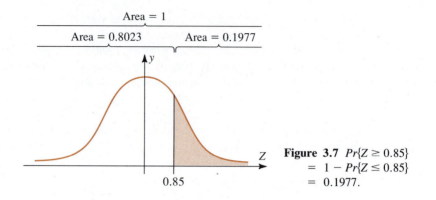

Figure 3.7 $Pr\{Z \geq 0.85\}$
$= 1 - Pr\{Z \leq 0.85\}$
$= 0.1977.$

Solution: As Figure 3.7 indicates, the probability $Pr\{Z \geq 0.85\}$ corresponds to the area of the region under the standard normal curve lying to the *right* of $a = 0.85$. Since the values of the distribution function $F(a)$, hence the entries of Table 4, correspond to the area under this curve lying to the *left* of $Z = a$, we use the fact that the area of the *entire* region lying under the graph of the curve must be 1:

$$Pr\{Z \geq 0.85\} + Pr\{Z \leq 0.85\} = 1$$
$$Pr\{Z \geq 0.85\} = 1 - Pr\{Z \leq 0.85\}$$
$$= 1 - 0.8023 \quad \text{(value from Table 4)}$$
$$= 0.1977. \qquad \square$$

Example 3 Find $Pr\{0.6 \leq Z \leq 2.25\}$ for the standard normal random variable Z.

Solution: Since both $a = 0.6$ and $b = 2.25$ are positive numbers, we may simply find the entries of Table 4 corresponding to these numbers and subtract:

$$Pr\{0.6 \leq Z \leq 2.25\} = Pr\{Z \leq 2.25\} - Pr\{Z \leq 0.6\}$$
$$= 0.9878 - 0.7257$$
$$= 0.2621.$$

(See Figure 3.8.) $\qquad \square$

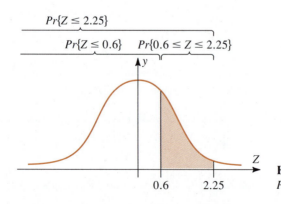

Figure 3.8 $Pr\{0.6 \leq Z \leq 2.25\} =$
$Pr\{Z \leq 2.25\} - Pr\{Z \leq 0.6\}.$

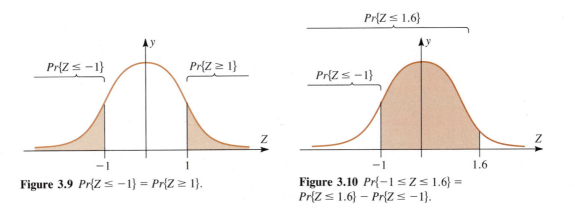

Figure 3.9 $Pr\{Z \le -1\} = Pr\{Z \ge 1\}$.

Figure 3.10 $Pr\{-1 \le Z \le 1.6\} =$
$Pr\{Z \le 1.6\} - Pr\{Z \le -1\}$.

Example 4 For the $N(0, 1)$ variable Z find $Pr\{-1 \le Z \le 1.6\}$.

Solution: Since $a = -1$ is less than zero, we cannot find the probability $Pr\{Z \le -1\}$ directly from the table. However, as Figure 3.9 shows, we can use the symmetry of the standard normal curve to calculate $Pr\{Z \le -1\}$ as

$$Pr\{Z \le -1\} = Pr\{Z \ge 1\}$$
$$= 1 - Pr\{Z \le 1\}$$
$$= 1 - 0.8413$$
$$= 0.1587$$

where we have obtained $Pr\{Z \le 1\} = 0.8413$ from Table 4.
    We may now proceed to calculate $Pr\{-1 \le Z \le 1.6\}$ as

$$Pr\{-1 \le Z \le 1.6\} = Pr\{Z \le 1.6\} - Pr\{Z \le -1\}$$
$$= 0.9452 - 0.1587$$
$$= 0.7865.$$

(See Figure 3.10.)   □

Probabilities for Other Normal Random Variables

We may use Table 4, Appendix II, to calculate probabilities for $N(\mu, \sigma)$ random variables X with expected value μ and standard deviation σ by first converting such variables to *standard* normal random variables using the equation

$$Z = \frac{X - \mu}{\sigma}. \tag{2}$$

It is a fact, which we shall not prove, that the variable Z in equation (2) has expected value 0 and standard deviation 1 (that is, it is a standard normal random variable). The endpoint a of the interval in which X may fall must be treated by the same conversion:

$$A = \frac{a - \mu}{\sigma}. \tag{3}$$

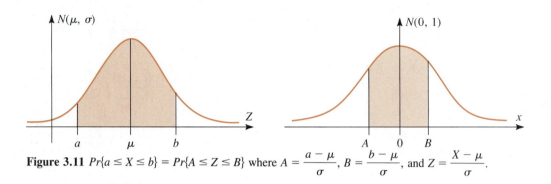

Figure 3.11 $Pr\{a \le X \le b\} = Pr\{A \le Z \le B\}$ where $A = \dfrac{a - \mu}{\sigma}$, $B = \dfrac{b - \mu}{\sigma}$, and $Z = \dfrac{X - \mu}{\sigma}$.

When this is done, then we know that

$$Pr\{X \le a\} = Pr\{Z \le A\}. \qquad (4)$$

We may therefore calculate the probability for the $N(\mu, \sigma)$ random variable X on the left-hand side of equation (4) by finding the probability on the right-hand side of the equation from Table 4 (Figure 3.11). The following examples illustrate this technique.

Example 5 For the normal random variable X having expected value $\mu = 5$ and standard deviation $\sigma = 2$, find $Pr\{6 \le X \le 9\}$. (See Figure 3.12.)

Solution: Using equation (3) with $\mu = 5$, and $\sigma = 2$, we find that the Z-value corresponding to the X-value $a = 6$ is

$$A = \frac{6 - 5}{2} = 0.5$$

and the Z-value corresponding to the X-value $b = 9$ is

$$B = \frac{9 - 5}{2} = 2.0.$$

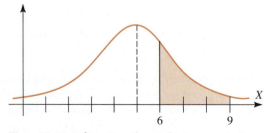

Figure 3.12 $Pr\{6 \le X \le 9\} = Pr\{0.5 \le Z \le 2.0\}$ for $N(5, 2)$ random variable X.

Thus

$$Pr\{6 \leq X \leq 9\} = Pr\{0.5 \leq Z \leq 2.0\}$$
$$= Pr\{Z \leq 2.0\} - Pr\{Z \leq 0.5\}$$
$$= 0.9772 - 0.6915 \qquad \text{(Table 4)}$$
$$= 0.2857. \qquad \qquad \square$$

Example 6 For the normal random variable X with $\mu = 60$ and $\sigma = 20$, find $Pr\{40 \leq X \leq 90\}$.

Solution: The Z-values corresponding to $a = 40$ and $b = 90$ are

$$A = \frac{40 - 60}{20} = \frac{-20}{20} = -1$$

and

$$B = \frac{90 - 60}{20} = \frac{30}{20} = 1.5.$$

Thus

$$Pr\{40 \leq X \leq 90\} = Pr\{-1 \leq Z \leq 1.5\}$$
$$= Pr\{Z \leq 1.5\} - Pr\{Z \leq -1\}$$
$$= Pr\{Z \leq 1.5\} - [1 - Pr\{Z \leq 1\}] \qquad \text{(see Example 4)}$$
$$= 0.9332 - (1 - 0.8413)$$
$$= 0.7745. \qquad \text{(See Figure 3.13.)} \qquad \square$$

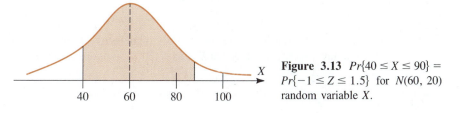

Figure 3.13 $Pr\{40 \leq X \leq 90\} = Pr\{-1 \leq Z \leq 1.5\}$ for $N(60, 20)$ random variable X.

Example 7 Biologists have determined that the variable representing the lifetime of a certain species of animals is normally distributed with expected value $\mu = 6$ years and standard deviation $\sigma = 1.5$ years. Find the probability that an individual animal of this species, selected at random, survives between 4 and 8 years.

Solution: We let X denote the lifetime for a member of this species. Then X is an $N(6, 1.5)$ random variable. The Z-values corresponding to the given X-values $a = 4$ and $b = 8$ are

$$A = \frac{4 - 6}{1.5} = \frac{-2}{1.5} = -1.33 \ldots$$

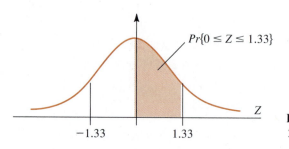

Figure 3.14 $Pr\{-1.33 \leq Z \leq 1.33\} = 2Pr\{0 \leq Z \leq 1.33\}$.

and

$$B = \frac{8 - 6}{1.5} = \frac{2}{1.5} = 1.33. \ldots$$

(In using Table 4, we round these two numbers to -1.33 and 1.33, respectively.) Then

$$
\begin{aligned}
Pr\{4 \leq X \leq 8\} &= Pr\{-1.33 \leq Z \leq 1.33\} \\
&= 2Pr\{0 \leq Z \leq 1.33\} \qquad \text{(see Figure 3.14)} \\
&= 2[Pr\{Z \leq 1.33\} - Pr\{Z \leq 0\}] \\
&= 2(0.9082 - 0.5000) \\
&= 0.8164.
\end{aligned}
$$

□

Our last example shows how knowledge of probability distributions is used in decision making.

Example 8 An automobile manufacturer purchases tires in bulk from a supplier. The supplier claims that the number of defective tires in any shipment of 1000 tires is a normally distributed random variable with expected value $\mu = 20$ and standard deviation $\sigma = 5$. If this claim is true, what is the probability that a shipment of 1000 tires contains 35 or more defective tires?

Solution: Let X denote the number of defective tires contained in the shipment. The manufacturer's claim is that X is an $N(20, 5)$ random variable. The Z-value corresponding to the X-value $b = 35$ is

$$B = \frac{35 - 20}{5} = 3.$$

Thus the probability of finding 35 or more defective tires would be

$$
\begin{aligned}
Pr\{X \geq 35\} &= Pr\{Z > 3\} \\
&= 1 - Pr\{Z \leq 3\} \\
&= 1 - 0.9987 \\
&= 0.0013.
\end{aligned}
$$

If the supplier's claim about μ and σ is correct, the probability of such an outcome is minuscule. If the manufacturer did find 35 defective tires in a shipment, there would be good reason to question the supplier's claim about quality!

□

Exercise Set 11.3

In Exercises 1–8 find the indicated probability for the standard $N(0, 1)$ normal random variable Z using Table 4, Appendix II.

1. $Pr\{Z \leq 2\}$

2. $Pr\{Z \leq 1.85\}$

3. $Pr\{1.2 \leq Z \leq 2.4\}$

4. $Pr\{0 \leq Z \leq 1.25\}$

5. $Pr\{-1 \leq Z \leq 2\}$

6. $Pr\{-2.15 \leq Z \leq 0\}$

7. $Pr\{-1.6 \leq Z \leq 1.35\}$

8. $Pr\{-1 \leq Z \leq -0.25\}$

In Exercises 9–16 find the indicated probability for the $N(\mu, \sigma)$ random variable by first converting to the standard $N(0, 1)$ random variable and then using Table 4.

9. $Pr\{X \leq 6\}$, $\mu = 4$, $\sigma = 2$

10. $Pr\{0 \leq X \leq 30\}$, $\mu = 20$, $\sigma = 20$

11. $Pr\{1 \leq X \leq 7\}$, $\mu = 2$, $\sigma = 2$

12. $Pr\{-3 \leq X \leq 5\}$, $\mu = 1$, $\sigma = 2$

13. $Pr\{35 \leq X \leq 65\}$, $\mu = 50$, $\sigma = 10$

14. $Pr\{2 \leq X \leq 3\}$, $\mu = 1$, $\sigma = 1$

15. $Pr\{-5 \leq X \leq 1\}$, $\mu = -1$, $\sigma = 4$

16. $Pr\{17 \leq X \leq 25\}$, $\mu = 22$, $\sigma = 5$

17. Heights in a certain population of adult males are found to be normally distributed with an expected value of $\mu = 70''$ and a standard deviation of $\sigma = 2''$. Find the probability that the height of an adult male selected at random from this population
   a. exceeds 6′ (that is, 72″)
   b. falls between 5′6″ and 6′2″ (that is, between 66″ and 74″).

18. An automobile leasing company finds that the annual mileage recorded for one of its cars is a normal random variable X with expected value $\mu = 18{,}000$ miles and standard deviation 2500 miles. Find
   a. $Pr\{16{,}000 \leq X \leq 20{,}000\}$
   b. $Pr\{X \geq 24{,}000\}$

19. A supplier of light bulbs claims that the number of defective light bulbs in any shipment of 1000 light bulbs is a normal random variable X with expected value 10 and standard deviation 3. If this is true, what is the probability of finding 15 or more defective bulbs in a randomly selected shipment?

20. The breaking strength of a brand of rope is a normal random variable with $\mu = 20$ lb and $\sigma = 10$ lb. What is the probability that a rope selected at random breaks at a force equal to or less than 115 lb?

21. The monthly rainfall in a region is a normal random variable with $\mu = 6''$ and $\sigma = 1.5''$. What is the probability that a month's rainfall will fall between 5″ and 7″?

22. If the combined SAT scores of entering freshmen at a college is a normal random variable with $\mu = 1080$ and $\sigma^2 = 1600$, find the probability that the SAT score for a freshman selected at random is greater than 1020.

23. A marketing firm determines that the annual household income of the consumers of a product is a normal random variable with $\mu = \$30{,}000$ and $\sigma = \$8000$. What percentage of these consumers have household incomes in excess of $42{,}000?

24. If the diameters of ball bearings produced by a properly functioning machine are normally distributed with expected value 3 cm and standard deviation $\sigma = 0.1$ cm, what is the probability that a randomly selected ball bearing will have diameter in excess of 3.24 cm?

25. Find the expected value of the $N(0, 1)$ random variable Z with probability density function $f(z) = \dfrac{1}{\sqrt{2\pi}} e^{-z^2/2}$ by evaluating the improper integral

$$E(Z) = \int_{-\infty}^{\infty} \frac{1}{\sqrt{2\pi}} z e^{-1/2(z^2)} \, dz$$

$$= \lim_{t \to -\infty} \int_{t}^{0} \frac{1}{\sqrt{2\pi}} z e^{-1/2(z^2)} \, dz$$

$$+ \lim_{t \to \infty} \int_{0}^{t} \frac{1}{\sqrt{2\pi}} z e^{-1/2(z^2)} \, dz.$$

26. Verify that properties (a) and (b) following Definition 5 hold for the general case of an $N(\mu, \sigma)$ normal random variable.

11.4 THE EXPONENTIAL PROBABILITY DISTRIBUTION

The *exponential* probability distribution is another example of a continuous random variable that occurs frequently in applications. Random variables that have exponential probability distributions typically represent the time required for an event to occur, such as the length of a telephone call, the time that elapses until a new automobile requires service, or the lifetime of a new small business.

The exponential probability distribution is determined by the following probability density function.

DEFINITION 6

Let $\alpha > 0$ be a positive real number. The continuous random variable X with probability density function

$$f(x) = \begin{cases} \alpha e^{-\alpha x}, & x \geq 0 \\ 0, & x < 0 \end{cases}$$

is called the **exponential random variable** with parameter α.

Note that the exponential random variable is determined by one parameter, while the normal random variable is determined by two parameters. Figure 4.1 shows the graph of the probability density function f for a typical exponential random variable X.

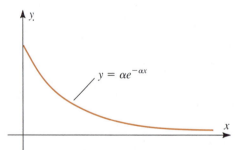

Figure 4.1 Graph of probability density function for a typical exponential random variable X.

Example 1 Verify that the function f in Definition 6 is a probability density function.

Solution: Since values of the exponential function $e^{-\alpha x}$ are positive for all numbers α and x, and since the number α is positive, values of the function $f(x) = \alpha e^{-\alpha x}$ are nonnegative for all values of x.

Since $f(x) = 0$ for $x < 0$, the integral of the function f over the entire sample space $(-\infty, \infty)$ is given by the improper integral

$$\int_0^\infty f(x)\, dx = \int_0^\infty \alpha e^{-\alpha x}\, dx$$

$$= \lim_{t \to \infty} \int_0^t \alpha e^{-\alpha x} \, dx$$

$$= \lim_{t \to \infty} -e^{-\alpha x} \Big]_0^t$$

$$= \lim_{t \to \infty} (1 - e^{-\alpha t})$$

$$= 1$$

as required for a probability density function. □

Example 2 For the exponential random variable X with probability density function $f(x) = 2e^{-2x}$ find $Pr\{1 \le X \le 2\}$.

Solution: From the definition of the probability determined by a probability density function we have

$$Pr\{1 \le X \le 2\} = \int_1^2 f(x) \, dx$$

$$= \int_1^2 2e^{-2x} \, dx$$

$$= -e^{-2x} \Big]_1^2$$

$$= -e^{-4} - (-e^{-2})$$

$$= \frac{1}{e^2} - \frac{1}{e^4}$$

$$\approx 0.117.$$ □

The Distribution Function

Recall that the *distribution* function F associated with a probability density function f is defined by

$$F(t) = Pr\{X \le t\}. \tag{1}$$

For the exponential probability density function f in Definition 6 equation (1) becomes, for $t > 0$

$$F(t) = Pr\{X \le t\} = \int_0^t f(x) \, dx$$

$$= \int_0^t \alpha e^{-\alpha x} \, dx$$

$$= -e^{-\alpha x} \Big]_0^t \tag{2}$$

$$= 1 - e^{-\alpha t}.$$

From equation (2) we may conclude that

$$Pr\{X \le t\} = 1 - e^{-\alpha t}, \qquad t \ge 0 \tag{3}$$

for the exponential random variable X with parameter α. Note that equation (3) allows probabilities to be calculated without integrating the probability density function. For example, the probability in Example 2 ($\alpha = 2$) can be calculated using equation (3) as

$$\begin{aligned} Pr\{1 \le X \le 2\} &= Pr\{X \le 2\} - Pr\{X \le 1\} \\ &= (1 - e^{-2\cdot 2}) - (1 - e^{-2\cdot 1}) \\ &= -e^{-4} + e^{-2}. \end{aligned}$$

Example 3 The telephone company determines that the length of time in seconds between incoming calls for directory assistance is an exponential random variable with $\alpha = 0.10$. If a call has just come in, what is the probability that the next call will occur within 5 seconds?

Solution: Using equation (3), we find that

$$\begin{aligned} Pr\{X \le 5\} &= 1 - e^{-(0.10)5} \\ &= 1 - e^{-0.5} \\ &= 1 - 0.607 \\ &= 0.393. \end{aligned} \qquad \square$$

Example 4 A company purchases a new delivery truck that comes with a 24-month warranty on all repairs. Experience has shown that the probability of such a truck requiring a major repair is an exponential random variable with parameter $\alpha = 0.05$ when the variable t (time) is measured in years. What is the probability of the company having to pay for the cost of a major repair to this truck within the first 3 years?

Solution: Since repairs during the first 2 years are covered under warranty, the probability of the company having to *pay* for a major repair in the first 3 years is just the probability that it occurs *between* the end of the second year and the end of the third year, or

$$\begin{aligned} Pr\{2 \le X \le 3\} &= Pr\{X \le 3\} - Pr\{X \le 2\} \\ &= (1 - e^{-(0.05)(3)}) - (1 - e^{-(0.05)(2)}) \\ &= e^{-0.10} - e^{-0.15} \\ &= 0.905 - 0.861 \\ &= 0.044. \end{aligned} \qquad \square$$

Expected Value   The expected value of an exponential random variable is closely related to its parameter α.

> The expected value of the exponential random variable X with parameter α is
> $$E(X) = \frac{1}{\alpha}. \tag{4}$$

To verify equation (4), we must evaluate the improper integral

$$E(X) = \int_0^\infty xf(x)\ dx = \int_0^\infty x \cdot \alpha e^{-\alpha x}\ dx$$

$$= \lim_{t \to \infty} \int_0^t \alpha x e^{-\alpha x}\ dx.$$

(5)

The integral $\int \alpha x e^{-\alpha x}\ dx = \alpha \int x e^{-\alpha x}\ dx$ requires integration by parts. We let

$$u = x \qquad dv = e^{-\alpha x}\ dx$$

$$du = dx \qquad v = -\frac{1}{\alpha} e^{-\alpha x}.$$

Then the integration by parts formula $\int u\ dv = uv - \int v\ du$ gives

$$\int \alpha x e^{-\alpha x}\ dx = \alpha \int x e^{-\alpha x}\ dx = \alpha \left[x \left(-\frac{1}{\alpha} e^{-\alpha x} \right) - \int -\frac{1}{\alpha} e^{-\alpha x}\ dx \right]$$

$$= x e^{-\alpha x} + \int e^{-\alpha x}\ dx$$

$$= -x e^{-\alpha x} - \frac{1}{\alpha} e^{-\alpha x} + C.$$

(6)

Combining equations (5) and (6) (with $C = 0$), we have

$$E(X) = \lim_{t \to \infty} \left(-x e^{-\alpha x} - \frac{1}{\alpha} e^{-\alpha x} \right) \Big|_0^t$$

$$= \lim_{t \to \infty} \left[\frac{1}{\alpha} - \left(t e^{-\alpha t} + \frac{1}{\alpha} e^{-\alpha t} \right) \right]$$

$$= \frac{1}{\alpha}$$

since $\lim_{t \to \infty} t e^{-\alpha t} = 0$ (see Exercise 20) and $\lim_{t \to \infty} \frac{1}{\alpha} e^{-\alpha t} = 0$.

Example 5 The number of hours that a certain type of lightbulb will burn until it expires is an exponential random variable X with expected value $E(X) = 100$ hours. Find the probability that

a. the light bulb expires within the first 80 hours.
b. the light bulb burns for more than 110 hours.

Solution: Since $E(x) = \dfrac{1}{\alpha} = 100$, the parameter for the exponential random variable X

is $\alpha = \dfrac{1}{100} = 0.01$. Using equation (3) with $\alpha = 0.01$, we find that

(a) $Pr\{X \leq 80\} = 1 - e^{-(0.01)(80)}$

$\qquad = 1 - e^{-0.8}$

$\qquad = 1 - 0.449$

$\qquad = 0.551.$

(b) $Pr\{X > 110\} = 1 - Pr\{X \leq 110\}$

$\qquad = 1 - (1 - e^{-(0.01)(110)})$

$\qquad = e^{-1.1}$

$\qquad = 0.333.$ □

Example 6 The lifetime of a computer component is determined to be an exponential random variable X with expected value $E(X) = 10,000$ hours. What is the probability that the component will fail between 8000 and 12,000 hours of use?

Solution: Since $E(X) = \dfrac{1}{\alpha} = 10,000$ we have $\alpha = \dfrac{1}{10,000} = 0.0001$. Thus

$Pr\{8000 \leq X \leq 12,000\} = Pr\{X \leq 12,000\} - Pr\{X \leq 8000\}$

$\qquad = (1 - e^{-(0.0001)(12,000)}) - (1 - e^{-(0.0001)(8000)})$

$\qquad = e^{-0.8} - e^{-1.2}$

$\qquad = 0.449 - 0.301$

$\qquad = 0.148.$ □

Exercise Set 11.4

In Exercises 1–5 find the indicated probability for the exponential random variable X with parameter α by integration.

1. $Pr\{0 \leq X \leq 4\}, \quad \alpha = 0.05$

2. $Pr\{0 \leq X \leq 10\}, \quad \alpha = 0.10$

3. $Pr\{2 \leq X \leq 5\}, \quad \alpha = 0.25$

4. $Pr\{1 \leq X \leq 9\}, \quad \alpha = 0.10$

5. $Pr\{2 \leq X \leq 3\}, \quad \alpha = 0.50$

In Exercises 6–10 find the indicated probability for the exponential random variable X with parameter α by using formula (3).

6. $Pr\{0 \leq X \leq 3\}, \quad \alpha = 0.2$

7. $Pr\{0 \leq X \leq 6\}, \quad \alpha = 0.1$

8. $Pr\{2 \leq X \leq 5\}, \quad \alpha = 0.4$

9. $Pr\{5 \leq X \leq 20\}, \quad \alpha = 0.02$

10. $Pr\{10 \leq X \leq 50\}, \quad \alpha = 0.01$

11. If X is an exponential random variable with expected value $E(X) = 20$, find $Pr\{5 \leq X \leq 30\}$.

12. For the exponential random variable X with expected value $E(X) = 10$ find $Pr\{5 \leq X \leq 15\}$.

13. Find α if X is an exponential random variable with parameter α for which $E(X) = 1$.

14. The random variable X representing the lifetime of a certain type of radial tire (in miles) is known to be an exponential random variable with expected value $E(X) = 40,000$ miles. Find the probability that one such tire will last for more than 50,000 miles.

15. The length of time between incoming telephone calls at a data-service company is determined to be an exponential random variable with expected value 4 minutes. Find the probability that the time between any two successive incoming calls is less than 2 minutes.

16. The length of time that a motorist must wait to cross a certain toll bridge is an exponential random variable with expected value 10 minutes. What is the probability that a motorist will have to wait more than 20 minutes at this bridge?

17. Let X be an exponential random variable. Find $Pr\{X > E(X)\}$. Why is the answer independent of α?

18. Find α if X is an exponential random variable with parameter α for which $Pr\{0 \le X \le 5\} = 0.6320$.

19. Find the variance of the exponential random variable X with parameter α using the facts that

$$\lim_{x \to \infty} x^2 e^{-\alpha x} = 0 \quad \text{and} \quad \lim_{x \to \infty} x e^{-\alpha x} = 0$$

when $\alpha > 0$.

20. Show that $\lim_{x \to \infty} x e^{-\alpha x} = 0$ for $a > 0$ as follows.

a. Expand the function $f(x) = e^{\alpha x}$ about $x = 0$ in a Taylor series to obtain

$$e^{\alpha x} = 1 + \alpha x + \frac{\alpha^2 x^2}{2} + \frac{\alpha^3 x^3}{3!} + \cdots$$
$$+ \frac{\alpha^n x^n}{n!} + \cdots.$$

b. Conclude from part a that $e^{\alpha x} > \dfrac{\alpha^2 x^2}{2}$ when $\alpha > 0$ and $x > 0$.

c. Conclude from part (b) that

$$xe^{-\alpha x} = \frac{x}{e^{\alpha x}} < \frac{x}{\left(\dfrac{\alpha^2 x^2}{2}\right)} = \frac{2}{\alpha^2 x}.$$

d. Conclude from part (c) that

$$\lim_{x \to \infty} xe^{-\alpha x} \le \lim_{x \to \infty} \frac{2}{\alpha^2 x} = 0 \qquad \text{for } \alpha > 0.$$

SUMMARY OUTLINE OF CHAPTER 11

- A **sample space** is the set of all possible simple outcomes for an experiment. (Page 589)

- An **event** is a subset of a sample space. (Page 589)

- A **discrete random variable** X is a variable that can equal any number in a finite sample space consisting only of numbers.

- A **probability distribution** for a discrete random variable X is a function defined on the sample space for X with two properties: (Page 593)

  (i) $0 \le Pr\{X = a\} \le 1$ for any number a in the sample space for X.
  (ii) The sum of the probabilities associated with the numbers in the sample space equals one.

- The **expected value** of the discrete random variable X is the number (Page 596)

  $$E(X) = x_1 p_1 + x_2 p_2 + \cdots + x_n p_n$$

  where $\{x_1, x_2, \ldots, x_n\}$ is the sample space for X and p_j is the probability $Pr\{X = x_j\}$, $j = 1, 2, \ldots, n$.

- The **variance** associated with this random variable is the number (Page 598)

  $$\text{Var}(X) = (x_1 - E(X))^2 p_1 + (x_2 - E(X))^2 p_2 + \cdots + (x_n - E(X))^2 p_n.$$

- The **standard deviation** σ of this random variable is the number $\sigma = \sqrt{\text{Var}(X)}$. (Page 599)

■ A **probability density function** for the continuous random variable X is a continuous nonnegative function f for which (Page 601)

$$Pr\{a \le X \le b\} = \int_a^b f(x) \, dx.$$

■ A **distribution function** for the continuous random variable X is a function F with the property that (Page 605)

$$F(x) = Pr\{X < x\}$$

for all x in the sample space for X.

■ If X is a continuous random variable with a continuous probability density function f and a probability distribution function F, we have (Page 606)

$$Pr\{a \le X \le b\} = \int_a^b f(x) \, dx = F(b) - F(a).$$

■ If X is a continuous random variable defined on the sample space $[a, b]$ with continuous probability density function f, the **expected value** of X is (Page 608)

$$E(X) = \int_a^b xf(x) \, dx$$

and the **variance** is

$$Var(X) = \int_a^b (x - E(X))^2 f(x) \, dx.$$

■ The **normal random variable** with parameters μ and σ, referred to as the $N(\mu, \sigma)$ variable, is the continuous random variable with probability density function (Page 612)

$$f(x) = \frac{1}{\sqrt{2\pi}\sigma} e^{-1/2(x - \mu/\sigma)^2}.$$

■ The $N(0, 1)$ random variable is called the **standard normal random variable**. (Page 613)

■ For the $N(\mu, \sigma)$ random variable X we have (Page 614)

$$E(X) = \mu, \qquad Var(X) = \sigma^2.$$

■ The **exponential random variable** with parameter α is the continuous random variable with probability density function (Page 622)

$$f(x) = \begin{cases} \alpha e^{-\alpha x}, & x \ge 0 \\ 0, & x < 0. \end{cases}$$

REVIEW EXERCISES—CHAPTER 11

1. Table 1 shows a probability distribution for the discrete random variable X. Find the probability of
   a. the event $\{1, 2, 5\}$
   b. the event $\{2, 3, 4\}$
   c. the event $\{1, 2, 3\}$

Table 1

x_i	1	2	3	4	5
p_i	0.1	0.1	0.2	0.3	0.3

2. Find the expected value of the random variable X in Exercise 1.

3. Find the variance of the random variable in Exercise 1.

4. Find the probability that a card drawn at random from an ordinary deck of playing cards is a red card numbered less than 10.

5. A lottery ticket costs $10, and 2000 tickets are sold. There is but one prize, a cash amount of $500. What is the expected value of the variable X representing the net winnings resulting from the purchase of a single ticket?

6. For the probability density function

$$f(x) = \begin{cases} \dfrac{x}{6}, & 0 \le x \le 2 \\[2mm] \dfrac{1}{3}, & 2 < x \le 4 \end{cases}$$

on the sample space [0, 4] find

a. $Pr\{0 \le X \le 2\}$
b. $Pr\{1 \le X \le 3\}$
c. $Pr\{1 \le X \le 4\}$

7. Find the number k so that the function $f(x) = kx$ is a probability density function on the sample space [0, 5].

8. For the probability density function in Exercise 7 find

a. $Pr\{0 \le X \le 2\}$
b. $Pr\{1 \le X \le 5\}$

9. For the probability density function $f(x) = k(4 - x^2)$ on the sample space [0, 2] find

a. k
b. $Pr\{0 \le X \le 1\}$
c. $Pr\{1 \le X \le 2\}$

10. Find the expected value of the random variable X with probability density function f in Exercise 7.

11. Find the variance, $Var(X)$, of the random variable X with probability density function f in Exercise 7.

12. Find the expected value of the random variable X with probability density function f in Exercise 9.

In Exercises 13–16 find the indicated probability for the standard $N(0, 1)$ normal random variable Z.

13. $Pr\{Z \le -0.5\}$ **14.** $Pr\{-2.1 \le Z \le 2.1\}$

15. $Pr\{Z \ge -0.6\}$ **16.** $Pr\{Z \le 1.7\}$

In Exercises 17–20 find the indicated probability for the $N(\mu, \sigma)$ normal random variable.

17. $Pr\{X \ge 7\}$, $\mu = 10$, $\sigma = 2$

18. $Pr\{6.5 \le X \le 10\}$, $\mu = 8.6$, $\sigma = 1.4$

19. $Pr\{-11 \le X \le 13\}$, $\mu = -2$, $\sigma = 6$

20. $Pr\{136 \le X \le 156\}$, $\mu = 140$, $\sigma = 8$.

21. The variable X representing a person's score on an intelligence test is a normal random variable with $\mu = 100$ and $\sigma = 10$. Find the probability that a score, selected at random, will

a. exceed 110.
b. exceed 120.

22. If the length of a goldfish's life is a normal random variable with expected value 15 months and variance 16 months, find the probability that a single goldfish selected at random will live at least nine months.

23. The annual snowfall in a certain region is found to be a normal random variable with expected value $\mu = 16$ inches and variance $\sigma^2 = 9$ inches. What is the probability that a single year's snowfall totals less than 10 inches?

24. If X is an $N(\mu, \sigma)$ random variable with $\mu = 6$ and $Pr\{4 \le X \le 8\} = 0.8664$, what is σ?

In Exercises 25–28 find the indicated probability for the exponential random variable X with parameter α.

25. $Pr\{X \ge 4\}$, $\alpha = 0.2$

26. $Pr\{0 \le X \le 70\}$, $\alpha = 0.2$

27. $Pr\{X \ge 12\}$, $\alpha = 0.10$

28. $Pr\{10 \le X \le 20\}$, $\alpha = 0.05$

29. Find $Pr\{2 \le X \le 6\}$ for the exponential random variable X with expected value $\mu = 4$.

30. The lifetime of a machine part is known to be an exponential random variable with expected value $\mu = 20$ months. Find the probability that such a part fails within the first year of use.

31. Find the expected value of the exponential random variable X if $Pr\{0 \le X \le 5\} = 0.221$.

32. The length of time between calls at an urban fire station is an exponential random variable with expected value $\mu = 2$ hours. If a call has just been received, what is the probability that the next call will occur within 30 minutes?

BASIC Computer Programs

Included here are four BASIC programs that are referred to in various examples and exercises in the text. They are presented as "bare-bones" prototypes, which can be used by those with access to computing facilities (personal computers, programmable calculators, or large computers) in designing programs that actually operate on particular machines. Notation appearing in these programs includes the following:

1. $a*b$ means the product ab.
2. $a \uparrow b$ means the exponentiation a^b.
3. a/b means the quotient $\dfrac{a}{b}$.

Proper development of computer software requires full documentation, as well as the inclusion of checks to ensure that the user does not attempt to supply inappropriate values to the program. (For example, in asking the user to specify the endpoints of an interval $[a, b]$, one should check to ensure that $a < b$.) We have made no attempt to do either, since we wish to highlight only the algorithm involved in the program.

Program 1: Values of the Function $f(x) = \dfrac{x^2 + x + 6}{x - 2}$

```
10   DEF FNF(X) = (X ↑ 2 + X + 6)/(X − 2)
20   PRINT "ENTER X"
30   INPUT X
40   LET Y = FNF(X)
50   PRINT "F(X) = "; Y
60   END
```

Comment: This program prints the function value $f(x) = \dfrac{x^2 + x + 6}{x - 2}$ for the user's choice of $x \neq 2$. To print values of a different function g, simply replace the right-hand side of the equation in line 10 with the expression for $g(x)$.

Program 2: Approximating the Number e

```
10  PRINT "ENTER N"
20  INPUT N
30  FOR M = 1 TO N
40  LET Y = (1 + 1/M) ↑ M
50  PRINT Y
60  NEXT M
70  END
```

Comment: This program prints the values $\left(1 + \dfrac{1}{M}\right)^M$ for integers M beginning with $M = 1$ and terminating with $M = N$.

Program 3: Calculating Compound Interest

```
10   PRINT "ENTER R"
20   INPUT R
30   PRINT "ENTER K"
40   INPUT K
50   PRINT "ENTER N"
60   INPUT N
70   PRINT "ENTER P"
80   INPUT P
90   LET Y = (1 + R/K) ↑ (N*K)
100  LET Z = Y*P
110  PRINT Z
120  END
```

Comment: This program prints the amount $\left(1 + \dfrac{r}{k}\right)^{nk} P$ on deposit n years after an initial deposit of P dollars has been subjected to compounding of interest k times per year at nominal rate r percent.

Program 4: Approximating Sum for $\displaystyle\int_a^b x^2\,dx$

```
10  DEF FNF(T) = T ↑ 2
20  PRINT "ENTER INTERVAL ENDPOINTS A,B"
30  INPUT A,B
40  PRINT "HOW MANY SUBINTERVALS?"
50  INPUT N
60  LET D = (B − A)/N
70  LET S = O
80  FOR I = 1 to N
```

```
 90    LET X = A + (I − 1)*D
100    LET S = S + FNF(X)*D
110   NEXT I
120   PRINT S
130   END
```

Comment: This program prints an approximating sum for the function $f(x) = x^2$ on the interval $[a, b]$ chosen by the user. The sum is associated with n subintervals of equal length, and the left endpoint of each subinterval is used as the "test point." To use the program for a different function, simply rewrite line 10 to correspond to the desired function.

Program 5:   Trapezoidal Rule Applied to $f(x) = \dfrac{1}{x}$

```
 10   DEF FNF(T) = 1/T
 20   PRINT "enter interval endpoints a,b with
        a > 0"
 30   INPUT A,B
 40   PRINT "how many subintervals?"
 50   INPUT N
 60   LET D = (B − A)/N
 70   LET S = FNF(A)
 80   FOR I = 1 TO (N − 1)
 90    LET X = A + I*D
100    LET S = S + 2*FNF(X)
110   NEXT I
120   LET S = S + FNF(B)
130   LET S = S*(D/2)
140   PRINT S
150   END
```

Comment: This program implements the Trapezoidal Rule for approximate integration for the function $f(x) = \dfrac{1}{x}$ on the interval $[a, b]$. The numbers a, b, and n (the number of subintervals) are supplied by the user.

Program 6:   Simpson's Rule Applied to the Function $f(x) = \dfrac{1}{x}$

```
 10   DEF FNF(T) = 1/T
 20   PRINT "enter interval endpoints a,b with a > 0"
 30   INPUT A,B
 40   PRINT "how many subintervals (an even
        number)?"
```

```
50   INPUT N
60   LET C = (B − A)/N
70   LET D = C/3
80   LET S = FNF(A)
90   FOR I = 1 TO (N − 1) STEP 2
100     LET S = S + 4*FNF(A + I*C)
110  NEXT I
120  FOR I = 2 TO (N − 2) STEP 2
130     LET S = S + 2*FNF(A + I*C)
140  NEXT I
150  LET S = S + FNF(B)
160  LET S = S*D
170  PRINT S
180  END
```

Comment: This program implements Simpson's Rule for approximate integration for the function $f(x) = \dfrac{1}{x}$ on the interval $[a, b]$. The numbers a, b, and n (the number of subintervals, *an even number*) are supplied by the user.

Program 7: Euler's Method Applied to the Initial Value Problem

$$\frac{dy}{dt} = e^t - 2v$$

$$y(0) = 1$$

```
10   LET Y = 1
20   LET T = 0
30   FOR I = 1 TO 4
40   T = T + 0.25
50   Y = Y + (EXP(T) − 2*Y)*(0.25)
60   PRINT T,Y
70   NEXT I
80   END
```

Comment: This program gives an approximation to the stated initial value problem on the interval $[0, 1]$ using $n = 4$ subintervals. The equation for Euler's method is implemented in step 50. Note that the right-hand side of this equation involves current values of both T and Y.

Mathematical Tables

Table 1. Exponential functions

x	e^x	e^{-x}	x	e^x	e^{-x}
0.00	1.00000	1.00000	.75	2.11700	.47237
.05	1.05127	.95123			
.10	1.10517	.90484	.80	2.22554	.44933
.15	1.16183	.86071	.85	2.33965	.42741
.20	1.22140	.81873	.90	2.45960	.40657
.25	1.28403	.77880	.95	2.58571	.38674
.30	1.34986	.74082	1.00	2.71828	.36788
.35	1.41907	.70469	2.00	7.38906	.13534
.40	1.49182	.67032	3.00	20.08554	.04979
.45	1.56831	.63763	4.00	54.59815	.01832
.50	1.64872	.60653	5.00	148.41316	.00674
			6.00	403.42879	.00248
.55	1.73325	.57695	7.00	1096.63316	.00091
.60	1.82212	.54881	8.00	2980.95799	.00034
.65	1.91554	.52205	9.00	8103.08393	.00012
.70	2.01375	.49659	10.00	22026.46579	.00005

Note: $e^{a+x} = e^a e^x$

Table 2. Natural logarithms

x	$\ln x$	x	$\ln x$	x	$\ln x$	x	$\ln x$
.1	−2.30258	2.6	.95551	5.1	1.62924	7.6	2.02815
.2	−1.60943	2.7	.99325	5.2	1.64866	7.7	2.04122
.3	−1.20396	2.8	1.02962	5.3	1.66771	7.8	2.05412
.4	−.91628	2.9	1.06471	5.4	1.68640	7.9	2.06686
.5	−.69314	3.0	1.09861	5.5	1.70475	8.0	2.07944
.6	−.51082	3.1	1.13140	5.6	1.72277	8.1	2.09186
.7	−.35666	3.2	1.16315	5.7	1.74047	8.2	2.10413
.8	−.22313	3.3	1.19392	5.8	1.75786	8.3	2.11626
.9	−.10535	3.4	1.22378	5.9	1.77495	8.4	2.12823
1.0	0.00000	3.5	1.25276	6.0	1.79176	8.5	2.14007
1.1	.09531	3.6	1.28093	6.1	1.80829	8.6	2.15176
1.2	.18232	3.7	1.30833	6.2	1.82455	8.7	2.16332
1.3	.26236	3.8	1.33500	6.3	1.84055	8.8	2.17475
1.4	.33647	3.9	1.36098	6.4	1.85630	8.9	2.18605
1.5	.40547	4.0	1.38629	6.5	1.87180	9.0	2.19722
1.6	.47000	4.1	1.41099	6.6	1.88707	9.1	2.20827
1.7	.53063	4.2	1.43508	6.7	1.90211	9.2	2.21920
1.8	.58779	4.3	1.45862	6.8	1.91692	9.3	2.23001
1.9	.64185	4.4	1.48160	6.9	1.93152	9.4	2.24071
2.0	.69315	4.5	1.50408	7.0	1.94591	9.5	2.25129
2.1	.74194	4.6	1.52606	7.1	1.96009	9.6	2.26176
2.2	.78846	4.7	1.54756	7.2	1.97408	9.7	2.27213
2.3	.83291	4.8	1.56862	7.3	1.98787	9.8	2.28238
2.4	.87547	4.9	1.58924	7.4	2.00148	9.9	2.29253
2.5	.91629	5.0	1.60944	7.5	2.01490	10.0	2.30259

Note: $\ln 10x = \ln x + \ln 10$

Table 3. Trigonometric functions (x in radians)

x	$\sin x$	$\cos x$	$\tan x$	x	$\sin x$	$\cos x$	$\tan x$
0.0	.00000	1.00000	0.00000	3.4	$-.25554$	$-.96680$	.26432
0.1	.09983	.99500	.10033	3.5	$-.35078$	$-.93646$	.37459
0.2	.19867	.98007	.20271				
0.3	.29552	.95534	.30934	3.6	$-.44252$	$-.89676$	.49347
0.4	.38942	.92106	.42279	3.7	$-.52984$	$-.84810$	.62473
0.5	.47943	.87758	.54630	3.8	$-.61186$	$-.79097$	.77356
				3.9	$-.68777$	$-.72593$	.94742
0.6	.56464	.82534	.68414	4.0	$-.75680$	$-.65364$	1.15782
0.7	.64422	.76484	.84229				
0.8	.71736	.69671	1.02964	4.1	$-.81828$	$-.57482$	1.42353
0.9	.78333	.62161	1.26016	4.2	$-.87158$	$-.49026$	1.77778
1.0	.84147	.54030	1.55741	4.3	$-.91617$	$-.40080$	2.28585
				4.4	$-.95160$	$-.30733$	3.09632
1.1	.89121	.45360	1.96476	4.5	$-.97753$	$-.21080$	4.63733
1.2	.93204	.36236	2.57215				
1.3	.96356	.26750	3.60210	4.6	$-.99369$	$-.11215$	8.86017
1.4	.98545	.16997	5.79788	4.7	$-.99992$	$-.01239$	80.71271
1.5	.99749	.07074	14.10142	$3\pi/2$	-1.00000	.00000	$-\infty$
$\pi/2$	1.00000	.00000	∞	4.8	$-.99616$	.08750	-11.38487
1.6	.99957	$-.02920$	-34.23254	4.9	$-.98245$	.18651	-5.26749
1.7	.99166	$-.12884$	-7.69660	5.0	$-.95892$	.28366	-3.38051
1.8	.97385	$-.22720$	-4.28626				
1.9	.94630	$-.32329$	-2.92710	5.1	$-.92581$	.37798	-2.44939
2.0	.90930	$-.41615$	-2.18504	5.2	$-.88345$	.46852	-1.88564
				5.3	$-.83227$	.55437	-1.50127
2.1	.86321	$-.50485$	-1.70985	5.4	$-.77276$	.63469	-1.21754
2.2	.80850	$-.58850$	-1.37382	5.5	$-.70554$	.70867	$-.99558$
2.3	.74571	$-.66628$	-1.11921				
2.4	.67546	$-.73739$	$-.91601$	5.6	$-.63127$	.77557	$-.81394$
2.5	.59847	$-.80114$	$-.74702$	5.7	$-.55069$	.83471	$-.65973$
				5.8	$-.46460$	.88552	$-.52467$
2.6	.51550	$-.85689$	$-.60160$	5.9	$-.37388$	.92748	$-.40311$
2.7	.42738	$-.90407$	$-.47273$	6.0	$-.27942$	.96017	$-.29101$
2.8	.33499	$-.94222$	$-.35553$				
2.9	.23925	$-.97096$	$-.24641$	6.1	$-.18216$	.98327	$-.18526$
3.0	.14112	$-.98999$	$-.14255$	6.2	$-.08309$	.99654	$-.08338$
3.1	.04158	$-.99914$	$-.04162$	2π	.00000	1.00000	.00000
π	.00000	-1.00000	.00000	6.3	.01681	.99986	.01682
3.2	$-.05837$	$-.99829$	.05847	6.4	.11655	.99318	.11735
3.3	$-.15775$	$-.98748$	.15975	6.5	.21512	.97659	.22028

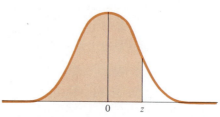

The entries in Table 4 are the probabilities that a random variable having the standard normal distribution takes on a value less than z: they are given by the area under the curve shaded in the above diagram.

Table 4. The standard normal distribution

z	.00	.01	.02	.03	.04	.05	.06	.07	.08	.09
0.0	.5000	.5040	.5080	.5120	.5160	.5199	.5239	.5279	.5319	.5359
0.1	.5398	.5438	.5478	.5517	.5557	.5596	.5636	.5675	.5714	.5753
0.2	.5793	.5832	.5871	.5910	.5948	.5987	.6026	.6064	.6103	.6141
0.3	.6179	.6217	.6255	.6293	.6331	.6368	.6406	.6443	.6480	.6517
0.4	.6554	.6591	.6628	.6664	.6700	.6736	.6772	.6808	.6844	.6879
0.5	.6915	.6950	.6985	.7019	.7054	.7088	.7123	.7157	.7190	.7224
0.6	.7257	.7291	.7324	.7357	.7389	.7422	.7454	.7486	.7517	.7549
0.7	.7580	.7611	.7642	.7673	.7704	.7734	.7764	.7794	.7823	.7852
0.8	.7881	.7910	.7939	.7967	.7995	.8023	.8051	.8078	.8106	.8133
0.9	.8159	.8186	.8212	.8238	.8264	.8289	.8315	.8340	.8365	.8389
1.0	.8413	.8438	.8461	.8485	.8508	.8531	.8554	.8577	.8599	.8621
1.1	.8643	.8665	.8686	.8708	.8729	.8749	.8770	.8790	.8810	.8830
1.2	.8849	.8869	.8888	.8907	.8925	.8944	.8962	.8980	.8997	.9015
1.3	.9032	.9049	.9066	.9082	.9099	.9115	.9131	.9147	.9162	.9177
1.4	.9192	.9207	.9222	.9236	.9251	.9265	.9279	.9292	.9306	.9319
1.5	.9332	.9345	.9357	.9370	.9382	.9394	.9406	.9418	.9429	.9441
1.6	.9452	.9463	.9474	.9484	.9495	.9505	.9515	.9525	.9535	.9545
1.7	.9554	.9564	.9573	.9582	.9591	.9599	.9608	.9616	.9625	.9633
1.8	.9641	.9649	.9656	.9664	.9671	.9678	.9686	.9693	.9699	.9706
1.9	.9713	.9719	.9726	.9732	.9738	.9744	.9750	.9756	.9761	.9767
2.0	.9772	.9778	.9783	.9788	.9793	.9798	.9803	.9808	.9812	.9817
2.1	.9821	.9826	.9830	.9834	.9838	.9842	.9846	.9850	.9854	.9857
2.2	.9861	.9864	.9868	.9871	.9875	.9878	.9881	.9884	.9887	.9890
2.3	.9893	.9896	.9898	.9901	.9904	.9906	.9909	.9911	.9913	.9916
2.4	.9918	.9920	.9922	.9925	.9927	.9929	.9931	.9932	.9934	.9936
2.5	.9938	.9940	.9941	.9943	.9945	.9946	.9948	.9949	.9951	.9952
2.6	.9953	.9955	.9956	.9957	.9959	.9960	.9961	.9962	.9963	.9964
2.7	.9965	.9966	.9967	.9968	.9969	.9970	.9971	.9972	.9973	.9974
2.8	.9974	.9975	.9976	.9977	.9977	.9978	.9979	.9979	.9980	.9981
2.9	.9981	.9982	.9982	.9983	.9984	.9984	.9985	.9985	.9986	.9986
3.0	.9987	.9987	.9987	.9988	.9988	.9989	.9989	.9989	.9990	.9990

Answers to Odd-Numbered Exercises

CHAPTER 1

Exercise Set 1.1

1. any real number

3. integers

5. integers

7. integers

9. any real number

11. d

13. a

15. b

17. g

19. f

21. True

23. Two (1 and -1)

25. $(-\infty, 3]$

27. $[-4, \infty)$

29. $[11, \infty]$

31. $\left[\dfrac{1}{2}, \infty\right)$

33. $\left(-\infty, \dfrac{4}{7}\right]$

35. $200P + 250B \leq 10,000$

37. 20

39. $24 < w < 30$

41. a. $1100 + 400x > 600x$
   b. $x \geq 6$

43. 15 ounces

Exercise Set 1.2

1.

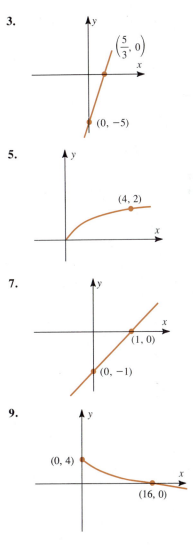

3.

5.

7.

9.

11.

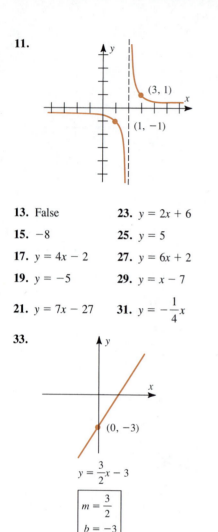

13. False

15. -8

17. $y = 4x - 2$

19. $y = -5$

21. $y = 7x - 27$

23. $y = 2x + 6$

25. $y = 5$

27. $y = 6x + 2$

29. $y = x - 7$

31. $y = -\dfrac{1}{4}x$

33.

$y = \dfrac{3}{2}x - 3$

$\boxed{m = \dfrac{3}{2}}$

$\boxed{b = -3}$

35.

$\left(0, \dfrac{10}{3}\right)$

$y = -\dfrac{2}{3}x + \dfrac{10}{3}$

$\boxed{m = -\dfrac{2}{3}}$

$\boxed{b = \dfrac{10}{3}}$

37.

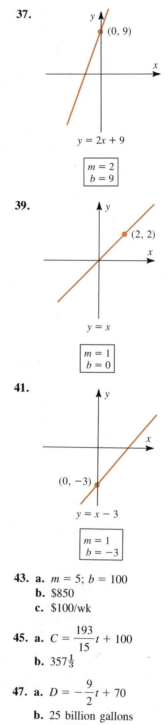

$(0, 9)$

$y = 2x + 9$

$\boxed{\begin{array}{l} m = 2 \\ b = 9 \end{array}}$

39.

$(2, 2)$

$y = x$

$\boxed{\begin{array}{l} m = 1 \\ b = 0 \end{array}}$

41.

$(0, -3)$

$y = x - 3$

$\boxed{\begin{array}{l} m = 1 \\ b = -3 \end{array}}$

43. a. $m = 5$; $b = 100$
b. \$850
c. \$100/wk

45. a. $C = \dfrac{193}{15}t + 100$
b. $357\frac{1}{3}$

47. a. $D = -\dfrac{9}{2}t + 70$
b. 25 billion gallons
c. by 1993 (using $t = 15.5$)

Exercise Set 1.3

1. $\sqrt{13}$ **5.** $\sqrt{8}$

3. $\sqrt{122}$ **7.** $\sqrt{50}$

11. $x^2 + y^2 = 9$ **13.** $x^2 - 8x + y^2 - 6y + 21 = 0$

15. $x^2 - 12x + y^2 - 8y - 48 = 0$

17. $(x - 5)^2 - 25$ **25.** center $= (-7, 5)$
                                  radius $= 2$

19. $16 - (x - 4)^2$ **27.** center $= (1, 3)$
                                  radius $= \sqrt{7}$

21. $\dfrac{7}{2} - 2\left(t + \dfrac{1}{2}\right)^2$ **29.** inside

23. center $= (1, -3)$
      radius $= \sqrt{22}$

Exercise Set 1.4

1. function   **5.** function

3. not a function   **7.** not a function

9. a. 7   **b.** -2
    c. 19   **d.** $7 - 3a$

11. a. $\dfrac{3}{2}$ **b.** 0

    c. 2   **d.** $\dfrac{1}{2}$

13. not a function   **17.** function

15. not a function   **19.** function

21. $(-\infty, 7)$ and $(7, \infty)$

23. $(-\infty, -1)$ and $(-1, \infty)$

25. $[-1, 1]$ **31.** $C(x) = 2x + 0.5$

27. $(-\infty, \infty)$ **33.** $C(x) = 70x + 1000$

29. $[-2, \infty)$ **35. a.** 600   **b.** 0   **c.** 20

37. $R(p) = \begin{cases} 400p, & p \le 20 \\ 600p - 10p^2 & p > 20 \end{cases}$

39. a.   **b.** 20

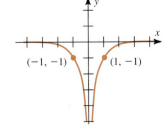

41. a. $P(x) = 75 + 0.005x$ **b.** \$275

43. a. $C(t) = \dfrac{7}{8}t + 315$

    b. 343 parts per million

Exercise Set 1.5

1. 9   **7.** $27x^{11/2}y^{-1}$

3. $\dfrac{1}{8}$ **9.** $6x^{7/3}y^{9/2}$

5. $27x^2$ **11.** True

13.

15.

17.

19.

21.

23.

25.

27.

29.

31.

33.

35.

37.

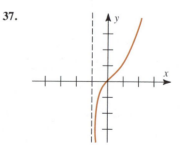

39. $w(\ell) = \dfrac{80}{\ell}$ Domain is $(0, \infty)$

41. a.

$u(x) = \sqrt{x + 2}$

b.

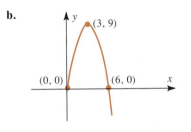

c. $U_2(x)$ **d.** $U_1(x)$

43. a. $T(3) = 30\left(1 + \dfrac{2}{\sqrt{3} + 1}\right) = 30(1 + 1) = 60$

seconds

b.

45. $C(x) = 400 + 40x + 0.2x^2$
$= 0.2(x^2 + 200x + 10{,}000) - 2{,}000 + 400$
$= 0.2(x + 100)^2 - 1{,}600$

a.

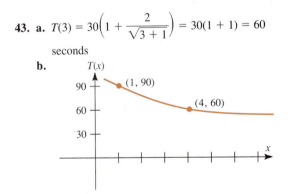

b. Fixed costs = \$400

x	$C(x)$
50	$400 + 2000 + 500 = 2900$
100	$400 + 4000 + 2000 = 6400$
150	$400 + 6000 + 4500 = 10{,}900$
200	$400 + 8000 + 8000 = 16{,}400$
250	$400 + 10{,}000 + 12{,}500 = 22{,}900$
300	$400 + 12{,}000 + 18{,}000 = 30{,}400$

Exercise Set 1.6

1. $x = -2, 2$

3. $x = -3, 2$

5. $x = -5, 3$

7. $x = 2$

9. $x = 0, 3$

11. $x = -\dfrac{1}{2} \pm \dfrac{\sqrt{5}}{2}$

13. $x = -\dfrac{3}{4} \pm \dfrac{\sqrt{17}}{4}$

15. $x = -1, 3$

17. $x = \dfrac{1}{6} \pm \dfrac{7}{6}$

19. $(2, 5)$

21. $(-1, 1), (2, 4)$

23. $(0, 0), (4, 2)$

25. $(-1, 1), (0, 2), (1, 3)$

27. $(4, 0), \left(\dfrac{5}{3}, -\dfrac{35}{3}\right)$

29. More items will be demanded than will be supplied; that is, $D(p_1) > S(p_1)$.

31. $x = \$3600$

33. a. $x = 10, 50$
    b. $10 < x < 50$

35. $t = 1, 5$ hours

37. $x = 10, 50$

Exercise Set 1.7

1. $x^2 + x - 5$

3. 1

5. $-\dfrac{11}{18}$

7. $\dfrac{1}{x + 2} - 2\sqrt{3 + x}$

9. 6

11. $-\dfrac{11}{3}$

13. $4 - x^2$

15. $3\sqrt{x} + 1$

17. $(3x + 1)^3$

19. $\dfrac{1}{\sqrt{x} + 1}$

21. $3x^{3/2} + 1$

23. $(x + 3)^2$

25. $u(x) = 2x$

27. a

29. b

31. d

33. a

35. d

37.

(0, 2)

39.

(2, 4)

41.

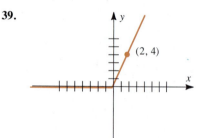

(−1, 0)   (2, 0)

43. a. $p(x) = \begin{cases} 500, & 0 \le x \le 5 \\ 500 - 10(x - 5), & 6 \le x \le 25 \\ 300, & 26 \le x \end{cases}$

b.

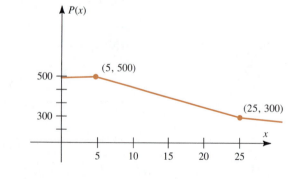

45. $f(c) = \sqrt{15 + 0.2c + 0.01c^2}$

47. a. $C(x) = (200x + 75)$ dollars

b. $R(x) = 1.4(200x + 75) = (280x + 105)$ dollars

c. $P(x) = \dfrac{280x + 105}{x} = \left(280 + \dfrac{105}{x}\right)$ dollars

Review Exercises—Chapter 1

1. a. $(-6, 3]$ **b.** $(-\infty, 4)$ **c.** $[2, \infty)$

3. $x \le -1$ **5.** $x \le \dfrac{2}{3}$

7.

(0, 1)

9.

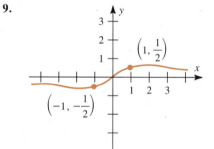

$\left(1, \dfrac{1}{2}\right)$

$\left(-1, -\dfrac{1}{2}\right)$

11. $\dfrac{1}{6}$ **15.** $b = 7$

13. $y = 5$

17.

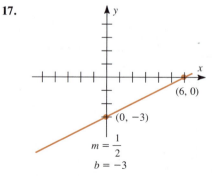

(6, 0)

(0, −3)

$m = \dfrac{1}{2}$

$b = -3$

19.

$m = -1$
$b = 4$

21. $(-\infty, 0)$ and $(0, \infty)$ **23.** $[-3, 3]$

25. a. $\dfrac{2}{9}$ **b.** $\dfrac{3}{14}$ **27.** 8   **29.** $x^3 y^6$

31.

33.

$r = \sqrt{6}$

35.

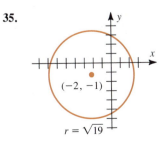

$r = \sqrt{19}$

37. 1, 2

39. 3

41. $-3, 3$

43. $(-8, -14)$

45. $(1, 1), (-1, 1)$

47. $x^4 + 4x^3 + 2x + 8$

49. -12

51. $\sqrt{x^2 - 4}$

53. $4x^2 - 28$

55.

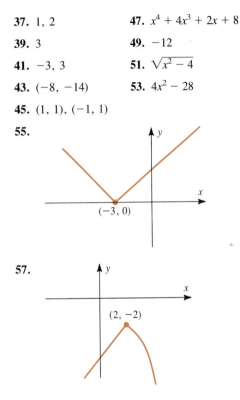

$(-3, 0)$

57.

$(2, -2)$

59. $200

61. b, c   **63.** False

65. $C(x) = \begin{cases} 20x, & 0 \le x \le 50 \\ 1000 + 15(x - 50), & 50 < x \end{cases}$

67. 98 cents per lb   **69.** 59.5°F

71. a. $21,000   **b.** 20%   **c.** $50,000

73. $15,000   **77.** 600

75. $Y(x) = x(400 - 2x)$

79. a. $75,000   **c.** 83,333

b. $V(n) = (100,000)\left(1 - \dfrac{n}{30}\right)$

CHAPTER 2

Exercise Set 2.1

1. $m = -2$ **5.** $m = -\dfrac{1}{2}$

3. $m = \dfrac{1}{4}$ **7.** $m = 2$

9. $m = -2$

11. $m = 3$

13. $m = 4$

15. $m = -1$

17. $y = 8x - 7$

19. $y = -\dfrac{1}{8}x + \dfrac{5}{8}$

21. $m = 2$

23. a. $m = 2$
b. $m = 2$

25. a. $p_0 = 4$
b. $m = 2$
c. $m = -2$

27. b. $m = 24$
c. $m = 25 - a$
d. $t = 25$

Exercise Set 2.2

1. 0

3. -3

5. -6

7. $\dfrac{1}{7}$

9. 3

11. 8

13. 4

15. -10

17. -2

19. $\dfrac{3}{10}$

21. $-\dfrac{1}{3}$

23. $\dfrac{1}{4}$

25. 0

27. 0

29. $-\infty$

Exercise Set 2.3

1. a. doesn't exist   **b.** 2   **c.** 2   **d.** 0

3. a. 2   **b.** doesn't exist   **c.** 0

5. -1

7. 0

9. 1

11. $x = -1, 1$

13. $x = -3$

15. $x = -3, -1, 1$

17. $x = 2$

19. $k = 3$

21. $k = -1$

23. $q = 50, 100$

25. $A = \$2865$

27. $a = 2$

Exercise Set 2.4

1. 2

3. $2x$

5. $4x^3$

7. $6x + 4$

9. $12x^3 + 4x - 4$

11. $3x^2 - 5x^4$

13. $12x^3 - 6$

15. $10x^4 + 6x^{-3}$

17. $\dfrac{4}{3}x^{-1/3} + 5x^{2/3}$

19. $-18x^{-3} + \dfrac{3}{2\sqrt{x}}$

21. $180x^8 + x^{-2/3}$

23. $-\dfrac{8}{3}x^{-5/3} + \dfrac{7}{4}x^{-5/4}$

25. 40

27. -284

29. $-\dfrac{3}{2\sqrt{2}} + \dfrac{5}{3}$

31. $y - 3 = 6(x - 1)$

33. $\dfrac{5}{16}$

35. $a = 5$

Exercise Set 2.5

1. a. $R'(x) = 10$
b. $C'(x) = 6$
c. $P(x) = 4x - 50$
d. $P'(x) = 4$

3. a. $R'(x) = 100 - 4x$
b. $C'(x) = 20 + 3x^2$
c. $P(x) = 80x - 2x^2 - x^3 - 400$
d. $P'(x) = 80 - 4x - 3x^2$

5. a. $R'(x) = 40 + \dfrac{25}{\sqrt{x}}$
b. $C'(x) = 20 + \dfrac{2}{3}x^{-1/3}$
c. $P(x) = 20x + 50\sqrt{x} - x^{2/3} - 150$
d. $P'(x) = 20 + \dfrac{25}{\sqrt{x}} - \dfrac{2}{3}x^{-1/3}$

7. a. $R'(x) = 400 - \dfrac{20}{3}x^{-1/3}$
b. $C'(x) = -\dfrac{5000}{x^2}$
c. $P(x) = 400x - 10x^{2/3} - \dfrac{5000}{x} - 40$
d. $P'(x) = 400 - \dfrac{20}{3}x^{-1/3} + \dfrac{5000}{x^2}$

9. a. $v(t) = 2t - 6$ **b.** $v(2) = -2$

11. a. $v(t) = -\dfrac{1}{2\sqrt{t}} + \dfrac{9}{2}t^{1/2}$
b. $v(4) = \dfrac{35}{4}$

13. a. $v(t) = \dfrac{3}{\sqrt{t}} + 9t^2 + \dfrac{1}{2}t^{-3/2}$

   b. $v(4) = \dfrac{2329}{16}$

15. a. $v(t) = 1$ **17.** $x = 20$
   b. $v(5) = 1$

19. a. $R(x) = 500x - x^2$
   b. $P(x) = 496x - 2x^2 - 150$
   c. $MR(x) = 500 - 2x$
       $MC(x) = 4 + 2x$
   d. $x = 124$
   e. Decreased to $x = 124$

21. a. $\dfrac{5}{2\sqrt{t}} - \dfrac{1}{4}$ **b.** Yes. For $t > 100$

23. $C'(x) = \dfrac{d}{dx}[F + V(x)]$
         $= V'(x)$

25. a. $MU(x) = 6 - 2x$ **27. a.** $v(0) = 6$
   b. $x = 3$ **b.** $t = 3$
   c. No   **c.** $v(6) = -6$

29. a. no   **b.** yes   **c.** $x_0 = 25$
   d. $P(16) = -160$; $P(25) = 0$; $P(49) = 440$;
       $P(100) = 1400$

Exercise Set 2.6

1. $f'(x) = 2x$

3. $f'(x) = 12x^3 - 24x^2 + 12x - 16$

5. $f'(x) = 6x^5 - 8x^3 + 2x$

7. $f'(x) = \dfrac{7}{2}x^{5/2} - \dfrac{1}{2}x^{-3/2}$

9. $f'(x) = x^{-2} + 24x^{-3}$

11. $f'(x) = \dfrac{-4}{(x-2)^2}$

13. $f'(x) = \dfrac{8x^2 - 48x - 32}{(x-3)^2}$

15. $f'(x) = 1$

17. $f'(x) = \dfrac{x-3}{(1+x)^3}$

19. $f'(x) = \dfrac{\dfrac{3}{2}x^{3/2} + 2x - \dfrac{1}{6}x^{1/6} - \dfrac{2}{3}x^{-1/3}}{(\sqrt{x}+1)^2}$

21. $f'(x) = \left(\dfrac{4}{3}\right)x^{1/3} + \left(\dfrac{2}{3}\right)x^{-2/3}$

23. $f'(x) = \dfrac{6x + 3 - x^2}{x^2(x+1)^2}$

25. $f'(x) = 3acx^2 + 2bcx + ad$

27. $\left(1, -\dfrac{3}{2}\right), \left(3, \dfrac{9}{2}\right)$

31. $f'(x) = -5x^4 + 16x^3 - 6x^2 - 8x + 3$

33. $f'(x) = (x - x^2)(x^3 - x^{-2})(-x^{-2} + 3x^{-4})$
       $+ (x - x^2)(x^{-1} - x^{-3})(3x^2 + 2x^{-3})$
       $+ (x^3 - x^{-2})(x^{-1} - x^{-3})(1 - 2x)$

35. a. $MR(x) = \dfrac{10{,}000(3\sqrt{x} + x)}{(2 + \sqrt{x})^2}$
   b. $MR(9) = 7200$

37. a. $MC(x) = 20 + 2x$
   b. $c(x) = \dfrac{400}{x} + 20 + x$
   c. $c'(x) = -\dfrac{400}{x^2} + 1$

39. $W'(t) = \dfrac{1000t}{(10 + t^2)^2}$ **41. a.** $MU(x) = \dfrac{9 - x^2}{(x^2 + 9)^2}$
                                        b. $x = 3$

Exercise Set 2.7

1. $f'(x) = 3(x + 4)^2$

3. $f'(x) = 6(2x - 7)(x^2 - 7x)^5$

5. $f'(x) = (x^4 - 5)^2(13x^4 - 5)$

7. $f'(x) = \dfrac{-6x}{(x^2 - 9)^4}$

9. $f'(x) = \dfrac{6}{\sqrt{x}}(3\sqrt{x} - 2)^3$

11. $f'(x) = \left(\dfrac{8}{3}x^{-1/3} - x^{-3/4}\right)(x^{2/3} - x^{1/4})^3$

13. $f'(x) = \dfrac{24(x - 3)^3}{(x + 3)^5}$

15. $f'(x) = 2(x + 2)^2(x^{-1} - x^{-2})$

17. $f'(x) = 3x^{-5/4}(4x^{-1/4} + 6)^{-4}$

19. $f'(x) = \dfrac{(9x^2 - 8x^3 + 1)(x^3 + 1)^2 + 1}{(1 - x)^2}$

21. $f'(x) = \dfrac{-12x - 6}{(x^2 + x + 1)^7}$

23. $f'(x) = \dfrac{4x + x^{3/2}}{2(1 + \sqrt{x})^4}$

25. $f'(x) = \dfrac{3}{2}(x^{1/2} - x^{-1/2} - x^{-3/2} + x^{-5/2})$

27. $f'(x) = \dfrac{5(x - 4)^4}{\sqrt{x^3 + 6}} - \dfrac{3x^2(x - 4)^5}{2(x^3 + 6)^{3/2}}$

29. $f'(x) = x^{-2/3}(x^{1/3} - 4)^2(x - \sqrt{x})^{-2/3}$
$\qquad - \dfrac{2}{3}(x - \sqrt{x})^{-5/3}\left(1 - \dfrac{2}{\sqrt{x}}\right)(x^{1/3} - 4)^3$

31. $y = 0$ **33.** $y = 2x - 1$

35. $MC(x) = \dfrac{16x}{\sqrt{40 + 16x^2}}$

37. a. $MP(x) = \dfrac{1}{3}(3x^2 + 10)(x^3 + 10x + 125)^{-2/3}$
  b. Yes

39. a. $N'(t) = \dfrac{30t}{(9 + t^2)^{3/2}}$
  b. No. $N'(t) > 0$ for all t.

41. $c'(x) = \dfrac{-(500\sqrt{40 + 16x^2} + 40)}{x^2\sqrt{40 + 16x^2}}$

Review Exercises—Chapter 2

1. 5

3. 77

5. 6

7. -3

9. $\dfrac{4}{3}$

11. $\sqrt{2}$

13. 4

15. $x = 2$

17. $x = -2, 3$

19. no x

21. $f'(x) = 2x - 1$

23. $f'(x) = \dfrac{2}{3}x^{-1/3}$

25. $f'(x) = \dfrac{-2}{x^3}$

27. $f'(x) = \dfrac{-1}{(x - 2)^2}$

29. $f'(x) = m$

31. $f'(x) = \dfrac{-7}{(3x - 7)^2}$

33. $f'(x) = \dfrac{4}{(x + 3)^2}$

35. $f'(x) = \dfrac{-2x}{(x^2 + 6)^2}$

37. $f'(x) = (6x + 12)(x^2 + 4x + 4)^2$

39. $f'(x) = \dfrac{-1}{3\sqrt{x}(\sqrt{x} + 3)^{5/3}}$

41. $f'(x) = \dfrac{-x}{(1 + x^2)^{3/2}}$ **43.** $f'(x) = \dfrac{1 + 2x^2}{\sqrt{1 + x^2}}$

45. $f'(x) = -6(1 - x^2)^3(6 + 2x)^{-4}(1 + 6x + x^2)$

47. $-\dfrac{5}{4}$ **49.** $3y + 2x - 14 = 0$

51. a. $p = 7$ **b.** $D'(7) = -\dfrac{1}{3}$ **c.** $S'(7) = 1$

53. $p = 200, 400$

55. a. $MC(x) = 30 + x$ **57. a.** $MC(x) = 40 + x^2$
  b. $x = 30$ **b.** $x = 6$

59. a. $p(x) = 250 - \dfrac{1}{2}x$ **61. a.** $c(x) = \dfrac{400}{x} + 50$

  b. $R(x) = 250x - \dfrac{1}{2}x^2$ **b.** $c'(x) = -\dfrac{400}{x^2}$

  c. $MR(x) = 250 - x$
  d. $x = 250$

CHAPTER 3

Exercise Set 3.1

	Increasing on	*Decreasing on*
1.	$(-\infty, \infty)$	
3.	$(-\infty, 0)$	$(0, \infty)$
5.	$\left(-\dfrac{\sqrt{3}}{3}, \dfrac{\sqrt{3}}{3}\right)$	$\left(-\infty, -\dfrac{\sqrt{3}}{3}\right)$ $\left(\dfrac{\sqrt{3}}{3}, \infty\right)$
7.	$(-\infty, -1)$ $(1, \infty)$	$(-1, 1)$

Increasing on	*Decreasing on*
9. $(-\infty, -2)$	$(-2, 2)$
$(2, \infty)$	
11. $(-\infty, -2)$	$(-2, 3)$
$(3, \infty)$	
13. $(-2, 0)$	$(-\infty, -2)$
$(1, \infty)$	$(0, 1)$
15. $(0, \infty)$	$(-\infty, 0)$
17. $(-\infty, -1)$	
$(-1, \infty)$	
19.	$(-\infty, 0)$
	$(0, 2)$
	$(2, \infty)$

21. $a = 3$

23. $x < 25$

25. a. For $x > 10$
    b. For $0 < x < 10$

27. a. $0 < x < 10$
    b. For all $x > 0$

29. a. $R(x) = 500x - \dfrac{x^2}{2}$

    b. For $0 < x < 500$

    c. $P(x) = 498x - \dfrac{x^2}{2} - 400$

    d. For $0 < x < 498$
    e. For $x > 498$

3. $x = 1$; rel. min.

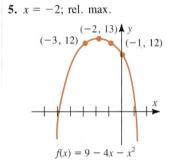

$f(x) = 3x^2 - 6x + 3$

5. $x = -2$; rel. max.

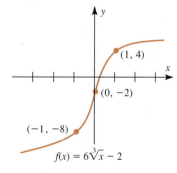

$f(x) = 9 - 4x - x^2$

Exercise Set 3.2

1. $x = 1$; rel. min.

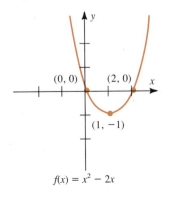

$f(x) = x^2 - 2x$

7. $x = 0$; neither

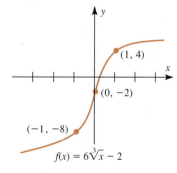

$f(x) = 6\sqrt[3]{x} - 2$

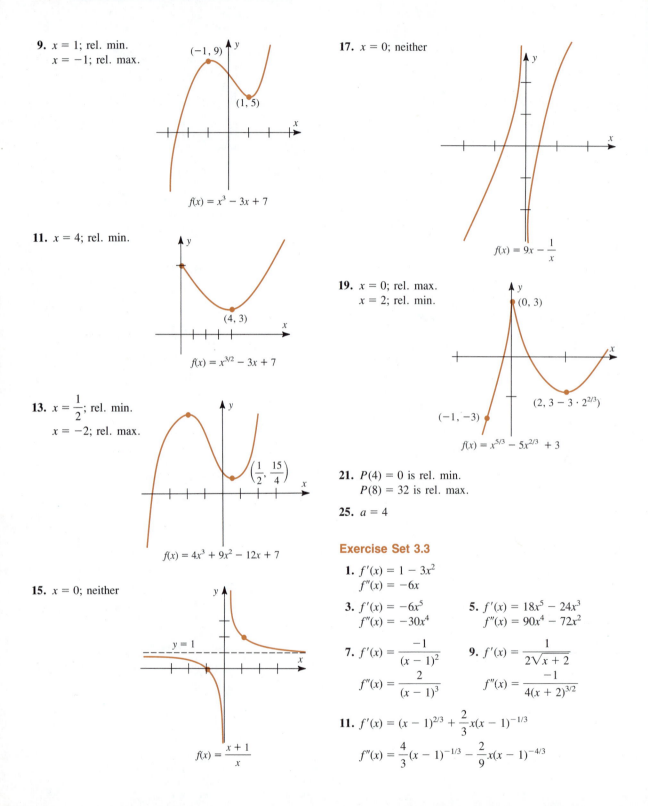

9. $x = 1$; rel. min.
$x = -1$; rel. max.

$(-1, 9)$
$(1, 5)$

$f(x) = x^3 - 3x + 7$

11. $x = 4$; rel. min.

$(4, 3)$

$f(x) = x^{3/2} - 3x + 7$

13. $x = \dfrac{1}{2}$; rel. min.

$x = -2$; rel. max.

$\left(\dfrac{1}{2}, \dfrac{15}{4}\right)$

$f(x) = 4x^3 + 9x^2 - 12x + 7$

15. $x = 0$; neither

$y = 1$

$f(x) = \dfrac{x + 1}{x}$

17. $x = 0$; neither

$f(x) = 9x - \dfrac{1}{x}$

19. $x = 0$; rel. max.
$x = 2$; rel. min.

$(0, 3)$
$(-1, -3)$
$(2, 3 - 3 \cdot 2^{2/3})$

$f(x) = x^{5/3} - 5x^{2/3} + 3$

21. $P(4) = 0$ is rel. min.
$P(8) = 32$ is rel. max.

25. $a = 4$

Exercise Set 3.3

1. $f'(x) = 1 - 3x^2$
$f''(x) = -6x$

3. $f'(x) = -6x^5$
$f''(x) = -30x^4$

5. $f'(x) = 18x^5 - 24x^3$
$f''(x) = 90x^4 - 72x^2$

7. $f'(x) = \dfrac{-1}{(x - 1)^2}$
$f''(x) = \dfrac{2}{(x - 1)^3}$

9. $f'(x) = \dfrac{1}{2\sqrt{x + 2}}$
$f''(x) = \dfrac{-1}{4(x + 2)^{3/2}}$

11. $f'(x) = (x - 1)^{2/3} + \dfrac{2}{3}x(x - 1)^{-1/3}$

$f''(x) = \dfrac{4}{3}(x - 1)^{-1/3} - \dfrac{2}{9}x(x - 1)^{-4/3}$

13. $f'(x) = -6x(1 - x^2)^2$
$f''(x) = 36x^2 - 6 - 30x^4$

15. concave up on $(-\infty, \infty)$
no inflection points

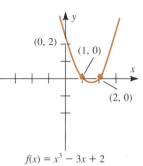

$f(x) = x^3 - 3x + 2$

17. concave down on $(-\infty, 3)$
concave up on $(3, \infty)$
$(3, -24)$ inflection point

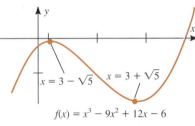

$f(x) = x^3 - 9x^2 + 12x - 6$

19. concave down on $(-\infty, 0)$
concave up on $(0, \infty)$
no inflection points

21. concave up on $(-\infty, -1)$
concave down on $(-1, \infty)$
no inflection points

$f(x) = \dfrac{x}{x + 1}$

23. concave down on $\left(-\infty, -\dfrac{1}{2}\right)$

concave up on $\left(-\dfrac{1}{2}, \infty\right)$

$\left(-\dfrac{1}{2}, 0\right)$ inflection point

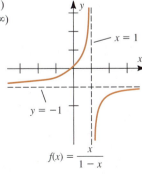

$f(x) = (2x + 1)^3$

25. concave down on $\left(-\infty, \dfrac{1}{2}\right)$

concave up on $\left(\dfrac{1}{2}, \infty\right)$

$\left(\dfrac{1}{2}, -\dfrac{7}{2}\right)$ inflection point

$f(x) = 2x^3 - 3x^2 + 18x - 12$

27. concave up on $(-\infty, 2)$
concave down on $(2, \infty)$
$(2, 0)$ inflection point

$f(x) = (x - 2)^{1/3}$

29. concave up on $(-\infty, 1)$
concave down on $(1, \infty)$
no inflection points

$f(x) = \dfrac{x}{1 - x}$

31. rel. min.   **35.** neither

33. rel. min.   **37.** concave up

39. a. $U_1(x)$
   b. investor 1
   c. (i) investor 2
      (ii) investor 1

47.

49.

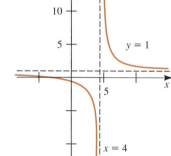

Exercise Set 3.4

1. -3 **15.** $+\infty$ **27.** $y = 7$

3. -1 **17.** $+\infty$ **29.** $x = 4$

5. $\dfrac{1}{7}$ **19.** $-\infty$ **31.** $x = -2$
                              $x = 2$

7. 0   **21.** $y = 0$ **33.** $x = -4$
                              $x = -1$

9. 0   **23.** $y = 2$ **35.** $x = 2$

11. 0   **25.** $y = 0$ **37.** $x = 2, 3$

13. $+\infty$

39. a. $C(x) = 2500 + 80x$
   b. $c(x) = \dfrac{2500}{x} + 80$
   c. 80

41. $A = 35$

43. yes; $y = 16$

45.

51.

53.

55.

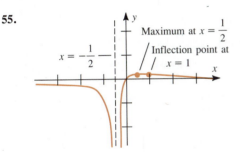

Maximum at $x = \dfrac{1}{2}$

Inflection point at $x = 1$

$x = -\dfrac{1}{2}$

57.

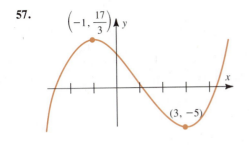

$\left(-1, \dfrac{17}{3}\right)$

$(3, -5)$

59.

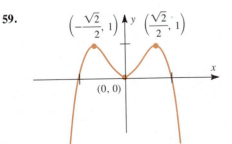

$\left(-\dfrac{\sqrt{2}}{2}, 1\right)$ $\left(\dfrac{\sqrt{2}}{2}, 1\right)$

$(0, 0)$

61.

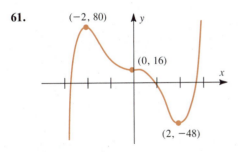

$(-2, 80)$

$(0, 16)$

$(2, -48)$

Exercise Set 3.5

1. Max. is $f(0) = 9$; min. is $f(-3) = f(3) = 0$

3. Max. is $f(-1) = f(3) = 6$; min. is $f(1) = 2$

5. Min. is $f\left(\dfrac{1}{2}\right) = \dfrac{11}{4}$; there is no maximum

7. Min. is $f\left(\dfrac{2}{3}\right) = -\dfrac{4}{27}$; max. is $f(3) = 18$

9. Min. is $f(-1) = -3$; max. is $f(0) = f(2) = 0$

11. Min. is $f(1) = 2$; max. is $f\left(\dfrac{1}{2}\right) = f(2) = \dfrac{5}{2}$

13. Min. is $f(0) = 2$; no max.

15. Max. is $f(2) = -\dfrac{1}{4}$; min. is $f(1) = f(3) = -\dfrac{1}{3}$

17. No min.; max. is $f(2) = 3$

19. Min. is $f(4) = -6$; max. is $f\left(\dfrac{1}{3}\right) = \dfrac{1}{\sqrt{3}} - \dfrac{1}{3^{3/2}} = \dfrac{2}{3\sqrt{3}}$

21. 2066

23. $P(0) = 160$ is max.
$P(4) = P(10) = 0$ is min.

25. $p = 50$

Exercise Set 3.6

1. 1990   **3.** $t = 2$ **5.** $(0, 36)$

7. When $\dfrac{N}{2}$ are infected

9. Max. is -8 at $(1, -7)$
Min. is -20 at $(3, -39)$

11. $p = 10$ **13.** \$40   **15.** $\ell = 4$, $w = 3$

17. $w = \dfrac{20}{4 + \pi}$, $r = \dfrac{10}{4 + \pi}$, $h = \dfrac{10}{4 + \pi}$

19. After $t = 1$ minute

21. $x = 8$ **23.** $x = 15$ trees

25. $v(5) = 0.062$ km/hr is min.
$v(7) = v(3) = 100$ km/hr is max.

27. Schedule maximum overtime possible.

29. Lay cable under water to point $x = 450$ meters from second boathouse.

31. $x = \dfrac{10}{3}$

Exercise Set 3.7

1. $x = 192$

3. $x = 5$

5. Companies 1, 2, and 4

7. 6 times per year

9. approximately 8 times per year

11. $x = 20$

13. a. $R(x) = x(600 - 3x)$
 b. $P(x) = 450x - 3.5x^2 - 400$
 c. $MR(x) = 600 - 6x$, $MC(x) = 150 + x$
 d. $x = 64.3$; $x = 64$ or 65, max. profit for $x = 64$
 e. they are equal

15. $MC(x) = c(x)$

19. 69 dishwashers

21. $E = \dfrac{1}{2}$

23. $E = 1$

25. $E = \dfrac{1}{3}$

27. a. $Q(101) = 99$
 b. $E(101) = \dfrac{101}{99}$ (elastic)

29. Elastic for all $p > 0$

31. Inelastic for all $p > 1$

Exercise Set 3.8

1. $\dfrac{dy}{dx} = -\dfrac{x}{y}$

3. $\dfrac{dy}{dx} = -\dfrac{y}{2x}$

5. $\dfrac{dy}{dx} = -1$

7. $\dfrac{dy}{dx} = \dfrac{1}{2y - 1}$

9. $\dfrac{dy}{dx} = \dfrac{5x^4}{4y^3}$

11. $\dfrac{dy}{dx} = \dfrac{y - \dfrac{1}{2}x^{-1/2}}{-x - \dfrac{1}{3}y^{-2/3}}$

13. $\dfrac{dy}{dx} = -1$

15. $\dfrac{dy}{dx} = -1$

17. $\dfrac{dy}{dx} = \dfrac{3}{5}$

19. $x + y = 8$

21. $\dfrac{dy}{dx} = \dfrac{1}{6}$

23. 40π m²/sec

25. 4 persons per day

27. \$117 million per year (approx.)

29. $12\sqrt{10}$ dollars per day

Exercise Set 3.9

1. No. $f(x) = |x|$ is not differentiable at $x = 0$

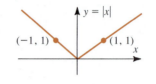

3. $c = 1$ **5.** $c = \dfrac{1}{4}$ **7.** $c = -\dfrac{3}{2}$

Review Exercises—Chapter 3

1. inc. on $(-\infty, 2)$ $f(2) = 4$, rel. max.
 dec. on $(2, \infty)$

3. dec. on $(-\infty, 1)$ $f(1) = 2$, rel. min.
 inc. on $(1, \infty)$

5. inc. on $(-\infty, -3)$ no rel. extrema
 inc. on $(-3, \infty)$

7. dec. on $(-4, -2\sqrt{2})$ $f(-2\sqrt{2}) = -8$, rel. min.
 inc. on $(-2\sqrt{2}, 2\sqrt{2})$ $f(2\sqrt{2}) = 8$, rel. max.
 dec. on $(2\sqrt{2}, 4)$

9. inc. on $(-\infty, 0)$ $f(0) = 2$, rel. max.
 dec. on $(0, 2)$ $f(2) = -2$, rel. min.
 inc. on $(2, \infty)$

	Minimum	*Maximum*
11.	$f(-1) = f(1) = -\dfrac{1}{3}$	$f(0) = -\dfrac{1}{4}$
13.	$f(-3) = -6$	$f(0) = 0$
15.	$f\left(-\dfrac{1}{\sqrt{2}}\right) = -\sqrt{2}$	$f(1) = 1$
17.	$f(-1) = -5$	$f(-2) = -3$
19.	$f(-1) = f(1) = -1$	$f(-2) = f(2) = 8$

21. Vertical asymptote: $x = 7$
 Horizontal asymptote: $y = 0$

23. No vertical asymptote
 Horizontal asymptote: $y = 1$

25. Vertical asymptote: $x = \pm 3$
Horizontal asymptote: $y = 1$

27. No vertical asymptote
Horizontal asymptote: $y = 1$ and $y = -1$

29. Vertical asymptote: $x = 0$
Horizontal asymptote: $y = 4$

31. -2

33. $\dfrac{7}{3}$ **37.** $-\infty$

35. $-\infty$ **39.** $+\infty$

41.

43.

45.

47.

49.

51. 40 cm²/sec

53. $-\dfrac{V}{324a^2\pi}$ m/sec   **59.** 7 times/yr

55. $x = 1000$ **61.** 127 items/week (5 fewer)

57. 60 km/hr   **63.** $p = \sqrt{50}$

65. a. $v(t) = 96 - 32t$
   b. after $t = 3$ sec, at $s(3) = 144$ ft
   c. at $t = 0$ when $s(0) = 0$
   d. $v(6) = -96$ ft/sec; speed $= 96$ ft/sec

67. a. $c(x) = 30 + \dfrac{200}{x} + 0.5x$

   b. $x = 20$

69. $\dfrac{dy}{dx} = \dfrac{-2xy - y^2}{x^2 + 2xy}$ **71.** $\dfrac{dy}{dx} = \dfrac{3}{8}$

CHAPTER 4

Exercise Set 4.1

1. a. 27   **c.** 343
   b. 2   **d.** $\dfrac{1}{8}$

3. a. $\dfrac{1}{8}$ **c.** 8
   b. $\dfrac{9}{4}$ **d.** $\dfrac{8}{27}$

5.

7.

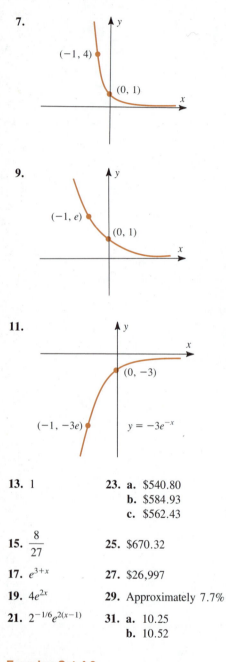

(−1, 4)

(0, 1)

9.

(−1, e)

(0, 1)

11.

(0, −3)

(−1, −3e) $y = -3e^{-x}$

13. 1

15. $\dfrac{8}{27}$

17. e^{3+x}

19. $4e^{2x}$

21. $2^{-1/6}e^{2(x-1)}$

Exercise Set 4.2

1. a. 2   **d.** 3
   b. 1   **e.** 4
   c. 4   **f.** 2

3. a. 1   **d.** −1.20396

b. 0   **e.** 3
c. 0.78846   **f.** −2

7. $x = 0$ **13.** $x = \pm e^4$

9. $x = \pm 2$ **15.** $x = 2$

11. $x = 0$

17. $\ln P = \ln 10 \left[\dfrac{-A}{t + C} + B \right]$

19. $x = \ln 100 - \ln p$

21. $T = 4.0547$ years   **25.** 35 items

23. 6.9315 years   **27.** $a = 15; b = 5$

Exercise Set 4.3

1. $\dfrac{1}{x}$ **9.** $6\left(\dfrac{x - 1}{x^2 - 2x}\right)[(\ln(x^2 - 2x)]^2$

3. $\dfrac{18}{3x - 2}$ **11.** $\dfrac{1}{t \ln t}$

5. $\ln x + 1$ **13.** $\dfrac{\ln x}{(1 + \ln x)^2}$

7. $\dfrac{3x^2 - 1}{2(x^3 - x)}$ **15.** $\dfrac{6}{x}(3 \ln \sqrt{x})^3$

17. $2x \ln(3x - 6) + \dfrac{x^2}{x - 2}$

19. $\dfrac{\ln x}{x}$ **21.** $\dfrac{4x^2}{x^3 + 3}$

23. $\dfrac{2}{3(x - 6)} - \dfrac{1}{2 + 2x}$

25. $\dfrac{dy}{dx} = (x + 3)^x \left[\ln(x + 3) + \dfrac{x}{x + 3} \right]$

27. $\dfrac{dy}{dx} = x^{\sqrt{x+1}} \left[\dfrac{\ln x}{2\sqrt{x + 1}} + \dfrac{\sqrt{x + 1}}{x} \right]$

29. $\dfrac{dy}{dx} = \dfrac{x(x + 1)(x + 2)}{(x + 3)(x + 4)} \left[\dfrac{1}{x} + \dfrac{1}{x + 1} + \dfrac{1}{x + 2} - \dfrac{1}{x + 3} - \dfrac{1}{x + 4} \right]$

31. $\dfrac{dy}{dx} = \left[\dfrac{(x^2 + 3)(1 - x)^4}{x\sqrt{1 + x}} \right] \left[\dfrac{2x}{x^2 + 3} - \dfrac{4}{1 - x} - \dfrac{1}{x} - \dfrac{1}{2 + 2x} \right]$

23. a. $540.80
   b. $584.93
   c. $562.43

25. $670.32

27. $26,997

29. Approximately 7.7%

31. a. 10.25
   b. 10.52

33. $y' = \dfrac{xy - y}{x - xy}$ **37.** $f(1) = 1$, rel. min.

35. $y = 3x - e$ **39.** no rel. extrema

41.

43. a. $\dfrac{1}{35} \approx 2.86\%$ **b.** $\dfrac{1}{40} = 2.5\%$ **c.** 0

Exercise Set 4.4

1. $6e^{6x}$

3. $2xe^{x^2-4}$

5. $\dfrac{(x-1)e^x}{x^2}$

7. $\dfrac{-e^{-x}}{2\sqrt{1 + e^{-x}}}$

9. $\left(2x - \dfrac{1}{2\sqrt{x}}\right)e^{x^2 - \sqrt{x}}$

11. $\dfrac{xe^x - 1}{(x+1)(e^x + 1)}$

13. $\dfrac{1}{2}(e^x - e^{-x})$

15. $4(x - e^{-2x})^3(1 + 2e^{-2x})$

31. $f\left(\dfrac{1}{\sqrt[3]{3}}\right) = \dfrac{e^{2/3}}{\sqrt[3]{3}}$, rel. max.

33. $p = \dfrac{100}{e}$ **37.** $\dfrac{\ln 2}{4}$

35. $x = 19.8$ **39.** after 0.83 yr

17. $\left(1 - \dfrac{2}{x^2}\right)e^{1/x^2}$

19. $\dfrac{e^x(1 + x - x^2 + x^3)}{(1 + x^2)^2}$

21. $\dfrac{dy}{dx} = \dfrac{1 - xy}{x^2}$

23. $2^x + x2^x \ln 2$

25. $\dfrac{4^x \ln 4}{2\sqrt{1 + 4^x}}$

27. $\dfrac{\log_2\sqrt{x}}{x \ln 2}$

29. $f(1) = 2e$

Exercise Set 4.5

1. $N(t) = Ce^{4t}$ **3.** $f(x) = Ce^{-2x}$

5. $y = Ce^{-3t}$

7. $N(t) = 3e^{4t}$

9. $N = 6e^{-3t}$

11. \$1123.32

15. 6.9315 years

17. a. $50\sqrt{2}$ mg
  b. 25 mg

19. a. 4 hrs
  b. 800

21. 275,568

23. $r = 0.112$

25. 5.49 mg

27. \$2,250,000

29. 23.3 hrs

31. 8.1 micrograms

Review Exercises—Chapter 4

1. a. $\dfrac{1}{343}$ **b.** 4   **c.** $\dfrac{4}{9}$ **d.** $\dfrac{27}{8}$

3. \$477.62   **5.** \$6065.31

7. a. $x = 0$ **c.** $x = 2$
  b. $x = 1, -2$ **d.** $x = \ln 2$

9. \$18.39

11. $\dfrac{1}{x}$

13. $-2xe^{6-x^2}$

15. $(3t^2 - 3)e^{t^3 - 3t + 2}$

17. $\dfrac{1}{2(x+1)}$

19. $\left(2x + \dfrac{1}{2}x^{3/2}\right)e^{\sqrt{x}}$

21. $\dfrac{x-2}{2x(x+2)}$

23. $\dfrac{1}{2x \ln \sqrt{x}}$

25. $e^{x-\sqrt{x}}\left(1 + x - \dfrac{\sqrt{x}}{2}\right)$

27. $\dfrac{2e^{-2x}}{1 - e^{-2x}}$

29. $e^{\sqrt{t} - \ln t}\left(\dfrac{1}{2\sqrt{t}} - \dfrac{1}{t}\right)$

31. $\dfrac{dy}{dx} = \dfrac{-y^2 - xy \ln y}{x^2 + xy \ln x}$

33. $y' = -\dfrac{y}{x}$

35. $\left(\dfrac{1}{\sqrt{e}}, -\dfrac{1}{2e}\right)$, rel. min.

37. $(-1, 1)$

39. $r = \ln 2$

41. $\dfrac{dh}{dt} = \dfrac{\ln t^2}{36\pi}$

43. $y = Ce^{-x}$

45. 254 million

47. 190 flies

CHAPTER 5

Exercise Set 5.1

1. $5x + C$

3. $\frac{1}{2}x^4 + \frac{5}{2}x^2 + C$

5. $\frac{1}{4}x^4 - 2x^3 + x^2 - x + C$

7. $\frac{3}{5}x^{5/3} - 2x^{3/2} + C$

9. $4 \ln x + C, \quad x > 0$

11. $\frac{3}{4}x^{4/3} - 6x^{1/3} - \frac{3}{2}x^{-2/3} + C$

13. $-4e^{-x} - \ln x + \frac{2}{3}x^{3/2} + C, \quad x > 0$

15. $-\frac{4}{5}x^{-5} + 2x^{-3} + C$

17. $\frac{1}{2}x^2 - \frac{12}{11}x^{11/6} + \frac{3}{5}x^{5/3} + C$

19. $\ln (x + 3) + C, \quad x > -3$

Exercise Set 5.2

1. ii

3. iv

5. $F(x) = 4x + 3$

7. $F(x) = \frac{1}{3}x^3 - x^2 - 3$

9. $F(x) = \ln x + 4$

11. $F(x) = \frac{3}{5}x^{5/3} - \frac{3}{2}x^{2/3} + 5$

13. $F(x) = \frac{5}{2}e^{2x} + 4x + \frac{15}{2}$

15. $C(x) = 700 + 250x$ dollars

17. a. $1187.50

   b. $C(x) = 1187.50 + 400x + \frac{1}{8}x^2$ dollars

19. $U(x) = \frac{2}{3}x^{3/2}$

21. a. $s(t) = t^3 + 3t^2 + 2t + 4$

   b. $s(4) = 124$

23. a. $N(t) = 20t + 16t^{3/2} + 2$

   b. $N(16) = 1346$

Exercise Set 5.3

1. $\frac{2}{3}(x + 2)^{3/2} + C$

3. $\frac{2}{3}(9 + x^2)^{3/2} + C$

5. $\frac{1}{2}e^{x^2} + C$

7. $\frac{1}{15}(x^3 + 5)^5 + C$

9. $-\frac{1}{6} \ln |1 - 3x^2| + C$

11. $\frac{1}{8}(1 + e^{2x})^4 + C$

13. $2\sqrt{e^x + 1} + C$

15. $-\frac{1}{2}(x^2 + 3x + 6)^{-2} + C$

17. $\frac{1}{4}\left(1 - \frac{1}{x}\right)^4 + C$

19. $\frac{1}{2}\sqrt{5 + x^4} + C$

21. $2[\ln x]^2 + C$

23. $\frac{9}{10}(x^{2/3} - 5)^{5/3} + C$

25. $(1 + e^{\sqrt{x}})^2 + C$

27. a. 40

   b. $C(x) = 500 + 40x + 10 \ln (1 + x^2)$

29. a. 0

   b. $R(x) = 25{,}000 \ln (100 + 0.2x^2) - 25{,}000 \ln 100$

31. $P(t) = 40{,}000 + 5[\ln(1 + e^{20t}) - \ln 2]$

Exercise Set 5.4

1. 32

3. $\frac{21}{2}$

5. $\frac{31}{20}$

7. $\frac{39}{4}$

9. $\frac{319}{420}$

11. 2

13. 16

15. $\frac{81}{2}$

17. $\frac{5}{2}$

19. 4

21. 3.4418 $(n = 100)$

23. 6.3086 $(n = 100)$

25. 12.4809 $(n = 100)$

Exercise Set 5.5

1. 8

3. $\frac{52}{3}$

5. $\dfrac{170}{3}$

7. $\dfrac{1}{4}(e^4 - 1)$

9. $-\dfrac{4}{3}$

11. $-\dfrac{4}{5}$

13. $\dfrac{32}{3}$

15. 0

17. $\dfrac{1}{2}\ln\left(\dfrac{28}{13}\right)$

19. $\dfrac{22}{3} - \dfrac{4}{3}\sqrt{2}$

21. $\dfrac{1}{4}$

23. 2

25. $2\sqrt{5} - 4$

27. 9

29. $\dfrac{3^{7/3}}{8}$

Exercise Set 5.6

1. 14

3. $\ln 3$

5. $\ln\sqrt{\dfrac{8}{3}}$

7. $\dfrac{52}{3}$

9. $\dfrac{116}{3}$

11. 6

13. $\dfrac{80}{3}$

15. 27

17. $\dfrac{125}{6}$

19. $\dfrac{1}{2}$

21. $\dfrac{125}{6}$

23. $\dfrac{8}{15}$

25. 3

27. $4^{2/3}$

Exercise Set 5.7

1. 25

3. $\dfrac{25{,}000}{3}$

5. 64

7. $\$1.44$

9. $15e^4 + 5$

11. $\$2180$

13. $\$9140$

15. 7600

17. $19{,}500$

19. $5000\left(\dfrac{1}{e} - \dfrac{1}{e^2}\right)$

21. $50{,}000\left(1 - \dfrac{1}{e}\right)$

Exercise Set 5.8

1. $\dfrac{351\pi}{3} = 117\pi$

3. 3π

5. $\dfrac{7\pi}{8}$

7. $\dfrac{128\pi}{15}$

9. $\dfrac{64\pi}{3}$

11. -1

13. $\dfrac{4^{3/2}}{6} = \dfrac{4}{3}$

15. $\dfrac{4}{3}$

17. $\dfrac{3\pi}{4}$

19. $\dfrac{1400}{3}$

Review Exercises—Chapter 5

1. $2x^3 - x^2 + x + C$ **3.** $\dfrac{3}{4}x^4 + C$

5. $\dfrac{t^3}{3} + \dfrac{6}{7}t^{7/3} + \dfrac{3}{5}t^{5/3} + C$

7. $\dfrac{x^3}{3} - \dfrac{7}{2}x^2 + 6\ln x + C$

9. $\dfrac{4}{3}x^3 - x + C$ **13.** $2e^{x/2} + C$

11. $-\dfrac{1}{2}\ln|1 - x^2| + C$ **15.** $2\sqrt{\ln x} + C$

17. $\dfrac{x^3}{3} - \dfrac{x^2}{2} + x - 2\ln|x + 1| + C$

19. $-\dfrac{1}{8}(1 - e^{2x})^4 + C$

21. **a.** $s(t) = t^2 + \dfrac{1}{t + 1} - 1$
  b. $a(t) = 2 + 2(t + 1)^{-3}$

23. **a.** $C(x) = 500 + 120x + 3x^2$
  b. $C(20) = 4100$

25. **a.** $P(x) = 130x - 3x^2 - 500$
  b. Yes; $P(20) = 900$

27. **a.** $V(t) = 20{,}000 + \dfrac{80{,}000}{t + 1}$
  b. $V(3) = 40{,}000$
  c. $20{,}000$

29. a. $C(x) = 70x + x^2 + 200$
    b. $C(0) = 200$

31. -3

33. $\dfrac{322}{5}$

35. $\dfrac{2}{3}(7^{3/2} - 8)$

37. $3(1 - \ln 2)$

39. $1 - 4 \ln 2$

41. $\dfrac{16}{3}$

43. 8

45. $\dfrac{675}{8}$

47. $\dfrac{3}{4} \ln\left(\dfrac{7}{2}\right)$

49. $\dfrac{9}{2} + \dfrac{6}{5}(8^{5/2} - 1)$

51. 8π

53. 63

55. $\dfrac{16}{5}$

57. 14

59. $1100 + 50 \ln\left(\dfrac{6}{5}\right)$

61. 5760

63. $\dfrac{7}{4}e^4 + 2e$

65. $500[e^2 - 1]$

67. $10{,}000\left(\dfrac{26^{3/2} - 1}{3}\right)$

69. $\dfrac{1}{6} \ln \dfrac{16}{7}$

e. $-\dfrac{13\pi}{6}$ **g.** $-\dfrac{11\pi}{6}$

f. $-\dfrac{3\pi}{4}$ **h.** -3π

7. a. $-\pi$ **c.** $-\dfrac{5\pi}{2}$

    b. -2π **d.** $-\dfrac{10\pi}{3}$

9. $-\dfrac{9\pi}{2}$ **11.** 3π **13.** $-\dfrac{9\pi}{4}$

15.

17.

$-\dfrac{\sqrt{2}}{2}, -\dfrac{\sqrt{2}}{2}$

19.

$(0, -1)$

CHAPTER 6

Exercise Set 6.1

1. a. $\dfrac{\pi}{2}$ **e.** $\dfrac{\pi}{3}$

   b. $\dfrac{\pi}{4}$ **f.** $-\dfrac{5\pi}{6}$

   c. $-\dfrac{3\pi}{4}$ **g.** π

   d. $\dfrac{\pi}{6}$ **h.** $\dfrac{7\pi}{6}$

3. a. $45°$ **e.** $157.5°$
   b. $270°$ **f.** $-150°$
   c. $-15°$ **g.** $330°$
   d. $210°$ **h.** $-135°$

5. a. $-\dfrac{\pi}{4}$ **c.** $-\dfrac{\pi}{6}$

   b. $-\dfrac{3\pi}{2}$ **d.** $-\dfrac{5\pi}{6}$

Exercise Set 6.2

1. a. $\dfrac{3}{5}$ **b.** $\dfrac{4}{5}$

3. a. $\dfrac{2}{\sqrt{13}}$ **b.** $\dfrac{3}{\sqrt{13}}$

5. a. $\dfrac{7}{\sqrt{65}}$ **b.** $\dfrac{4}{\sqrt{65}}$

7. $\sin t = \dfrac{\sqrt{2}}{2}$

$\cos t = -\dfrac{\sqrt{2}}{2}$

9. $\sin t = \dfrac{\sqrt{3}}{2}$

$\cos t = \dfrac{1}{2}$

11. $-\dfrac{\sqrt{2}}{2}$

13. 0

15. $\dfrac{\sqrt{3}}{2}$

17. $\dfrac{-\sqrt{2}}{2}$

19.

21.

23.

25. $x = 5\sqrt{2}$

$y = 5\sqrt{2}$

27. $x = \dfrac{5\sqrt{3}}{2}$

$y = \dfrac{5}{2}$

29. 0.2588

31. 0.9239

33. 0.9511

35. 0.2588

37. a. 12 hours
 b. 15 hours
 c. 9 hours

39. a. B
 b. $4 + B$
 c. $5 + B$

Exercise Set 6.3

1. $f'(x) = 3\cos 3x$

3. $\dfrac{dy}{dx} = \sin x + x\cos x$

5. $f'(x) = 3x^2\cos 2x - 2x^3\sin 2x$

7. $\dfrac{dy}{dx} = \cos^2 x - \sin^2 x$

9. $f'(x) = \dfrac{x\cos\sqrt{1 + x^2}}{\sqrt{1 + x^2}}$

13. $f'(x) = 0$

11. $\dfrac{dy}{dt} = -\dfrac{\sin\ln t}{t}$

15. $f'(t) = \dfrac{1 + \cos t}{t + \sin t}$

17. $\dfrac{dy}{dx} = \dfrac{3\cos x + x^2\cos x - 2x\sin x - 2x}{(3 + x^2)^2}$

19. $f'(x) = \sin\left(\dfrac{1}{x}\right) - \dfrac{1}{x}\cos\left(\dfrac{1}{x}\right)$

21. $y = x - \dfrac{\pi}{6} + \dfrac{\sqrt{3}}{2}$

23. $\dfrac{4}{49}$

25. Because $\sin^2 x + \cos^2 x = 1$

27. $f''(t) = \dfrac{-\sin\ln t - \cos\ln t}{t^2}$

29. $\dfrac{dy}{dx} = -\dfrac{1 + y\sin xy}{x\sin xy}$

31. $\dfrac{dy}{dx} = \dfrac{\cos y}{x\sin y + \cos y}$

33. max is $f\left(\dfrac{\pi}{4}\right) = \sqrt{2}$

min is $f\left(\dfrac{5\pi}{4}\right) = -\sqrt{2}$

35. max is $f\left(\dfrac{3\pi}{2}\right) = \dfrac{3\pi}{2} + 1$

min is $f\left(\dfrac{\pi}{2}\right) = \dfrac{\pi}{2} - 1$

37.

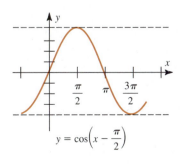

$y = \cos\left(x - \dfrac{\pi}{2}\right)$

39.

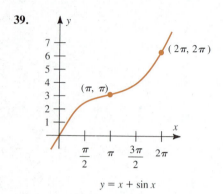

$y = x + \sin x$

41.

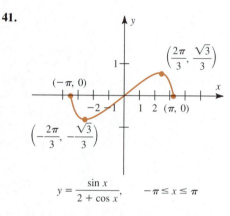

$y = \dfrac{\sin x}{2 + \cos x}$, $\quad -\pi \le x \le \pi$

43. max is $1 + \dfrac{\pi}{4}$ at 0 and 8 months

min is $1 - \dfrac{\pi}{4}$ at 4, 12 months.

45. a. $t = 11/2$ mo. (June 15)
b. $t = 23/2$ mo. (Dec. 15)

Exercise Set 6.4

1. $\dfrac{1}{6} \sin 6x + C$

3. 6

5. $-\dfrac{1}{3} \cos(1 + x^3) + C$

7. $-\dfrac{1}{2} \sin(\pi - t^2) + C$

9. 0

11. $-\dfrac{2}{3}(1 - \sin x)^{3/2} + C$

13. $-\dfrac{1}{2} \cos(x^2 + 2x) + C$

15. $\sin(\ln x) + C$

17. $e - 1$

19. $\dfrac{x}{2} - \dfrac{1}{4} \sin 2x + C$

21. $\dfrac{1}{6} \sin^3(x^2) + C$

23. $\dfrac{2}{\pi}$

25. 0

27. $2 + \dfrac{\pi^2}{4}$

29. $\dfrac{\pi^2}{4}$

Exercise Set 6.5

1. a. $\dfrac{1}{\sqrt{3}}$ **b.** $\sqrt{3}$ **c.** $\dfrac{2}{\sqrt{3}}$ **d.** 2

3. a. $\dfrac{3}{4}$ **b.** $\dfrac{4}{3}$ **c.** $\dfrac{5}{4}$ **d.** $\dfrac{5}{3}$

5. $\dfrac{50}{\sqrt{3}}$ **7.** $\dfrac{30}{\sqrt{3}}$ meters

9. $\sec x + x \sec x \tan x$ **15.** $-\sin x$

11. $-6 \csc^2(6x)$ **17.** $-6 \cot(3x) \csc^2(3x)$

13. $\sec x \tan^2 x + \sec^3 x$ **19.** $3x^2 \sec^2 x^3$

21. $\dfrac{x \sec \sqrt{1 + x^2} \tan \sqrt{1 + x^2}}{\sqrt{1 + x^2}}$

23. $\dfrac{-\csc^2 \ln \sqrt{x}}{2x}$ **29.** $\ln(\sqrt{2} - 1)$

25. $\dfrac{1}{4} \tan 4x + C$ **31.** $\dfrac{1}{2}x^2 + \cot x + C$

27. $\dfrac{1}{2} \tan x^2 + C$ **33.** $-\dfrac{1}{\pi} \csc(\pi x) + C$

35. $-\dfrac{1}{\pi} \cot \pi x - \dfrac{1}{\pi} \ln |\sin \pi x| + C$

37. $\dfrac{\pi}{4} - \ln 2$ **39.** 2π

41. $x = \dfrac{\pi}{2} \pm n\pi$, $n = 0, 1, 2, \ldots$

Review Exercises—Chapter 6

1. a. $\dfrac{\pi}{3}$ **c.** $\dfrac{2\pi}{3}$ **e.** $\dfrac{\pi}{18}$
   b. $\dfrac{7\pi}{36}$ **d.** $\dfrac{7\pi}{6}$ **f.** $\dfrac{5\pi}{12}$

3. a. $\dfrac{3\pi}{2}$ **c.** $\dfrac{3\pi}{4}$ **e.** π
   b. $\dfrac{3\pi}{2}$ **d.** $\dfrac{\pi}{3}$ **f.** π

5. $\dfrac{dy}{dx} = -\cos(\pi - x)$

7. $f'(x) = 3 \sec^2 3x$

9. $\dfrac{dy}{dx} = -\sin x \cdot \cos(\cos x)$ **11.** $\dfrac{dy}{dx} = \dfrac{\sec^2 x}{\tan x}$

13. $f'(x) = \pi \sec(\pi x)\tan(\pi x)e^{\sec \pi x}$

15. $\dfrac{dy}{dx} = 3(x + \sec x)^2(1 + \sec x \tan x)$

17. $f'(x) = \dfrac{-\sin x(1 + \sin x)e^{\cos x} - \cos x \cdot e^{\cos x}}{(1 + \sin x)^2}$

19. $\dfrac{dy}{dx} = \dfrac{-\sin x}{2(4 + \cos x)}$ **27.** $\dfrac{1}{3}e^{\tan 3x} + C$

21. $-\dfrac{1}{4}\cos 4x + C$ **29.** $-\dfrac{1}{5}\cos^5 x + C$

23. $\dfrac{1}{\pi}\sec \pi x + C$ **31.** $\dfrac{\sqrt{2}}{2} - 1$

25. $\dfrac{2}{3}(\tan x)^{3/2} + C$ **33.** $e^{\sqrt{2}} - e$

35. 0

37. a.

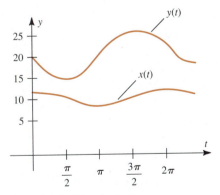

b. max. value is 12; min. is 8
c. max. value is 25; min. is 15

39. $m = 2$ **43.** $\dfrac{6}{\pi}\left[\ln\left(\dfrac{1}{\sqrt{3}}\right) - \ln(2 - \sqrt{3})\right]$

41. $\dfrac{1}{2}\ln(2 + \sqrt{3})$

CHAPTER 7

Exercise Set 7.1

1. $\dfrac{2}{3}(4 + \sin x)^{3/2} + C$ **19.** $\dfrac{1}{4}\sin^4 x + C$

3. $\dfrac{2}{9}(28)^{3/2}$ **21.** $\tan x + C$

5. $\dfrac{1}{2}(e - 1)$ **23.** $\dfrac{1}{6}$

7. $\dfrac{1}{5}\ln|1 + x^5| + C$ **25.** $\ln|\sec(1 + e^x)| + C$

9. $\ln(1 + e^x) + C$ **27.** $x + C$

11. $\dfrac{1}{4}$ **29.** $\ln 2$

13. $\dfrac{1}{8}(x^2 + 6x + 4)^4 + C$ **31.** $2\ln(1 + \sqrt{x}) + C$

15. $e^{\sec x} + C$ **33.** $\dfrac{1}{4}(2 - \sqrt{2})$

17. $2e(e - 1)$

35. $4000\ln 5 - 3200 \approx 3238.$

Exercise Set 7.2

1. $-xe^{-x} - e^{-x} + C$ **5.** $\dfrac{116}{15}$

3. $\dfrac{x^2}{2}\ln x - \dfrac{x^2}{4} + C$ **7.** $-x\cos x + \sin x + C$

9. $\dfrac{x}{12}(2x + 1)^6 - \dfrac{1}{168}(2x + 1)^7 + C$

11. $x^2e^x - 2xe^x + 2e^x + C$

13. $\dfrac{\pi}{4} - \ln\sqrt{2}$ **15.** $2 - \dfrac{5}{e}$

17. $x^2\sqrt{1 + x^2} - \dfrac{2}{3}(1 + x^2)^{3/2} + C$

19. $\dfrac{e^{2x}}{5}(\sin x + 2\cos x) + C$

21. $-\dfrac{1}{2}e^{-x^2}(x^2 + 1) + C$ **23.** $\dfrac{3}{2} - 2\ln 2$

25. $902

27. $\dfrac{1}{e-1}$

9. -0.29600 **21.** 0.6649

11. 24 **23.** 4.9501

13. 2.005 **25.** 1.2440

Exercise Set 7.3

15. 1.3239 **27.** 0.6658

1. $\dfrac{1}{18}(4+3x)^6+C$ **3.** $\ln|5+x|+\dfrac{5}{5+x}+C$

17. 0 **29.** 4.9471

19. -0.2939 **31.** 1.2738

5. $-\dfrac{1}{15}(7-3x)(7+2x)^{3/2}+C$

35. a. 2.4566 **39. a.** 1.6108
      b. 1.1713 **b.** $0.03\overline{3}$

7. $-\dfrac{1}{3}(3-x)\sqrt{3+2x}+C$

9. $\ln|x+\sqrt{x^2+16}|+C$

Review Exercises—Chapter 7

11. $\dfrac{-1}{2(5+x^2)}+C$ **17.** $2\ln\left(\dfrac{e^{x^2}}{1+e^{x^2}}\right)+C$

1. $\dfrac{2}{3}(x+1)^{3/2}+C$ **17.** $\dfrac{1}{\pi}$

13. $\dfrac{1}{24}\ln\left|\dfrac{3+4x}{3-4x}\right|+C$ **19.** $2\ln\left(\dfrac{e^{\sqrt{x}}}{1+e^{\sqrt{x}}}\right)+C$

3. $\dfrac{1}{2}\ln(1+x^2)+C$ **19.** $\dfrac{1}{3}x^3\ln x-\dfrac{1}{9}x^3+C$

15. $-\dfrac{1}{6}\ln\left|\dfrac{3+\sqrt{x^4+9}}{x^2}\right|+C$

5. $-\dfrac{22}{3}$ **21.** $-x\cot x+\ln|\sin x|+C$

21. $-\dfrac{1}{2}\cos x^2+\dfrac{\cos^3 x^2}{6}+C$

7. $\dfrac{1}{64}$ **23.** 1

9. $\dfrac{1}{3}(x^2+9)^{3/2}+C$ **25.** diverges

Exercise Set 7.4

11. $\dfrac{2}{3}$ **27.** $\sqrt{15}$

1. $\dfrac{1}{2}$ **17.** diverges

13. $\dfrac{1}{2}x^2\ln x-\dfrac{1}{4}x^2+C$ **29.** $\ln|x^2+4x+3|+C$

3. 1 **19.** diverges

5. diverges   **21.** diverges

15. $\dfrac{3}{2}(4^{2/3}-1)$

7. diverges   **23.** $1,000

9. diverges   **25. a.** $10\left(1-\dfrac{1}{e}\right)$ million dollars
               b. $10 million

CHAPTER 8

11. $\dfrac{1}{2}$ **27.** $\dfrac{\pi}{2}$

Exercise Set 8.1

13. diverges   **29.** $p<1$

1. a. 27 **5. a.** -4
  b. -42 **b.** 27

15. 2

  c. 32 **c.** -61

3. a. 0 **7. a.** 1

Exercise Set 7.5

  b. $\dfrac{7}{4}$ **b.** 3

1. 28 **5.** 1.3253

  c. $\dfrac{26}{9}$ **c.** 5

3. 1.8961 **7.** 0

9.

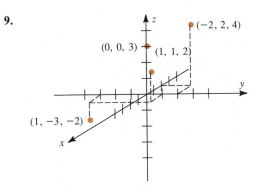

19. $z_x = \dfrac{y + ye^{xy}}{2\sqrt{xy + e^{xy}}}$; $z_y = \dfrac{x + xe^{xy}}{2\sqrt{xy + e^{xy}}}$

21. -4 **25.** 74

23. $\dfrac{1}{3}$ **27.** 0

29. a. 2 **b.** -2 **c.** -2 **d.** 2

31. a. $\dfrac{2y^2 - 2x^2}{(x^2 + y^2)^2}$ **b.** $\dfrac{-4xy}{(x^2 + y^2)^2}$

  c. $\dfrac{-4xy}{(x^2 + y^2)^2}$ **d.** $\dfrac{2x^2 - 2y^2}{(x^2 + y^2)^2}$

11. all (x, y) **17.** $R(x, y) = 10x + 15y$

33. $\dfrac{160}{27}$

13. $\{(x, y)|y \neq x\}$ **19.** $P(x, y) = 6x + 7y - 20$

15. $\{(x, y)|y > 0\}$ **21.** $P_0(r, T) = 10,000e^{-rT}$

35. a. $-\dfrac{44}{3}$ **b.** 2

23. a. $R(x, y) = 200x - 2x^2 + 150y - 3y^2$
  b. $P(x, y) = 196x - 2x^2 + 145y - 3y^2 - 500$

37. a. 150 **b.** $\dfrac{50}{3}$

Exercise Set 8.2

1. $\dfrac{\partial f}{\partial x} = 3$; $\dfrac{\partial f}{\partial y} = -6$

3. $\dfrac{\partial f}{\partial x} = 4y^2 - 9x^2y$; $\dfrac{\partial f}{\partial y} = 8xy - 3x^3 + 5y^4$

5. $\dfrac{\partial f}{\partial x} = \dfrac{1}{2\sqrt{x + y}}$; $\dfrac{\partial f}{\partial y} = \dfrac{1}{2\sqrt{x + y}}$

7. $\dfrac{\partial f}{\partial x} = \dfrac{-2y}{(x - y)^2}$; $\dfrac{\partial f}{\partial y} = \dfrac{2x}{(x - y)^2}$

9. $\dfrac{\partial f}{\partial x} = \dfrac{x}{x^2 + y^2}$; $\dfrac{\partial f}{\partial y} = \dfrac{y}{x^2 + y^2}$

11. $\dfrac{\partial f}{\partial x} = \dfrac{2}{3}x^{-1/3}y^{-1/3} - \dfrac{1}{2}\sqrt{y}\, x^{-3/2}$;

  $\dfrac{\partial f}{\partial y} = -\dfrac{1}{3}x^{2/3}y^{-4/3} + \dfrac{1}{2\sqrt{xy}}$

13. $\dfrac{\partial f}{\partial x} = \dfrac{1}{2\sqrt{x + 3e^y}}$; $\dfrac{\partial f}{\partial y} = \dfrac{3e^y}{2\sqrt{x + 3e^y}}$

15. $z_x = \dfrac{1}{y^2} + \dfrac{2y}{x^3}$; $z_y = \dfrac{-2x}{y^3} - \dfrac{1}{x^2}$

17. $z_x = \dfrac{y}{2\sqrt{x}}e^{\sqrt{x}-1}$; $z_y = e^{\sqrt{x}-1}$

Exercise Set 8.3

1. $(0, -2)$, rel. min.   **7.** $(0, 0)$, saddle point

3. $(-2, 1)$, rel. min.   **9.** $(1, 3)$, rel. min.

5. $(-3, 2)$, saddle point

11. $(0, 0)$, no conclusion by Second Derivative Test. Saddle by inspection.

13. $(0, 0)$, saddle point
  $\left(-\dfrac{4}{3}, -\dfrac{4}{3}\right)$, rel. max.

15. $\left(-\dfrac{4}{3}, \dfrac{5}{3}\right)$, rel. min.

17. $x = 10$, $y = 20$ **23.** $w = 2$, $\ell = 2$, $h = 4$

19. $x = 7.5$, $y = 2.5$ **25.** $w = 14$, $h = 14$, $\ell = 28$

21. $x = 4$, $y = 8$

Exercise Set 8.4

1. max. $= 2$ **5.** min. $= -16$
  no min.   max. $= 16$

3. min. $= 19$ **7.** min. $= 2$
  no max.   no max.

9. max. = 8
min. = -8

11. $x = 1$
$y = 6$

13. $x = 20$
$y = 20$

15. $x = 15$
$y = 25$

17. $x = 14$
$y = 14$
$h = 28$

19. $r = \dfrac{10}{\sqrt{6\pi}}$ cm

$h = \dfrac{20}{\sqrt{6\pi}}$ cm

21. $r = \pi^{-1/3}$
$h = 2\pi^{-1/3}$

Exercise Set 8.5

1. $y = 0.545x + 5.182$ **5.** $y = 0.429x + 0.714$

3. $y = 0.949x + 29.31$

7. a. $y = -2.14x + 12.2$
b. -0.64 (= 640 cars returned)

9. a. $y = -6.04x + 77.5$
b. 23.14

Exercise Set 8.6

1. -1

3. 0

5. 0

7. $\dfrac{e}{2} + \dfrac{1}{2e} - 1$

9. $\dfrac{3}{4} - \ln 2$

11. $\dfrac{11}{3}$

13. $\dfrac{1}{3}8^{3/2} - \dfrac{2}{3}5^{3/2} + \dfrac{1}{3}2^{3/2}$

15. $\dfrac{1}{24}$

17. 2

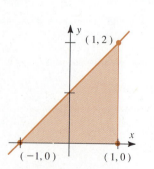

$(1,2)$
$(-1,0)$ $(1,0)$

19. $\dfrac{1}{10}$

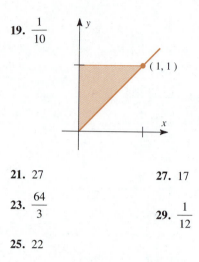

$(1,1)$

21. 27

23. $\dfrac{64}{3}$

25. 22

27. 17

29. $\dfrac{1}{12}$

Review Exercises—Chapter 8

1. $\{(x, y)\,|\,x^2 + y^2 \le 25\}$

3. a. -3 **b.** 3

5. $\{(x, y)\,|\,xy = 1\}$

7. $\dfrac{\partial f}{\partial x} = 3(x - y)^2 + \dfrac{1}{x}$; $\dfrac{\partial f}{\partial y} = -3(x - y)^2 + \dfrac{1}{y}$

9. $\dfrac{\partial f}{\partial x} = \dfrac{y^2}{(x + y)^2}$; $\dfrac{\partial f}{\partial y} = \dfrac{x^2}{(x + y)^2}$

11. $\dfrac{\partial f}{\partial x} = \dfrac{2}{3}x^{-1/3}y^{1/3} - \dfrac{3}{4}x^{-1/4}y^{-1/4}$;

$\dfrac{\partial f}{\partial y} = \dfrac{1}{3}x^{2/3}y^{-2/3} + \dfrac{1}{4}x^{3/4}y^{-5/4}$

13. $\dfrac{\partial f}{\partial x} = y\sqrt{y^2 - x^2} - \dfrac{x^2 y}{\sqrt{y^2 - x^2}}$;

$\dfrac{\partial f}{\partial y} = x\sqrt{y^2 - x^2} + \dfrac{xy^2}{\sqrt{y^2 - x^2}}$

15. $\dfrac{\partial f}{\partial x} = \dfrac{2}{3}(xy^2 - x^2y)^{-1/3}(y^2 - 2xy)$;

$\dfrac{\partial f}{\partial y} = \dfrac{2}{3}(xy^2 - x^2y)^{-1/3}(2xy - x^2)$

17. $\dfrac{\partial f}{\partial x} = \dfrac{x}{(x^2 + 4y^2 + 2z^2)}$; $\dfrac{\partial f}{\partial y} = \dfrac{4y}{(x^2 + 4y^2 + 2z^2)}$;

$\dfrac{\partial f}{\partial z} = \dfrac{2z}{(x^2 + 4y^2 + 2z^2)}$

19. $\dfrac{\partial^2 f}{\partial x^2} = \dfrac{-y^2}{(y^2 - x^2)^{3/2}};$

$\dfrac{\partial^2 f}{\partial y \partial x} = \dfrac{\partial^2 f}{\partial x \partial y} = \dfrac{xy}{(y^2 - x^2)^{3/2}};$

$\dfrac{\partial^2 f}{\partial y^2} = \dfrac{-x^2}{(y^2 - x^2)^{3/2}}$

21. $(0, 0)$, saddle   **27.** 8

23. $(0, 0)$, saddle   **29.** $\dfrac{1}{15}(2^{3/2} - 1)$

25. $(0, 0)$, saddle   **31.** $\dfrac{11}{12}$

33. a. $R(x, y) = 80x - 2x^2 + 120y - 4y^2$
  b. $P(x, y) = 60x - 2x^2 + 100y - 4y^2 - 1000$
  c. $x = 20, y = 15$ **d.** $x = 15, y = 12.5$

35. a. 1600   **37.** $x = 5$
  b. 200 $y = 15$

39. $f(1, 2, -3) = 15$

CHAPTER 9
Exercise Set 9.1

1. $\{5, 7, 9, 11, 13\}$

3. $\left\{\dfrac{1}{6}, \dfrac{3}{7}, \dfrac{5}{8}, \dfrac{7}{9}, \dfrac{9}{10}, \dots\right\}$

5. $\left\{-\dfrac{1}{e}, \dfrac{1}{e^2}, -\dfrac{1}{e^3}, \dfrac{1}{e^4}, -\dfrac{1}{e^5}\right\}$

7. $\left\{0, \dfrac{3}{14}, \dfrac{4}{13}, \dfrac{5}{14}, \dfrac{12}{31}\right\}$

9. $\left\{\dfrac{\sqrt{2}}{2}, -1, \dfrac{\sqrt{2}}{2}, 0, -\dfrac{\sqrt{2}}{2}\right\}$

11. $\dfrac{1}{3}$ **23.** 1

13. $\sqrt{5}$ **25.** 0

15. 1   **27.** e

17. 1   **29.** 1

19. 0   **31.** $\dfrac{A}{B}$

21. 0

33. a. $1,080   **35.** no
  b. $1166.40
  c. $1000(1.08)^n$

Exercise Set 9.2

1. $5 + 7 + 9 + 11$ **17.** $\dfrac{16}{3}$

3. $1 + \dfrac{1}{6} + \dfrac{1}{36} + \dfrac{1}{216}$ **19.** -2

5. $-\dfrac{1}{2} + \dfrac{1}{5} - \dfrac{1}{10} + \dfrac{1}{17}$ **21.** $-\dfrac{15}{4}$

7. $-\dfrac{1}{2} + \dfrac{1}{3} - \dfrac{1}{4} + \dfrac{1}{5}$ **25. a.** $1,937.50
  b. $2,000

9. $\dfrac{4}{3}$ **27.** $4 million

11. $\dfrac{7}{4}$ **29. b.** $1000\left(\dfrac{1 - e^{-rn}}{e^r - 1}\right)$

  e. $1000\left(\dfrac{1}{e^r - 1}\right)$

13. $\dfrac{1}{2}$ **31.** $80

15. diverges

Exercise Set 9.3

1. diverges   **25.** converges
3. converges   **27.** diverges
5. diverges   **29.** converges
7. converges   **31.** diverges
9. diverges   **33.** diverges
11. converges   **35.** converges
13. diverges   **37.** diverges
15. converges   **39.** diverges
17. diverges   **41.** diverges
19. converges   **43.** converges
21. diverges   **45.** diverges
23. converges   **47.** diverges

Exercise Set 9.4

1. diverges

3. converges

5. converges

7. diverges

9. diverges

11. converges

13. converges

15. converges

17. diverges

19. converges absolutely

21. converges absolutely

23. converges conditionally

25. converges conditionally

27. converges absolutely

29. diverges

Exercise Set 9.5

1. $P_3(x) = 1 - x + \dfrac{x^2}{2} - \dfrac{x^3}{3!}$

3. $P_4(x) = 1 - \dfrac{x^2}{2} + \dfrac{x^4}{4!}$

5. $P_2(x) = 1 - x + x^2$

7. $P_2(x) = \sqrt{2} + \dfrac{x}{2\sqrt{2}} - \dfrac{x^2}{16\sqrt{2}}$

9. $P_2(x) = \dfrac{\sqrt{2}}{2} + \dfrac{\sqrt{2}}{2}\left(x - \dfrac{\pi}{4}\right) - \dfrac{\sqrt{2}}{4}\left(x - \dfrac{\dot{\pi}}{4}\right)^2$

11. $P_4(x) = e\left[1 + (x - 1) + \dfrac{1}{2}(x - 1)^2 + \dfrac{1}{3!}(x - 1)^3 + \dfrac{1}{4!}(x - 1)^4\right]$

13. $P_3(x) = 3 + \dfrac{1}{6}(x - 9) - \dfrac{1}{216}(x - 9)^2 + \dfrac{1}{3888}(x - 9)^3$

15. $P_2(x) = 1 + x^2$

17. $P_2(x) = \dfrac{\sqrt{2}\pi}{8} + \left(\dfrac{\sqrt{2}}{2} - \dfrac{\sqrt{2}\pi}{8}\right)\left(x - \dfrac{\pi}{4}\right) - \dfrac{1}{2}\left(\sqrt{2} + \dfrac{\sqrt{2}\pi}{8}\right)\left(x - \dfrac{\pi}{4}\right)^2$

19. 0.3333 **21.** $\dfrac{12295}{3888}$

Exercise Set 9.6

1. $\ln(1.4) \approx 0.3413$
$|R_3(1.4)| \le 0.0064$

3. $\sqrt{3.91} \approx 1.9774$
$|R_2(3.91)| \le \dfrac{3}{8}(3.91)^{-5/2}(0.09)^3$

5. $\sin 48° \approx \dfrac{\sqrt{2}}{2}\left[1 + \dfrac{\pi}{60} - \dfrac{1}{2}\left(\dfrac{\pi}{60}\right)^2 - \dfrac{1}{6}\left(\dfrac{\pi}{60}\right)^3\right]$
$|R_4(48°)| \le \dfrac{1}{5!}\left(\dfrac{\pi}{60}\right)^5$

7. $\tan\left(\dfrac{\pi}{12}\right) \approx \dfrac{\pi}{12} + \dfrac{1}{3}\left(\dfrac{\pi}{12}\right)^3$

Using $\sec\dfrac{\pi}{12} < \sec\dfrac{\pi}{6} = \dfrac{2}{\sqrt{3}}$

$\tan\dfrac{\pi}{12} < \tan\dfrac{\pi}{6} = \dfrac{1}{\sqrt{3}}$,

$\left|R_3\left(\dfrac{\pi}{12}\right)\right| \le \dfrac{1}{4!}\left(\dfrac{298}{9\sqrt{3}}\right)\left(\dfrac{\pi}{12}\right)^4$

9. $|R_2(x)| \le \dfrac{1}{2}$ **11.** $n = 3$

Exercise Set 9.7

1. $1 + 2x + 2x^2 + \dfrac{4}{3}x^3 + \cdots + \dfrac{2^k x^k}{k!} + \cdots = \displaystyle\sum_{k=0}^{\infty} \dfrac{2^k x^k}{k!}$

3. $1 - x + x^2 - x^3 + \cdots = \displaystyle\sum_{k=0}^{\infty} (-1)^k x^k$

5. $(x - 1) - \dfrac{1}{2}(x - 1)^2 + \dfrac{1}{3}(x - 1)^3 - \dfrac{1}{4}(x - 1)^4 + \cdots = \displaystyle\sum_{k=1}^{\infty} \dfrac{(-1)^{k+1}}{k}(x - 1)^k$

7. $1 - \dfrac{(x - \pi/2)^2}{2} + \dfrac{(x - \pi/2)^4}{4!} - \cdots = \displaystyle\sum_{k=0}^{\infty} (-1)^k \dfrac{(x - \pi/2)^{2k}}{(2k)!}$

9. $\dfrac{\sqrt{3}}{2} + \dfrac{1}{2}\left(x - \dfrac{\pi}{3}\right) - \dfrac{\sqrt{3}}{2}\dfrac{(x - \pi/3)^2}{2} -$

$\dfrac{1}{2}\dfrac{(x - \pi/3)^3}{3!} + \cdots$

11. $(-\infty, \infty)$ **13.** $(0, 2]$

15. $-\left(x - \dfrac{\pi}{2}\right) + \dfrac{(x - \pi/2)^3}{3!} - \cdots$

Exercise Set 9.8

1. 0.2293 **7.** -1.495

3. 0.618 **9.** $(2.303, 2.303)$

5. 2.303

Review Exercises—Chapter 9

1. diverges   **7.** 0

3. diverges   **9.** $\dfrac{3}{2}$

5. $\dfrac{1}{2}$ **11.** 2

13. converges; p-series with $p = 2$

15. diverges; ratio test

17. converges; ratio test

19. diverges; $\lim\limits_{k \to \infty} a_k \neq 0$

21. diverges; integral test

23. converges; comparison with p-series

25. diverges; $\lim\limits_{k \to \infty} a_k \neq 0$

27. converges; ratio test

29. converges; comparison with p-series

31. $P_3(x) = 1 - 3x + \dfrac{9x^2}{2} - \dfrac{9x^3}{2}$

33. $P_2(x) = x$ **35.** $P_2(x) = x^2$

37. $P_3(x) = 3 + \dfrac{1}{6}(x - 9) - \dfrac{1}{216}(x - 9)^2 +$

$\dfrac{1}{3888}(x - 9)^3$

39. $\displaystyle\sum_{k=0}^{\infty} \dfrac{x^{k+1}}{k!}$

41. $\displaystyle\sum_{k=1}^{\infty} x^k = x + x^2 + x^3 + x^4 + \cdots$

43. a. $x^3 + 2x^2 - x + 3$
    b. $5 + 6(x - 1) + 5(x - 1)^2 + (x - 1)^3$

45. $\$750$ **47.** 0.8603 **49.** $f(2.0288) = 1.8197$

CHAPTER 10

Exercise Set 10.1

11. $y = 9x - \dfrac{x^2}{2} + C$

13. $y = te^t - e^t + C$

15. $y = \dfrac{4}{15}t^{5/2} + \dfrac{4}{3}t^{3/2} + C_1 t + C_2$

17. $y = \dfrac{1}{3}(1 + x^2)^{3/2} + \dfrac{2}{3}$

19. $y = \dfrac{1}{2} + \dfrac{5}{2}e^{-4t}$

25. $\dfrac{dV}{dt} = \sqrt{t + 1};\ V(0) = 5$

$V(t) = \dfrac{2}{3}(t + 1)^{3/2} + \dfrac{13}{3}$

27. $U'(x) = 6 - 2x;\ U(0) = 0$

Exercise Set 10.2

1. 72.1% **7. a.** 1306
                 b. $80,000$

3. After $T = 18$ weeks   **9.** $t = \ln 100 \approx 4.6$ days

5. $x = 0$ **11. a.** 19
                       b. after 22 minutes

Exercise Set 10.3

1. $y = Cx$ **7.** $y = \ln\left(\dfrac{1}{3}t^3 + C\right)$

3. $y = \dfrac{1}{2x^2 + C}$ **9.** $y = Ce^{x^2/2 + x} - 1$

5. $y = \ln(e^x + C)$ **11.** $y = 4e^{5x}$

13. $y = e^{x^2/2 - 2}$ **17.** $y \equiv 1$

15. $y = \pi e^{\sin x}$ **19.** $y = -2 + 4e^{3t}$

21. $P = \dfrac{100}{1 + Me^{-2t}}$

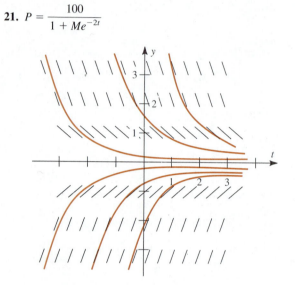

23. $k = -\dfrac{\ln 2}{4.5}$ billion

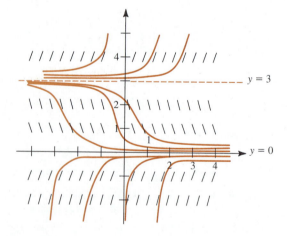

25. a. $\dfrac{dp}{dt} = -0.015p$
  b. $p(20) = 100e^{-0.3}$
  c. $p = 0$

27. a. $P'(t) = 0.10P(t) + 100$
  b. $P(7) = 1100e^{0.7} - 1000 = \$1215.13$

29. $P(5) = \dfrac{1}{1 + 999e^{(5/2)\,\ln\,(199/999)}} = 5350$

31. $y(10) = 30 + 70e^{(10/3)\,\ln\,(5/7)} = 58.2°C$

33. $y(t) = 3000 - 2000e^{-0.02t}$

35. $10 \ln 2 = 6.93$ years   **37.** $y = Cs^k$

Exercise Set 10.4

1. $y = Ae^{-3t} + Be^{-2t}$ **7.** $y = Ae^{-t} + Bte^{-t}$

3. $y = Ae^t + Be^{-t}$ **9.** $y = Ae^{-3t} + Bte^{-3t}$

5. $y = Ae^{2t} + Be^{-5t}$ **11.** $y = A \sin 2t + B \cos 2t$

13. $y = A \sin \sqrt{5}t + B \cos \sqrt{5}t$

15. $y = e^{2t}(A \sin 3t + B \cos 3t)$

17. $y = e^{-t/2}\left(A \sin \dfrac{t}{2} + B \cos \dfrac{t}{2}\right)$

19. $y = e^{3t} - e^{-2t}$ **21.** $y = 3e^{-t} + 4te^{-t}$

23. $y = -\sin 3t + 3 \cos 3t$

25. $x = A \sin 2t + B \cos 2t$
  $y = -2A \cos 2t + 2B \sin 2t$

27. $x(t) = 20 - 15\sqrt{2} \sin \sqrt{2}t + 80 \cos \sqrt{2}t$
  $y(t) = 10 + 30 \cos \sqrt{2}t + 80\sqrt{2} \sin \sqrt{2}t$

29. $x(t) \equiv 25$
  $y(t) \equiv 16$

Exercise Set 10.5

1.

3.

9.

5.

11.

7.

13.

15.

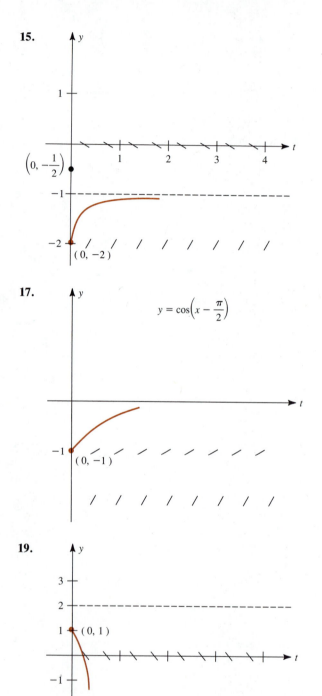

17.

$$y = \cos\left(x - \frac{\pi}{2}\right)$$

19.

21. become closer together

23. decrease   **27.** $y = 15$

25. $c = 1$ **29.** k is negative

Exercise Set 10.6

1.

t_j	0	0.25	0.5	0.75	1.0
y_j	0	1.5	2.25	3.375	5.0625
$y(t_j)$	1	1.6487	2.7183	4.4817	7.3891

$$y = e^{2t}$$

3.

t_j	0	0.25	0.50	0.75	1.0
y_j	1	1.5	1.75	1.875	1.9375
$y(t_j)$	1	1.3934	1.6321	1.7769	1.8647

$$y = 2 - e^{-2t}$$

5.

t_j	0	0.25	0.5	0.75	1.0
y_j	1	1	1.125	1.3906	1.8389
$y(t_j)$	1	1.0635	1.2663	1.6496	2.2974

$$y = 2e^{x^2/2} - 1$$

7.

t_j	0	0.25	0.50	0.75	1.0
y_j	1.5	1.5	1.375	1.1562	0.9102
$y(t_j)$	1.5	1.500	1.3750	1.1563	0.9102

$$y = e^{-t^2} + \frac{1}{2}$$

9.

t_j	0	0.25	0.5	0.75	1.0
y_j	1.5	2.25	3.25	4.625	6.5625
$y(t_j)$	1.5	2.3987	3.7183	5.7317	8.8890

$$y = t + \frac{1}{2} + e^{2t}$$

Review Exercises—Chapter 10

1. $y = 3 + Ce^t$ **3.** $y = 2 \sec \sqrt{x} + C$

5. $y = \dfrac{-2}{3t^2 + C}$

7. $y = Ce^{t^2/2+t} - 2$

9. $y = 2 + Ce^{-2t}$

11. $y = Ce^{-\sin t}$

13. $y = t + C$

15. $y = \dfrac{2}{1 + Me^{-2x}}$

17. $y^2 = t^2 + 2t + C$

19. $y = Ae^{-2t} + Be^{-3t}$

21. $y = Ae^{7x} + Be^{-2x}$

23. $y = Ae^{-4t} + Bte^{-4t}$

25. $y = 2 + \sqrt{1 + x^2}$

27. $y = 3e^{9t}$

29. $y = \pi e^{1-\cos x}$

31. $y = 2 + 50t - 5t^2$

33. $y = 5 \cos 4x$

35. $y = 6te^{3t}$

37.

39.

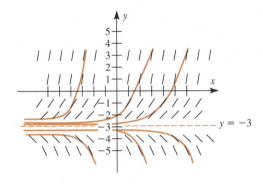

41. $V(t) = 199{,}856 + \dfrac{2}{3}(36 + t)^{3/2}$ thousand dollars

43. a. $\dfrac{dy}{dt} = 2y$ **b.** $\dfrac{d^2y}{dt^2} = y$

45. $y = e^{-t/5}$

47. $y(10) = 80 - 40e^{2(\ln 3 - \ln 4)} = 57.5°\text{F}$

CHAPTER 11

Exercise Set 11.1

1. a. $\dfrac{1}{2}$ **b.** $\dfrac{1}{2}$ **c.** $\dfrac{7}{12}$

3.

5. a. $\dfrac{1}{6}$ **b.** $\dfrac{5}{12}$ **c.** $\dfrac{31}{36}$

7. a. $\dfrac{3}{5}$ **b.** $\dfrac{2}{5}$

9. a. $\dfrac{8}{15}$ **b.** $\dfrac{7}{15}$ **c.** $\dfrac{3}{15}$

11. $\dfrac{1}{16}$

13. a. 1.7 **b.** 1.21 $\sigma = 1.1$

15. \$12 **17.** \$1750

Exercise Set 11.2

1. a. continuous **e.** continuous
 b. discrete **f.** discrete
 c. continuous **g.** discrete
 d. continuous

3. a. $\dfrac{1}{8}$ **b.** 0.65 or $\dfrac{1}{8}(8 - 2\sqrt{2})$ **c.** 1

5. a. $\dfrac{2}{9}$ **b.** $\dfrac{8}{9}$ **c.** $\dfrac{7}{9}$ **7.** $a = 2$

9. $\dfrac{8}{3}$ **11.** $\dfrac{25}{12}$ **13.** $\dfrac{4}{5}$

15. a. $\dfrac{1}{2}$ **b.** $\dfrac{1}{2}$ **c.** $6x - 6x^2$

17. $\dfrac{3}{16}x^2 - \dfrac{1}{32}x^3$ **19. a.** $\dfrac{5}{6}$ **b.** $\dfrac{3}{8}$

21. a. $k = 1$ **b.** $\dfrac{9}{10}$

17. 0.3679   **19.** $\dfrac{1}{\alpha^2}$

Exercise Set 11.3

1. 0.9772   **15.** 0.5328

3. 0.1069   **17. a.** 0.1587   **b.** 0.9544

5. 0.8185   **19.** 0.0475

7. 0.8567   **21.** 0.4972

9. 0.8413   **23.** 0.0668

11. 0.6853   **25.** 0

13. 0.8664

Exercise Set 11.4

1. 0.1813   **9.** 0.2345

3. 0.3200   **11.** 0.5557

5. 0.1447   **13.** $\alpha = 1$

7. 0.4512   **15.** 0.3935

Review Exercises—Chapter 11

1. a. 0.5   **b.** 0.6   **c.** 0.4

3. 1.64   **5.** $-9.75   **7.** $k = \dfrac{2}{25}$

9. a. $\dfrac{3}{16}$ **b.** $\dfrac{11}{16}$ **c.** $\dfrac{5}{16}$

11. $\dfrac{25}{18}$ **23.** 0.0228

13. 0.3085   **25.** 0.4493

15. 0.7257   **27.** 0.3012

17. 0.9332   **29.** 0.3834

19. 0.9270   **31.** $(20.020482) \approx 20.02$

21. a. 0.1587
     b. 0.0228

Index

II. Geometry Formulas

Triangle

$$A = \frac{1}{2}bh$$

Rectangle

$$A = bh$$

Circle

$$A = \pi r^2$$

$$C = \pi d = 2\pi r$$

Parallelogram

$$A = bh$$

19. $\displaystyle\int e^u \, du = e^u + C$

20. $\displaystyle\int a^u \, du = \frac{1}{\ln a} a^u + C, \; a \neq 1$

21. $\displaystyle\int ue^{au} \, du = \frac{e^{au}}{a^2} (au - 1) + C$

22. $\displaystyle\int \frac{du}{1 + e^u} = \ln\left(\frac{e^u}{1 + e^u}\right) + C$

23. $\displaystyle\int \frac{du}{a + be^{cu}} = \frac{u}{a} - \frac{1}{ac} \ln|a + be^{cu}| + C$

24. $\displaystyle\int \ln u \, du = u \ln u - u + C$

25. $\displaystyle\int u \ln u \, du = \frac{u^2}{2} \ln u - \frac{u^2}{4} + C$

26. $\displaystyle\int \frac{du}{u \ln u} = \ln|\ln u| + C$

27. $\displaystyle\int u^m \ln u \, du = u^{m+1}\left[\frac{\ln u}{m + 1} - \frac{1}{(m + 1)^2}\right] + C$

28. $\displaystyle\int \sin^2 u \, du = \frac{u}{2} - \frac{1}{4} \sin 2u + C$

29. $\displaystyle\int \sin^3 u \, du = -\frac{1}{3} \cos u(\sin^2 u + 2) + C$

30. $\displaystyle\int \cos^2 u \, du = \frac{1}{2} u + \frac{1}{4} \sin 2u + C$

31. $\displaystyle\int \cos^3 u \, du = \frac{1}{3} \sin u \, (\cos^2 u + 2) + C$

32. $\displaystyle\int \frac{du}{1 \pm \sin u} = \mp \tan\left(\frac{\pi}{4} \mp \frac{u}{2}\right) + C$

33. $\displaystyle\int \frac{du}{1 + \cos u} = \tan\left(\frac{u}{2}\right) + C$

34. $\displaystyle\int \frac{du}{1 - \cos u} = -\cot\left(\frac{u}{2}\right) + C$

APPLIED CALCULUS

Second edition